Stochastic Optimization Models in Finance

ECONOMIC THEORY AND MATHEMATICAL ECONOMICS

Consulting Editor: Karl Shell

UNIVERSITY OF PENNSYLVANIA
PHILADELPHIA, PENNSYLVANIA

Franklin M. Fisher and Karl Shell. **The Economic Theory of Price Indices:** ***Two Essays on the Effects of Taste, Quality, and Technological Change***

Luis Eugenio Di Marco (Ed.). **International Economics and Development:** ***Essays in Honor of Raúl Presbisch***

Erwin Klein. **Mathematical Methods in Theoretical Economics:** ***Topological and Vector Space Foundations of Equilibrium Analysis***

Paul Zarembka (Ed.). **Frontiers in Econometrics**

George Horwich and Paul A. Samuelson (Eds.). **Trade, Stability, and Macroeconomics:** ***Essays in Honor of Lloyd A. Metzler***

W. T. Ziemba and R. G. Vickson (Eds.). **Stochastic Optimization Models in Finance**

Stochastic Optimization Models in Finance

Edited by

W. T. ZIEMBA

Faculty of Commerce and Business Administration
The University of British Columbia
Vancouver, Canada

R. G. VICKSON

Department of Management Sciences
University of Waterloo
Waterloo, Ontario, Canada

ACADEMIC PRESS New York San Francisco London
A Subsidiary of Harcourt Brace Jovanovich, Publishers

ACADEMIC PRESS, INC.
111 Fifth Avenue, New York, New York 10003

United Kingdom Edition published by
ACADEMIC PRESS, INC. (LONDON) LTD.
24/28 Oval Road, London NW1

Library of Congress Cataloging in Publication Data

Ziemba, W T comp.
Stochastic optimization models in finance.

(Economic theory and mathematical economics series)
Bibliography: p.
Includes index.
1. Finance–Addresses, essays, lectures. 2. Mathematical optimization–Addresses, essays, lectures.
3. Stochastic processes–Addresses, essays, lectures.
I. Vickson, R. G., joint comp. II. Title.
HG174.Z54 1975 332'.01'84 75-2322
ISBN 0–12–780850–7

PRINTED IN THE UNITED STATES OF AMERICA

Dedicated to the memory of my father,
William Ziemba, 1910–1967,
from whom I inherit my interest in
gaming and financial problems.

W.T.Z.

To Lynne, who displayed infinite patience.

R.G.V.

CONTENTS

PART II. QUALITATIVE ECONOMIC RESULTS

PART III. STATIC PORTFOLIO SELECTION MODELS

PREFACE

There is no adequate book for an advanced course concerned with optimizing models of financial problems that involve uncertainty. The numerous texts and edited collections of articles related to quantitative business finance are largely concerned with deterministic models or stochastic models that are directly reducible to deterministic models. Our intention is that this present book of readings will partially fill this gap by providing a source that gives a reasonably thorough account of the mathematical theory and economic results relating to these problems. It is hoped that the volume can serve the dual role of text and research reference. The literature in this area is large and expanding rapidly. Hence to make the volume manageable we had to place severe constraints on our coverage. First and foremost we have only considered material relevant for optimizing models that explicitly involve uncertainty. As a result no material on statistical estimation procedures is included. Secondly, the concern is with the mathematical and economic theories involved and not with their application in practice and associated institutional aspects. However, our coverage does place emphasis on results and methods that can and have been utilized in the analysis of real financial problems. With some reluctance we also limited ourselves to models involving a single decision maker. Hence, material relating to gaming problems, market equilibrium, and other multiperson or multifirm problems is not included here. There is also very little material concerned with specific numerical algorithms or ad hoc solution approaches and that included is intended to be representative and not comprehensive.

The major criterion for inclusion of articles in the collection is that they present a significant methodological advance that is of lasting interest for teaching and research in finance. Since a major goal of the volume is to provide a source for students to "get quickly to the frontiers," preference was given to papers in areas where research is currently active. We have also tried to choose contributions that were well written and did not appear to have major gaps or errors in their presentation. Papers were also given lower priority if they have appeared in other books of readings or if similar material is available in such collections. This accounts for the very limited coverage of capital budgeting models. In several areas we felt it appropriate to provide new papers that summarize and extend existing results in the literature. Credit and blame for biases introduced in the selection of articles and their layout are due to WTZ.

The five parts of this work present material that is intended to be read in a sequential fashion part by part.

Part I is concerned with mathematical tools and expected utility theory. The treatment focuses on convexity and the Kuhn–Tucker conditions and the methods of dynamic programming. The next part is concerned with qualitative economic results, and particular attention is given to results relating to stochastic dominance, measures of risk aversion, and portfolio separation theorems. Part III is concerned with static models of portfolio selection. Particular attention is placed on the mean-variance and safety-first approaches and their extensions along with their relation to the expected utility approach. The questions of existence and diversifications of optimal portfolio policies and the effects of taxes on risk taking are also dealt with here. The fourth part is concerned with dynamic models that are in some sense reducible to static models. Part V is concerned with dynamic models that are most properly analyzed by dynamic methods. Particular attention is placed on models of portfolio revision, optimal capital accumulation, option strategies, and portfolio problems in continuous time.

Each part begins with an introduction that attempts to summarize and make cohesive the material that follows. This is followed by several articles that present major methodological advances in that area. At the end of each part there are numerous problems which form an integral part of the text and are subdivided into computational and review, and mind-expanding exercises. The computational and review exercises are intended to test the understanding of the preceding readings and how they relate to previous material. Some exercises present, hopefully, straightforward extensions, to new problems, of the methodological material in the readings. Other exercises fill gaps in the presentation of the reprinted articles. The mind-expanding exercises, on the other hand, are intended to traverse new problem areas whose analysis requires different techniques or nonstraightforward extensions of the material in the reprinted articles. Many of these exercises are quite difficult, and some present unsolved problems and conjectures. We have starred all exercises whose solution is unknown or in doubt to us. Some mind-expanding exercises describe previously unpublished problems and solutions. They also serve to survey, to some extent, many important papers that could not be included because of space limitations. Indeed, many exercises present major results from one or more published papers. The sources from which such exercises were adapted are indicated in the "Exercise Source Notes" at the end of each exercise section. We have tried to present the exercises in such a way that the reader can obtain maximum benefit per unit input of time spent in their solution. In particular results are generally stated and the reader is asked to verify them. Hence, the student should be able to understand the major results and conclusions of a particular exercise even if he cannot solve all its parts.

This collection of articles and exercises has not undergone extensive classroom testing. However, one of us (WTZ) has used some of the materials in courses at the University of British Columbia (in 1971) and at the University of California, Berkeley (in 1972). The typical students who would take a course based on this book are masters and doctoral level students in management science, operations research, and economics and doctoral level finance students. The ideal prerequisites consist of a course in elementary business finance using say Lusztig and Schwab (1973), Mao (1969), or Van Horne (1968), a course in nonlinear programming, using say Luenberger (1973) or Zangwill (1969), a course in probability theory using say Feller (1962) or Thomasion (1969), and a course in microeconomic theory using say Henderson and Quandt (1958) or Samuelson (1965). A course in intermediate capital theory using Fama and Miller (1972) or Mossin (1973) would also be of great value to the reader. Minimum prerequisites consist of a careful reading of Part I and the working of the accompanying computational and review exercises (perhaps supplemented by reading the first three chapters of Zangwill (1969) and Chapter Two of Arrow (1971), or by working the mind-expanding exercises in Part I); a calculus course using say Thomas (1953) or Allen (1971) and a course in intermediate probability theory using say Lippman (1971) or Hadley (1967). These minimum prerequisites should be adequate for the understanding of most of the articles in this volume if the reader has developed sufficient mathematical maturity. The book is intended for a two-semester course. It can also be used for a one-semester course related to static models (Parts I (1, 2), II, and III) or to dynamic models (Parts I (3), II (2), IV, and V).

ACKNOWLEDGMENTS

We would first like to thank the respective authors and publishers for their permission to reproduce the articles in this collection. Thanks are due to Professors Nils H. Hakansson, William F. Sharpe, and Karl Shell and Dr. Gordon B. Pye for initial encouragement to undertake the development of the book in its present form. Professor Shelby L. Brumelle participated in many discussions with us during the 1971–1972 academic year. Due to his appointment to a major administrative post at the University of British Columbia, he was unable to be a co-editor with us. In addition to his co-authored article with Vickson in Part II we would like to thank Professor Brumelle for contributing several problems. We would also like to thank the following colleagues and friends who read and commented on various parts of the book and contributed valuable problems: Professors Michael J. Brennan, W. Erwin Diewert, Richard C. Grinold, Loring G. Mitten, James A. Ohlson, and Cary A. Swoveland and Dr. Gordon B. Pye and Mr. Stig Larsson. Various portions of the manuscript were ably typed by Ms. Marlies Adamo, Ms. Sandra L. Schwartz, and Mrs. Ikuko Workman. The clerical assistance of Mr. Jupian Leung and Mr. Johnny Yu is gratefully acknowledged. Special thanks go to Mr. Amin Amershi, Ms. Trudy A. Cameron, Mr. Martin I. Kusy, and Mr. Katsushige Sawaki for their help in reading the proofs. Mr. Kusy also helped in the preparation of the index. Thanks are also due to the National Research Council of Canada for their continuing support of Professor Ziemba's research on stochastic programming and its applications under contract NRC-67-7147.

Part I
Mathematical Tools

INTRODUCTION

The first part of this book is devoted to technical prerequisites for the study of stochastic optimization models. We have selected articles and included exercises that appear to provide the necessary background for the study of the specific financial models discussed in the remainder of the book. The treatment in this part is, however, necessarily brief because of space limitations; hence the reader may wish to consult some of the noted additional references on some points.

The prerequisites for an in-depth study of stochastic optimization models are expected utility theory, convex functions and nonlinear optimization methods, and the embedded concepts of dynamic programming.

I. Expected Utility Theory

Fishburn's article presents a concise proof of a general expected utility theorem. He considers a set of outcomes and probability distributions over outcomes. The latter are termed horse lotteries. Given a set of reasonable assumptions concerning the decision-maker's preferences over alternative horse lotteries, Fishburn demonstrates the existence of a utility function and a subjective probability measure over the states of the world such that the decision-maker acts as if he maximized expected utility. The utility function is continuous and is uniquely defined up to a positive linear transformation (see Exercise CR-2 for examples). In some cases the utility function is bounded. However, most utility functions are unbounded in at least one direction. For such utility functions horse lotteries can be constructed that have arbitrarily large expected utility and an arbitrarily small probability of receiving a positive return. Such examples are called St. Petersburg paradoxes. They are illustrated in Exercises CR-4 and ME-2. Further restrictions on the utility function result if one makes additional assumptions concerning the decision-maker's behavior. For example, an investor who never prefers a fair gamble to the status quo must have a concave utility function (Exercise CR-1). Such investors will not simultaneously gamble and purchase insurance. Exercise CR-6 is concerned with this classic Friedman–Savage paradox; see also the papers by Yaari (1965a) and Hakansson (1970).[1] If the investor's preferences for horse lotteries are independent of his initial wealth level, then the utility function must be linear or exponential (Exercise CR-5). Similarly, if the investor's preferences for proportional gambles are independent of initial

[1] Throughout this book references cited by date will be found in the Bibliography at the end of the book.

wealth, then his utility function must be a logarithmic or power function (Exercise CR-17). Exercise ME-3 investigates the behavioral significance of assuming that a utility function defined over several commodities is separable or that its logarithm is separable.

The calculation of an allocation vector that maximizes expected utility may proceed via a stochastic programming or nonlinear programming algorithm as discussed in Part III. However, when the utility function is concave one can easily determine bounds on the maximum value of expected utility (Exercise ME-1). The lower bound is a consequence of Jensen's inequality (Exercise CR-7) which states that the expected value of a concave function is never exceeded by the value of the function evaluated at its mean. Some sharper Jensen-like bounds applicable under more restrictive assumptions appear in Ben-Tal and Hochman (1972) and Ben-Tal, Huang, and Ziemba (1974). Exercise CR-11 introduces the notion of a certainty equivalent, that is, a fixed vector whose utility equals the expected utility.

The expected utility approach provides a natural framework for the analysis of financial decision problems. However, there are several alternative approaches which are of particular interest in certain instances. Most notable are the approaches that trade off risk and return, and various safety-first and related chance-constrained formulations. These approaches and their relations with the expected utility approach are considered in some detail in Part III. See also Exercise CR-18 for an approach that augments a decision-maker's returns and costs in such a way that gambling and insurance conclusions that normally hinge on the shape of the utility function may be obtained for linear utility functions.

Fishburn's article provides a succinct but brief presentation of an expected utility theory. The mathematical level in this article is perhaps higher than in nearly all of the other material in this book. For this reason some readers may wish to consult the less general but more lucid developments by Arrow (1971), Jensen (1967a, b), and Pratt *et al.* (1964). A much fuller treatment (and comparison) of many expected utility theories may be found in the work of Fishburn (1970). See also Fishburn's paper (1968) for a concise survey of the broad area of utility theory.

II. Convexity and the Kuhn–Tucker Conditions

Many of the functions involved in financial optimization problems are convex (or concave, the negative of a convex function). Convex functions are defined on convex sets, which are sets such that the entire closed line segment joining any two points in the set is also in the set. Exercise ME-15 outlines some important properties of convex sets. Convex functions have the property that linear interpolations never underestimate the functions. Alternatively, linear supports never overestimate the functions. These and related characterizations are illustrated in Exercise ME-8. There are two important properties possessed by problems of minimizing convex functions subject to constraints having the form that a set of convex functions not exceed zero. First, the set of feasible points determined by the constraints is a convex set. Second, local minima are always global minima. For applications, certain generalizations of the convexity concept are needed. Mangasarian's first article introduces pseudo-convex functions. These functions are differentiable and possess both of the properties mentioned above, yet are not necessarily convex. They are essentially defined by the property that if a directional derivative is positive (points up), then the function continues to increase in the given direction. An even wider class of functions, called quasi-convex, is defined by retaining only the first property. There are several alternative ways to define quasi-convex functions and they are explored in Exercise ME-11. These generalized functions along with other related functions are discussed in Exercises CR-15 and ME-9. In Exercises CR-8 and 9 the reader is asked to investigate the convexity and generalized convexity properties of simple functions in one and several dimensions, respectively. Exercise ME-12 shows how one can verify whether or not a given function is a convex or generalized convex function by examining a Hessian matrix of second partial derivatives or a Hessian matrix bordered by first partials.

The most useful results concerning minimization problems are surely the Kuhn–Tucker necessary and sufficient conditions. In Mangasarian's article it is shown that the Kuhn–Tucker conditions are sufficient if the objective is pseudo-convex and the constraints are quasi-concave. The conditions are necessary if a mild constraint qualification is met. This is discussed in Exercise ME-13, where a proof of their necessity that utilizes Farkas' lemma is outlined. Exercises CR-10 and 13 illustrate how the Kuhn–Tucker conditions may be used to solve simple mean-variance tradeoff models and other optimization problems, respectively. The determination of the correct sign of the multipliers is considered in Exercise CR-16. The Kuhn–Tucker conditions are intimately related to the Lagrange function. Exercise ME-10 illustrates the relationship between saddle points of the Lagrangian and solutions of the minimization problem. The primal problem may be considered as a minimax of the Lagrangian where the min is with respect to original (primal) variables, and the max is with respect to Lagrange (dual) variables. A dual problem results when one maximins. When the functions are convex and differentiable one obtains the Wolfe dual. Mangasarian presents some results related to this dual problem in his first paper; other results are developed in Exercise ME-14. The reader is asked to determine the actual dual problems in some special cases in Exercise CR-12. The relationship between the dual variables and right-hand-side perturbations of the constraints as they relate to the optimal primal objective value is developed in Exercise ME-6.

Many of the functions that arise in financial optimization problems are composites of two or more simpler functions. In Mangasarian's second paper he presents a general theorem that relates the convexity and generalized convexity of composite functions to those of the simpler functions. In particular, convex functions of linear functions are convex, and convex nondecreasing functions of convex functions are convex. Such results are also useful in the analysis of the convexity properties of products, ratios, reciprocals, and so forth, of convex and related functions. Exercise ME-9 illustrates the use of such results on a class of convex-linear composite functions. The convexity and pseudo-convexity relationships between functions and their logarithms are considered in Exercise ME-4. Exercise ME-18 concerns the strict concavity of composite functions. Exercises CR-14 and ME-17 discuss properties of sums and integrals of convex and related functions. The monotonicity and convexity properties of functions defined by a minimization operation are considered in Exercise ME-7. Exercise ME-5 discusses a useful linearization result that provides the basis for the Frank Wolfe algorithm (Exercise III-ME-14).

The material presented related to convexity and minimization of constrained functions provides a basis for further study and contains most of the results needed in the sequel. The following sources provide additional results, methods, and references. The standard reference work on convexity is by Rockafellar (1970). Additional material may be found in the work of Berge (1963), Mangasarian (1969), Newman (1969), and Stoer and Witzgall (1970). The relationships between quasi-convex and convex functions are explored by Greenberg and Pierskalla (1971) (see also Ginsberg, 1973). For a simplified presentation of the Wolfe duality theory and its economic interpretation consult Balinsky and Baumal (1968). See the work of Mangasarian (1970) for the most general presentation of the Wolfe duality theory. A more general duality theory based on earlier work by Fenchel (1953) is given by Rockafellar (1970). A simplified presentation of the Rockafellar duality theory and its economic interpretation appears in Williams' paper (1970) (see also Cass, 1974). An extremely lucid presentation and comparison of several duality theories has been given by Geoffrion (1971). For a fuller treatment of the theory of nonlinear programming the reader is referred to textbooks by Lasdon (1970), Luenberger (1969, 1973), Mangasarian (1969), and Zangwill (1969) and the edited conference proceedings by Abadie (1967, 1970), Fletcher (1969b), Hammer and Zoutendijk (1974), and Rosen *et al.* (1970). The last five of these books provide excellent source material and reference for the study of specific algorithms.

III. Dynamic Programming

Many of the financial problems considered in this book may be best analyzed in a dynamic context. For example, a consumer's consumption is best thought of as occurring either continuously in time or during certain intervals of time. In such problems the decision maker will generally make a sequence of decisions, each one in turn dependent upon past data and its impact on the future. Naturally today's decision must be made in a way that reflects the fact that tomorrow's decision will be chosen to be the best one possible given the data available tomorrow. Thus intuitively it must be true that "an optimal set of decisions has the property that whatever the first decision is, the remaining decisions must be optimal with respect to the outcome which results from the first decision." This is Bellman's famous principle of optimality which may be proved via the contradiction argument that if this statement were not true, then certainly the policy under consideration could not be optimal. Given that there is an optimal policy, the principle of optimality provides a scheme for the numerical determination of such a policy. Moving forward in time the policy indicates the optimal decisions to make in all periods given all possible past data. That is, regardless of the state in which the decision maker finds himself, the policy must indicate his optimal action. Moving backward in time generally provides a mechanism for determining an optimal policy. Suppose there are n time periods and that all the possible states that might obtain in period $n-1$ are known. Then it is a relatively simple calculation to determine the best decision to make given that the individual finds himself in any one of these states. Suppose the states for period $n-2$ are given and that the payoffs are additive. Given that the decision maker is in any particular state in period $n-2$, an optimal decision is one which maximizes the sum of the immediate payoff in period $n-2$ plus the payoff that results from the optimal action made in period $n-1$. Now the optimal actions in period $n-1$ were previously calculated. Hence continuing period by period an optimal decision in any state in period t is one which maximizes the sum of the immediate return in period t plus the return from the optimal policy used in period $t+1$. One thus obtains a functional equation that may be solved sequentially to determine an optimal decision in each state in each period. The calculations are generally much simpler than an exhaustive enumeration because only those policies that involve optimal policies for periods $t+1, \ldots, n$ need be considered in determining optimal decisions for period t.

The article by Ziemba presents an introduction to the methods and uses of dynamic programming. His presentation is an adaptation of an earlier article by Denardo and Mitten; it is primarily concerned with discrete time terminating processes which include problems having finite time horizons.

However, some extensions to infinite horizon and continuous-time problems are made in the article and in Exercises ME-19 and 20. The basic elements of all dynamic programming problems are identified as the stages, states, decisions, transitions, and returns. Identification of these elements facilitates the formulation and solution of any particular problem. For terminating processes it is easy to prove directly the existence of an optimal policy under a mild monotonicity assumption using rather simple mathematics. The monotonicity assumption requires that if two policies a and b contain the same decision at some state x and if policy a produces at least as large a return as policy b for each state reachable from x, then it must follow that policy a produces at least as large a return as b starting from x. The reader is asked to investigate in Exercise CR-19 whether or not some common preference functions satisfy the monotonicity assumption. This approach to the study of sequential decision problems leads to an algorithm based on the functional equation of dynamic programming.

There are a number of important financial problems in which it is appropriate to consider a nonterminating model, such as problems whose horizon is either subject to chance or is infinite. If it is *assumed* that there exists an optimal policy, then, under the monotonicity assumption, it follows that the functional equation exists and can be utilized to compute an optimal policy. Some particular nonterminating models are considered in Exercises ME-19 and 20. Exercise ME-19 is concerned with the development of necessary and sufficient conditions for the validity of the functional equation for Markovian processes. In general the functional equation is valid if one is minimizing nonnegative costs or maximizing nonnegative returns. The functional equation is not generally valid in the reverse cases, i.e., when one is minimizing nonpositive costs or maximizing nonpositive returns. However, validity does obtain in some special cases. Exercise ME-20 extends these results to the case of bounded costs in a discounted expected cost minimization problem. This exercise also provides justifications for Howaıd's policy improvement and Manne's linear programming solution approach to Markovian sequential decision problems. Exercises CR-20 and 21 illustrate how one determines such a functional equation and performs the recursive calculations on simple one- and two-dimensional allocation problems, respectively. Exercise CR-26 considers an allocation problem over time and the effects that different constraints and horizons have on the optimal solution. Exercise CR-21 also illustrates how a Lagrange multiplier technique may be utilized to reduce memory requirements in the recursive scheme. Exercise CR-22 is concerned with an economic planning model and illustrates how the calculations and analysis may be modified to handle uncertainty. Exercise CR-23 concerns itself with a production problem that may be solved as a linear program as well as a dynamic program. It also involves uncertainty and the reader is asked to comment on the relative merits of the two solution approaches for various versions of the problem. Exercise CR-24 illustrates a simple way of deducing the functional equation via a decomposition scheme based on certain separability and monotonicity assumptions. The reader is asked in Exercise CR-25 to verify two results stated in the "Introduction to Dynamic Programming" by Ziemba.

The reader may consult Beckmann (1968), Bellman (1961), Bellman and Dreyfus (1962), Gluss (1972), Howard (1960), Jacobs (1967), and Nemhauser (1966) for introductory treatments of dynamic programming. More advanced treatments are given by Bellman (1957), Blackwell (1962), Dreyfus (1965), Howard (1972), Kaufmann and Cruon (1967), and Ross (1970). Exercises ME-19 and 20 present some of the pertinent results for the study of Markovian decision problems when the objective is the maximization of discounted expected return. A lucid presentation of additional results, including the study of alternative preference functions, may be found in the work of Ross (1970). See also Blackwell (1970) for a presentation of weak assumptions that lead to the optimality of stationary strategies. Veinott (1969) has given a unified presentation of the current state of this field. Many interesting gambling problems, whose solution and spirit of formulation are not unlike many of the problems in Part V of this book, may be posed as dynamic programming problems. The interested reader may consult Breiman (1964), Dubins and Savage (1965), Epstein (1967), Ross (1972), and Thorp (1969). The Epstein book provides an extremely lucid introduction to gambling problems. It is often advisable to develop special computational schemes to reduce the memory requirements needed to solve actual dynamic programming problems. An elementary treatment of some of these schemes may be found in the work of Bellman and Dreyfus (1962), Nemhauser (1966), and Wilde and Beightler (1967). For an advanced treatment of one such scheme consult Larson (1968). Continuous-time models pose special difficulties because one must generally utilize discrete time approximations to develop a dynamic programming functional equation. Such an approach is discussed and utilized in Merton's paper in Part V. See Kushner (1967) or Dreyfus (1965) for general discussions of such problems.

1. EXPECTED UTILITY THEORY

The Annals of Mathematical Statistics
1969, Vol. 40, No. 4, 1419–1429

A GENERAL THEORY OF SUBJECTIVE PROBABILITIES AND EXPECTED UTILITIES

By Peter C. Fishburn

Research Analysis Corporation

1. Introduction. The purpose of this paper is to present a general theory for the usual subjective expected utility model for decision under uncertainty. With a set S of states of the world and a set X of consequences let F be a set of functions on S to X. F is the set of acts. Under a set of axioms based on extraneous measurement probabilities, a device that is used by Rubin [14], Chernoff [3], Luce and Raiffa [9, Ch. 13], Anscombe and Aumann [1], Pratt, Raiffa, and Schlaifer [11], Arrow [2], and Fishburn [5], we shall prove that there is a real-valued function u on X and a finitely-additive probability measure P^* on the set of all subsets of S such that, for all $f, g \in F$,

$$f \precsim g \text{ if and only if } E[u(f(s)), P^*] \leqq E[u(g(s)), P^*]. \tag{1}$$

In (1), $\precsim$ ("is not preferred to") is the decision-maker's binary preference-indifference relation and $E(y, z)$ is the mathematical expectation of y with respect to the probability measure z.

Because we shall use extraneous measurement probabilities, (1) will be extracted from the more involved expression (2) that is presented in the next section. The axioms we shall use to derive (2) imply that P^* is uniquely determined and that u is unique up to a positive linear transformation. u may or may not be bounded: however, it is bounded if there is a denumerable partition of S each element of which has positive probability under P^*. Our theory places no restrictions on S and X except that they be nonempty sets with X containing at least two elements. X may or may not have a least (most) preferred consequence. In addition, no special restrictions are placed on P^*. For example, if S is infinite, it may or may not be true that $P^*(A) = 1$ for some finite subset $A \subseteq S$, and if $P^*(A) < 1$ for every finite $A \subseteq S$ it may or may not be true that S can be partitioned into an arbitrary finite number n of subsets such that $P^* = 1/n$ for each subset. Finally, no special properties will be implied for u apart from its uniqueness and its boundedness in the case noted above.

To indicate briefly how this differs from other theories, we note first that the theories of Chernoff [3], Luce and Raiffa [9], Anscombe and Aumann [1], Pratt, Raiffa, and Schlaifer [11], and Fishburn [5] assume that S is finite. The theory presented here is a generalization of a theory in Fishburn [5]. The theory of Davidson and Suppes [4] assumes that X is finite and implies that, if $x \prec y$ and $z \prec w$ and there is no consequence between x and y or between z and w then $u(y) - u(x) = u(w) - u(z)$. The theories of Ramsey [13] and Suppes [16]

Received 19 August 1968; revised 6 March 1969.

place no special restrictions on S but they imply that X is infinite and that if $u(x) < u(y)$ then there is a $z \varepsilon X$ such that $u(z) = .5u(x) + .5u(y)$. On the other hand Savage [15] does not restrict X in any unusual way, but his theory requires S to be infinite and implies that, for any positive integer n, there is an n-part partition of S such that $P^* = 1/n$ on each part of the partition. Arrow [2] also assumes this property for P^*.

2. Definitions and notation. $\mathcal{P}$ is the set of all simple probability measures (gambles) on X, so that if $P \varepsilon \mathcal{P}$ then $P(Y) = 1$ for some finite Y included in X. The probabilities used in $\mathcal{P}$ are extraneous measurement probabilities. They can be associated with outcomes of chance devices such as dice and roulette wheels.

With $P, Q \varepsilon \mathcal{P}$ and $\alpha \varepsilon [0, 1]$, $\alpha P + (1 - \alpha)Q$ is the direct linear combination of P and Q so that $\alpha P + (1 - \alpha)Q$ is in $\mathcal{P}$. Under this interpretation, $\mathcal{P}$ is a mixture set. By Herstein and Milnor's [7] definition, a *mixture set* is a set M and an operation that assigns an element $\alpha a + (1 - \alpha)b$ in M to $(a, b) \varepsilon M \times M$ and $\alpha \varepsilon [0, 1]$ in such a way that

$$(1)a + (0)b = a$$

$$\alpha a + (1 - \alpha)b = (1 - \alpha)b + \alpha a$$

$$\alpha(\beta a + (1 - \beta)b) + (1 - \alpha)b = (\alpha\beta)a + (1 - \alpha\beta)b$$

for all $a, b \varepsilon M$ and $\alpha, \beta \varepsilon [0, 1]$.

$\mathcal{H}$ is the set of all functions on S to $\mathcal{P}$. With $\mathbf{P} \varepsilon \mathcal{H}$, $\mathbf{P}(s)$ is the gamble in $\mathcal{P}$ assigned to $s \varepsilon S$ by $\mathbf{P}$. We shall call $\mathcal{H}$ the set of *horse lotteries*, after Anscombe and Aumann [1]. A pseudo-operational interpretation for $\mathbf{P}$ goes as follows. If $\mathbf{P}$ is selected and if state s obtains (to use Savage's term) then the gamble $\mathbf{P}(s)$ is used to select a consequence in X.

With $\mathbf{P}, \mathbf{Q} \varepsilon \mathcal{H}$ and $\alpha \varepsilon [0, 1]$, $\alpha \mathbf{P} + (1 - \alpha)\mathbf{Q}$ is the direct linear combination of $\mathbf{P}$ and $\mathbf{Q}$ so that $(\alpha\mathbf{P} + (1 - \alpha)\mathbf{Q})(s) = \alpha\mathbf{P}(s) + (1 - \alpha)\ \mathbf{Q}(s)$. Clearly, $\alpha\mathbf{P} + (1 - \alpha)\mathbf{Q}$ is in $\mathcal{H}$ when $\mathbf{P}, \mathbf{Q} \varepsilon \mathcal{H}$ and $\alpha \varepsilon [0, 1]$. It follows easily that $\mathcal{H}$ is a mixture set.

An *event* is any subset of S. A *partition* of S is a set of non-empty, disjoint events whose union equals S. $\mathbf{P}$ in $\mathcal{H}$ is *constant* on event A if and only if $\mathbf{P}(s) = \mathbf{P}(s')$ for all $s, s' \varepsilon A$, in which case we write $\mathbf{P} = P$ on A when $\mathbf{P}(s) = P$ for all $s \varepsilon A$. $\mathbf{P}$ and $\mathbf{Q}$ *agree* on event A if and only if $\mathbf{P}(s) = \mathbf{Q}(s)$ for all $s \varepsilon A$, in which case we write $\mathbf{P} = \mathbf{Q}$ on A.

We shall apply the binary relation $\leqslant$ to $\mathcal{H}$. $\mathbf{P} \sim \mathbf{Q}$ if and only if $\mathbf{P} \leqslant \mathbf{Q}$ and $\mathbf{Q} \leqslant \mathbf{P}$, and $\mathbf{P} < \mathbf{Q}$ if and only if $\mathbf{P} \leqslant \mathbf{Q}$ and not $\mathbf{Q} \leqslant \mathbf{P}$. With $P, Q \varepsilon \mathcal{P}$, we define $\leqslant$ on $\mathcal{P}$ in terms of $\leqslant$ on $\mathcal{H}$ thus:

$$P \leqslant Q \text{ if and only if } \mathbf{P} \leqslant \mathbf{Q} \text{ when } \mathbf{P} = P \text{ and } \mathbf{Q} = Q \text{ on } S.$$

With $P \varepsilon \mathcal{P}$ and $\mathbf{Q} \varepsilon \mathcal{H}$, $P \leqslant \mathbf{Q}$ means that $\mathbf{P} \leqslant \mathbf{Q}$ when $\mathbf{P} = P$ on S. $\mathbf{P} \lessdot Q$, $P < Q$, $P < \mathbf{Q}$, $\cdots$ are defined in similar ways.

A^c is the *complement* in S of event A, so that $A \cap A^c = \emptyset$ and $A \cup A^c = S$.

Event A is *null* if and only if $\boldsymbol{P} \sim \boldsymbol{Q}$ whenever $\boldsymbol{P}$, $\boldsymbol{Q}$ in $\mathcal{H}$ agree on A^c.

With these definitions at hand we shall turn to a set of axioms that implies the existence of a real function u on X and a probability measure P^* on (the set of all subsets of) S such that, for all $\boldsymbol{P}, \boldsymbol{Q} \in \mathcal{H}$,

$$(2) \qquad \boldsymbol{P} \preccurlyeq \boldsymbol{Q} \text{ if and only if } E[E(u, \boldsymbol{P}(s)), P^*] \leqq E[E(u, \boldsymbol{Q}(s)), P^*].$$

Expression (1) results from (2) on defining $f \preccurlyeq g$ if and only if $\boldsymbol{P} \preccurlyeq \boldsymbol{Q}$ when $\boldsymbol{P}(s)(f(s)) = \boldsymbol{Q}(s)(g(s)) = 1$ for all $s \in S$. When $\boldsymbol{P}(s)(f(s)) = 1$ for all $s \in S$, $E[E(u, \boldsymbol{P}(s)), P^*] = E[u(f(s)), P^*]$.

3. Axioms and summary theorem. In addition to the structural assumptions of the preceding section ($\mathcal{P}$ is the set of simple probability measures on X, $\mathcal{H}$ is the set of all functions on S to $\mathcal{P}$), we shall use the following six axioms. For all $\boldsymbol{P}, \boldsymbol{Q}, \boldsymbol{R} \in \mathcal{H}$:

A1. $\preccurlyeq$ is a weak order (transitive and connected, or complete) on $\mathcal{H}$;

A2. If $\boldsymbol{P} < \boldsymbol{Q}$ and $\alpha \in (0, 1)$ then $\alpha\boldsymbol{P} + (1 - \alpha)\boldsymbol{R} < \alpha\boldsymbol{Q} + (1 - \alpha)\boldsymbol{R}$;

A3. If $\boldsymbol{P} < \boldsymbol{Q}$ and $\boldsymbol{Q} < \boldsymbol{R}$ then $\alpha\boldsymbol{P} + (1 - \alpha)\boldsymbol{R} < \boldsymbol{Q}$ and $\boldsymbol{Q} < \beta\boldsymbol{P} + (1 - \beta)\boldsymbol{R}$ for some $\alpha, \beta \in (0, 1)$;

A4. $P < Q$ for some $P, Q \in \mathcal{P}$;

A5. If event A is not null, if $\boldsymbol{P} = P$ and $\boldsymbol{Q} = Q$ on A and if $\boldsymbol{P} = \boldsymbol{Q}$ on A^c then $\boldsymbol{P} \preccurlyeq \boldsymbol{Q}$ if and only if $P \preccurlyeq Q$;

A6. If $\boldsymbol{P}(s) < \boldsymbol{R}$ for all $s \in S$ then $\boldsymbol{P} \preccurlyeq \boldsymbol{R}$, and if $\boldsymbol{R} < \boldsymbol{Q}(s)$ for all $s \in S$ then $\boldsymbol{R} \preccurlyeq \boldsymbol{Q}$.

Axioms A1, A2, and A3 are fairly standard axioms in von Neumann-Morgenstern or Bernoullian expected-utility theory. A2 is an independence or sure-thing axiom, and a defense for its adoption is similar to the defenses made by Friedman and Savage [6], Savage [15], Raiffa [12], and others. In particular, it can be noted that a two-stage procedure can be associated with $\alpha\boldsymbol{P} + (1 - \alpha)\boldsymbol{R}$ whereby $\boldsymbol{P}$ is selected with probability α and $\boldsymbol{R}$ is selected with probability $1 - \alpha$ at the first stage. If $\boldsymbol{P}$ is selected at the first stage and if s obtains then $\boldsymbol{P}(s)$ is used to select an x at the second stage. A similar remark holds for $\boldsymbol{R}$ and $\boldsymbol{R}(s)$. If s obtains, the total probability that x will result is $\alpha\boldsymbol{P}(s)(x) + (1 - \alpha)\boldsymbol{R}(s)(x)$, which equals the probability $(\alpha\boldsymbol{P}(s) + (1 - \alpha)\boldsymbol{R}(s))(x)$ assigned to x by $\alpha\boldsymbol{P} + (1 - \alpha)\boldsymbol{R}$ when s obtains.

A3 is a typical Archimedean axiom, and A4 is similar to Savage's P5. It forecloses the uninteresting possibility that all consequences are indifferent to each other. A5 in the $\mathcal{H}$ context is similar in intent and structure to Savage's P3, which is part of his sure-thing principle. A5 says that preferences for constant horse lotteries are the same as preferences for corresponding horse lotteries that are constant on nonnull events and agree with one another on the complements of these events.

The final axiom A6 is another aspect of the sure-thing principle. It is similar to Savage's P7 extended to horse lotteries, except that Savage's P7 uses $\preccurlyeq$ where $<$ is used in A6.

One final definition will be needed for our summary theorem. We shall call $\boldsymbol{P} \varepsilon \mathcal{H}$ bounded if and only if there are real numbers a and b such that

$$P^*\{a \leqq E(u, \boldsymbol{P}(s)) \leqq b\} = 1.$$

According to common usage, $P^*\{\cdots\}$ is an abbreviation for $P^*(\{s \mid a \leqq E(u, \boldsymbol{P}(s)) \leqq b\})$.

SUMMARY THEOREM. A1-6 *imply that there is a finitely additive probability measure* P^* *on the set of all subsets of* S *and a real function* u *on* X *such that* (2) *holds for all* $\boldsymbol{P}, \boldsymbol{Q} \varepsilon \mathcal{H}$. *Moreover, when* A1-6 *hold and* P^* *and* u *satisfy* (2) *for all* $\boldsymbol{P}, \boldsymbol{Q} \varepsilon \mathcal{H}$, *then*:

(a) *Every horse lottery is bounded for* P^* *and* u;

(b) *For all* $A \subseteq S$, $P^*(A) = 0$ *if and only if event* A *is null*;

(c) u *is bounded if there is a denumerable partition of* S *such that each event in the partition has positive probability under* P^*;

(d) *A probability measure* Q^* *on the set of all subsets of* S *and a real function* u' *on* X *satisfy* (2) *in place of* P^* *and* u *for all* $\boldsymbol{P}, \boldsymbol{Q} \varepsilon \mathcal{H}$ *if and only if* $Q^* = P^*$ *and there are real numbers* $a > 0$ *and* b *such that* $u'(x) = au(x) + b$ *for all* $x \varepsilon X$.

4. Theorems. To establish the Summary Theorem we shall consider a series of theorems, several of which will be proved in the ensuing sections. We shall not prove the first theorem since it follows immediately from the expected-utility theorem proved by Jensen [8] supplemented by Theorem 18 in Luce and Suppes [10, p. 288].

THEOREM 1. *If* M *is a mixture set and* $\preccurlyeq$ *on* M *is a binary relation that satisfies the direct analogues of* A1, A2, *and* A3, *then there is a real-valued function* w *on* M *that satisfies* (3) *and* (4) *for all* $a, b \varepsilon M$ *and* $\alpha \varepsilon [0, 1]$ *and is unique up to a positive linear transformation when it has these two properties*:

$$a \preccurlyeq b \text{ if and only if } w(a) \leqq w(b) \tag{3}$$

$$w(\alpha a + (1 - \alpha)b) = \alpha w(a) + (1 - \alpha)w(b). \tag{4}$$

The next theorem, which uses Theorem 1 in its proof, amounts to a restatement of Theorems 2 and 3 in Fishburn [5]. Its proof is similar to the proofs of those theorems and will be omitted.

THEOREM 2. *If* A1-5 *hold and* $\{B_1, \cdots, B_n\}$ *is a finite partition of* S *then there are nonnegative numbers* $P_B^*(B_1), \cdots, P_B^*(B_n)$ *that sum to one, and there is a real function* u_B *on* X *such that, for all* $P_i, Q_i \varepsilon \mathcal{P}$, *if*

$$\boldsymbol{P} = P_i \text{ and } \boldsymbol{Q} = Q_i \text{ on } B_i; \qquad i = 1, \cdots, n$$

then

$$\boldsymbol{P} \preccurlyeq \boldsymbol{Q} \text{ if and only if } \textstyle\sum_{i=1}^n E(u_B, P_i)P_B^*(B_i) \leqq \sum_{i=1}^n E(u_B, Q_i)P_B^*(B_i). \tag{5}$$

Moreover, when A1–5 *hold and* P_B^* *and* u_B *have the stated properties then*:

(a) $P_B^*(B_i) = 0$ *if and only if* B_i *is null*;

(b) *Nonnegative numbers* $Q_B^*(B_1), \cdots, Q_B^*(B_n)$ *that sum to one and a real* *function* u_B' *on* X *satisfy* (5) *in place of* P_B^* *and* u_B *for all indicated* $\boldsymbol{P}$ *and* $\boldsymbol{Q}$ *if and only if* $Q_B^* = P_B^*$ *and* u_B' *is a positive linear transformation of* u_B.

Using this theorem we shall establish the following part of the Summary Theorem in the next section. With

$$\mathcal{H}_0 = \{\boldsymbol{P} \mid \boldsymbol{P} \in \mathcal{H} \text{ and } \boldsymbol{P} \text{ is constant on each event in some finite partition of } S\},$$

a horse lottery is in $\mathcal{H}_0$ only if it assigns no more than a finite number of $P \in \mathcal{P}$ to the states in S.

THEOREM 3. A1-5 *imply that there is a probability measure* P^* *on* S *and a real function* u *on* X *such that* (2) *holds for all* $\boldsymbol{P}, \boldsymbol{Q} \in \mathcal{H}_0$. *Moreover, when* A1-5 *hold and* P^* *and* u *satisfy* (2) *for all* $\boldsymbol{P}, \boldsymbol{Q} \in \mathcal{H}_0$, *then*:

(a) *For all* $A \subseteq S$, $P^*(A) = 0$ *if and only if event* A *is null;*

(b) *A probability measure* Q^* *on* S *and a real function* u' *on* X *satisfy* (2) *in place of* P^* *and* u *for all* $\boldsymbol{P}, \boldsymbol{Q} \in \mathcal{H}_0$ *if and only if* $Q^* = P^*$ *and* u' *is a positive linear transformation of* u.

Henceforth in this section u and P^* are assumed to satisfy the conditions of Theorem 3. Thus far we have not needed A6, which is required for the next result, to be proved in Section 6.

THEOREM 4. A1-6 *imply that* (2) *holds on the set of all bounded horse lotteries.*

We shall prove that all horse lotteries are bounded in two steps, the first of which is accomplished by

THEOREM 5. *If* A1-6 *hold and there is a denumerable partition of* S *such that each event in the partition has positive probability under* P^*, *then* u *on* X *is bounded.*

Clearly, if u on X is bounded then all horse lotteries are bounded. Theorem 5 will be proved in Section 7. To complete the second step of the proof that all horse lotteries are bounded, we consider first the following

PROPOSITION 1. *For each positive integer* n *there is an* n*-event partition of* S *such that each event in the partition has positive probability under* P^*

If Proposition 1 is true then there is a denumerable partition of S for which each event has positive probability. This is probably a well-known fact and we shall omit an explicit proof. The second step for horse-lottery boundedness and the final step in our proof of the Summary Theorem is noted in the following theorem, to be proved in Section 8.

THEOREM 6. *If* A1-6 *hold and if Proposition* 1 *is false then all horse lotteries are bounded.*

5. Proof of Theorem 3. Let A1–5 hold. By considering constant horse lotteries in $\mathcal{H}$, it follows from (5) that if $P, Q \in \mathcal{P}$ and if $\{B_1, \cdots, B_n\}$ and $\{C_1, \cdots, C_m\}$ are partitions of S then

$$E(u_B, P) \leqq E(u_B, Q) \text{ if and only if } E(u_C, P) \leqq E(u_C, Q).$$

Since $\mathcal{P}$ is a mixture set and A1, A2, and A3 imply that analogues of these axioms hold on $\mathcal{P}$ (or the set of constant horse lotteries), it follows from Theorem 1 that u_C on X is a positive linear transformation of u_B. Since a positive linear transformation of u_B in (5) will not affect the validity of (5), we can therefore delete the partition-specific subscript on u and have, in place of (5) when $\boldsymbol{P} = P_i$ and $\boldsymbol{Q} = Q_i$ on B_i,

(6) $\boldsymbol{P} \preccurlyeq \boldsymbol{Q}$ if and only if $\sum_{i=1}^{n} E(u, P_i)P_B^*(B_i) \leqq \sum_{i=1}^{n} E(u, Q_i)P_B^*(B_i)$.

For an event $A \subseteq S$ let

$$\mathcal{H}_A = \{\boldsymbol{P} \mid \boldsymbol{P} \in \mathcal{H} \text{ and } \boldsymbol{P} \text{ is constant on } A \text{ and on } A^c\}.$$

$\mathcal{H}_A$ is a mixture set. Suppose each of partitions $\{B_1, \cdots, B_n\}$ and $\{C_1, \cdots, C_m\}$ contains A. Then, if $\boldsymbol{P}$ in $\mathcal{H}_A$ equals P_A on A and P_{A^c} on A^c, (6) implies that, for all $\boldsymbol{P}, \boldsymbol{Q} \in \mathcal{H}_A$,

$$P_B^*(A)E(u, P_A) + [1 - P_B^*(A)]E(u, P_{A^c}) \leqq P_B^*(A)E(u, Q_A) + [1 - P_B^*(A)]E(u, Q_{A^c})$$

if and only if

$$P_C^*(A)E(u, P_A) + [1 - P_C^*(A)]E(u, P_{A^c}) \leqq P_C^*(A)E(u, Q_A) + [1 - P_C^*(A)]E(u, Q_{A^c}).$$

It then follows easily from Theorem 1 that $P_B^*(A) = P_C^*(A)$. Hence we can drop the partition-specific subscript on P^* and rewrite (6) as

(7) $\boldsymbol{P} \preccurlyeq \boldsymbol{Q}$ if and only if $\sum_{i=1}^{n} E(u, P_i)P^*(B_i) \leqq \sum_{i=1}^{n} E(u, Q_i)P^*(B_i)$.

It follows directly from Theorem 2 that P^* is uniquely determined, that $P^*(A) = 0$ only if A is null, and that u is unique up to a positive linear transformation. Finite additivity for P^* is easily demonstrated using partitions $\{A, B, (A \cup B)^c\}$ and $\{A \cup B, (A \cup B)^c\}$ in an analysis like that leading to (7) with $A \cap B = \varnothing$.

Finally, to obtain (2) for all $\boldsymbol{P}, \boldsymbol{Q} \in \mathcal{H}_0$, let $\boldsymbol{P} = P_i$ on B_i and $\boldsymbol{Q} = Q_j$ on C_j for the partitions $\{B_1, \cdots, B_n\}$ and $\{C_1, \cdots, C_m\}$. Applying (7) to the partition

$$\{B_i \cap C_j \mid i = 1, \cdots, n; j = 1, \cdots, m; B_i \cap C_j \neq \varnothing\}$$

we obtain

$$\boldsymbol{P} \preccurlyeq \boldsymbol{Q} \text{ if and only if } \textstyle\sum_i \sum_j E(u, P_i)P^*(B_i \cap C_j) \leqq \sum_i \sum_j E(u, Q_j)P^*(B_i \cap C_j)$$

which, by finite additivity for P^*, is the same as

$$\boldsymbol{P} \preccurlyeq \boldsymbol{Q} \text{ if and only if } \textstyle\sum_i E(u, P_i)P^*(B_i) \leqq \sum_j E(u, Q_j)P^*(C_j).$$

6. Proof of Theorem 4. Since $\mathcal{H}$ is a mixture set, Theorem 1 implies that there is a real function v on $\mathcal{H}$ such that, for all $\boldsymbol{P}, \boldsymbol{Q} \in \mathcal{H}$ and $\alpha \in [0, 1]$,

(8) $\boldsymbol{P} \preccurlyeq \boldsymbol{Q}$ if and only if $v(\boldsymbol{P}) \leqq v(\boldsymbol{Q})$

(9) $v(\alpha \boldsymbol{P} + (1 - \alpha)\boldsymbol{Q}) = \alpha v(\boldsymbol{P}) + (1 - \alpha)v(\boldsymbol{Q})$.

Expressions (8) and (9) hold also for all $\boldsymbol{P}$, $\boldsymbol{Q}$ in $\mathcal{H}_0$. Letting

$$w(\boldsymbol{P}) = E[E(u, \boldsymbol{P}(s)), P^*] \quad \text{for all } \boldsymbol{P} \in \mathcal{H}_0$$

it follows from Theorem 3 and

$$\begin{aligned} E[E(u, \alpha\boldsymbol{P}(s) + (1-\alpha)\boldsymbol{Q}(s)), P^*] &= E[\alpha E(u, \boldsymbol{P}(s)) + (1-\alpha)E(u, \boldsymbol{Q}(s)), P^*] \\ &= \alpha E[E(u, \boldsymbol{P}(s)), P^*] + (1-\alpha)E[E(u, \boldsymbol{Q}(s)), P^*] \end{aligned}$$

that, for all $\boldsymbol{P}, \boldsymbol{Q} \in \mathcal{H}_0$ and $\alpha \in [0, 1]$,

$$\boldsymbol{P} \preccurlyeq \boldsymbol{Q} \text{ if and only if } w(\boldsymbol{P}) \leqq w(\boldsymbol{Q})$$

$$w(\alpha\boldsymbol{P} + (1-\alpha)\boldsymbol{Q}) = \alpha w(\boldsymbol{P}) + (1-\alpha)w(\boldsymbol{Q}).$$

Then, by Theorem 1, w on $\mathcal{H}_0$ is a positive linear transformation of the restriction of v on $\mathcal{H}_0$. By an appropriate transformation we can, with no loss in generality, specify that

$$v(\boldsymbol{P}) = E[E(u, \boldsymbol{P}(s)), P^*] \tag{10}$$

for all $\boldsymbol{P} \in \mathcal{H}_0$. According to (8) the proof of Theorem 4 can be completed by proving that (10) holds for all bounded horse lotteries.

Our first step in this direction will be to prove that if $P^*(A) = 1$ and if c and d defined in the following expression are finite then

$$c = \inf\{E(u, \boldsymbol{P}(s)) \mid s \in A\} \leqq v(\boldsymbol{P}) \leqq \sup\{E(u, \boldsymbol{P}(s)) \mid s \in A\} = d. \tag{11}$$

Let $\boldsymbol{Q} = \boldsymbol{P}$ on A and $c \leqq \boldsymbol{Q}(s) \leqq d$ on A^c. Since A^c is null, $\boldsymbol{Q} \sim \boldsymbol{P}$ and hence $v(\boldsymbol{P}) = v(\boldsymbol{Q})$ by (8). To show that $c \leqq v(\boldsymbol{Q}) \leqq d$ when c and d are finite suppose to the contrary that $d < v(\boldsymbol{Q})$. With $c \leqq E(u, Q') \leqq d$ and $\boldsymbol{Q}' = Q'$ on S let $\boldsymbol{R} = \alpha\boldsymbol{Q} + (1-\alpha)\boldsymbol{Q}'$ with $\alpha < 1$ near enough to one so that

$$d < v(\boldsymbol{R}) = \alpha v(\boldsymbol{Q}) + (1-\alpha)v(\boldsymbol{Q}') < v(\boldsymbol{Q}).$$

Then $\boldsymbol{R} \prec \boldsymbol{Q}$ by (8). But since $E(u, \boldsymbol{Q}(s)) \leqq d < v(\boldsymbol{R})$ it follows from (8) that $\boldsymbol{Q}(s) \prec \boldsymbol{R}$ for all $s \in S$ so that axiom $A6$ implies $\boldsymbol{Q} \preccurlyeq \boldsymbol{R}$, a contradiction. Hence $d < v(\boldsymbol{Q})$ is false. By a symmetric proof, $v(\boldsymbol{Q}) < c$ is false. Therefore $c \leqq v(\boldsymbol{Q}) \leqq d$.

With $\boldsymbol{P}$ bounded let A with $P^*(A) = 1$ be an event on which $E(u, \boldsymbol{P}(s))$ is bounded and let c and d be defined as in (11). If $c = d$ then (10) is immediate. Henceforth assume that $c < d$. For notational convenience we shall take

$$c = 0, \qquad d = 1.$$

Let $\boldsymbol{Q}$ be as defined following (11) so that $v(\boldsymbol{Q}) = v(\boldsymbol{P})$ and

$$E[E(u, \boldsymbol{Q}(s)), P^*] = E[E(u, \boldsymbol{P}(s)), P^*].$$

To prove that

$$v(\boldsymbol{Q}) = E[E(u, \boldsymbol{Q}(s)), P^*]$$

let $\{A_1, \cdots, A_n\}$ be the partition (ignoring empty sets) of S defined by

$$A_1 = \{s \mid 0 \leqq E(u, \boldsymbol{Q}(s)) \leqq 1/n\}$$

$$A_i = \{s \mid (i-1)/n < E(u, \boldsymbol{Q}(s)) \leqq i/n\} \qquad i = 2, \cdots, n.$$

Let $P_i \varepsilon \mathcal{P}$ be such that

$$(12) \qquad (i-1)/n \leqq E(u, P_i) \leqq i/n \quad \text{for} \quad i = 1, \cdots, n.$$

The existence of such P_i is guaranteed by (11). Then let

$$(13) \qquad \boldsymbol{P}_i = \begin{cases} \boldsymbol{Q} & \text{on} \quad A_i \\ P_i & \text{on} \quad A_i^c \end{cases} \qquad i = 1, \cdots, n$$

$$(14) \qquad \boldsymbol{P}_0 = \sum_{i=1}^n n^{-1} \boldsymbol{P}_i$$

$$(15) \qquad \boldsymbol{R} = \sum_{j \neq i} (n-1)^{-1} P_j \quad \text{on} \quad A_i \quad \text{for} \quad i = 1, \cdots, n.$$

Since $\boldsymbol{P}_0(s) = \sum_i n^{-1} \boldsymbol{P}_i(s) = n^{-1}\boldsymbol{Q}(s) + (n-1)n^{-1} \sum_{j \neq i} (n-1)^{-1} P_j$ when $s \varepsilon A_i$,

$$\boldsymbol{P}_0 = n^{-1}\boldsymbol{Q} + (n-1)n^{-1}\boldsymbol{R},$$

so that, by (9) and (14),

$$(16) \qquad v(\boldsymbol{Q}) = \sum_{i=1}^n v(\boldsymbol{P}_i) - (n-1)v(\boldsymbol{R}).$$

Since $\boldsymbol{R} \varepsilon \mathcal{H}_0$, (10) implies that

$$v(\boldsymbol{R}) = \sum_{i=1}^n E(u, \sum_{j \neq i} (n-1)^{-1} P_j) P^*(A_i)$$
$$= (n-1)^{-1} \sum_{i=1}^n [\sum_{j \neq i} E(u, P_j)] P^*(A_i).$$

Substituting this in (16) we have

$$(17) \qquad v(\boldsymbol{Q}) = \sum_{i=1}^n v(\boldsymbol{P}_i) - \sum_{i=1}^n \sum_{j \neq i} E(u, P_j) P^*(A_i).$$

By (12), (13), and (11),

$$(18) \qquad (i-1)/n \leqq v(\boldsymbol{P}_i) \leqq i/n \qquad i = 1, \cdots, n.$$

This is true regardless of how the P_i are selected so long as they satisfy (12). In particular, since $0 = \inf \{E(u, \boldsymbol{Q}(s)) \mid s \varepsilon S\}$ and $1 = \sup \{E(u, \boldsymbol{Q}(s)) \mid s \varepsilon S\}$, we can select the P_i so that either

$$(19) \qquad E(u, P_1) = 1/n, \quad \text{and} \quad E(u, P_i) = (i-1)/n \quad \text{for} \quad i > 1$$

or

$$(20) \qquad E(u, P_i) = i/n \quad \text{for} \quad i < n, \quad \text{and} \quad E(u, P_n) = (n-1)/n.$$

Applying (19) and the left side of (18) to (17) we get

$$v(\boldsymbol{Q}) \geqq \sum_{i=1}^n (i-1)/n - \tfrac{1}{2}(n-1)P^*(A_1)$$
$$- \sum_{i=2}^n [\tfrac{1}{2}(n-1) - (i-1)/n + 1/n] P^*(A_i)$$
$$= \tfrac{1}{2}(n-1) - \tfrac{1}{2}(n-1) + \sum_{i=2}^n (i-1)n^{-1}P^*(A_i) - n^{-1}[1 - P^*(A_1)]$$
$$\geqq \sum_{i=1}^n (i-1)n^{-1}P^*(A_i) - 1/n.$$

Expressions (8) and (9) hold also for all $\boldsymbol{P}$, $\boldsymbol{Q}$ in $\mathcal{H}_0$. Letting

$$w(\boldsymbol{P}) = E[E(u, \boldsymbol{P}(s)), P^*] \quad \text{for all } \boldsymbol{P} \in \mathcal{H}_0$$

it follows from Theorem 3 and

$$\begin{aligned} E[E(u, \alpha\boldsymbol{P}(s) + (1-\alpha)\boldsymbol{Q}(s)), P^*] \\ &= E[\alpha E(u, \boldsymbol{P}(s)) + (1-\alpha)E(u, \boldsymbol{Q}(s)), P^*] \\ &= \alpha E[E(u, \boldsymbol{P}(s)), P^*] + (1-\alpha)E[E(u, \boldsymbol{Q}(s)), P^*] \end{aligned}$$

that, for all $\boldsymbol{P}, \boldsymbol{Q} \in \mathcal{H}_0$ and $\alpha \in [0, 1]$,

$$\boldsymbol{P} \preccurlyeq \boldsymbol{Q} \text{ if and only if } w(\boldsymbol{P}) \leqq w(\boldsymbol{Q})$$

$$w(\alpha\boldsymbol{P} + (1-\alpha)\boldsymbol{Q}) = \alpha w(\boldsymbol{P}) + (1-\alpha)w(\boldsymbol{Q}).$$

Then, by Theorem 1, w on $\mathcal{H}_0$ is a positive linear transformation of the restriction of v on $\mathcal{H}_0$. By an appropriate transformation we can, with no loss in generality, specify that

$$v(\boldsymbol{P}) = E[E(u, \boldsymbol{P}(s)), P^*] \tag{10}$$

for all $\boldsymbol{P} \in \mathcal{H}_0$. According to (8) the proof of Theorem 4 can be completed by proving that (10) holds for all bounded horse lotteries.

Our first step in this direction will be to prove that if $P^*(A) = 1$ and if c and d defined in the following expression are finite then

$$c = \inf\{E(u, \boldsymbol{P}(s)) \mid s \in A\} \leqq v(\boldsymbol{P}) \leqq \sup\{E(u, \boldsymbol{P}(s)) \mid s \in A\} = d. \tag{11}$$

Let $\boldsymbol{Q} = \boldsymbol{P}$ on A and $c \leqq \boldsymbol{Q}(s) \leqq d$ on A^c. Since A^c is null, $\boldsymbol{Q} \sim \boldsymbol{P}$ and hence $v(\boldsymbol{P}) = v(\boldsymbol{Q})$ by (8). To show that $c \leqq v(\boldsymbol{Q}) \leqq d$ when c and d are finite suppose to the contrary that $d < v(\boldsymbol{Q})$. With $c \leqq E(u, Q') \leqq d$ and $\boldsymbol{Q}' = Q'$ on S let $\boldsymbol{R} = \alpha\boldsymbol{Q} + (1-\alpha)\boldsymbol{Q}'$ with $\alpha < 1$ near enough to one so that

$$d < v(\boldsymbol{R}) = \alpha v(\boldsymbol{Q}) + (1-\alpha)v(\boldsymbol{Q}') < v(\boldsymbol{Q}).$$

Then $\boldsymbol{R} \prec \boldsymbol{Q}$ by (8). But since $E(u, \boldsymbol{Q}(s)) \leqq d < v(\boldsymbol{R})$ it follows from (8) that $\boldsymbol{Q}(s) \prec \boldsymbol{R}$ for all $s \in S$ so that axiom $A6$ implies $\boldsymbol{Q} \preccurlyeq \boldsymbol{R}$, a contradiction. Hence $d < v(\boldsymbol{Q})$ is false. By a symmetric proof, $v(\boldsymbol{Q}) < c$ is false. Therefore $c \leqq v(\boldsymbol{Q}) \leqq d$.

With $\boldsymbol{P}$ bounded let A with $P^*(A) = 1$ be an event on which $E(u, \boldsymbol{P}(s))$ is bounded and let c and d be defined as in (11). If $c = d$ then (10) is immediate. Henceforth assume that $c < d$. For notational convenience we shall take

$$c = 0, \qquad d = 1.$$

Let $\boldsymbol{Q}$ be as defined following (11) so that $v(\boldsymbol{Q}) = v(\boldsymbol{P})$ and

$$E[E(u, \boldsymbol{Q}(s)), P^*] = E[E(u, \boldsymbol{P}(s)), P^*].$$

To prove that

$$v(\boldsymbol{Q}) = E[E(u, \boldsymbol{Q}(s)), P^*]$$

let $\{A_1, \cdots, A_n\}$ be the partition (ignoring empty sets) of S defined by

$$A_1 = \{s \mid 0 \leqq E(u, \boldsymbol{Q}(s)) \leqq 1/n\}$$

$$A_i = \{s \mid (i-1)/n < E(u, \boldsymbol{Q}(s)) \leqq i/n\} \qquad i = 2, \cdots, n.$$

Let $P_i \, \varepsilon \, \mathcal{P}$ be such that

$$(12) \qquad (i-1)/n \leqq E(u, P_i) \leqq i/n \quad \text{for} \quad i = 1, \cdots, n.$$

The existence of such P_i is guaranteed by (11). Then let

$$(13) \qquad \boldsymbol{P}_i = \begin{cases} \boldsymbol{Q} & \text{on } A_i \\ P_i & \text{on } A_i^c \end{cases} \qquad i = 1, \cdots, n$$

$$(14) \qquad \boldsymbol{P}_0 = \textstyle\sum_{i=1}^{n} n^{-1}\boldsymbol{P}_i$$

$$(15) \qquad \boldsymbol{R} = \textstyle\sum_{j \neq i} (n-1)^{-1} P_j \quad \text{on} \quad A_i \quad \text{for} \quad i = 1, \cdots, n.$$

Since $\boldsymbol{P}_0(s) = \sum_i n^{-1}\boldsymbol{P}_i(s) = n^{-1}\boldsymbol{Q}(s) + (n-1)n^{-1} \sum_{j \neq i} (n-1)^{-1}P_j$ when $s \, \varepsilon \, A_i$,

$$\boldsymbol{P}_0 = n^{-1}\boldsymbol{Q} + (n-1)n^{-1}\boldsymbol{R},$$

so that, by (9) and (14),

$$(16) \qquad v(\boldsymbol{Q}) = \textstyle\sum_{i=1}^{n} v(\boldsymbol{P}_i) - (n-1)v(\boldsymbol{R}).$$

Since $\boldsymbol{R} \, \varepsilon \, \mathcal{H}_0$, (10) implies that

$$\begin{aligned} v(\boldsymbol{R}) &= \textstyle\sum_{i=1}^{n} E(u, \sum_{j \neq i} (n-1)^{-1}P_j)P^*(A_i) \\ &= (n-1)^{-1} \textstyle\sum_{i=1}^{n} [\sum_{j \neq i} E(u, P_j)]P^*(A_i). \end{aligned}$$

Substituting this in (16) we have

$$(17) \qquad v(\boldsymbol{Q}) = \textstyle\sum_{i=1}^{n} v(\boldsymbol{P}_i) - \sum_{i=1}^{n} \sum_{j \neq i} E(u, P_j)P^*(A_i).$$

By (12), (13), and (11),

$$(18) \qquad (i-1)/n \leqq v(\boldsymbol{P}_i) \leqq i/n \qquad i = 1, \cdots, n.$$

This is true regardless of how the P_i are selected so long as they satisfy (12). In particular, since $0 = \inf \{E(u, \boldsymbol{Q}(s)) \mid s \, \varepsilon \, S\}$ and $1 = \sup \{E(u, \boldsymbol{Q}(s)) \mid s \, \varepsilon \, S\}$, we can select the P_i so that either

$$(19) \quad E(u, P_1) = 1/n, \quad \text{and} \quad E(u, P_i) = (i-1)/n \quad \text{for} \quad i > 1$$

or

$$(20) \quad E(u, P_i) = i/n \quad \text{for} \quad i < n, \quad \text{and} \quad E(u, P_n) = (n-1)/n.$$

Applying (19) and the left side of (18) to (17) we get

$$\begin{aligned} v(\boldsymbol{Q}) &\geqq \textstyle\sum_{i=1}^{n} (i-1)/n - \frac{1}{2}(n-1)P^*(A_1) \\ &\quad - \textstyle\sum_{i=2}^{n} [\frac{1}{2}(n-1) - (i-1)/n + 1/n]P^*(A_i) \\ &= \tfrac{1}{2}(n-1) - \tfrac{1}{2}(n-1) + \textstyle\sum_{i=2}^{n} (i-1)n^{-1}P^*(A_i) - n^{-1}[1 - P^*(A_1)] \\ &\geqq \textstyle\sum_{i=1}^{n} (i-1)n^{-1}P^*(A_i) - 1/n. \end{aligned}$$

Applying (20) and the right side of (18) to (17) we get

$$v(Q) \leqq \sum_{i=1}^{n} i/n - \sum_{i=1}^{n-1} [\tfrac{1}{2}(n-1) - i/n + (n-1)/n]P^*(A_i)$$
$$- \tfrac{1}{2}(n-1)P^*(A_n)$$
$$= \sum_{i=1}^{n} in^{-1}P^*(A_i) + n^{-1}[1 - P^*(A_n)]$$
$$\leqq \sum_{i=1}^{n} in^{-1}P^*(A_i) + 1/n.$$

By the definition of E, $\sum (i-1)n^{-1}P^*(A_i) \leqq E[E(u, \boldsymbol{Q}(s)), P^*] \leqq \sum in^{-1}P^*(A_i)$, so that

$$|v(\boldsymbol{Q}) - E[E(u, \boldsymbol{Q}(s)), P^*]| \leqq 2/n \quad \text{for} \quad n = 1, 2, \cdots.$$

Therefore, $v(\boldsymbol{Q}) = E[E(u, \boldsymbol{Q}(s)), P^*]$.

7. Proof of Theorem 5. Let A1–6 hold and let $\bar{A}$ be a denumerable partition of S wth $P^*(A) > 0$ for all $A \varepsilon \bar{A}$. $\{P^*(A) | A \varepsilon \bar{A}\}$ must have a largest element, say $P^*(A_1)$. Then $\{P^*(A) | A \varepsilon \bar{A} - \{A_1\}\}$ must have a largest element, say $P^*(A_2)$. Continuing this process we get a sequence

$$A_1, A_2, \cdots \quad \text{with} \quad \{A_1, A_2, \cdots\} = \bar{A}$$

and

$$(21) \quad P^*(A_1) \geqq P^*(A_2) \geqq P^*(A_3) \geqq \cdots; \qquad P^*(A_i) > 0 \quad \text{for all} \quad i.$$

For definiteness suppose that u is unbounded above. By a linear transformation of u we can assume that $[0, \infty) \subseteq \{E(u, P) | P \varepsilon \mathcal{P}\}$. Let $P_i \varepsilon \mathcal{P}$ and $\boldsymbol{P} \varepsilon \mathcal{H}$ be such that

$$(22) \qquad E(u, P_i) = P^*(A_i)^{-1} \qquad i = 1, 2, \cdots$$
$$\boldsymbol{P} = P_i \quad \text{on} \quad A_i \qquad i = 1, 2, \cdots.$$

Also let $\boldsymbol{Q}_n \varepsilon \mathcal{H}$ be constant on each A_i $(i \leqq n)$ and constant on $\bigcup_{i=n+1}^{\infty} A_i$ with

$$(23) \quad E(u, \boldsymbol{Q}_n(s)) = P^*(A_n)^{-1} - P^*(A_i)^{-1} \qquad s \varepsilon A_i; \qquad i = 1, \cdots, n$$
$$E(u, \boldsymbol{Q}_n(s)) = 0 \qquad s \varepsilon \bigcup_{i=n+1}^{\infty} A_i.$$

Letting v on $\mathcal{H}$ be as given in (8) and (9) and satisfying (10) on $\mathcal{H}_0$,

$$(24) \qquad v(\boldsymbol{Q}_n) = \sum_{i=1}^{n} [P^*(A_n)^{-1} - P^*(A_i)^{-1}]P^*(A_i)$$
$$= P^*(A_n)^{-1} \sum_{i=1}^{n} P^*(A_i) - n \quad \text{for} \quad n = 1, 2, \cdots$$

since $\boldsymbol{Q}_n \varepsilon \mathcal{H}_0$ for each n.

By (22) and (23),

$$E(u, \tfrac{1}{2}\boldsymbol{P}(s) + \tfrac{1}{2}\boldsymbol{Q}_n(s)) = \tfrac{1}{2}P^*(A_i)^{-1} + \tfrac{1}{2}[P^*(A_n)^{-1} - P^*(A_i)^{-1}]$$
$$= \tfrac{1}{2}P^*(A_n)^{-1} \quad \text{for all} \quad s \varepsilon \bigcup_{1}^{n} A_i$$

and, by (21) and (23),

$$E(u, \tfrac{1}{2}\boldsymbol{P}(s) + \tfrac{1}{2}\boldsymbol{Q}_n(s)) \geqq \tfrac{1}{2}P^*(A_n)^{-1} \quad \text{for all} \quad s \varepsilon \bigcup_{n+1}^{\infty} A_i.$$

Therefore

$$\tfrac{1}{2}P^*(A_n)^{-1} = \inf\{E(u, \tfrac{1}{2}\boldsymbol{P}(s) + \tfrac{1}{2}\boldsymbol{Q}_n(s)) \mid s \,\varepsilon\, S\},$$

and hence, by (11), $v(\tfrac{1}{2}\boldsymbol{P} + \tfrac{1}{2}\boldsymbol{Q}_n) \geqq \tfrac{1}{2}P^*(A_n)^{-1}$, which, using (9) and (24) implies that

$$v(\boldsymbol{P}) \geqq P^*(A_n)^{-1} - P^*(A_n)^{-1} \textstyle\sum_{i=1}^{n} P^*(A_i) + n$$

$$\geqq n \quad \text{for} \quad n = 1, 2, \cdots.$$

Since this requires $v(\boldsymbol{P})$ to be infinite we have obtained a contradiction and conclude that u is not unbounded above. By a symmetric proof, u is not unbounded below. Hence, the hypotheses of Theorem 5 imply that u is bounded.

8. Proof of Theorem 6. Let A1–6 hold and assume that Proposition 1 is false. Then there must be a (unique) positive m for which there is an m-event partition of S that has positive probability for each event and such that every partition of S has at most m events that have positive probability.

For convenience assume that $u(y) = 0$ for a $y \,\varepsilon\, X$. Suppose then that $\boldsymbol{Q}$ in $\mathcal{H}$ is unbounded. For definiteness, assume that $\boldsymbol{Q}$ is unbounded above. Let $\boldsymbol{P}$ be obtained from $\boldsymbol{Q}$ by replacing each x for which $\boldsymbol{Q}(s)(x) > 0$ and $u(x) < 0$ by y with $u(y) = 0$, for all $s \,\varepsilon\, S$. Then $E(u, \boldsymbol{P}(s)) \geqq 0$ for all $s \,\varepsilon\, S$ and $\boldsymbol{P}$ is unbounded above so that, for every positive integer n,

$$P^*\{E(u, \boldsymbol{P}(s)) \geqq n\} > 0.$$

Because no partition has more than m events with positive probability, $P^*\{E(u, \boldsymbol{P}(s)) \geqq n\}$ can change no more than m times as n increases. Hence, there is a positive integer N and an $\alpha > 0$ such that

$$P^*\{E(u, \boldsymbol{P}(s)) \geqq n\} = \alpha \quad \text{for all} \quad n \geqq N. \tag{25}$$

Let

$$E(u, P_i) = i \quad \text{for} \quad i = 1, 2, \cdots$$

and

$$\boldsymbol{Q}_n = \begin{cases} \boldsymbol{P} & \text{on} \quad \{s \mid E(u, \boldsymbol{P}(s)) \geqq n\} \\ P_n & \text{on} \quad \{s \mid E(u, \boldsymbol{P}(s)) < n\} \end{cases}$$

$$\boldsymbol{R}_n = \begin{cases} P_n & \text{on} \quad \{s \mid E(u, \boldsymbol{P}(s)) \geqq n\} \\ \boldsymbol{P} & \text{on} \quad \{s \mid E(u, \boldsymbol{P}(s)) < n\}. \end{cases}$$

Then, with $\boldsymbol{P}_n = P_n$ on S, $\tfrac{1}{2}\boldsymbol{P} + \tfrac{1}{2}\boldsymbol{P}_n = \tfrac{1}{2}\boldsymbol{Q}_n + \tfrac{1}{2}\boldsymbol{R}_n$, so that with v on $\mathcal{H}$ as defined by (8) and (9) and satisfying (10) for all bounded horse lotteries,

$$v(\boldsymbol{P}) + n = v(\boldsymbol{Q}_n) + v(\boldsymbol{R}_n) \qquad n = 1, 2, \cdots. \tag{26}$$

Since $\boldsymbol{R}_n$ is bounded, its definition, (10), and (25) imply

$$v(\boldsymbol{R}_n) = E[E(u, \boldsymbol{R}_n(s)), P^*] \geqq n\alpha \quad \text{for all} \quad n \geqq N.$$

Since $P_{n-1} \prec Q_n(s)$ for all $s \, \varepsilon \, S$, $A6$ implies that $P_{n-1} \preccurlyeq Q_n$ so that

$$v(Q_n) \geqq n - 1 \quad \text{for all} \quad n.$$

These inequalities and (26) yield

$$v(P) \geqq n\alpha - 1 \quad \text{for all} \quad n \geqq N,$$

which requires $v(P)$ to be infinite, a contradiction. Therefore Q cannot be unbounded above. A symmetric proof shows that Q cannot be unbounded below. Hence, when $A1$–6 hold and Proposition 1 is false, every horse lottery is bounded.

REFERENCES

[1] ANSCOMBE, F. J. and AUMANN, R. J. (1963). A definition of subjective probability. *Ann. Math. Statist.* **34** 199–205.
[2] ARROW, K. J. (1966). Exposition of a theory of choice under uncertainty. *Synthese* **16** 253–269.
[3] CHERNOFF, H. (1954). Rational selection of decision functions. *Econometrica* **22** 422–443.
[4] DAVIDSON, D. and SUPPES, P. (1956). A finitistic axiomatization of subjective probability and utility. *Econometrica* **24** 264–275.
[5] FISHBURN, P. C. (1967). Preference-based definitions of subjective probability. *Ann. Math. Statist.* **38** 1605–1617.
[6] FRIEDMAN, M. and SAVAGE, L. J. (1952). The expected-utility hypothesis and the measurability of utility. *J. Political Economy* **60** 463–474.
[7] HERSTEIN, I. N. and MILNOR, J. (1953). An axiomatic approach to measurable utility. *Econometrica* **21** 291–297.
[8] JENSEN, N. E.(1967). An introduction to Bernoullian utility theory. I. Utility functions. *Swedish J. Economics* **69** 163–183.
[9] LUCE, R. D. and RAIFFA, H. (1957). *Games and Decisions.* Wiley. New York.
[10] LUCE, R. D. and SUPPES, P. (1965). Preference, utility, and subjective probability. *Handbook of Mathematical Psychology.* **3** 249–410. Wiley, New York.
[11] PRATT, J. W., RAIFFA, H. and SCHLAIFER, R. (1964). The foundations of decision under uncertainty: an elementary exposition. *J. Amer. Statist. Assoc.* **59** 353–375.
[12] RAIFFA, H. (1961). Risk, ambiguity, and the Savage axioms: comment. *Quarterly J. Economics* **75** 690–694.
[13] RAMSEY, F. P. (1931). *The Foundations of Mathematics and Other Logical Essays.* Harcourt, Brace, and Co., New York. [Reprinted in H. E. Kyburg and H. E. Smokler (Eds.), *Studies in Subjective Probability.* Wiley, New York, 1964.]
[14] RUBIN, H. (1949). Postulates for the existence of measurable utility and psychological probability (abstract). *Bull. Amer. Math. Soc.* **55** 1050–1051.
[15] SAVAGE, L. J. (1954). *The Foundations of Statistics.* Wiley, New York.
[16] SUPPES, P. (1956). The role of subjective probability and utility in decision making. *Proc. Third Berkeley Symp. Math. Statist. Prob.* **5** 61–73. Univ. of California Press.

1.   EXPECTED UTILITY THEORY   21

2. CONVEXITY AND THE KUHN-TUCKER CONDITIONS

J.SIAM Control
Ser. A, Vol. 3, No. 2
Printed in U.S.A., 1965

PSEUDO-CONVEX FUNCTIONS*

O. L. MANGASARIAN†

Abstract. The purpose of this work is to introduce pseudo-convex functions and to describe some of their properties and applications. The class of all pseudo-convex functions over a convex set C includes the class of all differentiable convex functions on C and is included in the class of all differentiable quasi-convex functions on C. An interesting property of pseudo-convex functions is that a local condition, such as the vanishing of the gradient, is a global optimality condition. One of the main results of this work consists of showing that the Kuhn-Tucker differential conditions are sufficient for optimality when the objective function is pseudo-convex and the constraints are quasi-convex. Other results of this work are a strict converse duality theorem for mathematical programming and a stability criterion for ordinary differential equations.

1. Introduction. Throughout this work, we shall be concerned with the real, scalar, single-valued, differentiable function $\theta(x)$ defined on the non-empty open set D in the m-dimensional Euclidean space E^m. We let C be a subset of D and let ∇_x denote the $m \times 1$ partial differential operator

$$\nabla_x = \left[\frac{\partial}{\partial x_1}, \cdots, \frac{\partial}{\partial x_m}\right]',$$

where the prime denotes the transpose. We say that $\theta(x)$ is *pseudo-convex* on C if for every x^1 and x^2 in C,

$$(1.1) \qquad (x^2 - x^1)'\nabla_x\theta(x^1) \geqq 0 \quad \text{implies} \quad \theta(x^2) \geqq \theta(x^1).$$

We say that $\theta(x)$ is *pseudo-concave* on C if for every x^1 and x^2 in C,

$$(1.2) \qquad (x^2 - x^1)'\nabla_x\theta(x^1) \leqq 0 \quad \text{implies} \quad \theta(x^2) \leqq \theta(x^1).$$

Thus $\theta(x)$ is pseudo-concave if and only if $-\theta(x)$ is pseudo-convex. In the subsequent paragraphs we shall confine our remarks to pseudo-convex functions. Analogous results hold for pseudo-concave functions by the appropriate multiplication by -1.

We shall relate the pseudo-convexity concept to the previously established notions of convexity, quasi-convexity [1], [2] and strict quasi-convexity [3], [5].

The function $\theta(x)$ is said to be *convex* on C, [2], if C is convex and if for every x^1 and x^2 in C,

$$(1.3) \qquad \theta(\lambda x^1 + (1 - \lambda)x^2) \leqq \lambda\theta(x^1) + (1 - \lambda)\theta(x^2)$$

* Received by the editors March 4, 1965.
† Shell Development Company, Emeryville, California.

for every λ such that $0 \leqq \lambda \leqq 1$. Equivalently, $\theta(x)$ is convex on C if

$$\theta(x^2) - \theta(x^1) \geqq (x^2 - x^1)'\nabla_x\theta(x^1) \tag{1.4}$$

for every x^1 and x^2 in C.

The function $\theta(x)$ is said to be *quasi-convex* on C, [1], [2], if C is convex and if for every x^1 and x^2 in C,

$$\theta(x^2) \leqq \theta(x^1) \quad \text{implies} \quad \theta(\lambda x^1 + (1 - \lambda)x^2) \leqq \theta(x^1) \tag{1.5}$$

for every λ such that $0 \leqq \lambda \leqq 1$. Equivalently, $\theta(x)$ is quasi-convex on C if

$$\theta(x^2) \leqq \theta(x^1) \quad \text{implies} \quad (x^2 - x^1)'\nabla_x\theta(x^1) \leqq 0. \tag{1.6}$$

The function $\theta(x)$ is said to be *strictly quasi-convex* on C, [3], [5], if C is convex and if for every x^1 and x^2 in C, $x^1 \neq x^2$,

$$\theta(x^2) < \theta(x^1) \quad \text{implies} \quad \theta(\lambda x^1 + (1 - \lambda)x^2) < \theta(x^1) \tag{1.7}$$

for every λ such that $0 < \lambda < 1$. It has been shown [5] that every lower semicontinuous strictly quasi-convex function is quasi-convex but not conversely.

In the next section we shall give some properties of pseudo-convex functions and show how these properties can be used to generalize some previous results of mathematical programming, duality theory and stability theory of ordinary differential equations. Theorem 1 generalizes the Arrow-Enthoven version [1, Theorem 1] of the Kuhn-Tucker differential sufficient optimality conditions for a mathematical programming problem. Theorem 2 gives a generalization of Huard's converse duality theorem of mathematical programming [4, Theorem 2] and Theorem 3 generalizes a stability criterion for equilibrium points of nonlinear ordinary differential equations [8, Theorem 1].

2. Properties of pseudo-convex functions and applications. In this section we shall give some properties of pseudo-convex functions and some extensions of the results of mathematical programming and ordinary differential equations.

PROPERTY 0. *Let $\theta(x)$ be pseudo-convex on C. If $\nabla_x\theta(x^0) = 0$, then x^0 is a global minimum over C.*

Proof. For any x in C,

$$(x - x^0)'\nabla_x\theta(x^0) = 0,$$

and hence by (1.1),

$$\theta(x) \geqq \theta(x^0),$$

which establishes the property.

PROPERTY 1. *Let C be convex. If $\theta(x)$ is convex on C, then $\theta(x)$ is pseudo-convex in C, but not conversely.*

Proof. If $\theta(x)$ is convex on C, then by (1.4),

$$(x^2 - x^1)'\nabla_x\theta(x^1) \geqq 0 \qquad \text{implies} \qquad \theta(x^2) \geqq \theta(x^1),$$

which is precisely (1.1). That the converse is not necessarily true can be seen from the example

$$\theta(x) \equiv x + x^3, \qquad x \in E^1,$$

which is pseudo-convex on E^1 but not convex.[1]

PROPERTY 2. *Let C be convex. If $\theta(x)$ is pseudo-convex on C, then $\theta(x)$ is strictly quasi-convex (and hence quasi-convex) on C, but not conversely.*

Proof. Let $\theta(x)$ be pseudo-convex on C. We shall assume that $\theta(x)$ is not strictly quasi-convex on C and show that this leads to a contradiction. If $\theta(x)$ is not strictly quasi-convex on C, then it follows from (1.7) that there exist $x^1 \neq x^2$ in C such that

$$\theta(x^2) < \theta(x^1), \tag{2.1}$$

and

$$\theta(x) \geqq \theta(x^1), \tag{2.2}$$

for some $x \in L$, where

$$L = \{x \mid x = \lambda x^1 + (1 - \lambda)x^2, 0 < \lambda < 1\}. \tag{2.3}$$

Hence there exists an $\bar{x} \in L$ such that

$$\theta(\bar{x}) = \max_{x \in \bar{L}} \theta(x), \tag{2.4}$$

where

$$\bar{L} = L \cup \{x^1, x^2\}. \tag{2.5}$$

Now define

$$f(\lambda) = \theta((1 - \lambda)x^1 + \lambda x^2), \qquad 0 \leqq \lambda \leqq 1. \tag{2.6}$$

Hence

$$\theta(\bar{x}) = f(\bar{\lambda}), \tag{2.7}$$

where

$$\bar{x} = (1 - \bar{\lambda})x^1 + \bar{\lambda}x^2, \qquad 0 < \bar{\lambda} < 1. \tag{2.8}$$

[1] To see that $x + x^3$ is pseudo-convex, note that $\nabla_x\theta(x) = 1 + 3x^2 > 0$. Hence $(x - x^0)'\nabla_x\theta(x^0) \geqq 0$ implies that $x \geqq x^0$ and $x^3 \geqq (x^0)^3$, and thus

$$\theta(x) - \theta(x^0) = (x + x^3) - (x^0 + (x^0)^3) \geqq 0.$$

We have from (2.4) through (2.7) that $f(\lambda)$ achieves its maximum at $\bar{\lambda}$. Hence it follows by the differentiability of $\theta(x)$ and the chain rule that

$$(2.9) \qquad (x^2 - x^1)'\nabla_x \theta(\bar{x}) = \frac{df(\bar{\lambda})}{d\lambda} = 0.$$

Since

$$(2.10) \qquad x^2 - \bar{x} = x^2 - (1 - \bar{\lambda})x^1 - \bar{\lambda}x^2 = (1 - \bar{\lambda})(x^2 - x^1),$$

it follows from (2.9) and (2.10) and the fact that $\bar{\lambda} < 1$, that

$$(2.11) \qquad (x^2 - \bar{x})'\nabla_x\theta(\bar{x}) = 0.$$

But by the pseudo-convexity of $\theta(x)$, (2.11) implies that

$$(2.12) \qquad \theta(x^2) \geqq \theta(\bar{x}).$$

Hence from (2.1) and (2.12),

$$\theta(x^1) > \theta(\bar{x}),$$

which contradicts (2.4). Hence $\theta(x)$ must be strictly quasi-convex on C.

That the converse is not necessarily true can be seen from the example

$$\theta(x) \equiv x^3, \qquad x \in E^1,$$

which is strictly quasi-convex on E^1, but not pseudo-convex.

PROPERTY 3. *Let C be convex. If $\theta(x)$ is pseudo-convex on C, then every local minimum*[2] *is a global minimum.*

Proof. By Property 2, $\theta(x)$ is strictly quasi-convex on C. Now if $\bar{x}$ is a local minimum, then

$$(2.13) \qquad \theta(\bar{x}) \leqq \theta(x) \qquad \text{for every} \quad x \in N(\bar{x}) \cap C,$$

where $N(\bar{x})$ is some neighborhood of $\bar{x}$. Let x be any point in C, but not in $N(\bar{x}) \cap C$. Then there exists a $\bar{\lambda}$, $0 < \bar{\lambda} < 1$, such that

$$\tilde{x} = ((1 - \bar{\lambda})\bar{x} + \bar{\lambda}x) \in N(\bar{x}) \cap C.$$

Now if $\theta(x) < \theta(\bar{x})$, then by the strict quasi-convexity of $\theta(x)$,

$$\theta(\bar{x}) > \theta(\tilde{x}),$$

which contradicts (2.13). Hence $\theta(x) \geqq \theta(\bar{x})$, which proves Property 3.

THEOREM 1. *Let $\theta(x)$, $g_1(x)$, $\cdots$, $g_n(x)$ be differentiable functions on E^m. Let C be a convex set in E^m and $\theta(x)$ be pseudo-convex on C and $g_1(x)$, $\cdots$, $g_n(x)$ be quasi-convex on C. If there exist an $x^0 \in C$ and $y^0 \in E^n$ satisfy-*

[2] A local minimum is an $\bar{x} \in C$ such that $\theta(\bar{x}) \leqq \theta(x)$ for all $x \in N(\bar{x}) \cap C$, where $N(\bar{x})$ is some neighborhood of $\bar{x}$.

ing the Kuhn-Tucker differential conditions [7], *namely,*

(2.14) $$\nabla_x\theta(x^0) + \nabla_x\sum_{i=1}^{n} y_i^0 g_i(x^0) = 0,$$

(2.15) $$\sum_{i=1}^{n} y_i^0 g_i(x^0) = 0,$$

(2.16) $$g_i(x^0) \leqq 0, \qquad i = 1, \cdots, n,$$

(2.17) $$y_i^0 \geqq 0, \qquad i = 1, \cdots, n,$$

then

(2.18) $$\theta(x^0) = \min_{x \in C} \{\theta(x) \mid g_i(x) \leqq 0, i = 1, \cdots, n\}.$$

Proof. The proof is similar to part of the proof of [1, Theorem 1]. Let

$$I = \{i \mid g_i(x^0) < 0\}.$$

Hence $g_i(x^0) = 0$ for $i \notin I$. From (2.15), (2.16) and (2.17) it follows that

(2.19) $$y_i^0 = 0 \qquad \text{for} \qquad i \in I.$$

Let

$$R = \{x \mid g_i(x) \leqq 0, i = 1, 2, \cdots, n, x \in C\}.$$

Then $g_i(x) \leqq g_i(x^0)$ for $i \notin I$, $x \in R$. Hence by the quasi-convexity of the g_i's on R it follows from (1.6) that

(2.20) $$(x - x^0)'\nabla_x g_i(x^0) \leqq 0 \qquad \text{for} \qquad i \notin I, x \in R.$$

Hence by (2.20) and (2.17) we have that

(2.21) $$(x - x^0)'\nabla_x \sum_{i \notin I} y_i^0 g_i(x^0) \leqq 0 \qquad \text{for} \qquad x \in R,$$

and from (2.19) we have

(2.22) $$(x - x^0)'\nabla_x \sum_{i \in I} y_i^0 g_i(x^0) = 0 \qquad \text{for} \qquad x \in R.$$

Hence (2.21) and (2.22) imply

$$(x - x^0)'\nabla_x \sum_{i=1}^{n} y_i^0 g_i(x^0) \leqq 0 \qquad \text{for} \qquad x \in R,$$

which in turn implies, by (2.14), that

(2.23) $$(x - x^0)'\nabla_x\theta(x^0) \geqq 0 \qquad \text{for} \qquad x \in R.$$

But by the pseudo-convexity of $\theta(x)$ on R, (2.23) implies that

$$\theta(x) \geqq \theta(x^0) \qquad \text{for} \qquad x \in R.$$

For the case when the set I is empty, the above proof is modified by deleting (2.19), (2.22) and references thereto. For the case when $I = \{1, 2, \cdots, n\}$ the above proof is modified by deleting that part of the proof *between* (2.19) and (2.22) and references thereto.

It should be noted here that the above theorem is indeed a generalization of Arrow and Enthoven's result [1, Theorem 1]. Every case covered there is covered by the above theorem, but not conversely. An example of a case not covered by Arrow and Enthoven is the following one:

$$\min_{x \in E^1} \{-e^{-x^2} \mid -x \leqq 0\}.$$

Another application of pseudo-convex functions may be found in duality theory. Consider the primal problem

$$\text{(PP)} \qquad \min_{x \in E^m} \{\theta(x) \mid g(x) \leqq 0\},$$

where $\theta(x)$ is a scalar function on E^m and $g(x)$ is an $n \times 1$ vector function on E^m. For the above problem Wolfe [10] has defined the dual problem as

$$\text{(DP)} \qquad \max_{x \in E^m, y \in E^n} \{\psi(x, y) \mid \nabla_x \psi(x, y) = 0, y \geqq 0\},$$

where

$$\psi(x, y) \equiv \theta(x) + y' g(x).$$

Under appropriate conditions Wolfe has shown [10, Theorem 2] that if x^0 solves (PP), then x^0 and some y^0 solve (DP). Conversely, under somewhat stronger conditions, Huard [4, Theorem 2] showed that if (x^0, y^0) solves (DP), then x^0 solves (PP). Both Wolfe and Huard required, among other things, that $\theta(x)$ and the components of $g(x)$ be convex. We will now show that Huard's theorem can be extended to the case where $\theta(x)$ is pseudo-convex and the components of $g(x)$ are quasi-convex, and that Wolfe's theorem is not amenable to such an extension.

THEOREM 2. (*Strict converse duality theorem*)[3]. *Let $\theta(x)$ be a pseudo-convex function on E^m and let the components of $g(x)$ be differentiable quasi-convex functions on E^m.*

(a) *If (x^0, y^0) solves* (DP) *and $\psi(x, y^0)$ is twice continuously differentiable with respect to x in a neighborhood of x^0, and if the Hessian of $\psi(x, y^0)$ with respect to x is nonzero at x^0, then x^0 solves* PP.

(b) *Let x^0 solve* (PP) *and let $g(x) \leqq 0$ satisfy the Kuhn-Tucker constraint qualification* [7]. *It does not necessarily follow that x^0 and some y^0 solve* (DP).

Proof. (a) The assumption that the Hessian of $\psi(x, y^0)$ with respect to x is nonzero at x^0 insures the validity of the following Kuhn-Tucker neces-

[3] For the difference between "duality" and "strict duality," the reader is referred to [9].

sary conditions for some $v^0 \in E^m$:

$$\nabla_x\psi(x^0, y^0) + \nabla_x v^{0\prime}\nabla_z\psi(x^0, y^0) = 0,$$

$$\nabla_y\psi(x^0, y^0) + \nabla_y v^{0\prime}\nabla_z\psi(x^0, y^0) \leqq 0,$$

$$y^{0\prime}\nabla_y\psi(x^0, y^0) + y^{0\prime}\nabla_y v^{0\prime}\nabla_z\psi(x^0, y^0) = 0,$$

$$y^0 \geqq 0,$$

$$\nabla_z\psi(x^0, y^0) = 0.$$

The first and last equations above, together with the assumption that the Hessian of $\psi(x, y^0)$ is nonzero at x^0, imply that $v^0 = 0$. Hence the above necessary conditions become:

$$\nabla_z\psi(x^0, y^0) = 0,$$

$$\nabla_y\psi(x^0, y^0) = g(x^0) \leqq 0,$$

$$y^{0\prime}\nabla_y\psi(x^0, y^0) = y^{0\prime}g(x^0) = 0,$$

$$y^0 \geqq 0.$$

But from Theorem 1, with $C = E^m$, these conditions are sufficient for x^0 to be a solution of (PP).

(b) This part of the theorem will be established by means of the following counter-example:

(PP1) $$\min_{x \in E^1} \{-e^{-x^2} \mid -x + 1 \leqq 0\},$$

(DP1) $$\max_{x \in E^1, y \in E^1} \{-e^{-x^2} - yx + y \mid 2xe^{-x^2} - y = 0, y \geqq 0\}.$$

The solution of (PP1) is obviously $x^0 = 1$, whereas (DP1) has no maximum solution but has a zero supremum.

Finally, we give an application of pseudo-concavity outside the realm of of mathematical programming. In particular, we extend a stability criterion for equilibrium points of ordinary differential equations [8, Theorem 1].

THEOREM 3. (*Stability criterion*). *Let*

$$\dot{x} = f(t, x)$$

be a system of ordinary differential equations, where x and f are m-dimensional vectors and $0 \leqq t < \infty$. Let $f(t, x)$ be continuous in the (x, t) space and let $f(t, 0) = 0$ for $0 \leqq t < \infty$, so that $x = 0$ is an equilibrium point. If $x'f(t, x)$ is a pseudo-concave function of x on E^m for $0 \leqq t < \infty$, then $x = 0$ is a stable equilibrium point.

Proof. Consider the Lyapunov function

$$V(x, t) = \tfrac{1}{2}x'x,$$

which is obviously positive definite. It follows that for $0 \leqq t < \infty$,

$$\dot{V} = x'\dot{x} = x'f(t, x) \leqq 0,$$

where the last inequality follows from the pseudo-concavity of $x'f(t, x)$ in x and the fact that $f(t, 0) = 0$. Hence by Lyapunov's stability theorem [6], $x = 0$ is a stable equilibrium point.

It should be noted that the above proof would not go through had we merely required that $x'f(t, x)$ be quasi-concave instead of pseudo-concave.

3. Remarks on pseudo-convex functions. Properties 1 and 2 and the fact that every differentiable strictly quasi-convex function is also quasi-convex [5] establish a hierarchy among differentiable functions that is depicted in Fig. 1. In other words, if we let S_1, S_2, S_3, and S_4 represent the sets of all differentiable functions defined on a *convex* set C in E^m that are, respectively, convex, pseudo-convex, strictly quasi-convex, and quasi-convex, then

$$S_1 \subset S_2 \subset S_3 \subset S_4.$$

Functions belonging to S_1, S_2, or S_3 share the property that a local minimum is a global minimum. Functions belonging to S_4 do not necessarily have this property. The Kuhn-Tucker differential conditions are sufficient for optimality, (see (2.18)), provided that $g_i(x)$, $i = 1, \cdots, n$ belong to S_4 and $\theta(x)$ belongs to S_1 or S_2, but not if $\theta(x)$ belongs to S_3 or S_4. It seems that the pseudo-convexity of $\theta(x)$ and the quasi-convexity of $g_i(x)$ are the weakest conditions that can be imposed so that relations (2.14) to (2.17) are sufficient for optimality.

There does not seem to be a simple extension of the concept of pseudo-convexity to nondifferentiable functions. This may be due to the fact that pseudo-convexity eliminates inflection points, and such points are easily described by derivatives, but not otherwise.

Finally, it should be remarked that the convexity of the set C is inherent in the definition of quasi-convexity. In contrast, the convexity of C is not needed in the definition of pseudo-convexity. Thus, without the convexity of C, we may have a pseudo-convex function that is not quasi-convex. For example, over the nonconvex set

$$C = \{x \mid x \in E^1, x \neq 0\},$$

the function

$$\theta(x) = \begin{cases} x & \text{for } x < 0, \\ x + 1 & \text{for } x > 0, \end{cases}$$

is pseudo-convex but obviously not quasi-convex, since C is nonconvex.

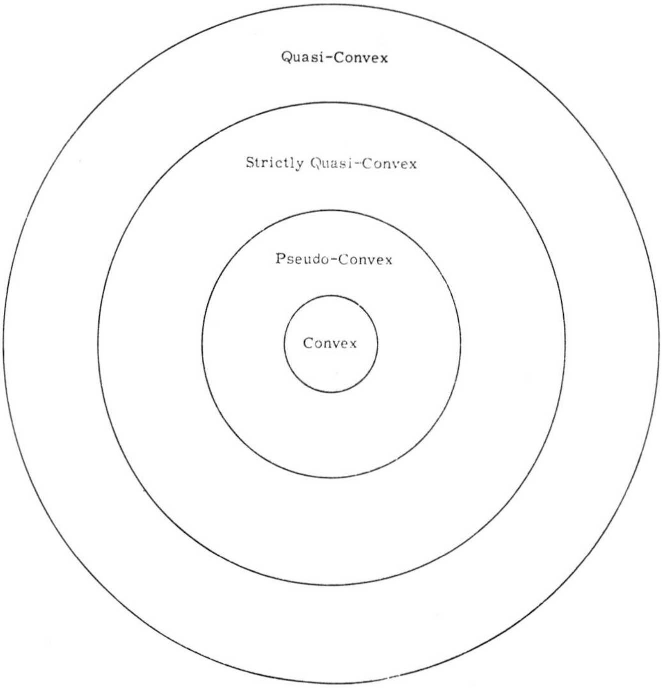

FIG. 1

4. Acknowledgement. I am indebted to my colleagues, S. Karamardian and J. Ponstein, for stimulating discussions on this paper.

REFERENCES

[1] K. J. ARROW AND A. C. ENTHOVEN, *Quasi-concave programming*, Econometrica, 29 (1961), pp. 779–800.

[2] C. BERGE, *Topological Spaces*, Macmillan, New York, 1963.

[3] M. A. HANSON, *Bounds for functionally convex optimal control problems*, J. Math. Anal. Appl., 8 (1964), pp. 84–89.

[4] P. HUARD, *Dual programs*, IBM J. Res. Develop., 6 (1962), pp. 137–139.

[5] S. KARAMARDIAN, *Duality in mathematical programming*, forthcoming.

[6] N. N. KRASOVSKII, *Stability of Motion*, Stanford University Press, Stanford, California, 1963.

[7] H. W. KUHN AND A. W. TUCKER, *Nonlinear programming*, Proceedings of the

Second Berkeley Symposium on Mathematical Statistics and Probability, University of California Press, Berkeley, 1951, pp. 481–492.
[8] O. L. MANGASARIAN, *Stability criteria for nonlinear ordinary differential equations*, this Journal, 1 (1963), pp. 311–318.
[9] O. L. MANGASARIAN AND J. PONSTEIN, *Minmax and duality in nonlinear programming*, International Symposium on Mathematical Programming, London, 1964; J. Math. Anal. Appl., to appear.
[10] P. WOLFE, *A duality theorem for nonlinear programming*, Quart. Appl. Math., 19 (1961), pp. 239–244.

Reprinted from *Cahiers du Centre d'Études de Recherche Opérationelle* **12**, 114-122 (1970).

CONVEXITY, PSEUDO-CONVEXITY AND QUASI-CONVEXITY OF COMPOSITE FUNCTIONS

O.L. MANGASARIAN
Mathematics Research Center
University of Wisconsin, Madison (U.S.A.)

ABSTRACT

A number of recent results which establish the convexity, pseudo-convexity or quasi-convexity of certain functions are shown to be special cases of the fact that under suitable conditions a composite function is convex, pseudo-convex or quasi-convex.

Many of the theoretical results and computational algorithms of nonlinear programming require that the functions entering the problem be convex, pseudo-convex or quasi-convex [8, 1, 14, 9, 7, 11, 6]. Accordingly, a number of papers have been recently published which establish the convexity, pseudo-convexity or quasi-convexity of certain *specific* functions [13, 2, 10, 12] and the *convexity* of a *general* composite function [3]. All these results are shown to be simple consequences of the fact that under suitable conditions a *general* composite function is convex, pseudo-convex or quasi-convex.

Preliminaries

R^n denotes the n-dimensional Euclidean space, small Latin letters denote vectors or vector functions, and small Greek letters denote real numbers or numerical functions. Subscripts denote components of vectors, superscripts denote specific vectors. If x and y are in R^n then xy denotes their scalar product $x_1 y_1 + \ldots + x_n y_n$. If θ is a differentiable numerical function on R^n that is $\theta: R^n \to R$, then $\nabla\theta(x)$ denotes the n-dimensional vector of the first partial derivatives of θ, $\partial\theta(x)/\partial x_i$, $i = 1, \ldots, n$. If f is a differentiable m-dimensional vector function on R^n, that is $f: R^n \to R^m$, then $\nabla f(x)$ denotes the $m \times n$ Jacobian matrix of the first partial derivatives $\partial f_i(x)/\partial x_j$, $i = 1, \ldots, m$, $j = 1, \ldots, n$. If φ is a differentiable numerical function defined on $R^m \times R^k$, then $\nabla_1 \varphi(y, z)$

denotes the m-dimensional vector of the m first partial derivatives $\partial\varphi(y, z)/\partial y_i$, $i = 1, \ldots, m$, and $\nabla_2\varphi(y, z)$ denotes the k-dimensional vector of the k first partial derivatives $\partial\varphi(y, z)/\partial z_i$, $i = 1, \ldots, k$.

A set Γ in R^n is said to be *convex* if and only if for each x^1 and x^2 in Γ, $(1-\lambda)x^1 + \lambda x^2$ is also in Γ for each λ satisfying $0 \leqslant \lambda \leqslant 1$.

Let θ be a numerical function defined on some set Γ in R^n.

If Γ is convex, then θ is said to be *convex* on Γ if and only if for each x^1 and x^2 in Γ and $0 \leqslant \lambda \leqslant 1$

$$\theta((1-\lambda)x^1 + \lambda x^2) \leqslant (1-\lambda)\theta(x^1) + \lambda\theta(x^2).$$

If Γ is open and convex and θ is differentiable on Γ, then θ is convex on Γ if and only if for each x^1 and x^2 in Γ

$$\theta(x^2) - \theta(x^1) \geqslant \nabla\theta(x^1)(x^2 - x^1).$$

If Γ is open and θ is differentiable on Γ, then θ is said to be *pseudo-convex* [9] on Γ if and only if for each x^1 and x^2 in Γ

$$\nabla\theta(x^1)(x^2 - x^1) \geqslant 0 \Rightarrow \theta(x^2) \geqslant \theta(x^1).$$

(If Γ is also convex, then each differentiable convex function on Γ is a pseudo-convex function on Γ but not conversely [9].)

If Γ is convex, then θ is said to be *quasi-convex* [1, 4] on Γ if and only if the set $\{x \mid x \in \Gamma, \theta(x) \leqslant \alpha\}$ is convex for *each* real α. Equivalently θ is quasi-convex on Γ if and only if for each x^1 and x^2 in Γ

$$\theta(x^2) \leqslant \theta(x^1) \Rightarrow \theta((1-\lambda)x^1 + \lambda x^2) \leqslant \theta(x^1) \quad \text{for} \quad 0 \leqslant \lambda \leqslant 1.$$

(If Γ is convex, then each pseudo-convex function on Γ is a differentiable quasi-convex function but not conversely [9].)

Let φ be a numerical function defined on some set Λ in $R^m \times R^k$. φ is said to be *increasing-decreasing* (or incr-decr) on Λ if and only if for each (y^1, z^1) and (y^2, z^2) in Λ

$$y^2 \geqslant y^1 \quad \text{and} \quad z^2 \leqslant z^1 \Rightarrow \varphi(y^2, z^2) \geqslant \varphi(y^1, z^1)\,^*.$$

φ is said to be *increasing*-——— (or incr-———) on Λ if and only if for each (y^1, z) and (y^2, z) in Λ

$$y^2 \geqslant y^1 \Rightarrow \varphi(y^2, z) \geqslant \varphi(y^1, z).$$

φ is said to be *decreasing*-——— (or decr-———) on Λ if and only

* $y^2 \geqslant y^1$ denotes $y_i^2 \geqslant y_i^1$ for $i = 1, \ldots, m$.

if for each (y^1, z) and (y^2, z) in Λ

$$y^2 \leqslant y^1 \Rightarrow \varphi(y^2, z) \geqslant \varphi(y^1, z)\,. \quad .$$

The following lemma follows directly from the definition of differentiability and the mean value theorem.

Lemma: Let φ be a differentiable numerical function defined on an open convex set Λ in $R^m \times R^k$. Then

$$\varphi \text{ incr-decr on } \Lambda \Leftrightarrow \nabla_1 \varphi(y, z) \geqslant 0 \text{ and } \nabla_2 \varphi(y, z) \leqslant 0$$
$$\text{for all } (y, z) \text{ in } \Lambda.$$

(The convexity of Λ is needed only in the backward implication of the above lemma.)

Principal result

Theorem: Let Γ be a convex set in R^n, let f be an m-dimensional vector function defined on Γ, let g be a k-dimensional vector function defined on Γ, and let φ be a numerical function defined on $R^m \times R^k$, let $\theta(x) = \varphi(f(x), g(x))$, and let any one of the following four assumptions hold

	f (on Γ)	g (on Γ)	φ (on $R^m \times R^k$)		
(i)	*convex*	*concave*	*incr-decr*;	or	$\nabla_1 \varphi(y, z) \geqslant 0$, $\nabla_2 \varphi(y, z) \leqslant 0$
(ii)	*linear*	*linear*	———		
(iii)	*convex*	*linear*	*incr*-——;	*or*	$\nabla_1 \varphi(y, z) \geqslant 0$
(iv)	*concave*	*linear*	decr-——;	*or*	$\nabla_1 \varphi(y, z) \leqslant 0$.

Then the following implications hold

(I) φ *convex on* $R^m \times R^k \Rightarrow \theta$ *convex on* Γ

(II) Γ *open, f and g differentiable on* Γ—
θ *pseudo-convex on* $R^m \times R^k$ — $\Rightarrow \theta$ pseudo-convex on Γ

(III) φ *quasi-convex on* $R^m \times R^k \Rightarrow \theta$ *quasi-convex on* Γ.

(To keep the statement of the theorem and subsequent results simple we require that φ have certain properties on $R^m \times R^k$ when we actually need these properties only on the convex hull of the product $\Gamma_1 \times \Gamma_2$ of the image sets

$$\Gamma_1 = \{y \mid y = f(x), x \in \Gamma\} \quad \text{and} \quad \Gamma_2 = \{z \mid z = g(x), x \in \Gamma\}.)$$

Proof

(I) We shall prove this part of the theorem under assumption (i) first.

Let x^1, $x^2 \in \Gamma$ and let $0 \leqslant \lambda \leqslant 1$. Then

$$\begin{aligned}\theta((1-\lambda)x^1 + \lambda x^2) &= \varphi(f((1-\lambda)x^1 + \lambda x^2), g((1-\lambda)x^1 + \lambda x^2)) \\ &\leqslant \varphi((1-\lambda)f(x^1) + \lambda f(x^2), (1-\lambda)g(x^1) + \lambda g(x^2)) \\ &\quad \text{(since } f \text{ is convex, } g \text{ concave and } \varphi \text{ increasing-decreasing)} \\ &\leqslant (1-\lambda)\varphi(f(x^1), g(x^1)) + \lambda\varphi(f(x^2), g(x^2)) \\ &\quad \text{(since } \varphi \text{ is convex)} \\ &= (1-\lambda)\theta(x^1) + \lambda\theta(x^2)\,,\end{aligned}$$

and hence θ is convex. Under assumption (ii) the first inequality above is an equality, and under assumption (iii) or (iv) it remains an inequality.

(II) We again prove this part of the theorem under assumption (i) first. Let x^1, $x^2 \in \Gamma$. Then

$$\begin{aligned}\nabla\theta(x^1)(x^2-x^1) &= (\nabla_1\varphi(f(x^1), g(x^1))\nabla f(x^1) + \\ &\quad + \nabla_2\varphi(f(x^1), g(x^1))\nabla g(x^1))(x^2-x^1) \\ &\quad \text{(by the chain rule)} \\ &\leqslant \nabla_1\varphi(f(x^1), g(x^1))(f(x^2) - f(x^1)) + \\ &\quad + \nabla_2\varphi(f(x^1), g(x^1))(g(x^2) - g(x^1)) \\ &\quad \text{(by the convexity of } f \text{, concavity of } g \text{, increasing-decreasing property of } \varphi \text{ and the Lemma).}\end{aligned}$$

Hence

$$\begin{aligned}\nabla\theta(x^1)(x^2-x^1) \geqslant 0 &\Rightarrow \nabla_1\varphi(f(x^1), g(x^1))(f(x^2) - f(x^1)) + \\ &\quad + \nabla_2\varphi(f(x^1), g(x^1))(g(x^2) - g(x^1)) \geqslant 0 \\ &\quad \text{(by the above inequality)} \\ &\Rightarrow \varphi(f(x^2), g(x^2)) \geqslant \varphi(f(x^1), g(x^1)) \\ &\quad \text{(by the pseudo-convexity of } \varphi\text{)} \\ &\Rightarrow \theta(x^2) \geqslant \theta(x^1)\,,\end{aligned}$$

and hence θ is pseudo-convex. Under assumption (ii) the first inequality in the above proof becomes an equality, and under assumptions (iii) or (iv) it remains an inequality.

(III) Let assumption (i) hold, let x^1, $x^2 \in \Gamma$ and let $0 \leqslant \lambda \leqslant 1$. Then

$$\theta(x^2) \leqslant \theta(x^1) \Rightarrow \varphi(f(x^2), g(x^2)) \leqslant \varphi(f(x^1), g(x^1))$$

$$\Rightarrow \varphi((1-\lambda)f(x^1) + \lambda f(x^2), (1-\lambda)g(x^1) + + \lambda g(x^2)) \leqslant \varphi(f(x^1), g(x^1))$$

(since φ is quasi-convex)

$$\Rightarrow \phi(f((1-\lambda)x^1 + \lambda x^2), g((1-\lambda)x^1 + \lambda x^2)) \leqslant \varphi(f(x^1), g(x^1))$$

(since f is convex, g concave and φ increasing-decreasing)

$$\Rightarrow \theta((1-\lambda)x^1 + \lambda x^2) \leqslant \theta(x^1),$$

and hence θ is quasi-convex. The above implications again hold if we replace assumption (i) by assumption (ii), (iii), or (iv). Q.E.D.

Corollary: If any one of the following four assumptions holds

	f (on Γ)	g (on Γ)	φ (on $R^m \times R^k$)
(i′)	*concave*	*convex*	*incr-decr*; *or* $\nabla_1\varphi(y,z) \geqslant 0$, $\nabla_2\varphi(y,z) \leqslant 0$
(ii′)	*linear*	*linear*	———
(iii′)	*concave*	*linear*	*incr-*——; *or* $\nabla_1\varphi(y,z) \geqslant 0$
(iv′)	*convex*	*linear*	*decr-*——; *or* $\nabla_1\varphi(y,z) \leqslant 0$,

then the following implications hold

(I′) *φ concave on $R^m \times R^k$ $\Rightarrow$ θ concave on Γ*

(II′) *Γ open, f and g differentiable on Γ— φ pseudo-concave on $R^m \times R^k$ — $\Rightarrow$ θ pseudo-concave on Γ*

(III′) *φ quasi-concave on $R^m \times R^k$ $\Rightarrow$ θ quasi-concave on Γ.*

Remark 1. When we set $k = 0$, that is $\theta(x) = \varphi(f(x))$, the result (Ii) above is a well known convexity result of a composite function [4, p. 191, Theorem 4], and (IIIi) is similarly a known quasi-convexity result of a composite function [4, p. 207, Theorem 1]. The case (Ii) was recently established by Bereanu [3]. The remaining cases, (Iii), (Iiii), (Iiv), (II), and

(III) have not, to the author's knowledge, been given before. These cases, together with (Ii) subsume all the interesting recent results in the literature [2, 10, 12, 13] in which convexity, pseudo-convexity or quasi-convexity of certain *specific* functions were established. (The pseudo-convex case is of particular interest since many of the theoretical and computational results of nonlinear programming that were established for convex functions also hold for pseudo-convex functions.)

Remark 2. Most of the computational algorithms of nonlinear programming [7, 11, 6] require the convexity of the functions defining the problem in order to guarantee convergence to a global minimum. Some of these algorithms however will also converge to a global minimum if we merely have pseudo-convex functions instead. For example the Frank-Wolfe algorithm [7] which finds a global minimum of a continously differentiable function θ on a polyhedron will also converge to a global minimum if we merely require θ to be pseudo-convex on the polyhedron rather than convex. (See the convergence proof of the algorithm on page 90 of [5].) Result II of the theorem then extends the Frank-Wolfe algorithm to functions θ defined by $\theta(x) = \varphi(f(x), g(x))$ where assumptions (i), (ii), (iii), or (iv) are satisfied and φ is pseudo-convex. Some such functions are nonlinear fractional functions and bi-nonlinear functions which are described in the next section.

Applications

We indicate now how our principal result includes the recent results of [13, 2, 10, 12] which establish the convexity, pseudo-convexity or quasi-convexity of certain specific functions.

(A) *Nonlinear fractional functions* [13, 10, 12]. Let Γ be a convex set in R^n, let ρ and σ be numerical functions defined on Γ, let $\theta(x) = \rho(x)/\sigma(x)$, and let any one of the following assumptions hold on Γ

	ρ	σ
(1)	convex $\geqslant 0$	concave > 0
(2)	concave $\leqslant 0$	convex < 0
(3)	convex $\leqslant 0$	convex > 0
(4)	concave $\geqslant 0$	concave < 0
(5)	linear	linear $\neq 0$
(6)	linear $\leqslant 0$	convex $\neq 0$

(7)	linear $\geqslant 0$	concave $\neq 0$
(8)	convex	linear > 0
(9)	concave	linear < 0

Then the following hold

(II) θ is pseudo-convex on Γ if Γ is open and ρ and σ are differentiable on Γ.

(III) θ is quasi-convex on Γ.

If any one of the following assumptions hold on Γ

	ρ	σ
(1′)	concave $\geqslant 0$	convex > 0
(2′)	convex $\leqslant 0$	concave < 0
(3′)	concave $\leqslant 0$	concave > 0
(4′)	convex $\geqslant 0$	convex < 0
(5′)	linear	linear $\neq 0$
(6′)	linear $\leqslant 0$	concave $\neq 0$
(7′)	linear $\geqslant 0$	convex $\neq 0$
(8′)	concave	linear > 0
(9′)	convex	linear < 0,

then the following hold

(II′) θ is pseudo-concave on Γ if Γ is open and ρ and σ are differentiable on Γ.

(III′) θ is quasi-concave on Γ.

The above results follow from our principal result by observing that the numerical function $\varphi(y, z) = y/z$, $(y, z) \in R \times R$ is pseudo-convex and hence also quasi-convex [9] on either of the convex sets $\{(y, z) \mid (y, z) \in R \times R, z > 0\}$ or $\{(y, z) \mid (y, z) \in R \times R, z < 0\}$, and by making the identifications

$$f(x) = \rho(x), \quad g(x) = \sigma(x)$$

for the cases (1), (5), (8), and (9); and

$$f(x) = \sigma(x), \quad g(x) = \rho(x)$$

for the cases (2), (6), and (7); and

$$f(x) = \begin{bmatrix} \rho(x) \\ \\ \sigma(x) \end{bmatrix}$$

for the case (3); and

$$g(x) = \begin{bmatrix} \rho(x) \\ \\ \sigma(x) \end{bmatrix}$$

for the case (4).

(B) *Reciprocal functions* [6, 2, 10]. Let Γ be a convex set in R^n, let σ be a numerical function defined on Γ, and let $\theta(x) = 1/\sigma(x)$. Then

(1) σ concave > 0 on Γ $\Rightarrow$ θ convex > 0 on Γ,
(2) σ convex < 0 on Γ $\Rightarrow$ θ concave < 0 on Γ.

The above follows from (Iiv) and (I'iv') respectively if we define $\varphi(y, z) = 1/y$, $(y, z) \in R \times R$ and note that $\nabla_1 \varphi(y, z) = -1/(y)^2$, $\nabla_2 \varphi(y, z) = 0$ and that $\varphi(y, z)$ is convex for $y > 0$ and concave for $y < 0$.

(C) *Bi-nonlinear functions* [10]. Let Γ be an open convex set in R^n, let ρ and μ be differentiable functions defined on Γ, and let $\theta(x) = \rho(x)\mu(x)$. Then the following implications hold on Γ

(1) ρ convex $\leqslant 0$, μ concave > 0 $\Rightarrow$ θ pseudo-convex $\leqslant 0$
(2) ρ convex < 0, μ concave $\geqslant 0$ $\Rightarrow$ θ pseudo-convex $\leqslant 0$
(3) ρ convex < 0, μ convex $\leqslant 0$ $\Rightarrow$ θ pseudo-concave $\geqslant 0$
(4) ρ concave $\geqslant 0$, μ concave > 0 $\Rightarrow$ θ pseudo-concave $\geqslant 0$.

If we write $\theta(x) = \rho(x)/[1/\mu(x)]$, then (C1) follows from (B1) and (A3); and (C4) follows from (B1) and (A1'). If we write $\theta(x) = \mu(x)/[1/\rho(x)]$, then (C2) follows from (B2) and (A4); and (C3) follows from (B2) and (A2').

(D) *Special fractional function* [2]. Let Γ be a convex set in R^n, let α and β be numerical functions on Γ, and let $\theta(x) = (\alpha(x))^2/\beta(x)$. Then the following implications hold on Γ.

(1) α convex $\geqslant 0$, β concave > 0 $\Rightarrow$ θ convex > 0
(2) α concave $\leqslant 0$, β concave > 0 $\Rightarrow$ θ convex $\geqslant 0$
(3) α linear, β concave > 0 $\Rightarrow$ θ convex $\geqslant 0$
(1') α concave $\leqslant 0$, β convex < 0 $\Rightarrow$ θ concave $\leqslant 0$
(2') α convex $\geqslant 0$, β convex < 0 $\Rightarrow$ θ concave $\leqslant 0$
(3') α linear, β convex < 0 $\Rightarrow$ θ concave $\leqslant 0$.

If we observe that the function φ defined by

$$\varphi(y, z) = \frac{(y)^2}{z}, \quad (y, z) \in R \times R, \quad z \neq 0,$$

is convex on the convex set $\{(y, z) \mid (y, z) \in R \times R,\, z > 0\}$, and is concave on the convex set $\{(y, z) \mid (y, z) \in R \times R,\, z < 0\}$, then (D1) follows from (Ii), and (D′1′) from (I′i′) if we make the identification $f(x) = \alpha(x)$, $g(x) = \beta(x)$; (D3) follows from (Iiv), and (D′3′) from (I′iv′) if we make the identification $f(x) = \beta(x)$, $g(x) = \alpha(x)$; (D2) follows from (Ii) and (D′2′) from (I′i′) if we make the identification $g(x) = \begin{bmatrix} \alpha(x) \\ \beta(x) \end{bmatrix}$.

REFERENCES

[1] Arrow, K.J. and Enthoven, A.C., "Quasi-concave programming," *Econometrica*, **29**, 1961, 779-800.

[2] Bector, C.R., "Programming problems with convex fractional functions," *Operations Research*, **16**, 1968, 383-391.

[3] Bereanu, B., "On the composition of convex functions," *Dicussion paper*, No. 6801, Center for Operations Research and Econometrics, Catholic University of Louvain, January 1968.

[4] Berge, C., *Topological spaces*, McMillan, New York, 1963.

[5] Berge, C. and Ghouila-Houri, A., *Programming, games, and transportation networks*, Wiley, New York, 1965.

[6] Fiacco, A.V. and McCormick, G.P., "The sequential unconstrained minimization technique for nonlinear programming, a primal dual method," *Management Science*, **10**, 1965, 360-364.

[7] Frank, M. and Wolfe, P., "An algorithm for quadratic programming," *Naval Research Logisiics Quarterly*, **3**, 1956, 95-110.

[8] Kuhn, H.W. and Tucker, A.W., "Nonlinear programming," in *Proceeding of the Second Berkeley Symposium on Mathematical Statistics and Probability*, ed. J. Neyman, University of California Press, 1951, 481-492.

[9] Mangasarian, O.L., "Pseudo-convex functions," *SIAM J. on Control*, **3**, 1965, 281-290.

[10] Mangasarian, O.L., "Nonlinear fractional programming," *Mathematics Research Center Technical Summary*, Report #819, University of Wisconsin, Madison, 1967.

[11] Rosen, J.B., "The gradient projection method for nonlinear programming," Part I, *J. SIAM*, **8**, 1960, 181-217, Part II, *J. SIAM*, **9**, 1961, 514-532.

[12] Swarup, K., "Indefinite quadratic programming," *Cahiers du Centre d'Études de Recherche Opérationnelle*, **8**, 1966, 217-222.

[13] Tuy, H., "Sur les inégalités linéaires," *Colloquium Mathematicum*, **13**, 1964, 107-123.

[14] Wolfe, P., "A duality theorem for nonlinear programming," *Quart. of Applied Math.*, **19**, 1961, 239-244.

3. DYNAMIC PROGRAMMING

*Introduction to Dynamic Programming**

W. T. Ziemba

THE UNIVERSITY OF BRITISH COLUMBIA

I. Introduction

Dynamic programming is a strategy for the solution and analysis of certain types of optimization problems. The technique has been utilized in many diverse fields such as production and inventory planning, transportation, reliability, control theory, economic and physical planning, and design. The method has also been used in the analysis of many problems in the realm of recreational mathematics such as puzzles, chess, checkers, blackjack, roulette, and other games of chance.

The technique of dynamic programming will be of paramount use in the analysis of the multiperiod financial problems that are considered in Parts IV and V of this book. This article presents an introduction to the fundamental ideas, uses, and limitations of dynamic programming. The reading of this article along with the solution of the associated problems should provide sufficient background for the study of the material in the sequel.

Generally speaking, dynamic programming is concerned with problems where an individual must make a sequence of decisions in a specified order. Problems amenable to analysis by dynamic programming methods must generally be cast into such a framework even though their original formulation may not appear to have a sequential nature. Most financial problems, however, do have a sequential nature where periods of time separate the various decision points. However, this need not be the case for the application of the dynamic programming technique.

The underlying mathematical structure of problems amenable to dynamic programming has been explored by many authors since Richard Bellman developed the technique in the late 1940s. The basic ideas are even older, dating at least to Massé in the 1920s. The literature concerned with dynamic programming is positively voluminous—Bellman himself has authored several hundred papers and more than ten books on the subject. Our intent here is not to survey this literature[1] but to present a self-contained introduction to

* Adapted from E. V. Denardo and L. G. Mitten, "Elements of Sequential Decision Processes," *Journal of Industrial Engineering* **18** (1967), 106–112.

[1] For additional information and depth the reader may consult the references listed in the introduction.

the use of the dynamic programming approach. The paper presents a general framework and a mathematical analysis that attempts to isolate those problems for which the dynamic programming approach is valid. The exposition is an adaptation of Denardo and Mitten (1967) that is based in part on some of their earlier work (see the literature [4, 5, 8]).[2]

We will utilize the term sequential decision processes (SDPs) to describe the phenomenon under study. It will be assumed that the mechanism governing the evolution of the process is explicit and that the SDPs satisfy certain termination and monotonicity assumptions. The termination assumption guarantees that the process is completed in a (fixed) finite number of transitions. The basic result for such processes is that there exists an optimal policy. The development also yields an algorithm for determining an optimal policy and its return function.

The analysis of terminating SDPs requires less formidable mathematical tools than is required for the analysis of general SDPs. However, the more complex models, such as infinite horizon models, share many features of the terminating models. The language used to describe terminating models may be used for general SDPs. Also the monotonicity property plays a central role in the analysis of these models as well as of the terminating models. Furthermore the property of evolution dependence, discussed in Section V, facilitates the analysis of a large class of infinite horizon models using the methods discussed here. The development illustrates the central role in dynamic programming played by the notion of an optimal policy. This notion may be accepted as a premise called the "optimality postulate" in which it is assumed that an optimal policy does exist. Utilizing this postulate and the monotonicity assumption, the functional equations are derived.

II. Sequential Decision Processes

A review of the wide variety of problems that have been treated by dynamic programming reveals that the vast majority (if not all) may be characterized as processes that pass through a set of states in response to a sequence of choices of decisions. The values associated with the process typically depend on both the states traversed and the decisions made. The following five factors may be identified as the basic elements of dynamic programming problems: stages, states, decisions, transitions, and returns. These terms will be given

[2] Throughout this book, references cited by numbers in square brackets will be found in the reference lists at the ends of the articles in which they occur.

more precise definitions. The monotonicity assumption[3] discussed below and the notions of stages, states, decisions, transitions, and returns jointly constitute a sequential decision process.

Points or epochs in the SDP at which a decision must be made divide the process into stages which may be indexed as $T = \{1, 2, \ldots\}$. The state space at stage $t \in T$ is denoted by the set Ω_t, a subset of k-dimensional Euclidean space. An element $\omega \in \Omega_t$ is called the state at stage t and may be interpreted as a description of one of the conditions or situations in which the process may exist. Hence Ω_t is the set of all possible states that might obtain at stage t.

To specify the status of the process both the state and stage index must be identified. Let the set $\Omega \equiv \{(\omega, t) \mid \omega \in \Omega_t,\ t \in T\}$ denote the space–time set. An element $x \in \Omega$ is called a state. Sufficient conditions will be presented concerning the process to conclude that the states have the

Markovian Regeneration Property A state is a regeneration point for the process in the sense that the costs associated with current and future decisions depend on the current state, but not on the prior states. Current and future costs may depend on the current state, but they do not depend on the history of prior states and decisions leading to this state.

Associated with each state $x \in \Omega$ is an arbitrary nonempty set D_x called the decision set for x. An element $d_x \in D_x$ is called a decision and represents one of the choices available when the process is in state x. The process moves through space–time in response to the decisions made at each stage. The random variable X_t which takes values in Ω_t denotes the position or state of the process at time t.

We now describe the mechanism governing the trajectory of the process $\{X_t \mid t \in T\}$ in response to the decisions chosen. When the process is in state $x = (\omega, t)$, selection of decision d_x determines a (possibly defective)[4] probability distribution function $P(\cdot; x, d_x)$ on Ω_t, where $P(z; x, d_x) \equiv$

[3] The monotonicity assumption is quite mild and is satisfied by most properly posed dynamic programming problems. See Exercises CR-19 and 24 for some examples.

[4] A probability distribution function is defective when its total mass is strictly less than 1, i.e., $\lim_{z \to \infty} P(z; x, d_x) < 1$. The defect $1 - P(\infty; x, d_x)$ is the probability that the process terminates at state x. It is useful to define the set of states $T(x, d_x)$ to which the process might possibly move from state x if decision d_x is selected. More formally $z \in T(x, d_x)$ iff $P[z+\varepsilon; x, d_x] - P[z-\varepsilon; x, d_x] > 0$ for each $\varepsilon > 0$, i.e., $z \in T(x, d_x)$ iff each open neighborhood of z contains positive mass. If the process moved to a particular state with certainty, then $T(x, d_x)$ would contain exactly one element. If the defect of the transition probability is 1 at state x wherever d_x is selected, then $T(x, d_x)$ is empty and the process terminates with certainty.

$\text{Prob}(X_{t+1} \leqq z \mid (X_t, t) = x, d_x)$ is called the probability transition function of the process and governs its evolution. The transition from stage t to stage $t+1$ may be viewed as shown in Fig. 1. The mechanism inside the box may be

Fig. 1. Transition from stage t to $t+1$.

thought of as that which chooses a new state from Ω_{t+1} via the inputs and the probability distribution over Ω_{t+1}. Implicit in this process is the assumption that given $x = (X_t, t)$ and d_x, the random variable X_{t+1} is independent of the random variables $X_s, s < t$, and their corresponding decisions. Note that the deterministic case is that special instance when all probability mass in each stage t falls upon one member of Ω_t.

A policy δ is an ordered collection of decisions containing one decision for each state in Ω. The policy space Δ is the collection of all such policies. Thus, a policy $\delta \in \Delta$ prescribes a particular decision for each and every state $x \in \Omega$ and the policy space consists of all possible combinations of decisions at the various states. The policy space is the Cartesian product of all the decision sets; that is, $\Delta = \prod_{x \in \Omega} D_x$. The symbol δ_x will frequently be used to denote the decision in δ that applies to state x. Given a particular policy δ, the process $\{(X_t, t \mid t \in T\}$ is Markovian. That is, given the present state, the future evolution of the process is independent of the past. Each policy δ is assumed to have a real-valued return function $V_\delta(x)$ which reflects the net reward that would accrue if the process were started in state x and the specified decisions in δ were applied at each of the states through which the process evolves. The optimal return from x is $f(x) \equiv \max_{\delta \in \Delta} V_\delta(x)$. It is assumed throughout that the maximum is attained by some $\delta \in \Delta$. It will prove convenient to have notation to describe those states in Ω that are accessible from state x in any finite number of transitions. Let $T(x)$ be the set of points that can be reached from state x in one transition; more generally, for $A \subset \Omega$, let $T(A)$ be the set of points that can be reached in one transition from at least one state $x \in A$. Then, one has

$$T(x) = \bigcup_{d_x \in D_x} T(x, d_x), \qquad T(A) = \bigcup_{x \in A} T(x).$$

Next, define $T^n(x)$ as the set of all points that can be reached from state x in exactly n transitions. Then $T^1(x) = T(x)$ and $T^{n+1}(x) = T[T^n(x)]$ for $n \geqq 1$. Finally, define Ω_x as the set of points that are accessible from x in any positive number of transitions, with

$$\Omega_x = \bigcup_{n=1}^{\infty} T^n(x).$$

An important assumption concerning the structure of the process is the

Monotonicity Assumption[5] Suppose δ and γ are two policies such that for some state x, $\delta_x = \gamma_x$ and $v_\delta(y) \geqq v_\gamma(y)$ for all $y \in \Omega_x$. Then $v_\delta(x) \geqq v_\gamma(x)$.

This assumption states, roughly, that if two policies δ and γ contain the same decision at x, i.e., $\delta_x = \gamma_x$, and if policy δ produces at least as large a return as γ for each state $y \in \Omega_x$, then policy δ produces at least as large a return as γ at x.

When $\Omega_x = \varnothing$, the hypothesis is interpreted to reduce to $\delta_x = \gamma_x$, and the conclusion becomes $v_\delta(x) = v_\gamma(x)$, since $v_\delta(x) \geqq v_\gamma(x)$ and interchanging the roles of δ and γ yields $v_\gamma(x) \geqq v_\delta(x)$.

As an example let us consider a multiperiod consumption–investment model. In each period (stage) the consumer must allocate his current wealth between consumption and investment. Suppose that the consumer's state in period t is completely characterized by his wealth level w. A policy must specify the level of consumption at each state (w_t, t). Consumption cannot exceed wealth, if borrowing is not allowed; hence $D_{(w_t,t)} \equiv \{c \mid 0 \leqq c \leqq w\}$. The distribution function of wealth in period $t+1$ given current wealth and the decision must be specified. A common assumption is that wealth in period $t+1$ is proportional to that invested in period t, the constant of proportionality being a random variable ξ_t with distribution function $F_t(z) = \text{Prob}[\xi_t \leqq z]$. That is, there are stochastic constant returns to scale. Suppose that the ξ_t are independent random variables. Then $P[z;(\omega,t),c] = \text{Prob}[\xi_t(\omega - c \leqq z] = F_t(z/(\omega - c))$. It may be noted that the model can be made more realistic if one is willing to expand and complicate the state space. For example, the state might specify not only wealth level but also various economic indicators, or the liquidity characteristics of various assets that comprise this wealth.

The transition function and policy δ together determine the probability distribution of the process $W_1, W_2, W_3, \ldots$, where W_t is the wealth at time t. Thus the process $\{C_t\colon\ t \in T\}$, where $C_t = \delta_{(W_t,t)}$, is also determined.

[5] A monotonicity property was first explicitly used in the study of sequential decision processes by Mitten [8]. Denardo developed it [4] and used it in the analysis of a broad class of nonterminating processes [5].

Finally we must specify the return function. Suppose that the utility of consumption, at stage t, discounted to the present takes the additive form $\beta^{t-1}u_t(c_t)$. Then

$$\begin{aligned}
v_\delta(w,1) &= E\left\{\sum_{t\in T} \beta^{t-1}u_t(\delta_{(w_t,t)})\,|\,W_1 = w, \delta\right\} \\
&= u_1(\delta_{(w,1)}) + E\left\{\sum_{t>1} \beta^{t-1}u_t(\delta_{(w_t,t)})\,|W_1 = w, \delta\right\} \\
&= u_1(c_1) + \beta E\{v_\delta(W_2,2)\,|\,W_1 = w, \delta\} \\
&= u_1(c_1) + \beta\int_{-\infty}^{+\infty} v_\delta(z,2)\,dP[z;w,1,c_1]
\end{aligned}$$

is the expected value of following the policy δ if the consumer has initial wealth w. The reader is asked in Exercise CR-19 to verify that the monotonicity assumption is satisfied by this return function if the u_t are monotone nondecreasing.

If $|u_t|$ is bounded, say, by $M < \infty$, then the return function is well defined because

$$\left|\sum_{t=1}^{\infty} \beta^{t-1}u_t(c_t)\right| \leqq M/(1-\beta) < \infty, \qquad 0<\beta<1.$$

If one considers a finite planning period of, say, T periods, then the return function might be $v_\delta(w) = E\{\sum_{t=1}^{T} u_t(c_t)\}$ and the transition functions such that $P(z;\omega,N,d) = 0$ for each $z \in \mathbf{R}$, $\omega \in \Omega_N$, and $d \in D_{(\omega,N)}$. Thus the process terminates in period T. Note that a discount factor is no longer needed to ensure that v_δ is well defined, since there are only T terms in the sum. If it is further assumed that the process does not terminate previous to period T, then the model satisfies the termination assumption presented in the following section.

III. Terminating Process

Terminating processes are those for which the process is completed in a fixed finite number of transitions n_x for any initial state x. More formally we may state the

Termination Assumption For each $x \in \Omega$ and $d_x \in D_x$, the defect $1-P(\infty;x,D_x)$ is either 0 or 1. Moreover for each $x \in \Omega$ there exists a $n_x < \infty$

such that $T^{n_x}(x) = \varnothing$. Naturally n_x may vary with x. The termination condition implies that for each $x \in \Omega$, n_x is the smallest positive integer such that $T^n(x) = \varnothing$ for all $n \geqq n_x$.

For $m \geqq 1$, let S_m be the subset of Ω containing those points from which termination always occurs in m transitions or less but does not always occur in fewer than m transitions. That is, let $S_m = \{x \in \Omega \mid n_x = m\}$. The termination assumption implies that $\Omega = \bigcup_{m=1}^{\infty} S_m$. Similarly, let S^m be the set of all states in Ω from which termination always occurs in m transitions or less; that is, let $S^m = \bigcup_{i=1}^{m} S_i$. Should there exist an integer N such that $n_x \leqq N$ for every $x \in \Omega$ and $n_x = N$ for at least one x, then the process is called an *N-stage* sequential decision process. *N-stage* processes cover a wide variety of important applications.

The term "regeneration point" arises since the process is Markovian. Hence the probability law governing transitions from a given state is independent of what states preceded it. Here, the return $v_\delta(x)$ does not depend on how state x was attained. Note that $\{x\} \cup \Omega_x$ contains x and all states which are accessible from x. The following lemma demonstrates that $v_\delta(x)$ depends only on those decisions in δ that apply to these states. The procedure of proof will be to perform induction from the "end" of the process; that is, starting with S^1, the proof is inductive on n. The following simple consequence of the termination assumption is used in the proof: If $x \in S_n$, then $\Omega_x \subset S^{n-1}$, a fact that follows from the observation that if $z \in \Omega_x$, then $n_z \leqq n-1$.

Lemma Suppose the monotonicity and termination assumptions are satisfied. Let δ and γ be any two policies such that $\delta_z = \gamma_z$ for $z = x$ and all $z \in \Omega_x$. Then $v_\delta(x) = v_\gamma(x)$.

Proof The proof is inductive on S^n. First, if $x \in S^1$, then $v_\delta(x) = v_\gamma(x)$, by the monotonicity assumption. Assume from the inductive hypothesis that the lemma is true for every $z \in S^n$. Pick $x \in S_{n+1}$. The hypothesis of the lemma is that $\delta_x = \gamma_x$ and $\delta_y = \gamma_y$ for all $y \in \Omega_x$. Since $\Omega_x \subset S^n$, the inductive hypothesis states that $v_\delta(z) = v_\gamma(z)$ for every $z \in \Omega_x$. Hence, by the monotonicity assumption, $v_\delta(x) = v_\gamma(x)$. The lemma is then true for every x in $S_{n+1} \cup S^n = S^{n+1}$, completing the inductive argument. Hence, the lemma is true for every $x \in \Omega_x$.

While the hypothesis of the lemma could have been stated as a readily verifiable assumption about the structure of the process, the usual route of getting along with the minimum number of assumptions has been followed. The user who is trying to decide whether a problem he has in mind fits this model might wish to assure himself that each state is a regeneration point, as well as verifying the monotonicity and termination assumptions.

IV. The Main Theorem and an Algorithm

In the optimization problems being considered, the objective is to find a policy (say, γ) that maximizes the total return available at some state x; that is, to find a $\gamma \in \Delta$ such that $v_\gamma(x) = f(x) = \max_{\delta \in \Delta} v_\delta(x)$. If $v_\gamma(x) = f(x)$, then the policy γ is called "optimal for x," and if $v_\gamma(z) = f(z)$ for all $z \in B$ (an arbitrary subset of Ω), then γ is called "optimal for B." There is no inherent reason why a policy that is optimal for x need be optimal for any other state, say $y \neq x$. Although intuition might suggest that in most interesting optimization problems the optimal policy for one state would not be optimal for other states, it will be shown below that for problems satisfying the assumptions of the previous section there exist policies that are optimal for *every* state in Ω.

The main result, expressed in the form of the theorem that follows, makes use of the following notation. Let $D_x^* = \{\delta_x \mid v_\delta(x) = f(x)\}$, and let $\Delta^* = \times_{x \in \Omega} D_x^*$. It is noted that $\Delta^* \neq \varnothing$, since each set D_x^* contains at least one decision. By way of interpretation, note that if $d_x \in D_x^*$, then d_x is the decision at x for a policy δ which is optimal for x; that is, $d_x = \delta_x$ and $v_\delta(x) = f(x)$. The policies in Δ^* then consist of all combinations of the decisions from the sets D_x^*.

Theorem If the termination and monotonicity assumptions are satisfied, then every policy $\delta \in \Delta^*$ is optimal for Ω.

Proof Consider any $\delta \in \Delta^*$. First it is shown that δ is optimal for S_1. With arbitrary $x \in S_1$, the monotonicity assumption implies $v_\delta(x) = v_\gamma(x)$ for any policy γ such that $\gamma_x = \delta_x$. Since, by definition of Δ^*, some such γ satisfies $v_\gamma(x) = f(x)$, policy δ is optimal for x. Since choice of x was arbitrary, δ is optimal for S_1.

For an inductive hypothesis, suppose δ is optimal for S^n. Let x be an arbitrary element of S_{n+1}, and let γ be a policy such that $v_\gamma(x) = f(x)$ and $\gamma_x = \delta_x$. (Such a policy exists by definition of Δ^*.) Since $\Omega_x \subset S^n$, the inductive hypothesis states that $v_\delta(z) = f(z)$ for all $z \in \Omega_x$. Hence, $v_\delta(z) \geqq v_\gamma(z)$ for all $z \in \Omega_x$; then the monotonicity assumption assures $v_\delta(x) \geqq v_\gamma(x)$. Hence, $f(x) = v_\delta(x)$, implying that δ is optimal for S_{n+1}. Since $\Omega = \bigcup_{n=1}^\infty S_n$, the inductive argument proves that δ is optimal for Ω.

The theorem asserts the existence of a (nonempty) set of policies, each of which is optimal for Ω, and it characterizes the decisions making up these policies. In addition, the theorem states that ties between decisions that are all optimal for a particular point can be broken arbitrarily.

Inherent in the proof of the theorem is a procedure for constructing a policy that is optimal for Ω. Indeed, much of the merit of dynamic programming lies in the efficiency of this algorithm. For the algorithm, it will be convenient to describe the return function in slightly different notation. Define the function $h(x, d_x, v_\delta)$ by

$$h(x, d_x, v_\delta) = v_\gamma(x) \qquad \text{where} \quad \gamma_x = d_x \quad \text{and} \quad \gamma_z = \delta_z \quad \text{for all} \quad z \neq x.$$

Two notions are present in this definition. First, the dependence on δ has been suppressed; only v_δ remains. This is justified by the lemma, which assures that if δ and γ are two policies such that $v_\delta = v_\gamma$, then $v_\pi(x) = v_\lambda(x)$ where π and λ are defined by $\pi_x = \lambda_x = d_x$ and, for all $z \neq x$, $\pi_z = \delta_z$ and $\lambda_z = \gamma_z$. Second, the dependence on d_x in $h(x, d_x, v_\delta)$ has been made explicit. Note that $h(x, d_x, v_\delta)$ is not a function of δ_x—that decision δ_x is immaterial, and it is not required that $\delta_x = d_x$. In economic terms, $h(x, d_x, v_\delta)$ might be interpreted as the cumulative return obtained by starting at state x and choosing decision d_x with the prospect of receiving the terminating reward $v_\delta(z)$ if transition occurs to state z. The algorithm is now displayed in the form of a corollary, whose proof replicates that of the theorem and is left as an exercise for the reader (Exercise CR-25).

Corollary 1 Suppose the monotonicity and termination assumptions are satisfied. Then for each n, the policy δ^n, constructed by the following procedure, is optimal for S^n.

1. With arbitrary δ°, set $f_0 = v(\delta^\circ)$ and $n = 1$.
2. For each $x \in S_n$, pick any d_x^* such that

$$h(x, d_x^*, f_{n-1}) = \max_{d_x} h(x, d_x, f_{n-1}).$$

3. Define δ^n and f_n by

$$\delta_z^n = \begin{cases} d_z^* & \text{if} \quad z \in S_n \\ \delta_z^{n-1} & \text{if} \quad z \notin S_n, \end{cases}$$

$$f_n = v(\delta^n).$$

4. If $\Omega = S^n$, stop. Otherwise, increment n by 1 and return to step 2.

The total number of policies whose returns must be evaluated in step 2 of the algorithm is $\sum_{x \in \Omega} \#(D_x)$, where $\#(A)$ is the number of elements in the set A. In contrast, the total number of possible policies is $\prod_{x \in \Omega} \#(D_x)$. By reducing the product ($\prod$) to a sum ($\sum$), the algorithm often effects a substantial reduction in the number of policies that need to be evaluated. Frequently, this reduction brings optimization problems within the range of tractable computation.

The preceding results also imply

Corollary 2 Suppose the monotonicity and termination assumptions are satisfied. Then

$$f(x) = \max_{d_x} h(x, d_x, f) \qquad \text{for each} \quad x \in \Omega.$$

Furthermore, if

$$v_\delta(x) = \max_{d_x} h(x, d_x, v_\delta) \qquad \text{for each} \quad x \in \Omega,$$

then $v_\delta = f$.

The preceding is the standard functional equation of dynamic programming, and Corollary 2 attests that f is its unique solution.

The functional equation for the investment–consumption model is

$$v_\delta(w, t) = \max_{0 \leqq c \leqq w} \left[u_t(c) + \beta \int_{-\infty}^{\infty} v_\delta(z, t+1)\, dP(z; w, t, \delta) \right].$$

If $t = T$, then the equation simplifies to

$$v_\delta(w, T) = \max_{0 \leqq c \leqq w} [u_T(c)] = u_T(w)$$

provided u_T is nondecreasing. In this context u_T may be interpreted as the utility of a bequest.

An implication of the terminating process concept is that the defect of each transition probability is 0 or 1. For a consumption model, this might be overly restrictive, since the consumer might die in any state. For example, assume that $1 - P[\infty, x, d_x] > \varepsilon$ for each $x \in \Omega$. So the process has at least probability ε of terminating in each state. This assumption ensures that the return function is well defined, since

$$v_\delta(w, t) \leqq E \sum_{n \geqq t} |u(c_n)| \leqq M/\varepsilon < \infty \qquad \text{provided} \quad |u| \leqq M.$$

V. Nonterminating Processes

The main result of Section IV is that a policy exists that is optimal for Ω.

While this result might seem unintuitive at first, it has in fact been verified for a wide variety of more complex settings. These more complex settings include infinite horizon problems for which a discount factor is applied at each epoch of the evolution of the process [3, 5, 7] and infinite horizon problems in which the average rate of return per unit time is maximized [1, 2, 7]. Also included are processes whose duration is subject to chance [6, 9, 10] and those found in sequential analysis [11]. These models fail to satisfy the termination assumption. They do not admit of induction from the end, and their analysis necessarily relies on more difficult mathematical tools.

Rather than attempting to outline the analysis required for these more complex cases, one can start with the main conclusion generally obtained in their analysis, termed the

Optimality Postulate There exists a policy δ that is optimal for Ω.

It shall be shown that the standard functional equation follows readily from the optimality postulate and the monotonicity assumption.

Experience with a wide range of dynamic programming problems might convince the practitioner that the optimality postulate is satisfied by a particular problem he has in mind. Since the mathematics that is otherwise required might be quite complex, there is some merit to adopting this conclusion as a postulate and proceeding heuristically. Indeed, Bellman's [1] classical approach to dynamic programming is heuristic in that it centers around the principle of optimality—a property of the optimal policy which one must either verify or accept for his particular problem. The optimality postulate also serves as an alternative starting point to the principle of optimality.

Preference for the optimality postulate or the principle of optimality as a starting point is, it is felt, largely a matter of personal taste. Furthermore, they are so closely interrelated that the optimality postulate might justifiably be thought of as a minor variant of the principle of optimality. The optimality postulate is perhaps more elementary and more readily established as the conclusion of a theorem. On the other hand, the principle of optimality leads somewhat more directly to the functional equations of dynamic programming.

The central issue in applying this heuristic approach to a particular problem is determining whether the principle of optimality or optimality postulate is satisfied. Experience in the principle of optimality has not been flawless; it is not difficult to convince oneself incorrectly that it applies to a particular problem. In fact, the literature abounds with misapplication of the principle of optimality. Every problem known to the authors for which some policy is optimal for Ω also satisfies the version of the monotonicity property stated below.

Since the question of whether a particular return function satisfies the monotonicity property is readily decided, doing so helps one determine whether to apply either the ptinciple of optimality or the optimality postulate.

Monotonicity Property Let δ and γ be any two policies such that $v_\delta \geqq v_\gamma$. Then $h(x, d_x, v_\delta) \geqq h(x, d_x, v_\gamma)$ for every $x \in \Omega$ and $d_x \in D_x$.

With one exception, the definitions and notation introduced in the preceding sections are unchanged here. The state space Ω, the decision set D_x, the policy space Δ, the return $v_\delta(x)$, and the optimal return $f(x)$ are exactly as defined before. However, the interpretation of $h(x, d_x, v_\delta)$ is changed slightly. The current interpretation of $h(x, d_x, v_\delta)$ is as the return for starting at x, choosing decision d_x for x, and thereafter choosing decision δ_z whenever state z is visited. Should state x be revisited, decision δ_x (not d_x) will be selected for it. Clearly, if state x cannot be revisited, then $h(x, d_x, v_\delta) = v_\gamma(x)$ for that policy γ defined by $\gamma_x = d_x$ and $\gamma_z = \delta_z$ for all $z \neq x$—just as before. If state x is revisited, then $h(x, d_x, v_\delta)$ might not be the return $v_\gamma(x)$ for any policy γ, since in the case that $\delta_x \neq d_x$ no policy γ corresponds to the decision procedure used for determining $h(x, d_x, v_\delta)$.

Unidirectional Processes

Next, the notion that a process evolves unidirectionally is made specific. Each point x in Ω has a real number $t(x)$ associated with it, where $t(x)$ might be thought of as the time at which x occurs. For processes that evolve unidirectionally, it shall be assumed roughly that the return v_δ depends only on δ_x and on the decisions made in states occurring in future time.

Evolution-Dependence Assumption Consider any two policies δ and γ such that $\delta_x = \gamma_z$ and $\delta_z = \gamma_z$ for all z such that $t(z) > t(x)$. Then $v_\delta(x) = v_\gamma(x)$.

In this sense, a state is thought of as a regeneration point for the process. The assumption is readily tested for a particular problem. Of course, the lemma in the preceding section verifies this assumption for terminating processes. The evolution-dependence assumption assures that $h(x, d_x, v_\delta)$ is independent of δ_z for every z satisfying $t(z) \leqq t(x)$. Hence, $h(x, d_x, v_\delta) = v_\gamma(x)$ for that policy γ defined by $\gamma_x = d_x$ and $\gamma_z = \delta_z$ for all $z \neq x$.

With ξ as a policy that is optimal for Ω, one has $v_\xi = f$ and

$$f(x) = \max_{d_x \in D_x} h(x, d_x, f) \qquad \text{for each} \quad x \in \Omega.$$

Cyclic Processes

Often, as in a Markov chain model [7], states may be revisited many or infinitely many times. In this situation, one can remember the history, redefine the state space, and generate a model that evolves unidirectionally. This generated model has many more states than the original problem, as well as a considerable amount of symmetry. With Ω' as the state space for the generated problem, the essential question is whether a policy that is optimal for Ω' need remember the history of the process. If not, it is called "stationary" and is equivalent to a policy that is optimal for the original problem. See Exercises ME-17 and 18 in this regard.

The issue shall not be dealt with here of imposing enough structure on the original problem that its history-remembering version has a stationary policy that is optimal for Ω'. Rather, the alternative route of introducing an assumption that allows one to deal directly with the original problem shall be followed.

Continuation Assumption If $h(x, d_x, v_\delta) \geqq v_\delta(x)$, then $v_\gamma(x) \geqq h(x, d_x, v_\delta)$, where γ is the policy determined by $\gamma_x = d_x$ and $\gamma_z = \delta_z$ for $z \neq x$.

The continuation assumption states simply that if the return is improved by substituting d_x for δ_x the first time state x is encountered, then substituting d_x for δ_x indefinitely also improves it. This assumption is more difficult to verify than the monotonicity assumption but it has intuitive appeal and has been verified for many processes (see, for instance, Denardo [5, Lemma 2]) With ξ as a stationary policy that is optimal for Ω, the continuation assumption implies $h(x, d_x, v_\xi) \leqq h(x, \xi_x, v_\xi) = f(x)$. Hence,

$$f(x) = \max_{d_x \in D_x} h(x, d_x, f) \qquad \text{for each} \quad x \in \Omega.$$

For cyclic and unidirectional processes, the standard functional equation of dynamic programming has been obtained as an outgrowth of the optimality postulate and supplementary assumptions: f is optimal over stationary policies—this is no restriction if there exists a stationary optimal policy.

ACKNOWLEDGMENT

This article was adapted from the paper by Denardo and Mitten (1967). It is intended to provide an introduction to dynamic programming for readers of this book. I would like to thank Professors E. V. Denardo and L. G. Mitten for permission to include this adaptation of their article and for their helpful comments. Errors and misconceptions in this presentation are the responsibility of the adapter and not Denardo and Mitten. Thanks are also due to S. L. Brumelle who prepared the consumption–investment example and some notes that became part of an earlier version of this adaptation, and to R. G. Vickson for helpful comments and suggestions.

References

1. BELLMAN, R. E., *Dynamic Programming*. Princeton Univ. Press, Princeton, New Jersey, 1957.
2. BLACKWELL, D., "Discrete Dynamic Programming." *Annals of Mathematical Statistics* **33** (1962), 719–726.
3. BLACKWELL, D., "Discounted Dynamic Programming." *Annals of Mathematical Statistics* **36**, (1965), 226–235.
4. DENARDO, E. V., *Sequential Decision Processes*. Ph.D. Dissertation. Northwestern Univ., Evanston, Illinois, June 1965.
5. DENARDO, E. V., "*Contraction Mappings in the Theory Underlying Dynamic Programming*." *SIAM Review* **9** (1967), 165–177.
6. DUBINS, L. E., AND SAVAGE, L. J., *How To Gamble If You Must*. McGraw-Hill, New York, 1965.
7. HOWARD, R. A., *Dynamic Programming and Markov Processes. Technol.* MIT Press, Cambridge, Massachusetts, 1960.
8. MITTEN, L. G., "Composition Principles for Synthesis of Optimal Multi-Stage Processes." *Journal of the Operations Research Society of America* **12**, (1964), 610–619.
9. SHAPLEY, L. S., "Stochastic Games." *Proceedings of the National Academy Sciences of the United States* **39**, (1953), 1095–1100.
10. STRAUCH, R. E., "Negative Dynamic Programming." *Annals of Mathematical Statistics* **37**, (1966), 871–890.
11. WALD, A., *Sequential Analysis*. Wiley, New York, 1947.

COMPUTATIONAL AND REVIEW EXERCISES

1. Consider an investor having an initial wealth of \$16 whose utility function for wealth w is $w^{1/2}$.

(a) Show that he is indifferent between the status quo and the following two gambles:
(i) lose \$7 or gain \$9 with equal probability, and
(ii) lose \$16 with probability 22/80, gain \$48 with probability 1/8 and gain \$9 with probability 3/5.

(b) Will he prefer to accept one or more of these gambles if his initial wealth is \$25?

(c) Change the odds on these gambles so that he is indifferent between each of the gambles and the status quo if his initial wealth is \$30.

(d) Consider the gambles in (a) and let the utility function be w^{α}. For what values of α strictly between 0 and 1 does the investor prefer the status quo, gamble (i), and gamble (ii)?

(e) Suppose the utility function is a general concave function u. Will the investor ever prefer or be indifferent between the status quo and the fair gamble, gain or lose (a) with equal probability?

(f) What happens if u is strictly concave or convex?

2. A consequence of the expected utility theorem is that cardinal utility functions are unique only up to positive linear transformation. Thus $u(w) = a + bv(w)$ for $b > 0$ represents the same preferences as v. To specify the utility function uniquely one must specify two boundary conditions, such as requiring that u go through the origin with slope 1, i.e., $u(0) = 0$, $u'(0) = 1$. Determine what a and b must be in this case if $v(w)$ is

(a) $\log(w + w_0)$, where $w_0 > 0$;
(b) $(w + w_0)^{\alpha}$, where $w_0 > 0$ and $0 < \alpha < 1$;
(c) $w - \beta w^2$, $\beta > 0$; (d) $-e^{-\delta w}$, $\delta > 0$.

3. Prove that each of the utility functions in Exercise 2 is strictly concave and nondecreasing on suitably restricted domains of w.

4. Consider a gamble where a fair coin is repeatedly tossed until a head is obtained. If a head is obtained on the first toss, the payoff is \$2, \$4 if the head is obtained on the second toss, \$8 on the third, and so on, so that with each additional toss the payoff doubles.

(a) Show that the expected return from this gamble is infinite.

(b) Let x be the amount that one is willing to pay for the gamble. What is the probability that a profit is made if $x = 10, 20, 50, 100, 1000$?

(c) What would you be willing to pay for the gamble? Illustrate the St. Petersburg paradox.

(d) Illustrate how the paradox may be resolved by choosing a particular concave utility function (such as $\log w$ or $w^{1/2}$, where w is wealth) and by showing that there exists an optimal value of the gamble at all (finite) initial wealth levels.

(e) Show that it is always possible to create a gamble that leads to a paradox if the gambler's utility function is unbounded.

5. The preference ordering of an investor generally depends on his wealth level w.

(a) Show that if the utility functions $u(w)$ and $u(w+\beta)$ represent the same preference ordering for all values of β, then these utility functions satisfy

$$u(w+\beta) = a(\beta)\,u(w) + b(\beta),$$

where it is noted that the constants a and b may depend on β.

(b) Show that the only utility functions that have the property that their preference orderings are independent of wealth are $u(w) = aw$ or $u(w) = be^{cw}$ where a, b, and c are real numbers. [*Hint*: Differentiate the equation in (a) with respect to w and β, combine these, and solve the resulting differential equation with $da(\beta)/d\beta =$ or $\neq 0$.]

6. Suppose that an investor's utility function is $u(w) = w^3$.
(a) Construct a fair gamble, an actuarily unfair insurance situation, and a wealth level for which the investor simultaneously wishes to gamble and purchase insurance.

For the example used in (a), show that at different wealth levels the investor is
(b) unwilling to purchase the given insurance, and
(c) unwilling to take the given gamble.
(d) Is there a wealth level at which the investor is unwilling to purchase the given insurance or to take the given gamble?
(e) Is the result of (d) general in the sense that its answer remains the same for all gambles that illustrate (a)?

Illustrate the wealth ranges and sets of gambles for which the investor is
(f) a risk averter, or
(g) a risk taker.

Show that on E, u is
(h) quasi-concave, and
(i) not pseudo-concave.

7. (Jensen's inequality) Suppose $\Lambda \subset E^n$ and $H \subset E^m$ are convex sets.
(a) Show that the function $f(x)\colon \Lambda \to (-\infty, +\infty]$ is convex on Ω if and only if

$$\sum_{i=1}^{n} \lambda_i f(x^i) \geqq f\left(\sum_{i=1}^{n} \lambda_i x^i\right), \qquad \text{where} \quad \lambda_i \geqq 0, \quad \sum_{i=1}^{n} \lambda_i = 1.$$

(b) Show that the function $f(\xi)\colon H \to (-\infty, +\infty]$ is convex on H if and only if

$$E_\xi f(\xi) \geqq f(\bar{\xi})$$

provided $\bar{\xi}$ exists.
(c) Show that the inequalities are reversed if f is concave.
(d) Show that these inequalities become equalities if f is linear.
(e) Show that the inequality in (b) becomes an equality for all f if and only if all of the mass of ξ is at $\bar{\xi}$.

8. Determine whether the following functions of x defined on E^1 are pseudo-convex, convex, strictly convex, quasi-concave, pseudo-concave, concave, strictly concave, or none of these.
(a) x^3; (b) $x^3 + x$; (c) $\exp(-x^2)$;
(d) $\exp(-x^3)$; (e) e^{2x}; (f) $\exp(2^x)$;
(g) $|x^{2m+1}|$ for $m = 0, 1, \ldots, T$; (h) $|x^{2m}|$ for $m = 0, 1, \ldots, T$;
(i) $\sin x$; (j) $\exp(x^2) + x^3$.

9. Determine whether the following functions of $(x_1, \ldots, x_n)$ defined on E^n are quasi-convex, pseudo-convex, convex, strictly convex, quasi-concave, pseudo-concave, concave, strictly concave, or none of these.
(a) $\sum c_i x_i$.
(b) $\sum c_i x_i + \frac{1}{2} \sum \sum C_{ij} x_i x_j$, where $C = [C_{ij}]$ is negative definite.
(c) $(a_0 + \sum a_i x_i)/(b_0 + \sum b_i x_i)$ on the subset of E^n, where the denominator is nonzero.

(d) $x_1 x_2$ on $E_+{}^2 \equiv \{x \mid x_1 \geqq 0,\ x_2 \geqq 0\}$.
(e) $(x_1 x_2)^{1/2}$ on $E_+{}^2$. (f) $x_1 x_2 \cdots x_n$ on $E_+{}^n$.
(g) $\sum \log [1+\log(1+x_i)^{b_i}]^{a_i},\ a_i, b_i > 0$ on $E_+{}^n$.

10. Suppose an investor wishes to allocate \$100 between stocks and bonds. Bonds yield 1.05 per dollar invested whereas stocks yield 1.10. The investor trades off higher yields versus higher risks according to the preference function {mean−(0.3) variance}. Suppose bonds are risk free and stocks have a variance of 0.5. Find the optimal allocations using the Kuhn–Tucker conditions.

11. Let $x = 0$ represent your present wealth. If P is a probability measure on amounts of money that represent potential incremental additions to your present wealth and if $P \sim x$ (where x is considered as a sure-thing addition to your present wealth), then x is a certainty equivalent for P; i.e., $P \sim x$ means that you are indifferent between gambling with P and receiving x as an outright gift. Estimate your certainty equivalent for P when
(a) $P(0) = 0.5,\ P(10{,}000) = 0.5$.
(b) $P(0) = 0.1,\ P(1{,}000{,}000) = 0.9$.
(c) $P(0) = \frac{1}{3},\ P(1000) = \frac{1}{3},\ P(3000) = \frac{1}{3}$.

Estimate your certainty equivalent for
(d) $P(0) = 0.01,\ P(5000) = 0.99,\ x \sim P$.
(e) $Q(0) = 0.99,\ Q(5000) = 0.01,\ y \sim Q$.
(f) $R(0) = 0.5,\ R(5000) = 0.5,\ z \sim R$.
(g) Show that the expected utility theory implies that $R \sim \frac{1}{2}P + \frac{1}{2}Q$.
(h) Does this mean that $z = \frac{1}{2}(x+y)$?
(i) Does it mean that z is indifferent to a 50–50 gamble between x and y?

12. Find and examine the concavity properties of the Wolfe duals of the following problems.
(a) $\{\min a'x + \frac{1}{2}x'Qx \mid Ax \leqq b,\ x \geqq 0\}$, where Q is positive semidefinite;
(b) $\{\min a'x \mid Ax \geqq b,\ x \geqq 0\}$; show that the dual is equivalent to the ordinary linear programming dual $\{\max \pi'b \mid A'\pi \leqq a,\ \pi \geqq 0\}$; and
(c) $\{\min a'x + \frac{1}{2}x'Qx \mid c'x + \frac{1}{2}x'Bx \leqq b\}$, where Q and B are positive semidefinite.

13. Utilize the Kuhn–Tucker conditions to show that:
(a) $x^* = (1, 0)$ solves
$$\min\{[(x_1+1)^3/3] + x_2 \mid x_1 \geqq 1,\ x_2 \geqq 0\};$$
(b) $x^* = (1, 1)$ solves
$$\max\{10x_1 - 2x_1{}^2 - x_1{}^3 + 8x_2 - x_2{}^2 \mid x_1 + x_2 \leqq 2,\ x_1 \geqq 0,\ x_2 \geqq 0\};$$
(c) $x^* = (5)$ solves $\{\min -e^x \mid 3 \leqq x \leqq 5\}$;
(d) $x^* = (\sqrt[5]{2/3}, 0)$ solves
$$\min\{x_1{}^3 + 1/x_1{}^2 \mid x_1 + x_2{}^2 \leqq 3,\ x_1 \geqq 0,\ x_2 \geqq 0\};$$
(e) $x^* = (5, 3)$ solves
$$\max \exp[-(x_1{}^2 + x_2{}^2)^{-1/k}] \mid 0 \leqq x_1 \leqq 5,\ 0 \leqq x_2 \leqq 3\} \quad \text{where} \quad k > 0.$$
Be sure to verify that the sufficiency assumptions are satisfied.

14. Suppose $f(x)$ and $g(x)$ are defined on the convex set $C \subset R^n$.
(a) Show that f is quasi-concave (pseudo-concave) iff $\alpha f(x) + \beta$ is quasi-concave (pseudo-concave) for arbitrary scalars $\alpha \geqq 0$ and β.
(b) Show that $\alpha f(x) + \beta g(x)$ is not necessarily quasi-concave (pseudo-concave) if f and g are quasi-concave (pseudo-concave) and $\alpha > 0$ and $\beta > 0$ are arbitrary scalars.
(c) Show that $\alpha f(x) + \beta g(x)$ is (strictly) concave if (at least one of) f and g are (strictly) concave and $\alpha > 0$ and $\beta > 0$ are arbitrary scalars.

15. Besides functions that are convex (C), strictly convex (SC), quasi-convex (QC), and pseudo-convex (PC), it is of interest to consider the related functions that are strictly quasi-convex, strictly pseudo-convex, and XC convex. Let $f(x)$ be a numerical function defined on the convex set $C \subset R^n$. Then f is strictly quasi-convex (SQC) if

$$f(x^2) < f(x^1) \quad \Rightarrow \quad f[\lambda x^1 + (1-\lambda)x^2] < f(x^1), \qquad \forall 0 < \lambda < 1, \quad x^1, x^2 \in C;$$

f is strictly pseudo-convex (SPC) if f is differentiable and

$$f(x^2) \leqq f(x^1) \quad \Rightarrow \quad (x^2 - x^1)' \nabla f(x^1) < 0, \qquad \forall x^1, x^2 \in C, \quad x^1 \neq x^2;$$

and f is XC convex if

$$f(x^2) \leqq f(x^1) \quad \Rightarrow \quad f[\lambda x^1 + (1-\lambda)x^2] < f(x^1),$$
$$\forall x^1, x^2 \in C, \quad x^1 \neq x^2, \quad \text{and} \quad 0 < \lambda < 1.$$

(a) Show that a local minimum of a quasi-convex function is not necessarily a global minimum by considering the function

$$f(x) = \begin{cases} a & \text{if } 0 \leqq x < \frac{1}{2} \\ b & \text{if } \frac{1}{2} \leqq x \leqq 1 \end{cases} \qquad \text{where } a \neq b.$$

(b) Suppose a local minimum x^1 of a quasi-convex function f is not a global minimum. Then f is constant in the intersection of x_1 and the line segment between x^1 and any global minimum x^*.

(c) Show that a local minimum of a strictly quasi-convex or a strictly pseudo-convex function is a global minimum.

(d) Show that $D \equiv \{x \mid g_j(x) \geqq 0,\ i = 1, \ldots, m\}$ is convex if the g_j are quasi-concave.

(e) Show that the seven kinds of convexity may be related as follows: SC $\Rightarrow$ C, C $\Rightarrow$ PC, PC $\Rightarrow$ SQC, and SQC $\Rightarrow$ QC; and SC $\Rightarrow$ XC, XC $\Rightarrow$ SQC, and PC + XC $\Rightarrow$ SPC. Assume differentiability as needed, and that the SQC functions are lower semicontinuous, otherwise SQC $\not\Rightarrow$ QC.

(f) Show that the seven kinds of convexity are indeed different by verifying the results in the following table, where a plus indicates that the function has the stated property.

Function	Region R	SC	C	SPC	PC	SQC	QC	XC
$f_1(x) = \begin{cases} 0 \\ -(x-1)^2 \end{cases}$	$0 \leqq x \leqq 1$ $1 \leqq x \leqq 2$	−	−	−	−	−	+	−
$f_2(x, y) = -x^2$	$0 \leqq x, y \leqq 1$	−	−	−	−	+	+	−
$f_3(x, y) = -x^2 - x$	$0 \leqq x, y \leqq 1$	−	−	−	+	+	+	−
$f_4(x) = 0$	$0 \leqq x \leqq 1$	−	+	−	+	+	+	−
$f_5(x) = -x$	$0 \leqq x \leqq 1$	−	+	+	+	+	+	+
$f_6(x) = x^2$	$0 \leqq x \leqq 1$	+	+	+	+	+	+	+
$f_7(x) = -x^2 - x$	$0 \leqq x \leqq 1$	−	−	+	+	+	+	+
$f_8(x) = -x^2$	$0 \leqq x \leqq 1$	−	−	−	−	+	+	+

(g) Suppose f is either SC, SPC, or SQC. Show that f either attains a unique minimum on the closed convex set C or has no minimum.

16. The determination of the correct sign of the Lagrange multiplier is of paramount importance in the use of the Kuhn–Tucker conditions.

(a) Suppose that the problem is

$$\{\min f(x) \mid g(x) \leqq 0\}.$$

Show that the Kuhn–Tucker conditions are $\nabla f(x) + \lambda' \nabla g(x) = 0$, $g(x) \leqq 0$, and $\lambda \geqq 0$. [*Hint*: Investigate the direction gradients for the problem $\{\min x^2 \mid -x \leqq 0\}$.]
(b) Utilize the result in (a) to verify the following chart:

	$\max f$	$\min f$
$g \geqq 0$	$\lambda \geqq 0$	$\lambda \leqq 0$
$g \leqq 0$	$\lambda \leqq 0$	$\lambda \geqq 0$

giving the sign of the multipliers in the Kuhn–Tucker conditions: $\nabla f(x) + \lambda' \nabla g(x) = 0$, $g(x) \leqq 0$ or $\geqq 0$.

17. Suppose that an investor's preference ordering for proportional gambles on wealth is independent of his wealth level.

(a) Show that

$$p_1 u[w(1+r_1)] + p_2 u[w(1+r_2)] = q_1 u[w(1+s_1)] + q_2 u[w(1+s_2)],$$

if and only if

$$p_1 u(1+r_1) + p_2 u(1+r_2) = q_1 u(1+s_1) + q_2 u(1+s_2)$$

for all p_i, q_i, r_i, s_i, and w, with $w > 0$, $p_i, q_i \geqq 0$, $p_1 + p_2 = q_1 + q_2 = 1$.
(b) Show that u must have the form w^α or $\log w$.

18. This problem presents an alternative to the expected utility approach based on the assumption that the decision maker is an augmented income maximizer. What is altered in this approach is not the shape of the utility function (assumed to be linear) but the measure of gains and losses.

Let c be the cost of a lottery ticket or an insurance policy and r_c the opportunity rate on c. Suppose that the possible monetary outcomes are $\xi_1, \ldots, \xi_m$ having probabilities $p_i > 0$ and that the opportunity interest rate on each ξ_i is r_i. Over a T-period horizon the augmented cost is $c_a = c(1+r_c)^T$ and the expected augmented gain (or loss) is $y_a = \sum_i \xi_i p_i (1+r_i)^T$. Under the suggested approach, choices under uncertainty are dictated by the relative sizes of c_a and y_a.

(a) When does this approach reduce to the expected utility approach?
(b) Show that a gamble is fair if $c = \sum p_i \xi_i$.
(c) Show that for fair gambles if

$$r_i \gtreqqless r_c \qquad \text{for all} \quad i, \qquad \text{then} \quad y_a \gtreqqless c_a.$$

(d) What can you conclude from (c)? What can be said if

$$r_i > r_c \qquad \text{and} \qquad r_j < r_c?$$

(e) Show that an augmented income maximizer will prefer a ticket in a fair lottery over the sure income c if lending rates are increasing functions of the amount lent—assuming that each $\xi_i > c > 0$, except ξ_n, which is zero.
(f) Suppose that borrowing rates are increasing functions of the amount borrowed. Suppose further that in the fair insurance case the investor is able to pay for the premium without outside borrowing but will have to borrow if adverse outcomes occur without insurance. Show that it is optimal to purchase insurance as long as the lending rate is less than the borrowing rate.
(g) Show that the augmented income approach can explain the phenomenon of simultaneous insurance and gambling activities.

19. Refer to the "Introduction to Dynamic Programming." Verify that the following return functions satisfy the monotonicity assumption.

(a) $v_\delta(x) = r(x, \delta_x) \cdot v_\delta(z)$, where $\{z\} = T(x, \delta_x)$ and $r(x, \delta_x)$ and $v_\delta(z)$ are nonnegative.

(b) $v_\delta(x) = r(x, \delta_x) + \sum_{z \in T(x, \delta_x)} v_\delta(z)$, where the cardinality of $T(x, \delta_x)$ is finite for each x.

(c) $v_\delta(x) = r(x, \delta_x) + \int_{z \in T(x, \delta_x)} p(z: x, \delta_x) v_\delta(z)$, where $p(\cdot) \geqq 0$.

(d) $v_\delta(x) = \min_{u \in U_x}\{r(x, \delta_x, u) + \sum_{z \in T(x, \delta_x)} p[z: x, \delta_x, u] v_\delta(z)\}$, where $p(\cdot) \geqq 0$.

(e) When would one obtain these return functions?

(f) Show that (a)–(d) may be generalized by replacing "+" and multiplication wherever they appear by $\oplus$, where $\oplus$ is any commutative, symmetric binary operator that preserves inequalities on a subset H of the real numbers; that is, $(a \oplus b) \oplus c = a \oplus (b \oplus c)$, $a \oplus b = b \oplus a$, and if $a \geqq b$, then $a \oplus c \geqq b \oplus c$.

(g) Verify that the following are $\oplus$ operators:

$$a \oplus b = a + b, \qquad a \oplus b = \max\{a, b\}, \qquad a \oplus b = \min\{a, b\};$$
$$a \oplus b = e^{a+b} \qquad \text{and} \qquad a \oplus b = ab,$$

the last if H is the nonnegative orthant.

(h) Show that the monotonicity assumption is satisfied in the consumption–investment model if the u_t are monotone nondecreasing.

(i) Suppose $u_t(c_t) = \alpha_t + \beta_t c_t + \gamma_t c_t^2 + \gamma_t c_t^3$. Determine conditions on the parameters $\alpha_t, \beta_t, \gamma_t$, and δ_t so that the monotonicity assumption is satisfied for all $c_t \in [0, M]$, where $M < \infty$.

20. Consider the allocation problem

$$f_N(b) \equiv \max R(x_1, \ldots, x_N) \equiv g_1(x_1) + \cdots + g_N(x_N)$$
$$\text{s.t.} \quad x_1 + \cdots + x_N = b, \quad \text{all } x_n \geqq 0.$$

Hence $f_N(b)$ is the optimal return from an allocation of resources b to N activities.

(a) Derive the functional equation

$$f_n(b) = \max_{0 \leqq x_n \leqq b}\{g_n(x_n) + f_{n-1}(b - x_n)\} \qquad \text{for } n = 2, 3, \ldots, N \text{ and } b \geqq 0.$$

(b) Interpret the meaning of the functional equation in (a).

The recurrence relation in (a) yields a theoretical method for obtaining the sequence $\{f_n(b)\}$ inductively once $f_1(b)$ is determined.

(c) Show that $f_1(b) = g_1(b)$.

(d) Develop an algorithm to solve the functional equation in (a) based on the grid $0, \Delta, 2\Delta, \ldots, T\Delta = b$.

(e) Apply the algorithm in (d) when $N = 3$, $b = 10$, $g_1(x_1) = x_1^3$, $g_2(x_2) = 4x_2^2$, and $g_3(x_3) = 8x_3$.

Suppose now that the x_n are subject to the constraints $a_n \leqq x_n \leqq b_n$.

(f) Derive the new functional equation.

(g) Modity the algorithm in (d) to solve the functional equation in (f).

(h) Solve the problem in (e) when $(a_1, a_2, a_3) = (0, 1, 0)$ and $(b_1, b_2, b_3) = (1, 2, 4)$.

21. This problem is a generalization of Exercise 20 to the two-dimensional constraint case and illustrates the use of Lagrange multipliers in dynamic programming problems. Consider the two-dimensional resource allocation problem.

$$f_N(b, c) \equiv \max R(x_1, \ldots, x_N, y_1, \ldots, y_N) = \sum_{i=1}^{N} g_i(x_i, y_i),$$
$$\text{s.t.} \quad \sum x_i = b, \quad \text{all } x_i \geqq 0,$$
$$\sum y_i = c, \quad y_i \geqq 0.$$

(a) Show that $f_1(b,c) = g_1(b,c)$.
(b) Develop the functional equation

$$f_n(b,c) = \max_{0 \leqq x_n \leqq b} \max_{0 \leqq y_n \leqq c} [g_n(x_n, y_n) + f_{n-1}(b-x_n, c-y_n)] \text{ for } n = 2,3,\dots,N.$$

(c) Develop a grid algorithm to solve the problem in (b).
(d) How much more complicated is the two-dimensional case as opposed to the one-dimensional case considered in Exercise 20?
(e) Apply the algorithm when $N = 3$, $b = 10$, $c = 15$, $g_1(x_1, y_1) = x_1^3 + y_1^2$, $g_2(x_2, y_2) = 4x_2^2 + y_2$, and $g_3(x_3, y_3) = 4x_2^2$.

Consider the problem of maximizing the modified function

$$g_1(x_1, y_1) + \cdots + g_N(x_N, y_N) - \lambda[y_1 + \cdots + y_N]$$
$$\text{s.t.} \quad x_1 + \cdots + x_N = b, \tag{*}$$

and all $x_i, y_i \geqq 0$, where λ is a fixed parameter. Assume that

$$\lim_{y_n \to \infty} [g_n(x_n, y_n)/y_n] = 0.$$

(f) Show that the maximization over y_n and x_n may be done independently.
(g) Show that

$$r_n(x_n, \lambda) \equiv h_n(x_n) \equiv \max_{y_n \geqq 0} [g_n(x_n, y_n) - \lambda y_n].$$

(h) Show that the original problem is then equivalent to maximizing $h_1(x_1) + \cdots + h_N(x_N)$ subject to (*) if λ is varied until the restriction $\sum y_n(\lambda) = c$ is met.
(i) Modify the algortihm in Exercise 20 to solve this problem.

22. Suppose an economy has a productive capcaity of M in year 1 and we wish to plan production for the years $1, \dots, N$. Productive capacity can be used either for the production of consumer goods or for the development of additional production capacity. Consumers receive $f(x)$ dollars worth of goods if x units of the productive capacity are allocated (for one year) to the manufacture of consumer goods. At the end of the year, some of the productive capacity allocated to the production of consumer goods will be worn out, and only a quantity αx, $0 < \alpha < 1$, will be usable in the following year. If y units of productive capacity are allocated to the formation of capital goods, then at the end of the year, we have a productive capacity of δy, $\delta > 1$, corresponding to these y units. Suppose that α, δ, and f are stationary in time but that consumer goods obtained at a later time are not as valuable as those available immediately. The one-period discount factor is $0 < r < 1$. It is of interest to determine how the economy should divide its productive capacity between consumption and investment goods in each of the N years so as to maximize the present value of the economy's consumption.
(a) Formulate the problem as a dynamic program.
(b) Develop a flow chart indicating the calculations involved to solve the problem for general f.

Suppose $f(x) = a + bx$, where $a, b > 0$.
(c) Find the optimal policy.
(d) Let $M = 10$, $N = 5$, $r = 0.9$, $\alpha = \frac{1}{2}$, $\delta = 1\frac{1}{2}$, $a = 1$, and $b = \frac{1}{2}$. Find the optimal decisions.

Suppose now that δ is a random variable and that $p_r\{\delta = \delta_1\} = p_1 > 0$ and $p_r\{\delta = \delta_2\} = 1 - p_1$.
(e) Formulate a dynamic programming model of the economy's problem assuming that the goal is to maximize the present value of the economy's expected consumption.
(f) Let $p_1 = p_2 = \frac{1}{2}$, $\delta_1 = 1\frac{1}{4}$, and $\delta_2 = 1\frac{3}{4}$. Compute the optimal policy using the data in (d).

(g) Compare the results in (f) with those in (d).

(h) What happens if x and/or y are discrete variables? [*Hint*: See Nemhauser (1966, p. 126).]

23. A firm estimates that their sales in the next 6 months will be

Month:	1	2	3	4	5	6
Sales:	100	110	95	90	120	130

The production level is 95 units per month and the intial inventory level is 25 units. Changes in the level of production from one month to the next cost the firm $5\,|x_t - x_{t-1}|$ dollars, where x_t is the production level in period t. It costs the firm one dollar for each unit held in inventory for one month. The firm wishes to determine production levels in each month in order to minimize total cost. It is supposed that the firm must always meet the demand quotas and that items in the final inventory are worthless.

(a) Formulate this problem as a dynamic program.

(b) Formulate this problem as a linear program.

(c) Which is the best solution technique?

Suppose the sales estimates are subject to the following source of error. Let d and f be the actual demand and the forecasted demand, respectively, and suppose that

$$p_r\{d=f\} = \tfrac{1}{2}, \qquad p_r\{d=f+5\} = \tfrac{1}{4}, \qquad \text{and} \qquad p_r\{d=f-5\} = \tfrac{1}{4}.$$

Suppose demand must always be met and that the objective is to minimize expected total costs.

(d) Formulate the problem as a dynamic program.

(e) Formulate the problem as a linear program.

(f) Compare these two solution approaches.

Alternatively suppose that unsatisfied demand is subject to a penalty of \$10 per unit but that no more than 5 units of the demand may be unsatisfied.

(g) Formulate this problem as a dynamic program.

(h) Formulate this problem as a linear program.

(i) Compare the two solution approaches.

24. Consider an N-period dynamic problem where the states are $x_1, \dots, x_N$. Suppose that the system will be in state $x_n = r_{n-1}(x_{n-1}, d_{n-1})$ if decision d_{n-1} is made when the system is in state x_{n-1}. Let g be a preference function over states $x_2, \dots, x_{N+1}$. Then the problem is to determine a set of decisions $d_1, \dots, d_N$ that maximize

$$g(x_{N+1}, \dots, x_2) = g\,[r_N(x_N, d_N), \dots, r_1(x_1, d_1)].$$

Assume that an optimal policy exists. The function g is said to be separable if there exist functions g_1 and g_2 such that

$$g(x_{N+1}, \dots, x_2) = g_1\,[x_{N+1}, g_2(x_N, \dots, x_2)].$$

(a) Show that g is always separable if $N = 2$.

(b) Interpret the separability assumption.

The function g_1 is said to be monotonic if g_1 is a nondecreasing function of g_2 for all fixed x_{N+1}. The dynamic problem is said to decompose if

$$\max_{d_N, \dots, d_1} g\,[r_N(x_N, d_N), \dots, r_1(x_1, d_1)]$$

$$= \max_{d_N} g_1 \left[r_N(x_N, d_N), \max_{d_{N-1}, \dots, d_1} g_2\{r_{N-1}(x_{N-1}, d_{N-1}), \dots, r_1(x_1, d_1)\} \right].$$

(c) Show that the monotonicity assumption guarantees decomposition when $N = 2$.
(d) Show that the monotonicity and separability assumptions guarantee decomposition for general N.
(e) Show that decomposition is always possible if $g = r_N + \cdots + r_1$.
(f) Is decomposition always possible if $g = \sum_{j=2}^{N+1} h_j(x_j)$?
(g) What does decomposition imply regarding the functional equation of dynamic programming.
(h) Develop conditions on h_j, $j = 2, \ldots, N+1$, so that decomposition is possible when $g = \prod_{j=2}^{N+1} h_j$.

25. Refer to the "Introduction to Dynamic Programming."
(a) Verify Corollary 1.
(b) Verify Corollary 2.

26. M tons of ore are produced in an N-stage operation. The revenue and costs related to the extraction at stage i are $a_i x_i$ and $b_i x_i^2$, where x_i denotes the number of tons extracted at state i. It is desired to find the optimal amount to be extracted at each stage to maximize profits given that all the ore must be processed.
(a) Solve the problem when $M = 10$, $N = 3$, $(a_1, a_2, a_3) = (4, 5, 4)$, and $(b_1, b_2, b_3) = (1, 3, 6)$.
(b) Solve the problem with the data in (a) if the x_i are restricted to integers.
(c) Suppose $M = 10$, $N = \infty$, $a_i = (1.10)^i$, $b_i = (1.08)^i$, and that net profits are discounted at the rate of 6% per year. Find the optimal x_i. Is there a stationary policy?
(d) Consider the problem in (c) when $N = 10$. Solve the problem via linear programming.
(e) What happens in the solution of (d) and (e) if each x_i is restricted to 0, 1, or 2?

Exercise Source Notes

Exercise 4 was adapted from Bernoulli (1954); Exercise 5 was adapted from Borch (1968); Exercise 7 illustrates Jensen's inequality (Jensen, 1906); Exercise 15 was adapted from Ponstein (1967); Exercise 18 was adapted from Kim (1973); portions of Exercise 19 were adapted from Denardo and Mitten (1967); Exercises 20 and 21 were adapted from Bellman and Dreyfus (1962); Exercise 22 was inspired by a problem in the work of Hadley (1964); Exercise 24 was adapted from Nemhauser (1966) and Mitten (1964); and Exercise 26 was inspired by a problem written by Professor L. G. Mitten that illustrates Bellman's gold-mining problem [see Bellman and Dreyfus (1962)].

MIND-EXPANDING EXERCISES

1. Suppose that $u(w)$ is concave where wealth $w = \sum_{i=1}^{n} \xi_i x_i \equiv \xi' x$, where ξ_i is the return per dollar invested in i and x_i is the level of investment in i. Assume that the choice of $x \equiv (x_1, \ldots, x_n)$ is limited to the convex set $K \subset E^n$, and let $\bar{\xi}$ be the expected value of ξ.

(a) Prove that the following upper and lower bounds may be obtained on the optimal objective value of the maximum expected utility

$$\overset{\text{(i)}}{E_\xi u(\xi'\bar{x}) \leqq \max_{x \in K} E_\xi u(\xi' x)} \overset{\text{(ii)}}{\leqq \max_{x \in K} u(\bar{\xi}' x)},$$

where E_ξ represents mathematical expectation with respect to the random vector ξ and $\bar{x}$ is a solution (assumed to exist) to the problem on the far right.

(b) What assumptions are needed to prove inequality (i)?

(c) Prove that if u is monotone nondecreasing, then x, a solution to $\{\max \bar{\xi}' x \mid x \in K\}$, may be used in place of $\bar{x}$ for inequality (ii).

(d) Prove that if $u(0) = 0$ and $u'(0) = 1$, that one has the easier to compute upper bound

$$\max_{x \in K} u(\bar{\xi}' x) \leqq \max_{x \in K} \bar{\xi}' x.$$

2. Suppose u is a nondecreasing utility function with $u(0) = 0$ unbounded from above.

(a) Find a sequence of random variables $x_1, x_2, \ldots$ such that $Eu(x_1) \leqq Eu(x_2) \leqq \cdots$ and $Eu(x_n) \to \infty$ for which $P[u(x_n) = 0] \to 1$. That is, x_n is preferred to x_{n-1} on the basis of expected utility, but in the limit with probability 1, the investor receives a zero return.

(b) Show that the result of (a) cannot happen if u is bounded from above.

(c) Comment on the significance of (a) and (b).

3. Suppose $u(x_1, \ldots, x_T) \equiv u(x)$ is a von Neumann–Morgenstern cardinal utility function over consumption $(x_1, \ldots, x_T)$. Assume that u possesses the following regularity assumptions: (i) $\{x \mid u(x) \geqq \alpha\}$ is strictly convex; (ii) $u(x^a) > u(x^b)$ if $x_t^a > 0$ for all t and $x_t^b = 0$ for some t; (iii) $\partial u / \partial x_t > 0$; and (iv) u is twice continuously differentiable. We say that u is additive if there exist T functions $u_t(x_t)$ such that $u(x) = \sum u_t(x_t)$; ordinally additive if there exists an F, $F > 0$ such that $F(u(x)) = \sum u_t(x_t)$; and log additive if there exist constants a and $b > 0$ such that $\log[a + bu(x)] = \sum u_t(x_t)$.

(a) Illustrate some common utility functions that are additive, ordinally additive, and log additive, respectively.

Let X^a and Y^a represent two T-dimensional consumption paths and let $\gamma_a \in [0,1]$. A lottery ticket L_a denoted by (γ_a, X^a, Y^a) provides consumptions X^a and Y^a with probabilities γ_a and $1 - \gamma_a$, respectively. Note that $u(L_a) = \gamma_a u(X^a) + (1 - \gamma_a) u(Y^a)$. Two lottery tickets L_a and L_b are a pair of t-standard lottery tickets if $\gamma_a = \gamma_b$ and $x_t^a = x_t^b$ for a given t.

(b) Interpret the concept of a t-standard lottery.

An individual's preferences are said to satisfy the strong additivity axiom if his preference between two t-standard lottery tickets in a given pair is independent of the level of x_t for all pairs of t-standard lottery tickets and all t.

(c) Interpret the strong additivity assumption.

(d) Show that an individual's preferences satisfy the strong additivity axiom if and only if u is additive.

We say that two simple lottery tickets are a pair of t-normal lottery tickets if $\gamma_a = \gamma_b = \frac{1}{2}$ and $x_t^a = y_t^a = x_t^b = y_t^b \equiv z_t$ for a given t.

(e) Interpret the concept of a t-normal lottery ticket.

An individual's preferences are said to satisfy the weak additivity axiom if his preference between two t-normal lottery tickets in a given pair is independent of the level of z_t for all pairs of t-normal lottery tickets and all t.

(f) Interpret the weak additivity axiom.

(g) Show that if an individual's preferences satisfy the weak additivity axiom, then u is ordinally additive.

(h) Show that an individual's preferences satisfy the weak additivity axiom if and only if u is additive or log additive.

Pollak (1967) summarizes his findings as follows:

> The strong additivity axiom is satisfied if and only if the von Neumann–Morgenstern utility function is the sum of functions each of which depends on the level of consumption in a single period. The weak additivity axiom is satisfied in this case and is also satisfied if the von Neumann–Morgenstern utility function can be written as the product of functions each of which depends on the level of consumption in a single period. Furthermore, these are the only cases in which the weak additivity axiom is satisfied.
>
> The assumption that an individual's von Neumann–Morgenstern utility function is ordinally additive is a severe restriction on his preferences; it is equivalent to the assumption that his ordinal utility function is additive. The assumption that his von Neumann–Morgenstern utility function is additive or log additive implies a more severe restriction on preferences, namely, that they satisfy the weak additivity axiom. The assumption that his von Neumann–Morgenstern utility function is additive implies a still more severe restriction, namely, the strong additivity axiom. Simplifications resulting from special assumptions about the form of the von Neumann–Morgenstern utility function are paid for in lost generality and applicability. By expressing these assumptions in terms of preferences, it may be easier to assess their cost.

4. Suppose $f(x)$ is positive and is defined on the open convex set C in E^n.

(a) Show that f is convex on C if $\log f$ is convex on C.

(b) Show that f is pseudo-concave on C if $\log f$ is concave on C.

5. Suppose $f, g_1, \ldots, g_m$: $R^n \to R$, where f is pseudo-concave and the g_j are quasi-concave. Show that x^* solves $\{\max f(x) \mid g_i(x) \geqq 0, i=1, \ldots, n\}$ if and only if it solves the problem with a linear objective function $\{\max \nabla f(x^*)'x \mid g_i(x) \geqq 0, i=1, \ldots, n\}$.

6. Consider the problem

$$Z(a,b) \equiv \min f(a,x),$$
$$\text{s.t.} \quad g(x) \leqq b, \quad x \geqq 0,$$

where $x \in E^n$, $b \in E^m$, $g\colon E^n \to E^m$, $a \in A \subset E^s$, f is a convex function of x and a concave function of a, and the g_j are convex functions of x. Suppose that a minimizing point x^* exists $\forall a \in A$ and b. Show that

(a) Z is convex in b, and

(b) Z is concave in a.

(c) Interpret these perturbation results.

(d) Let y_j be an optimal dual variable corresponding to $g_j(x) \leqq b_j$. Show that

$$\left[\frac{\partial Z(\cdot)}{\partial b_j}\right]_+ \leqq y_j \leqq \left[\frac{\partial Z(\cdot)}{\partial b_j}\right]_-,$$

where + and − indicate right- and left-directional derivatives.

(e) When is Z strictly convex in b? [*Hint*: Begin by assuming that $b^1 \neq b^2$; then $x^1 \neq x^2$, where x^i is a minimizing vector when $b = b^i$].

7. Prove the following results concerning the monotonicity and convexity of functions defined by a minimization operation. Let $E_0 \equiv (-\infty, +\infty]$ and $h: C \to E_0$ be said to be nondecreasing (strictly increasing) iff $x \geqq y$, $x, y \in C$ $(x \neq y)$ implies that $h(x) \geqq (>) h(y)$.

Suppose $g: C \times D \to E_0$, $f(x) \equiv \min_{z \in D} g(x, z)$, and that for each $x \in C$ the minimum is attained at a $z^*(x) \in D$.

(a) Suppose g is nondecreasing (strictly increasing) in $x \in C$ for each fixed $z \in D$; then f is nondecreasing (strictly increasing).

Suppose $e: C \times D(x) \to E_0$, $d(x) \equiv \min_{z \in D(x)} e(x, z)$, and that for each $x \in C$, the minimum is attained at a $z^*(x) \in D(x)$.

(b) Suppose e is (strictly) convex on the convex set $\{(x, z) \mid x \in C,\ z \in D(x)\}$; then d is (strictly) convex.

8. Show that $f(x)$ is a concave function on the convex set $C \subset R^n$ iff:

(a) $\{(y, x) \mid y \leqq f(x),\ x \in C\}$ is a convex set for all y;

(b) $f(x+b)$ is concave on C for any $b \in R^n$;

(c) $f(x^2) - f(x^1) \leqq \nabla f(x^1)\,(x^2 - x^1)$ for all $x^1, x^2 \in C$, assuming f is differentiable; and

(d) $[\nabla f(x^2) - \nabla f(x^1)]'(x^2 - x^1) \leqq 0$ for all $x^1, x^2 \in C$, assuming f is differentiable.

(e) Illustrate these results graphically in R on the function $\log x$.

(f) Devise similar results when f is strictly concave.

9. Let $f(x)$ be defined on the convex set $C \subset E^n$. Show that f is, respectively, convex, concave, strictly convex, or strictly concave on C if and only if for each $x^1, x^2 \in C$ $g(\lambda) \equiv f[(1-\lambda)x^1 + \lambda x^2]$ is, respectively, convex, concave, strictly convex, or strictly concave on the line segment $[0, 1]$.

10. Suppose (x^0, λ^0) is a saddle point of the Lagrange function $L(x, \lambda) \equiv f(x) + \lambda' g(x)$, that is,

$$L(x, \lambda^0) \leqq L(x^0, \lambda^0) \leqq L(x^0, \lambda) \qquad \forall x \in R^n, \quad \lambda \geqq 0, \quad \lambda \in R^m.$$

(a) Prove that x^0 is a solution of $\{\max f(x) \mid g(x) \geqq 0\}$.

(b) Show that $\max_x \min_{\lambda \geqq 0} L(x, \lambda) = \max_{\lambda \geqq 0} \min_x L(x, \lambda) = L(x^0, \lambda^0)$.

(c) Suppose $m = n = 1$, $f(x) = -x^2$, $g(x) = x - 2$. Show that L has a saddle point.

(d) Consider the problem $\{\max x^2 \mid 0 \leqq x \leqq 1\}$. Show that the corresponding Lagrangian does not have a saddle point.

(e) Suppose x^* solves the problem in (a), the Lagrangian has a saddle point, and λ^* are the dual multipliers. Show that $\lambda_i^* g_i(x^*) = 0$ for each i.

(f) Part (a) indicates that if a Lagrangian has a saddle point, then the corresponding x vector is a maximizing point. Show that the converse is not generally true by considering the case when $f(x) = x$, $n = m = 1$, and $g_1(x) = -x^2$.

(g) Suppose that f and g are concave and a constraint qualification, such as the existence of $\hat{x}$ such that $g(\hat{x}) > 0$, is satisfied (see Exercise 13 below). Show that if $\tilde{x}$ solves the problem in (a), then there exists a $\tilde{\lambda} \geqq 0$ such that $(\tilde{x}, \tilde{\lambda})$ is a saddle point and $\tilde{\lambda}' g(\tilde{x}) = 0$.

11. Suppose a numerical function $f(x)$ is defined on a convex set $C \subset E^n$. Show that alternative definitions of quasi-concavity are

(a) $f(x)$ is quasi-concave iff

$$f\{\lambda x^1 + (1-\lambda)x^2\} \geqq \min\{f(x^1), f(x^2)\} \qquad \forall 0 \leqq \lambda \leqq 1, \quad x^1, x^2 \in C;$$

and

(b) $f(x)$ is quasi-concave and differentiable iff

$$f(x^1) \leqq f(x^2) \quad \Rightarrow \quad \nabla f(x^1)'(x^2 - x^1) \geqq 0 \qquad \forall x^1, x^2 \in C.$$

12. (Hessians, bordered Hessians, convex, and generalized convex functions) Suppose the numerical function $f(x)$ is defined on the open convex set $C \subset E^n$, and is twice continuously differentiable.

(a) Let

$$D_r(x) \equiv \begin{bmatrix} 0 & \dfrac{\partial f(x)}{\partial x_1} & \cdots & \dfrac{\partial f(x)}{\partial x_r} \\ \dfrac{\partial f(x)}{\partial x_1} & \dfrac{\partial^2 f(x)}{\partial x_1^2} & \cdots & \dfrac{\partial^2 f(x)}{\partial x_1\,\partial x_r} \\ \vdots & \vdots & & \vdots \\ \dfrac{\partial f(x)}{\partial x_r} & \dfrac{\partial^2 f(x)}{\partial x_r\,\partial x_1} & \cdots & \dfrac{\partial^2 f(x)}{\partial x_r^2} \end{bmatrix} \qquad \text{for} \quad r = 1, \dots, n.$$

Show that f is pseudo-convex if $|D_r(x)| < 0$ for $r = 1, \dots, n$ and $\forall x \in C$.

(b) The condition in (a) is sufficient to determine that f is quasi-concave. Show that this condition is not necessary by examining the quadratic function $c'x + x'Dx$ defined on the convex cone $\{x \mid x'Dx \leqq 0,\ c'x \geqq 0\}$ for

$$D = \begin{pmatrix} 0 & 0 & 0 \\ 0 & -1 & 0 \\ 0 & 0 & 1 \end{pmatrix}.$$

(c) Show that a necessary condition that f be concave on C is that

$$|D_r(x)| \leqq 0 \qquad \text{for} \quad r = 1, \dots, n \quad \text{and} \quad \forall x \in C.$$

(d) Show that the condition in (c) is not sufficient by examining the quadratic function $-x_2^2 + x_3^2$ on $E_+^3 \equiv \{(x_1, x_2, x_3) \mid x_i \geqq 0,\ i = 1, 2, 3\}$.

The conditions for quasi-convexity are similar to the criterion for the positive definiteness of a symmetric matrix and hence the convexity of f.

(e) Let

$$H_r(x) \equiv \begin{bmatrix} \dfrac{\partial^2 f(x)}{\partial x_1^2} & \cdots & \dfrac{\partial^2 f(x)}{\partial x_1\,\partial x_r} \\ \vdots & & \vdots \\ \dfrac{\partial^2 f(x)}{\partial x_r\,\partial x_1} & \cdots & \dfrac{\partial^2 f(x)}{\partial x_r^2} \end{bmatrix} \qquad \text{for} \quad r = 1, \dots, n.$$

Show that f is strictly convex if $|H_r(x)| > 0$ for $r = 1, \dots, n$ and $\forall x \in C$.

(f) Utilize the function x^4 to show that the condition in (e) is not necessary.

(g) Show that $|H_r(x)| \geqq 0$ for $r = 1, \dots, n$ and $\forall x \in C$ does *not* imply that f is convex.

(h) Show that f is convex if *all* the principal minors of $H_n(x)$ have nonnegative determinants $\forall x \in C$.

(i) The results in (e)–(h) suggest the conjecture that nonpositivity of all principal minors of $D_n(x)$ $\forall x \in C$ is a sufficient condition for the quasi-convexity of f. However, it is possible for a non-quasi-convex function to have all principal minors of $D_2(x)$ nonpositive as does the concave function $-x_1^2 - 2x_1x_2 - x_2^2$ on E^2.

(j) Show that the condition in (h) is not necessary either by examining $x_1^2 - x_2^2$ on E_+^2.

(k) It is tempting to believe that $\forall x \in C$ either: (i) $\nabla f(x) = 0$ and all principal minors of $H_n(x)$ have nonnegative determinants or (ii) $\nabla f(x) \neq 0$ and all principal minors of $D_n(x)$ have nonpositive minors if and only if f is quasi-convex on C. Show, however, that this conjecture is false by considering $f(x) = -x^4$.

The conditions in (k) are close to those which are necessary and sufficient but they are not stringent enough. Necessary and sufficient conditions can be derived using the concept of a semistrict local maximum (SSLM). Let g be a function of one variable defined over a nonempty (possibly infinite) open interval (a,b). We say that g attains a SSLM at $t_0 \in (a,b)$ if there exist $\delta > 0$, $\delta_1 \geqq 0$, $\delta_2 \geqq 0$ such that $t_0-\delta_1-\delta \geqq a$, $t_0+\delta_2+\delta \leqq b$, and (i) $g(t) = g(t_0)$ for $t_0-\delta_1 \leqq t \leqq t_0+\delta_2$; (ii) $g(t_0) > g(t)$ for $t_0-\delta_1-\delta < t < t_0-\delta_1$; and (iii) $g(t_0) > g(t)$ for $t_0+\delta_2 < t < t_0+\delta_2+\delta$.

(l) Graphically illustrate functions that possess a local maximum that are and are not SSLM.

(m) Show that a SSLM is a strict local minimum if $\delta_1 = \delta_2 = 0$.

(n) Show that a local minimum is not necessarily a SSLM.

Necessary and sufficient conditions for quasi-convexity may be stated in terms of normalized vectors orthogonal to the gradient.

(o) Show that f is quasi-convex over C if and only if for all $x \in C$ and all vectors d such that $d^t d = 1$ and $d^t \nabla f(x) = 0$, either $d^t H_n(x) d > 0$ or $d^t H_n(x) d = 0$ and the function of one variable $g(t) = f(x^0 + td)$ does *not* attain a SSLM at $t = 0$.

(p) Show that the condition in (o) will correctly determine that $f(x) = -x^4$ is not quasi-convex.

For specialized functions the necessary and sufficient conditions for quasi-convexity are often simplified.

(q) Suppose $\partial f(x)/\partial x_i < 0$ for all $x \in C$. Show that f is quasi-convex if and only if for every $x \in C$ and all vectors d such that $d^t d = 1$ and $d^t\, \nabla f(x) = 0$ we have $d^t H_n(x) d \geqq 0$.

(r) Suppose $f(x) = \sum_{i=1}^n f_i(x_i)$ and $C = E_n{}^+$ and $\partial^2 f(x)/\partial x_i{}^2 \neq 0$ for all i. Show that

$$\sum_{i=1}^{n} \frac{(\partial f(x)/\partial x_i)^2}{\partial^2 f(x)/\partial x_i{}^2} < 0.$$

(s) Let C be a convex cone; i.e., C is a convex set and $\lambda x \in C$ for all $\lambda \geqq 0$ if $x \in C$. Suppose f is nonnegative and positively homogeneous on C; i.e., $f(x) \geqq 0$ and $f(\lambda x) = \lambda f(x)$ for all $\lambda > 0$ and $x \in C$. Show that f is quasi-convex if and only if it is convex. Show that the result may not obtain if f is negative or f is nonnegatively homogeneous.

(t) Utilize the condition in (o) on the examples in (i) and (j).

(u) Suppose f is twice continuously differentiable and quasi-convex on C. Show that f is pseudo-convex at any point $x \in C$, where $\nabla f(x) \neq 0$.

13. Under appropriate concavity assumptions, as outlined in Mangasarian's first paper, the Kuhn–Tucker conditions are sufficient for x^* to be an optimal solution of

$$\{\max f(x) \,|\, g_j(x) \geqq 0,\ j = 1, \ldots, m\}.$$

However, the conditions are not always necessary.

(a) Show that the Kuhn–Tucker conditions do not hold at $x^* = 0$, the optimal solution to $\{\max x \,|\, x^2 \leqq 0\}$.

The feasible region in (a) is not convex so it might be conjectured that the conditions would be necessary if the feasible region was convex. This is not the case, however. Consider the constraint set

$$K \equiv \{(x_1, x_2) \,|\, x_1 \geqq 0,\ x_2 \geqq 0,\ (1-x_1-x_2)^3 \geqq 0\}.$$

(b) Show that K is a convex polyhedron.

(c) Show that the Kuhn–Tucker necessary conditions are not satisfied at the point $x = (\frac{1}{2}, \frac{1}{2})'$.

The problem in (b)–(c) hints that the functions involved determine whether the Kuhn–Tucker conditions are necessary, not the shape of the feasible region. Suppose d is a direction vector in E^n, $A(x)$ is the set of binding constraints at x, i.e., those $g_i(x) = 0$, and F is the feasible region. Let $D(x) \equiv [d \mid \exists\sigma > 0 \colon \sigma \geqq \tau \geqq 0 \Rightarrow x+\tau d \in F\}$, i.e., the set of feasible directions at x and $\mathscr{D}(x) \equiv \{d \mid \nabla g_j(x)'d \geqq 0,\ i \in A(x)\}$, the set of outward-pointing gradients of the active constraints.

(d) Show that if x^* is optimal for the nonlinear programming problem, then $\nabla f(x^*)'d \leqq 0\ \forall d \in D(x^*)$ or $\bar{D}(x^*)$ (the closure of D).

(e) Show that $\bar{D}(x) \subset \mathscr{D}(x)$.

Then a constraint qualification is $\mathscr{D}(x^*) = \bar{D}(x^*)$, i.e., the set of outward-pointing gradients of the active constraints equals the set of feasible directions.

(f) Show that the Kuhn–Tucker conditions are necessary under the constraint qualification. [*Hint*: Utilize Farkas' lemma, i.e., the statement $q'x \leqq 0$ for all x such that $Ax \geqq 0$ is equivalent to the statement that there exist $u \geqq 0$ such that $q + A'u = 0$.]

(g) Show the following assumptions are sufficient to guarantee that the constraint qualification is satisfied:

(i) all constraint functions are linear; or

(ii) all constraint functions are convex or pseudo-convex and the constraint set has a nonempty interior; or

(iii) the gradients of all binding constraints are linearly independent.

14. Suppose the primal problem is $\{\min f(x) \mid g_j(x) \leqq 0,\ j = 1, \ldots, m\}$; then the Wolfe dual problem is

$$\{\max h(x,\lambda) \equiv f(x) + \textstyle\sum \lambda_j g_j(x) \mid \lambda \geqq 0,\ \nabla_x f(x) + \sum \lambda_j \nabla_x g_j(x) = 0\},$$

where f and the g_j are differentiable functions.

(a) (Weak duality) Suppose x^1 is feasible for the primal problem, (x^2, λ^2) is feasible for the dual problem, and f and the g_j are convex at x^2. Show that $f(x^1) \geqq h(x^2, \lambda^2)$.

(b) (Wolfe's duality theorem) Suppose f and the g_j are convex functions. Let x^1 solve the primal problem and suppose the g_j satisfy a constraint qualification. Show that there exists a $\lambda^1 \in E^m$ such that (x^1, λ^1) solve the dual problem and $f(x^1) = h(x^1, \lambda^1)$.

(c) (Mangasarian's converse duality theorem) Suppose f and the g_j are convex functions. Let x^1 solve the primal problem and suppose the g_j satisfy a constraint qualification. Show that if (x^2, λ^2) solve the dual problem and $h(x, \lambda^2)$ is strictly convex at x^2, then $x^2 = x^1$ and $f(x^1) = h(x^2, \lambda^2)$.

(d) (Huard's converse duality theorem) Suppose (x^2, λ^2) solve the dual problem and f and the g_j are convex at x^2. Show that if $h(x, \lambda^2)$ is twice continuously differentiable at x^2 and the $n \times n$ Hessian matrix $\nabla_x \nabla_x h(x^2, \lambda^2)$ is nonsingular, then x^2 solves the primal problem and $f(x^2) = h(x^2, \lambda^2)$.

15. Prove the following properties of convex sets.

(a) Show that the intersection of a finite or infinite number of convex sets in E^n is a convex set.

(b) A necessary and sufficient condition for a set C to be convex is that for each integer $m \geqq 1$:

$$\left.\begin{aligned} x^1, \ldots, x^m &\in C \\ \lambda_1, \ldots, \lambda_m &\geqq 0 \\ \lambda_1 + \cdots + \lambda_m &= 1 \end{aligned}\right\} \quad \Rightarrow \quad \lambda_1 x^1 + \cdots + \lambda_m x^m \in C;$$

that is, all convex combinations of points of C belong to C.

(c) (Carathéodory's theorem) Let $D \subset E^n$. If x is a convex combination of points of D, then x is a convex combination of $n+1$ or fewer points of D.
(d) (Separation theorem) Let C and D be nonempty disjoint convex sets in E^n. Then there exists a plane $\{x \mid c'x = \alpha,\ c \neq 0,\ c, x \in E^n\}$ which separates them so that $c'x \leqq \alpha$ if $x \in C$ and $c'x \geqq \alpha$ if $x \in D$.
(e) (Strict separation theorem) Let C and D be two closed nonempty convex sets in E^n, where D is bounded. If C and D are disjoint (i.e., $C \cap D = \emptyset$), then there exists a plane $\{x \mid c'x = \alpha,\ c \neq 0,\ c, x \in E^n\}$ which strictly separates them so that $c'x < \alpha$ if $x \in C$ and $c'x > \alpha$ if $x \in D$.

16. It is often of interest to be able to transform generalized convex functions into convex functions. One searches for a $\phi\colon R \to R$ such that $g(x) \equiv \phi[f(x)]$ is convex when f is, say, quasi- or pseudo-convex, where K is a convex subset of E^n and $f, g\colon K \to E$.

Suppose $x^1, x^2 \in K$; then f is said to be r-convex if

$$f\{\lambda x^1 + (1-\lambda)x^2\} \leqq \begin{cases} \log\{\lambda \exp[rf(x^1)] + (1-\lambda)\exp[rf(x^2)]\}^{1/r} & \text{if } r \neq 0, \\ \lambda f(x^1) + (1-\lambda) f(x^2) & \text{if } r = 0, \end{cases}$$

so that 0-convexity is ordinary convexity. Similarly, f is r-concave if the inequality is reversed.
(a) Show that $\log x$ is 1-convex and 1-concave in addition to being 0-concave.
(b) Show that f is s-convex (s-concave) for every $s > r$ ($s < r$) if f is r-convex (r-concave).
(c) Show that all r-convex (r-concave) functions are quasi-convex (quasi-concave). [*Hint*: Investigate $r = \pm\infty$.]
(d) Show that

$$f(x) = \begin{cases} 1 & \text{if } x \leqq 2, \\ 2 & \text{if } x > 2 \end{cases}$$

is quasi-convex but not r-convex for any r.
(e) Suppose $g(x) \equiv \exp(rf(x))$. Show that f is r-convex (r-concave) with $r \neq 0$ iff g is convex (concave) whenever $r > 0$ and concave (convex) whenever $r < 0$. [*Hint*: Exponentiate r times the r-convex definition.]
(f) Show that f is r-convex iff $-f$ is $-r$-concave.
(g) Suppose f is r-convex (r-concave). Show that $f+\alpha$ is r-convex (r-concave) for all $\alpha \in E$. Show that kf is (r/k)-convex [(r/k)-concave] if $k > 0$.
(h) Suppose f and g are r-convex (r-concave) and $\alpha_1, \alpha_2 > 0$. Show that

$$h(x) \equiv \begin{cases} \log[\alpha_1 \exp[rf(x)] + \alpha_2 \exp[rg(x)]\}^{1/r} & \text{if } r \neq 0, \\ \alpha_1 f(x) + \alpha_2 g(x) & \text{if } r = 0, \end{cases}$$

is r-convex (r-concave).
(i) Suppose f is r-convex, $r \leqq 0$ ($r \geqq 0$), and g is s-convex and nondecreasing on E. Show that $h(x) \equiv g[f(x)]$ is s-convex (s-concave).
(j) Show that f is r-convex (r-concave) iff for all $x^1, x^2 \in K$ the function $l(\lambda) \equiv f[\lambda x^1 + (1-\lambda)x^2]$ is r-convex (r-concave) on $[0, 1]$.
(k) Show that f is r-convex and differentiable iff for all $x^1, x^2 \in K$,

$$\begin{aligned} (1/r)\exp[rf(x^2)] &\geqq (1/r)\exp[rf(x^1)]\{1 + r(x^2 - x^1)' \nabla f(x^1)\} && \text{if } r \neq 0, \\ f(x^2) &\geqq f(x^1) + (x^2 - x^1)' \nabla f(x^1) && \text{if } r = 0, \end{aligned}$$

where ∇ denotes the gradient operator, i.e., $(\partial/\partial x_1, \ldots, \partial/\partial x_n)$.

(l) Suppose f is twice continuously differentiable on (a,b). Show that f is r-convex iff

$$r\left[\frac{df(x)}{dx}\right]^2 + \frac{d^2f(x)}{dx^2} \geqq 0 \qquad \text{for all} \quad x \in (a,b).$$

(m) Suppose f is twice continuously differentiable on K (assumed open). Show that f is r-convex (r-concave) iff the matrix

$$Q(x) \equiv r\,\nabla f(x)\,(\nabla f(x))' + \nabla(\nabla f(x))$$

is positive (negative) semidefinite for all $x \in K$.

(n) Show that all differentiable r-convex (r-concave) functions are pseudo-convex (pseudo-concave), for $r = \pm\infty$.

(o) Suppose f is twice continuously differentiable and quasi-convex on K (assumed open). It can be shown that, if there exists a real number

$$r^* = \sup_{\substack{x \in K \\ |z|=1}} \frac{-z'\,\nabla(\nabla f(x))\,z}{[z'\,\nabla f(x)]^2},$$

whenever the denominator is not zero, then f is r^*-convex. The r-concave analog is obtained by replacing supremum by infimum. Show that $r^* = 1$ if $f(x) = \log(x)$. Show that no r^* exists if $f(x) = x^3$.

17. Exercise CR-14 explores the convexity properties of the addition of convex and related functions. It is of interest to explore such results when a function is defined by an integral. Let $\Xi \subset E^k$, $C \subset E^n$, where C is convex and $E_0 = (-\infty, +\infty]$.

(a) Suppose $f\colon C \times \Xi \to E_0$ is convex in $x \in C$ for all fixed $\rho \in \Xi$ and $\int f(x,\rho)\,dF(\rho) > -\infty$. Show that $g(x) \equiv \int f(x,\rho)\,dF(\rho)$ is convex on C.

(b) Suppose f is strictly convex in $x \in C$ for all fixed $\rho \in \Xi$ and $|g(x)| < \infty$. Show that g is strictly convex on C.

Suppose that f takes the univariate form $f(x,\rho) = u(\rho'x)$, where $n = k$.

(c) Show that g is convex in x if u is convex in the scalar w as long as $g > -\infty$.

(d) Show that strict convexity of u and $|g| < \infty$ does not imply that g is strictly concave.

(e) Show that the additional regularity hypothesis that $\rho'x^2 = \rho'x^2$ $\forall \rho \in \Xi$ implies that $x^1 \equiv x^2$ along with the hypotheses in (d) will guarantee that g is strictly concave.

(f) Show that the assumption in (e) amounts to assuming that the variance–covariance matrix of ρ is nonsingular if the ρ_i have a joint normal distribution.

It is also of interest to explore the monotonicity properties of functions defined by integrals. Let f and C be general and not necessarily convex. A function $h\colon C \to E$ is said to be nondecreasing (strictly increasing) iff $x \geqq y$, $x, y \in C$ $(x \neq y)$ implies that $h(x) \geqq (>)\, h(y)$.

(g) Suppose f is nondecreasing in $x \in C$ for all fixed $\rho \in \Xi$. Show that g is nondecreasing in $x \in C$.

(h) Suppose f is strictly increasing in $x \in C$ for all fixed $\rho \in \Xi$ and $|g| < \infty$. Show that g is strictly increasing in $x \in C$.

18. (Strict concavity of composite functions) Suppose $f(x) = \log(a_1 x_1 + a_2 x_2)$ is defined on the unit simplex $X = \{(x_1, x_2) \mid x_1 \geqq 0,\ x_2 \geqq 0,\ x_1 + x_2 = 1\}$, $a_1, a_2 > 0$ and $a_1 \neq a_2$.

(a) Show that the Hessian matrix of f is singular at every point in X.

(b) Show, however, that f is strictly concave on X. [*Hint*: Utilize the inequality between the geometric and arithmetic means.]

(c) Is f strictly concave on $Y = \{(x_1, x_2) \mid x_1 > 0,\ x_2 > 0\}$?

(d) Show that the result in (b) does not generalize to the case when $X \subset E^n$ for $n \geqq 3$ by considering $f(x) = \log(a_1 x_1 + a_2 x_2 + a_3 x_3)$, $X = \{[x_1, x_2, x_3) \mid x_i \geqq 0, \sum x_i = 1\}$, where the a_i are unequal positive constants.

(e) Show that the result in (b) does not generalize to arbitrary strictly concave functions g and convex sets $X \subset E^2$ by devising a counterexample to the conjecture: Suppose $f(x) = g[a_1 x_1 + a_2 x_2 + b]$, where g is strictly concave and strictly increasing, $a_1 \neq a_2$, $a_i \neq 0$, and $|f(x)| < \infty$ for $x \in X$. Then f is strictly concave.

The basic difficulty is that vectors x^1, x^2, and $x^\lambda = \lambda x^1 + (1-\lambda) x^2$ can be constructed so that $h(x^1) = h(x^2) = h(x^\lambda)$ where $h(x) = a_1 x_1 + a_2 x_2 + b$ and $0 < \lambda < 1$. Hence $f(x^\lambda) = \lambda f(x^1) + (1-\lambda) f(x^2)$. One way to rule out this possibility is to assume that the function h is one-to-one on X, that is, there is a unique vector $x \in X$ such that $h(x) = y$ for any y.

(f) Show that the function in (b) is one-to-one whereas the functions in (c), (d), and (e) are not.

Let $f(x) = g[h_1(x), \ldots, h_m(x)]$ where $x \in E^n$, each $h_i\colon E^n \to E^1$, $g\colon E^m \to E^1$, and $f\colon E^n \to E^1$.

(g) Suppose g is strictly concave and nondecreasing, each h_i is concave, and at least one h_i is one-to-one. Show that f is strictly concave.

Show that the result in (g) has the following variants.

(h) Suppose g is strictly concave and nonincreasing, each h_i is convex, and at least one h_i is one-to-one. Show that f is strictly concave.

(i) Suppose $f(x) = g[h_1(x), h_2(x)]$, g is strictly concave and nondecreasing in its first argument and nonincreasing in its second argument, h_1 is concave, h_2 is convex, and at least one of the h_i is one-to-one. Show that f is strictly concave.

The strict concavity of f is also guaranteed if one of the h_i is strictly concave.

(j) Suppose each h_i is concave and some h_i, say h_k, is strictly concave, g is concave, nondecreasing, and strictly increasing in its kth component. Show that f is strictly concave.

(k) Develop results along the lines of (h) and (i) for the case when some h_k is strictly concave.

19. A process is observed at discrete points in time in one of the possible states $0, 1, 2, \ldots$. After each state observation an action $a \in A$ is chosen where A is finite. The payoff is $R(i,a)$ if action a is chosen in state i. Furthermore, the next state of the process is determined according to the Markov transition probabilities $p_{ij} \geqq 0$, $\sum_j p_{ij} = 1$, $i, j \geqq 0$. Let $V_f(i)$ be the total expected payoff if the initial state is i and policy f is employed, and $V(i) \equiv \sup_f V_f(i)$. Suppose $R(i,a) \geqq 0$.

(a) Show that the functional equation is

$$V(i) = \max_i \{R(i,a) + \sum_j p_{ij}(a) V_j\}.$$

(b) Show that a policy f is optimal, that is, $V_f(i) = V(i)$ for $i = 0, 1, 2, \ldots$, if and only if its return function satisfies the functional equation. [*Hint*: Consider using some policy in stage 1 and switching to f.]

(c) Show that a weaker sufficient condition for the result in (b) is that $V_f(i) \geqq 0$.

(d) Show that the result in (b) is not true when $R(i,a) \leqq 0$. [*Hint*: Suppose there are two states and two actions and $R(1,1) = 0$, $R(1,2) = -1$, $R(2,i) = 0$, $i = 1, 2$, $P_{1,1}(1) = 1$, $P_{1,2}(2) = 1$, $P_{2,2}(i) = 1$, $i = 1, 2$. Show that the policy that always chooses action 2 is not optimal yet it satisfies the functional equation.]

(e) Suppose for each initial state i and policy f there is a stopping state s. The state s has the property that once the process enters s it remains in s with zero return. Show that if the process enters state s with probability 1 in a bounded number of transitions, then the result in (b) is valid if returns are nonpositive.

A policy is said to be stationary if the action it chooses at any time is a deterministic function of the state at that time.

(f) Assume that $R(i,a) \leqq 0$ and the state space (as well as the action space) is finite. If policy f is stationary, then it is optimal if

$$V_f(i) > R(i,a) + \sum_j p_{ij}(a) V_f(j)$$

for all $a \neq f(i)$ and all i. [*Hint*: Consider the discount factor $0 < \beta < 1$. Establish the result for all β sufficiently near 1, then let $\beta \to 1$.]

Suppose now that $R(i,a) \leqq 0$. For convenience let $C(i,a) \equiv -R(i,a)$ and suppose that the decision maker wishes to minimize nonnegative costs rather than maximize nonpositive payoffs. Let $W_f(i)$ be the total expected costs if the initial state is i and policy f is employed, and $W(i) \equiv \sup_f W_f(i)$.

(g) Show that the functional equation is now

$$W(i) = \min_a \left\{ C(i,a) + \sum_j p_{ij}(a) W(j) \right\}.$$

(h) Policy f is optimal if it consists of actions that minimize the right-hand side of the functional equation for each state i. [*Hint*: Note that if f is determined by the functional equation, the decision maker can get within $\varepsilon/2$ of W if f is used for one state and then a policy is used that is within $\varepsilon/2$ of W.]

(i) Show that a weaker sufficient condition for the result in (h) is that $W(i) \geqq 0$.

(j) Show that the result in (h) is not true in the positive case, i.e., when $R(i,a) \geqq 0$. [*Hint*: Suppose the states are the positive integers and when in state i the decision maker may either accept a terminal reward $1 - 1/i$ or else receive no reward and go to state $i+1$. Show that no optimal policy exists.]

(k) Attempt to find conditions similar to those in (e) and (f) that are sufficient for the validity of the result in (h) when $R(i,a) \geqq 0$.

(l) Are any of the results true for general $R(i,a)$?

20. This exercise considers a broad class of Markovian decision problems for which a stationary policy is always optimal. An optimal policy can then be found using Howard's policy improvement routine based on the functional equation. The details are considered in the finite state space case in which Manne's equivalent linear program provides an alternative computational scheme.

The process is supposed to be observed at discrete points $t = 0, 1, 2, \ldots$ to be in one of the states $X_t = 0, 1, 2, \ldots$. In each time period t in each state i the decision maker chooses an action $a \in A$ where A is finite. This results in a cost $C(i,a)$ and a transformation $p_{ij}(a) \equiv \text{Prob}\{X_{t+1} = j \mid X_t = i, a_t = a\}$. Hence both costs and transition probabilities are Markovian in the sense that they depend only on the last state and action. Assume that costs are bounded by $M < \infty$ so that $|C(i,a)| < M$. A policy is a rule for choosing actions. The policy is stationary if the action it chooses at time t is a deterministic function of the state X_t.

(a) Show that the sequence of states $\{X_t\}$ forms a Markov chain with transition probabilities $p_{ij}[f(i)]$ if a stationary policy f is employed in each state i.

Let $V_\pi(i)$ be the expected total discounted cost incurred using policy π. Hence

$$V_\pi(i) = E_\pi\left[\sum_{t=0}^{\infty} \alpha^t C(X_t, a_t) \mid X_0 = i\right], \qquad i \geqq 0,$$

where E_π represents mathematical expectation conditional on the fact that policy π is used and $0 < \alpha < 1$ is the discount factor. Let $V_\alpha(i) = \inf_\pi V_\pi(i)$. Policy π is α-optimal if its expected cost is minimal for every initial state i, i.e., $V_\pi(i) = V_\alpha(i)$.

(b) Show that the functional equation is

$$V_\alpha(i) = \min_a \left\{ C(i,a) + \alpha \sum_{j=0}^{\infty} p_{ij}(a) V_\alpha(j) \right\}, \qquad i \geq 0.$$

Let $B(I)$ be the set of all bounded (real-valued) functions on the state space.

(c) Show that $V_\pi \in B(I)$ for all policies π. If f is a stationary policy, define the mapping $T_f: B(I) \to B(I)$ by

$$(T_f u)(i) = C[i, f(i)] + \alpha \sum_{j=0}^{\infty} p_{ij}[f(i)] u(j).$$

(d) Interpret this expression and show that $T_f u \in B(I)$.

For notational convenience let $T_f^1 = T_f$ and $T_f^n = T_f(T_f^{n-1})$ for $n > 1$. For two functions $u, v \in B(I)$ we write $u \leqq (=) V$ if $u(i) \leqq (=) v(i)$ for all $i \geqq 0$, and $u_n \to u$ if $u_n(i) \to u(i)$ uniformly in i for $i \geqq 0$.

(e) Suppose $u, v \in B(I)$ and that f is a stationary policy. Show that (i) $u \leqq v \Rightarrow T_f u \leqq T_f v$, (ii) $T_f V_f = V_f$, and (iii) $T_f^n u \to V_f$ for all $u \in B(I)$.

(f) Use the properties in (e) to show that a stationary policy f is α-optimal if in each state i it chooses an action that minimizes the right-hand side of the functional equation in (b). Hence if an α-optimal policy exists, it may be taken as stationary and determined by the functional equation.

Suppose the expected cost function V_f of a stationary policy f has been evaluated. Let f^* be a stationary policy which when in every state i chooses an action that minimizes $C(i,a) + \alpha \sum_{i=0}^{\infty} p_{ij}(a) V_f(j)$.

(g) Show that f^* is at least as good as f, that is, $V_{f^*}(i) \leqq V_f(i)$ for all $i \geqq 0$.

Using the theory of contraction mappings it can be shown [see, e.g., Ross (1970)] that f^* is strictly better for some initial state i if f is not α-optimal. This follows because V_α is the unique solution of

$$V_\alpha(i) = \min_a \{C(i,a) + \alpha \sum_{j=0}^{\infty} p_{ij}(a) V_\alpha(j)\}, \qquad i \geqq 0.$$

Furthermore for any $u \in B(I)$, $T_\alpha^n \to V_\alpha$ as $n \to \infty$. Suppose now that the state space as well as the action space is finite. Denote the states by $0, 1, \ldots, m$. Note that for any stationary policy f, V_f is the unique solution to

$$V_f(i) = C[i, f(i)] + \alpha \sum_{j=0}^{m} p_{ij}[f(i)] V_f(i), \qquad i = 0, \ldots, m.$$

(h) Show that $V_f(i)$ may be easily calculated because the expression above amounts to a linear system of $m+1$ equations in $m+1$ unknowns.

Thus f may be improved by choosing an action that minimizes $C(i,a) + \alpha \sum_{j=0}^{\infty} p_{ij}(a) V_f(j)$ for each i. If the new policy is different from f, then it is strictly better than f for at least one initial state and not worse in any state. If the new policy is identical to f or has the same minimum cost in each initial state as f, then f and the new policy are both α-optimal.

(i) Show that

$$T_\alpha u \geqq u \quad \Rightarrow \quad V_\alpha \geqq u,$$

where

$$(T_\alpha u)(i) = \min_a \left\{ C(i,a) + \alpha \sum_{j=0}^{\infty} p_{ij}(a) u(j) \right\}.$$

(j) Show that an optimal policy V_α may be obtained by maximizing u:

$$\text{s.t.} \quad T_\alpha u \geqq u.$$

[*Hint*: Show that $T_\alpha V_\alpha = V_\alpha$ using the functional equation in (b).]

(k) Show that the problem in (j) is equivalent to the linear program

$$\max \sum_{i=0}^{m} u(i)$$

$$\text{s.t.} \quad C(i,a) + \alpha \sum_{j=0}^{m} p_{ij}(a)\, u(j) \geqq u(i) \qquad \text{for all} \quad a \in A \quad \text{and} \quad i = 0, \ldots, m.$$

Exercise Source Notes

Portions of Exercise 1 were adapted from Hillier (1969); Exercise 2 was written by Professor S. L. Brumelle; Exercise 3 was adapted from Pollak (1967); Exercise 4 was adapted from Klinger and Mangasarian (1968); portions of Exercise 7 were adapted from Iglehart (1965); Exercise 9 was adapted from Mangasarian (1969); portions of Exercise 10 were adapted from Karlin (1958) and Zangwill (1969); Exercise 12 was adapted from Ferland (1971), Diewert (1973), Katzner (1970), and Arrow and Enthoven (1961); Exercise 13 was adapted from Zangwill (1969), Fiacco and McCormick (1968), and Kortanek and Evans (1968); Exercises 14 and 15 were adapted from Mangasarian (1969); Exercise 16 was adapted from Avriel (1972); Exercise 19 was adapted from Ross (1972) who extended and clarified some results due to Blackwell (1967) and Strauch (1966); and Exercise 20 was adapted from Ross (1970) and is based in part on results due to Blackwell (1962), Howard (1960), and Manne (1960).

Part II

Qualitative Economic Results

INTRODUCTION

In the second part of the book we are concerned with the qualitative analysis of portfolio choice. The material covered in this section is single-period portfolio analysis, in which case the investor's goal is to maximize the expected utility of terminal (end-of-period) wealth. The investor will normally have a number of options available from which he can freely choose. Typically, he can invest his wealth in a risk-free asset with known nonnegative rate of return, and he can invest in a number of risky assets with random rates of return. It is assumed that the investor knows the joint probability distribution of returns on the risky assets. The investor can control the fraction of his wealth to be invested in each of the available assets. In a realistic case there may be borrowing or short-sale constraints that limit the investor's choice of control variables. For any given feasible allocation, the return on the investment will be a random variable, and the investor's problem is to choose an allocation which maximizes the expected utility of gross portfolio return.

The relevant mathematical tools included in this part are concerned with stochastic ordering (comparison of random variables) and risk-aversion measures. Both of these tools will continue to be useful throughout the remainder of the book. The qualitative analysis of portfolio choice is limited in this part primarily to the question of the aggregation of risky assets which enable decentralized decision-making via the separation theorems. This topic is discussed using both the expected utility and the mean–variance (or safety-first) criteria. A deeper treatment of the quantitative and computational aspects of portfolio choice is presented in Part III.

I. Stochastic Dominance

The Hanoch and Levy article introduces the idea of stochastic ordering of random variables. The paper is concerned with the following question: Given two random variables, when can it be said that the expected utility of one is at least as great as the expected utility of the other for all utility functions in some general class? The paper discusses this question for two utility function classes of relevance to finance: the class of nondecreasing utility functions, and the class of nondecreasing concave utility functions. It is shown that a necessary and sufficient condition for ordering with respect to the class of nondecreasing utility functions is that the cumulative distribution functions of the random variables be ordered. That is, a random variable X has consistently larger expected utility than a random variable Y if and only if the distribution function of X is everywhere less than or equal to that of Y. This type of ordering is commonly referred to as first-degree stochastic dominance. The result stated above is known in the statistics literature, and was proved by Lehmann (1955).

Its usefulness for single-period portfolio analysis is rather limited, since the required ordering of cumulative distributions is a severe restriction on the structure of investment returns. However, the result is useful in later sections of the book, because the restriction is often satisfied by the successive rows of Markov transition matrices. Hanoch and Levy also show that a necessary and sufficient condition for ordering with respect to the class of nondecreasing concave utility functions is that the first integrals of the cumulative distributions be ordered. This ordering, commonly called second-degree stochastic dominance, is also well known in the statistics literature, and the result stated above was originally proved by Blackwell (1951) (see also Sherman, 1951). Since a concave utility function is a natural mathematical expression of risk aversion, this result is of considerable importance in clarifying the qualitative analysis of choice behavior for risk averters. Tesfatsion (1974) has indicated that the Hanoch and Levy proofs are not strictly valid, since finiteness of certain integrals is implicitly assumed but does not follow from the stated assumptions.

The original Markowitz theory of portfolio allocation was based on the plausible hypothesis of mean–variance efficiency. In this approach the assumption is made that investor preferences can be ranked completely in terms of expected value and variance of portfolio return. In this case, any investor will choose a portfolio which lies somewhere on the mean–variance efficiency frontier, that is, in the set of portfolios that minimize variance for any given mean return. Various forms of mean–variance portfolio analysis have dominated the finance literature until very recently, although such analysis is generally not consistent with expected utility theory. The concepts of first- and second-degree stochastic dominance are useful in clarifying this issue. Hanoch and Levy analyze some simple special cases of the dominance relations. When the graphs of the cumulative distributions cross precisely once, it is shown that second-degree dominance is equivalent to a simple ordering of the means. The mean–variance rules are also examined, and are shown to be valid for random variables which belong to a two-parameter family such that the "standardized" random variables (centered about the mean and divided by standard deviation) have a common distribution. In particular, mean–variance rules are valid for normally distributed random variables. For random variables depending on more than two parameters, examples are given which show that mean–variance analysis is invalid.

The reader is asked to examine the dominance relations explicitly in Exercises CR-1 and 3. The application of stochastic dominance to discounted returns is considered in Exercise CR-2. Mean–variance analysis for normally distributed random variables is examined in CR-4, while CR-5 presents an example which illustrates that mean-variance analysis is generally false for an arbitrary two-parameter family of random variables (further restrictions are needed). Exercises CR-6, 7, and 8 examine necessary conditions for ordering random variables when the utility functions are quadratic or cubic. In Exercise ME-5 another type of dominance relation of relevance to convex costs rather than concave utilities, is considered. Exercise ME-11 examines dominance relations for concave utility functions which are not assumed to be nondecreasing. Exercise ME-7 illustrates the caution which must be exercised when dealing with dominance relations, in that the addition of a random variable to each side of the relation may render it invalid. It should be noted that all such dominance relations are relevant only to the problem of choosing one random variable from a given set of random variables. In the portfolio problem, the possibility of investing partially in each of several assets will generally introduce all of the random assets into the optimal solution, even if one of them strictly dominates the others. This diversification behavior is illustrated in the extreme in the paradoxical Exercise CR-12. Stochastic dominance relations between different total portfolios are what matters in the portfolio allocation problem. Exercise ME-22 utilizes stochastic dominance results in the analysis of a reinsurance problem.

The Brumelle and Vickson paper examines stochastic ordering relations from a unified view, based on recent work in the field of measure theory. The emphasis is on simple tools which permit better understanding of the underlying structure of such dominance relations, and which allow some new dominance relations to be discovered in a straightforward manner. The "technical" level of the paper is not unduly high, but the general tone is perhaps somewhat abstract. The paper emphasizes the fact that most of the interesting classes of utility functions are convex sets in a function space, in the sense that they are closed under addition and multiplication by nonnegative constants. If the sets are represented as closed convex hulls of their extreme points (just as for convex sets in ordinary Euclidean space), then necessary and sufficient conditions for dominance are obtained by restricting one's attention to the extreme points. Since the extreme points can often be determined by inspection, the stochastic dominance tests drop out immediately. The paper reexamines first- and second-degree dominance from this viewpoint, and then treats the interesting case of third-degree dominance. This is the ordering relation relevant to the class of concave nondecreasing utility functions having nonnegative third derivative (i.e., having convex first derivative). The interest in this class arises from the fact that it includes many of the economically relevant utility functions for wealth. This ordering relation was introduced by Whitmore (1970). In Exercise ME-6 it is shown that third-degree dominance is equivalent to an ordering of double integrals of cumulative distributions. Exercise ME-20 extends the theory to nth-degree stochastic dominance.

The paper also discusses an alternative formulation of second-degree dominance which emphasizes the underlying probabilistic content of the ordering. The ordering is shown to be equivalent to the existence of a bivariate distribution with respect to which the given random variables form a submartingale or martingale sequence; see Exercise V-CR-19 for elementary properties of martingales. Equivalently, one of the random variables must have the same distribution as the other plus nonpositive "noise." Exercise ME-12 presents another approach to the proof of this result. Exercise ME-13 shows how dominance for random variables having unequal means can be reduced to the equal mean case. Exercise ME-15 asks for an algorithm to construct the random noise variable mentioned above.

II. Measures of Risk Aversion

The Pratt paper shows how to relate risk preferences of an individual to specific properties of his utility function. Pratt introduces two important risk-aversion indices which are often called Arrow–Pratt risk-aversion indices because they were independently studied by Arrow [see Arrow (1965, 1971)] and Pratt. These are (1) the absolute risk-aversion index, and (2) the relative risk-aversion index. Since an individual's preference orderings are invariant under positive linear transformations of the utility functions, any meaningful risk-aversion measures must also be invariant under such transformations. The absolute and relative risk-aversion indices both satisfy this requirement. Pratt shows that a natural restriction on the utility function is that its absolute risk-aversion index be a nonincreasing function. This assumption implies that the insurance premium an investor is willing to pay to cover a fixed risk is a nonincreasing function of wealth. Several useful properties of such utility functions are presented. First, the property of nonincreasing absolute risk aversion is shown to remain valid for positive linear transformations of the wealth level. Composition of one such function with another is shown to yield a function of the same type. Finally, the set of such functions is shown to be convex. These properties can be utilized to generate a large number of utility functions having nonincreasing absolute risk aversion by starting with a small sample of such functions. Pratt also discusses investor behavior with respect to risks which are proportional to wealth. He shows that the assumption of nondecreasing relative risk aversion is equivalent to the requirement that the investor's proportional insurance premium increases with wealth.

Additional material regarding the risk-aversion measures and their application is found in the exercises. Exercise CR-10 shows that risky assets are not inferior goods if the absolute risk-aversion index is nonincreasing. Exercise CR-24 considers the case when the risky asset has a two-point distribution. Exercise CR-25 shows how the results apply to simple insurance and foreign exchange problems. Exercise ME-19 is concerned with plunging (avoidance of a secure asset) in portfolio selection. Exercise ME-21 presents a more detailed analysis of the qualitative behavior of portfolios consisting of a blend of one safe and one risky asset. Several ways to characterize portfolios are seen to relate to the absolute and relative risk-aversion indices. In Exercise CR-11, the relevance of absolute risk aversion to the problem of optimal insurance policies is outlined. Effects of changes in mean and variance of risky assets are examined in Exercises ME-3 and 4, in connection with the problem of optimal foreign exchange holdings. Exercise CR-17 presents an alternative approach to risk aversion, based on mean-variance indifference curves. In Exercises ME-8 and 9, additional results are presented regarding the effects of risk-aversion indices on the qualitative behavior of an investor's insurance premium as a function of mean and variance of risky return (see also Zeckhauser and Keeler, 1970). In Exercise ME-10 it is shown that a bounded concave utility function must have a relative risk-aversion index which increases "on the average," from values less than 1 for small arguments to values greater than 1 for large arguments. Exercise ME-23 presents a not commonly assumed utility function that has the desirable properties that it is increasing, bounded, strictly concave, and has decreasing absolute and increasing relative risk aversion. See Sankar (1973) for another desirable utility function. In Exercise ME-1 the reader is asked to devise stochastic dominance tests which would be appropriate to the class of utility functions exhibiting decreasing absolute or increasing relative risk aversion.

III. Separation Theorems

The Lintner paper discusses the application of mean-variance and safety-first analysis to portfolio selection. Under the assumptions that there exists a risk-free asset which can be lent or borrowed in unlimited amounts at a common positive rate of interest, the optimal portfolio problem becomes considerably simplified. Lintner proves a very important separation theorem due to Tobin (1958), which splits the computation into two parts. First, a single optimal mutual fund, the same for all investors, is calculated by solving a fractional programming problem. Then the optimal portfolio for any particular investor is calculated by means of a univariate unconstrained optimization. The Tobin–Lintner separation theorem implies that all investors will purchase different amounts of a common mutual fund, and this reduces the many-risk-asset problem to an effective single-risk-asset problem. If borrowing of the risk-free asset is limited, the separation theorem need not hold. In this case the computation of an optimal portfolio becomes considerably more difficult, and would generally need to be done separately for each investor's utility function. This is also true if the interest rate of borrowing exceeds that of lending. Some of the computational aspects of similar problems are treated in the Ziemba paper in Part III. The Lintner paper along with

Sharpe (1964) proved also to be seminal building blocks in the modern theory of capital asset pricing; see Jensen (1972) for an assessment of the current state of the development and application of this theory.

Exercise CR-9 extends the Tobin–Lintner separation theorem to the case of homogeneous risk measures more general than variance. Exercise CR-19 examines an equivalent quadratic programming problem for determining the optimal mutual fund. Exercise CR-20 examines the Lintner separation theorem when the variance–covariance matrix of the risky assets is not positive definite. In Exercises CR-22 and 23, optimal portfolios are obtained for a single risky asset under unlimited and limited borrowing. Exercise ME-18 is concerned with the existence of solutions to the linear complementary problem arising from the Lintner paper. Exercise ME-16 examines substitution and complementary effects in the choice of risky assets, for the case of a quadratic utility function.

The Vickson paper discusses restrictions on the investor's utility function which are necessary and sufficient for separation of the general portfolio problem. The paper is an extension of a recent paper of Cass and Stiglitz (1970) in which the separation property was examined for the case when unlimited borrowing and short sales are allowed. The problem is to determine conditions on the utility function which will ensure that an arbitrary portfolio problem in many assets (one of which may be riskless) reduces to an equivalent two-asset problem, involving two mutual funds which are independent of initial wealth. When such a property holds, determination of the optimal portfolio is reduced to a two-stage procedure: (1) calculation of the optimal mutual funds and (2) determination of actual investment in each mutual fund. If there exists a riskless asset, it may or may not be one of the "mutual funds" in the optimal portfolio. If it is one of the mutual funds, then the optimal risky asset proportions are independent of wealth; otherwise the risky asset proportions will generally be wealth dependent. The type of separation property discussed in the Vickson paper is totally different from that in the Lintner paper. In the latter, the optimal mutual fund is determined solely on the basis of the means, variances, and covariances of the risky assets, and the rate of return on the riskless asset; it is independent of wealth and the utility function. In the separation property of the Vickson paper, however, the mutual funds are independent of wealth but dependent on the parameters of the utility function. In both types of "separation," the actual investment in the mutual funds is dependent on both the wealth level and the utility function.

Necessary conditions on marginal utility are derived for separation in general markets, including markets with or without a risk-free asset. These conditions on marginal utility are also sufficient for separation in general markets with unlimited borrowing and short sales. When borrowing constraints are introduced, these conditions are essentially still sufficient for separation in a local sense (for all wealth in some interval). Under certain circumstances they are also sufficient for separation in a global sense (for all wealth). This distinction between local and global separation, which arises because of constraints, was introduced explicitly by Hakansson (1969a), but was also present implicitly in the original mean-variance theory of Markowitz (1952).

Exercise CR-21 asks for some proof details which were omitted in the paper. Exercise ME-14 examines the separation property from another viewpoint. Exercise ME-4 applies some of the results of the paper to qualitative analysis of optimal foreign exchange holdings.

IV. Additional Reading Material

There are a number of articles which the reader may find helpful. Quirk and Saposnik (1962) presented an early treatment of stochastic dominance concepts. The paper of Hadar and Russell (1969) discusses first- and second-degree stochastic dominance in a simple way for bounded random variables. Rothschild and Stiglitz (1970) present a simple proof of Theorem 2.1 of the Brumelle and Vickson paper; their paper is the basis of Exercise ME-12. They also introduce the idea of one random variable having "more weight in the tails" than another, and they demonstrate its equivalence to second-degree dominance. In a second paper (1971) they apply stochastic dominance to the qualitative analysis of several economic problems. Diamond and Stiglitz (1974) review and extend some of the Rothschild–Stiglitz analysis and develop results concerned with the effects of increasing risk when expected utility is held constant (see also Diamond and Yaari, 1972). Vickson (1974) presents necessary and sufficient stochastic dominance conditions for increasing concave utility functions having nonincreasing absolute risk aversion in the case when the random variables are discrete. The stochastic dominance tests may be performed using a dynamic programming scheme. In Vickson (1975), he develops necessary conditions and (different) sufficient conditions for general random variables. In addition, he shows that for general random variables with equal means, necessary and sufficient conditions are precisely Whitmore's (1970) conditions for third degree stochastic dominance.

Fishburn (1974a, b) develops theorems for first and second degree stochastic dominance involving convex combinations of probability distribution functions. Levy and Hanoch (1970) and Joy and Porter (1974) test the stochastic dominance rules on a set of mutual funds and compare the resulting efficient sets with those obtained using mean-variance analysis [see also Porter and Gaumnitz (1972), Levy and Sarnat (1970), and Porter (1974)]. Porter (1972) examines the use of stochastic dominance rules in capital budgeting decisions under risk. Porter *et al.* (1973) have constructed special algorithms for efficiently performing stochastic dominance tests when there are many assets. Levy (1973) has considered the stochastic dominance problem when the investment returns have log-normal distributions. Meyer (1966) and Phelps (1966) present very general treatments of orderings of measures on topological spaces. The book by Arrow (1971) discusses the role of risk-aversion measures in qualitative investor behavior, and is highly recommended as supplementary reading. Yaari (1969) presents an axiomatic approach to risk-aversion measures. Shell (1973) presents an elementary treatment of behavior toward risk and its consequences for the elasticity of demand for money. The Tobin (1958) paper elucidates the Tobin–Lintner separation theorem, and discusses the role of risk in liquidity preference. The classic paper of Markowitz (1952) introduced a mean-variance analysis in portfolio allocation. Stiglitz (1973) presents a simple discussion of portfolio separation theorems and their applications to monetary theory.

1. STOCHASTIC DOMINANCE

G. Hanoch and H. Levy
Reprinted from *The Review of Economic Studies* **36**, 335-346 (1969).

The Efficiency Analysis of Choices Involving Risk [1]

I. INTRODUCTION

The choice of an individual decision-maker among alternative risky ventures, may be regarded as a two-step procedure. Firstly, he chooses an efficient set among all available portfolios, independently of his tastes or preferences. Secondly, he applies his individual preferences to this set, in order to choose the desired portfolio. The subject of this paper is the analysis of the first step. That is, it deals with optimal selection rules, which minimise the efficient set, by discarding any portfolio that is inefficient, in the sense that it is inferior to a member of the efficient set, from point of view of each and every individual, when all individuals' utility-functions are assumed to be of a given general class of admissible functions.

It is assumed throughout that any utility function is of the Von-Neumann-Morgenstern [21] type, i.e., it is determined up to a linear transformation, and is non-decreasing, and that individuals maximize expected utility.[2] The risks considered are random variables X with given (cumulative) probability distributions $F(x) = \Pr\{X \leqq x\}$. $F(x)$ is thus a non-decreasing function, continuous on the right, with $F(-\infty) = 0$, $F(\infty) = 1$. It may be discrete, continuous or mixed, with finite or infinite range. The analysis is not influenced by whether these probabilities are " objective " or " subjective ". However, the distributions are assumed to be fully specified, with no vagueness or uncertainty attached to their specification.

The present analysis is carried out in terms of a single dimension (e.g., money), both for the utility functions and for the probability distributions. However, the results may easily be extended, with minor changes in the theorems and the proofs, to the multivariate case.[3]

Section II gives a necessary and sufficient condition for efficiency, when no further restrictions are imposed on the utility functions.

Section III states and proves the optimal efficiency criterion in the presence of general risk aversion, i.e., for concave utility functions.

Section IV analyzes the conditions under which the well-known and widely used mean-variance criterion (μ, σ) is a valid efficiency criterion.

Finally, Section V gives some conclusions, remarks, and suggestions for further theoretical and empirical analysis related to the present results.

II. UNRESTRICTED UTILITY—THE GENERAL EFFICIENCY CRITERION

Given two risks (random variables), X and Y, with the respective (known, cumulative) probability distribution functions $F(x)$, $G(y)$, we say that XDY, or FDG—X dominates Y,

[1] The authors are grateful to A. Beja and S. Kaniel for valuable comments. The referees commented on an early draft, that similar results appeared in other works, unpublished and unknown to us at that time: a paper by Hadar and Russel [7], a Thesis by J. Hammond [8], and a book by Pratt, Raiffa and Schlaifer [17]. There is also some overlap with Quirk and Saposnik [18], and Feldstein [5]. However, the present paper gives a more general treatment and some significant modifications to most of these results.

[2] The axiomatic basis for this assumption is given in: Herstein and Milnor [10], Marschak [15], Von-Neumann and Morgenstern [21].

[3] This, however, raises many special and interesting problems, and will be dealt with more extensively in a forthcoming paper by the first author.

or F dominates G—if:

$$\Delta Eu = E_F u(x) - E_G u(y) \geqq 0,$$

(i.e., the expected utility of X is greater than or equal to the expected utility of Y) for every utility function u, in the class U of all admissible utility functions, and if $\Delta Eu > 0$ for some u in U. In this section, U is the class of all non-decreasing functions (assumed to have finite values for any finite value of x).[1] The variables X and Y[2] are defined here as the money payoffs of a given venture, which are additions (or reductions) to the individual's (constant) wealth, so that no restriction to positive values of X or Y is necessary. (The utility function u depends, of course, on the individual's initial wealth position).

A sufficient condition for FDG (or XDY) is an efficiency criterion.[3] The set of all risks which are not dominated by another risk according to the given criterion, is an efficient set. The weaker the sufficient condition, the smaller the efficient set.

The minimal efficient set E is generated when the criterion is optimal, that is, when the condition for FDG is both sufficient and necessary.

If the class of admissible utility functions is restricted to a proper subset of U, the minimal efficient set may be reduced to a subset of E, and the corresponding optimal efficiency criterion may be weakened.

Before proceeding to state and prove the general criterion, we need the following:

Lemma 1. *Let G, F, be two (cumulative) distributions, and $u(x)$ a non-decreasing function, with finite values for any finite x; then*

$$\Delta Eu = E_F u(x) - E_G u(x) = \int [G(x) - F(x)] du(x).^4$$

Proof. By definition,[5]

$$\Delta Eu = \int u dF - \int u dG. \qquad \text{...(1)}$$

Integrating (1) by parts gives:

$$\Delta Eu = \left[\int d(u\,.\,F) - \int F du\right] - \left[\int d(u\,.\,G) - \int G du\right]$$

$$= \int d[u\,.\,(F-G)] + \int (G-F) du.$$

In order to show that the first term on the right vanishes, we define a sequence of functions $u_n(x)$, converging to $u(x)$:

$$u_n(x) = \begin{cases} u(-n) & \text{for } x < -n \\ u(x) & -n \leqq x \leqq n \\ u(n) & x > n. \end{cases}$$

[1] Quirk and Saposnik [18] proved a similar criterion for the discrete, finite case, and assumed a strictly monotone utility. They also sketched a proof for the (pure) continuous case, requiring that $u(x)$ be bounded and piecewise differentiable (p. 144). All these requirements are not necessary. Similar comments apply to Hadar and Russel [7], who assumed $u(x)$ to be twice continuously differentiable, and restricted to a finite range. Cf. also Hammond [8].

[2] In much of the following discussion, we do not distinguish in notation between the variables X and Y, restricting our attention to the two distributions, $F(x)$ and $G(x)$.

[3] An additional requirement is the transitivity of the sufficient condition; but this is obviously assured by the given definition of dominance, since $E_F u - E_G u \geqq 0$ and $E_G u - E_H u \geqq 0$, imply $E_F u - E_H u \geqq 0$, and FDH.

[4] The integrals throughout are Stieltjes-Lebesgues integrals, ranging on all real values of x, unless specified otherwise; cf. Cramér [4], p. 62.

[5] The arguments in the functions appearing in integrals are omitted in cases where no misunderstanding should arise.

But now,

$$\int d[u(F-G)] = \lim_{n\to\infty} \int d[u_n(F-G)]$$
$$= \lim_{n\to\infty} \{u(n)[F(\infty)-G(\infty)]-u(-n)[F(-\infty)-G(-\infty)]\}$$
$$= \lim_{n\to\infty} \{u(n).0-u(-n).0\} = 0.$$

Hence $\Delta Eu = \int (G-F)du$, if the integral exists.[1] Q.E.D.

The optimal criterion for *FDG* is given in Theorem 1.

Theorem 1. *Let F, G and u be as in Lemma* 1. *A necessary and sufficient condition for FDG is:* $F(x) \leqq G(x)$ *for every x, and* $F(x_0)<G(x_0)$ *for some* x_0.

Proof. (*a*) The sufficiency follows immediately from Lemma 1.

$$G-F \geqq 0 \Rightarrow \int (G-F)du \geqq 0$$

when u is non-decreasing. If $G(x_0)-F(x_0)>0$, due to the right-continuity of F and G, there is an interval $x_0 \leqq x< x_0+\beta$ where $G(x)-F(x)>0$. To show that there exist some u_0 for which $\Delta Eu_0>0$, choose $u_0(x)$ as follows:

$$u_0(x) = \begin{cases} x_0 & x \leqq x_0 \\ x & x_0 \leqq x \leqq x_0+\beta \\ x_0+\beta & x \geqq x_0+\beta. \end{cases}$$

Then $u_0 \in U$, and $\int (G-F)du = \int_{x_0}^{x_0+\beta} (G-F)dx>0.$

(*b*) The necessity is proved similarly. If, for some x_1, $G(x_1)-F(x_1)<0$, there is an interval $[x_1, x_1+\varepsilon)$, where $G(x)-F(x)<0$; choose:

$$u_1(x) = \begin{cases} x_1 & x \leqq x_1 \\ x & x_1 \leqq x \leqq x_1+\varepsilon \\ x_1+\varepsilon & x \geqq x_1+\varepsilon. \end{cases}$$

Then,
$$\int (G-F)du_1 = \int_{x_1}^{x_1+\varepsilon} (G-F)dx<0.$$

In addition, if $F(x) = G(x)$ for all x, then

$$\int (G-F)du = 0 \text{ for all } u. \quad \text{Q.E.D.}$$

The interpretation of this criterion is straightforward: Given that F and G are distinct, F dominates G, if and only if, for every value x, the probability of getting x or less is not larger with F than with G. This also means, that $1-F \geqq 1-G$, or that the probability of getting more than x is not smaller with F than with G, for every x. Hence, the cumulative probability distribution F is a shift downward (or to the right) of the distribution G.[2]

[1] The formal equality holds even if the integral diverges, since ΔEu is then infinite.
[2] Thus F is " stochastically larger " than G. Cf. Quirk and Saposnik [18].

Whenever the two cumulative distributions intersect, they cannot dominate one another; that is: one can find one utility function u where $\Delta Eu>0$, and another function v where $\Delta Ev<0$ (where both u and v are non-decreasing).

It follows immediately (since $u(x) = x$ is admissible) that $E_F X > E_G Y$, i.e., a larger mean value of X, is a necessary condition for dominance of X. The variance, however, plays no direct role in the efficiency criterion. For example, consider two random variables with rectangular distributions, i.e. X has a constant probability density in the range $x_1 \leqq x \leqq x_2$, and Y a constant density in the range $y_1 \leqq y \leqq y_2$.[1] If $x_1 > y_1$ and $x_2 > y_2$, X is preferred to Y with every possible utility function. The variance of X (which equals $\frac{1}{12}(x_2-x_1)^2$) may be much larger than var Y, and the degree of risk aversion implied by $u(x)$ may be as high as one wants to assume, still X is better than Y, if individuals do not prefer less money to more money (i.e., if $u(x)$ is non-decreasing).

If one wishes to consider utility functions which vary only in a bounded range,[2] it can easily be verified that the optimal criterion for dominance of F over G, is that $F(x) \leqq G(x)$ for all x and $F(x_0) < G(x_0)$ for some x_0 *in the given range*. F and G may intersect outside the range, but this would not affect $\int (G-F)du = \Delta Eu$.

III. EFFICIENCY IN THE FACE OF RISK AVERSION

It is sometimes argued,[3] both on theoretical and on empirical grounds, that risk aversion, or concave utility functions, is the only case to consider. We do not wish to take sides in this controversy, but to analyze the implications of such an assumption for our efficiency analysis. In Theorem 2 we present an optimal criterion, when U_1 is the class of all non-decreasing, concave utility functions, i.e. when

$$u(\alpha x_1+(1-\alpha)x_2) \geqq \alpha u(x_1)+(1-\alpha)u(x_2)$$

for any pair x_1, x_2, and a positive fraction α.[4]

Theorem 2. *Let F and G be two (cumulative) distributions. A necessary and sufficient condition for FDG, for every u(x) which is non decreasing and concave, is*

$$\int_{-\infty}^{x} [G(t)-F(t)]dt \geqq 0$$

for every x, and $G \neq F$ *for some* x_0.

That is, the accumulated area under G should not be less than the area under F, below any real value of x (where F and G are distinct).

Proof. (*a*) The necessity of the condition follows again from Lemma 1. Suppose for some x_0, $\int_{-\infty}^{x_0} (G-F)dt<0$. Now, define $u_2(x)$ as follows:

$$u_2(x) = \begin{cases} x & x \leqq x_0 \\ x_0 & x \geqq x_0. \end{cases}$$

1 The function $F(x)$ is:

$$F(x) = \begin{cases} 0 & x \leqq x_1 \\ \dfrac{x-x_1}{x_2-x_1} & x_1 \leqq x \leqq x_2 \\ 1 & x \geqq x_2; \end{cases}$$

$G(y)$ is defined similarly.

2 E.g., if an individual cannot be worse off than losing everything, this means that $u(x) = u(-w)$, for $x \leqq -w$, when w is the initial wealth. The range is thus bounded from below.

3 See Arrow [1], and Yaari [22].

4 Cf. Hammond [8], and Hadar and Russel [7], for similar theorems.

Observe that $u_2(x)$ is in U_1, since it is non-decreasing and concave; but

$$\Delta Eu_2 = \int (G-F)du_2 = \int_{-\infty}^{x_0} [G(t)-F(t)]dt<0,$$

by assumption. Again, $F = G$ for all x, implies $\Delta Eu = 0$ for all u. Therefore, the condition is indeed necessary.

(*b*) The proof of sufficiency is somewhat more involved. Define two characteristic functions:

$$I_A(x) = 1;\ I_B(x) = 0, \quad \text{when } G(x) \geqq F(x)$$

$$I_A(x) = 0;\ I_B(x) = 1, \quad \text{when } G(x)<F(x);$$

and define a transformation Tx by the following equation:

$$\int_{-\infty}^{Tx} I_A(t)|\,G-F\,|\,dt = \int_{-\infty}^{x} I_B(t)|\,G-F\,|\,dt. \qquad \text{...(2)}$$

Since $\lim_{x\to-\infty} \int_{-\infty}^{x} |\,G-F\,|\,dt = 0$, we have:

$$\int_{-\infty}^{x} (G-F)dt = \int_{-\infty}^{x} I_A(t)|\,G(t)-F(t)|\,dt - \int_{-\infty}^{x} I_B(t)|\,G(t)-F(t)|\,dt$$

which is non-negative, by assumption. The first term on the right is a non-decreasing function of x, hence the equality (2) may only be maintained if $Tx \leqq x$, for all x. One can verify that Tx is almost everywhere continuous and differentiable, and that $T'(x) \geqq 0$ (that is, Tx is non-decreasing).

Differentiating equation (2), which holds for all x, one gets:

$$I_A(Tx)|\,G(Tx)-F(Tx)|\,T'(x) = I_B(x)|\,G(x)-F(x)| \text{ for almost all } x. \qquad \text{...(3)}$$

We shall now prove that $\int_{t\leqq x} [G(t)-F(t)]du(t) \geqq 0$ for all x.

$$\int_{t\leqq x} (G-F)du(t) = \int_{t\leqq x} I_A\,|\,G-F\,|\,du(t) - \int_{t\leqq x} I_B\,|\,G-F\,|\,du(t). \qquad \text{...(4)}$$

Substituting (3) in the integrand of the second term, we get:

$$\int_{t\leqq x} (G-F)du(t) = \int_{t\leqq x} I_A\,|\,G-F\,|\,du(t) - \int_{t\leqq x} I_A(Tt)|\,G(Tt)-F(Tt)|\,T'(t)du(t).$$

Consider now the second term on the right. Remembering that the integrand is non-negative, we get

$$\int_{t\leqq x} I_A(Tt)|\,G(Tt)-F(Tt)|\,T'(t)du(t) \leqq \int_{t\leqq x} I_A(Tt)|\,G(Tt)-F(Tt)|\,T'(t)du(Tt),$$

since $Tt \leqq t$, and $u(t)$ is concave.

A change of variable in the last term, $Z = Tt$ gives:

$$\int_{Z\leqq Tx} I_A(Z)|\,G(Z)-F(Z)|\,du(Z),$$

so that, collecting this result back in (4):

$$\int_{t \leqq x} (G-F)du \geqq \int_{t \leqq x} I_A \,|\, G-F \,|\, du - \int_{t \leqq Tx} I_A \,|\, G-F \,|\, du$$

$$= \int_{Tx \leqq t \leqq x} I_A \,|\, G-F \,|\, du(t) \geqq 0, \text{ since } Tx \leqq x.$$

Now, the non-negativity of ΔEu follows, since

$$\Delta Eu = \int (G-F)du = \lim_{x \to \infty} \int_{t \leqq x} [G(t)-F(t)]du(t) \geqq 0$$

for all concave u. And since $F \equiv G \Rightarrow \int (G-F)du = 0$, we need $F \neq G$ for some x_0, to assure that $\Delta Eu > 0$ for some u. Q.E.D.

One may verbalize the criterion and the proof given above as follows: The two cumulative distributions may intersect many times, as long as the negative areas between them (where $F > G$) to the left of any x, remain smaller, in absolute value, than the accumulated positive areas (where $F < G$). This assures that $\int_{-\infty}^{x} (G-F)dt$ is non-negative Hence $\Delta Eu = \int_{-\infty}^{x} (G-F)du$ is non-negative for any linear utility $u(t) = t$, or $u(t) = Kt$. Therefore it must be true *a fortiori* for any concave utility, since the negative area to the right of any positive area, will weigh less with a concave $u(t)$ than with a linear one, in the total area $\int_{-\infty}^{x} (G-F)du(t)$.

Here, $E_F(X) \geqq E_G(Y)$ (where $F \neq G$) is a necessary condition for dominance of X over Y. But here, too, the variance of X need not be smaller than the variance of Y. The following example should make this clear:

Example 1

x, y	Pr (x)	Pr (y)	$F(x)$	$G(y)$	$G-F$	$\int_0^x (G-F)dt$
0	0·65	0·86	0·65	0·86	0·21	0
1	0·25	0·03	0·90	0·89	−0·01	0·21
2	0·10	0·11	1·00	1·00	0	0·20

In this example $E_F(X) = 0{\cdot}45 > E_G(Y) = 0{\cdot}25$,

$$\text{var}_F\,(X) = 0{\cdot}4475 > \text{var}_G\,(Y) = 0{\cdot}4075,$$

but X is preferred to Y for any concave utility function. Since $F(1) > G(1)$, the general criterion of theorem 1 is not satisfied. Indeed, choosing a convex $u(x) = 10^{x^2}$, we get Y preferred to X: $\Delta Eu = E_F u(x) - E_G u(y) \cong 1003 - 1101 < 0$.

Applying the criterion of Theorem 2 to discrete distributions, one gets the following criterion:

Let $x_1, x_2, \ldots, x_k, \ldots$ be (an increasing) sequence of values where F and/or G have a non-zero probability f_i or g_i.

The optimal criterion for *FDG* is:

$$\sum_{i=1}^{k-1} (x_{i+1}-x_i) \sum_{j=1}^{i} (g_j-f_j) = \sum_{i=1}^{k-1} (x_k-x_i)(g_i-f_i)$$
$$= x_k \sum_{i=1}^{k-1} (g_i-f_i) - \sum_{i=1}^{k-1} x_i(g_i-f_i) \geqq 0, \text{ for all } k \geqq 2.$$

It should be evident, that the criteria of Theorems 1 and 2 are valid optimal efficiency criteria, since they are transitive (see footnote 3 on page 336 above), and independent of individual wealth positions (assumed constant) or tastes, apart from the tastes implied by the definition of the class U.

An important special case of Theorem 2 is when the two distributions intersect only once.[1] In this case, a necessary and sufficient condition for (concavity) dominance of F, is that $F<G$ *below* the intersection point, when $E_F \geqq E_G$. In this case, the area between F and G to the left of the intersection [where $(G-F)>0$], exceeds the area to the right, where $(G-F)<0$. This is stated and proved as follows:

Theorem 3. *Let F, G be two distributions with mean values μ_1, μ_2, respectively, such that for some $x_0<\infty$, $F \leqq G$ for $x<x_0$ (and $F<G$ for some $x_1<x_0$) and $F \geqq G$ for $x \geqq x_0$; then FDG (for concave utility functions) if and only if $\mu_1 \geqq \mu_2$.*

Proof. (*a*) If $\mu_1 \geqq \mu_2$, we have, by applying Lemma 1 to the case $u(t) = t$:

$$\mu_1-\mu_2 = \int_{-\infty}^{\infty} (G-F)dt = \int_{-\infty}^{x_0} (G-F)dt - \int_{x_0}^{\infty} |G-F|\, dt \geqq 0.$$

Therefore $\int_{-\infty}^{x} (G-F)dt \geqq 0$ for all $x<\infty$, and by Theorem 2, *FDG*.

(*b*) If $\mu_1<\mu_2$,

$$\mu_1-\mu_2 = \int_{-\infty}^{x_0} (G-F)dt - \int_{x_0}^{\infty} |G-F|\, dt<0.$$

Hence, for some finite $x_2>x_0$,

$$\int_{-\infty}^{x_2} (G-F)dt = \int_{-\infty}^{x_0} (G-F)dt - \int_{x_0}^{x_2} |G-F|\, dt<0,$$

whereas for x_0 we have $\int_{-\infty}^{x_0} (G-F)dt>0$, and the condition of Theorem 2 does not hold for all x. That is, F or G do not dominate one another for all concave utilities. Q.E.D.

IV. THE LIMITATIONS OF THE MEAN-VARIANCE EFFICIENCY CRITERION

Since the famous analysis of behaviour towards risk by Tobin [20], and the subsequent developments of the subject of portfolio selection by Markowitz [13], [14] and others,[2] one is accustomed to identify the *variance* of a distribution of returns with the degree of its riskiness. Assuming a general aversion to risk seems to imply, therefore, a general aversion to variance. Hence the implication which is in the basis of Markowitz's mean-variance criterion, that whenever the mean and the variance are both higher in one distribution than in the other, it is not possible to judge between the two in terms of efficiency,

[1] This is termed " simply intertwined distributions " by Hammond [8].
[2] E.g., Baumol [2], Friedman and Savage [6], Lintner [12] and Sharpe [19].

but one has to resort to individual tastes, in order to determine the preferred item. However, this is well known to be generally invalid. (Example 1 above shows this for a case where the criterion of Theorem 1 is not satisfied).

The converse conclusion of the mean variance approach, that a larger mean and a smaller variance ought to be preferred by all people with aversion to risk, is also unsound. This could be best made clear by another example:

Example 2

x	Pr (x)	y	Pr (y)
1	0·80	10	0·99
100	0·20	1000	0·01

$$EX = 20{\cdot}8 > EY = 19{\cdot}9$$
$$\text{var } X \doteq 1468 < \text{var } Y = 9703$$

Hence the Markowitz criterion is satisfied for x. But suppose the utility function is $u(Z) = \log_{(10)}Z$, which is a well-behaved function for all positive values, displaying risk-aversion everywhere.[1] With this utility, however, $Eu(x) = 0{\cdot}4$, $Eu(y) = 1{\cdot}02$, and Y is preferred to X!

Examples (1) and (2) thus show that Markowitz's (μ, σ) criterion is neither necessary nor sufficient for dominance, if the expected utility maxim is adhered to.[2] (See also the following Example 3.)

Under what condition, then, is the (μ, σ) criterion valid? Tobin [20] claimed that the mean vs. variance analysis is relevant in two cases:

(*a*) When the utility function is quadratic;

(*b*) When the distributions of the portfolios are all members of a two-parameter family. Let us examine these two cases.

(*a*) *Quadratic Utility*. There are two main objections to the use of quadratic utility: [3]

(1) that it is relevant for a bounded range only (the rising portion); and (2) that it displays increasing absolute risk aversion everywhere.[4] Disregarding the second one, at this stage, one may deal with the first objection by imposing appropriate bounds on the range of relevant values, or on the coefficients of u, so that $u(x)$ is an increasing function throughout the relevant range. In this case, the mean-variance criterion is well known to be a sufficient condition for dominance. However, as has been shown, it is not a necessary condition (e.g. in Example 1, X is dominant for all concave utilities, hence for quadratic utility; but var $X >$ var Y). We have given elsewhere some conditions for efficiency with quadratic utility,[5] which are much weaker sufficient conditions than Markowitz's.[6]

(*b*) *Two-parameter Distributions*. If instead of restricting the admissible utility functions, one is willing to consider only a restricted class of two-parameter distributions,[7]

[1] $u(Z) = \log Z$ also shows decreasing risk-aversion everywhere, in accordance with the requirement stated by Arrow [1] and by Pratt [16].

[2] A different " mean-variance approach ", which is often confused with this in the literature, is where μ and σ are assumed to be *direct arguments* in the utility function, with *assumed* given signs of their partial effects (cf. Borch [3]; Markowitz [14], p. 209). As proved by Example 2, this approach is inconsistent with the axioms of expected utility. Its logical basis thus seems to be questionable, or at least vague.

[3] Cf. Arrow [1], Hirshleifer [11], Pratt [16].

[4] That is, $\frac{-u''(x)}{u'(x)}$ increases with x. This measure of risk aversion was developed independently by Arrow [1] and Pratt [16].

[5] Hanoch and Levy [9]. These conditions are " optimal " relative to the given information about the distributions.

[6] E.g., $\mu_1 > \mu_2$, and $(\sigma_1^2 - \sigma_2^2) < (\mu_1 - \mu_2)^2$ is sufficient for dominance of (μ_1, σ_1), for any quadratic utility; *loc. cit.*

[7] See also Pratt, Raiffa and Schlaifer [17], sec. 23.22.2. Feldstein [5] discusses some counter-examples, for general two-parameter distributions (e.g. log-normal).

the mean-variance criterion becomes a sufficient condition for efficiency. However, by Theorem 1 and the discussion of Section II, it is evident that even under these assumptions the (μ, σ) is not a necessary condition, since if the two cumulative distributions do not intersect, one dominates the other for any set of preferences, whether or not the variance is smaller. But for two such distributions that do intersect, the mean-variance criterion is equivalent to the condition of Theorem 2, and thus it is both necessary and sufficient in the face of risk-aversion. This is summarized and formalized in the following theorem.

Theorem 4. *Let $F(x)$ and $G(y)$ be two distinct distributions with means μ_1 and μ_2, and variances σ_1^2, σ_2^2, respectively, such that $F(x) = G(y)$, for all x and y which satisfy $\frac{x-\mu_1}{\sigma_1} = \frac{y-\mu_2}{\sigma_2}$. Let $\mu_1 \geqq \mu_2$, and $F(x_1) > G(x_1)$ for some x_1 (i.e., $F(x)$ and $G(x)$ intersect). Then F dominates G for all concave $u(x)$, if and only if $\sigma_1^2 \leqq \sigma_2^2$.*

That is, F and G belong to *the same family* of distributions, with two *parameters which are independent functions of μ and σ^2, respectively.*[1] In that case, $Z = \frac{x-\mu}{\sigma}$ is distributed identically with mean 0 and variance 1 (according to some $H(Z;\ 0, 1)$), for both F and G. In these limited cases, when the two distributions $F(x)$ and $G(x)$ intersect, the mean-variance criterion is both necessary and sufficient for efficiency. The proof is an immediate application of Theorem 3.

Proof. If $\sigma_1 = \sigma_2 = \sigma$; and $\mu_1 > \mu_2$; For all x, $\frac{x-\mu_1}{\sigma} < \frac{x-\mu_2}{\sigma}$; thus $F(x) \leqq G(x)$ for all x, and there is no intersection point. F dominates G by Theorem 1. If $\mu_1 = \mu_2$, F and G are identical.

If $\sigma_1 \neq \sigma_2$, we have an intersection point at x_0, where

$$\frac{x_0-\mu_1}{\sigma_1} = \frac{x_0-\mu_2}{\sigma_2} \quad \text{or:} \quad x_0 = \frac{\mu_2\sigma_1-\mu_1\sigma_2}{\sigma_1-\sigma_2}.$$

(*a*) If $\sigma_1 > \sigma_2$, then for $x < x_0$, $\frac{x-\mu_1}{\sigma_1} > \frac{x-\mu_2}{\sigma_2}$, and $F(x) \geqq G(x)$ ($F > G$ for x_1); and for $x > x_0$, $\frac{x-\mu_1}{\sigma_1} \leqq \frac{x-\mu_2}{\sigma_2}$, and $F(x) \leqq G(x)$. Thus, the condition of Theorem 3 is not satisfied, and F cannot dominate G.

(*b*) If $\sigma_1 < \sigma_2$, then for $x < x_0$ we have $\frac{x-\mu_2}{\sigma_1} < \frac{x-\mu_2}{\sigma_2}$, and $F \leqq G$; for $x > x_0$ $\frac{x-\mu_1}{\sigma_1} > \frac{x-\mu_2}{\sigma_2}$, and $F \geqq G$. Since $F < G$ for some x (given that $\mu_1 \geqq \mu_2$), F dominates G by Theorem 3. Q.E.D.

Obviously, the (μ, σ) criterion is optimal, when the distributions considered are all Gaussian Normal. But the symmetric nature of this distribution seems to deny its usefulness as a good approximation to reality, for at least some types of risky portfolios.

Even for symmetric distributions, the mean-variance criterion is not valid, when the distribution has more than two parameters. The following example should make this clear.

[1] Feldstein [5] shows examples where this requirement (of the two parameters being independent functions of μ or σ) is not satisfied, leading to invalidity of the (μ, σ) criterion. Tobin [20] does not discuss this basic requirement.

Y

Example 3

We consider the following 3-parameters (μ, h, p) family of symmetric, discrete, distributions of the random variable X $(p<\frac{1}{2})$:

x	$\Pr(x)$	$F(x)$	$\int_{-\infty}^{x} F(t)dt$
$\mu-h$	p	p	0
μ	$1-2p$	$1-p$	hp
$\mu+h$	p	1	h

Here $EX = \mu$; $\text{var } X = 2h^2p$. Choose: $\mu_1 = \mu_2 = 0$; $h_1 = 1$, $h_2 = 2$; $p_1 = 0{\cdot}25$, $p_2 = 0{\cdot}1$. Then: $EX_1 = EX_2 = 0$; $\text{var } X_1 = 0{\cdot}5 < \text{var } X_2 = 0{\cdot}8$;

$$\int_{-\infty}^{-1} (F_2 - F_1)dt = 0{\cdot}1 > 0,$$

but

$$\int_{-\infty}^{0} (F_2 - F_1)dt = h_2p_2 - h_1p_1 = -0{\cdot}05 < 0,$$

and by Theorem 2, X_1 does not dominate X_2 for all concave utilities.

V. CONCLUSION

The main conclusions to be derived from this analysis, are negative as well as positive.

On the negative side, one has to be very critical of using automatically the mean-variance criterion for choice among risky ventures. In fact, any variation of such a criterion, based on the mean and the variance alone, is generally invalid.[1]

The identification of riskiness with variance, or with any other single measure of dispersion, is clearly unsound. There are many obvious cases, where more dispersion is desirable, if it is accompanied by an upward shift in the location of the distribution, or by an increasing positive asymmetry.

A more detailed analysis of the relation between skewness and risk is a desirable route to follow, if one is trying to restrict the information about distributions to a small number of parameters.[2]

On the positive side, we have given optimal criteria for both the general case and the risk-aversion case. These criteria are not hard to comprehend, and are quite general, being independent of restrictions on either the form of utility or on the type of distributions. However, a complete specification of the distribution is a necessary condition for their application. When the distributions are of the finite, discrete type (i.e., a finite number of distinct outcomes with given probabilities), the computations needed for efficiency comparisons are relatively simple. Approximating any distribution of a risky item by such a finite, discrete probability distribution might be feasible, and might furnish relatively good results.

It should be an interesting empirical study, to try and select an efficient set of stocks, basing the computations on available data of stock returns over time, and to compare these with results obtained by using different sub-optimal criteria.[3]

[1] For example, a criterion offered by Baumol [2], is not valid, since it is a weaker condition than Markowitz's. In cases where (μ, σ) is not sufficient, Baumol's criterion $(\mu - k\sigma)$ is not sufficient *a fortiori*. If (μ, σ) is optimal, Baumol's condition reduces the " efficient set " still further, and is therefore invalid.

[2] Cf., however, Hanoch and Levy [9].

[3] Some results along these lines, based on Israel stock-market data, are included in a thesis by Haim Levy.

Further analysis on the theoretical side should be devoted to the translation of these general criteria into equivalent criteria for special families of probability distributions, of a more flexible nature than the simple, symmetric, two-parameter normal distributions.

Finally, all this should be tied up with the theory of consumption under uncertainty, by generalization to multivariate utilities and distributions; with the theory of the financial structure of the firm and its cost of capital; and with a general efficiency analysis of interdependent risks, with applications to the construction of optimal portfolios, rather than the choice among alternative, independent, portfolios.

The Hebrew University,
Jerusalem

G. Hanoch
H. Levy.

First version received 15.6.68; *final version received* 9.1.69

REFERENCES

[1] Arrow, J. K. *Aspects of the Theory of Risk Bearing* (Helsinki, 1965).

[2] Baumol, W. J. " An Expected Gain-Confidence Limit Criterion for Portfolio Selection ", *Management Science*, **10**, No. 1 (October 1963), 174-182.

[3] Borch, K. " A Note on Utility and Attitudes to Risk ", *Management Science* (July 1963).

[4] Cramér, H. *Mathematical Methods of Statistics* (Princeton University Press, Princeton, New Jersey, 1946).

[5] Feldstein M. " Mean Variance Analysis in the Theory of Liquidity Preference and Portfolio Selection ", *Review of Economic Studies*, **36** (January 1968).

[6] Friedman, M. and Savage, L. J. " The Utility Analysis of Choices Involving Risk ", *The Journal of Political Economy* (August 1958), pp. 279-304.

[7] Hadar, J. and Russel, W. R. " Rules for Ordering Uncertain Prospects ", *American Economic Review* (forthcoming).

[8] Hammond, J. " Towards Simplifying the Analysis of Decisions under Uncertainty where Preference is Non-Linear ", (Unpublished Thesis, Harvard Business School, 1968).

[9] Hanoch, G. and Levy, H. " Efficient Portfolio Selection with Quadratic and Cubic Utility ", *Journal of Business* (forthcoming).

[10] Herstein, I. N. and Milnor, J. " An Axiomatic Approach to Measurable Utility ", *Econometrica*, **21**, No. 2, April (1953); reprinted as *Cowles Commission Papers*, New Series, No. 65.

[11] Hirshleifer, J. " Investment Decision under Uncertainty: Choice-Theoretic Approaches ", *The Quarterly Journal of Economics*, **79** (Nov. 1965) No. 4, pp. 509-536.

[12] Lintner, J. " Security Price, Risk and Maximal Gains from Diversification ", *The Journal of Finance*, **20** (December 1965), 587-615.

[13] Markowitz, H. M. " Portfolio Selection ", *The Journal of Finance*, **6**, No. 1 (March 1952), 77-91.

[14] Markowitz, H. M. *Portfolio Selection* (New York, Wiley, 1959).

[15] Marschak, J. "Rational Behavior, Uncertain Prospects, and Measurable Utility", *Econometrica*, **18**, No. 2 (April 1950); reprinted as *Cowles Commission Papers*, New Series, No. 43.

[16] Pratt, J. W. " Risk Aversion in the Small and Large ", *Econometrica* (January-April 1964), pp. 122-136.

[17] Pratt, Raiffa and Schlaifer. *Introduction to Statistical Decision Theory* (McGraw-Hill (Preliminary Edition) 1968).

[18] Quirk, J. P. and Saposnik, R. " Admissibility and Measurable Utility Functions ", *Review of Economic Studies* (1962).

[19] Sharpe, W. F. " Capital Assets Prices: A Theory of Market Equilibrium under Conditions of Risk ", *Journal of Finance* (1964), pp. 425-442.

[20] Tobin, J. " Liquidity Preference as Behaviour Towards Risk ", *Review of Economic Studies*, **25** (1957-58), 65-68.

[21] Von Neumann, J. and Morgenstern, O. *Theory of Games and Economic Behaviour* (Princeton University Press, Princeton, N.J., Third Edition 1953).

[22] Yaari, M. E. " Convexity in the Theory of Choice under Risk ", *The Quarterly Journal of Economics*, **79** (May 1965), No. 2, 278-290.

A Unified Approach to Stochastic Dominance

S. L. Brumelle
UNIVERSITY OF BRITISH COLUMBIA

and

R. G. Vickson
UNIVERSITY OF WATERLOO

I. Introduction to Stochastic Dominance

The fundamental problem facing an investor may be viewed as that of choice between two investment alternatives with random returns X and Y. For an expected utility maximizer with known utility function u, X is preferred to Y if $Eu(X) \geqq Eu(Y)$. However, it is difficult in practice to find an investor's utility function. Accordingly, it would be most useful to know whether X is preferred to Y by *all* investors with utility functions u in some general class U. This approach to the problem of choice seeks to characterize the investor's utility function only in terms of its general properties (e.g., monotonicity, concavity) rather than its specific numerical values. For purposes of portfolio theory, the class of increasing concave utility functions is of special importance, since it characterizes risk-averse investors.

Given a class U of utility functions, X is said to (stochastically) *dominate* Y if $Eu(X) \geqq Eu(Y)$ for all $u \in U$. The stochastic dominance relation over U thus partially orders the set of random variables.

With each random variable X, there is a distribution function $F(x) = P[X \leqq x]$, and a probability measure μ defined by $\mu\{[X \leqq x]\} = F(x)$. It is convenient to use the notation $\mu(f)$ to denote $Ef(X)$. That is, $\mu(f) = Ef(X) = \int_{\mathbf{R}} f(x)\, dF(x)$.

The distribution function of the random variable Y will be denoted by G and the corresponding measure by ν so that

$$\nu(f) = Ef(Y) = \int_{\mathbf{R}} f(y)\, dG(y).$$

The previous definition can be restated in the new notation: X dominates Y (with respect to U) if and only if $\mu(f) \geqq \nu(f)$ for each $f \in U$ for which $\mu(f)$ and $\nu(f)$ exist. One advantage of the notation $\mu(f)$ over $Ef(X)$ is the resulting similarity with pointwise ordering of functions.

Since integration is a linear operation, $\lambda(af+bg) = a\lambda(f)+b\lambda(g)$ for $a, b \in \mathbf{R}$ and any probability measure λ. Thus, $\mu(f) \geqq \nu(f)$ for each $f \in U$ if and only if $\mu(f) \geqq \nu(f)$ for each $f \in c(U)$ where $c(U)$ denotes all nonnegative linear combinations of functions in U, that is, the convex cone generated by U.

Also, if $f_n \to f$ in an appropriate manner so that $\mu(f_n) \to \mu(f)$ and $\nu(f_n) \to \nu(f)$, and if $f_n \in U$ for each n, then $\mu(f) \geqq \nu(f)$. Thus, in some cases $c(U)$ can be further extended to its closure under limits. For example, if all

$f \in U$ are continuous and vanish outside of some compact set K, then $\mu(f) \geqq \nu(f)$ for each $f \in U$ if an only if $\mu(f) \geqq \nu(f)$ for each $f \in \overline{c(U)}$, where the bar denotes closure under pointwise limits.

In characterizing investor utility functions by means of general properties, the resulting classes U are often automatically (closed) convex cones. This is true, in particular, for the classes consisting of increasing functions or of increasing concave functions. Given a (closed) convex cone U of functions, it will often be relatively simple to find a proper subset G of U such that U is the closed convex cone generated by G (i.e., such that any function f in U is a limit of positive weighted sums of functions in G). In such a case a necessary condition for stochastic dominance over U is that $\mu(g) \geqq \nu(g)$ for all $g \in G$, since $G \subset U$. Because of the additivity of stochastic dominance, it is usually quite simple to establish by limiting arguments the converse statement, that is, that dominance over G implies dominance over U. In this situation, a simplified test for stochastic dominance is obtained, since the ordering need only be checked within the subset G of U. It is clearly of interest to find the smallest set G which generates U. Since U is convex, the "best" G will generally be the set of extreme points of U (*cf.* Krein–Milman theorem [8]). All of the stochastic dominance tests which have appeared in the finance literature can be understood in terms of the ideas just given. Identification of the appropriate generating class G leads to explicit tests for dominance in terms of simply computed properties of the distribution functions. These "practical" tests are really nothing more than the requirement of dominance over G, translated into more specific terms.

In Section II, three examples of stochastic dominance relations are treated from this unifying point of view. Tests are derived for stochastic dominance over the classes of increasing functions, of increasing concave functions, and of increasing concave functions with nonnegative third derivative. These examples are drawn from the finance literature, and are of interest because of their implications for investor behavior. In Section III, stochastic dominance over the class of concave functions is reexamined in order to bring out its "probabilistic" content. If X dominates Y with respect to all concave utility functions, X is less "risky" than Y in an economic sense since all risk-averse investors prefer it. An interesting question is this: In what probabilistic, as opposed to economic sense, is X less risky than Y? It is shown that Y has the same distribution as $X+Z$, where $E[Z \mid X] \leqq 0$. Thus Y may be interpreted as X plus "noise" Z, and is therefore more "variable" than X. The derivation in Section III is given for the general case of multidimensional random variables, since this is no more difficult than the univariate case of direct interest to finance.

II. Examples of Stochastic Dominance Relations

In the following examples, "increasing" should be interpreted in the weak sense, i.e., as "nondecreasing," and is denoted by the symbol $\nearrow$. An analogous convention is for "decreasing". The operators $\wedge$ and $\vee$ are used for "minimum" and "maximum," respectively. For example, $\min[a, b] = a \wedge b$.

EXAMPLE 1

The first example is referred to as first-order stochastic dominance by Hadar and Russell [4] and appeared in the work of Lehmann [6] (also see Hanoch and Levy [5]). This example shows that an investor with an increasing utility function prefers X to Y if $P[Y \geqq w] \leqq P[X \geqq w]$ for all values of w, i.e., $1-G(w) \leqq 1-F(w)$ for each w.

Define the function $I_{[x \geqq w]}$ to be 1 if $x \geqq w$ and 0 if $x < w$. Then $P[Y \geqq w] = EI_{[Y \geqq w]}$ and $P[X \geqq w] = EI_{[X \geqq w]}$. Given a function f, the function f^+ is called the positive part of f and is defined by $f^+ = \max[f, 0]$. Similarly, $f^- = \min[f, 0]$. Note that $f = f^+ + f^-$.

Theorem 2.1 Suppose that f is an increasing function defined on $R = (-\infty, \infty)$ with $\nu(f^-) > -\infty$. Let $G = \{I_{[\cdot \geqq w]},\ w \in R\}$ and suppose that $\nu(g) \leqq \mu(g)$ for each $g \in G$. Then $\nu(f) \leqq \mu(f)$. Note that $\nu(g) \leqq \mu(g)$ for all $g \in G$ if and only if $1-F(w) \geqq 1-G(w)$ for all w in R.

Proof Since $\lambda(f+b) = \lambda(f)+b$ for any probability measure λ and $b \in R$, we can without loss of generality assume that $f(0) = 0$.

It is convenient to consider f^+ and f^- separately. Define

$$f_n^+(x) = \begin{cases} f(i/2^n) & \text{if } \ i/2^n \leqq x \leqq (i+1)/2^n, \quad i = 0, 1, 2, \ldots, n2^n \\ f(n) & \text{if } \ x \geqq n. \end{cases}$$

Clearly, each f_n^+ is a positive linear combination of functions of the form $I_{[\cdot \geqq w]}$. Therefore, $\nu(f_n^+) \leqq \mu(f_n^+)$ for each n.

Also, $f_n^+ \nearrow f^+$. Thus by the monotone convergence theorem [7, p. 124],

$$\nu(\lim f_n^+) = \lim \nu(f_n^+) \qquad \text{and} \qquad \mu(\lim f_n^+) = \lim \mu(f_n^+).$$

Consequently, $\nu(f^+) \leqq \mu(f^+)$. Similarly, define f_n^- by

$$f_n^-(x) = \begin{cases} f(i/2^n) & \text{if } \ (i-1)/2^n \leqq x \leqq i/2^n, \quad i = -n2^n, \ldots, -2, -1, 0, \\ f(-n) & \text{if } \ x \leqq -n. \end{cases}$$

Each f_n^- is a positive linear combination of step functions of the form $-I_{[\cdot \leqq w]}$. Since $-I_{[\cdot \leqq w]} = I_{[\cdot \geqq w]} - 1$, $\nu(-I_{[\cdot \leqq w]}) \leqq \mu(-I_{[\cdot \leqq w]})$. Therefore, $\nu(f_n^-) \leqq (f_n^-)$ for each n.

Since $f_n^- \searrow f^-$ and $\nu(f^-) > -\infty$, it follows by the dominated convergence theorem [7, p. 125] that $-\infty < \nu(f^-) = \lim_n \nu(f_n^-) \leqq \lim_n \mu(f_n^-) = \mu(f^-)$. Since $f = f^+ + f^-$, $\nu(f) \leqq \mu(f)$, which concludes the proof.

The condition $\nu(f^-) > -\infty$ was imposed to ensure that $\nu(f)$ and $\mu(f)$ are defined. The condition $\mu(f^+) < \infty$ would do as well.

The converse of Theorem 2.1 is very easy.

Theorem 2.2 If $\nu(f) \leqq \mu(f)$ for all increasing functions f for which the inequality is defined, then $1-F(w) \geqq 1-G(w)$ for each w.

Proof $I_{[\cdot \geqq w]}$ is an increasing function for each w. Let $f(x) = I_{[x \geqq w]}$. Since $\mu(f)$ and $\nu(f)$ both exist, the hypothesis implies that $\nu(f) \leqq \mu(f)$. But $\mu(f) = 1-F(w)$ and $\nu(f) = 1-G(w)$, so that $1-F(w) \geqq 1-G(w)$.

EXAMPLE 2

This example is called second-degree stochastic dominance by Hadar and Russell [4]. It is closely related to the ordering in mean proposed by Veinott (see Bessler and Vienott [1]; also see Hanoch and Levy [5]). This example shows that an investor with a concave increasing utility function (i.e., a risk-averse investor) prefers X to Y if and only if $E[X \wedge w] \geqq E[Y \wedge w]$ for each w.

Theorem 2.3 $\nu(f) \leqq \mu(f)$ for all increasing concave f defined on $\mathbf{R} = (-\infty, +\infty)$ with $\mu(f^+) < \infty$, if and only if $\nu(g) \leqq \mu(g)$ for each $g \in G$, where G is the set of functions of the form $g(x) = \min(x, w) = x \wedge w$.

Note that if X and Y are bounded below by a, then

$$\mu(g) = E(X \wedge w) = \int_a^w x\, dF(x) + w(1-F(w)) = a + \int_a^w (1-F(x))\, dx,$$

where the last equality follows from an integration by parts. Similarly, $\nu(g) = a + \int_0^w (1-G(x))\, dx$. In this case, $\nu(g) \leqq \mu(g)$ $\forall g \in G$ becomes $\int_a^w [1-F(y)]\, dy \geqq \int_a^w [1-G(y)]\, dy$ for $w \geqq a$.

Proof As in Example 1, $f(0)$ can be assumed to be 0 without loss of generality.

Since f is concave increasing, there exists some nonnegative and decreasing g such that $f(x) = \int_0^x g(y)\, dy$.

Using the approximations

$$g_n(x) = \begin{cases} g\left(\dfrac{i+1}{2^n}\right) & \text{if } \dfrac{i}{2^n} < x \leqq \dfrac{i+1}{2^n}, \quad i = -n2^n, \ldots, -2, -1, 0, 1, 2, \ldots, n2^n \\ g(-n) & \text{if } x \leqq -n, \\ 0 & \text{if } x > n, \end{cases}$$

we have $g_n \nearrow g$. Each g_n is a positive linear combination of functions of the form $I_{[\cdot \leqq w]}$.

Define $f_n(x) = \int_0^x g_n(y)\,dy$. Since each g_n is a positive linear combination of functions of the form $I_{[\cdot \leqq w]}$, each f_n is a positive linear combination of functions of the form

$$\int_0^x I_{[y \leqq w]}\,dy = (x \wedge w) - w^-.$$

Note that

$$f_n^{+}(x) \nearrow \int_0^x g(y)\,dy = f^{+}(x) \qquad \text{for} \quad x \geqq 0.$$

Let

$$h_n(x) = \begin{cases} g(o) \cdot x & \text{for} \quad x \leqq 0 \\ f_n^{+}(x) & \text{for} \quad x \geqq 0 \end{cases}$$

and

$$h(x) = \begin{cases} g(o) \cdot x & \text{for} \quad x \leqq 0, \\ f^{+}(x) & \text{for} \quad x \geqq 0. \end{cases}$$

By the monotone convergence theorem, $\mu(h_n) \nearrow \mu(h)$ and $v(h_n) \nearrow v(h)$. Since $v(h_n) \leqq \mu(h_n)$ for each n, it follows that $v(h) \leqq \mu(h) \leqq \mu(f^{+})$, which is $<\infty$ by hypothesis. Similarly, $f_n^{-}(x) \searrow \int_0^x g(y)\,dy = f^{-}(x)$ for $x \leqq 0$. The function $e_n(x) = f_n^{-}(x) - g(o)x^{-}$ is a positive linear combination of functions of the form $x \wedge w$. Let $e(x) = f^{-}(x) - g(0)x^{-}$.

By the monotone convergence theorem, since $0 \geqq e_n \searrow e$ it follows that

$$v(e_n) \searrow v(e) \geqq v(f^{-}) \qquad \text{and} \qquad \mu(e_n) \searrow \mu(e) \geqq v(e).$$

Since $f(x) = e(x) + h(x)$, it follows that $v(f) \leqq \mu(f)$.

To prove the converse, let $g(x) = x \wedge w$. Then $\mu(g^{+}) < \infty$ and g is concave increasing. Consequently, by the hypothesis it follows that $v(g) \leqq \mu(g)$.

Example 3

This example is called third-degree stochastic dominance, and was introduced by Whitmore [14]. It shows that an investor with a concave increasing utility function f with a nonnegative third derivative (provided f is thrice differentiable) prefers X to Y if $E[(X-w)^2 I_{[X \leqq w]}] \leqq E[(Y-w)^2 I_{[Y \leqq w]}]$ for each w.

This example is of interest because of its relation to the Pratt–Arrow index of risk aversion, $r = -f''/f'$. If f is a concave increasing function such that r is decreasing, then necessarily $f''' \geqq 0$. However, the converse does not hold. Note that $f''' \geqq 0$ is equivalent to convexity of f'. Thus, let U be the set of functions f defined on $(-\infty, +\infty)$ such that there exists a convex decreasing nonnegative function h with $f(x) = \int_0^x h(y)\,dy$.

Theorem 2.4 $v(f) \leqq \mu(f)$ for all $f \in U$ with $\mu(f^+) < \infty$, if and only if $EX \geqq EY$ and $v(g) \leqq \mu(g)$ for each $g \in G$, where G is the set of functions of the form $-(w-x)^2 I_{[x \leqq w]}$.

Note that if X is bounded below by a, then

$$E[(X-w)^2 I_{[X \leqq w]}] = a - w + 2\int_a^w (1-F(x))(x-w)\,dx$$

follows by an integration by parts. Thus if X and Y are bounded below by a, then $v(g) \leqq \mu(g)$ for each $g \in G$ if and only if

$$\int_a^w (1-F(x))(w-x)\,dx \geqq \int_a^w (1-G(x))(w-x)\,dx$$

for each $w \geqq a$, and $EX \geqq EY$.

Proof Without loss of generality, it can be assumed that $f(0) = 0$. If $f(x) = c \cdot x$, then $v(f) \leqq \mu(f)$ if and only if $EX \geqq EY$.

If $f \in U$ is not linear, then there exists some convex, nonnegative, decreasing function h which is not identically constant, such that $f(x) = \int_0^x h(y)\,dy$. As in Example 2, h can be arbitrarily well approximated by a positive weighted sum of functions of the form $(w-x)^+$. Formally, there exists some function p such that $h(x) = \int_0^x p(y)\,dy$. Let $L = \sup\{x: p(x) < 0\}$. Since h is not constant, $L > -\infty$. If $L = \infty$, then define

$$h_{ni}(x) = [p(K_i)(x-K_i) + h(K_i)]^+$$

where

$$K_i = \frac{2^n n - i + 1}{2^n}, \qquad i = 1, 2, \ldots, 1 + 2^{n+1} n.$$

If $L < \infty$, then define

$$h_{n1}(x) = \left[-\frac{1}{2^n}(x - h(L)) + h(L)\right]^+ \qquad \text{and}$$

$$h_{ni} = [p(K_i)(x-K_i) + h(K_i)]^+,$$

where

$$K_i = L - \frac{i-1}{2^n}, \qquad i = 2, 3, \ldots, n2^n + 1.$$

In either case, define $h_n(x) = \max_i\{h_{ni}(x)\}$. Then $h_n \nearrow h$ and each $h_n(x)$ is a positive linear combination of functions of the form $(w-x)^+$. Define $f_n(x) = \int_o^x h_n(y)\,dy$.

Then f_n is a positive linear combination of functions of the form $\int_o^x (w-y)^+\,dy = ag(x)+b$ for some $g \in G$ and some constants a and b with $a \geqq 0$. Since $f_n \nearrow f$ on $[0,\infty)$ and $f_n \searrow f$ on $(-\infty,0)$, it follows that $v(f) \leqq \mu(f)$ if $v(g) \leqq \mu(g)\ \forall g \in G$, by applying the monotone and dominated convergence theorems. The converse is trivial since $G \subseteq U$.

III. Probabilistic Content of Stochastic Dominance

In this section an alternative formulation of second-degree stochastic dominance is presented. This formulation emphasizes the underlying probabilistic content of the dominance idea. Results are proved only for the case of bounded random variables.

Suppose that $\Omega \subset \mathbf{R}^r$ is a compact convex set in r-dimensional Euclidean space (e.g., the unit cube). Suppose that X and Y are random variables such that $P[X\in\Omega] = P[Y\in\Omega] = 1$. If $v = (v^1, v^2, \ldots, v^r)$ and $v' = (v'^1, v'^2, \ldots, v'^r)$ are vectors in $\mathbf{R}^r$, the notation $v \leqq v'$ will be used to denote the r-dimensional ordering $v^i \leqq v'^i$ $(1 \leqq i \leqq r)$. A real-valued function f on Ω is nondecreasing if $f(v) \leqq f(v')$ whenever $v \leqq v'$. Let C be the set of all real-valued continuous functions on Ω, and let $\mathscr{S}$ be the set of concave nondecreasing functions on Ω. The basic result on stochastic dominance is contained in the following.

Theorem 3.1 The following conditions are equivalent:

(a) $Eu(X) \geqq Eu(Y)$ $(u \in \mathscr{S})$.

(b) There exists a random variable Z such that $P[Y \leqq \xi] = P[X+Z \leqq \xi]$ $(\xi \in \Omega)$, and $E(Z\,|\,X) \leqq 0$ almost surely (i.e., except on a set of probability 0). The theorem states that X dominates Y if an only if Y is "noisier" than X; for Y has the same distribution function as $X+Z$, and Z is random even if a specific realization of X is given. It should be emphasized that the random variables Y and $X+Z$ need not be *equal* even though they have the same distribution function.

Theorem 3.1 was established for a compact interval $\Omega = [a, b] \subset \mathbf{R}^1$ by Blackwell [2][1] and for discrete, finite multidimensional X and Y by Sherman [12]. The technically complex algebraic arguments of the original proofs were later supplanted by technically simpler, but somewhat "abstract" functional analysis arguments [8, 13]. The present treatment is a compromise between the two approaches. The "modern" proofs are followed for the discrete case, and standard limiting arguments are then used to obtain the general case.

III.1 Some Notions from Functional Analysis

Recall that the set C of continuous functions on Ω is a vector space over the real numbers $\mathbf{R}$:

(i) $f, g \in C \;\Rightarrow\; f+g \in C$;
(ii) $f \in C,\ c \in \mathbf{R} \;\Rightarrow\; cf \in C$.

The vector space C becomes a separable metric space when endowed with the sup norm topology, defined by the metric

$$d(f,g) = \sup_{x \in \Omega} |f(x) - g(x)| = \max_{x \in \Omega} |f(x) - g(x)|.$$

Thus C has many of the properties of ordinary Euclidean space $\mathbf{R}^m$, except that it has infinite dimension. A *linear functional* on C is a real-valued linear function on C, that is, a linear map from the functions $f \in C$ into the real numbers. A random variable X may be considered to be a nonnegative linear functional on C. For if μ is the probability measure corresponding to X $[\mu(A) = \int_A dF_X(x)]$, the functional $\mu(f) = Ef(X)$ $(f \in C)$ has the following obvious properties:

(i) $\mu(f) \in \mathbf{R}$ $(f \in C)$;
(ii) $\mu(f+g) = \mu(f) + \mu(g)$ $(f, g \in C)$;
(iii) $\mu(cf) = c \cdot \mu(f)$ $(f \in C,\ c \in \mathbf{R})$;
(iv) $\mu(f) \geqq 0$ $(f \in C, f \geqq 0)$;
(v) $\mu(e) = 1$, where $e \in C$ is the unit constant function $(e(x) = 1$ for all $x \in \Omega)$.

The converse result is also true: Any linear functional μ on C which satisfies (i)–(v) corresponds to a random variable X such that $\mu(f) = Ef(X)$ [3, Chapter IV, p. 120].

Another important class of functionals on C is the class H of *sublinear functionals* $h \in H$, satisfying

[1] For the case $EX = EY$, this result was recently rediscovered by Rothschild and Stiglitz [10], and was applied to the qualitative analysis of several economic problems in a companion paper [11].

(i) $h(f) \in \mathbf{R}$ $(h \in H, f \in C)$;
(ii) $h(f+g) \leqq h(f)+h(g)$ $(h \in H; f, g \in C)$;
(iii) $h(cf) = ch(f)$ $(h \in H; f \in C;\ c \geqq 0)$.

Note that a sublinear functional is simply a positively homogeneous convex function on the space C. The relevance of sublinear functionals is contained in the following very important theorem.

Hahn–Banach Theorem Let E be a linear space, F a linear subspace of E. Let l be a linear functional on F, h a sublinear functional on E. If $l(f) \leqq h(f)$ for all $f \in F$, then l can be extended to a linear functional $\bar{l}$ defined on all of E, such that $\bar{l}(f) = l(f)$ for $f \in F$, and $\bar{l}(f) \leqq h(f)$ for all $f \in E$.

The proof of this theorem is not technically difficult, but uses somewhat abstract arguments involving the axiom of choice. The proof is omitted since it can be found in most books on analysis (e.g., Munroe [9]).

III.2 The Discrete Case

To prove Theorem 3.1, it suffices to prove the implication (a) $\Rightarrow$ (b), since the converse (b) $\Rightarrow$ (a) follows trivially from Jensen's inequality. Let X and Y be discrete random variables concentrated at the points $\Omega = \{x_1, x_2, \ldots, x_k\} \subset \mathbf{R}^r$. Assume without loss of generality that $\Omega \subset I^r = [0,1]^r$, the r-dimensional unit cube. Let C (respectively, $\mathscr{S}$) denote the continuous (respectively, concave nondecreasing) functions on I^r. Let $P[X = x_i] = p_i \geqq 0$, $P[Y = x_i] = q_i \geqq 0$, with $\sum_{i=1}^k p_i = \sum_{i=1}^k q_i = 1$. The random variables X, Y can be viewed as nonnegative linear functionals μ and ν on C:

$$\mu(f) = \sum_{i=1}^{k} p_i f(x_i), \qquad \nu(f) = \sum_{i=1}^{k} q_i f(x_i).$$

Let H be the set of sublinear functionals on C. The key result is

Lemma 1 Let $h \in H$, and suppose that $h = \sum_{i=1}^k p_i h_i$, where $h_i \in H$ $(1 \leqq i \leqq k)$. If $\nu(f) \leqq h(f)$ for all $f \in C$, then there exists a decomposition $\nu = \sum_{i=1}^k p_i \nu_i$ such that

(i) ν_i is a linear functional on C, and
(ii) $\nu_i(f) \leqq h_i(f)$ for all $f \in C$.

Proof (Meyer [8]) Let $E = C^k$ be the linear space consisting of k-tuples $(f_1, f_2, \ldots, f_k)$ with $f_i \in C$ $(1 \leqq i \leqq k)$. Let F be the linear subspace of E consisting of "diagonal" elements $(f, f, \ldots, f)$, $f \in C$. Define a linear functional $\bar{\nu}$ on F and a sublinear functional $\bar{h}$ on E:

$$\bar{\nu}(f, f, \ldots, f) = \nu(f), \qquad \bar{h}(f_1, f_2, \ldots, f_k) = \sum_{i=1}^{k} p_i h_i(f_i).$$

By hypothesis, $\bar{v} \leqq \bar{h}$ on F. Thus, by the Hahn-Banach theorem, $\bar{v}$ can be extended to a linear functional $\bar{\bar{v}}$ on E, with $\bar{\bar{v}} \leqq \bar{h}$ on E. Clearly, the functional $\bar{\bar{v}}$ may be written as $\bar{\bar{v}}(f_1, f_2, \ldots, f_n) = \sum_{i=1}^{k} p_i v_i(f_i)$. Now $\bar{\bar{v}} \leqq \bar{h}$ on $E \Rightarrow v_i(f) \leqq h_i(f)$ for all $f \in C$. Since $\bar{\bar{v}} = \bar{v}$ on F one has $v(f) = \sum_{i=1}^{k} p_i v_i(f)$.

Returning to the proof of Theorem 3.1, define for each $f \in C$ the upper envelope $\bar{f}$:

$$\bar{f}(x) = \inf\{g(x) \mid g \in \mathscr{S}, g \geqq f\}.$$

The following properties are evident.

(1) $f \leqq \bar{f} \leqq \sup\{f(x) \mid x \in I^r\}$.
(2) $\bar{f} \in \mathscr{S}$ for all $f \in C$.
(3) $\overline{(f+g)} \leqq \bar{f} + \bar{g} \quad (f, g \in C)$.
(4) $\overline{(cf)} = c \cdot \bar{f} \quad (f \in C;\ c \geqq 0)$.

The upper envelope operation thus produces a sublinear functional $h_i(f) = \bar{f}(x_i)$ for each $x_i \in \Omega$. Define

$$h(f) = \sum_{i=1}^{k} p_i h_i(f) = \sum_{i=1}^{k} p_i \bar{f}(x_i).$$

From properties (1) and (2) of the upper envelope, and from hypothesis (a) of Theorem 3.1, it follows that

$$v(f) \leqq v(\bar{f}) \leqq \mu(\bar{f}) = \sum_{i=1}^{k} p_i \bar{f}(x_i) = h(f)$$

for all $f \in C$. By Lemma 1, $v = \sum_{i=1}^{k} p_i v_i$, with $v_i \leqq h_i$. One has

(1′) v_i is a nonnegative linear functional. For $f \leqq 0 \Rightarrow \bar{f} \leqq 0 \Rightarrow v_i(f) \leqq h_i(f) = \bar{f}(x_i) \leqq 0$.

(2′) $v_i(e = 1$. For $0 \leqq v_i(e) \leqq \bar{e}(x_i) \leqq 1$, and $1 = v(e) = \sum_{i=1}^{k} p_i v_i(e) \leqq \sum_{i=1}^{k} p_i = 1$.

Thus v_i is a probability measure on I^r. Furthermore, one has

(3′) v_i is a discrete probability measure, concentrated on Ω.

For if $f \geqq 0$ is any continuous function which vanishes on $\Omega = \{x_1, x_2, \ldots, x_k\}$, then

$$0 = \sum_{i=1}^{k} q_i f(x_i) = v(f) = \sum_{i=1}^{k} p_i v_i(f) \quad \Rightarrow \quad v_i(f) = 0, \qquad 1 \leqq i \leqq k.$$

Define $d_{ij} = v_i(\{x_j\}) = P[Y_i = x_j]$, where Y_i is the random variable corresponding to v_i. Then

(1″) $\sum_{j=1}^{k} d_{ij} = 1$, $d_{ij} \geqq 0$ for all i, j.

(2″) $q_j = \sum_{i=1}^{k} p_i d_{ij}$. For let $f \geqq 0$ be any continuous function such that $f(x_j) = 1, f(x_i) = 0$ $(i \neq j)$. Then

$$q_j = v(f) = \sum_{i=1}^{k} p_i v_i(f) = \sum_{i=1}^{k} p_i d_{ij}.$$

(3″) $\sum_{j=1}^{k} d_{ij} x_j \leqq x_i$ (vector inequality). For let $\phi^m(x) = x^m$ be the mth coordinate function on I^r $(1 \leqq m \leqq r)$. Clearly $\phi^m \in \mathscr{S}$, so $\bar{\phi}^m = \phi^m$. Thus

$$\sum_j d_{ij} x_j^{\ m} = v_i(\phi^m) \leqq \bar{\phi}^m(x_i) = \phi^m(x_i) = x_i^{\ m}.$$

The proof of Theorem 3.1 is now completed by defining

$$Z\colon P[Z = x_j - x_i \mid X = x_i] = d_{ij}.$$

The properties $P[X+Z \leqq \xi] = P[Y \leqq \xi]$ and $E(Z \mid X) \leqq 0$ follow immediately from (2″) and (3″).

III.3 The General Case

Suppose now that X and Y are concentrated in Ω, a compact convex subset of the interior of I^r. Since $Eu(X) \geqq Eu(Y)$ for all concave nondecreasing u on Ω, this is true also for all $u \in \mathscr{S}$, the set of concave nondecreasing functions on I^r. For each $n = 1, 2, \ldots$, partition I^r into n^r disjoint half-open cubes K_α of side $1/n$. These cubes are of the form $\prod_{i=1}^{r} (a_i, a_i + n^{-1}]$ and are obtained by cutting the edges of I^r into halves, thirds, quarters, etc. They contain their upper right-hand boundaries, but not their lower left-hand boundaries. Clearly the supports of μ and v (the probability measures corresponding to X and Y) are contained in the disjoint union $\bigcup_{\alpha=1}^{n^r} K_\alpha$. Since any function $u \in \mathscr{S}$ is nondecreasing,

$$v(u) = \sum_{\alpha=1}^{n^r} \int_{K_\alpha} u(\xi)\, dF_Y(\xi) \geqq \sum_{\alpha=1}^{n^r} u(\underline{\xi}_\alpha) \int_{K_\alpha} dF_Y(\xi),$$

where $\underline{\xi}_\alpha$ is the lower left-hand corner of K_α. Define $q_\alpha = \int_{K_\alpha} dF_Y(\xi)$. Then $v(u) \geqq \sum_{\alpha=1}^{n^r} q_\alpha u(\underline{\xi}_\alpha)$ $(u \in \mathscr{S})$. Similarly, $\mu(u) \leqq \sum_{\alpha=1}^{n^r} p_\alpha u(\bar{\xi}_\alpha)$ where $\bar{\xi}_\alpha$ is the upper right-hand corner of K_α, and $p_\alpha = \int_{K_\alpha} dF_X(\xi)$. Defining the discrete random variables X_n and Y_n such that $P[X_n = \bar{\xi}_\alpha] = p_\alpha$, $P[Y_n = \underline{\xi}_\alpha] = q_\alpha$, it follows that $Eu(X_n) \geqq Eu(Y_n)$ $(u \in \mathscr{S})$. Thus, by the discrete version of Theorem 3.1, there exists Z_n taking values in $S = [-1, 1]^r$, such that

$$P[Y_n \leqq \xi] = P[X_n + Z_n \leqq \xi] \qquad \text{and} \qquad E(Z_n \mid X_n) \leqq 0.$$

Since X_n and Z_n have uniformly bounded range, the sequence of bivariate c.d.f.'s $H_n(x,z)$ of (X_n, Z_n) on $I^r \times S$ contains a subsequence $H.$ which converges to the c.d.f. H of a bivariate random variable (X,Z) on $I^r \times S$ [3, Chapter VIII, p. 267]. Similarly, the sequence $\{n'\}$ contains a subsequence $\{n''\}$ such that $Y_{n''} \to Y$. Since $EZ = \int_{I^r \times S} z\, d^2H(x,z)$ exists, the conditional expectation $E(Z \mid X)$ exists μ-almost surely [3, Chapter V, pp. 162–165]. Let $A \subset I^r$ be any μ-measurable set. Recalling that $E(Z_{n''} \mid X_{n''}) \leqq 0$, it follows that

$$\begin{aligned} 0 &\geqq \lim_{n'' \to \infty} \int_A E(Z_{n''} \mid X_{n''} = x)\, dF_{X_{n''}}(x) \\ &= \lim_{n'' \to \infty} \int_{A \times S} z\, d^2 H_{n''}(x,z) = \int_{A \times S} z\, d^2 H(x,z) \\ &= \int_A E(Z \mid X = x)\, dF_X(x). \end{aligned}$$

Thus $E(Z \mid X) \leqq 0$ μ-almost surely (and therefore also P-almost surely, where P is the "underlying" probability measure). A simple argument shows that $P[Y \leqq \xi] = P[X+Z \leqq \xi]$, $\xi \in I^r$, and this completes the proof.

III.4 Extension

In Theorem 3.1 it is assumed that X and Y are bounded random variables. In applications one is often interested in unbounded random variables, and a generalization of Theorem 3.1 can be proved for these situations under additional assumptions. The following results due to Strassen [13] are stated without proof.

Theorem 3.2 If X and Y are random variables in $\mathbf{R}^1$ having finite means EX and EY, the following conditions are equivalent:

(a) $Eu(X) \geqq Eu(Y)$ for all concave nondecreasing u.

(b) There exists Z such that $P[Y \leqq \xi] = P[X+Z \leqq \xi]$ for all $\xi \in \mathbf{R}^1$, and $E(Z \mid X) \leqq 0$ almost surely.

Theorem 3.3 If X and Y are random variables in E^r having finite means EX and EY (with $EX = EY$), the following conditions are equivalent:

(a) $Eu(X) \geqq Eu(Y)$ for all concave u (whether nondecreasing or not).

(b) There exists Z such that $P[Y \leqq \xi] = P[X+Z \leqq \xi]$ for all $\xi \in \mathbf{R}^1$, and $E(Z \mid X) = 0$ almost surely.

The generalization of Theorem 3.3 to the case of unequal means and concave nondecreasing u in (a), appears to be an unsolved problem.

References

1. Bessler, S. A., and Vienott, Jr., A. F., "Optimal Policy for Dynamic Multi-Echelon Inventory Model". *Naval Research Logistics Quarterly* **13** (1966), 355–389.
2. Blackwell, D., "Comparison of Experiments", *Proceedings of Symposium on Mathematical Statistics and Probability, 2nd, Berkeley, 1951* (J. Neyman and L. M. Lecam, eds.), pp. 93–102. Univ. of California Press, Berkeley, 1951.
3. Feller, W., *An Introduction to Probability Theory and Its Applications*, Second edition, Vol. 2. Wiley, New York, 1971.
4. Hadar, J., and Russell, W. R., "Rules for Ordering Uncertain Prospects." *American Economic Review* **49** (1969), 25–34.
5. Hanoch, G., and Levy, H., "Efficiency Analysis of Choices Involving Risk." *Review of Economic Studies* **36** (1969), 335–346.
6. Lehmann, E. L., "Ordered Families of Distributions." *Annals of Mathematical Statistics* **26** (1955), 399–419.
7. Loève, M., *Probability Theory*, Third edition. Van Nostrand-Reinhold, Princeton, New Jersey, 1963.
8. Meyer, P. A., *Probability and Potentials*. Ginn (Blaisdell), Boston, Massachusetts, 1966.
9. Munroe, M. E., *Introduction to Measure and Integration*. Addison-Wesley, Reading, Massachusetts, 1953.
10. Rothschild, M., and Stiglitz, J. E., "Increasing Risk: I. A Definition." *Journal of Economic Theory* **2**, (1970), 225–243.
11. Rothschild, M., and Stiglitz, J. E., "Increasing Risk: II. Its Economic Consequences." *Journal of Economic Theory* **3**, (1971), 66–84.
12. Sherman, S., "On a Theorem of Hardy, Littlewood, Polya, and Blackwell." *Proceedings of the National Academy of Sciences of the United States* **37** (1951), 826–831.
13. Strassen, V., "The Existence of Probability Measures with Given Marginals." *Annals of Mathematical Statistics* **36** (1965), 423–439.
14. Whitmore, G. A., "Third-Degree Stochastic Dominance." *American Economic Review* **60** (1970), 457–459.

2. MEASURES OF RISK AVERSION

Econometrica, Vol. 32, No. 1–2 (January–April, 1964)

RISK AVERSION IN THE SMALL AND IN THE LARGE[1]

BY JOHN W. PRATT

This paper concerns utility functions for money. A measure of risk aversion in the small, the risk premium or insurance premium for an arbitrary risk, and a natural concept of decreasing risk aversion are discussed and related to one another. Risks are also considered as a proportion of total assets.

1. SUMMARY AND INTRODUCTION

LET $u(x)$ BE a utility function for money. The function $r(x) = -u''(x)/u'(x)$ will be interpreted in various ways as a measure of local risk aversion (risk aversion in the small); neither $u''(x)$ nor the curvature of the graph of u is an appropriate measure. No simple measure of risk aversion in the large will be introduced. Global risks will, however, be considered, and it will be shown that one decision maker has greater local risk aversion $r(x)$ than another at all x if and only if he is globally more risk-averse in the sense that, for every risk, his cash equivalent (the amount for which he would exchange the risk) is smaller than for the other decision maker. Equivalently, his risk premium (expected monetary value minus cash equivalent) is always larger, and he would be willing to pay more for insurance in any situation. From this it will be shown that a decision maker's local risk aversion $r(x)$ is a decreasing function of x if and only if, for every risk, his cash equivalent is larger the larger his assets, and his risk premium and what he would be willing to pay for insurance are smaller. This condition, which many decision makers would subscribe to, involves the third derivative of u, as $r' \leqq 0$ is equivalent to $u'''u' \geqq u''^2$. It is not satisfied by quadratic utilities in any region. All this means that some natural ways of thinking casually about utility functions may be misleading. Except for one family, convenient utility functions for which $r(x)$ is decreasing are not so very easy to find. Help in this regard is given by some theorems showing that certain combinations of utility functions, in particular linear combinations with positive weights, have decreasing $r(x)$ if all the functions in the combination have decreasing $r(x)$.

The related function $r^*(x) = xr(x)$ will be interpreted as a local measure of aversion to risks measured as a proportion of assets, and monotonicity of $r^*(x)$ will be proved to be equivalent to monotonicity of every risk's cash equivalent measured as a proportion of assets, and similarly for the risk premium and insurance.

These results have both descriptive and normative implications. Utility functions for which $r(x)$ is decreasing are logical candidates to use when trying to describe the behavior of people who, one feels, might generally pay less for insurance against

[1] This research was supported by the National Science Foundation (grant NSF-G24035).

a given risk the greater their assets. And consideration of the yield and riskiness per investment dollar of investors' portfolios may suggest, at least in some contexts, description by utility functions for which $r^*(x)$ is first decreasing and then increasing.

Normatively, it seems likely that many decision makers would feel they ought to pay less for insurance against a given risk the greater their assets. Such a decision maker will want to choose a utility function for which $r(x)$ is decreasing, adding this condition to the others he must already consider (consistency and probably concavity) in forging a satisfactory utility from more or less malleable preliminary preferences. He may wish to add a further condition on $r^*(x)$.

We do not assume or assert that utility may not change with time. Strictly speaking, we are concerned with utility at a specified time (when a decision must be made) for money at a (possibly later) specified time. Of course, our results pertain also to behavior at different times if utility does not change with time. For instance, a decision maker whose utility for total assets is unchanging and whose assets are increasing would be willing to pay less and less for insurance against a given risk as time progresses if his $r(x)$ is a decreasing function of x. Notice that his actual expenditure for insurance might nevertheless increase if his risks are increasing along with his assets.

The risk premium, cash equivalent, and insurance premium are defined and related to one another in Section 2. The local risk aversion function $r(x)$ is introduced and interpreted in Sections 3 and 4. In Section 5, inequalities concerning global risks are obtained from inequalities between local risk aversion functions. Section 6 deals with constant risk aversion, and Section 7 demonstrates the equivalence of local and global definitions of decreasing (and increasing) risk aversion. Section 8 shows that certain operations preserve the property of decreasing risk aversion. Some examples are given in Section 9. Aversion to proportional risk is discussed in Sections 10 to 12. Section 13 concerns some related work of Kenneth J. Arrow.[2]

Throughout this paper, the utility $u(x)$ is regarded as a function of total assets rather than of changes which may result from a certain decision, so that $x=0$ is equivalent to ruin, or perhaps to loss of all readily disposable assets. (This is essential only in connection with proportional risk aversion.) The symbol $\sim$ indicates that two functions are equivalent as utilities, that is, $u_1(x) \sim u_2(x)$ means there exist constants a and b (with $b>0$) such that $u_1(x)=a+bu_2(x)$ for all x. The utility functions discussed may, but need not, be bounded. It is assumed, however, that they are sufficiently regular to justify the proofs; generally it is enough that they be twice continuously differentiable with positive first derivative, which is already re-

[2] The importance of the function $r(x)$ was discovered independently by Kenneth J. Arrow and by Robert Schlaifer, in different contexts. The work presented here was, unfortunately, essentially completed before I learned of Arrow's related work. It is, however, a pleasure to acknowledge Schlaifer's stimulation and participation throughout, as well as that of John Bishop at certain points.

quired for $r(x)$ to be defined and continuous. A variable with a tilde over it, such as $\tilde{z}$, is a random variable. The risks $\tilde{z}$ considered may, but need not, have "objective" probability distributions. In formal statements, $\tilde{z}$ refers only to risks which are not degenerate, that is, not constant with probability one, and interval refers only to an interval with more than one point. Also, increasing and decreasing mean nondecreasing and nonincreasing respectively; if we mean strictly increasing or decreasing we will say so.

2. THE RISK PREMIUM

Consider a decision maker with assets x and utility function u. We shall be interested in the *risk premium* π such that he would be indifferent between receiving a risk $\tilde{z}$ and receiving the non-random amount $E(\tilde{z})-\pi$, that is, π less than the actuarial value $E(\tilde{z})$. If u is concave, then $\pi \geqq 0$, but we don't require this. The risk premium depends on x and on the distribution of $\tilde{z}$, and will be denoted $\pi(x,\tilde{z})$. (It is not, as this notation might suggest, a function $\pi(x,z)$ evaluated at a randomly selected value of z, which would be random.) By the properties of utility,

$$u(x+E(\tilde{z})-\pi(x,\tilde{z}))=E\{u(x+\tilde{z})\}\,. \tag{1}$$

We shall consider only situations where $E\{u(x+\tilde{z})\}$ exists and is finite. Then $\pi(x,\tilde{z})$ exists and is uniquely defined by (1), since $u(x+E(\tilde{z})-\pi)$ is a strictly decreasing, continuous function of π ranging over all possible values of u. It follows immediately from (1) that, for any constant μ,

$$\pi(x,\tilde{z})=\pi(x+\mu,\,\tilde{z}-\mu)\,. \tag{2}$$

By choosing $\mu=E(\tilde{z})$ (assuming it exists and is finite), we may thus reduce consideration to a risk $\tilde{z}-\mu$ which is actuarially neutral, that is, $E(\tilde{z}-\mu)=0$.

Since the decision maker is indifferent between receiving the risk $\tilde{z}$ and receiving for sure the amount $\pi_a(x,\tilde{z})=E(\tilde{z})-\pi(x,\tilde{z})$, this amount is sometimes called the cash equivalent or value of $\tilde{z}$. It is also the asking price for $\tilde{z}$, the smallest amount for which the decision maker would willingly sell $\tilde{z}$ if he had it. It is given by

$$u(x+\pi_a(x,\tilde{z}))=E\{u(x+\tilde{z})\}\,. \tag{3a}$$

It is to be distinguished from the bid price $\pi_b(x,\tilde{z})$, the largest amount the decision maker would willingly pay to obtain $\tilde{z}$, which is given by

$$u(x)=E\{u(x+\tilde{z}-\pi_b(x,\tilde{z}))\}\,. \tag{3b}$$

For an unfavorable risk $\tilde{z}$, it is natural to consider the insurance premium $\pi_I(x,\tilde{z})$ such that the decision maker is indifferent between facing the risk $\tilde{z}$ and paying the non-random amount $\pi_I(x,\tilde{z})$. Since paying π_I is equivalent to receiving $-\pi_I$, we have

$$\pi_I(x,\tilde{z})=-\pi_a(x,\tilde{z})=\pi(x,\tilde{z})-E(\tilde{z})\,. \tag{3c}$$

If $\tilde{z}$ is actuarially neutral, the risk premium and insurance premium coincide.

The results of this paper will be stated in terms of the risk premium π, but could equally easily and meaningfully be stated in terms of the cash equivalent or insurance premium.

3. LOCAL RISK AVERSION

To measure a decision maker's local aversion to risk, it is natural to consider his risk premium for a small, actuarially neutral risk $\tilde{z}$. We therefore consider $\pi(x, \tilde{z})$ for a risk $\tilde{z}$ with $E(\tilde{z})=0$ and small variance σ_z^2; that is, we consider the behavior of $\pi(x,\tilde{z})$ as $\sigma_z^2 \to 0$. We assume the third absolute central moment of $\tilde{z}$ is of smaller order than σ_z^2. (Ordinarily it is of order σ_z^3.) Expanding u around x on both sides of (1), we obtain under suitable regularity conditions[3]

$$u(x-\pi)=u(x)-\pi u'(x)+O(\pi^2)\,, \tag{4a}$$

$$\begin{aligned} E\{u(x+\tilde{z})\} &= E\{u(x)+\tilde{z}u'(x)+\tfrac{1}{2}\tilde{z}^2u''(x)+O(\tilde{z}^3)\} \\ &= u(x)+\tfrac{1}{2}\sigma_z^2 u''(x)+o(\sigma_z^2)\,. \end{aligned} \tag{4b}$$

Setting these expressions equal, as required by (1), then gives

$$\pi(x,\tilde{z})=\tfrac{1}{2}\sigma_z^2 r(x)+o(\sigma_z^2)\,, \tag{5}$$

where

$$r(x) = -\frac{u''(x)}{u'(x)} = -\frac{d}{dx}\log u'(x)\,. \tag{6}$$

Thus the decision maker's risk premium for a small, actuarially neutral risk $\tilde{z}$ is approximately $r(x)$ times half the variance of $\tilde{z}$; that is, $r(x)$ is twice the risk premium per unit of variance for infinitesimal risks. A sufficient regularity condition for (5) is that u have a third derivative which is continuous and bounded over the range of all $\tilde{z}$ under discussion. The theorems to follow will not actually depend on (5), however.

If $\tilde{z}$ is not actuarially neutral, we have by (2), with $\mu=E(\tilde{z})$, and (5):

$$\pi(x,\tilde{z})=\tfrac{1}{2}\sigma_z^2 r(x+E(\tilde{z}))+o(\sigma_z^2)\,. \tag{7}$$

Thus the risk premium for a risk $\tilde{z}$ with arbitrary mean $E(\tilde{z})$ but small variance is approximately $r(x+E(\tilde{z}))$ times half the variance of $\tilde{z}$. It follows also that the risk premium will just equal and hence offset the actuarial value $E(\tilde{z})$ of a small risk $(\tilde{z})$; that is, the decision maker will be indifferent between having $\tilde{z}$ and not having it when the actuarial value is approximately r(x) times half the variance of $\tilde{z}$. Thus $r(x)$

[3] In expansions, $O(\;)$ means "terms of order at most" and $o(\;)$ means "terms of smaller order than."

may also be interpreted as twice the actuarial value the decision maker requires per unit of variance for infinitesimal risks.

Notice that it is the variance, not the standard deviation, that enters these formulas. To first order any (differentiable) utility is linear in small gambles. In this sense, these are second order formulas.

Still another interpretation of $r(x)$ arises in the special case $\tilde{z} = \pm h$, that is, where the risk is to gain or lose a fixed amount $h > 0$. Such a risk is actuarially neutral if $+h$ and $-h$ are equally probable, so $P(\tilde{z}=h) - P(\tilde{z}=-h)$ measures the *probability premium* of $\tilde{z}$. Let $p(x,h)$ be the probability premium such that the decision maker is indifferent between the status quo and a risk $\tilde{z} = \pm h$ with

$$(8) \qquad P(\tilde{z}=h) - P(\tilde{z}=-h) = p(x,h)\,.$$

Then $P(\tilde{z}=h) = \frac{1}{2}[1+p(x,h)]$, $P(\tilde{z}=-h) = \frac{1}{2}[1-p(x,h)]$, and $p(x,h)$ is defined by

$$(9) \qquad u(x) = E\{u(x+\tilde{z})\} = \tfrac{1}{2}[1+p(x,h)]\,u(x+h) + \tfrac{1}{2}[1-p(x,h)]u(x-h)\,.$$

When u is expanded around x as before, (9) becomes

$$(10) \qquad u(x) = u(x) + hp(x,h)u'(x) + \tfrac{1}{2}h^2u''(x) + O(h^3)\,.$$

Solving for $p(x,h)$, we find

$$(11) \qquad p(x,h) = \tfrac{1}{2}hr(x) + O(h^2)\,.$$

Thus for small h the decision maker is indifferent between the status quo and a risk of $\pm h$ with a probability premium of $r(x)$ times $\frac{1}{2}h$; that is, $r(x)$ is twice the probability premium he requires per unit risked for small risks.

In these ways we may interpret $r(x)$ as a measure of the *local risk aversion* or *local propensity to insure* at the point x under the utility function u; $-r(x)$ would measure locally liking for risk or propensity to gamble. Notice that we have not introduced any measure of risk aversion in the large. Aversion to ordinary (as opposed to infinitesimal) risks might be considered measured by $\pi(x,\tilde{z})$, but π is a much more complicated function than r. Despite the absence of any simple measure of risk aversion in the large, we shall see that comparisons of aversion to risk can be made simply in the large as well as in the small.

By (6), integrating $-r(x)$ gives $\log u'(x) + c$; exponentiating and integrating again then gives $e^c u(x) + d$. The constants of integration are immaterial because $e^c u(x) + d \sim u(x)$. (Note $e^c > 0$.) Thus we may write

$$(12) \qquad u \sim \int e^{-\int r}\,,$$

and we observe that the local risk aversion function r associated with any utility function u contains all essential information about u while eliminating everything arbitrary about u. However, decisions about ordinary (as opposed to "small") risks are determined by r only through u as given by (12), so it is not convenient entirely to eliminate u from consideration in favor of r.

4. CONCAVITY

The aversion to risk implied by a utility function u seems to be a form of concavity, and one might set out to measure concavity as representing aversion to risk. It is clear from the foregoing that for this purpose $r(x) = -u''(x)/u'(x)$ can be considered a measure of the concavity of u at the point x. A case might perhaps be made for using instead some one-to-one function of $r(x)$, but it should be noted that $u''(x)$ or $-u''(x)$ is not in itself a meaningful measure of concavity in utility theory, nor is the curvature (reciprocal of the signed radius of the tangent circle) $u''(x)(1 + [u'(x)]^2)^{-3/2}$. Multiplying u by a positive constant, for example, does not alter behavior but does alter u'' and the curvature.

A more striking and instructive example is provided by the function $u(x) = -e^{-x}$. As x increases, this function approaches the asymptote $u = 0$ and looks graphically less and less concave and more and more like a horizontal straight line, in accordance with the fact that $u'(x) = e^{-x}$ and $u''(x) = -e^{-x}$ both approach 0. As a utility function, however, it does not change at all with the level of assets x, that is, the behavior implied by $u(x)$ is the same for all x, since $u(k+x) = -e^{-k-x} \sim u(x)$. In particular, the risk premium $\pi(x, \tilde{z})$ for any risk $\tilde{z}$ and the probability premium $p(x, h)$ for any h remain absolutely constant as x varies. Thus, regardless of the appearance of its graph, $u(x) = -e^{-x}$ is just as far from implying linear behavior at $x = \infty$ as at $x = 0$ or $x = -\infty$. All this is duly reflected in $r(x)$, which is constant: $r(x) = -u''(x)/u'(x) = 1$ for all x.

One feature of $u''(x)$ does have a meaning, namely its sign, which equals that of $-r(x)$. A negative (positive) sign at x implies unwillingness (willingness) to accept small, actuarially neutral risks with assets x. Furthermore, a negative (positive) sign for all x implies strict concavity (convexity) and hence unwillingness (willingness) to accept any actuarially neutral risk with any assets. The absolute magnitude of $u''(x)$ does not in itself have any meaning in utility theory, however.

5. COMPARATIVE RISK AVERSION

Let u_1 and u_2 be utility functions with local risk aversion functions r_1 and r_2, respectively. If, at a point x, $r_1(x) > r_2(x)$, then u_1 is locally more risk-averse than u_2 at the point x; that is, the corresponding risk premiums satisfy $\pi_1(x, \tilde{z}) > \pi_2(x, \tilde{z})$ for sufficiently small risks $\tilde{z}$, and the corresponding probability premiums satisfy $p_1(x, h) > p_2(x, h)$ for sufficiently small $h > 0$. The main point of the theorem we are about to prove is that the corresponding global properties also hold. For instance, if $r_1(x) > r_2(x)$ for all x, that is, u_1 has greater local risk aversion than u_2 everywhere, then $\pi_1(x, \tilde{z}) > \pi_2(x, \tilde{z})$ for every risk $\tilde{z}$, so that u_1 is also globally more risk-averse in a natural sense.

It is to be understood in this section that the probability distribution of $\tilde{z}$, which determines $\pi_1(x, \tilde{z})$ and $\pi_2(x, \tilde{z})$, is the same in each. We are comparing the risk premiums for the same probability distribution of risk but for two different utilities. This does not mean that when Theorem 1 is applied to two decision makers, they must have the same personal probability distributions, but only that the notation is imprecise. The theorem could be stated in terms of $\pi_1(x,\tilde{z}_1)$ and $\pi_2(x,\tilde{z}_2)$ where the distribution assigned to $\tilde{z}_1$ by the first decision maker is the same as that assigned to $\tilde{z}_2$ by the second decision maker. This would be less misleading, but also less convenient and less suggestive, especially for later use. More precise notation would be, for instance, $\pi_1(x,F)$ and $\pi_2(x,F)$, where F is a cumulative distribution function.

THEOREM 1: *Let $r_i(x)$, $\pi_i(x,\tilde{z})$, and $p_i(x)$ be the local risk aversion, risk premium, and probability premium corresponding to the utility function u_i, $i=1,2$. Then the following conditions are equivalent, in either the strong form (indicated in brackets), or the weak form (with the bracketed material omitted).*

(a) $r_1(x) \geqq r_2(x)$ *for all* x [*and* $>$ *for at least one* x *in every interval*].

(b) $\pi_1(x,\tilde{z}) \geqq [>]\, \pi_2(x,\tilde{z})$ *for all* x *and* $\tilde{z}$.

(c) $p_1(x,h) \geqq [>]\, p_2(x,h)$ *for all* x *and all* $h>0$.

(d) $u_1(u_2^{-1}(t))$ *is a* [*strictly*] *concave function of* t.

(e) $\dfrac{u_1(y)-u_1(x)}{u_1(w)-u_1(v)} \leqq [<] \dfrac{u_2(y)-u_2(x)}{u_2(w)-u_2(v)}$ *for all* v, w, x, y *with* $v<w \leqq x<y$.

The same equivalences hold if attention is restricted throughout to an interval, that is, if the requirement is added that $x, x+\tilde{z}, x+h, x-h, u_2^{-1}(t), v, w$, *and* y, *all lie in a specified interval.*

PROOF: We shall prove things in an order indicating somewhat how one might discover that (a) implies (b) and (c).

To show that (b) follows from (d), solve (1) to obtain

$$\pi_i(x,\tilde{z}) = x + E(\tilde{z}) - u_i^{-1}(E\{u_i(x+\tilde{z})\}) . \tag{13}$$

Then

$$\begin{aligned} \pi_1(x,\tilde{z}) - \pi_2(x,\tilde{z}) &= u_2^{-1}(E\{u_2(x+\tilde{z})\}) - u_1^{-1}(E\{u_1(x+\tilde{z}(\}) \\ &= u_2^{-1}(E\{\tilde{t}\}) - u_1^{-1}(E\{u_1(u_2^{-1}(\tilde{t}))\}) , \end{aligned} \tag{14}$$

where $\tilde{t} = u_2(x+\tilde{z})$. If $u_1(u_2^{-1}(t))$ is [strictly] concave, then (by Jensen's inequality)

$$E\{u_1(u_2^{-1}(\tilde{t}))\} \leqq [<]\, u_1(u_2^{-1}(E\{\tilde{t}\})) . \tag{15}$$

Substituting (15) in (14), we obtain (b).

To show that (a) implies (d), note that

$$(16)\qquad \frac{d}{dt}u_1(u_2^{-1}(t)) = \frac{u_1'(u_2^{-1}(t))}{u_2'(u_2^{-1}(t))},$$

which is [strictly] decreasing if (and only if) $\log u_1'(x)/u_2'(x)$ is. The latter follows from (a) and

$$(17)\qquad \frac{d}{dx}\log\frac{u_1'(x)}{u_2'(x)} = r_2(x)-r_1(x)\,.$$

That (c) is implied by (e) follows immediately upon writing (9) in the form

$$(18)\qquad \frac{1-p_i(x,h)}{1+p_i(x,h)} = \frac{u_i(x+h)-u_i(x)}{u_i(x)-u_i(x-h)}\,.$$

To show that (a) implies (e), integrate (a) from w to x, obtaining

$$(19)\qquad -\log\frac{u_1'(x)}{u_1'(w)} \geqq [>] -\log\frac{u_2'(x)}{u_2'(w)} \quad \text{for} \quad w<x\,,$$

which is equivalent to

$$(20)\qquad \frac{u_1'(x)}{u_1'(w)} \leqq [<] \frac{u_2'(x)}{u_2'(w)} \quad \text{for} \quad w<x\,.$$

This implies

$$(21)\qquad \frac{u_1(y)-u_1(x)}{u_1'(w)} \leqq [<] \frac{u_2(y)-u_2(x)}{u_2'(w)} \quad \text{for} \quad w\leqq x<y\,,$$

as may be seen by applying the Mean Value Theorem of differential calculus to the difference of the two sides of (21) regarded as a function of y. Condition (e) follows from (21) upon application of the Mean Value Theorem to the difference of the reciprocals of the two sides of (e) regarded as a function of w.

We have now proved that (a) implies (d) implies (b), and (a) implies (e) implies (c). The equivalence of (a)–(e) will follow if we can prove that (b) implies (a), and (c) implies (a), or equivalently that not (a) implies not (b) and not (c). But this follows from what has already been proved, for if the weak [strong] form of (a) does not hold, then the strong [weak] form of (a) holds on some interval with u_1 and u_2 interchanged. Then the strong [weak] forms of (b) and (c) also hold on this interval with u_1 and u_2 interchanged, so the weak [strong] forms of (b) and (c) do not hold. This completes the proof.

We observe that (e) is equivalent to (20), (21), and

$$(22)\qquad \frac{u_1(w)-u_1(v)}{u_1'(x)} \geqq [>] \frac{u_2(w)-u_2(v)}{u_2'(x)} \quad \text{for} \quad v<w\leqq x\,.$$

6. CONSTANT RISK AVERSION

If the local risk aversion function is constant, say $r(x)=c$, then by (12):

(23) $u(x)\sim x$ if $r(x)=0$;

(24) $u(x)\sim -e^{-cx}$ if $r(x)=c>0$;

(25) $u(x)\sim e^{-cx}$ if $r(x)=c<0$.

These utilities are, respectively, linear, strictly concave, and strictly convex.

If the risk aversion is constant locally, then it is also constant globally, that is, a change in assets makes no change in preference among risks. In fact, for any k, $u(k+x)\sim u(x)$ in each of the cases above, as is easily verified. Therefore it makes sense to speak of "constant risk aversion" without the qualification "local" or "global."

Similar remarks apply to constant risk aversion on an interval, except that global consideration must be restricted to assets x and risks $\tilde{z}$ such that $x+\tilde{z}$ is certain to stay within the interval.

7. INCREASING AND DECREASING RISK AVERSION

Consider a decision maker who (i) attaches a positive risk premium to any risk, but (ii) attaches a smaller risk premium to any given risk the greater his assets x. Formally this means

(i) $\pi(x,\tilde{z})>0$ for all x and $\tilde{z}$;

(ii) $\pi(x,\tilde{z})$ is a strictly decreasing function of x for all $\tilde{z}$.

Restricting $\tilde{z}$ to be actuarially neutral would not affect (i) or (ii), by (2) with $\mu=E(\tilde{z})$.

We shall call a utility function (or a decision maker possessing it) *risk-averse* if the weak form of (i) holds, that is, if $\pi(x,\tilde{z})\geqq 0$ for all x and $\tilde{z}$; it is well known that this is equivalent to concavity of u, and hence to $u''\leqq 0$ and to $r\geqq 0$. A utility function is *strictly risk-averse* if (i) holds as stated; this is equivalent to strict concavity of u and hence to the existence in every interval of at least one point where $u''<0, r>0$.

We turn now to (ii). Notice that it amounts to a definition of strictly decreasing risk aversion in a global (as opposed to local) sense. On would hope that decreasing global risk aversion would be equivalent to decreasing local risk aversion $r(x)$. The following theorem asserts that this is indeed so. Therefore it makes sense to speak of "decreasing risk aversion" without the qualification "local" or "global." What is nontrivial is that $r(x)$ decreasing implies $\pi(x,\tilde{z})$ decreasing, inasmuch as $r(x)$ pertains directly only to infinitesimal gambles. Similar considerations apply to the probability premium $p(x,h)$.

THEOREM 2: *The following conditions are equivalent.*

(a′) *The local risk aversion function* $r(x)$ *is* [*strictly*] *decreasing.*

(b′) *The risk premium* $\pi(x,\tilde{z})$ *is a [strictly] decreasing function of* x *for all* $\tilde{z}$.

(c′) *The probability premium* $p(x,h)$ *is a [strictly] decreasing function of* x *for all* $h>0$.

The same equivalences hold if "increasing" is substituted for "decreasing" throughout and/or attention is restricted throughout to an interval, that is, the requirement is added that x, $x+\tilde{z}$, $x+h$, *and* $x-h$ *all lie in a specified interval.*

PROOF: This theorem follows upon application of Theorem 1 to $u_1(x)=u(x)$ and $u_2(x)=u(x+k)$ for arbitrary x and k.

It is easily verified that (a′) and hence also (b′) and (c′) are equivalent to

(d′) $u'(u^{-1}(t))$ is a [strictly] convex function of t.

This corresponds to (d) of Theorem 1. Corresponding to (e) of Theorem 1 and (20)–(22) is

(e′) $u'(x)u'''(x) \geqq (u''(x))^2$ [and $>$ for at least one x in every interval].

The equivalence of this to (a′)–(c′) follows from the fact that the sign of $r'(x)$ is the same as that of $(u''(x))^2 - u'(x)u'''(x)$. Theorem 2 can be and originally was proved by way of (d′) and (e′), essentially as Theorem 1 is proved in the present paper.

8. OPERATIONS WHICH PRESERVE DECREASING RISK AVERSION

We have just seen that a utility function evinces decreasing risk aversion in a global sense if an only if its local risk aversion function $r(x)$ is decreasing. Such a utility function seems of interest mainly if it is also risk-averse (concave, $r \geqq 0$). Accordingly, we shall now formally define a utility function to be [*strictly*] *decreasingly risk-averse* if its local risk aversion function r is [strictly] decreasing and nonnegative. Then by Theorem 2, conditions (i) and (ii) of Section 7 are equivalent to the utility's being strictly decreasingly risk-averse.

In this section we shall show that certain operations yield decreasingly risk-averse utility functions if applied to such functions. This facilitates proving that functions are decreasingly risk-averse and finding functions which have this property and also have reasonably simple formulas. In the proofs, $r(x)$, $r_1(x)$, etc., are the local risk aversion functions belonging to $u(x)$, $u_1(x)$, etc.

THEOREM 3: *Suppose* $a>0$: $u_1(x)=u(ax+b)$ *is [strictly] decreasingly risk-averse for* $x_0 \leqq x \leqq x_1$ *if and only if* $u(x)$ *is [strictly] decreasingly risk-averse for* $ax_0+b \leqq x \leqq ax_1+b$.

PROOF: This follows directly from the easily verified formula:

(26) $$r_1(x)=ar(ax+b)\,.$$

THEOREM 4: *If* $u_1(x)$ *is decreasingly risk-averse for* $x_0 \leqq x \leqq x_1$, *and* $u_2(x)$ *is decreasingly risk-averse for* $u_1(x_0) \leqq x \leqq u_1(x_1)$, *then* $u(x)=u_2(u_1(x))$ *is decreasingly* *risk-averse for $x_0 \leqq x \leqq x_1$, and strictly so unless one of u_1 and u_2 is linear from some x on and the other has constant risk aversion in some interval.*

PROOF: We have $\log u'(x) = \log u_2'(u_1'(x)) + \log u_1'(x)$, and therefore

$$(27) \qquad r(x) = r_2(u_1(x))u_1'(x) + r_1(x) .$$

The functions $r_2(u_1(x))$, $u_1'(x)$, and $r_1(x)$ are $\geqq 0$ and decreasing, and therefore so is $r(x)$. Furthermore, $u_1'(x)$ is strictly decreasing as long as $r_1(x) > 0$, so $r(x)$ is strictly decreasing as long as $r_1(x)$ and $r_2(u_1(x))$ are both >0. If one of them is 0 for some x, then it is 0 for all larger x, but if the other is strictly decreasing, then so is r.

THEOREM 5: *If $u_1, \ldots, u_n$ are decreasingly risk-averse on an interval $[x_0, x_1]$, and $c_1, \ldots, c_n$ are positive constants, then $u = \Sigma_1^n c_i u_i$ is decreasingly risk-averse on $[x_0, x_1]$, and strictly so except on subintervals (if any) where all u_i have equal and constant risk aversion.*

PROOF: The general statement follows from the case $u = u_1 + u_2$. For this case

$$(28) \qquad r = -\frac{u_1'' + u_2''}{u_1' + u_2'} = \frac{u_1'}{u_1' + u_2'} r_1 + \frac{u_2'}{u_1' + u_2'} r_2 ;$$

$$(29) \qquad r' = \frac{u_1'}{u_1' + u_2'} r_1' + \frac{u_2'}{u_1' + u_2'} r_2' + \frac{u_1'' u_2' - u_1' u_2''}{(u_1' + u_2')^2} (r_1 - r_2)$$

$$= \frac{u_1' r_1' + u_2' r_2'}{u_1' + u_2'} - \frac{u_1' u_2'}{(u_1' + u_2')^2} (r_1 - r_2)^2 .$$

We have $u_1' > 0$, $u_2' > 0$, $r_1' \leqq 0$, and $r_2' \leqq 0$. Therefore $r' \leqq 0$, and $r' < 0$ unless $r_1 = r_2$ and $r_1' = r_2' = 0$. The conclusion follows.

9. EXAMPLES

9.1. *Example 1.* The utility $u(x) = -(b-x)^c$ for $x \leqq b$ and $c > 1$ is strictly increasing and strictly concave, but it also has strictly *increasing* risk aversion: $r(x) = (c-1)/(b-x)$. Notice that the most general concave quadratic utility $u(x) = \alpha + \beta x - \gamma x^2$, $\beta > 0$, $\gamma > 0$, is equivalent as a utility to $-(b-x)^c$ with $c = 2$ and $b = \frac{1}{2}\beta/\gamma$. Therefore a quadratic utility cannot be decreasingly risk-averse on any interval whatever. This severely limits the usefulness of quadratic utility, however nice it would be to have expected utility depend only on the mean and variance of the probability distribution. Arguing "in the small" is no help: decreasing risk aversion is a local property as well as a global one.

9.2. *Example 2.* If

(30) $u'(x)=(x^a+b)^{-c}$ with $a>0, c>0$,

then $u(x)$ is strictly decreasingly risk-averse in the region

(31) $x>[\max\{0, -b, b(a-1)\}]^{1/a}$.

To prove this, note

(32) $$r(x)= -\frac{d}{dx}\log u'(x) = \frac{ac}{x+bx^{1-a}},$$

which is $\geqq 0$ and strictly decreasing in the region where the denominator $x+bx^{1-a}$ is $\geqq 0$ and strictly increasing, which is the region (30). (The condition $x\geqq 0$ is included to insure that x^a is defined; for $a\geqq 1$ it follows from the other conditions.)

By Theorem 3, one can obtain a utility function that is strictly decreasingly risk-averse for $x>0$ by substituting $x+d$ for x above, where d is at least the right-hand side of (31). Multiplying x by a positive factor, as in Theorem 3, is equivalent to multiplying b by a positive factor.

Given below are all the strictly decreasingly risk-averse utility functions $u(x)$ on $x>0$ which can be obtained by applying Theorem 3 to (30) with the indicated choices of the parameters a and c:

(33) $a=1, 0<c<1$: $u(x)\sim(x+d)^q$ with $d\geqq 0, 0<q<1$;

(34) $a=1, c=1$: $u(x)\sim\log(x+d)$ with $d\geqq 0$;

(35) $a=1, c>1$: $u(x)\sim -(x+d)^{-q}$ with $d\geqq 0, q>0$;

(36) $a=2, c=.5$: $u(x)\sim\log(x+d+[(x+d)^2+b])$ with $d\geqq |b|^{\frac{1}{2}}$;

(37) $a=2, c=1$: $u(x)\sim\arctan(\alpha x+\beta)$ or $\log(1-(\alpha x+\beta)^{-1})$ with $\alpha>0, \beta\geqq 1$;

(38) $a=2, c=1.5$: $u(x)\sim[1+(\alpha x+\beta)^{-2}]^{-\frac{1}{2}}$ or $-[1-(\alpha x+\beta)^{-2}]^{-\frac{1}{2}}$ with $\alpha>0, \beta\geqq 1$.

9.3. *Example 3.* Applying Theorems 4 and 5 to the utilities of Example 2 and Section 6 gives a very wide class of utilities which are strictly decreasingly risk-averse for $x>0$, such as

(39) $u(x)\sim -c_1 e^{-cx}-c_2 e^{-dx}$ with $c_1>0, c_2>0, c>0, d>0$.

(40) $u(x)\sim\log(d_1+\log(x+d_2))$ with $d_1\geqq 0, d_2\geqq 0, d_1+\log d_2\geqq 0$.

10. PROPORTIONAL RISK AVERSION

So far we have been concerned with risks that remained fixed while assets varied. Let us now view everything as a proportion of assets. Specifically, let $\pi^*(x,\tilde{z})$ be the *proportional risk premium* corresponding to a proportional risk $\tilde{z}$; that is, a decision maker with assets x and utility function u would be indifferent between receiving a risk $x\tilde{z}$ and receiving the non-random amount $E(x\tilde{z})-x\pi^*(x,\tilde{z})$. Then $x\pi^*(x,\tilde{z})$ equals the risk premium $\pi(x,x\tilde{z})$, so

$$\pi^*(x,\tilde{z}) = \frac{1}{x}\,\pi(x,x\tilde{z})\ . \tag{41}$$

For a small, actuarially neutral, proportional risk $\tilde{z}$ we have, by (5),

$$\pi^*(x,\tilde{z})=\tfrac{1}{2}\sigma_z^2 r^*(x)+o(\sigma_z^2)\ , \tag{42}$$

where

$$r^*(x)=xr(x)\ . \tag{43}$$

If $\tilde{z}$ is not actuarially neutral, we have, by (7),

$$\pi^*(x,\tilde{z})=\tfrac{1}{2}\sigma_z^2 r^*(x+xE(\tilde{z}))+o(\sigma_z^2)\ . \tag{44}$$

We will call r^* the *local proportional risk aversion* at the point x under the utility function u. Its interpretation by (42) and (44) is like that of r by (5) and (7).

Similarly, we may define the *proportional probability premium* $p^*(x,h)$, corresponding to a risk of gaining or losing a proportional amount h, namely

$$p^*(x,h)=p(x,xh)\ . \tag{45}$$

Then another interpretation of $r^*(x)$ is provided by

$$p^*(x,h)=\tfrac{1}{2}hr^*(x)+O(h^2)\ , \tag{46}$$

which follows from (45) and (11).

11. CONSTANT PROPORTIONAL RISK AVERSION

If the local proportional risk aversion function is constant, say $r^*(x)=c$, then $r(x)=c/x$, so the utility is strictly decreasingly risk-averse for $c>0$ and has negative, strictly increasing risk aversion for $c<0$. By (12), the possibilities are:

$$u(x)\sim x^{1-c} \quad \text{if} \quad r^*(x)=c<1\ , \tag{47}$$

$$u(x)\sim \log x \quad \text{if} \quad r^*(x)=1\ , \tag{48}$$

$$u(x)\sim -x^{-(c-1)} \quad \text{if} \quad r^*(x)=c>1\ . \tag{49}$$

If the proportional risk aversion is constant locally, then it is constant globally, that is, a change in assets makes no change in preferences among proportional risks. This follows immediately from the fact that $u(kx)\sim u(x)$ in each of the cases above. Therefore it makes sense to speak of "constant proportional risk aversion" without the qualification "local" or "global." Similar remarks apply to constant proportional risk aversion on an interval.

12. INCREASING AND DECREASING PROPORTIONAL RISK AVERSION

We will call a utility function [strictly] increasingly or decreasingly proportionally risk-averse if it has a [strictly] increasing or decreasing local proportional risk aversion function. Again the corresponding local and global properties are equivalent, as the next theorem states.

THEOREM 6: *The following conditions are equivalent.*

(a″) *The local proportional risk aversion function* $r^*(x)$ *is [strictly] decreasing.*

(b″) *The proportional risk premium* $\pi^*(x,\tilde{z})$ *is a [strictly] decreasing function of* x *for all* $\tilde{z}$.

(c″) *The proportional probability premium* $p^*(x,h)$ *is a [strictly] decreasing function of* x *for all* $h>0$.

The same equivalences hold if "increasing" is substituted for "decreasing" throughout and/or attention is restricted throughout to an interval, that is, if the requirement is added that x, $x+x\tilde{z}$, $x+xh$, *and* $x-xh$ *all lie in a specified interval.*

PROOF: This theorem follows upon application of Theorem 1 to $u_1(x)=u(x)$ and $u_2(x)=u(kx)$ for arbitrary x and k.

A decreasingly risk-averse utility function may be increasingly or decreasingly proportionally risk-averse or neither. For instance, $u(x)\sim -\exp[-q^{-1}(x+b)^q]$, with $b\geqq 0$, $q<1$, $q\neq 0$, is strictly decreasingly risk-averse for $x>0$ while its local proportional risk aversion function $r^*(x)=x(x+b)^{-1}[(x+b)^q+1-q]$ is strictly increasing if $0<q<1$, strictly decreasing if $q<0$ and $b=0$, and neither if $q<0$ and $b>0$.

13. RELATED WORK OF ARROW

Arrow[4] has discussed the optimum amount to invest when part of the assets x are to be held as cash and the rest invested in a specified, actuarially favorable risk. If $\tilde{\imath}$ is the return per unit invested, then investing the amount a will result in assets $x+a\tilde{\imath}$. Suppose $a(x,\tilde{\imath})$ is the optimum amount to invest, that is $a(x,\tilde{\imath})$ maximizes $E\{u(x+a\tilde{\imath})\}$. Arrow proves that if $r(x)$ is [strictly] decreasing, increasing, or constant for all x, then $a(x,\tilde{\imath})$ is [strictly] increasing, decreasing, or constant, respectively, except that $a(x,\tilde{\imath})=x$ for all x below a certain value (depending on $\tilde{\imath}$). He also proves a theorem about the asset elasticity of the demand for cash which is equivalent to the statement that if $r^*(x)$ is [strictly] decreasing, increasing, or constant for all x, then the optimum proportional investment $a^*(x,\tilde{\imath})=a(x,\tilde{\imath})/x$ is [strictly] increasing, decreasing, or constant, respectively, except that $a^*(x,\tilde{\imath})=1$ for all x below a certain value. In the present framework it is natural to deduce these re-

[4] Kenneth J. Arrow, "Liquidity Preference," Lecture VI in "Lecture Notes for Economics 285, The Economics of Uncertainty," pp. 33-53, undated, Stanford University.

sults from the following theorem, whose proof bears essentially the same relation to Arrow's proofs as the proof of Theorem 1 to direct proofs of Theorems 2 and 6. For convenience we assume that $a_1(x,\tilde{\imath})$ and $a_2(x,\tilde{\imath})$ are unique.

THEOREM 7: *Condition* (a) *of Theorem* 1 *is equivalent to*

(f) $a_1(x,\tilde{\imath}) \leqq a_2(x,\tilde{\imath})$ *for all* x *and* $\tilde{\imath}$ [*and* $<$ *if* $0 < a_1(x,\tilde{\imath}) < x$].

The same equivalence holds if attention is restricted throughout to an interval, that is, if the requirement is added that x *and* $x+\tilde{\imath}x$ *lie in a specified interval.*

PROOF: To show that (a) implies (f), note that $a_j(x,\tilde{\imath})$ maximizes

$$(50) \qquad v_j(a) = \frac{1}{u_j'(x)} E\{u_j(x+a\tilde{\imath})\}, \qquad j=1,2.$$

Therefore (f) follows from

$$(51) \qquad \frac{d}{da}\{v_1(a)-v_2(a)\} = E\left\{\tilde{\imath}\left(\frac{u_1'(x+a\tilde{\imath})}{u_1'(x)} - \frac{u_2'(x+a\tilde{\imath})}{u_2'(x)}\right)\right\} \leqq [<] 0,$$

which follows from (a) by (20).

If, conversely, the weak [strong] form of (a) does not hold, then its strong [weak] form holds on some interval with u_1 and u_2 interchanged, in which case the weak [strong] form of (f) cannot hold, so (f) implies (a). (The fact must be used that the strong form of (f) is actually stronger than the weak form, even when x and $x+\tilde{\imath}x$ are restricted to a specified interval. This is easily shown.)

Assuming u is bounded, Arrow proves that (i) it is impossible that $r^*(x) \leqq 1$ for all $x > x_0$, and he implies that (ii) $r^*(0) \leqq 1$. It follows, as he points out, that if u is bounded and r^* is monotonic, then r^* is increasing. (i) and (ii) can be deduced naturally from the following theorem, which is an immediate consequence of Theorem 1 (a) and (e).

THEOREM 8: *If* $r_1(x) \geqq r_2(x)$ *for all* $x > x_0$ *and* $u_1(\infty) = \infty$, *then* $u_2(\infty) = \infty$. *If* $r_1(x) \geqq r_2(x)$ *for all* $x < \varepsilon$, $\varepsilon > 0$, *and* $u_2(0) = -\infty$, *then* $u_1(0) = -\infty$.

This gives (i) when $r_1(x) = 1/x$, $r_2(x) = r(x)$, $u_1(x) = \log x$, $u_2(x) = u(x)$. It gives (ii) when $r_1(x) = r(x)$, $r_2(x) = c/x$, $c > 1$, $u_1(x) = u(x)$, $u_2(x) = -x^{1-c}$.

This section is not intended to summarize Arrow's work,[4] but only to indicate its relation to the present paper. The main points of overlap are that Arrow introduces essentially the functions r and r^* (actually their negatives) and uses them in significant ways, in particular those mentioned already, and that he introduces essentially $p^*(x,h)$, proves an equation like (46) in order to interpret decreasing r^*, and mentions the possibility of a similar analysis for r.

Harvard University

Addendum

In retrospect, I wish footnote 2 had made clear that Robert Schlaifer's contribution included formulating originally the concept of decreasing risk aversion in terms of the probability premium and proving that it implies $r(x)$ is decreasing; i.e., that (c′) implies (a′) in Theorem 2.

In addition, John Lintner found a discrepancy in using the results that turned out to be due to a more significant error: The first term on the right side of Eq. (44) should be divided by $1+E(\tilde{z})$.

3. SEPARATION THEOREMS

Reprinted from *The Review of Economics and Statistics* 47, 13-37 (1965).

THE VALUATION OF RISK ASSETS AND THE SELECTION OF RISKY INVESTMENTS IN STOCK PORTFOLIOS AND CAPITAL BUDGETS*

John Lintner

Introduction and Preview of Some Conclusions

THE effects of risk and uncertainty upon asset prices, upon rational decision rules for individuals and institutions to use in selecting security portfolios, and upon the proper selection of projects to include in corporate capital budgets, have increasingly engaged the attention of professional economists and other students of the capital markets and of business finance in recent years. The essential purpose of the present paper is to push back the frontiers of our knowledge of the logical structure of these related issues, albeit under idealized conditions. The immediately following text describes the contents of the paper and summarizes some of the principal results.

The first two sections of this paper deal with the *problem of selecting* optimal security portfolios by risk-averse investors who have the alternative of investing in risk-free securities with a positive return (or borrowing at the same rate of interest) and who can sell short if they wish. The first gives alternative and hopefully more transparent proofs (under these more general market conditions) for Tobin's important "separation theorem" that ". . . the proportionate composition of the non-cash assets is independent of their aggregate share of the investment balance . . . " (and hence of the optimal holding of cash) for risk averters in purely competitive markets when utility functions are quadratic *or* rates of return are multivariate normal.[1] We then note that the same conclusion follows from an earlier theorem of Roy's [19] without dependence on quadratic utilities or normality. The second section shows that *if short sales are permitted*, the best portfolio-mix of risk assets can be determined by the solution of a single simple set of simultaneous equations without recourse to programming methods, and when covariances are zero, a still simpler ratio scheme gives the optimum, whether or not short sales are permitted. When covariances are not all zero and short sales are excluded, a single quadratic programming solution is required, but sufficient.

Following these extensions of Tobin's classic work, we concentrate on the set of risk assets held in risk averters' portfolios. In section III we develop various significant *equilibrium properties within* the risk asset portfolio. In particular, we establish conditions under which stocks will be held long (short) in optimal portfolios even when "risk premiums" are negative (positive). We also develop expressions for different combinations of expected rate of return on a given security, and its standard deviation, variance, and/or covariances which will result in the same relative holding of a stock, *ceteris paribus*. These "indifference functions" provide direct evidence on the moot issue of the appropriate functional relationships between "required rates of return" and relevant risk parameter(s) — and on the related issue of how "risk classes" of securities may best be delineated (if they are to be used).[2]

*This paper is another in a series of interrelated theoretical and statistical studies of corporate financial and investment policies being made under grants from the Rockefeller Foundation, and more recently the Ford Foundation, to the Harvard Business School. The generous support for this work is most gratefully acknowledged. The author is also much indebted to his colleagues Professors Bishop, Christenson, Kahr, Raiffa, and (especially) Schlaifer, for extensive discussion and commentary on an earlier draft of this paper; but responsibility for any errors or imperfections remains strictly his own.

[Professor Sharpe's paper, "Capital Asset Prices: A Theory of Market Equilibrium Under Conditions of Risk" (*Journal of Finance*, September 1964) appeared after this paper was in final form and on its way to the printers. My first section, which parallels the first half of his paper (with corresponding conclusions), sets the algebraic framework for sections II, III and VI, (which have no counterpart in his paper) and for section IV on the equilibrium prices of risk assets, concerning which our results differ significantly for reasons which will be explored elsewhere. Sharpe does not take up the capital budgeting problem developed in section V below.]

[1]Tobin [21, especially pp. 82–85]. Tobin assumed that funds are to be a allocated only over "monetary assets" (risk-free cash and default-free bonds of uncertain resale price) and allowed no short sales or borrowing. See also footnote 24 below. Other approaches are reviewed in Farrar [38].

[2]It should be noted that the classic paper by Modigliani and Miller [16] was silent on these issues. Corporations were assumed to be divided into homogeneous classes having the property that all shares of all corporations in any given class differed (at most) by a "scale factor," and hence (*a*) were perfectly correlated with each other and (*b*) were perfect substitutes for each other in perfect markets (p. 266). No comment was made on the measure of risk or uncertainty (or other attributes) relevant to the identification of different "equiva-

There seems to be a general presumption among economists that relative risks are best measured by the standard deviation (or coefficient of variation) of the rate of return,[3] but in the simplest cases considered — specifically when all covariances are considered to be invariant (or zero) — the indifference functions are shown to be linear between expected rates of return and their *variance*, not standard deviation.[4] (With variances fixed, the indifference function between the ith expected rate of return and its pooled covariance with other stocks is hyperbolic.) There is no simple relation between the expected rate of return required to maintain an investor's relative holding of a stock and its standard deviation. Specifically, when covariances are non-zero and variable, the indifference functions are complex and non-linear *even if* it is assumed that the *correlations* between rates of return on different securities are invariant.

To this point we follow Tobin [21] and Markowitz [14] in assuming that current security prices are given, and that each investor acts on his own (perhaps unique) probability distribution over rates of return given these market prices. In the rest of the paper, we assume that investors' joint probability distributions pertain to dollar returns rather than rates of return[5], and for simplicity we assume that all investors assign identical sets of means, variances, and covariances to the distribution of these dollar returns. However unrealisic the latter assumption may be, it enables us, in section IV, to derive a set of (stable) equilibrium market prices which at least fully and explicitly reflect the presence of uncertainty *per se* (as distinct from the effects of diverse expectations), and to derive further implications of such uncertainty. In particular, the aggregate market value of any company's equity is equal to the capitalization at the risk-free interest rate of a uniquely defined *certainty-equivalent* of the probability distribution of the aggregate dollar returns to all holders of its stock. For each company, this certainty equivalent is the expected value of these uncertain returns less an adjustment term which is proportional to their aggregate risk. The factor of proportionality is the *same for all companies* in equilibibrum, and may be regarded as a *market price of dollar risk*. The relevant risk of each company's stock is measured, moreover, not by the standard deviation of its dollar returns, but by the *sum* of the *variance* of its own aggregate dollar returns *and* their *total covariance* with those of all other stocks.

The next section considers some of the implications of these results for the normative aspects of the capital budgeting decisions of a company whose stock is traded in the market. For simplicity, we impose further assumptions required to make capital budgeting decisions independent of decisions on how the budget is financed.[6] The capital budgeting problem becomes a quadratic programming problem analogous to that introduced earlier for the individual investor. This capital budgeting-portfolio problem is formulated, its solution is given and some of its more important properties examined. Specifically, the minimum expected return (in dollars of expected present value) required to justify the allocation of funds to a given risky project is shown to be an increasing function of each of the following factors: (*i*) the risk-free rate of return; (*ii*) the "market price of (dollar) risk"; (*iii*) the variance in the project's own present value return; (*iv*) the project's aggregate present value return-*co*variance with assets already held by the company, and (*v*) its total covariance with other projects concurrently included in the capital budget. *All five* factors are involved explicitly in the corresponding (derived) formula for the minimum acceptable *expected rate* of return on an investment project. In this model, all means

lent return" classes. Both Propositions I (market value of *firm* independent of capital structure) and II (the linear relation between the expected return on equity shares and the debt-equity ratio for firms within a given class) are derived from the above assumptions (and the further assumption that corporate bonds are riskless securities); they involve no inter-class comparisons, ". . . nor do they involve any assertion as to what is an adequate compensation to investors for assuming a given degree of risk. . . ." (p. 279).

[3] This is, for instance, the presumption of Hirschleifer [8, p. 113], although he was careful not to commit himself to this measure alone in a paper primarily focussed on other issues. For an inductive argument in favor of the standard deviation of the rate of return as the best measure of risk, see Gordon [5, especially pp. 69 and 761. See also Dorfman in [3, p. 129 ff.] and Baumol [2].

[4] Except in dominantly "short" portfolios, the constant term will be larger, and the slope lower, the higher the (fixed) level of covariances of the given stocks with other stocks.

[5] The dollar return in the period is the sum of the cash dividend and the increase in market price during the period.

[6] We also assume that common stock portfolios are not "inferior goods," that the value of *all other* common stocks is invariant, and any effect of changes in capital budgets on the *covariances* between the values of different companies' *stocks* is ignored.

and (co)variances of present values must be calculated at the riskless rate r^*. We also show that *there can be no "risk-discount" rate* to be used in computing present values to accept or reject individual projects. In particular, the *"cost of capital"* as defined (for uncertainty) anywhere in the literature *is not the appropriate rate* to use in these decisions *even if* all new projects have the same "risk" as existing assets.

The final section of the paper briefly examines the complications introduced by institutional limits on amounts which either individuals or corporations may borrow at given rates, by rising costs of borrowed funds, and certain other "real world" complications. It is emphasized that the results of this paper are not being presented as directly applicable to practical decisions, because many of the factors which matter very siginificantly in practice have had to be ignored or assumed away. The function of these simplifying assumptions has been to permit a rigorous development of theoretical relationships and theorems which reorient much current theory (especially on capital budgeting) and provide a basis for further work.[7] More detailed conclusions will be found emphasized at numerous points in the text.

I — Portfolio Selection for an Individual Investor: The Separation Theorem

Market Assumptions

We assume that (1) *each individual* investor can invest any part of his capital in certain *risk-free assets* (e. g. deposits in insured savings accounts[8]) all of which pay interest at a common positive rate, exogeneously determined; and that (2) he can invest *any fraction* of his capital *in any* or all of a given finite set of *risky* securities which are (3) traded in a single *purely competitive market*, free of transactions costs and taxes, at given market prices,[9] which consequently do not depend on his investments or transactions. We also assume that (4) any investor may, if he wishes, borrow funds to invest in risk assets. Except in the final section, we assume that the *interest rate paid* on such loans is the same as he would have received had he invested in risk-free savings accounts, and that there is *no limit* on the amount he can borrow at this rate. Finally (5) he makes all purchases and sales of securities and all deposits and loans at discrete points in time, so that in selecting his portfolio at any "transaction point," each investor will consider only (*i*) the cash throw-off (typically interest payments and dividends received) within the period to the next transaction point and (*ii*) changes in the market prices of stocks during this same period. The *return* on any common stock is defined to be the sum of the cash dividends received plus the change in its market price. The return on any portfolio is measured in exactly the same way, including interest received or paid.

Assumptions Regarding Investors

(1) Since we posit the existence of assets yielding *positive risk-free* returns, we assume that each investor has already decided the fraction of his total capital he wishes to hold in cash and non-interest bearing deposits for reasons of liquidity or transactions requirements.[10] Henceforth, we will speak of *an investor's capital* as the stock of funds he has available for profitable investment *after* optimal cash holdings have been deducted. We also assume that (2) each investor will have assigned a *joint probability distribution* incorporating his best judgments regarding the returns on all *individual stocks*, or at least will have specified an expected value and variance to every return and a covariance or correlation to every pair of returns. All expected values of returns are finite, all variances are non-zero and finite, and all correlations of returns are less than one in absolute value (i. e. the covariance matrix is positive-definite). The investor computes the expected value and variance of the total return on any possible *portfolio*, or mix of any specified amounts of any or all of the individual stocks, by forming the appropriately weighted average or sum of these components expected returns, variances and covariances.

[7] The relation between the results of this paper and the models which were used in [11] and [12] is indicated at the end of section V.

[8] Government bonds of appropriate maturity provide another important example when their "yield" is substituted for the word "interest."

[9] Solely for convenience, we shall usually refer to all these investments as common stocks, although the analysis is of course quite general.

[10] These latter decisions are independent of the decisions regarding the allocation of remaining funds between risk-free assets with positive return and risky stocks, which are of direct concern in this paper, because the risk-free assets with positive returns clearly dominate those with no return once liquidity and transactions requirements are satisfied at the margin.

With respect to an investor's *criterion for choices* among different attainable combinations of assets, we assume that (3) if any two mixtures of assets have the *same expected return*, the investor will prefer the one having the *smaller variance* of return, and if any two mixtures of assets have the *same variance* of returns, he will prefer the one having the *greater expected value*. Tobin [21, pp. 75–76] has shown that such preferences are implied by maximization of the expected value of a von Neumann-Morgenstern utility function if *either* (*a*) the investor's *utility* function is *concave* and *quadratic or* (*b*) the investor's *utility* function is *concave, and* he has assigned probability distributions such that the *returns* on *all possible portfolios differ at most by a location and scale parameter*, (which will be the case if the joint distribution of all individual stocks is multivariate normal).

Alternative Proofs of the Separation Theorem

Since the interest rates on riskless savings bank deposits ("loans to the bank") and on borrowed funds are being assumed to be the same, we can treat borrowing as negative lending. Any portfolio can then be described in terms of (i) the *gross* amount invested in stocks, (ii) the fraction of this amount invested in each individual stock, and (iii) the *net* amount invested in loans (a negative value showing that the investor has borrowed rather than lent). But since the *total net* investment (the algebraic sum of stocks plus loans) is a given amount, the problem simply requires finding the jointly optimal values for (1) the ratio of the gross investment in stocks to the total net investment, and (2) the ratio of the gross investment in each individual stock to the total gross investment in stocks. It turns out that although the solution of (1) depends upon that of (2), in our context the latter is independent of the former. Specifically, the *separation theorem* asserts that:

Given the assumptions about borrowing, lending, and investor preferences stated earlier in this section, *the optimal proportionate composition of the stock* (*risk-asset*) *portfolio* (i.e. the solution to sub-problem 2 above) *is independent of the ratio of the gross investment in stocks to the total net investment.*

Tobin proved this important separation theorem by deriving the detailed solution for the optimal mix of risk assets *conditional* on a given gross investment in this portfolio, and then formally proving the critical invariance property stated in the theorem. Tobin used more restrictive assumptions that we do regarding the available investment opportunities and he permitted no borrowing.[11] Under our somewhat broadened assumptions in these respects, the problem fits neatly into a traditional Fisher framework, with different available combinations of expected values and standard deviations of return on alternative *stock portfolios* taking the place of the original "production opportunity" set and with the alternative investment choices being concurrent rather than between time periods. Within this framework, alternative and more transparent proofs of the separation theorem are available which do not involve the actual calculation of the best allocation in stocks over individual stock issues. As did Fisher, we shall present a simple algebraic proof[12], set out the logic of the argument leading to the theorem, and depict the essential geometry of the problem.[13]

As a preliminary step, we need to establish the relation between the investor's total investment in *any* arbitrary mixture or portfolio of individual stocks, his total net return from all his investments (including riskless assets and any borrowing), and the risk parameters of his investment position. Let the *interest rate* on riskless assets or borrowing be r^*, and the *uncertain return* (dividends plus price appreciation) *per dollar invested in the given portfolio of stocks* be r. Let w represent the *ratio* of gross investment in stocks to

[11] Tobin considered the special case where cash with no return was the only riskless asset available. While he formally required that all assets be held in non-negative quantities (thereby ruling out short sales), and that the total value of risk assets held not be greater than the investment balance available without borrowing, these non-negativity and maximum value constraints were not introduced into his formal solution of the optimal investment mix, which in turn was used in proving the invariance property stated in the theorem. Our proof of the theorem is independent of the programming constraints neglected in Tobin's proof. Later in this section we show that when short sales are properly and explicitly introduced into the set of possible portfolios, the resulting equations for the optimum portfolio mix are identical to those derived by Tobin, but that insistence on no short sales results in a somewhat more complex programming problem (when covariances are non-zero), which may however, be readily handled with computer programs now available.

[12] An alternative algebraic proof using utility functions explicitly is presented in the appendix, note I.

[13] Lockwood Rainhard, Jr. has also independently developed and presented a similar proof of the theorem in an unpublished seminar paper.

total *net* investment (stock plus riskless assets minus borrowing). Then the investor's net return per dollar of total net investment will be

(1) $\bar{y} = (1-w)r^* + w\bar{r} = r^* + w(\bar{r} - r^*)$; $0 \leq w < \infty$,

where a value of $w < 1$ indicates that the investor holds some of his capital in riskless assets and receives interest amounting to $(1-w)r^*$; while $w > 1$ indicates that the investor borrows to buy stocks on margin and pays interest amounting to the absolute value of $(1-w)r^*$. From (1) we determine the mean and variance of the net return per dollar of total net investment to be:

(2a) $\bar{y} = r^* + w(\bar{r} - r^*)$, and

(2b) $\sigma^2_y = w^2 \sigma^2_r$.

Finally, after eliminating w between these two equations, we find that the direct relation between the expected value of the investor's net return per dollar of his total net investment and the risk parameters of his investment position is:

(3a) $\bar{y} = r^* + \theta \sigma_y$, where

(3b) $\theta = (\bar{r} - r^*)/\sigma_r$.

In terms of *any* arbitrarily selected *stock* portfolio, therefore, the investor's *net* expected rate of return on his total net investment is related *linearly* to the *risk* of return on his total net investment as *measured* by *the standard deviation* of his return. Given *any* selected stock portfolio, this linear function corresponds to Fisher's "market opportunity line"; its intercept is the risk-free rate r^* and its slope is given by θ, which is determined by the parameters $\bar{r}$ and σ_r of the particular stock portfolio being considered. We also see from (2a) that, by a suitable choice of w, the investor can use *any* stock mix (and *its* associated "market opportunity line") to obtain an expected return, $\bar{y}$, as high as he likes; but that, because of (2b) and (3b), as he increases his investment w in the (tentatively chosen) mix, the standard deviation σ_y (and hence the variance σ^2_y) of the return on his total investment also becomes proportionately greater.

Now consider all possible stock portfolios. Those portfolios having the same θ value will lie on the same "market opportunity line," but those having different θ values *will offer different "market opportunity lines"* (between expected return and risk) for the investor to use. The investor's problem is to choose which stock portfolio-mix (or market opportunity line *or* θ value) to use *and* how intensively to use it (the proper value of w). Since *any* expected return $\bar{y}$ can be obtained from *any* stock mix, an investor adhering to our choice criterion will minimize the variance of his over-all return σ^2_y associated with *any* expected return he may choose *by confining all his investment in stocks to the mix with the largest θ value. This portfolio minimizes the variance associated with any $\bar{y}$ (and hence any w value)* the investor may prefer, and *consequently, is independent* of $\bar{y}$ and w. This establishes the separation theorem[14], once we note that our assumptions regarding available portfolios[15] insure the existence of a maximum θ.

It is equally apparent that *after* determining the optimal stock portfolio (mix) by maximizing θ, the investor can complete his choice of an over-all investment position by substituting the θ of this optimal mix in (3) and decide which over-all investment position by substituting of the available $(\bar{y}, \sigma_y)$ pairs he prefers by referring to his own utility function. Substitution of this best $\bar{y}$ value in (2a) determines a unique best value of the ratio w of gross investment in the optimal stock portfolio to his total net investment, and hence, the optimal amount of investments in riskless savings deposits or the optimal amount of borrowing as well.

This separation theorem thus has four immediate *corrolaries* which can be stated:

(*i*) *Given* the assumptions about borrowing and lending stated above, any investor whose choices maximize the expectation of any particular utility function consistent with these conditions will make *identical decisions regarding the proportionate composition of his stock* (risk-asset) *portfolio. This is true regardless of the particular utility function*[16] whose expectation he maximizes.

(*ii*) Under these assumptions, only a *single point* on the Markowitz "Efficient Frontier" *is relevant* to the investor's decision regarding his investments in *risk* assets.[17] (The next section

[14] See also the appendix, note I for a different form of proof.

[15] Specifically, that the amount invested in any stock in any stock mix is infinitely divisible, that all expected returns on individual stocks are finite, that all variances are positive and finite, and that the variance-covariance matrix is positive-definite.

[16] When probability assessments are multivariate normal, the utility function may be polynomial, exponential, etc. Even in the "non-normal" case when utility functions *are* quadratic, they may vary in its parameters. See also the reference to Roy's work in the text below.

[17] When the above conditions hold (see also final para-

shows this point can be obtained directly without calculating the remainder of the efficient set.)

Given the same assumptions, (*iii*) the parameters of the investor's particular utility within the relevant set determine *only* the ratio of his total gross investment in stocks to his total *net* investment (including riskless assets and borrowing); and (*iv*) the investor's wealth is also, consequently, relevant to determining the *absolute size* of his investment in individual stocks, but *not* to the *relative distribution* of his gross investment in stocks *among individual issues*.

The Geometry of the Separation Theorem and Its Corrolaries

The algebraic derivations given above can be represented graphically as in chart 1. Any given available stock portfolio is characterized by a pair of values $(\sigma_r, \bar{r})$ which can be represented as a point in a plane with axes σ_y and $\bar{y}$. Our assumptions insure that the points representing all available stock mixes lie in a finite region, all parts of which lie to the right of the vertical axis, and that this region is bounded by a closed curve.[18] The contours of the investor's utility function are concave upward, and any movement in a north and or west direction denotes contours of greater utility. Equation (3) shows that all the (σ_y, y) pairs attainable by combining, borrowing, or lending with *any* particular stock portfolio lie on a ray from the point $(0, r^*)$ though the point corresponding to the stock mix in question. Each possible stock portfolio thus determines a unique "market opportunity line". Given the properties of the utility function, it is obvious that shifts from one possible mix to another which *rotate* the associated market opportunity line *counter colckwise* will *move the investor to preferred positions regardless of the point on the line* he had tentatively chosen. The slope of this market-opportunity line given by (3) is θ, and the limit of the favorable rotation is given by the maximum attainable θ, which identifies the optimal mix M.[19] Once this best mix, M, has been determined, the investor completes the optimization of his total investment position by selecting the point on the ray through M which is tangent to a utility contour in the standard manner. If his utility contours are as in the U_i set in chart 1, he uses savings accounts and does not borrow. If his utility contours are as in U_j set, he borrows in order to have a gross investment in his best stock mix greater than his net investment balance.

Risk Aversion, Normality and the Separation Theorem

The above analysis has been based on the assumptions regarding markets and investors stated at the beginning of this section. One crucial premise was investor *risk-aversion* in the form of *preference for expected return* and *preference against return-variance, ceteris paribus*. We noted that Tobin has shown that *either* concave-*quadratic* utility functions *or* multivariate *normality* (of probability assessments) *and any concave* utility were *sufficient* conditions to validate this premise, but they were *not* shown (or alleged) to be *necessary* conditions. This is probably fortunate because the quadratic utility of income (or wealth!) function, in spite of its popularity in theoretical work, has several undesirably restrictive and implausible properties,[20] and, despite

---

graph of this section), the modest narrowing of the relevant range of Markowitz' Efficient Set suggested by Baumol [2] is still larger than needed by a factor strictly proportionate to the number of portfolios he retains in his truncated set! This is true since the relevant set is a single portfolio under these conditions.

[18] See Markowitz [14] as cited in the appendix, note I.

[19] The analogy with the standard Fisher two-period production-opportunity case in perfect markets with equal borrowing and lending rates is clear. The optimal set of production opportunities available is found by moving along the envelope function of efficient combinations of projects onto ever higher present value lines to the highest attainable. This best set of production opportunities is independent of the investor's particular utility function which determines only whether he then lends or borrows in the market (and by how much in either case) to reach hi best over-all position. The only differences between this case and ours lie in the concurrent nature of the comparisons (instead of inter-period), and the rotation of the market opportunity lines around the common pivot of the riskless return (instead of parallel shifts in present value lines). See Fisher [4] and also Hirschlaifer [7], figure 1 and section 1a.

[20] In brief, not only does the quadratic function imply negative marginal utilities of income or wealth much "too soon" in empirical work unless the risk-aversion parameter is very small — in which case it cannot account for the degree of risk-aversion empirically found,— it also implies that, over a major part of the range of empirical data, common stocks, like potatoes in Ireland, are "inferior" goods. Offering more return at the same risk would so sate investors that they would reduce their risk-investments *because* they were more attractive. (Thereby, as Tobin [21] noted, denying the negatively sloped demand curves for *riskless* assets which are standard doctrine in "liquidity preference theory" — a conclusion which cannot, incidentally, be avoided by "limit arguments" on quadratic utilities such as he used, once borrowing and leverage are admitted.)

its mathematical convenience, multivariate normality is doubtless also suspect, especially perhaps in considering common stocks.

It is, consequently, very relevant to note that by using the Bienaymé-Tchebycheff inequality, Roy [19] has shown that investors operating on his "Safety First" principle (i.e. make risky investments so as to minimize the upper bound of the probability that the realized outcome will fall below a pre-assigned "disaster level") should maximize the ratio of the *excess* expected portfolio return (over the disaster level) to the standard deviation of the return on the portfolio[21] — which is precisely our criterion of max θ when his disaster level is equated to the risk-free rate r^*. This result, of course, does not depend on multivariate normality, and uses a different argument and form of utility function.

The *Separation Theorem*, and its Corrolaries (*i*) and (*ii*) above — and all the rest of our following analysis which depends on the maximization

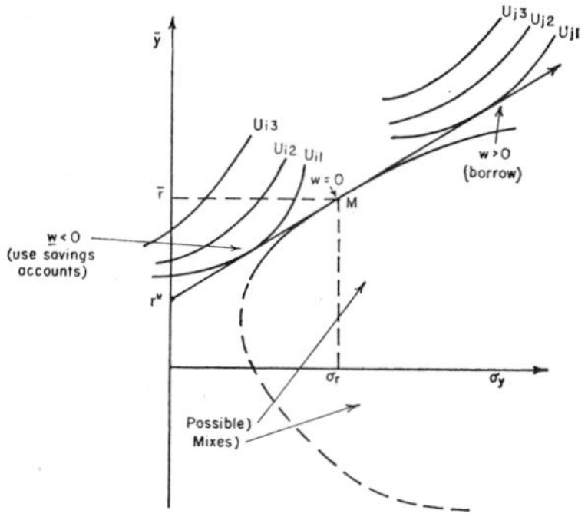

FIGURE I

of θ — is thus rigorously appropriate in the non-multivariate normal case for Safety-Firsters who minimax the stated upper bound of the chance of doing less well on portfolios including risk assets than they can do on riskless investments, just as it is for concave-expected utility maximizers in the "normal" case. On the basis of the same probability judgments, these Safety-Firsters will use the same proximate criterion function (max θ) and will choose proportionately the same risk asset portfolios as the more orothodox "utility maximizers" we have hitherto considered.

II — Portfolio Selection: The Optimal Stock Mix

Before finding the optimal stock mix — the mix which maximizes θ in (3*b*) above — it is necessary to express the return on any arbitrary mix in terms of the returns on individual stocks included in the portfolio. Although short sales are excluded by assumption in most of the writings on portfolio optimization, this restrictive assumption is arbitrary for some purposes at least, and we therefore broaden the analysis in this paper to include short sales whenever they are permitted.

Computation of Returns on a Stock Mix, When Short Sales are Permitted

We assume that there are m different stocks in the market, denoted by $i = 1, 2, \ldots, m$, and treat short sales as negative purchases. We shall use the following basic notation:

- $|h_i|$ – The ratio of the gross investment in the ith stock (the market value of the amount bought *or* sold) to the gross investment in all stocks. A positive value of h_i indicates a *purchase*, while a negative value indicates a *short sale*.
- $\bar{r}_i$ – The return per dollar invested in a *purchase* of the ith stock (cash dividends plus price appreciation)
- $\bar{r}$ – As above, the return per dollar invested in a particular *mix* or *portfolio* of stocks.

Consider now a gross investment in the entire mix, so that the actual investment in the ith stock is equal to $|h_i|$. The returns on purchases and short sales need to be considered separately.

First, we see that if $|h_i|$ is invested in a *pur-*

This function also implausibly implies, as Pratt [17] and Arrow [1] have noted, that the insurance premiums which people would be willing to pay to hedge *given* risks *rise* progressively with wealth or income. For a related result, see Hicks [6, p. 802].

[21] Roy also notes that when judgmental distributions *are* multivariate normal, maximization of this criterion *minimizes* the probability of "disaster" (failure to do better in stocks than savings deposits or government bonds held to maturity). It should be noted, however, minimization of the probability of short falls from "disaster" levels in this "normal" case is strictly *equivalent* to expected utility maximization under *all* risk-averters' utility functions. The equivalence is *not* restricted to the utility function of the form (0, 1) (zero if "disaster" occurs, one if it doesn't), as claimed by Roy [19, p. 432] and Markowitz [14, p. 293 and following.].

chase ($h_i > 0$), the return will be simply $h_i\bar{r}_i$. For reasons which will be clear immediately however, we write this in the form:

$$(4a)\quad h_i\bar{r}_i = h(\bar{r}_i - r^*) + |h_i|\, r^*.$$

Now suppose that $|h_i|$ is invested in a *short sale* ($h_i < 0$), this gross investment being equal to the price received for the stock. (The price received must be deposited in escrow, and in addition, an amount equal to margin requirements on the current price of the stock sold must be remitted or loaned to the actual owner of the securities borrowed to effect the short sale.) In computing the *return* on a short sale, we know that the short seller must pay to the person who lends him the stock any dividends which accrue while the stock is sold short (and hence borrowed), and his capital gain (or loss) is the negative of any price appreciation during this period. In addition, the short seller will receive interest at the riskless rate r^* on the sales price placed in escrow, and he may or may not *also* receive interest at the same rate on his cash remittance to the lender of the stock. To facilitate the formal analysis, we *assume* that *both interest components* are *always received* by the short seller, and that margin requirements are 100%. In this case, the short seller's *return* per dollar of his gross investment will be $(2r^* - \bar{r}_i)$, and if he invests $|h_i|$ in the short sale ($h_i < 0$), its contribution to his portfolio return will be:

$$(4b)\quad |h_i|\,(2r^* - \bar{r}_i) = h_i(\bar{r}_i - r^*) + |h_i|\, r^*.$$

Since the right-hand sides of (4*a*) and (4*b*) are identical, the total return per dollar invested in *any* stock mix can be written as:

$$(5)\quad \bar{r} = \Sigma_i\,[h_i(\bar{r}_i - r^*) + |h_i|\, r^*] = r^* + \Sigma_i h_i(\bar{r}_i - r^*)$$

because $\Sigma_i\,|h_i| = 1$ by the definition of $|h_i|$.

The expectation and variance of the return on any stock mix is consequently

$$(6a)\quad \bar{r} = r^* + \Sigma_i h_i(\bar{r}_i - r^*) = r^* + \Sigma_i h_i \bar{x}_i,$$

$$(6b)\quad \check{r} = \Sigma_{ij} h_i h_j \check{r}_{ij} = \Sigma_{ij} h_i h_j \check{x}_{ij}$$

where $\check{r}_{ij}$ represents the variance $\sigma_{rij}{}^2$ when $i = j$, and covariances when $i \neq j$. The notation has been further simplified in the right-hand expressions by defining:

$$(7)\quad \bar{x}_i = \bar{r}_i - r^*,$$

and making appropriate substitutions in the middle expressions. The quantity θ defined in (3*b*) can thus be written:

$$(8)\quad \theta = \frac{\bar{r} - r^*}{(\check{r})^{1/2}} = \frac{\bar{x}}{(\check{x})^{1/2}} = \frac{\Sigma_i h_i \bar{x}_i}{(\Sigma_{ij} h_i h_j \check{x}_{ij})^{1/2}}.$$

Since h_i may be either positive or negative, equation (6*a*) shows that a portfolio with $\bar{r} < r^*$ and hence with $\theta > 0$ exists if there is one or more stocks with $\bar{r}_i$ not exactly equal to r^*. We assume throughout the rest of the paper that such a portfolio exists.

Determination of the Optimal Stock Portfolio

As shown in the proof of the Separation Theorem above, the optimal stock portfolio is the one which maximizes θ as defined in equation (8). We, of course, wish to maximize this value subject to the constraint

$$(9)\quad \Sigma_i\,|h_i| = 1,$$

which follows from the definition of $|h_i|$. But we observe from equation (8) that θ is a *homogeneous function of order zero* in the h_i: the value of θ is *unchanged by any proportionate* change in all h_i. Our problem thus reduces to the simpler one of finding a vector of values yielding the *unconstrained* maximum of θ in equation (8), after which we may scale these initial solution values to satisfy the constraint.

The Optimum Portfolio When Short Sales are Permitted

We first examine the partial derivatives of (8) with respect to the h_i and find:

$$(10)\quad \frac{\partial\,\theta}{\partial h_i} = (\sigma_x)^{-1}\,[\bar{x}_i - \lambda(h_i\check{x}_{ii} + \Sigma_j h_j \check{x}_{ij})\,],$$

where,

$$(11)\quad \lambda = \bar{x}/\sigma_x{}^2 = \Sigma_i h_i \bar{x}_i / \Sigma_i \Sigma_j h_i h_j \check{x}_{ij}.$$

[22] In recent years, it has become increasingly common for the short seller to waive interest on his deposit with the lender of the security — in market parlance, for the borrowers of stock to obtain it "flat"— and when the demand for borrowing stock is large relative to the supply available for this purpose, the borrower may pay a cash premium to the lender of the stock. See Sidney M. Robbins, [18, pp. 58–59]. It will be noted that these practices reduce the expected return of short sales without changing the variance. The formal procedures developed below permit the identification of the appropriate stocks for short sale assuming the expected return is $(2r^* - \bar{r}_i)$. If these stocks were to be borrowed "flat" or a premium paid, it would be *simply necessary to iterate the solution after replacing* $(\bar{r}_i - r^*)$ in (4*b*) and (5) *for these stocks* with the value $(\bar{r}_i)$ — and *if*, in addition, a premium p_i is paid, the term $(\bar{r}_i + p_i)$ should be substituted (where $p_i \geqq 0$ is the premium (if any) per dollar of sales price of the stock to be paid to lender of the stock). With equal lending and borrowing rates, changes in margin requirements will not affect the calculations. (I am indebted to Prof. Schlaifer for suggesting the use of absolute values in analyzing short sales.)

The *necessary and sufficient conditions* on the *relative* values of the h_i for a stationary *and the unique (global) maximum*[23] are obtained by setting the derivatives in (10) equal to zero, which give the set of equations

(12) $z_i\check{x}_{ii} + \Sigma_j z_j \check{x}_{ij} = \bar{x}_i, \quad i = 1, 2, \ldots, m;$

where we write

(13) $z_i = \lambda h_i.$

It will be noted the set of equations (12)—which are identical to those Tobin derived by a different route[24]—are *linear* in the own-*variances*, pooled *covariances*, and *excess returns* of the respective securities; and since the covariance matrix $\check{\underline{x}}$ is positive definite and hence non-singular, this system of equations has a unique solution

(14) $z_i^0 = \Sigma_j \check{x}^{ij} \bar{x}_j$

where $\check{x}^{ij}$ represents the ij^{th} element of $(\check{\underline{x}})^{-1}$, the inverse of the covariance matrix. Using (13), (7), and (6*b*), this solution may also be written in terms of the primary variables of the problem in the form

(15) $h_i^0 = (\lambda^0)^{-1}\Sigma_j \check{r}^{ij}(\bar{r}_j - r^*), \quad$ all i.

Moreover, since (13) implies

(16) $\Sigma_i |z_i| = \lambda \Sigma_i |h_i|,$

λ^0 may readily be evaluated, after introducing the constraint (9) as

(17) $\Sigma_i |z_i^0| = \lambda^0 \Sigma_i |h_i^0| = \lambda^0$

The optimal *relative* investments z_i^0 can consequently be scaled to the optimal proportions of the stock portfolio h_i^0, by dividing each z_i^0 by the sum of their absolute values. A comparison of equations (16) and (11) shows further that:

(18) $\Sigma_i |z_i^0| = \lambda^0 = \bar{x}^0/\sigma_{x^0}^2;$

i.e. the sum of the absolute values of the z_i^0 yields, as a byproduct, the value of the ratio of the expected excess rate of return on the optimal portfolio to the variance of the return on this best portfolio.

It is also of interest to note that if we form the corresponding λ-ratio of the expected excess return to its variance for each ith stock, we have at the optimum:

(19) $h_i^0 = (\lambda_i/\lambda^0) - \Sigma_{j \neq i} h_j^0 \check{x}_{ij}/\check{x}_{ii}$ where
$\lambda_i = \bar{x}_i/\check{x}_{ii}.$

The optimal fraction of each security in the best portfolio is equal to the ratio of *its* λ_i to that of the entire portfolio, *less* the ratio of its pooled covariance with other securities to its own variance. Consequently, *if* the investor were to act on the assumption that all covariances were zero, he could pick his optimal portfolio mix very simply by determining the λ_i ratio of the expected excess return $\bar{x}_i = \bar{r}_i - r^*$ of each stock to its variance $\check{x}_{ii} = \check{r}_{ii}$, and setting each $h_i = \lambda_i/\Sigma\lambda_i$; for with no covariances,[25] $\Sigma\lambda_i = \lambda^0 = \bar{x}^0/\sigma_{x^0}^2$. With this simplifying assumption, the λ_i ratios of each stock suffice to determine the optimal mix by simple arithmetic;[26] in the more general case with non-zero covariances, a single set[27] of linear equations must be solved in the usual way, but no (linear or non-linear) programming is required and no more than one point on the "efficient frontier" need ever be computed, given the assumptions under which we are working.

The Optimum Portfolio When Short Sales are not Permitted

The exclusion of short sales does not complicate the above analysis *if* the investor is willing to act on an assumption of no correlations between the returns on different stocks. In this case, he finds his best portfolio of "long" holding by merely eliminating all securities whose λ_i-

[23]It is clear from a comparison of equations (8) and (11), showing that sgn θ = sgn λ, that only the vectors of h_i values corresponding to $\lambda > 0$ are relevant to the maximization of θ. Moreover, since θ as given in (8) and all its first partials shown in (10) are continuous functions of the h_i, it follows that when short sales are permitted, any maximum of θ must be a stationary value, and any stationary value is a maximum (rather than a minimum) when $\lambda > 0$ because θ is a convex function with a positive-definite quadratic form in its denominator. For the same reason, any maximum of θ is a unique (global) maximum.

[24]See Tobin, [21], equation (3.22), p. 83. Tobin had, however, formally required no short selling or borrowing, implying that this set of equations is valid under these constraints [so long as there is a single riskless asset (pp. 84–85)]; but the constraints were ignored in his derivation. We have shown that this set of equations *is* valid *when short sales* are properly included in the portfolio *and borrowing* is available in perfect markets in unlimited amounts. The alternative set of equilibrium conditions required when short sales are ruled out is given immediately below. The complications introduced by borrowing restrictions are examined in the final section of the paper.

[25] With no covariances, the set of equations (12) reduces to $\lambda h_i = \bar{x}_i/\check{x}_{ii} = \lambda_i$, and after summing over all $i = 1, 2 \ldots m$, and using the constraint (9), we have immediately that $|\lambda^0| = \Sigma_i |\lambda_i|$, and $\lambda^0 > 0$ for max θ (instead of min θ).

[26] Using a more restricted market setting, Hicks [6, p. 801] has also reached an equivalent result when covariances are zero (as he assumed throughout).

[27] See, however, footnote 22, above.

ratio is negative, and investing in the remaining issues in the proportions $h_i = \lambda_i/\Sigma\lambda_i$ in accordance with the preceding paragraph.

But in the more generally realistic cases when covariances are nonzero *and* short sales are not admitted, the solution of a single bilinear or quadratic programming problem is required to determine the optimal portfolio. (All other points on the "efficient frontier," of course, continue to be irrelevant so long as there is a riskless asset and a "perfect" borrowing market.) The optimal portfolio mix is now given by the set of h_i^0 which maximize θ in equation (8) subject to the constraint that all $h_i \geqq 0$. As before, the (further) constraint that the sum of the h_i be unity (equation 9) may be ignored in the initial solution for the *relative* values of the h_i [because θ in (8) is homogeneous of order zero]. To find this optimum, we form the Lagrangian function

$$(20) \quad \phi(\underline{h}, \underline{u}) = \theta + \Sigma_i u_i h_i$$

which is to be maximized subject to $h_i \geqq 0$ and $u_i \geqq 0$. Using (11), we have immediately

$$(21) \quad \frac{\partial\phi}{\partial h_i} \gtreqless 0 \leftrightarrow \bar{x}_i - \lambda(h_i \check{x}_{ii} + \Sigma_j h_j \check{x}_{ij}) + \alpha u_i \gtreqless 0.$$

As in the previous cases, we also must have $\lambda > 0$ for a maximum (rather than a minimum) of ϕ, and we shall write $z_i = \lambda h_i$ and $v_i = \alpha u_i$. The necssary and sufficient conditions for the vector of *relative* holdings z_i^0 which maximizes θ in (20) are consequently,[28] using the Kuhn-Tucker theorem [9],

$$(22a) \quad z_i^0 \check{x}_{ii} + \Sigma_j z_j^0 \check{x}_{ij} - v_i^0 = \bar{x}_i, \; i = 1, 2, \ldots m;$$

where

$$(22b\text{-}d) \quad z_i^0 \geqq 0, \; v_i^0 \geqq 0, \; z_i^0 v_i^0 = 0.$$

This system of equations can be expeditiously solved by the Wilson Simplicial Algorithm [23].

Now let m' denote the number of stocks with strictly positive holdings $z_i^\circ > 0$ in (22b), and renumber the entire set of stocks so that the subset satisfying this strict inequality [and, hence also, by (22d) $v_i^0 = 0$] are denoted 1, 2, . . . , m'. *Within this m' subset of stocks found to belong in the optimal portfolio with positive holdings,* we consequently have, using the constraint (19),

$$(17a) \quad \Sigma_{i=1}^{m'} z_i^0 = \lambda^0 \Sigma_{i=1}^{m'} h_i^0 = \lambda^0$$

so that the *fraction* of the optimal portfolio invested in the ith stock (where $i = 1, 2 \ldots m'$) is

$$(23) \quad h_i^0 = z_i^0/\lambda^0 = z_i^0/\Sigma_{i=1}^{m'} z_i^0 .$$

Once again, using (17a) and (11), the sum of the z_i^0 within this set of stocks held yields as a by-product the ratio of the expected excess rate of return on the optimal *portfolio* to the variance of the return on this best portfolio:

$$(18a) \quad \Sigma_{i=1}^{m'} z_i^0 = \lambda^0 = \bar{x}^0/\sigma^2_{x^0}.$$

Moreover, since $z_i^0 > 0$ in (22a and 22b) strictly implies $v_i^0 = 0$ by virtue of (22c), equation (22a) *for the subset of positively held stocks* $i = 1, 2 \ldots m'$ is formally identical to equation (12). We can, consequently, use these equations to bring out certain significant properties of the security portfolios which will be held by risk-averse investors trading in perfect markets.[29] *In the rest of this paper, all statements with respect to "other stocks" will refer to other stocks included within the portfolio.*

III Risk Premiums and Other Properties of Stocks Held Long or Short in Optimal Portfolios

Since the covariances between most pairs of stocks will be positive, it is clear from equation (19) that stocks held long ($h_i^0 > 0$) in a portfolio will generally be those whose expected

---

[28] Equation (22a–22d) can readily be shown to satisfy the six necessary and two further sufficient conditions of the Kuhn-Tucker theorem. Apart from the constraints $\underline{h} \geqq 0$ and $\underline{u} \geqq 0$ which are automatically satisfied by the computing algorithm [conditions (22b and 22c)] the four *necessary* conditions are:

1) $\left[\frac{\partial\phi}{\partial h_i}\right]^0 \leqq 0$. This condition is satisfied *as a strict equality* in our solutions by virtue of equation (22a) [See equation (21)]. This strict equality also shows that,

2) $h_i^0 \left[\frac{\partial\phi}{\partial h_i}\right]^0 = 0$, the first complementary slackness condition is also satisfied.

3) $\left[\frac{\partial\phi}{\partial u_i}\right]^0 \geqq 0$. This condition is satisfied because from equation (20), $\left[\frac{\partial\phi}{\partial u_i}\right]^0 = h_i^0 \geqq 0$ by virtue of equation (22b). This same equation shows that the second complementary slackness condition,

4) $u_i^0 \left[\frac{\partial\phi}{\partial u_i}\right]^0 = 0$, may be written $u_i^0 h_i^0 = 0$ which is also satisfied because of equation (22c) since $\alpha \neq 0$.

The two additional *sufficiency* conditions are of course satisfied because the variance-covariance matrix $\underline{\check{X}}$ is positive definite, making $\phi(\underline{h}, \underline{u}^0)$ a concave function on $\underline{h}$ and $\phi(\underline{h}^0, \underline{u})$ a convex function of $\underline{u}$.

[29] More precisely, the properties of portfolios when both the investors and the markets satisfy the conditions stated at the outset of section I or, alternatively, when investors satisfy Roy's premises as noted previously.

return is enough greater than the risk-free rate to offset the disutility, so to speak, of the contribution of their variance and pooled covariance to the risk of the entire portfolio. This much is standard doctrine. Positive covariances with other securities held long in the portfolio raise the minimum level of $\bar{x}_i > 0$ which will lead to the inclusion of the ith stock as a positive holding in the optimal portfolio. But equation (19) shows that stocks whose expected returns are *less* than the riskless rate (i.e. $\bar{x}_i < 0$ or $\bar{r}_i < r^*$) will *also* be held *long* ($h_i^0 > 0$) *provided* that *either* (*a*) they are *negatively correlated* in sufficient degree with other important stocks *held long* in the portfolio, or (*b*) that they are *positively correlated* in sufficient degree with other important stocks *held short* in the portfolio. The precise condition for $h_i^0 > 0$ when $\bar{x}_i < 0$ is that the weighted sum of the ith covariances be sufficiently negative to satisfy

(19*a*) $h_i^0 > 0 \leftrightarrow |\Sigma_{j\neq i} h_j^0 \check{x}_{ij}| > |\bar{x}_i/\lambda^0|$,

which follows from (19) since $\check{x}_{ii} > 0$.

Since our $\bar{x}_i$ is precisely what is usually called the "risk premium" in the literature, we have just shown that *the "risk premiums" required on risky securities* (i.e. those with σ_i and $\sigma_i^2 > 0$) for them *to be held long by optimizing risk-averse investors in perfect markets need not always be positive*, as generally presumed. *They will in fact be negative* under either of the conditions stated in (*a*) and (*b*) above, summarized in (19*a*). The explanation is, of course, that a long holding of a security which is negatively correlated with other long holdings tends to reduce the variance of the whole portfolio by offsetting some of the variance contributed by the other securities in the portfolio, and this "variance-offsetting" effect may dominate the security's own-variance and even a negative expected excess return $\bar{x}_i < 0$.

Positive correlations with other securities held short in the portfolio have a similar variance-offsetting effect.[30]

Correspondingly, it is apparent from (19) itself that any stock with *positive* excess returns or risk premiums ($\bar{x}_i > 0$) will be held *short* ($h_i^0 < 0$) in the portfolio *provided that either* (*a*) it is *positively correlated* in sufficient degree with other stocks *held long* in the portfolio, or (*b*) it is *negatively correlated* in sufficient degree with other stocks *held short* in the portfolio. *Positive* (negative) *risk premiums are neither a sufficient nor a necessary condition for a stock to be held long* (short).

Indifference Contours

Equation (12) (and the equivalent set (22*a*) restricted to stocks actually held in portfolios) also enables us to examine the *indifference contours* between expected excess returns, variances, or standard deviations and covariances of securities which will result in the *same fraction* h_i^0 of the investor's portfolio being held in a given security. The general presumption in the literature, as noted in our introduction,[31] is that the market values of risk assets are adjusted in perfect markets to maintain a *linear* relation between expected rates of return (our $\bar{r}_i = \bar{x}_i + r^*$) *and risk* as measured by the *standard deviation* of return σ_i on the security in question. This presumption probably arises from the fact that this relation *is* valid for trade offs *between* a riskless security *and* a single risk asset (or a *given mix* of risk assets to be held in fixed proportions). But it can *not* be validly attributed to indifferent trade offs *between* risk assets *within* optimizing risk-asset portfolios. In point of fact, it can easily be shown that there is a *strictly linear indifference contour* between the *expected return* $\bar{r}_i$ (or the expected excess return $\bar{x}_i$) *and the variance* σ_i^2 (*not* the standard deviation σ_i) of the individual security, and this linear function has very straightforward properties. The assumption made in this derivation that the covariances σ_{ij} with other securities are invariant is a more reasonable one than is perhaps readily apparent.[32] Subject to the acceptability

[30] Stocks with negative expected excess returns or "risk premiums" ($\bar{x}_i < 0$) will, of course, enter into portfolios only as short sales (provided these are permitted) when the inequality in (19*a*) is reversed, i.e.

$h_i^0 < 0 \leftrightarrow \Sigma_{j\neq i} h_j^0 \check{x}_{ij} + \bar{x}_i/\lambda^0 < 0$. When short sales are not permitted, and (19*a*) is not satisfied, stocks with $\bar{x}_i < 0$ simply do not appear in the portfolio at all.

[31] See footnote 3 for references and quotations.

[32] Fixed covariances are directly implied by the assumption that every pair of ith and j^{th} stocks are related by a one-common-factor model (e.g. the general state of the economy or the general level of the stock market), so that, letting $\bar{\mu}$ represent the general exogenous factor and $\bar{\omega}$ the random outcome of endogenous factors under management's control, we have

$$\bar{x}_i = a_i + b_i \bar{\mu} + \bar{\omega}_i$$
$$\bar{x}_j = a_j + b_j \bar{\mu} + \bar{\omega}_j$$

with $\bar{\mu}$, $\bar{\omega}_i$, and $\bar{\omega}_j$ mutually independent. This model implies

$$\sigma_i^2 = b_i^2 \sigma_\mu^2 + \sigma_\omega^2, \text{ and } \sigma_{ij} = b_i b_j \sigma_\mu^2,$$

so that if management, say, varies the part under its control, of this latter assumption, it follows that *risk classes of securities should be scaled in terms of variances* of returns rather than standard deviations (with the level of covariances reflected in the parameters of the linear function). The complexities involved when indifference contours are scaled on covariances or standard deviations are indicated below.

The conclusion that the indifference contour between $\bar{x}_i$ and the variance σ_i^2 is *linear* in the general case when all covariances σ_{ij} are held constant is established in the appendix, note II, by totally differentiating the equilibrium conditions (12) [or the equivalent set (22*a*) restricted to the m' stocks held in the portfolio]. But *all* pairs of values of $\bar{x}_i$ and σ_i^2 along the linear indifference coutour which holds h_i^0 fixed at some given level also rigorously imply that the proportionate mix of *all other* stocks in the portfolio is *also unchanged*. Consequently, we may proceed to derive other properties of this indifference contour by examining a simple "two security" portfolio. (The ith security is renumbered "1," and "all other" securities are called the second security.) If we then solve the equilibrium conditions[33] (12) in this two-stock case and hold $K = h_1^0/h_2^0$ constant, we have

(24) $K = h_1^0/h_2^0 = \text{constant} = (\bar{x}_1\sigma_2^2 - \bar{x}_2\sigma_{12})/(\bar{x}_2\sigma_1^2 - \bar{x}_1\sigma_{12})$

which leads to the desired explicit expression, using $\bar{r}_1 = \bar{x}_1 + r^*$,

(25) $\bar{r}_1 = r^* + W\sigma_{12} + WK\sigma_1^2,$

where

(25*a*) $W = \bar{x}_2/(\sigma_2^2 + K\sigma_{12}).$

Since[34] $WK = \lambda^0 h_1^0$ and $\lambda^0 > 0$, the *slope* of this indifference contour between $\bar{x}_1$ and σ_1^2 will always be positive when $h^0_1 > 0$ (as would be expected, because when σ_{12} is held constant, increased variance requires added return to justify any given positive holding[35]); but when the first stock is held short, its expected (or excess) return and its variance along the contour vary inversely (as they should since "shorts" profit from price declines). Moreover, if we regard σ_{12} as an exogenous "shift" parameter, the *constant term* (or intercept) of this indifference contour varies directly[36] with σ_{12}, and the slope of $\bar{x}_1$ on σ_1^2 varies inversely[37] with σ_{12} in the usual case, when $\bar{x}_2 > 0$.

Now note that (25) and (25*a*) can be written

(25*b*) $\bar{r}_1 = r^* + \bar{x}_2(\sigma_{12} + K\sigma_1^2)/(\sigma_2^2 + K\sigma_{12}),$

which clearly depicts a hyperbolic (rather than linear) indifference contour on σ_{12} if σ_1^2 is regarded as fixed, and a more complex function between $\bar{r}_1$ (or x_1) and the standard deviation σ_1, which may be written (using $\sigma_{12} = \sigma_1\sigma_2\rho$),

$$(25b') \quad \bar{x}_1 = \frac{\bar{x}_2 K\sigma_1^2\,[1 + \rho(K\sigma_1/\sigma_2)^{-1}]}{\sigma_2^2\,(1 + \rho K\sigma_1/\sigma_2)}$$

The *slope* of the indifference contour between $\bar{x}_1$ and σ_1 is a still more involved function, which may be written most simply as

$$(25c) \quad \frac{\partial\bar{x}_1}{\partial\sigma_1} = \frac{\bar{x}_2\,[2K\sigma_1\sigma_2^2 + (K^2\sigma_1^2\sigma_2 + \sigma_2^3)\rho]}{(\sigma_2^2 + K\sigma_{12})^2}$$

$$= 2K\sigma_1\bar{x}_2\frac{1+(\rho/2)\,[(K\sigma_1/\sigma_2) + (\sigma_2/K\sigma_1)].}{\sigma^2(1 + \rho K\sigma_1/\sigma_2)^2}$$

It is true, in the usual situation with $K > 0$, $\bar{x}_2 > 0$, and $\rho > 0$, that $\bar{x}_1$ $(= \bar{r}_1 - r^*)$ and $\partial x_1/\partial\sigma_1$ are necessarily positive as common doctrine presumes, *but* the complex non-linearity is evident even in this "normal case" restricted to two stocks — and the *positive risk premium* $\bar{x}_1$ *and positive slope* on σ_1, of course, *cannot be generalized*. For instance, in the admittedly less usual but important case with $\bar{x}_2 > 0$ and the intercorrelation $\rho < 0$, *both* $\bar{x}_1$ and $\partial\bar{x}_1/\partial\sigma_1$ are *alternatively negative and positive* over different ranges[38] of σ_1 for any fixed h_1^0 or $K > 0$.

$\bar{\omega}$ and σ_ω^2, the covariance will be unchanged. (This single-common-factor model is essentially the same as what Sharpe [20] calls the "diagonal" model.)

[33] The explicit solution is $z_1^0 = \lambda^0 h_1^0 = (\bar{x}_1\sigma_2^2 - \bar{x}_2\sigma_{12})/(\sigma_1^2\sigma_2^2 - \sigma_{12}^2)$; and $z_2^0 = \lambda^0 h_2^0 = (\bar{x}_2\sigma_1^2 - \bar{x}_1\sigma_{12})/(\sigma_1^2\sigma_1^2 - \sigma_{12}^2)$; where $\lambda^0 = z_1^0 + z_2^0$.

[34] Upon substituting (24) in (25) and using the preceding footnote, we have $W = \lambda^0 h_2^0 = z_2^0$, from which it follows that $WK = \lambda^0 h_2^0 h_1^0/h_2^0 = \lambda^0 h_1^0$.
As noted earlier, we have $\lambda^0 > 0$ (because the investor maximizes and does not minimize θ). [It may be noted that W is used instead of z_2^0 in (25) in order to incorporate the restriction on the indifference contours that K is constant, and thereby to obtain an expression (25*a*) which does not contain $\bar{x}_1$ and σ_1^2 (as does z_2^0 without the constraint of constant K).]

[35] Note that this is true whether the "other security" is held long or short.

[36] Let the constant term in (25) be $C = r^* + W\sigma_{12}$. Then

$$\frac{\partial C}{\partial\sigma_{12}} = \frac{(\sigma_2^2 + K\sigma_{12})\,\bar{x}_2 - \bar{x}_2\sigma_{12}K}{(\sigma_2^2 + K\sigma_{12})^2} = \frac{\bar{x}_2\,\sigma_2^2}{(\sigma_2^2 + K\sigma_{12})^2}$$

which has the same sign as $\bar{x}_2$, independent of the sign of K, σ_{12}, or $\bar{x}_1$.

[37] We have $\partial WK/\partial\sigma_{12} = -K^2\bar{x}_2/(\sigma_2^2 + K\sigma_{12})^2$, which has a sign opposite to that of $\bar{x}_2$.

[38] With $K > 0$, $\bar{x}_2 > 0$, and $\rho < 0$, we have from (25*b'*)

$\bar{x}_1 < 0$ if $0 < K\sigma_1/\sigma_2 < |\rho|$, and

$\bar{x}_1 > 0$ if $|\rho| < K\sigma_1/\sigma_2 < |\rho^{-1}|$.

On the other hand, from (25*c*) we have

Moreover, *in contrast* to the $\bar{x}_i - \sigma_1^2$ contour examined above, the pairs of values along the $\bar{x}_i - \sigma_i$ contour which hold h_i^0 constant do *not* imply an unchanged mix[39] of the other stocks in the optimizing portfolio when $m' > 2$; nor is λ^0 invariant along an $\bar{x}_i - \sigma_i$ contour, as it is along the $\bar{x}_1 - \sigma_1^2$ contour with covariances constant. For both reasons, the indifference contour between $\bar{x}_1$ and σ_1 for portfolios of $m' > 2$ stocks is very much more complex than for the two-stock case, whereas the "two-stock" contour (3) between $\bar{x}_1$ and σ_1^2 is exact for any number of stocks (when "all other" stocks are pooled in fixed proportions, as we have seen they can validly be). We should also observe that there does not seem to be an easy set of economically interesting assumptions which lead to *fixed correlations* as σ_1 varies (as assumed in deriving $\bar{x}_1 - \sigma_1$ indifference contours) in marked contrast to the quite interesting and plausible "single-factor" model (see footnote 32 above) which directly validates the assumption of fixed covariances used in deriving the $\bar{x}_1 - \sigma_1^2$ indifference contours.

In sum, we conclude that — however natural or plausible it may have seemed to relate risk premiums to standard deviations of return *within* portfolios of risk assets, and to scale risk classes of securities on this same basis — risk premiums can most simply *and* plausibly be related directly to *variances* of returns (with the level of covariances reflected in the parameters of the linear function). Since the principal function of the concept of "risk class" has been to delineate a required level of risk premium, we conclude further that risk classes should also be delineated in the same units (variances) if, indeed, the concept of risk class should be used at all.[40]

IV — Market Prices of Shares Implied by Shareholder Optimization in Purely Competitive Markets Under Idealized Uncertainty

Our analysis to this point has followed Tobin [21] and Markowitz [14] in assuming that current security prices are *exogenous data*, and that each investor acts on his own (doubtless unique) probability distribution over rates of return, *given* these market prices. I shall continue to make the same assumptions concerning markets and investors introduced in section I. In particular, it is assumed that security markets are purely competitive, transactions costs and taxes are zero, and *all* investors prefer a greater mean rate of return for a given variance and a lesser rate of return variance for any given mean return rate. But in this and the following section, I shall *assume* (1) that investors' joint probability distributions pertain to *dollar returns rather than rates* of return — the dollar return in the period being the sum of the cash dividend and the increase of market price during the period. Also, for simplicity, assume that (2) for *any* given set of market prices for all stocks, *all* investors assign *identical* sets of means, variances, and convariances to the joint distribution of these dollar returns (and hence for *any* set of prices, to the vector of means and the variance-covariances matrix of the rates of return $\tilde{r}_i$ of all stocks), and that all correlations between stocks are < 1.

This assumption of identical probability beliefs or judgments by all investors in the market restricts the applicability of the analysis of this and the following section to what I have elsewhere characterized as *idealized uncertainty* [10, pp. 246–247]. But however unrealistic this latter assumption may be, it does enable us to derive a set of (stable) equilibrium market prices — and an important theorem concerning the properties of these prices — which at least fully and explicitly reflect the presence of uncertainty *per se* (as distinct from the effects of diverse judgmental distributions among investors).

Note first that the assumption of identical probability judgments means that (1) *the same stock mix will be optimal for every investor* (although the actual dollar gross investment in this mix — and the ratio, w, of gross investment in this mix to his net investment balance — will vary from one investor to the next). It consequently follows that, when the market is in equilibrium, (2) the h_i^0 given by equation (15) or (12) can be interpreted as the ratio of the *aggregate market* value of the ith stock to the total aggregate market value of all stocks, and hence, (3) *all h_i will be strictly positive.*

In order to develop further results, define

$$\partial\bar{x}_1/\partial\sigma_1 < 0 \text{ if } 0 < K\sigma_1/\sigma_2 < |\rho^{-1}| - \sqrt{\rho^{-2} - 1},$$

and

$$\partial\bar{x}_1/\partial\sigma_1 > 0 \text{ if } |\rho^{-1}| - \sqrt{\rho^{-2} - 1} < K\sigma_1/\sigma_2 < |\rho^{-1}|.$$

[39] See appendix, note 11(*b*).

[40] However, see below, especially the "fifth" through "seventh" points enumerated near the end of Section V.

V_{0i} — the aggregate market value of the ith stock at time zero,
$\bar{R}_i$ — the aggregate return on the ith stock (the sum of aggregate cash dividends paid and appreciation in aggregate market value over the transaction period); and
$T \equiv \Sigma_i V_{0i}$, the aggregate market value of *all* stock in the market at time zero.

The original economic definitions of the variables in the portfolio optimization problem give

$$(26a) \quad h_i = V_{0i}/T,$$

$$(26b) \quad \bar{r}_i = \bar{R}_i/V_{0i},$$

$$(26c) \quad \bar{x}_i = \bar{r}_i - r^* = (\bar{R}_i - r^* V_{0i})/V_{0i},$$

$$(26d) \quad \check{x}_{ij} = \check{r}_{ij} = \check{R}_{ij}/V_{0i} V_{0j},$$

where $\check{R}_{ij}$ is the covariance of the aggregate dollar returns of the ith and j^{th} stocks (and $\check{R}_{ii}$ is the ith stock's aggregate return variance). The equilibrium conditions (12) may now be written

$$(12a) \quad \frac{\bar{R}_i - r^* V_i^0}{V_{0i}} = \lambda \frac{V_i^0}{T} \frac{\check{R}_{ii}}{(V_{0i})^2} + \lambda \Sigma_{j \neq i} \frac{V_{0i}}{T} \frac{\check{R}_{ij}}{V_{0i} V_{0j}},$$

which reduces to

$$(27) \quad \bar{R}_i - r^* V_{0i} = (\lambda/T)\,[\check{R}_{ii} + \Sigma_{j \neq i} \check{R}_{ij}] = (\lambda/T)\, \Sigma_j \check{R}_{ij}.$$

Now $\bar{R}_i - r^* V_{0i}$ represents the *expected* excess of the aggregate dollar return on the ith security over earnings at the riskless rate on its aggregate market value, and $\Sigma_j \check{R}_{ij}$ represents the aggregate risk (direct dollar return variance and total covariance) entailed in holding the stock. Equation (27) consequently establishes the following:

Theorem: Under Idealized Uncertainty, equilibrium in purely competitive markets of risk-averse investors requires that the values of all stocks will have adjusted themselves so that the *ratio* of the expected excess aggregate dollar returns of each stock to the aggregate dollar risk of holding the stock will be *the same for all* stocks (and equal to λ/T), when the risk of each stock is measured by the variance of its own dollar return and its combined covariance with that of all other stocks.

But we seek an explicit equation[41] for V_{0i}, and to this end we note that partial summation of equation (27) over *all other* stocks gives us

$$(28) \quad \Sigma_{k \neq i} (\bar{R}_k - r^* V_{0k}) = (\lambda/T) \Sigma_{k \neq i} \Sigma_j \check{R}_{kj}.$$

After dividing each side of (27) by the corresponding side of (28), and solving for V_{0i}, we then find that the aggregate market value of the ith stock is related to the concurrent market values of the *other* $(m-1)$ stocks by

$$(29) \quad V_{0i} = (\bar{R}_i - W_i)/r^*$$

where

$$(29a) \quad W_i = \gamma_i \Sigma_j \check{R}_{ij} = \gamma_i (\check{R}_{ii} + \Sigma_{j \neq i} \check{R}_{ij})$$

and

$$(29b) \quad \gamma_i = \frac{\Sigma_{k \neq i} (\bar{R}_k - r^* V_{0k})}{\Sigma_{k \neq i} \Sigma_j \check{R}_{kj}} = \frac{\Sigma_{k \neq i} (\bar{R}_k - r^* V_{0k})}{\Sigma_{k \neq i} \Sigma_{j \neq i} \check{R}_{kj} + \Sigma_{j \neq i} \check{R}_{ij}}.$$

Since (29b) *appears* to make the slope coefficient γ_i unique to each company, we must note immediately that dividing each side of (27) by its summation over *all* stocks shows that the aggregate market value of the ith stock is *also* related to the concurrent market values of *all* (m) stocks[42] by equation (29) when W_i is written as

$$(29c) \quad W_i = (\lambda/T)\, \Sigma_j \check{R}_{ij},$$

and

$$(29d) \quad \lambda/T = \frac{\Sigma_i (\bar{R}_i - r^* V_{0i})}{\Sigma_i \Sigma_j \check{R}_{ij}}.$$

But from equations (28) and (29*b*), we see that

$$(29e) \quad \gamma_i = \gamma_j = \lambda/T,$$

a *common value* for *all companies in the market.* The values of W_i given by (29*a*) and (29*c*) are consequently *identical*, and *the subscripts on* γ *should henceforth be ignored.*

In words, equations (29) establish the following further

Theorem: Under Idealized Uncertainty, in purely competitive markets of risk-averse investors,

A) the total market value of any stock in equilibrium is equal to the *capitalization* at the *risk-free interest rate* r^* of the *certainty equivalent* $(R_i - W_i)$ of its uncertain *aggregate dollar return* $\bar{R}_i$;

B) *the difference* W_i between the expected value $\bar{R}_i$ of these returns and their certainty

[41] I do not simply rearrange equation (27) at this point since (λ/T) includes V_{0i} as one of its terms (see equation (29*d*) below).

[42] Alternatively, equations (29) and (29*c*) follow directly from (27), and (29*d*) may be established by substituting (26*a–d*) in (11).

equivalent is *proportional* for *each* company to *its aggregate risk* represented by the *sum* ($\Sigma_j \check{R}_{ij}$) of the *variance* of these returns and their total covariance with those of all other stocks; and

C) the factor of proportionality ($\gamma = \lambda/T$) is the *same* for *all* companies in the market.

Certain corrolaries are immediately apparent:

Corrolary I: Market values of securities are related to standard deviations of dollar returns by way of variances and covariances, *not directly* and *not linearly*.

Corrolary II: The aggregate risk ($\Sigma_j R_{ij}$) of the ith stock which is directly relevant to its aggregate market value V_{0i} is simply *its contribution* to the aggregate *variance* of the dollar returns (for *all* holders together) of *all* stocks (which is $\Sigma_i \Sigma_j \check{R}_{ij}$).

Corrolary III: The *ratio* $(\bar{R}_i - W_i)/\bar{R}_i$ of the *certainty-equivalent* of aggregate dollar returns to their expected value is, in general, *different for each* i^{th} company when the market is in equilibrium;[43] but for all companies, this certainty-equivalent to expected-dollar-return ratio is the *same linear function* $\{1-\gamma[\Sigma_j\check{R}_{ij}/\bar{R}_i]\}$ of total dollar risk ($\Sigma_j \check{R}_{ij}$) attributable to the ith stock deflated by its expected dollar return $\bar{R}_i$.

Several further implications also follow immediately. First, note that equation (29) can be written

$$(29') \quad V_{0i} = (\bar{R}_i - W_i)/r^* = (V_{0i} + \bar{R}_i - W_i)/(1 + r^*) = (\bar{H}_i - W_i)/(1 + r^*).$$

Since $\bar{R}_i$ was defined as the sum of the aggregate cash dividend and increase in value in the equity during the period, the *sum* $V_{0i} + \bar{R}_i$ is equal to the expected value of the sum (denoted $\bar{H}_i$) of the cash dividend and end-of-period aggregate market value of the equity, and the elements of the covariance matrix $\underline{\underline{H}}$ are identical to those in $\underline{\underline{\check{R}}}$. *All* equations (29) can consequently be validly rewritten substituting H for R throughout [and $(1 + r^*)$ for r^*], *thus explicitly determining all current values* V_{0i} *directly by the joint probability distributions over the end-of-period realizations*[44] $\tilde{H}_i$. (The value of W_i, incidentally, is not affected by these substitutions.) Our assumption that investors hold joint probability distributions over dollar returns $\tilde{R}_i$ is consequently *equivalent* to an assumption that they hold distributions over end-of-period realizations, and *our analysis applies equally under either assumption.*

Moreover, after the indicated substitutions, equation (29′) *shows that the current aggregate value of any equity is equal to the certainty-equivalent of the sum of its prospective cash receipts (to shareholders) and total market value at the end of the period, discounted at the riskless rate* r^*. Similarly, by an extension of the same lines of analysis, the certainty equivalent of the cash dividend and market value at the end of the first period clearly may be regarded as the then-present-values using riskless discount rates of the certainty-equivalents of random receipts still further in the future. *The analysis thus justifies viewing market values as riskless-rate present values of certainty-equivalents of random future receipts*, where certainty-equivalents are related to expected values by way of variances and covariances weighted by adjustment factors γ_{it}, which may or may not be the same for each future period t.

Still another implication of equation (29) is of a more negative character. Those who like (or hope) to find a "risk" discount rate k_r with which to discount expected values under uncertainty will find from (29) that, using a subscript i for the individual firm

$$(29'') \quad V_{0i} = \frac{\bar{R}_i}{k_{ri}} = \frac{\bar{R}_i}{r^*(1 - W_i/\bar{R}_i)^{-1}} = \frac{\bar{R}_i}{r^*(1 - \gamma\Sigma_j\check{R}_{ij}/\bar{R}_i)^{-1}}$$

so that

$$(30) \quad k_{ri} = r^*(1 - \gamma\Sigma_j\check{R}_{ij}/\bar{R}_i)^{-1}.$$

It is apparent that (*i*) the *appropriate "risk" discount rate* k_{ri} *is unique to each individual company in a competitive equilibrium* (because of the first half of corrolary III above); (*ii*) that efforts to derive it complicate rather than simplify the analysis, since (*iii*) it is a *derived* rather than a primary variable; and that (*iv*) it explicitly involves all the elements required for the determination of V_{0i} itself, and, (*v*) does so in a more

[43] From equations (27), (29), (29*a*), and (29*e*), this statement is true for all pairs of stocks having different aggregate market values, $V_{0i} \neq V_{0j}$.

[44] Because we are assuming only "idealized" uncertainty, the distribution of these end-of-period realizations will be independent of judgments regarding the dividend receipt and end-of-period market value separately. See Lintner [10] and Modigliani-Miller [16].

complex and non-linear fashion.[45] Having established these points, the rest of our analysis returns to the more direct and simpler relation of equation (29).

V — Corporate Capital Budgeting Under Idealized Uncertainty

Capital budgeting decisions within a corporation affect both the expected value and variances — and hence, the certainty-equivalents — of its prospective aggregate dollar returns to its owners. When the requisite conditions are satisfied, equation (29) thus provides a normative criterion for these decisions, derived from a competitive equilibrium in the securities market.

In developing these important implications of the results of the last section, I of course maintain the assumptions of idealized uncertainty in purely competitive markets of risk-averse investors with identical probability distributions, and I continue to assume, for simplicity, that there are no transactions costs or taxes. The identity of probability distributions over outcomes now covers corporate management as well as investors, and includes potential corporate investments in the capital budget as well as assets currently held by the company. Every corporate management, *ex ante*, assigns probability zero to default on its debt, and all investors also treat corporate debt as a riskless asset. I thus extend the riskless investment (or borrowing) alternative from individual investors to corporations. Each company can invest any amount of its capital budget in a perfectly safe security (savings deposit or certificate of deposit) at the riskless rate r^*, or it may borrow unlimited amounts at the *same* rate in the current or any future period.[46] I also assume that the investment opportunities available to the company in any time period are regarded as independent of the size and composition of the capital budget in any other time period.[47] I also assume there is no limited liability to corporate stock, nor any institutional or legal restriction on the investment purview of any investor, and that the riskless rate r^* is expected by everyone to remain constant over time.

Note that this set of assumptions is sufficient to validate the famous (taxless) Propositions I and II of Modigliani and Miller [15]. In particular, under these severely idealized conditions, for any given size and composition of corporate assets (investments), investors will be indifferent to the *financing* decisions of the company. Subject to these conditions, we can, consequently, derive valid decision rules for capital budgets which do not explicitly depend upon concurrent financing decisions. Moreover, these conditions make the present values of the cash flows *to any* company from its real (and financial) assets and operations equal to the total market value of investors' *claims* to these flows, i.e., to the sum of the aggregate market value of its common (and preferred) stock outstanding and its borrowings (debt)[48]. They also make any change in shareholders claims equal to the change in the present values of flows (before interest deductions) to the company less any change in debt service. The *changes* in the market value of the equity V_{0i} induced by capital budgeting decisions will consequently be precisely equal to

$$(31)\qquad \Delta V_{0i} = \Delta\,(\bar{R}_i - W_i)/(1 + r^*) \\ = \Delta\,(\bar{H}_i - W_i)/(1 + r^*),$$

where $\Delta\bar{H}_i$ is the net change induced in the *expected* present value at the end of the first period of the cash inflows (net of interest charges) to the ith company attributable to *its assets*[49] when all present values are computed at the riskless rate r^*.

These relationships may be further simplified in a useful way by making three additional assumptions: that (i) the *aggregate market value*

[45] It may also be noted that even when *covariances* between stocks are constant, the elasticity of k_{rt} with respect to the variance $\breve{R}_{ii}$ (and *a fortiori*) to the standard deviation of return) is a unique (to the company) multiple of a hyperbolic relation of a variance-expected-return ratio:

$$(30a)\qquad \frac{\breve{R}_{ii}}{k_{rt}}\;\frac{\partial k_{rt}}{\partial \breve{R}_{ii}} = \frac{\gamma\,\breve{R}_{ii}/\bar{R}_i}{1 - \gamma\,(\Sigma_{j\neq i}\breve{R}_{ij}/\bar{R}_i) - \gamma\,\breve{R}_{ii}/\bar{R}_i}\,.$$

[46] The effects of removing the latter assumption are considered briefly in the final section.

[47] This simplifying assumption specifies a (stochastic) comparative static framework which rules out the complications introduced by making investor expectations of future growth in a company's investment opportunities conditional on current investment decisions. I examine the latter complications in other papers [11], and [12].

[48] See Lintner [10]. Note that in [10, especially p. 265, top 1st column] I argued that additional assumptions were needed to validate the "entity theory" under uncertainty — the last sentence of the preceding paragraph, and the stipulation that corporate bonds are riskless meet the requirement. See, however, Modigliani-Miller [16].

[49] By definition, ΔH_i is the change in the expected sum of dividend payment and market value of the equity at the end of the period. This is made equal to the statement in the text by the assumptions under which we are operating.

of *all other* stocks—and (*ii*) the *covariances* $\check{R}_{ij}$ *with* all other *stocks* are invariant to the capital budgeting decisions of the ith company; while (*iii*) the (optimal) *portfolio* of risk assets is not an "inferior good" (in the classic Slutsky-Hicks sense) *vis a vis* riskless assets. The reasonableness of (*iii*) is obvious (especially in the context of a universe of risk-averse investors!), and given (*iii*), assumption (*i*) is a convenience which only involves ignoring (generally small) second-order feedback effects (which will not reverse signs); while the plausibility of (*ii*) as a good working first approximation was indicated above (footnote 32).[50]

In this context, we now show that capital budgeting decisions by the ith firm will raise the aggregate market value of its equity V_{0i}—and hence by common agreement be in the interest of its shareholders—so long as the induced change in expected dollar return is greater than the product of the market price γ of risk and the induced variance of dollar returns, i.e.,

$$(32)\quad \Delta\bar{R}_i - \gamma\Delta\check{R}_{ii} = \Delta\bar{H}_i - \gamma\Delta\check{H}_{ii} > 0.$$

This assertion (or theorem) can be proved as follows. The total differential of (29) is

$$(29f)\quad r^*\Delta V_{0i} - \Delta\bar{R}_i + \gamma\Delta\check{R}_{ii} + (\Sigma_j\check{R}_{ij})\Delta\gamma = 0$$

so that under the above assumptions

$$(29g)\quad \Delta\bar{R}_i \geqq \gamma\,\Delta\check{R}_{ii} + (\Sigma_j\check{R}_{ij})d\,\gamma \rightarrow \Delta V_{0i} \geqq 0 \rightarrow \Delta T \geqq 0.$$

But using (29*e*) and (29*d*), we have

$$(29h)\quad \Delta\gamma = (\Delta\bar{R}_i - \gamma\Delta\check{R}_{ii})/\Sigma_i\Sigma_j\check{R}_{ij}$$

so that

$$(29i)\quad \Delta\bar{R}_i = \gamma\Delta\check{R}_{ii} \rightarrow \Delta\gamma = 0 \rightarrow \Delta V_{0i} = 0 \rightarrow \Delta T = 0,$$

and the first equality in (29*i*) defines the relevant *indifference function*.[51] Moreover, using (29*h*) and the fact that $\Sigma_j\check{R}_{ij} < \Sigma_i\Sigma_j\check{R}_{ij}$, we have from (29*g*):

$$(29j)\quad \Delta\bar{R}_i \geqq \gamma\Delta\bar{R}_{ii} \rightarrow \Delta\bar{R}_i \geqq \gamma\Delta\check{R}_{ii} + (\Sigma_j\check{R}_{ij})\,\Delta\gamma,$$

and consequently

$$(29k)\quad \Delta\bar{R}_i \geqq \gamma\,\Delta\check{R}_{ii} \rightarrow \Delta V_{0i} \geqq 0 \rightarrow \Delta T \geqq 0,$$

from which (32) follows immediately.

In order to explore the implications of (32) further, it will now be convenient to consider in more detail the capital budgeting decisions of a company whose *existing assets* have a present value computed at the rate r^* (and measured at the *end* of the first period) of $\tilde{H}_0^{(1)}$, a random variable with expected value $\bar{H}_0^{(1)}$ and variance $\check{H}_{00}$.

The company may be provisionally holding any fraction of $\bar{H}_0$ in savings deposits or CD's yielding r^*, and it may use any such funds (or borrow unlimited amounts at the *same* rate) to make new "real" investments. We assume that the company has available a set of new projects $1, 2 \ldots j \ldots n$ which respectively involve *current* investment outlays of $H_j^{(0)}$, and which have present values of the relevant incremental cash flows (valued at the *end* of the first period) of $\tilde{H}_j^{(1)}$. Since any diversion (or borrowing) of funds to invest in any project involves an opportunity cost of $r^*H_j^{(0)}$, we also have the "excess" dollar end-of-period present value return

$$(33)\quad \tilde{X}_j^{(1)} = \tilde{H}_j^{(1)} - r^*H_{j0}^{(0)}.$$

Finally, we shall denote the $(n+1)$ 'th order covariance matrix (including the existing assets $\tilde{H}_0$) by $\underline{\underline{\check{H}}}$ or $\underline{\underline{\check{X}}}$ whose corresponding elements $\check{H}_{jk} = \check{X}_{jk}$.

Determination of the Optimal Corporate Capital-Budget-Portfolio

In this simplified context, it is entirely reasonable to expect that the corporation will seek to maximize the left side of equation[52] (32) as its capital budgeting criterion. At first blush, a very complex *integer* quadratic-programming solution would seem to be required, but fortunately we can break the problem down inductively and find a valid formulation which can be solved in essentially the same manner as an individual investor's portfolio decision.

[50] It is, however, necessary in general to redefine the variables in terms of *dollar* returns (rather than rates of return), but this seems equally reasonable.

[51] Note that this indifference function can also be derived by substituting equations (26*a–d*) directly into that found in section III above (equation 6*b*) in appendix Note II or equation (25) in the text) for the relevant case where covariances are invariant.

[52] Under our assumption that stock portfolios are not inferior goods, sgn ΔT = s gn $[\Delta\bar{R}_i - \gamma\Delta\check{R}_{ii}]$ so that (although generally small in terms of percentages) the induced change in aggregate values of all stocks will reinforce the induced change in the *relative* value of the ith stock; the fact that $\Delta\gamma$ also has the same sign introduces a countervailing feedback, but as shown above [note especially (29*g*)], this latter effect is of second order and cannot reverse the sign of the criterion we use. In view of the overwhelming informational requirements of determining the maximum of a fully inclusive criterion function which allowed formula induced adjustments external to the firm, *and* the fact our criterion is a monotone rising function of this ultimate ideal, the position in the text follows.

First, we note that if a single project j is added to an existing body of assets $\bar{H}_0{}^0$, we have

$$(34a)\quad \Delta\bar{R}_i - \gamma\Delta\check{\underline{R}}_{ii} = \bar{H}_j{}^{(1)} - r^*H_j{}^0 - \gamma\,[H_{jj} + 2\check{H}_{j0}] = X_j{}^{(1)} - \gamma\,[\check{X}_{jj} + 2\check{X}_{j0}].$$

Now suppose a project k is also added. The total change from j and k *together* is

$$(34b)\quad (\Delta R_i - \gamma\Delta\check{R}_{ii}) = X_j{}^{(1)} + X_k{}^{(1)} - \gamma\,[\check{X}_{jj} + \check{X}_{kk} + 2\check{X}_{j0} + 2\check{X}_{k0} + 2\check{X}_{jk}],$$

while the *increment* due to adding *k with j already in the budget* is

$$(34c)\quad (\Delta R_i - \gamma\check{R}_{ii}) = \bar{X}_k{}^{(1)} - \gamma\,[\check{X}_{kk} + 2\check{X}_{k0} + 2\check{X}_{ik}].$$

Given the goal of maximizing the left side of (32), the kth project should be added to the budget (already provisionally containing j) *if and only if* the *right* side of (34c) is > 0— and if this condition is satisfied, the same test expression written for j, given inclusion of k, will show whether j should stay in. Equation (34c) appropriately generalized to any number of projects, is thus a *necessary condition* to be satisfied by *each project in an optimal budget*, given the inclusion of all other projects simultaneously satisfying this condition.

The unstructured iterative or search procedure suggested by our two-project development can obviously be short-circuited by programming methods, and the integer aspect of the programming (in this situation) can conveniently be by-passed by assuming that the company may accept all or any fractional part a_j, $0 \leqq a_j \leqq 1$, of any project (since it turns out that all a_j in the final solution will take on *only* limiting values). Finally, thanks to this latter fact, the objective of maximizing the left side of (32) is equivalent[53] to maximizing

$$(32')\quad Z = H_0{}^{(1)} + \Sigma_j a_j \bar{H}_j{}^{(1)} - r^*\Sigma_j a_j H_j{}^{(0)} - \gamma[\Sigma_j a_j \check{H}_{jj} + 2\Sigma_j a_j \check{H}_{j0} + 2\Sigma_{j\neq k\neq 0} a_j a_k \check{H}_{jk}] = \bar{H}_0{}^{(1)} + \Sigma_j a_j \bar{X}_j - \gamma[\Sigma_j a_j \check{X}_{jj} + 2\Sigma_j a_j \check{X}_{j0} + 2\Sigma_{j\neq k\neq 0} a_j a_k \check{X}_{jk}],$$

subject to the constraints that $0 \leqq a_j \leqq 1$ for all a_j, $j = 1, 2 \ldots n$. Not only will all a_j be binary variables in the solution, but the generalized form of the necessary condition (34c) will be given by the solution [see equation (37) below].

In order to maximize Z in (32′) subject to the constraints on a_j, we let $q_j = 1 - a_j$ for convenience, and form the Lagrangian function

$$(35)\quad \psi\,(a, \mu, \eta) = Z + \Sigma_j \mu_j a_j + \Sigma_j \eta_j q_j$$

which is to be maximized subject to $a_j \geqq 0$, $q_j \geqq 0$, $\mu_j \geqq 0$, and $\eta_j \geqq 0$, where μ_j and η_j are the Lagrangian multipliers associated with the respective constraints $a_j \geqq 0$ and $q_j \geqq 0$. Using (33), we have immediately

$$(35')\quad \frac{\partial\psi}{\partial a_j} \gtrless 0 \leftrightarrow \bar{X}_j - \gamma\,[a_j\check{X}_{jj} + 2\Sigma_j a_j \check{X}_{j0} + 2\Sigma_{k\neq 0}\, a_k \check{X}_{jk}] + \mu_j - \eta_j \gtrless 0.$$

Using the Kuhn-Tucker Theorem [9], the necessary and sufficient conditions for the optimal vector of investments $a_j{}^0$ which maximize ψ in (35) are consequently[54]

$$(36a)\quad \gamma\,[a_j{}^0\,\check{X}_{jj} + 2a_j{}^0\check{X}_{j0} + 2\Sigma_{k\neq j\neq 0}\, a_k{}^0\check{X}_{jk}] - \mu_j{}^0 + \eta_j{}^0 = \bar{X}_j$$

when

$$(36b, c, d, e)\quad a_j{}^0 \geqq 0,\ q_j{}^0 \geqq 0,\ \mu_j{}^0 \geqq 0,\ \eta_j{}^0 \geqq 0$$

and

$$(36f, g)\quad \mu_j{}^0\, a_j{}^0 = 0,\ \eta_j{}^0\, q_j{}^0 = 0,$$

where

$$j = 1, 2 \ldots n$$

in each set (36a) – (36g).

Once again, these equations can be readily solved by the Wilson Simplicial Algorithm [23] on modern computing equipment. It may be observed that this formulation in terms of independent investment projects can readily be generalized to cover mutually exclusive, contingent, and compound projects[55] with no difficulty. It is also apparent that the absence of a financing constraint (due principally to our assumption

[53] For the reason given, the maximum of (32′) is the same as it would be if (32′) had been written in the more natural way using $a_i{}^2$ instead of a_j as the coefficient of $\check{H}_{jj}$; the use of a_j is required to make the form of (35′) and (37) satisfy the requirement of (34c).

[54] The proof that the indicated solution satisfies the Kuhn-Tucker conditions with respect to the variables $a_j{}^0$ and $\mu_j{}^0$ is identical to that given above footnote 28 upon the substitution of $\underline{X}$ for $\underline{x}$, a_j for h_i, and μ_j for u_i, and need not be repeated. The two additional *necessary* conditions are

(3′) $\left[\dfrac{\partial\psi}{\partial\eta_j}\right]^0 \geqq 0$, which is satisfied, since from (35) we have $\left[\dfrac{\partial\psi}{\partial\eta_j}\right]^0 = q_j{}^0 \geqq 0$ by virtue of (36c); and this latter relation shows that the corresponding complementary slackness condition,

(4′) $\mu^0{}_j \left[\dfrac{\partial\psi}{\partial\mu_j}\right]^0 = 0$, may be written $\mu_j{}^0\, q_j{}^0 = 0$, and is therefore satisfied because of (36g).

All three *sufficiency* conditions are also satisfied because the variance-covariance matrix $\underline{X}$ is positive definite, making $\psi\,(\underline{a}, \underline{u}^0, \underline{n}^0)$ a concave function on $\underline{a}$ and $\psi\,(\underline{a}, \underline{u}^0, \underline{n}^0)$ a convex function on both $\underline{u}$ and $\underline{n}$.

[55] See Weingartner [22], 11 and 32–34.

that new riskless debt is available in unlimited amounts at a fixed rate r^*) insures that all projects will either be accepted or rejected *in toto*. All a_j^0 will be either 0 or 1, and the troublesome problems associated with fractional projects or recourse to integer (non-linear) programming do not arise.

Consider now the set of *accepted* projects, and denote this subset with asterisks. We then have all $a_{j*}^0 = a_{k*}^0 = 1$; the corresponding $\mu_{j*}^0 = \mu_{k*}^0 = 0$; and for any project j^*, the corresponding $\eta_{j*}^0 > 0$ (i.e. *strictly positive*),[56] and the number η_{j*}^0 is the "dual evaluator" or "shadow price" registering the *net gain* to the company *and* its shareholders of accepting the project. Rewriting the corresponding equation from (36*a*), we have[57]

$$(37)\qquad \eta_{j*}^0 = \bar{H}_{j*}^{(1)} - r^* H_{j*}^{(0)} - \gamma\,[\check{H}_{j*j*} + 2\check{H}_{j*0} + 2\Sigma_{k*\neq j*\neq 0}\check{H}_{j*k*}] > 0.$$

Several important features and implications of these results should be emphasized. First of all, note that we have shown that *even* when uncertainty is admitted in only this highly simplified way, and when any effect of changes in capital budgets on the *covariances* between returns on different companies' *stocks* is ignored, the minimum expected return (in dollars of expected present value $\bar{H}_{j*}^{(1)}$) required to justify the allocation of funds to a given risky project costing a given sum $H_{j*}^{(0)}$ is an increasing function of each of the following factors: (*i*) the risk-free rate of return r^*; (*ii*) the "market price of dollar risk", γ; (*iii*) the variance $\check{H}_{j*j*}$ in the project's own present value return; (*iv*) the project's aggregate present value return-*co*variance $\check{H}_{j*0}$ with assets already held by the company, and (*v*) its total covariance $\Sigma_{k*\neq j*\neq 0}\check{H}_{j*k*}$ with other projects concurrently included in the capital budget.

Second, it follows from this analysis that, if uncertainty is recognized to be an important fact of life, and risk-aversion is a significant property of relevant utility functions, appropriate *risk-variables* must be introduced *explicitly* into the analytical framework used in analysis, and that these risk-variables will be *essential components* of any optimal decision rules developed. Important insights can be, and have been, derived from "certainty" models, including some *qualitative* notions of the *conditional* effects of changes in availability of funds due to fund-suppliers' reactions to uncertainty,[58] but such models ignore the decision-maker's problem of optimizing *his* investment decisions in the face of the stochastic character of the outcomes among which *he* must choose.

Third, it is clear that *stochastic considerations are a primary source of interdependencies among projects*, and these must *also enter explicitly* into optimal decision rules. In particular, note that, although own-variances are necessarily positive and subtracted in equation (37), the net gain n_{j*}^0 may still be positive and justify acceptance *even if* the expected end-of-period "excess" present-value return $(\check{X}_{j*}^{(1)} = \bar{H}_{j*}^{(1)} - r^* H_{j*}^{(0)})$ is negative[59]— so long as its total present-value-covariances $(\check{H}_{j*0} + \Sigma_{k*\neq j*\neq 0}\check{H}_{j*k*})$ are also negative and sufficiently large. *Sufficiently risk-reducing investments rationally belong in corporate capital budgets even at the expense of lowering expected present value returns* — an important (and realistic) feature of rational capital budgeting procedure not covered (nor even implied) in traditional analyses.

Fourth, note that, as would by now be expected, for any fixed r^* and γ, the net gain from a project is a *linear* function of its (present value) *variance* and *covariances* with existing company assets and concurrent projects. Standard deviations are not involved except as a component of (co)variances.

Fifth, the fact that the risk of a project involves all the elements in the bracketed term in (37), including covariances with other concurrent projects, indicates that in practice it will often be extremely difficult, if not impossible, to classify *projects* into respectively homogeneous "risk classes." The practice is convenient (and desirable where it does not introduce significant bias) but our analysis shows it is *not essential*, and the considerations which follow show it to be *a*

[56] We are of course here ignoring the very exceptional and coincidental case in which $\eta_{j*}^0 = 0$ which implies that a_{j*}^0 is indeterminate in the range $0 \leqq a_{0j} \leqq 1$, the company being *totally indifferent* whether or not all (or any part) of a project is undertaken.

[57] We use $\check{H}_{j*k*}$ to denote elements the original covariance matrix $\check{\underline{H}}$ *after* all rows and columns associated with rejected projects have been removed.

[58] See Weingartner [22] and works there cited. Weingartner would of course agree with the conclusion stated here, see pp. 193–194.

[59] Indeed, in extreme cases, a project should be accepted even if the expected end-of-period present value $\bar{H}_{j*}^{(1)}$ is less than cost $H_{j*}^{(0)}$, provided negative correlations with existing assets and other concurrent investments are sufficiently strong and negative.

dangerous expedient which is positively misleading as generally employed in the literature.

Sixth, it must be emphasized that — following the requirements of the market equilibrium conditions (29) from which equations (36), (37), and (38) were derived — *all means and (co)variances of present values have been calculated using the riskless rate* r^*. In this connection, recall the non-linear effect on present values of varying the discount rate used in their computation. Also remember the further facts that (*i*) the means and variances of the distributions of present values computed at different discount rates do not vary in proportion to each other when different discount rates are applied to the same set of future stochastic cash flow data, and that (*ii*) the changes induced in the means and variances of the present values of different projects having different patterns and durations of future cash flows will also differ greatly as discount rates are altered. From these considerations alone, it necessarily follows that *there can be no single "risk discount rate"* to use in computing present values for the purpose of deciding on the acceptance or rejection of *different individual projects* out of a subset of projects *even if all projects in the subset have the same degree of "risk."*[60] The same conclusion follows *a fortiori* among projects with different risks.

Seventh, the preceding considerations, again *a fortiori*, insure that *even if all new projects have the same degree of "risk" as existing assets, the "cost of capital"* (as defined for uncertainty *anywhere* in the literature) *is not the appropriate discount rate to use in accept-reject decisions on individual projects* for capital budgeting.[61] This is true whether the "cost of capital" is to be used as a "hurdle rate" (which the "expected return" must exceed) *or* as a discount rate in obtaining present values of net cash inflows and outflows.

Perhaps at this point the reader should be reminded of the rather heroic set of simplifying assumptions which were made at the beginning of this section. One consequence of the unreality of these assumptions is, clearly, that the results are not being presented as directly applicable to practical decisions at this stage. Too many factors that matter very significantly have been left out (or assumed away). But the very simplicity of the assumptions has enabled us to develop rigorous proofs of the above propositions which do differ substantially from current treatments of "capital budgeting under uncertainty." A little reflection should convince the reader that *all the above conclusions will still hold under more realistic (complex) conditions.*

Since we have shown that selection of individual projects to go in a capital budget under uncertainty by means of "risk-discount" rates (or by the so-called "cost of capital") is fundamentally in error, we should probably note that the decision criteria given by the solutions of equation (36) [and the acceptance condition (37)] — which directly involve the means and variances of present values computed at the riskless rate — do have a valid counterpart in the form of a "required expected rate of return." Specifically, if we let $[\Sigma\check{H}_{j*}]$ represent the entire bracket in equation (37), and divide through by the original cost of the project $H_{j*}^{(0)}$, we have

$$(38) \quad \bar{H}_{j*}^{(1)}/H_{j*}^{(0)} = r_{j*} > r^* + \gamma\,[\Sigma\check{H}_{j*}]/H_{j*}^{(0)}.$$

Now the ratio of the expected *end-of-period* present value $\bar{H}_{j*}^{(1)}$ to the initial cost $H_{j*}^{(0)}$ — i.e. the left side of (38), which we write r_{j*} — is precisely (the expected value of) what Lutz called the *net* short term marginal efficiency of the investment [13 p. 159]. We can thus say that the *minimum acceptable expected rate of return* on a project is a (positively sloped) linear function of the ratio of the project's *aggregate incremental* present-value-variance-covariance $(\Sigma\check{H}_{j*})$ to its cost $H_{j*}^{(0)}$. The slope coefficient is still the "market price of dollar risk", γ, and the intercept is the risk-free rate r^*. (It will be observed that our "accept-reject" rule for individual projects under uncertainty thus reduces to Lutz' rule under certainty — as it should — since with certainty the right-hand ratio term is zero.) To

[60] Note, as a corollary, it also follows that *even if* the world were simple enough that a single "as if" risk-discount rate could in principle be found, the same considerations insure that *there can be no simple function relating the appropriate "risk-discount" rate to the riskless rate* r^* *and "degree of risk," however measured.* But especially in this context, it must be emphasized that a single risk discount rate *would produce non-optimal choices* among projects *even if* (*i*) all projects could be assigned to meaningful risk-classes, *unless* it were also true that (*ii*) all projects had the same (actual) time-pattern of net cash flows and the same life (which is a condition having probability measure zero under uncertainty!).

[61] Note particularly that, even though we are operating under assumptions which validate Modigliani and Miller's propositions I and II, and the form of finance is *not* relevant to the choice of projects, we nevertheless cannot accept their use of their ρ_k — their cost of capital — as the relevant discount rate.

avoid misunderstanding and misuse of this relation, however, several further observations must be emphasized.

a) Equation (38) — like equation (37) from which it was derived — states a necessary condition of the (Kuhn-Tucker) optimum with respect to the projects selected. It may validly be *used to choose* the desirable projects out of the larger set of *possible* projects *if the covariances among potential projects* $\check{H}_{j \neq k \neq 0}$ *are all zero.*[62] *Otherwise, a programming solution of equation set* (36) *is required*[63] *to find which subset of projects* H_{j*} satisfy either (37) or (38), essentially because the total variance of any project $[\Sigma\check{H}_j]$ is dependent on *which* other projects are *concurrently included* in the budget.

b) Although the risk-free rate r^* enters equation (38) *explicitly* only as the intercept [or constant in the linear (in)equation form], it must be emphasized again that it *also enters implicitly as the discount rate used in computing the means and variances of all present values which appear* in the (in)equation. *In consequence,* (*i*) *any shift* in the value of r^* *changes every term* in the function. (*ii*) The changes in $\bar{H}^{(1)}{}_{j*}$ and $\Sigma\check{H}_{j*}$ are *non-linear and non-proportional* to each other.[64] Since (*iii*) any shift in the value of r^* changes *every covariance* in equation (36*a*) *non-proportionately,* (*iv*) the *optimal subset of projects* j^* is *not invariant* to a change in the risk-free rate r^*. Therefore (*v*), *in principal, any shift in the value of* r^* *requires a new programming solution of the entire set of equations* (36).

c) Even for a predetermined and fixed r^*, and even with respect only to *included* projects, the condition expressed in (38) is rigorously *valid only under the full set* of simplifying assumptions stated at the beginning of this section. In addition, the programming solution of equation (36), and its derivative property (38), *simultaneously determines both the optimal composition and the optimal size* of the capital budget *only under this full set* of simplifying assumptions. Indeed, even if the twin assumptions of a fixed riskless rate r^* and of formally unlimited borrowing opportunities at this rate are retained[65], *but* other assumptions are (realistically) generalized — specifically to permit expected returns on new investments at any time to depend in part on investments made in prior periods, and to make the "entity value" in part a function of the finance mix used — *then* the (set of) programming solutions merely determines the optimal *mix or composition* of the capital budget *conditional* on each possible aggregate budget size and risk.[66] Given the resulting "investment opportunity function" — which is the three-dimensional Markowitz-type envelope of efficient sets of projects — the optimal capital budget size and risk can be determined directly by market criteria (as developed in [11] and [12])[67] but will depend explicitly on concurrent financing decisions (e.g. retentions and leverage).[68]

VI — Some Implications of More Relaxed Assumptions

We have come a fairly long way under a progressively larger set of restrictive assumptions. The purpose of the exercise has not been to provide results *directly* applicable to practical decisions at this stage — too much (other than uncertainty *per se*) that matters greatly in prac-

[62] Note that covariances $\check{H}_{j0}$ with *existing* assets need not be zero since they are independent of other projects and may be combined with the own-variance $\check{H}_{jj}$.

[63] In strict theory, an iterative *exhaustive* search over *all possible* combinations *could* obviate the programming procedure, but the number of combinations would be very large in practical problems, and economy dictates programming methods.

[64] This statement is true *even if* the set of projects j^* were invariant to a change in r^* which in general will not be the case, as noted in the following text statement.

[65] If these assumptions are not retained, the position *and* composition of the investment opportunity function (defined immediately below in the text) are themselves dependent on the relevant discount rate, for the reasons given in the "sixth" point above and the preceding paragraph. (See also Lutz [13 p. 160].) Optimization then requires the solution of a much different and more complex set of (in)equations, simultaneously encompassing finance-mix *and* investment mix.

[66] This stage of the analysis corresponds, in the standard "theory of the firm," to the determination of the optimal mix of factors for each possible scale.

[67] I should note here, however, that on the basis of the above analysis, the correct *marginal expected* rate of return for the investment opportunity function should be the value of r_j^* [See left side equation (38) above] for the marginally included project at each budget size, i.e. the ratio of end-of-period present value computed at the riskless rate r^* to the project cost — rather than the different rate (generally used by other authors) stated in [12, p. 54 top]. Correspondingly, the relevant *average* expected return is the same ratio computed for the budget as a whole. Correspondingly, the relevant *variance* is the variance of this ratio. None of the subsequent analysis or results of [12] are affected by this corrected specification of the inputs to the investment opportunity function.

[68] This latter solution determines the optimal point on the investment opportunity function at which to operate. The optimal *mix* of projects to include in the capital budget is that which corresponds to the optimal point on the investment opportunity function.

tice has been assumed away — but rather to develop rigorously some of the fundamental implications of uncertainty *as such* for an important class of decisions about which there has been much confusion in the theoretical literature. The more negative conclusions reached — such as, for instance, the serious distortions inherently involved in the prevalent use of a "risk-discount rate" or a "company-risk-class" "cost-of-capital" for project selection in capital budgeting — clearly will hold under more general conditions, as will the primary role under uncertainty of the *risk-free rate* (whether used to calculate *distributions* of present values *or* to form *present values of certainty-equivalents*). But others of our more affirmative results, and especially the particular equations developed, are just as clearly inherently conditional on the simplifying assumptions which have been made. While it would be out of place to undertake any exhaustive inventory here, we should nevertheless note the impact of relaxing certain key assumptions upon some of these other conclusions.

The particular formulas in sections II–V depend *inter-alia* on the *Separation Theorem* and each investor's consequent preference for the stock *mix* which maximizes θ. Recall that in proving the Separation Theorem in section I we assumed that the investor could borrow unlimited amounts at the rate r^* equal to the rate on savings deposits. Four alternatives to this assumption may be considered briefly. (1) *Borrowing Limits*: The Theorem (and the subsequent development) holds *provided* that the margin requirements turn out *not* to be binding; but if the investor's utility function is such that, given the portfolio which maximizes θ, he prefers a w greater than is permitted, *then* the Theorem does not hold and the utility function must be used explicitly to determine the optimal stock mix.[69] (2) *Borrowing rate r^{**} greater* than "lending rate" r^*: (*a*) If the max θ using r^* implies a $w < 1$, the theorem holds in original form; (*b*) if the max θ using r^* implies $w > 1$ *and* (upon recomputation) the max θ using r^{**} in equations (3*b*), (7) and (8) implies $w > 1$, the theorem also holds but r^{**} (rather than r^*) *must be used* in sections II–V; (*c*) if max θ using r^* implies $w > 1$ *and* max θ using r^{**} implies $w < 1$, *then* there will be no borrowing *and* the utility function must be used explicitly to determine the optimal stock mix.[70] (3) *Borrowing rate an increasing function of leverage* $(w - 1)$: The theorem still holds under condition (2*a*) above, but if max θ using r^* implies $w > 1$ *then* the *optimal mix and the optimal financing* must *be determined simultaneously using the utility function* explicitly.[71] (4) The latter conclusion also follows immediately *if the borrowing rate is not independent of the stock mix.*

The *qualitative* conclusions of sections II and III hold even if the Separation Theorem does not, but the formulas would be much more complex. Similarly, the stock market equilibrium in section IV — and the parameters used for capital budgeting decisions in section V — will be altered if different investors in the market are affected differently by the "real world" considerations in the preceding paragraph (because of different utility functions, or probability assessments), or by differential tax rates. Note also that even if all our original assumptions through section IV are accepted for investors, the results in section V would have to be modified to allow for all real world complications in the cost and availability of debt and the tax treatment of debt interest versus other operating income. Finally, although explicitly ruled out in section V, it must be recalled that "limited liability," legal or other institutional restrictions or premiums, or the presence of "market risk" (as distinct from default risk) on corporate debt, *are sufficient both* to make the optimal *project mix* in the capital budget *conditional* on the finance mix (notably retentions and leverage), *and* the finance mix itself *also* something to be optimized.

Obviously, the need for further work on all these topics is great. The present paper will have succeeded in its essential purpose if it has rigorously pushed back the frontiers of theoretical understanding, and opened the doors to more fruitful theoretical and applied work.

[69] See appendix, note III.

[70] See appendix, note IV.

[71] See appendix, note V.

APPENDIX

Note I — Alternative Proof of Separation Theorem and Its Corrolaries

In this note, I present an alternative proof of the *Separation Theorem* and its corrolaries using utility functions explicity. Some readers may prefer this form, since it follows traditional theory more closely.

Let $\bar{y}$ and σ_y be the expected value and variance of the rate of return on any asset mixture and A_0 be the amount of the investor's total net investment. Given the assumptions regarding the market and the investor, stated in the text, the investor will seek to maximize the expected utility of a function which can be written in general form as

$$(1') \quad E[U(A_0\bar{y}, A_0\sigma_y)] = \bar{U}(A_0\bar{y}, A_0\sigma_y),$$

subject to his investment opportunities characterized by the risk-free rate r^*, at which he can invest in savings deposits or borrow any amount he desires, and by the set of all stock mixes available to him, each of which in turn is represented by a pair of values $(\bar{r}, \sigma_r)$. Our assumptions establish the following properties[72] of the utility function in (1′):

$$(1a') \quad \begin{cases} \partial\bar{U}/\partial\bar{y} = A_0\bar{U}_1 > 0; \ \partial\bar{U}/\partial\sigma_y = A_0\bar{U}_2 < 0; \\ \left.\dfrac{d\bar{y}}{d\sigma_y}\right|_{\bar{U}} = -\bar{U}_2/\bar{U}_1 > 0; \ \left.\dfrac{d^2\bar{y}}{d\sigma_y{}^2}\right|_{\bar{U}} > 0. \end{cases}$$

Also, with the assumptions we have made,[73] all available stock mixes will lie *in a finite region* all parts of which are strictly to the right of the vertical axis in the σ_r, r plane since all available mixes will have positive variance. The boundary of this region will be a closed curve[74] and the region is convex.[75] Moreover, since $\bar{U}_1 > 0$ and $\bar{U}_2 < 0$ in (1a′), all mixes within this region are dominated by those whose (σ_r, r) values lie on the part of the boundary associated with values of $\bar{r} > 0$, *and* for which changes in σ_r and $\bar{r}$ are positively associated. This is Markowitz' Efficient Set or "E-V" Frontier. We may write its equation[76] as

$$(2') \quad \bar{r} = f(\sigma_r), \ f'(\sigma_r) > 0, \ f''(\sigma_r) < 0.$$

Substituting (2′) in (2) and (3) in the text, we find the first order conditions for the maximization of (1) subject to (2), (3), and (2′) to be given by the equalities in

$$(3a') \quad \partial\bar{U}/\partial w = \bar{U}_1(\bar{r} - r^*) + \bar{U}_2\sigma_r \geqq 0.$$

$$(3b') \quad \partial\bar{U}/\partial\sigma_r = \bar{U}_1 w f'(\sigma) + \bar{U}_2 w \geqq 0.$$

which immediately reduce to the two equations [using (3*a*) from the text]

$$(4') \quad \theta = -\bar{U}_2/\bar{U}_1 = f'(\sigma_r).$$

Second order conditions for a maximum are satisfied because of the concavity of (1′) and (2′). The separation theorem follows immediately from (4′) when we note that the equation of the first and third members $\theta = f'(\sigma)$ *is precisely the condition for the maximization*[77] *of* θ, since

$$(5a') \quad \frac{\partial\theta}{\partial\sigma_r} = \frac{\sigma_r[f'(\sigma_r)] - [\bar{r} - r^*]}{\sigma_r{}^2} = \frac{f'(\sigma_r) - \theta}{\sigma_r}$$

$$(5b') \quad \frac{\partial^2\theta}{\partial(\sigma_r)^2} = \frac{\sigma_r f''(\sigma_r) + [f'(\sigma_r) - \theta] - [f'(\sigma_r) - \theta]}{\sigma_r{}^2}$$

$$= f''(\sigma_r)/\sigma_r < 0 \text{ for all } \sigma_r > 0.$$

A necessary condition for the maximization of (1′) is consequently the maximization of θ (as asserted), which is independent of w. The value of $(-\bar{U}_2/\bar{U}_1)$, however, directly depends on w (for *any* given value of θ), and a second necessary condition for the maximization of $\bar{U}$ is that w be adjusted to bring this value $(-\bar{U}_2/\bar{U}_1)$ into equality with θ, thereby satisfying the usual tangency condition between utility contours and the market opportunity function (3) in the text. These two necessary conditions are also *sufficient* because of the concavity of (1′) and the positive-definite property of the matrix of risk-investment opportunities. Q.E.D.

Note II

a) Indifference Contours Between x_i and σ^2_i When all σ_{ij} are Constant

The conclusion that the indifference contour between $\bar{x}_i$ and the variance $\sigma_i{}^2$ is *linear* in the general case when all covariances σ_{ij} are held constant can best be established by totally differentiating the equilibrium conditions (12) in the text [or the equivalent set (22*a*) restricted to the m' stocks held in the portfolio] which yields the set of equations

[72] For formal proof of these properties, see Tobin, [21], pp. 72–77.

[73] Specifically, that the amount invested in any stock in any stock mix is infinitely divisible, that all expected returns on individual stocks are finite, that all variances are positive and finite, and that the variance-covariance matrix is positive-definite.

[74] Markowitz [14] has shown that, in general, this closed curve will be made up of successive hyperbolic segments which are strictly tangent at points of overlap.

[75] Harry Markowitz, [14], chapter VII. The shape of the boundary follows from the fact that the point corresponding to any mix (in positive proportions summing to one) of any two points on the boundary lies to the left of the straight line joining those two points; and all points on and within the boundary belong to the set of available $(\sigma_r, \bar{r})$ pairs because any such point corresponds to an appropriate combination in positive proportions of at least one pair of points on the boundary.

[76] Note that the stated conditions on *the* derivatives in (2′) hold even in the exceptional cases of discontinuity. Markowitz [14], p. 153.

[77] This conclusion clearly holds even in the exceptional cases (noted in the preceding footnote) in which the derivatives of $r = f(\sigma_r)$ are not continuous. Equation (3*a*′) will hold as an exact equality because of the continuity of the utility function, giving $\theta = -\bar{U}_2/\bar{U}_1$. By equation (3*b*′), expected utility $\bar{U}$ increases with σ_r for all $f'(\sigma) \geqq -\bar{U}_2/\bar{U}_1 = \theta$, and the max σ_r consistent with $f'(\sigma) \geqq \theta$ maximizes θ by equation (5*a*′).

$$\lambda^0 \sigma_1^2 dh_1^0 + \lambda^0 \sigma_{12} dh_2^0 + \ldots . + \lambda^0 \sigma_{1i} dh_i^0 + \ldots . + \lambda^0 \sigma_{1m'} dh_{m'}^0 + \frac{\bar{x}_1}{\lambda^0} d\lambda^0 = 0$$

$$\vdots$$

$$(6) \quad \lambda^0 \sigma_{i1}^0 dh_1^0 + \lambda^0 \sigma_{i2} dh_2^0 + \ldots . + \lambda^0 \sigma_i^2 dh_i^0 + \ldots . + \lambda^0 \sigma_{im'} dh_{m'}^0 + \frac{\bar{x}_i}{\lambda^0} d\lambda^0 = dx_i - \lambda^0 h_i d\sigma_i^2$$

$$\vdots$$

$$\lambda^0 \sigma_{m'1} dh_1^0 + \lambda^0 \sigma_{m'2} dh_2^0 + \ldots . + \lambda^0 \sigma_{m'i} dh_i^0 + \ldots . + \lambda^0 \sigma^2_{m'} dh_{m'}^0 + \frac{\bar{x}_{m'}}{\lambda^0} d\lambda^0 = 0$$

$$dh_1^0 + dh_2^0 + \ldots . + dh_i^0 + \ldots . + dh_{m'}^0 = 0$$

Denoting the coefficient matrix on the left by $\underline{\underline{H}}$, and the i, j^{th} element of its inverse by H^{ij}, we have by Cramer's rule,

$$(6a') \quad dh_i^0 = (d\bar{x}_i - \lambda^0 h_i^0 \, d\sigma_i^2) \, H^{ii}.$$

Since $\underline{\underline{H}}$ is non-singular, h_i^0 will be constant along an indifference contour if and only if

$$(6b') \quad d\bar{x}_i = \lambda^0 h_i^0 \, d\sigma_i^2.$$

The indifference contour is strictly linear because the slope coefficient $\lambda^0 h_i^0$ is invariant to the absolute levels of $\bar{x}_i$ and σ_i^2 when h_i^0 is constant, as may be seen by noting that

$$(6c') \quad d\lambda_0^0 = (d\bar{x}_i - \lambda^0 h_i^0 \, d\sigma_i^2) \, H^{i\lambda 0}$$

so that

$$(6d') \quad dh_i^0 = 0 \longrightarrow d\lambda_i^0 = 0,$$

when only $\bar{x}_i$ and σ_i^2 are varied. Moreover, any pair of changes $d\bar{x}_i$ and $d\sigma_i^2$ which hold $dh_i^0 = 0$ by (6a' and b') imply *no change* in the relative holding h_j^0 of *any other* security, since $dh_j^0 = (d\bar{x}_i - \lambda^0 h_i^0 \, d\sigma_i^2) \, H^{ij} = 0$ for all $j \neq i$ when $dh_i^0 = 0$. Consequently, *all* pairs of values of $\bar{x}_i$ and σ_i^2 along the linear indifference contour which holds h_i^0 fixed at some given level rigorously imply that the proportionate mix of *all other* stocks in the portfolio is *also unchanged* — as was also to be shown.

b) Indifference Contours Between x_i and σ_i When ρ Constant

If the equilibrium conditions (12) are differentiated totally to determine the indifference contours between $\bar{x}_i$ and σ_i, the left-hand side of equations (6′) above will be unaffected, but the right side will be changed as follows: In the ith equation

$$d\bar{x}_i - \lambda^0 [2h_i^0 \sigma_i - \Sigma_{j \neq i} h_j^0 \sigma_j \rho_{ij}] \, d\sigma_i = d\bar{x}_i - \lambda^0 (h_i^0 \sigma_i - \bar{x}_i / \sigma_i) d\sigma_i$$

replaces $d\bar{x}_i - \lambda^0 h_i^0 d\sigma_i^2$; the last equation is unchanged; and in all other equations $-\lambda^0 h_i^0 \sigma_j \rho_{ij} d\sigma_i$ replaces 0. We then have

$$(7a') \quad dh_i^0 = [d\bar{x}_i - \lambda^0 (h_i^0 \sigma_i - \bar{x}_i/\sigma_i) \, d\sigma_i] \, H^{ii} - \lambda^0 h_i^0 \, \Sigma_{j \neq i} \sigma_j \, \rho_{ij} \, H^{ij} \, d\sigma_i;$$

$$(7b') \quad dh_j^0 = [d\bar{x}_i - \lambda^0 (h_i^0 \sigma_i - \bar{x}_i/\sigma_i) \, d\sigma_i] \, H^{ij} - \lambda^0 h_i^0 \, \Sigma_{K \neq i} \, \sigma_K \, \rho_{iK} \, H^{jK} \, d\sigma_i;$$

$$(7c') \quad d\lambda^0 = [d\bar{x}_i - \lambda^0 (h_i^0 \sigma_i - \bar{x}_i/\sigma_i) \, d\sigma_i] \, H^{i\lambda 0} - \lambda^0 h_i \, \Sigma_{K \neq i} \, \sigma_K \, \rho_{iK} \, H^{\lambda 0 K} \, d\sigma_i.$$

Clearly, in this case, $dh_i^0 = 0$ does *not* imply $dh_j^0 = 0$, *nor* does it imply $d\lambda^0 = 0$.

Note III — Borrowing Limits Effective

In principle, in this case the investor must compute all the Markowitz efficient boundary segment joining M (which maximizes θ in figure 1) to the point N corresponding to the greatest attainable $\bar{r}$. Given the fixed margin w, he must then project all points on this original (unlevered) efficient set (see equation 2′ above) to determine the new (levered) efficient set of (σ_y, $\bar{y}$) pairs attainable by using equations (2a, b) in the text; and he will then choose the (σ_y, $\bar{y}$) pair from this latter set which maximizes utility. With concave utility functions this optimum (σ_y, $\bar{y}$) pair will satisfy the standard optimizing tangency conditions between the (recomputed) efficient set and the utility function. The situation is illustrated in figure 2.

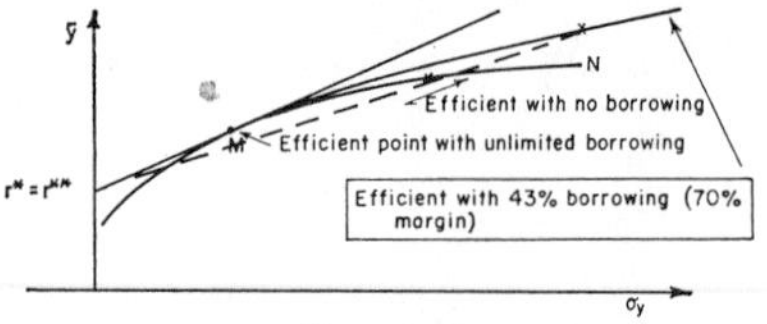

FIGURE II

Note IV — Borrowing Rate r^{**} is Higher than Lending Rate r^*

The conclusions stated in the text are obvious from the graph of this case (which incidentally is *formally* identical to Hirschleifer's treatment of the same case under certainty in [7].)

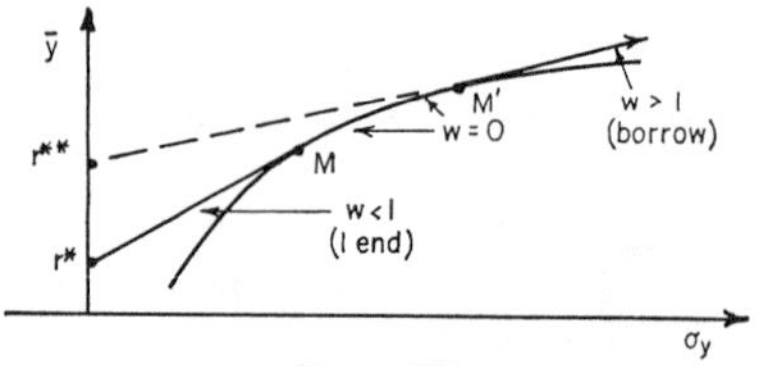

FIGURE III

The optimum depends uniquely upon the utility function if it is tangent to the efficient set with no borrowing in the range MM'.

Note V — Borrowing Rate is Dependent on Leverage

With $r^{**} = g(w)$, $g'(w) > 0$, and when the optimum $w > 1$ so that borrowing is undertaken, θ itself from equation (3) in the text becomes a function of w, which we will write $\theta\,(w)$. The optimizing equations, corresponding to (3′a, b) above in note I, then become

(6a′) $\partial \bar{U} / \partial w = \bar{U}_1 [(r - r^{**}) - wg'(w)] + \bar{U}_2 \sigma_r \geqq 0$

(6b′) $\partial \bar{U} / \partial \sigma_r = \bar{U}_1 wf'(\sigma) + \bar{U}_2 w \geqq 0$

which reduce to the two equations

(7′) $\theta(w) - wg'(w)/\sigma_r = -\bar{U}_2/\bar{U}_1 = f'(\sigma).$

The equation of the first and third members $\theta(w) - wg'(w)/\sigma_r = f'(\sigma)$ is *no longer* equal to the *maximization* of θ itself, *nor* is the solution of this equation *independent* of w which is required for the validity of the Separation Theorem. It follows that the selection of the optimal stock mix (indexed by θ) and of w *simultaneously depend upon the parameters of the utility function* (and, with normal distribution, *also* upon its *form*). Q.E.D.

[1] ARROW, KENNETH, "Comment on the Portfolio Approach to the Demand for Money and Other Assets," *The Review of Economics and Statistics, Supplement*, XLV (Feb., 1963), 24–27.

[2] BAUMOL, WILLIAM J., "An Expected Gain-Confidence Limit Criterion for Portfolio Selection," *Management Science*, X (Oct., 1963), 174–82.

[3a] DORFMAN, ROBERT, "Basic Economic and Technologic Concepts" In Arthur Maass, et. al., *Design of Water-Resource Systems* (Cambridge, Harvard University Press, 1962).

[3b] FARRAR, DONALD E., *The Investment Decision Under Uncertainty* (Englewood Cliffs, N.J., Prentice-Hall, 1962)

[4] FISHER, IRVING, *The Theory of Interest* (New York, 1930).

[5] GORDON, MYRON J., *The Investment, Financing and Valuation of the Corporation* (Homewood, Illinois: Richard D. Irwin, 1962).

[6] HICKS, J. R., "Liquidity", *The Economic Journal*, LXXII (Dec., 1962).

[7] HIRSCHLEIFER, JACK, "On the Theory of Optimal Investment Decision," *Journal of Political Economy*, LXVI (Aug., 1958).

[8] HIRSCHLEIFER, JACK, "Risk, the Discount Rate, and Investment Decisions," *American Economic Review*, LI (May, 1961).

[9] KUHN, H. W., and A. W. TUCKER, "Nonlinear Programming" in J. Neyman ed., *Proceedings of the Second Berkeley Symposium on Mathematical Statistics and Probability* (Berkeley: University of California Press, 1951), 481–492.

[10] LINTNER, JOHN, "Dividends, Earnings, Leverage, Stock Prices and the Supply of Capital to Corporations," *Review of Economics and Statistics*, XLIV (Aug., 1962).

[11] LINTNER, JOHN, "The Cost of Capital and Optimal Financing of Corporate Growth," *The Journal of Finance*, XVIII (May, 1963).

[12] LINTNER, JOHN, "Optimal Dividends and Corporate Growth Under Uncertainty," *The Quarterly Journal of Economics*, LXXVIII (Feb., 1964).

[13] LUTZ, FREDERICK and VERA, *The Theory of Investment of the Firm* (Princeton, 1951).

[14] MARKOWITZ, HARRY, *Portfolio Selection* (New York, 1959).

[15] MODIGLIANI, FRANCO and MILLER, MERTON, "The Cost of Capital, Corporation Finance and the Theory of Investment," *American Economic Review*, XLVIII (June, 1958).

[16] MODIGLIANI, FRANCO and MILLER, MERTON, "Dividend Policy, Growth and the Valuation of Shares," *Journal of Business*, XXXIV (Oct., 1961).

[17] PRATT, JOHN W., "Risk Aversion in the Small and in the Large," *Econometrica*, XXXIII (Jan.–April, 1964), 122–136.

[18] ROBBINS, SIDNEY M., *Managing Securities* (Boston: Houghton Mifflin Company, 1954).

[19] ROY, A. D., "Safety First and the Holding of Assets," *Econometrica*, XX (July, 1952), 431–449.

[20] SHARPE, WILLIAM F., "A Simplified Model for Portfolio Analysis," *Management Science*, IX (Jan., 1963), 277–293.

[21] TOBIN, JAMES, "Liquidity Preference as Behavior Toward Risk," *Review of Economic Studies*, XXVI, (Feb., 1958), 65–86.

[22] WEINGARTNER, H. MARTIN, *Mathematical Programming and the Analysis of Capital Budgeting Problems* (Englewood Cliffs, New Jersey: Prentice-Hall, 1963).

[23] WILSON, ROBERT B., *A Simplicial Algorithm for Concave Programming* (unpublished D.B.A. thesis, Harvard Business School, 1963).

Separation in Portfolio Analysis

R. G. Vickson

UNIVERSITY OF WATERLOO

I. Introduction

Most modern treatments of portfolio analysis are founded on the expected utility theory of von Neumann and Morgenstern [1]. When there is a single risky asset, expected utility maximization leads to a straightforward theory of demand for risk. In this case such questions as the demand for the risky asset as a function of wealth, of interest rate, and of taxation policies—in short, the microeconomics of risk—are all reasonably simple to analyze. This pleasant state of affairs generally breaks down if many additional risky assets are introduced into the problem. In the latter case the complexities of multidimensional nonlinear programming resist not only one's ability to calculate the optimal portfolio, but also one's ability to conceptualize the problem effectively. It is therefore important to know if the problem can be recast as a new problem involving only one or two "composite" assets, for if this is true the problem becomes effectively one or two dimensional. For example, if the optimal portfolio as a function of wealth leaves the risky asset proportions unchanged, the entire set of risky assets is effectively combined into a single mutual fund. The many-asset case is thus reduced to a single-asset case. In some circumstances, such a reduction to a single mutual fund might not be possible, but reduction to two distinct mutual funds might be. The many-asset case is thus reduced to a two-asset case—a significant reduction in problem size. A portfolio problem which collapses in this manner down to a problem involving one or two mutual funds is said to exhibit the *separation* property. Such problem behavior clearly has many advantages for both qualitative and quantitative purposes.

In a recent paper, Cass and Stiglitz [2] have derived restrictions on the forms of utility functions which lead to separation of the portfolio problem for general markets. They find that only a very small class of utility functions allows general separation, these being the functions $u(\cdot)$ such that $u'(w) = (a+bw)^c$ or $u'(w) = ae^{bw}$. It is well known that general (increasing, concave) utility functions lead to separation in the case of very special markets, having joint normal returns, in the presence of a risk-free asset which can be borrowed or lent without limit at a common interest rate (Tobin [3]; Lintner [4], reprinted in Part II, Chapter 3 of this volume). It is perhaps less well known that this is true also for markets with stable Paretian returns.[1] The main contribution of Cass and Stiglitz has been to prove that, for completely general markets, only very special utility functions can exhibit separation. Cass and Stiglitz gave results only for the case wherein borrowing and short sales were permitted, so that nonnegativity constraints on investments could be ignored. In this paper such constraints are included and are shown not to change the class of utility functions which yield separation. The extra constraints have the effect of forcing a distinction between global and local separation (Hakansson [6]) but otherwise lead to no significant changes in the theory. The utility functions leading to local versus global separation are determined and are found to agree with those of Hakansson. The entire development follows closely that of Cass and Stiglitz [2], but many of the proofs have been shortened or left as exercises for the reader.

II. The Separation Property

It is assumed throughout that the investor has a utility function for wealth $u(\cdot)$ with domain $(\mathbf{u}, \infty)$, where $-\infty \leqq \mathbf{u} \leqq 0$. It will also be assumed that u is twice continuously and piecewise three times continuously differentiable on $(\mathbf{u}, \infty)$, that u is strictly increasing $(u' > 0)$ on some interval $(0, L) \subset (\mathbf{u}, \infty)$, and that u is strictly concave $(u'' < 0)$ on $(\mathbf{u}, \infty)$. Finally, it will be assumed that $\mathbf{u}$ is a natural lower bound for the domain of u in the sense that $u'(\mathbf{u}) = \infty$ or $u(\mathbf{u}) = -\infty$, or both.[2] The investor having initial wealth $w > 0$ can invest in securities $1, \ldots, n$, with security i having the random gross return $\rho_i(\cdot) \geqq 0$ per dollar invested. This nonnegativity assumption implies limited liability: The investor can lose at most his total initial wealth. Note that normal or stable Paretian returns are ruled out by this assumption. The customary approach which treats the assets as "infinitely divisible" goods, available in continuous amounts, is also adhered to throughout this paper.

Let $y_1, \ldots, y_n$ be the amount invested in each security $1, \ldots, n$. Let

$$\rho = (\rho_1, \ldots, \rho_n)^{\mathrm{T}} \in E^n \tag{1}$$

[1] Optimal portfolios are given by mean-variance analysis in the normal case and by mean–dispersion analysis in the stable case. For both problems, homogeneity of the risk measure yields separation into mutual funds which are independent of the utility function (see Breen [5]; the Ziemba paper in Part III, Chapter 1 of this volume; and Exercise II-CR-9).

[2] If $u'(\mathbf{u}) < \infty$ and $u(\mathbf{u}) > -\infty$, then $u(w)$ can be extended as a concave function to values $w < \mathbf{u}$. In this case, the domain $(\mathbf{u}, \infty)$ would not correspond to a complete specification of the investor's preference ordering for gambles on wealth.

and

$$e = (1, \ldots, 1)^{\mathrm{T}} \in E^n \tag{2}$$

where T denotes transposition. The investor's problem is

$$\mathrm{Pk}: \max\{F(y) \mid y \in K_k^{\,n}(w)\}, \tag{3}$$

where

$$F(y) = Eu(\rho^{\mathrm{T}} y), \tag{4}$$

$$K_k^{\,n}(w) = \{x \in E^n \mid e^{\mathrm{T}} x = w, x_i \geqq 0 \text{ for } i \in I_+\}, \tag{5}$$

and

$$I_+ = \{i_1, \ldots, i_k\}, \tag{6}$$

The set I_+ labels the assets which cannot be sold short. The special case $I_+ = \varnothing$, problem $P0$, omits all nonnegativity constraints, and is the problem considered by Cass and Stiglitz [2]. The contraint set for wealth level unity will be denoted by $S_k^{\,n}$:

$$S_k^{\,n} = K_k^{\,n}(1). \tag{7}$$

Let $y^*(w) \in K_k^{\,n}(w)$ be the optimal solution to (3), and define $x^*(w) = (1/w)\, y^*(w) \in S_k^{\,n}$. Clearly, $x^*(w)$ is the vector of optimal asset proportions. Problem Pk is said to exhibit *global separation* if and only if there exist wealth-independent vectors $x^1, x^2 \in S_k^{\,n}$ such that

$$x^*(w) = m(w)\, x^1 + [1 - m(w)]\, x^2 \tag{8}$$

for all $w > 0$. Problem Pk is said to exhibit *local separation* if (8) holds only for $w \in (a, b) \neq (0, \infty)$. In both of these definitions, the mutual funds x^1, x^2 and the proportion $m(w)$ invested in x^1, may depend on the specific utility functions used, but this dependence will be suppressed in the notation.[3] The separation property splits the computation of optimal investment into two subproblems: (i) the determination of the wealth-independent mutual funds, and (ii) the determination of investment in each mutual fund as a function of wealth. Note that if $k > 0$, global separation of problem Pk might not be possible, since $x^*(w)$ in (8) might become infeasible for certain w. In such a case, local separation still holds.

[3] In the Tobin–Lintner [3, 4] separation, x^1 and x^2 would be independent of the utility function, but $m(w)$ would not be, since it is governed by mean-variance indifference curves.

II.1 Necessary Conditions for Separation

The Kuhn–Tucker necessary and sufficient conditions [7, 8] for optimality of y^* are

$$\frac{\partial}{\partial y_i} F(y^*) - \lambda - \mu_i = 0, \qquad i \in I_+, \tag{9a}$$

$$\frac{\partial}{\partial y_i} F(y^*) - \lambda = 0, \qquad i \notin I_+, \tag{9b}$$

$$e^{\mathrm{T}} y^* = w, \tag{10}$$

$$y_i^* \geqq 0, \qquad \mu_i \leqq 0, \qquad \mu_i y^* = 0, \qquad i \in I_+. \tag{11}$$

To understand the restrictions imposed by separation, it is convenient to begin with the case of an interior optimum y^*, or alternatively, the case $I_+ = \varnothing$. By continuity, if $y_i^*(w_0) > 0$ for $i \in I_+$, there exists an interval W containing w_0 such that $y_i^*(w) > 0$ for $i \in I_+$, $w \in W$. By (11), $\mu_i = 0$ for $i \in I_+$, $w \in W$; λ may thus be chosen as an independent variable, and $w = w(\lambda)$ determined by (10). The optimal solution $y^*(w)$ may be rewritten as $y(\lambda)$. From the separation property, Eq. (8), it follows that

$$y(\lambda) = a(\lambda)\,x^1 + b(\lambda)\,x^2 \tag{12}$$

where

$$a(\lambda) = w(\lambda)\,m(\lambda), \qquad b(\lambda) = w(\lambda)\,[1 - m(\lambda)]. \tag{13}$$

Differentiating (12) with respect to λ gives

$$y'(\lambda) = a'(\lambda)\,x^1 + b'(\lambda)\,x^2, \qquad y''(\lambda) = a''(\lambda)\,x^1 + b''(\lambda)\,x^2. \tag{14}$$

Let $Y = [y, y', y'']$ denote the $n \times 3$ matrix whose columns are y, y', and y''. Since the columns of Y are linear combinations of two fixed vectors x^1, x^2, it follows that Y must have rank $\leqq 2$.[4] Thus any 3×3 submatrix of Y must be singular, with vanishing determinant. This gives the following alternative characterization of separation:

$$\begin{vmatrix} y_i(\lambda) & y_i'(\lambda) & y_i''(\lambda) \\ y_j(\lambda) & y_j'(\lambda) & y_j''(\lambda) \\ y_k(\lambda) & y_k'(\lambda) & y_k''(\lambda) \end{vmatrix} = 0 \tag{15}$$

for any three distinct i, j, and k.

[4] Recall that the rank of a matrix is the maximum number of linearly independent columns or rows.

II.2 Restriction of the Utility Functions

Cass and Stiglitz [2] pointed out that separation can hold for *general* markets only if it holds for a convenient set of *special* markets. Once the utility functions have been suitably restricted by the special markets it becomes easier to examine the general market problem. It is convenient to examine first the separation property for discrete random returns. Thus, suppose that there are only finitely many states of nature, $\theta = 1, \ldots, N$, with $\Pr[\theta = i] = p_i$. The returns $\rho_i(\theta)$ are specified completely by an $n \times N$ nonnegative matrix ρ_{ij}, $i = 1, \ldots, n$, $j = 1, \ldots, N$. Assuming an interior solution (or $I_+ = \varnothing$), the following optimality conditions are obtained:

$$\sum_{j=1}^{N} p_j u'\left(\sum_{k=1} y_k \rho_{kj}\right) \rho_{ij} = \lambda, \qquad i = 1, \ldots, n. \tag{16}$$

Consider first the case $N = n$ and with nonsingular matrix ρ. Then (16) can be solved to give

$$u'\left(\sum_k y_k \rho_{ki}\right) = \frac{\lambda}{p_i} \sum_j R_{ij}, \tag{17}$$

where

$$R = \rho^{-1}. \tag{18}$$

Since u' is continuously differentiable and strictly decreasing, the inverse function $g(\cdot) = u'^{-1}(\cdot)$ is a well-defined, strictly decreasing, continuously differentiable function on its domain. Furthermore, since u' is piecewise twice continuously differentiable, so is g. Solving (18) gives

$$\sum_{j=1}^{n} y_j \rho_{ji} = g\left(\frac{\lambda}{p_i} \sum_{j=1}^{n} R_{ij}\right) \equiv g(\lambda k_i). \tag{19}$$

From the separation equation (8), it follows that

$$g(\lambda k_i) = a(\lambda) m_i + b(\lambda) n_i \tag{20}$$

where

$$m_i = \sum_j x_j^1 \rho_{ji}, \qquad n_i = \sum_j x_j^2 \rho_{ji}. \tag{21}$$

Clearly, the consideration which yielded Eq. (15) will apply to the vector $g(\lambda k_i)$, $i = 1, \ldots, n$, as well as to the vector $y(\lambda)$, so the following necessary condition obtains:

$$\begin{vmatrix} g(k_i \lambda) & g(k_j \lambda) & g(k_l \lambda) \\ k_i g'(k_i \lambda) & k_j g'(k_j \lambda) & k_l g'(k_l \lambda) \\ k_i^2 g''(k_i \lambda) & k_j^2 g''(k_j \lambda) & k_l^2 g''(k_l \lambda) \end{vmatrix} = 0 \tag{22}$$

for distinct i, j, and l. Assuming that $0 < k_i < u'(0)$ for $i = 1, \dots, n$ (which is always possible by suitable choice of ρ), it follows that $y(\lambda)$ will actually be an interior solution for all $\lambda \leqq 1$. Since (22) holds as an identity in λ and k_i, it is permissible to set $\lambda = 1$, to fix k_j and k_l at arbitrary values in the interval $(0, u'(0))$, and to let $z = k_i$ be a variable. Then (22) becomes

$$r_1 z^2 g''(z) + r_2 z g'(z) + r_3 g(z) = 0 \tag{23}$$

for real constants r_1, r_2, and r_3. It is easy to verify that, if α and β are the roots of the quadratic equation

$$r_1 x(x-1) + r_2 x + r_3 = 0, \tag{24}$$

then the general solution to (23) is

$$g(z) = \begin{cases} Az^{\alpha} + Bz^{\beta} & \text{if} \quad \alpha \neq \beta, \\ z^{\alpha}(A + B \log z) & \text{if} \quad \alpha = \beta, \end{cases} \tag{25}$$

with constant A and B. Note that $g(z)$ must have the form (25) in each interval within which g'' is continuous. Although it might seem that g could be made up from different "pieces" of the form (25), with constants A, B, α, and β changing, this is, in fact, not true. The reader can verify that separation, together with the assumed continuity of g and g', implies that the single set of constants A, B, α, β applies throughout the entire subdomain $\{z \mid z < u'(0)\}$. In terms of u, this means that the following functional representation holds:

$$\xi = \begin{cases} Au'(\xi)^{\alpha} + Bu'(\xi)^{\beta} & (\alpha \neq \beta), \qquad (26a) \\ u'(\xi)^{\alpha}[A + B \log u'(\xi)] & (\alpha = \beta) \qquad (26b) \end{cases}$$

for all $\xi \in (0, \infty)$. Note that separation forces u to be three times continuously differentiable on $(0, \infty)$, even though this was originally assumed to be true only in a piecewise sense.

For problem Pk with $0 \leqq k < n$, the reader can show that Eqs. (26a) and (26b) are valid throughout the entire domain $\xi \in (\mathbf{u}, \infty)$ of u, even if $\mathbf{u} < 0$. In fact, it is always possible to construct a nonsingular matrix ρ such that all relevant nonnegativity constraints are satisfied strictly, while the unconstrained assets are sold short, so that $g(z)$ is "probed" for $z \in (\mathbf{u}, 0)$. However, if $k = n$, the nonnegativity constraints form a natural boundary which always restricts ξ in (26) to the set $(0, \infty)$. In this case, even if $\mathbf{u} < 0$, no information about the behaviour of $u(\xi)$ for $\xi < 0$ can be obtained from the portfolio problem. The following fundamental theorem of Cass and Stiglitz has thus been proved.

Theorem 1 A necessary condition for local or global separation of problem *Pk* is that marginal utility satisfy either

$$Au'(\xi)^{\alpha} + Bu'(\xi)^{\beta} = \xi \qquad (\text{with} \quad \alpha \neq \beta)$$

or

$$u'(\xi)^{\alpha}[A + B \log u'(\xi)] = \xi$$

for all $\xi \in \Xi_k$, with

$$\Xi_k = \begin{cases} (\mathbf{u}, \infty) & \text{if} \quad k < n, \\ (0, \infty) & \text{if} \quad k = n. \end{cases} \tag{27}$$

For general discrete random returns with $N > n$, there are two cases which must be considered: (1) markets without money, that is, markets having no risk-free asset, and (2) markets with money. Each case is treated separately.

Case 1 To derive *necessary* conditions for separation in general markets without money, it is enough to derive these conditions for a special subclass of such markets. In particular, suppose that the matrix ρ has the structure [2]:

$$[\rho] = \left[\begin{array}{c|cc} & 0 & 0 \\ \tilde{\rho} & \vdots & \vdots \\ & 0 & 0 \\ \hline 0 \cdots 0 & \rho_n & \rho_n \sigma \end{array}\right], \tag{28}$$

where $\tilde{\rho}$ is a nonsingular $(n-1) \times (n-1)$ matrix, and $N = n+1$. The parameter $\sigma > 0$ is arbitary. Assuming an interior solution (which can always be obtained by suitable choice of ρ) it follows as in (19) that

$$y_i = \sum_{j=1}^{n-1} g\left(\frac{\lambda}{p_j} \sum_{k=1}^{n-1} R_{ik}\right) R_{ji}, \qquad i = 1, \ldots, n-1, \tag{29}$$

where $R = \tilde{\rho}^{-1}$. There are two subcases:

(i) $g(z) = Az^{\alpha} + Bz^{\beta}$. Equation (29) becomes

$$y_i = Ak_i \lambda^{\alpha} + Bl_i \lambda^{\beta}, \qquad i = 1, \ldots, n-1, \tag{30}$$

where

$$k_i = \sum_{j=1}^{n-1} R_{ji} \left(\frac{1}{p_j} \sum_{k=1}^{n-1} R_{ik}\right)^{\alpha} \tag{31}$$

and where l_i is given by (31) with α replaced by β. From the separation equation (12) it follows that y_i has the form (30) also for $i = n$. Thus there exist vectors $k, l \in E^n$ such that $y = Ak\lambda^\alpha + Bl\lambda^\beta$. Equation (16) gives, for $i = n$;

$$\rho_n[p_n u'(f(\lambda)) + p_{n+1}\sigma u'(\sigma f(\lambda))] = \lambda, \tag{32}$$

where

$$f(\lambda) = k_n \rho_n A\lambda^\alpha + l_n \rho_n B\lambda^\beta. \tag{33}$$

Since (32) is an identity in λ and σ, each term on the left-hand side must be separately linear in λ (see Exercise CR-21). Thus

$$p_n \rho_n u'(f(\lambda)) = c\lambda + d, \qquad p_{n+1}\rho_n \sigma u'(\sigma f(\lambda)) = (1-c)\lambda - d, \tag{34}$$

so

$$f(\lambda) = g\left[\frac{c\lambda + d}{p_n \rho_n}\right] = A\left[\frac{c\lambda + d}{p_n \rho_n}\right]^\alpha + B\left[\frac{c\lambda + d}{p_n \rho_n}\right]^\beta, \tag{35}$$

and

$$\sigma f(\lambda) = g\left[\frac{(1-c)\lambda - d}{\rho_n p_{n+1}\sigma}\right] = A\left[\frac{(1-c)\lambda - d}{p_{n+1}\rho_n \sigma}\right]^\alpha + B\left[\frac{(1-c)\lambda - d}{p_{n+1}\rho_n \sigma}\right]^\beta. \tag{36}$$

Equations (33), (35), and (36) can be true as identities in λ only if (a) $\alpha = 1$, $\beta = 0$ (or $\alpha = 0$, $\beta = 1$), or (b) $B = 0$, $A \neq 0$ (or $A = 0$, $B \neq 0$). (See Exercise CR–21.) In case (a) the utility function is quadratic, while in case (b) the marginal utility has the form $u'(\xi) = b\xi^c$.

(ii) $g(z) = z^\alpha(A + B\log z)$. Arguments like those above imply $B = 0$, giving $u'(\xi) = b\xi^c$. (See Exercise CR–21.) The following theorem has thus been proved [2]:

Theorem 2 A necessary condition for local or global separation of problem *Pk* in general markets without money is that u' have one of the forms:

$$u'(\xi) = a + b\xi \qquad \text{(quadratic utility)}$$

or

$$u'(\xi) = b\xi^c \qquad \text{(constant relative risk aversion)}$$

for $\xi \in \Xi_k$.

Case 2 To derive necessary conditions for separation in general markets with money, it is only necessary to consider matrices ρ of the form [2]

$$[\rho] = \left[\begin{array}{ccc|ccc} & & & 0 & 0 & 0 \\ & \tilde{\rho} & & \vdots & \vdots & \vdots \\ & & & 0 & 0 & \cdot \\ \hline 0 & \cdots & 0 & \rho_n & \rho_n \sigma & \rho_n \sigma^2 \\ r & \cdots & r & r & r & r \end{array}\right], \tag{37}$$

where $\tilde{\rho}$ is a nonsingular $(n-2)\times(n-2)$ matrix aud $\sigma > 0$ is arbitrary. Note that asset n is money, having the same rate of return $r > 0$ in all states of nature. By a simple modification of the proof of Theorem 2, the following result may be obtained. (See Exercise CR–21.)

Theorem 3 A necessary condition for local or global separation of problem *Pk* in general markets with money is that u' have one of the forms

$$u'(\xi) = (a+b\xi)^c$$

or

$$u'(\xi) = ae^{b\xi} \qquad \text{(constant absolute risk aversion)}$$

for $\xi \in \Xi_k$.

Note In the statement of Theorem 3 it is not assumed that money is one of the mutual funds in an optimal portfolio. The latter property, called *monetary separation* is, of course, also covered by the necessary conditions stated in the theorem.

III. Sufficient Conditions for Separation

In this section, the extent to which the necessary conditions stated in Theorems 2 and 3 are also sufficient for separation is determined. The first result is

Theorem 4 The following conditions are sufficient for local or global separation of problem *Pk* in general markets without money:

(i) $u'(\xi) = a+b\xi \quad (a < 0,\ b > 0)$, or
(ii) $u'(\xi) = b\xi^c \quad (b > 0,\ c < 0)$

for $\xi \in \Xi_k$. For case (i), separation is global if $k = 0$ (problem *P*0), and local if $0 < k \leqq n$; in this case, local separation will hold for all w, with possibly different mutual funds applying in different regions of w. For case (ii), separation is global for all problems Pk, $0 \leqq k \leqq n$.

Proof (i) $u'(\xi) = a + b\xi$. For problem $P0$ the solution can be written down explicitly:

$$x^*(w) = k(w)x^1 + [1 - k(w)]x^2, \tag{38}$$

where

$$x^1 = \sigma^{-1}e/e^{\mathrm{T}}\sigma^{-1}e, \qquad x^2 = \sigma^{-1}\mu/e^{\mathrm{T}}\sigma^{-1}\mu, \tag{38}$$

and

$$k = 1 + \frac{a}{bw}e^{\mathrm{T}}\sigma^{-1}\mu. \tag{40}$$

In (39) and (40),

$$\mu = (E\rho_1, \ldots, E\rho_n)^{\mathrm{T}} \tag{41}$$

and

$$\sigma = (E\rho_i\rho_j), \qquad 1 \leqq i, \ \ j \leqq n. \tag{42}$$

The solution as written displays global separation explicitly. For problem Pk with $k > 0$, (38) will be valid in an interval $w \in W$ if $x_i^*(w) > 0$ for $i \in I_+$. Hence, in this case also, local separation will hold. When none of these conditions is satisfied, local separation can still be proved by noting that the solution must lie on the Markowitz [9] mean-variance efficiency frontier, and by noting that any point on this frontier is either a single mutual fund or a convex combination of two neighbouring mutual funds.

(ii) $u'(\xi) = b\xi^c$. For problem $P0$ the solution $x^*(w)$ is identical to the solution of the following set of nonlinear equations.

$$E\left\{(\rho_i - \rho_n)\left[\rho_n + \sum_{j=1}^{n-1}(\rho_j - \rho_n)x_j^*\right]^c\right\} = 0, \qquad i = 1, \ldots, n-1, \tag{43}$$

and

$$x_n^* = 1 - \sum_{i=1}^{n-1} x_i^*. \tag{44}$$

Again, (43) and (44) are also valid for problem Pk with $k > 0$ if $x_i^*(w) \geqq 0$ for $i \in I_+$. Note that in both these cases $x^*(w)$ collapses onto a *single* mutual fund which is independent of w. When (43) and (44) are not valid the latter result is still true, as can be seen most easily by showing that the Kuhn–Tucker conditions for $x^*(w)$ are independent of w. Thus, global separation holds.

It is clear by comparing Theorems 2 and 4, that the question of separation in general markets *without* money has been completely solved. However, for general markets *with* money, the situation is considerably less clear-cut, as the following result shows.

Theorem 5 Let $\rho_1, \ldots, \rho_n$ be a general market with ρ_n having the certain (gross) rate of return $r > 0$, i.e., ρ_n is money. Then sufficient conditions for local or global separation of problem Pk are

(i) $u'(\xi) = (a+b\xi)^c$, with either
  (ia) $c = 1$, or
  (ib) $a > 0,\ b > 0,\ c < 0$ and $n \notin I_+$, or
  (ic) $a = 0,\ b > 0,\ c < 0$.
(ii) $u'(\xi) = ae^{b\xi} \quad (a > 0,\ b < 0)$.

Note that cases (ia) and (ic) are already covered by Theorem 4. For case (ib), global separation holds. However, if $n \in I_+$, extra conditions may be required[5] to guarantee even local separation if $a, b > 0$ and $c < 0$. In all cases (i), if $n \notin I_+$, global monetary separation obtains, but if $n \in I_+$, money may or may not be one of the mutual funds in an optimal portfolio. However, for case (ii) local monetary separation holds for large w.

Proof (i) $u'(\xi) = (a+b\xi)^c$. For $c = 1$ (quadratic utility), Theorem 4 applies. The fact that global monetary separation holds if $n \notin I_+$ follows from the work of Lintner [4], reprinted as the first article of this chapter. Reference to Lintner's paper also shows that monetary separation need not obtain if $n \in I_+$. The general problem may be written

$$\max \{F(\tilde{y}) \mid \tilde{y} \in Y(w)\},$$

where

$$Y(w) = \left\{\tilde{y} \in E^{n-1} \mid \tilde{y}_i \geqq 0,\ i \in I_+ - \{n\} \text{ and } \sum_{i=1}^{n-1} \tilde{y}_i \leqq [br + a/w]^{-1} \right.$$
$$\left. \text{if } n \in I_+,\ \sum_{i=1}^{n-1} \tilde{y}_i \text{ unconstrained if } n \notin I_+ \right\}, \tag{45}$$

and where

$$F(\tilde{y}) = \begin{cases} \pm E\left[1 + b\sum_{i=1}^{n-1} (\rho_i - r)\tilde{y}_i\right]^{c+1} & (c \gtrless -1) \\ E \log\left[1 + b\sum_{i=1}^{n-1} (\rho_i - r)\tilde{y}_i\right] & (c = -1). \end{cases} \tag{46}$$

[5] This is an unsolved problem.

The optimal solution $x^*(w)$ is

$$x_i^*(w) = \left(br + \frac{a}{w}\right)\tilde{y}_i, \qquad i = 1, \ldots, n-1, \tag{47}$$

$$x_n^*(w) = 1 - \sum_{i=1}^{n-1} x_i^*(w). \tag{48}$$

It follows, from (45)–(48), that global monetary separation holds if $n \notin I_+$, and that global separation (whether monetary or not) holds if $a = 0$. It is also evident that local monetary separation holds for all w sufficiently large if $n \in I_+$, and if $\sum_{i=1}^{n-1} y_i < 1/br$ in the solution for infinite wealth level w. Whether local separation, monetary or otherwise, holds at all if this condition is not satisfied, remains an unresolved question.

(ii) $u'(\xi) = ae^{b\xi}$. In this case the problem may be rewritten

$$\max\{F(y) \mid y \in Y'(w)\},$$

where

$$Y'(w) = \left\{y \in E^{n-1} \mid y_i \geqq 0,\ i \in I_+ - \{n\} \text{ and } \sum_{i=1}^{n-1} y_i \leqq w \right.$$
$$\left. \text{if } n \in I_+,\ \sum_{i=1}^{n-1} y_i \text{ unconstrained if } n \notin I_+ \right\}, \tag{49}$$

and where

$$F(y) = -E\exp\left\{b\sum_{i=1}^{n-1}(\rho_i - r)\,y_i\right\}. \tag{50}$$

The optimal solution $x^*(w)$ is

$$x_i^*(w) = (1/w)\,y_i, \qquad i = 1, \ldots, n-1, \tag{51}$$

$$x_n^*(w) = 1 - \sum_{i=1}^{n} x_i^*(w).$$

Evidently, global monetary separation holds if $n \notin I_+$, and local monetary separation holds if $n \in I_+$ and w is sufficiently large. This completes the proof of the theorem.

The reader will notice that, for both utility functions $u'(\xi) = ae^{b\xi}$ and $u'(\xi) = (a + b\xi)^c$ $(a, b > 0, c < 0)$, the investor holds a larger fraction of wealth in money for larger values of w, if he holds money at all. This investor behavior is easily understood on a less formal level through the following argument. Because of monetary separation, the problem is equivalent to that of money and a single risky asset. Both of the utility functions just given display nondecreasing relative risk aversion,[6] and Arrow [10] has shown, for

[6] Recall that the relative risk-aversion measure is given by $R(w) = -wu''(w)/u'(w)$.

the single risky asset problem, that the fraction of wealth devoted to money increases with wealth in such cases.

For convenience, a summary chart of Theorems 4 and 5 is given in Table I.

TABLE I

A. *General Markets Without Money*

Marginal utility	Constraints	Type of separation
$u'(\xi) = a + b\xi$	Unconstrained	Global
$a > 0,\ b < 0$	Constrained	Local
$u'(\xi) = b\xi^c$	Unconstrained	Global
$b > 0,\ c < 0$	Constrained	Global

B. *General Markets With Money*

Marginal utility	Constraints on money	Restrictions	Type of separation
$u'(\xi) = a + b\xi$	Unconstrained	—	Global monetary
$a > 0,\ b < 0$	No borrowing		Local, all w
$u'(\xi) = b\xi^c$	Unconstrained	—	Global monetary
$b > 0,\ c < 0$	No borrowing		Global
$u'(\xi) = ae^{b\xi}$	Unconstrained	—	Global monetary
$a > 0,\ b < 0$	No borrowing		Local monetary, large w
$u'(\xi) = (a+b\xi)^c$	Unconstrained		Global monetary
$a, b > 0,\ c < 0$	No borrowing	Positive fraction in money at infinite wealth	Local monetary, large w

The dependence of optimal mutual funds on utility function parameters can be ascertained from the results above. Indeed, a number of parametric properties follow immediately from the observation that investor preferences are invariant under utility scale changes [1]. For the case $u'(\xi) = (a+b\xi)^c$, this principle implies that the investor preference-orderings and optimal portfolios must depend on a and b only through the ratio a/b if this is nonzero. For the case $a = 0$, $u'(\xi) = b\xi^c$, the optimal portfolio must be independent of b. It is gratifying to note that these conclusions follow also from Eqs. (38)–(40),

(43)–(44), and (45)–(48). In all cases the optimal mutual funds x^1 and x^2 depend strongly on the "exponent" c and on the ratio a/b if this is nonzero. For quadratic utility with $c = 1$, x^1 and x^2 are independent of a/b in the unconstrained problem $P0$, and are piecewise constant functions of a/b in the constrained problems Pk ($k > 0$), varying only according to which segment of the Markowitz frontier is tangent to a mean–variance indifference curve. When $c \neq 1$ and $a/b \neq 0$, x^1 and x^2 depend on both c and a/b, for unconstrained as well as constrained problems. For utility functions of the constant absolute risk-aversion class, $u'(\xi) = ae^{b\xi}$, investor preference-orderings and optimal solutions must be independent of a. Again, this follows directly from Eqs. (49)–(51).

Throughout this paper the allowed type of investment in an asset has either been unconstrained or purely nonnegative. A more general case is that which allows limited borrowing or short sales. It is clear that such a generalization leads to no substantial changes in the theory. In particular, the conclusions of Theorems 4 and 5 remain valid in this more general situation. The utility functions giving local versus global separation in markets with money agree with those obtained by Hakansson [6] using different methods. (See also Exercise ME-14.)

References

1. von Neumann, J., and Morgenstern, O., *Theory of Games and Economic Behavior.* Princeton Univ. Press, Princeton, New Jersey, 1953.
2. Cass, D., and Stiglitz, J. E., "The Structure of Investor Preferences and Asset Returns, and Separability in Portfolio Allocations: A Contribution to the Pure Theory of Mutual Funds." *Journal of Economic Theory* **2** (1970), 122–160.
3. Tobin, J., "Liquidity Preference as Behavior Toward Risk." *Review of Economic Studies* **36** (1958), 65–86.
4. Lintner, J., "The Valuation of Risk Assets and the Selection of Risky Investments in Stock Portfolios and Capital Budgets." *Review of Economis and Statistics* **47** (1965), 13–37, reprinted in this volume.
5. Breen, W., "Homogeneous Risk Measures and the Construction of Composite Assets." *Journal of Financial and Quantitative Analysis* **3** (1968), 405–413.
6. Hakansson, N. H., "Risk Disposition and the Separation Property in Portfolio Selection." *Journal of Financial and Quantitative Analysis* **4** (1969), 401–416.
7. Kuhn, H. W., and Tucker, A. W., "Nonlinear Programming." *Proceedings of the Symposium on Mathematical Statistics and Probability, 2nd, Berkeley, 1951* (J. Neyman ed.). Univ. of California Press, Berkeley, 1951.
8. Zangwill, W., *Nonlinear Programming: A Unified Approach.* Prentice-Hall, Englewood Cliffs, New Jersey, 1969.
9. Markowitz, H. M., "Portfolio Selection." *Journal of Finance* **6** (1952), 77–91.
10. Arrow, K. J., *Aspects of the Theory of Risk-Bearing.* Yrjö Jahnsson Foundation, Helsinki, 1965.

COMPUTATIONAL AND REVIEW EXERCISES

1. Suppose that gross returns per $100 of investment for eight alternative investment opportunities were

Year \ Investment	A	B	C	D	E	F	G
1963	90	70	160	145	60	110	150
1964	110	90	170	145	60	140	170
1965	110	70	170	120	20	140	170
1966	110	90	170	180	40	160	160
1967	110	130	170	120	60	160	170
1968	90	70	170	145	70	160	170
1969	110	130	160	190	120	160	180
1970	110	70	170	145	70	160	170
Mean net return	5.00	−10.00	67.50	50.00	−37.50	48.75	67.50

Assume that the investor attaches equal probability to past outcomes as predictors of future returns.

(a) Plot the cumulative probability distributions.
(b) Utilize the plots in (a) to show that investments C, D, and G dominate the others for all monotone nondecreasing utility functions.
(c) Show that investment C is preferred if the investor has a logarithmic utility function.
(d) Show that investment G is preferred if the investor has the quadratic utility function $u(w) = w^2$.
(e) Construct a utility function for which investment D is preferred.

2. (a) Consider an investment problem in which it is desired to discount the wealth level before taking the expected utility. Show that the stochastic dominance theorems in the Hanoch–Levy paper apply to a reinterpreted discounted wealth.
(b) Show that the stochastic dominance theorems are independent of initial wealth, if the investment returns are independent of initial wealth.

3. Suppose investments X and Y have the following returns

Outcome (z)	$\Pr(X = z)$	$\Pr(Y = z)$
0	$\frac{1}{4}$	$\frac{1}{2}$
$\frac{1}{2}$	$\frac{1}{4}$	0
1	0	$\frac{1}{2}$
$1\frac{1}{2}$	$\frac{1}{4}$	0
2	$\frac{1}{4}$	0

(a) Show that investment X is preferred to investment Y for all monotone nondecreasing utility functions even though the variance of X is larger than the variance of Y.
(b) Suppose that $\Pr(X{=}z)$ has entries $(\frac{1}{4}, \frac{1}{2}, 0, 0, \frac{1}{4})$. Show that X is not preferred to Y for all monotone nondecreasing utility functions.
(c) Referring to (b) show that X is preferred to Y for all monotone nondecreasing concave utility functions, even though the variance of X is larger than the variance of Y.

4. Suppose investments X and Y have normal distributions $N(\mu_1, \sigma_1^2)$ and $N(\mu_2, \sigma_2^2)$, respectively. Suppose that $\mu_2 \leqq \mu_1$ and $\sigma_2^2 < \sigma_1^2$.

(a) Find a nondecreasing concave utility function such that Y is preferred to X.

(b) Find a nondecreasing utility function such that X is preferred to Y.

(c) Suppose that $\mu_2 \geqq \mu_1$ and $\sigma_2^2 \leqq \sigma_1^2$. Show that Y is preferred to X for all nondecreasing concave utility functions.

5. Consider the following return structure

Outcome z	Probability
a	p
$\frac{1}{2}$	q
1	$1-p-q$

where p, q, and a are given parameters such that $a < \frac{1}{2}$, $p \geqq 0$, $q \geqq 0$, $p+q \leqq 1$. Suppose for investments X and Y that $(a, p, q) = (\frac{1}{8}, \frac{7}{16}, \frac{7}{16})$ and $(0, \frac{1}{8}, \frac{3}{4})$, respectively.

(a) Show that $\mu_x \leqq \mu_y$, $\sigma_x^2 \geqq \sigma_y^2$.

(b) Find a nondecreasing concave utility function such that X is preferred to Y.

(c) Find a nondecreasing concave utility function such that Y is preferred to X.

(d) Interpret the results in (b) and (c).

(e) Suppose a is the same for both X and Y (but p and q remain as originally specified). Show that Y is preferred to X for all nondecreasing utility functions.

(f) Find values of p and q such that Y is preferred to X for all nondecreasing concave utility functions.

(g) Interpret the results of (f) and (g) in the light of (d).

6. Consider an investor having a quadratic utility for wealth,

$$u(w) = 2kw - w^2, \qquad \text{where} \quad k > 0 \quad \text{and} \quad 0 \leqq w \leqq k.$$

Suppose that random investments X_1 and X_2 with $0 \leqq X_1, X_2 \leqq k$, have means μ_1, μ_2 and variances σ_1^2, σ_2^2, respectively.

(a) Show that X_1 is preferred to X_2, i.e., has higher expected utility, if and only if $2\Delta\mu(k-\bar{\mu}) - \Delta\sigma^2 > 0$, where $\Delta\mu \equiv \mu_1 - \mu_2$, $\bar{\mu} \equiv \frac{1}{2}(\mu_1 + \mu_2)$, and $\Delta\sigma^2 \equiv \sigma_1^2 - \sigma_2^2$.

(b) Show that X_1 is preferred to X_2 if $\mu_1 > \mu_2$ and $\sigma_1 < \sigma_2$.

(c) Show that (b) is not necessarily true if the range of X_1 and X_2 is $[0, \infty)$.

(d) Find an X_1 and an X_2 such that X_1 is preferred to X_2 even though $\mu_1 < \mu_2$.

Suppose that random investments $X_1, X_2, \ldots, X_n$ have a known least upper bound α, where $\alpha \leqq k$.

(e) Show that X_1 is preferred to X_2 if $\Delta\mu \geqq 0$ and $2\,\Delta\mu(\alpha - \bar{\mu}) - \Delta\sigma^2 > 0$.

(f) Show that X_1 is preferred to X_2 if the relation in (e) holds with some particular number $\bar{x}$ (between $\bar{\mu}$ and α) substituted for α. In particular, if $\mu_m = \max_i \mu_i$, a sufficient condition for X_1 to be preferred to X_2 is that $2\,\Delta\mu(\mu_m - \bar{\mu}) - \Delta\sigma^2 > 0$.

(g) Show that the sufficient condition in (f) is weaker (hence, more general) than the (μ, σ^2) criterion. That is, construct an example in which $2\,\Delta\mu(\mu_m - \bar{\mu}) - \Delta\sigma^2 > 0$ (hence X_1 is preferred to X_2), but with $\Delta\sigma^2 > 0$ [so that X_1 and X_2 cannot be ranked by the (μ, σ^2) criterion].

(h) Show that the sufficient conditions in (a), (b), (e), and (f) are transitive. That is if X_1 is preferred to X_2 and X_2 is preferred to X_3 by any of the criteria, then X_1 is also preferred to X_3 by the same criterion.

7. Consider an investor having the quadratic utility function of Exercise 6. We are interested in preference criteria for the dominance of X_1 over X_2 when the only information known is that (i) $0 \leqq X_1,\ X_2 \leqq k$; (ii) the means and variances μ_1, μ_2 and σ_1, σ_2 are known with $\Delta\mu = \mu_2 - \mu_1 \geqq 0$; and (iii) X_1 and X_2 have symmetric distributions. Let R_1 and R_2 be the length of the ranges of X_1, X_2, i.e.,

$$R_i = \inf[b-a \mid P(X_i > b) = P(X_i < a) = 0], \qquad i = 1,2.$$

(a) Show that $R_i \leqq 2(k-\mu_i)$.
(b) Show that $R_i \geqq 2\sigma_i$.

Let $A \equiv \max\{\mu_1+\alpha_1,\ \mu_2+\alpha_2\}$.

(c) Show that X_1 is preferred to X_2 if $2\,\Delta\mu(A-\bar{\mu}) - \Delta\sigma^2 > 0$.
(d) Given symmetric returns X_1, X_2, X_3 as above, choose

$$A \equiv \max\{\mu_1+\sigma_1,\ \mu_2+\sigma_2,\ \mu_3+\sigma_3\}.$$

With this A, show that the sufficient condition in (c) is transitive.
(e) Show that condition (c) is weaker (i.e., more general) than that of part (e) in Exercise 6.

8. Consider an investor having a cubic utility function for wealth, $u(w) = w + bw^2 + cw^3$, where $0 \leqq b^2 < 3c$.

(a) Show that X_1 is preferred to X_2 if $EX_1 > EX_2$ and

$$3(EX_1{}^2 - EX_2{}^2)^2 < 4(EX_1 - EX_2)(EX_1{}^3 - EX_2{}^3).$$

(b) Show that the condition in (a) is transitive, that is, if X_1 is preferred to X_2 and X_2 is preferred to X_3, then X_1 is also preferred to X_3.

9. The Tobin separation theorem gives sufficient conditions for the optimal proportions held in the risky assets to be independent of the investor's utility function and the proportion of the investor's initial wealth that is invested in the risk-free asset. The theorem is proved and discussed in Lintner's paper in a mean variance analysis setting. It is of interest to develop similar separation theorems for wider classes of risk measures.

Suppose initial wealth is one dollar and that $x = (x_2, \ldots, x_n)$ are the investments made in the risky investments $i = 2, \ldots, n$ which have the known joint cumulative distribution function $F(\rho_2, \ldots, \rho_n)$. Let $-x_0$ be the level of borrowing in a risk free asset at rate $\bar{\rho}_0$, and x_1 be the level of lending in a risk free asset at rate $\bar{\rho}_1$. The return is then $\sum_{i=0}^{n} \rho_i x_i$. Consider a risk measure $R(x)$ and suppose the R is homogeneous of degree m, differentiable, and concave on the convex set K which represents constraints on the choice of x save the budget constraint. Assume that $\{(x_0, \ldots, x_n) \mid x \in K,\ x_0 \leqq 0,\ x_1 \geqq 0,\ \sum_{i=0}^{n} x_i = 1\}$ is compact.

(a) Suppose $\bar{\rho}_0 = \bar{\rho}_1$ and that $K = \{x \mid x \geqq 0\}$. Show that the ratios of the optimal investments in $x_2, \ldots, x_n$ are independent of the optimal investment in the risk free asset as long as this investment $(x_0 + x_1)$ is nonzero. {*Hint*: Develop the Kuhn–Tucker conditions for

$$\left\{\min R(x) \;\middle|\; \sum_{i=0}^{n} \bar{\rho}_i x_i \geqq \alpha,\ \sum_{i=0}^{n} x_i = 1,\ x_i \geqq 0,\ i = 1, \ldots, n,\ x_0 \leqq 0\right\}$$

and show that if x^* satisfies them, then so does $(\beta_1 x_0, \beta_1 x_1, \beta_2 x_2, \ldots, \beta_2 x_n)$ for all $\beta_1 > 0$, $\beta_2 > 0$.]
(b) Show that the separation property holds if only lending is allowed and $x_1 > 0$.
(c) Show that the separation property holds if only borrowing is allowed and $x_0 < 0$.

Suppose $\bar{\rho}_0 > \bar{\rho}_1$ and that $K = \{x \mid x \geqq 0\}$.

(d) Show that it is never optimal to borrow at $\bar{\rho}_0$ and lend at $\bar{\rho}_1$.
(e) Show that the separation property holds unless $x_0 = x_1 = 0$.

(f) Graphically illustrate the situation in (e) in mean dispersion space.
(g) Prove (a)–(c) and (e) when $K = E^n$, i.e., when short sales are allowed.
(h) Attempt to prove (a)–(c) and (e) when K is the convex set $\{x \mid g_i(x) \geqq 0,\ i = 1, \ldots, k\}$, where each g_i is differentiable and quasi-concave and K satisfies a constraint qualification. What must be assumed about the g_i? Is the separation global?

10. Consider a risk-averse investor with initial wealth w_0 who may invest any portion of w_0 in a risk-free asset with zero return and the remainder in a risky asset with rate of return X, a random variable. Let a be the amount invested in the risky asset, and $m = w_0 - a$ the amount invested in the risk-free asset. The individual chooses a to maximize expected utility of terminal wealth w, where $w = w_0 + aX$. Let $W(a) = Eu(w_0 + aX)$.

(a) Show that

$$W'(a) = E(u'(w)X) \qquad \text{and} \qquad W''(a) = E[u''(w)X^2] < 0.$$

(b) Show that $W(a)$ has its maximum at $a = 0$ if and only if $W'(0) \leqq 0$. Show that a necessary and sufficient condition for this to occur is that $EX \leqq 0$.
(c) Show that $a > 0$ if and only if $EX > 0$. This demonstrates that a risk-averter will always take some part of an actuarially favorable gamble.

Consider the case in which $W(a)$ is maximized at an interior point $a^* \neq 0$. It is of interest to study the behavior of a^* as a function of initial wealth w_0.

(d) Show that

$$\frac{da^*}{dw_0} = -\frac{E[u''(w^*)X]}{E[u''(w^*)X^2]} \qquad \text{where} \quad w^* = w_0 + a^*X.$$

Suppose that the investor exhibits decreasing absolute risk aversion, i.e., $R(w)$ is non-increasing in w.

(e) Show that $u''(w_0 + a^*X)X \geqq -R(w_0)u'(w_0 + a^*X)X$.
(f) Show that $E[u''(w_0 + a^*X)X] \geqq 0$. This demonstrates that $da^*/dw_0 \geqq 0$, so that the amount of risky investment does not decrease with increasing initial wealth.
(g) Show that $da^*/dw_0 < 0$ if $R(w)$ is strictly increasing.

11. (Optimal insurance policy) Assume that an individual owns two assets one of which is risky. The value of the safe asset is A, while the value of the risky asset is B. The individual may insure the risky asset during the period of risk for i. Thus if the individual insures, his wealth becomes $A + B - i$. If the individual does not insure, his wealth is the random variable

$$\xi = \begin{cases} A & \text{with probability } p, \\ A + B & \text{with probability } 1 - p, \end{cases}$$

where p is the probability of losing the risky asset during the period.

(a) Show that the maximum premium i that the individual would be willing to pay for insurance is given by the equation

$$pu(A) + (1-p)u(A+B) = u(A+B-i), \tag{1}$$

where u is the individual's utility function. Assume that the individual's utility function is monotonic, nondecreasing, and concave.
(b) Show that the maximum premium (i) must exceed the acturarial value of the loss (pB).
(c) Show that i is a nondecreasing function of both p and B.

Suppose that u is differentiable and exhibits decreasing absolute risk aversion.

(d) Show that the larger the safe portion of the portfolio, the smaller the value of the premium. [*Hint*: First show that

$$\frac{di}{dA} = -\frac{pu'(A) + (1-p)u'(A+B) - u'(A+B-i)}{u'(A+B-i)}. \tag{2}$$

Then eliminate p from (2) to obtain the numerator $N(i)$. Show that $N(0) = N(B) = 0$ and that N attains a maximum at some $i = \bar{i}$ and that $N'(i) > 0$ if $i < \bar{i}$ and $N'(i) < 0$ if $i > \bar{i}$, hence $N(i) \geqq 0$. Thus di/dA will be nonpositive.]

Suppose that the individual may specify the desired amount of coverage, say C, where $0 \leqq C \leqq B$. Let the premium rate be c. Then the premium is cC and the individual's final wealth is the random variable

$$\xi = A + B - x + (Cx/B) - cC,$$

where x is the random size of the damage that occurs.

Suppose the investor wishes to maximize $f(C) = Eu(\xi)$ subject to $0 \leqq C \leqq B$

(e) Show that

$$f'(C) = E\left\{u'(\xi)\left(\frac{x}{B} - c\right)\right\} \qquad \text{and that} \qquad f''(C) = E\left\{u''(\xi)\left(\frac{x}{B} - c\right)^2\right\} < 0.$$

(f) Show that it is never optimal to take full coverage (i.e., $C = B$) if the premium is actuarially unfavorable [i.e., $cB > E(x)$].

(g) When is it optimal to take no coverage?

(h) Show that the optimal coverage is a decreasing function of wealth if the individual displays decreasing risk aversion.

Consider an individual who may have to pay a claim in the random amount of x and who wishes to provide a deductible insurance cover against this contingency. If S is the deductible amount, the amount received by the individual is the random variable

$$\phi = \begin{cases} 0 & \text{if} \quad x \leqq S, \\ x - S & \text{otherwise.} \end{cases}$$

The premium is, say, $p(S) = (1+\lambda)E(\phi)$, where $\lambda E(\phi)$ is the insurance company's loading fee.

Assuming that the indivudial's present wealth is A, his final wealth is the random variable

$$\xi = A - p(S) - x + \phi.$$

(i) Show that the expected utility of final wealth is

$$Eu(\xi) = \int_0^S u(A - p(S) - x)\, dF(x) + u(A - p(S) - S)\int_S^\infty dF(x).$$

(j) When is the optimal S finite?

(k) Suppose that the optimal S is finite and that the individual has decreasing risk aversion. Show that the optimal deductible amount is a nondecreasing function of wealth.

(l) When is the optimal $S = 0$?

(m) What do the various models conclude regarding the statement that individuals with decreasing risk aversion assume more risk the larger their wealth levels?

12. Consider an expected utility maximizing investor with strictly concave twice continuously differentiable utility function $u(\cdot)$ over wealth w. Suppose there are two possible investments having random returns ρ_1 and ρ_2 (defined on the bounded range $[a,b]$) per dollar invested. Let initial wealth be w_0 and suppose that the investment allocation is the choice of k, where final wealth is defined to be

$$w = w_0[1 + k\rho_1 + (1-k)\rho_2].$$

Let $\rho_1(A) = \rho_1 + A$, for any constant A.

(a) Suppose ρ_1 and ρ_2 have symmetric distributions but are not positively correlated, and $A \in I \equiv [a-b, b-a]$. Show that $A > 0$ implies $k^* > \frac{1}{2}$, $A = 0$ implies $k^* = \frac{1}{2}$, and $A < 0$ implies $k^* < \frac{1}{2}$, where k^* maximizes expected utility. [*Hint*: Investigate the first-order conditions at $k = \frac{1}{2}$.]

(b) Show that $dk^*/dA|_{A=0} > 0$ in (a) but that dk^*/dA is not generally positive. However, show that dk^*/dA is strictly positive over I if u displays constant absolute risk aversion.

(c) The result in (a) indicates that for the case of an additive shift the investor will hold more than half his portfolio in the stochastically dominant asset. It is of interest to determine whether or not strict first-degree stochastic dominance will imply $k^* > \frac{1}{2}$. Show that this is not the case by considering the following example. Let $w_0 = 1$, $u(w) = \log(w - 0.2)$, and ρ_1 and ρ_2 have the joint probability distribution

ρ_2 \ ρ_1	-0.9	$1+B$
-0.9	0	0.5
1	0.5	0

,

where $B > 0$. Show that ρ_1 strictly dominates ρ_2 for all strictly increasing utility functions. Show that

$$k^* \cong \frac{1}{2}\left(\frac{3.61 + 1.8B}{3.61 + 1.9B}\right) < \frac{1}{2}.$$

(d) Show that essentially all the results above are valid if the assumption that the random returns are bounded is replaced by the assumption that k is bounded and expected utility is bounded.

(e) Suppose u displays constant relative risk aversion, where the constant $\in (0, 1]$. Suppose ρ_1 and ρ_2 have symmetric distributions, $a = -1$. Let $\rho_1(C) = (1+C)\rho_1 + C$, where $C \in (-1, \infty)$. Show that $dk^*/dC > 0$, $C > 0$, implies $k^* > \frac{1}{2}$, $C = 0$ implies $k^* = \frac{1}{2}$, and $C < 0$ implies $k^* < \frac{1}{2}$. (Note that strict inequalities require that ρ_1 and ρ_2 satisfy the regularity assumption that $\Pr[\rho_i = -1] < 1$.)

(f) The assumption that the relative risk-aversion constant $\in (0, 1]$ is crucial for dk^*/dC to be positive. Let $w_0 = \frac{1}{2}$ and $u(w) = -w^{-(c-1)}$, $c > 1$, where c is the risk-aversion constant. Suppose ρ_1 and ρ_2 have the joint probability distribution

ρ_2 \ ρ_1	0	2
0	0	0.5
2	0.5	0

Show that $dk^*/dA|_{A=0} < 0$ whenever $c > 4$. Investigate the behavior of k^* for this example.

13. (Concavity of indifference curves) Let $u(w)$ be an investor's utility function over wealth w and suppose $f(w; \mu, \sigma)$ is the distribution over wealth. Assume that f depends only on mean (μ) and standard deviation (σ) or independent functions of these parameters.

(a) Show that an indifference curve may be written as

$$I = \int u(\mu + z\sigma) f(z; 0, 1)\, dz, \qquad \text{where} \quad z = (w-\mu)/\sigma.$$

(b) Totally differentiate I to show that

$$\frac{d\mu}{d\sigma} = \frac{-\int z u'(w) f(z; 0, 1)\, dz}{\int u'(w) f(z; 0, 1)\, dz}.$$

(c) Use (b) to show that $d\mu/d\sigma$ is >0 if u is strictly concave and <0 if u is strictly convex.

(d) Show further that $d^2\mu/d\sigma^2 < 0$ if u is strictly concave so that the indifference curves are strictly concave.

(e) Show that the indifference curves are strictly convex if u is strictly convex.

Consider the quadratic utility function $u(w) = (1+b)w + bw^2$.

(f) Show that $0 < b < 1$ for risk lovers and $-1 < b < 0$ for risk averters.

(g) Show that

$$\frac{d\mu}{d\sigma} = \frac{-\sigma}{[(1+b)/2b] + \mu} \qquad \text{and} \qquad \frac{d^2\mu}{d\sigma^2} = \frac{-[1 + (d\mu/d\sigma)^2]}{[(1+b)/2b] + \mu}.$$

14. Implicit in the proof of the concavity of mean–variance indifference curves in Exercise 13 is that the means are independent of the variances. Consider a log-normal distribution over w, the wealth outcome of the investment. Let m and s^2 be the expected value and variance of $\log w$. Then the mean and variance of w are $\mu = \exp(m + s^2/2)$ and $\sigma^2 = \mu^2(\exp(s^2) - 1)$, respectively (see Exercise III-ME-20), which are dependent. Assume that the utility function is logarithmic.

(a) Show that $E_w u(w) = E_w[\log w] = m = \log\mu - \frac{1}{2}\log((\sigma^2/\mu^2) + 1)$.

(b) Show that $d\mu/d\sigma = \mu\sigma/(2\sigma^2 + \mu^2) > 0$.

(c) Interpret (b) and show that it implies that the indifference curve $\mu = h(\sigma)$ is pseudo-convex and pseudo-concave.

(d) Show that $d^2\mu/d\sigma^2$ has the same sign as $(1 - 2k^4 - k^2)$, where $k \equiv \sigma/\mu$.

(e) Show that the indifference curve is convex in the area of the μ–σ plane where $k \leqq 0.707$ and concave where $k \geqq 0.707$.

(f) Illustrate these results graphically.

15. (a) Show that an investor's utility function is logarithmic if his marginal utility of wealth is proportional to the reciprocal of his wealth level.

Consider an investor with a logarithmic utility function, $\log w$, faced with a choice between a riskless asset with zero net return and a risky asset having net returns r_1 and r_2 with probabilities p_1 and p_2. Let $0 \leqq x \leqq w_0$ be the optimal amount invested in the risky asset if the investors initial wealth is w_0.

(b) Show that $x = w_0$ if $\min(r_1, r_2) > 0$.

(c) Show that $x = 0$ if $\max(r_1, r_2) < 0$.

(d) Show that $x = -w_0\bar{r}/r_1 r_2 > 0$ if $\max(r_1, r_2) > 0$, $\min(r_1, r_2) < 0$, and $\bar{r} \equiv p_1 r_1 + p_2 r_2 > 0$.

(e) Interpret the results of (b)–(d) noting particularly how x and x/w (the fraction invested in the risky asset) change as w_0, $\bar{r}$, r_1, and r_2 change.

(f) Repeat the calculation in (d) when the return on the risk-free asset is $r_0 > 0$.

(g) Show that x/w_0 is independent of w_0 when there are n possible net realizations for the risky asset $r_1, \ldots, r_n$ having probabilities $p_1, \ldots, p_n$, respectively, and the net risk-free return is r_0 for the cases analogous to (b)–(d).

(h) Referring to (g) show that x/w_0 rises or falls with w if c, an exogenous source of wealth obtained at the same time that the risky asset return is realized, is negative or positive, respectively.

16. Suppose an investor's utility function is the quadratic function $u(w) = w - \alpha w^2$.

(a) Show that the investor prefers, is indifferent to, or does not prefer risks if α is negative, zero, or positive, respectively.

(b) Suppose w has a distribution having mean $\bar{w}$ and variance σ^2. Show that the expected utility is $\bar{w} - \alpha(\bar{w}^2 + \sigma^2)$.

(c) At what wealth level does the marginal utility of wealth become negative?

(d) Show that u has increasing absolute risk aversion and constant relative risk aversion, if $\alpha \neq 0$.

(e) Discuss the implications of the results in (c) and (d).

Consider the investment allocation of w_0 dollars between a risky asset having net return r which has mean $\bar{r}$ and variance σ_r^2 and a risk-free asset having net return $r_0 = 0$.

(f) Utilize (b) to show that the optimal investment in the risky asset is $x = \bar{r}(1 - \alpha w_0)/\alpha\sigma_r^2$, where $\alpha > 0$.

(g) When is x positive, 0, or w_0?

(h) Interpret the results of (f) and (g) noting particularly how x and x/w_0 (the fraction invested in the risky asset) change with $\bar{r}$, σ_r^2, α, and w_0.

(i) Repeat the calculation of (f) when $r_0 > 0$.

17. (Alternative definition of risk aversion) Consider a set of portfolios whose returns are random variables R having probability density functions $f(R; \mu, \sigma)$ all of which differ only by scale and location factors, i.e., $f(R; \mu, \sigma) = \phi[(R - \mu)/\sigma]$. Show that for an investor having a utility function u, the portfolios are characterized completely by indifference curves in the (μ, σ) plane, i.e., show that

(a) $E[u(R) \mid \mu, \sigma] = \int u(\mu + \rho\sigma)\phi(\rho)\, d\rho \equiv \bar{u}(\mu, \sigma)$.

(b) Show that the slope of an indifference curve is

$$\frac{d\mu}{d\sigma} = -\frac{\int \rho u'(\mu+\rho\sigma)\phi(\rho)\, d\rho}{\int u'(\mu+\rho\sigma)\phi(\rho)\, d\rho}.$$

(c) Introducing the Arrow–Pratt absolute risk-aversion index $A(w) = -(u''(w))/(u'(w))$, show that the second derivative of an indifference curve is

$$\frac{d^2\mu}{d\sigma^2} = \frac{\int A(\mu+\rho\sigma)(\rho + d\mu/d\sigma)^2 u'(\mu+\rho\sigma)\phi(\rho)\, d\rho}{\int u'(\mu+\rho\sigma)\phi(\rho)\, d\rho}.$$

(d) Suppose that the function $\phi(\rho)$ is symmetric about $\rho = 0$. Show that $d\mu/d\sigma \geqq 0\ (>0)$ if u is nondecreasing and concave (strictly concave). Interpret this result in terms of risk aversion.

(e) Show that an indifference curve $\mu(\sigma)$ is strictly increasing and convex (strictly convex) if $u(w)$ is strictly increasing and concave (strictly concave).

18. Consider an expected utility maximizer with a logarithmic utility function over wealth w. Suppose that his investment choice is between a sure asset with zero return (money) and a risky asset which is log-normally distributed with mean μ and variance σ^2. Let α be the proportion of risky assets held, ξ the value of each dollar invested in the risky asset and let the initial fortune be $F > 0$.

(a) Show that the expected utility is

$$Z(\alpha) \equiv E_\xi \log[(1-\alpha)F + \alpha\xi F] = \log F + E_\xi \log[1 + \alpha(\xi - 1)].$$

(b) Show that Z is concave in α.

(c) Show that it is optimal to invest in only the risky asset if $(\sigma/\mu)^2 \leqq \mu - 1$. [*Hint*: Show that $dZ(\alpha)/d\alpha > 0$ for $0 \leqq \alpha < 1$ and $dZ(1)/d\alpha \geqq 0$ when this condition obtains.]

(d) Develop a similar result to (c) when the risk-free asset has return $(1+r)$, $r > 0$, for each dollar invested.

19. In Lintner's paper it is shown how to compute the efficient frontier by solving a linear complementary problem of the form $w = \Sigma z - a$, $z'w = 0$, $z \geqq 0$, $w \geqq 0$, where Σ is the positive-definite variance–covariance matrix of returns and a is the mean return vector net of the risk-free rate of return.

(a) Show that solving the linear complementary problem is equivalent to solving the strictly concave quadratic program

$$\{\max a'z - \tfrac{1}{2}z'\Sigma z \mid z \geqq 0\}. \tag{1}$$

[*Hint*: Examine the Kuhn–Tucker conditions of (1).]

(b) Develop a modification of the simplex method to solve this quadratic program.

20. In Lintner's article it is shown that under the normality and existence of riskless asset assumptions, the investor's decision problem may be decomposed into a two-stage process. In stage 1, a choice is made of the optimal proportions of the risky assets, and in stage 2 optimal proportions of the risky asset and the riskless asset are chosen.

(a) Show that the proportions of the risky assets that are optimal are unique if the variance–covariance matrix Σ of the risky assets is positive definite. [*Hint*: Begin by showing that the efficient surface is strictly concave.]

(b) Illustrate graphically the various situations that may occur if Σ is not positive-definite.

(c) How does one calculate these "different" ratios when Σ is positive semidefinite?

(d) Is there a circumstance in which it would be useful to know more than one of these ratios?

21. Referring to the Vickson article, prove

(a) If $f(\lambda) = a\lambda^{\alpha} + b\lambda^{\beta}$ with a, b, α, and β constant ($\alpha \neq \beta$), and if

$$c_1(\sigma)\,u'(f(\lambda)) + c_2(\sigma)\,u'(\sigma f(\lambda)) = \lambda$$

for all $\sigma > 0$ and all λ, then $u'(f(\lambda))$ and $u'(\sigma f(\lambda))$ are linear in λ. [*Hint*: If $u'(f(\lambda))$ is not linear, there exists an interval L in which it is either strictly convex or strictly concave. Consider what happens near $\sigma = 1$.]

(b) Equations (33), (35), and (36) imply either $u'(\xi) = a + b\xi$ or $u'(\xi) = b\xi^c$.

(c) For $g(z) = z^{\alpha}(A + B\log z)$, Eq. (38) implies $B = 0$.

(d) Prove Theorem 3.

22. An investor having utility function $u(w) = -e^{-kw}$ and initial wealth w_0 can invest in a risk-free asset with net rate of return $r \geqq 0$ and a risky asset with normally distributed net rate of return ρ. Let $\mu = E\rho$, $\sigma^2 = \operatorname{var}\rho$, and let $R = w_0[1 + (1-a)r + a\rho]$ be terminal wealth, given an amount $aw_0 \geqq 0$ invested in the risky asset.

(a) Show that the investor's preference ordering for the random returns R can be expressed completely in terms of the mean μ_R and variance $\sigma_R{}^2$ of R.

(b) Show that the investor's mean–variance indifference curves are given by the equations $\frac{1}{2}k\sigma_R{}^2 - \mu_R = l =$ constant. Show that preference decreases with increasing l.

(c) Show that the optimal investment proportion is $a^* = (\mu - r)/kw_0\sigma^2$ if $\mu \geqq r$ and borrowing is unlimited.

Suppose the investor is indifferent between a certain return of \$1000 and a gamble which returns \$0 or \$5000 with equal probability.

(d) Show that $k = 0.657 \times 10^{-3}$.

Suppose that $r = 0.06$, $\mu = 0.12$, $\sigma = 0.04$, and $w_0 = \$1000$.

(e) Show that the optimal policy is to borrow \$56,000 and to invest \$57,000 in the risky asset.

(f) What is the optimal policy if $w_0 = \$5000$?

23. For the investor of Exercise 22 having $w_0 = \$1000$ initially, find the optimal policy under the following conditions.

(a) Borrowing is limited to \$1000.

(b) Borrowing is unlimited provided there is at least a 0.99 probability of repaying the loan plus interest.

(c) Borrowing is unlimited provided there is at least a 0.99 probability of repaying the loan, excluding interest.

(d) The loan, excluding interest, must be repaid with probability 1.

24. A risky investment returns $(1+X_1) > 1$ with probability $\pi > 0$ and $(1+X_2) < 1$ with probability $1-\pi$. Let $\bar{X} = \pi X_1 + (1-\pi) X_2 > 0$ and $\sigma^2 > 0$ be the variance of X. An investor saves part of his wealth $w_0 > 0$ and invests the balance, say a, in the risky investment. Let u be the investor's utility functions and suppose that $u' > 0$ and $u'' < 0$.

(a) Show that the optimal investment, say a^*, is the solution of

$$X_1 \pi u'(w_0 + X_1 a) = -X_2(1-\pi) u'(w_0 + X_2 a).$$

(b) Suppose $u(w) = \ln w$. Show that

$$a^* = -\frac{w_0 \bar{X}}{X_1 X_2} > 0 \quad \text{and that} \quad \frac{da^*}{dw_0} > 0.$$

(c) Suppose $u(w) = w - w^2/2\alpha$ for $\alpha \geqq w_0$. Show that

$$a^* = \frac{(\alpha - w_0)\bar{X}}{\sigma^2 + (\bar{X})^2} > 0 \quad \text{and that} \quad \frac{da^*}{dw_0} < 0.$$

(d) Suppose $u(w) = 1 - e^{-\lambda w}$ for $\lambda > 0$. Show that

$$a^* = \frac{1}{\lambda(X_1 - X_2)} \ln \frac{\pi X_1}{(\pi - 1) X_2} > 0 \quad \text{and that} \quad \frac{da^*}{dw_0} = 0.$$

(e) Let $R(w)$ denote the investor's risk-aversion function. Show that

$$\frac{da^*}{dw_0} \{ \gtreqless 0 \quad \text{if} \quad R' \{ \lesseqgtr 0.$$

25. Show that the model and hence the results in Exercise 24 apply to the following three investment situations.

(a) An investor has initial wealth $G > 0$. Wealth saved provides a net return of $r > 0$. A risky investment has net return of Y_1 $(>r)$ and Y_2 $(<r)$ with probabilities $1 > \pi > 0$ and $1-\pi$. [*Hint*: Let $w_0 = (1+r)G$, $X_1 = Y_1 - r$, and $X_2 = Y_2 - r$.]

(b) An investor has b and L dollars invested in cash and a risky asset, respectively. With probability $0 < 1-\pi < 1$ the risky asset is worthless; otherwise it has value L. The investor can insure against loss of any portion y of the risky asset at a cost of py, where $0 < p < 1$. [*Hint*: Let $w_0 = b + (1-p)L$, $X_1 = p$, and $X_2 = 1-p$.]

(c) A foreign exchange speculator owes g English pounds in 90 days. The forward rate of exchange relating dollars to pounds is f. The spot rate of exchange is s_1 $(>f)$ or s_2 $(<f)$ with probabilities π and $1-\pi$. Let V be the value of the speculator's other assets in 90 days and h be the amount hedged $(0 \leqq h \leqq g)$. [*Hint*: Let $w_0 = V - fg$, $X_1 = f - s_1$, and $X_2 = f - s_2$.]

Exercise Source Notes

Exercise 1 was adapted from Levy and Sarnat (1972a); Exercise 2 was adapted from Levy and Sarnat (1971); Exercises 6–8 were adapted from Hanoch and Levy (1971); Exercise 9 was adapted from Breen (1968); Exercise 10 was based on Arrow (1971); Exercise 11 was based on Mossin (1968a); Exercise 12 was adapted from Rentz and Westin (1972); Exercise 13 was adapted from Tobin (1958); portions of Exercise 14 were adapted from Feldstein (1969); Exercises 15 and 16 were adapted from Freimer and Gordon (1968); Exercise 17 was adapted from Adler (1969); portions of Exercise 18 were adapted from Feldstein (1969); Exercises 24 and 25 were adapted from notes written by Professor R. C. Grinold; they further illustrate the ideas in Exercise 10 due to Arrow (1971).

MIND EXPANDING EXERCISES

*1. Determine stochastic dominance theorems for utility functions that are strictly increasing and concave and for which:

(a) The absolute risk-aversion function is nonincreasing in w.

(b) The relative risk-aversion function is nondecreasing in w.

2. (a) Suppose that a utility function u over wealth w is the cubic function $u(w) = w^3 - 2kw^2 + (k^2+g^2)w$. Suppose $k^2 \geqq 3g^2$. Show that u is monotone increasing, strictly concave, and has decreasing absolute risk aversion in the range $0 \leqq w \leqq \frac{2}{3}k - \frac{1}{3}(k^2-3g^2)^{1/2}$.

*(b) For the cubic function $u(w) = w^3 - \beta w^2 + \gamma w$ determine restrictions on the range of w and the parameters β and γ so that u is monotone nondecreasing, concave, and has nondecreasing relative risk aversion.

3. It is useful in risky situations to know how the optimal asset proportions change as the distributions of the risky assets are altered. Such results indicate how robust the optimal decisions are. They are also useful for policy makers who wish to influence the investment in certain assets since imposition of taxes and other measures influence the return distributions of these assets. An interesting problem where such results are useful is the problem of determining optimal foreign exchange positions. In the simplest case, the investor, an expected utility maximizer of terminal wealth w, with utility function u $(u' > 0, u'' < 0)$ must decide the number of dollars x to invest in a particular currency. His initial wealth w_0 returns him $w_0(1+r)$, with certainty $(r \geqq 0)$, at the end of the investment period. For each dollar invested in the risky currency he makes a profit of ξ dollars ($\xi < 0$ indicates a loss), where ξ is the difference between the spot and forward exchange prices of the currency when the forward contract matures. For simplicity assume that the investor can undertake at zero initial cost any commitment of resources in the forward currency. The decision problem is then to choose the x that maximizes $Eu(w) \equiv Eu[(1+r)w_0 + \xi x]$.

(a) Show that the first- and second-order conditions for utility maximization are $E[\xi u'(w)] = 0$ and $E[\xi^2 u''(w)] < 0$, respectively, and that these conditions are sufficient for x to be the unique maximum.

Consider an upward shift in the mean of ξ, when all other moments about the mean remain constant. Denote the new random variable by $\delta = \xi + \alpha$.

(b) Utilize the first-order conditions to show that a small change in α with expected utility remaining maximized leads to the first-order condiiton

$$E[u'(w+\alpha x) + \delta u''(w+\alpha x)\{x + \delta\, \partial x/\partial\alpha\}] = 0.$$

(c) Show that the partial derivative of x with respect to α when evaluated at the initial position, namely, $\alpha = 0$, is

$$\left.\frac{\partial x}{\partial \alpha}\right|_{\alpha=0} = -\frac{\{E[u'(w)] + xE[\xi u''(w)]\}}{E[\xi^2 u''(w)]}.$$

(d) Utilize the first-order conditions to show that

$$\frac{\partial x}{\partial w_0} = -\frac{(1+r)E[\xi u''(w)]}{E[\xi^2 u''(w)]}.$$

(e) Combine the results in (b)–(d) to show that

$$\left.\frac{\partial x}{\partial \alpha}\right|_{\alpha=0} = k + \frac{[x\, \partial x/\partial w_0]}{(1+r)}, \qquad \text{where} \quad k \equiv -\frac{E[u'(w)]}{E[\xi^2 u''(w)]}.$$

(f) Show that $k > 0$.

[Note (Exercise CR-10) that the sign of $\partial x/\partial w_0$ is positive, zero, or negative as the absolute risk-aversion function $-u''(w)/u'(w)$ is decreasing, constant, or increasing in w.]

(g) Show that $x \geqq 0$, and decreasing absolute risk aversion implies that $\partial x/\partial \alpha|_{\alpha=0}$ is positive.

(h) Indicate when $\partial x/\partial \alpha|_{\alpha=0}$ is negative and interpret this result.

Consider a change in expectations such that all possible expectations of ξ are multiplied by a factor of λ.

(i) Show that $\tilde{x} = x/\lambda$ satisfies the first- and second-order conditions for a maximum.

(j) Interpret the result of (i) and indicate how the result is useful for policy purposes to encourage or discourage investment in the risky currency.

Consider a change in the dispersion of ξ about a constant expected value. Let $\Phi = a + b\xi$, where $a = 0$, $b = 1$ indicates the initial position. Increasing b tends to stretch the distribution of ξ since all outcomes are multiplied by a number larger than 1. To keep the mean constant a must be adjusted so that it satisfies

$$d\bar{\Phi} = da + db\,\bar{\xi} = 0 \qquad \text{or} \qquad da/db = -\bar{\xi}.$$

(k) Show that the first-order conditions become

$$E\left\{\frac{d}{db}(a+b\xi)\,u'[(1+r)\,w_0 + (a+b\xi)\,x]\right\} = 0 \qquad \text{when} \quad d\bar{\Phi} = 0.$$

(l) Utilize the results of (e) and (k) to show that

$$(dx/db)_{\bar{\xi}\ \mathrm{const}} = -x - \bar{\xi}(\partial x/\partial \alpha).$$

(m) Show that x and $\bar{\xi}$ have the same sign, and that if $\bar{\xi} > 0$ and u has decreasing absolute risk aversion, $(dx/db)_{\bar{\xi}\ \mathrm{const}}$ is negative.

(n) Indicate when $(dx/db)_{\bar{\xi}\ \mathrm{const}}$ is positive and interpret this result.

4. Referring to the introduction to Exercise 3 suppose that the speculator may invest in n forward exchange positions, $i = 1, \ldots, n$, where x_i refers to the investment level in currency i and ξ_i to the random profit made per dollar invested in i. The problem is then to choose $x \equiv (x_1, \ldots, x_n)$ to maximize $Eu(w) = Eu[(1+r)\,w_0 + \xi' x]$.

(a) Show that first- and second-order conditions for expected utility maximization are $E[\xi u'(w)] = 0$ (n equations) and $E[\xi\xi' u''(w)]$ negative definite. Show that these conditions are sufficient for x to be the unique maximum.

Consider a shift in the mean of the ith return from ξ_i to $\xi_i + \alpha_i$.

(b) Utilize the first-order conditions to show that a small change in α_i evaluated at $\alpha_i = 0$ with expected utility remaining maximized leads to the first-order condition

$$E[\xi\xi' u''(w)]\,(\partial x/\partial \alpha_i)_{\alpha_i=0} = E[\xi u''(w)]\,x_i + [m] \qquad (n \text{ equations}),$$

where $[m]$ is a column vector of zeros except for $-E[u'(w)]$ in the ith component.

(c) Utilize the first-order conditions to show that

$$E[\xi\xi' u''(w)]\,[\partial x/\partial w_0] = E[\xi u''(w)]\,(1+r).$$

(d) Show that the results in (b) and (c) imply that

$$(\partial x_j/\partial \alpha_i)_{\alpha_i=0} = S_{ij} + x_i(\partial x_j/\partial w_0)/(1+r),$$

where $S_{ij} \equiv -E[u'(w)]\,D_{ij}/D$, D is the determinant of the matrix $E[\xi\xi' u''(w)]$, and D_{ij} is the cofactor of the ijth element of D.

(e) Show that $S_{ii} > 0$ and that S_{ij} may be positive or negative (as forward markets are complements or substitutes).

(f) Show that decreasing absolute risk aversion does not imply that $x_i \, \partial x_i/\partial w_0$ is positive.

Consider the special class of utility functions that satisfy the differential equation $-u'(w)/u''(w) = a+bw$. Theorem 5 in Vickson's paper shows that for such utility functions the optimal risky asset ratios are independent of the initial wealth w_0. Hence the optimal speculative positions have the form $x = k[a+(1+r)bw_0]$, where the constant k depends on the distribution of ξ.

(g) For the special class of utility functions suppose that the absolute and relative risk-aversion functions are nonincreasing and nondecreasing, respectively. Show that $a \geqq 0$, $b \geqq 0$, and that $x_i(\partial x/\partial w_0) = k_i^2[a+(1+r)bw_0][(1+r)b] \geqq 0$.

(h) Interpret the result in (g).

Consider the investor's response if every possible return in the ith forward market is multiplied by the factor λ_i, $i = 1, \ldots, n$.

(i) Show that $\tilde{x}_i = x_i/\lambda_i$ satisfies the first- and second-order conditions for a maximum.

(j) Indicate the tax policy implications of (i), noting in particular that $\tilde{x}_j$ is independent of λ_i if $i \neq j$.

Consider a change in the dispersion of the returns in a particular currency when the expected return in that currency and the distribution of returns in all other markets remain constant (except for changes in the covariances between the return in the given currency and the returns from the other currencies).

Let $\Phi_i = a_i + b_i \xi_i$, with $\alpha_i = 0$, $\beta_i = 1$ initially. An increase in dispersion may be viewed as an increase in b_i coupled with a decrease in a_i so that $\overline{\Phi}_i = \bar{\xi}_i$ (i.e., $da_i/db_i = -\bar{\xi}_i$).

(k) Utilize the first-order conditions to show that

$$[\partial x_t/\partial b]_{\bar{\xi}\,\text{const}} = \begin{cases} -x_i - \bar{\xi}_i[\partial x_j/\partial \alpha_i]_{\alpha_i=0}, & i = j, \\ -\bar{\xi}_i[\partial x_j/\partial \alpha_i]_{\alpha_i=0}, & i \neq j, \end{cases}$$

where $\partial x_j/\partial \alpha_i$ is defined as in (d).

(l) Discuss the results of (k).

Consider the possibility that w_1, terminal wealth from sources other than foreign exchange speculation, is the random variable $p'y + (1+r)y_0$, where p_i is the return per dollar invested in asset i at level y_i and y_0 represents the dollars invested in a risk-free asset having return $(1+r)$, $r \geqq 0$. The budget constraint is $e'y + y_0 = w_0$, where e is a column vector of ones. The decision problem then is to choose (x, y) to maximize $Eu[(1+r)w_0 + \xi'x + (p-re)'y]$.

(m) Show that if $r = 0$, the problem is identical to that studied above, hence results (a)–(l) apply directly.

(n) Suppose $r > 0$. Show that all of the preceding results are valid if the changes in asset returns are restricted to those involved with forward exchange positions.

(o) Suppose $r > 0$. Illustrate how results analogous to the above may be obtained for changes in the asset returns of the other risky assets.

5. Suppose X and Y are random variables taking values in a closed convex subset of the real line. Define Y to be stochastically smaller in mean than X, written $Y \underset{M}{\prec} X$ if and only if $E(Y-t)^+ \leqq E(X-t)^+$ for all t, where $\theta^+ \equiv \max\{\theta, 0\}$.

(a) Show that $Y \underset{M}{\prec} X$ if and only if $\int_t^\infty (G(x)-F(x))\,dx \geqq 0$, where F and G are the cumulative distribution functions of X and Y, respectively.

(b) Show that $Y \underset{M}{\prec} X$ if and only if $Eu(Y) \leqq Eu(X)$ for all convex nondecreasing utility functions u.

(c) Interpret (b).
(d) Illustrate some problem types for which such utility functions would arise naturally.
(e) Recall from the Hanoch–Levy paper that X is preferred to Y by risk averters, say $Y \underset{2}{\precsim} X$, if $Eu(x) \geqq Eu(Y)$ for all concave nondecreasing utility functions u. Under what conditions are the statements $Y \underset{2}{\precsim} X$ and $Y \underset{M}{\precsim} X$ equivalent? [*Hint*: Investigate the case in which F and G cross.]
(f) Find a counter example to the following statement. If $Y \underset{M}{\precsim} X$, that is, X is preferred to Y by all risk lovers, then $X \underset{2}{\precsim} Y$; that is, Y is preferred to X by all risk averters.

6. Consider the class of utility functions U_3 defined on a finite interval $[a, b]$ such that u is three times continuously differentiable and $u' > 0$, $u'' \leqq 0$, and $u''' \geqq 0$.
(a) Referring to the paper by Pratt, show that the absolute risk-aversion index may be nondecreasing or nonincreasing for $u \in U_3$.
(b) Interpret the result in (a).

Recall from the Brumelle and Vickson paper that the concept of third-degree stochastic dominance $Y \underset{3}{\precsim} X$ is defined by the condition

$$\int_a^b u(x)\,[dF(x) - dG(x)] \geqq 0 \qquad \text{for all} \quad u \in U_3,$$

where F and G are the cumulative distribution functions of X and Y, respectively.
(c) Show that $Y \underset{3}{\precsim} X$ iff
 (i) $\int_a^x \int_a^y [G(z) - F(z)]\, dz\, dy \geqq 0$ for all $x \in [a, b]$,
 (ii) $\int_a^b [G(y) - F(y)]\, dy \geqq 0$.
(d) Graphically interpret condition (i) of (c).
(e) Show that $Y \underset{3}{\precsim} X$ iff

$$x^2[G(x) - F(x)]/2 - x\int_a^x y[dG(y) - dF(y)] + \int_a^x (y^2/2)\,[dG(y) - dF(y)] \geqq 0 \qquad \text{for all} \quad x \in [a, b],$$

and

$$\int_a^b y[dF(y) - dG(y)] \geqq 0.$$

(f) Denoting means and variances of F and G by $\mu_F, \mu_G, \sigma_F^2, \sigma_G^2$, respectively, show that $Y \underset{3}{\precsim} X$ only if

$$(\sigma_G^2 - \sigma_F^2) + (\mu_F - \mu_G)(2b - \mu_F - \mu_G) \geqq 0 \qquad \text{and} \qquad \mu_F - \mu_G \geqq 0.$$

(g) Give an example to show that these, the conditions in (f), are not sufficient for $Y \underset{3}{\precsim} X$.
(h) Develop an algorithmic flow chart suitable for computer programming to determine whether $Y \underset{3}{\precsim} X$ utilizing the results in (f). [*Hint*: Utilize a grid for the interval $[a, b]$.]

7. Define the symbol $\lesssim$ by $X \lesssim Y$ if and only if $P[X \geqq x] \leqq P[Y \geqq x]$ for all x. Suppose that $X \lesssim Y$ and show that
(a) $-Y \lesssim -X$,
(b) $rX \lesssim rY$ for all $r \geqq 0$, and
(c) $X + Z \lesssim Y + Z$ if Z is independent of X and of Y.

Suppose that $X + Z \lesssim Y + Z$.
(d) Find a counterexample to the statement that $X + 2Z \lesssim Y + 2Z$ if Z is independent of X and Y.

8. Consider an investor having a utility function for wealth $u(w)$ which is strictly increasing in w. Recall that the risk premium $\pi(w,z)$ of a random return z given initial wealth w is defined by the equation

$$u[w+Ez-\pi(w,z)] = Eu(w+z).$$

(a) Show that if h is any constant,

$$\pi(w+h,z) = \pi(w,z+h).$$

Let

$$A(t) = -\frac{u''(t)}{u'(t)},$$

$$P(t;w) = -\frac{tu''(t+w)}{u'(t+w)},$$

and

$$R(t) = -\frac{tu''(t)}{u'(t)}$$

be the absolute, partial relative, and relative measures of risk aversion, respectively.

(b) Show that $\Psi_A(t) = u'[u^{-1}(t)]$ is concave, linear, or convex according as $A(t)$ is nondecreasing, constant, or nonincreasing.

(c) Show that $\Psi_p(t) = u^{-1}(t)u'[u^{-1}(t)] - wu'[u^{-1}(t)]$ is concave, linear, or convex according as $P(t;w)$ is nondecreasing, constant, or nonincreasing in t.

(d) Show that $\Psi_R(t) = u^{-1}(t)u'[u^{-1}(t)]$ is concave, linear, or convex according as $R(t)$ is nondecreasing, constant, or nonincreasing.

(e) Show that statements (b)–(d) may be altered to strictly concave, linear, strictly convex as ... is strictly increasing, constant, or strictly decreasing.

Suppose that z is a random variable with distribution function F defined on an interval $[a,b]$. We are interested in the behavior of the risk premium $\pi(w,z)$ as z and w are subject to scale changes. If z is replaced by λz and $\lambda =$ constant, we restrict a and λ in such a way that $P[w+\lambda z < 0] = 0$, so that the individual can lose at most his total initial wealth.

(f) Show that $(\partial/\partial\lambda)[\pi(w,\lambda z)/\lambda] > 0$, $=0$, or <0 according as $P(t;w)$ is strictly increasing, constant, or strictly decreasing. [*Hint*: Show that $(\partial/\partial\lambda)[\pi(w,\lambda z)/\lambda] \gtreqless 0$ if and only if

$$u'[u^{-1}(Eu(w+\lambda z))][u^{-1}(Eu(w+\lambda z)) - w] \lesseqgtr E[u'(w+\lambda z)\lambda z].$$

Recall from Exercise I-CR-7 that $\Psi_p E(t) \gtreqless E\Psi_p(t)$ according as Ψ_p is concave, linear, or convex. Examine the variable $t = u(w+\lambda z)$.]

(g) Show that $(\partial/\partial w)[\pi(w,z)] \lesseqgtr 0$ according as $A(t)$ is strictly increasing, constant, or strictly decreasing.

(h) Show that $(\partial/\partial\lambda)[\pi(\lambda w,\lambda z)/\lambda] \lesseqgtr 0$ according as $R(t)$ is strictly increasing, constant, or strictly decreasing.

(i) Discuss the economic interpretation of the results, particularly regarding local versus global measures of risk aversion.

9. Let the partial relative risk-aversion function $P(t;w)$ and relative risk-aversion function $R(t)$ be defined as in Exercise 8.

(a) Let w be fixed. Show that if $P(t;w)$ is nonincreasing in t for t in some interval $(0,t_0)$ with $t_0 > 0$, then either

(i) $P(t;w) = 0$ [so $u''(t+w) = 0$ for $0 < t < t_0$], or

(ii) $w = 0$.

Let $w > 0$ be fixed, and suppose that $t_0 > 0$.

(b) Show that if $P(t;w)$ is monotone (strictly monotone) in t for $0 < t < t_0$, then it is nondecreasing (strictly increasing) for $0 < t < t_0$.

(c) Show that if $R(t)$ is nondecreasing, then either

(i) $u''(t) = 0$, or

(ii) $P(t;w)$ is strictly increasing in t for each w.

10. Consider a strictly increasing utility function $u(w)$ which is concave and bounded on $[0,\infty)$. It is of interest to examine the behavior of $R(w) = -wu''(w)/u'(w)$ for large and small values of w. Suppose there exist numbers r and $w_0 > 0$ such that $R(w) \leqq r$ for $w \geqq w_0$.

(a) Show that

$$u'(w) \geqq cw^{-r} \qquad \text{for} \quad w \geqq w_0, \qquad \text{where} \quad c \equiv u'(w_0)\, w_0{}^r > 0.$$

(b) Show that

$$u(w) \geqq \begin{cases} u(w_0) + \dfrac{c}{1-r}[w^{1-r} - w_0^{1-r}], & r \neq 1 \\ u(w_0) + c[\log w - \log w_0], & r = 1. \end{cases}$$

(c) Show that $R(w)$ cannot converge to a limit which is less than 1 as $w \to \infty$.

(d) Show that $R(w)$ cannot converge to a limit which is greater than 1 as $w \to 0$. This shows that $R(w)$ must be increasing on the average, from values somewhat less than 1 for small w to values somewhat greater than 1 for large w.

11. Suppose the random variables X and Y have distributions F and G, respectively, and that

(i) $\int_{-\infty}^{\infty} (G-F)\, du \geqq 0$ for all concave nondecreasing u, and

(ii) $\int_{-\infty}^{\infty} (G-F)\, du \leqq 0$ for all concave nonincreasing u.

(a) Show that $\int_{-\infty}^{x} [G(t)-F(t)]\, dt \geqq 0$ for all x and $\int_{-\infty}^{\infty} [G(t)-F(t)]\, dt = 0$.

(b)* Either verify the converse of (a), i.e., that the conditions in (a) imply (i)–(ii), or find a counterexample.

(c) Show that (i)–(ii) is equivalent to $Eu(X) \geqq Eu(Y)$ for all concave u, provided all the integrals involved exist and are finite.

12. Let X and Y be discrete random variables concentrated on the finite set of points $\Omega = \{\alpha_1, \alpha_2, \ldots, \alpha_n\} \subset [a,b]$. Let X_1 be a discrete random variable concentrated on Ω, and let $p_i = P[X = \alpha_i]$, $p_i' = P[X_1 = \alpha_i]$. X_1 is said to differ from X by a mean-preserving spread (MPS) if

(i) $p_i = p_i'$ except for four values i_1, i_2, i_3, and i_4, with $\alpha_{i_1} \leqq \alpha_{i_2} \leqq \alpha_{i_3} \leqq \alpha_{i_4}$;

(ii) $p_{i_1}' - p_{i_1} = p_{i_2} - p_{i_2}' \geqq 0$, $p_{i_3} - p_{i_3}' = p_{i_4}' - p_{i_4} \geqq 0$;

(iii) $\sum_{j=1}^{4} \alpha_{i_j}(p_{i_j}' - p_{i_j}) = 0$.

(a) Show that X_1 has less probability mass in the "middle" and more mass in the "tails" than X; that is, X_1 is more spread out than X.
(b) Show that $EX_1 = EX$.

Let F and F_1 be the cumulative distribution functions of X and X_1, respectively. Define $T_1(x) = \int_a^x [F_1(y) - F(y)]\,dy$.
(c) Show that $T_1(x) \geqq 0$ for all $x \in [a,b]$, and $T_1(b) = 0$.
(d) Construct a discrete random variable Z_1 such that $P[X_1 \leqq x] = P[X+Z_1 \leqq x]\ \forall x$, with $E[Z_1 \mid X] = 0$.

The random variable Y is said to differ from X by a sequence of MPSs if there exists a finite sequence of random variables

$$X = X_0, X_1, X_2, \ldots, X_N, X_{N+1} = Y$$

such that X_{i+1} differs from X_i by a MPS. Let G be the cumulative distribution function of Y; assume that $EX = EY$, and

$$\int_a^x [G(y) - F(y)]\,dy \geqq 0 \qquad \text{for all} \quad x \in [a,b].$$

(e) Show that Y differs from X by a sequence of MPSs. [*Hint*: Use induction on the number of steps of the function $G-F$.]
(f) Show by constructive arguments that there exists a random variable Z such that $P[Y \leqq x] = P[X+Z \leqq x]\ \forall x$, with $E[Z \mid X] = 0$.
*(g) Generalize the discussion to the case $EX > EY$.

13. Two random variables X_1 and X_2 are equal in distribution, written as $X_1 \overset{d}{=} X_2$, if $P[X_1 \leqq x] = P[X_2 \leqq x]\ \forall x$. Define two orderings of random variables:
(i) $Y \underset{N}{\precsim} X$ if $Y \overset{d}{=} X+Z$ with $E[Z \mid X] \leqq 0$, and
(ii) $Y \underset{E}{\precsim} X$ if $Y \overset{d}{=} X+Z$ with $E[Z \mid X] = 0$.

(a) Show that if $Y \underset{N}{\precsim} X$, then there exists a nonnegative function f such that $Y \underset{E}{\precsim} X - f(X)$. Conversely, if $Y \underset{E}{\precsim} X$ and f is any nonnegative function, then $Y \underset{N}{\precsim} X + f(X)$.
(b) Show that if $Eu(X) \geqq Eu(Y)$ for all concave nondecreasing u where X and Y are defined on finite ranges, then there exists $f(\cdot) \geqq 0$ such that $Eu[X - f(X)] \geqq Eu(Y)$ for all concave u (whether nondecreasing or not).
(c) If X and Y are defined on finite ranges, if $EX = EY$, and if $Eu(X) \geqq Eu(Y)$ for all concave nondecreasing u, show that $Eu(X) \geqq Eu(Y)$ for all concave u.

14. It is of interest to examine the question of separation in portfolio analysis when short sales are permitted and when the investor is required to remain solvent with certainty at the end of the period. Assume that the investor's utility function for wealth, $u(w)$, is twice continuously differentiable for $w > 0$, and that $u'(w_1) > 0$ for some $w_1 > 0$. Consider an investor having initial wealth $w > 0$, confronted by the following investment possibilities:
(i) An amount x_0 in a risk-free asset returning $(r-1) > 0$ per dollar; x_0 may be negative (indicating borrowing), but the debt must be repaid at the end of the period.
(ii) An amount x_i in a risky asset returning ρ_i per dollar, where $i = 1, \ldots, N$; there exists S, $0 \leqq S \leqq N$, such that the first S of the assets may be sold short, but must be covered at the end of the period. Assume that the distribution function of risky returns, $F(\xi) = P\{\rho \leqq \xi\}$, is known. Assume further that
(iii) $P\{\rho_i \in [0, B]\} = 1$, $i = 1, \ldots, N$, for some finite $B > 0$, and
(iv) $P\{\sum_{i=1}^N (\rho_i - r)\theta_i < 0\} > 0$ for all finite $\theta \neq 0$ which satisfy $\theta_i \geqq 0$ for $i > S$.

Condition (iv) is called the "no-easy-money" (NEM) condition; it states that *no* combination of risky assets yields with probability 1 a return exceeding the risk-free return (this will be clarified below).

The x_i satisfy $\sum_{i=0}^{N} x_i = w$, $x_i \geqq 0$, $i = S+1, \ldots, N$. The return on the investment is the random variable

$$R = x_0 r + \sum_{i=1}^{N} x_i \rho_i = \sum_{i=1}^{N} (\rho_i - r) x_i + rw.$$

Since any borrowing must be fully secured, we have the additional *solvency constraint*

(v) $P\{\sum_{i=1}^{N} (\rho_i - r) x_i + rw \geqq 0\} = 1.$

The investor's problem is

$$P1: \quad \max_x Eu\left[\sum_{i=1}^{N} (\rho_i - r) x_i + rw\right]$$

$$\text{s.t.} \quad P\left\{\sum_{i=1}^{N} (\rho_i - r) x_i + rw \geqq 0\right\} = 1 \quad \text{and} \quad x_i \geqq 0, \quad i = S+1, \ldots, N.$$

Define

$$b(x) = \inf\left\{b \mid P\left[\sum_{i=1}^{N} (\rho_i - r) x_i < b\right] > 0\right\}.$$

(a) Show that the NEM conditions imply $b(x) < 0$ for $x \neq 0$, and $x_i \geqq 0$, $i > S$.
(b) Show that $b(0) = 0$ and $b(\lambda x) = \lambda b(x)$ for $\lambda \geqq 0$.
(c) Show that the solvency constraint (v) becomes

$$b(x) + rw \geqq 0 \qquad \text{or} \qquad b(x/rw) + 1 \geqq 0.$$

(d) Let $X = \{x \mid b(x) + rw \geqq 0,\ x_i \geqq 0,\ i > S\}$ be the set of investments satisfying the solvency constraints. Show that X is a nonempty compact, convex set. (Therefore infinite borrowing or infinite short sales are ruled out.)
(e) Show that problem $P1$ has a solution having a finite maximum; show that the solution is unique if u is strictly concave. [*Hint*: Make appropriate assumptions along the lines indicated in problem I-ME-18.]

Denote the solution to problem $P1$, for wealth level w, by $x^*(w)$. In the presence of the solvency constraint it becomes necessary to distinguish between *local* and *global* separation. We say that the separation property holds

(vi) *locally*, if there exists $w_0 > 0$ such that (1) $x_j^*(w_0) \neq 0$, and (2) $x_j^*(w) \neq 0$ in a neighborhood M of w_0 implies (3) $x_i^*(w)/x_j^*(w) = c_i$ is constant for $w \in M$ and $i = 1, \ldots, N$ and (4) there exists a distribution function F such that at least two of the ratios above are nonzero; and it holds

(vii) *globally*, if (1) $x_j^*(w_0) \neq 0$ implies (2) $x_i^*(w)/x_j^*(w) = c_i$ is constant for all $w > 0$ and $i = 1, \ldots, N$, and (3) at least two of these ratios are nonzero for some F.

(f) Let $u(w) = -e^{-cw}$, $c > 0$. For this u, show that local, but not global separation holds. [*Hint*: Consider the associated problem

$$P2: \quad \max_x Eu\left[\sum_{i=1}^{N} (\rho_i - r) x_i + k\right] \quad \text{s.t.} \quad b(x/k) + 1 \geqq 0 \quad \text{and} \quad x_i \geqq 0,$$

$$i = S+1, \ldots, N, \qquad \text{where} \quad k > 0 \text{ is constant}.$$

Show that there exist F and k such that the solution x^0 of $P2$ has at least two nonzero components and satisfies the interior constraint conditions $b(x^0/k)+1 > 0$. Show that $P1$ becomes

$$\max[-u(rw-k)]\, Eu\left[\sum_{i=1}^{N}(\rho_i - r)x_i + k\right]$$

$$\text{s.t.}\quad x_i \geqq 0, \quad i = S+1,\ldots,N \quad\text{and}\quad b(x/rw)+1 \geqq 0.$$

Show that $x^*(w) = x^0$ for all w such that $b(x^0/rw)+1 \geqq 0$, and show this includes all $w \geqq (1/r)k$. Finally, show that this condition fails to hold for w in some neighborhood of $w = 0$, so global separation fails.]

(g) Given the utility functions $u(w) = w$, $u(w) = -w^c$ $(c < 0)$, or $u(w) = w^c$ $(0 < c < 1)$, show that global separation holds. [*Hint*: For each u, show that there exist functions g, h such that $u(x+y) = g((x/y)+1)h(y)$, and show that $P1$ becomes

$$P3: \qquad \max[u(rw/k)]\, Eu\left[\sum_{i=1}^{N}(\rho_i - r)x_i(k/rw) + k\right]$$

$$\text{s.t.}\quad b(x/rw)+1 \geqq 0 \quad\text{and}\quad x_i \geqq 0, \quad i \geqq S+1.$$

Referring to problem $P2$, show that the solution is $x^*(w) = (rw/k)x^0$ for all $w > 0$.]

(h) For $u(w) = w$, show that $x_i^*(w) \neq 0$ for at most one $i \geqq 1$ if the ρ_i are independently distributed. Show by an example that this conclusion is false if the ρ_i are not independent.

15. Given that $Eu(X) \geqq Eu(Y)$ for all concave nondecreasing u, find an algorithm for constructing a joint distribution function of X and Z such that $E[Z\,|\,X] \leqq 0$ and $P[Y \leqq x] = P[X+Z \leqq x]$ for all x.

16. It is of interest to study substitution and complementary effects in the choice of risky assets. Suppose the investor's utility function over wealth w is $u(w) = w - \frac{1}{2}\alpha w^2$, where $\alpha > 0$. Let $x = (x_1,\ldots,x_n) \geqq 0$ be the asset allocation vector and $\rho \equiv (\rho_1,\ldots,\rho_n)$ denote the random returns per dollar invested. Suppose the mean return vector is $\bar{\rho}$ and the variance–covariance matrix of ρ is Σ, which is assumed to be positive definite.

(a) Develop the Kuhn–Tucker conditions for the expected utility maximizing problem, assuming initial wealth is w_0.

(b) Suppose all $x_i^* > 0$. Let $m_{ij} \equiv -\alpha(\sigma_{ij} + \bar{\rho}_i\bar{\rho}_j)$. Show that the conditions in (a) reduce to $\sum_{j=1}^{n} m_{ij}x_j - \lambda = -\bar{\rho}_i$, $i = 1,\ldots,n$, and $e'x = w_0$, where $e \equiv (1,\ldots,1)'$.

(c) Let

$$H \equiv \begin{bmatrix} m_{11} & \cdots & m_{1n} & -1 \\ \vdots & & \vdots & \vdots \\ m_{n1} & & m_{nn} & -1 \\ -1 & & -1 & 0 \end{bmatrix},$$

and D be the determinant of H and D_{ij} the cofactor of the ijth element of H. We are interested in the demand response to changes in expected returns. Show that

$$\left(\frac{\partial x_j}{\partial w_0}\right)_{\mu_j=\text{const}} = \frac{D_{n+1,j}}{D}$$

and

$$\left(\frac{\partial x_j}{\partial \mu_i}\right)_{w_0=\text{const}} = \underbrace{-(1-\alpha\bar{\rho}'x)\frac{D_{ij}}{D}}_{=S_{ij}} + \alpha x_i \sum_{k=1}^{n} \bar{\rho}_k \frac{D_{kj}}{D}.$$

Then

$$S_{ij} = \left(\frac{\partial x_j}{\partial \mu_i}\right)_{w_0 = \text{ onst}} - \alpha x_i \sum_{k=1}^{n} \bar{p}_k \frac{D_{kj}}{D}.$$

[*Hint*: Totally differentiate the conditions in (*b*).]

(d) Consider the effect of a change in the total expected return to the investor without any change in risk. Then $E(w) = \bar{p}'x - \tau$ and $\mathrm{var}(w) = x'\Sigma x$, where τ is a tax if $\tau \geqq 0$ or a subsidy if $\tau \leqq 0$. Show that the Kuhn–Tucker conditions are

$$H \begin{bmatrix} x_1 \\ \vdots \\ x_n \\ -\lambda \end{bmatrix} = \begin{bmatrix} -(1+\alpha\tau)\bar{p}_1 \\ \vdots \\ -(1+\alpha\tau)\bar{p}_n \\ -w_0 \end{bmatrix}.$$

Differentiate both sides by τ to show that

$$-\left[\frac{\partial x_j}{\partial E(w)}\right]_{\bar{p}_i = \text{const}} = \left(\frac{\partial x_j}{\partial \tau}\right)_{\bar{p}_i = \text{const}} = \frac{-\alpha \sum_{k=1}^{n} \bar{p}_k D_{kj}}{D}.$$

(e) Utilize (d) to show that

$$S_{ij} = \left(\frac{\partial x_j}{\partial p_i}\right)_{w_0 = \text{const}} - X_i \left[\frac{\partial x_j}{\partial E(w)}\right]_{\bar{p}_k = \text{const}}.$$

Show how S_{ij} may be interpreted as the effect of the change in μ_i on x_j provided that the investor is compensated for the change in μ_i so as to enable him to enjoy the same expected wealth with the same risk—analogous to the Slutsky equation in consumer theory.

We may say that assets i and j $(i \neq j)$ are substitutes or complements if S_{ij} is negative or positive, respectively.

(f) Show that $S_{ij} = S_{ji}$ so that the substitution effect is symmetric.

(g) Show that $S_{ii} > 0$. Hence the rise in expected return on an asset will increase its own demand—if the wealth effect is neglected.

(h) Show that $\sum_j S_{ij} = 0$. Hence $\sum_{j \neq i} S_{ij} < 0$ so that substitution effects dominate the complementary effects. Show that

$$\sum_{i=1}^{m} \sum_{j=1}^{m} S_{ij} > 0 \quad \text{and} \quad \sum_{i=1}^{m} \sum_{j=m+1}^{n} S_{ij} < 0 \quad \text{if} \quad 1 \leqq m < n.$$

Hence if the n assets are divided into two arbitrary groups these groups can be treated as substitutes for each other—if the substitution effects within each group are negligible.

(i) Utilize the result in (d) to show that

$$\sum_j \bar{p}_j \left(\frac{\partial x_j}{\partial \tau}\right)_{\mu_i = \text{const}} = -\alpha \sum_j \sum_k \frac{D_{kj}}{D} \bar{p}_k \bar{p}_j > 0,$$

so that the investor will react to increases in τ (lump-sum taxes) by increasing the expected value of his wealth.

Let us now consider the effect of the change in risk on the demand for assets. Suppose only σ_{ij} is variable and all other parameters are constant. Differentiating the equations in (b) with respect to x_i and σ_{ij} gives

$$\sum_j m_{ij}\, dx_j - d\lambda = \alpha \sum_j x_j\, d\sigma_{ij}$$

and

$$\sum_j dx_j = dw_0.$$

Let $T_{ij}{}^k \equiv \partial x_k / \partial \sigma_{ij}$. Then

$$T_{ij}{}^k = \alpha \frac{(x_i D_{jk} + x_j D_{ik})}{D} \qquad \text{if} \quad i \neq j$$

and

$$T_{ii}{}^k = \alpha \frac{x_i D_{ik}}{D}.$$

(j) Show that $T_{ii}{}^i < 0$; that is, the demand for an asset is reduced when it becomes more risky.
(k) Show that $(i \neq k)$ $T_{ii}{}^k > 0$ (<0) if x_i and x_k are substitutes (complements). That is, the increase in risk of a particular asset increases the demand for all assets that are substitutes for it and reduces the demand for all assets that are complements to it.
(l) Show that if x_i and x_j are complements, the increase in σ_{ij} will reduce the demand for x_i and x_j.
(m) Suppose i, j, and k are all different. Show that if σ_{ij} increases, the risk of a composite asset consisting of x_i and x_j increases. Thus x_k increases (decreases) if x_k is a substitute (complement) with x_i and x_j.
(n) Suppose Y is a square nonsingular matrix and t is a scalar. Let $\partial Y/\partial t = |\partial Y_{ij}/\partial t|$.

Show that

$$\frac{\partial Y^{-1}}{\partial t} = -Y^{-1} \frac{\partial Y}{\partial t} Y^{-1}.$$

[*Hint*: Differentiate $Y^{-1}Y = I$.]
(o) Show that

$$\left(\frac{\partial S_{ij}}{\partial \sigma_{ij}}\right)_{\overline{\rho_k} = \text{const}} = -(1 - \alpha \bar{\rho}' x) \frac{\partial h^{ij}}{\partial \sigma_{ij}} + \alpha h^{ij} \left[x_i \frac{\partial x_j}{\partial E(w)} + x_j \frac{\partial x_i}{\partial E(w)} \right]$$

where h_{ij} is the ijth element of H^{-1}. [*Hint*: Show that $h^{ii} = D_{ii}/D < 0$ and $\partial h^{ij}/\partial \sigma_{ij} > 0$; then utilize the result in (n) along with the definition of $T_{ij}{}^k$ and the result in (d).] Show that $(\partial S_{ij}/\partial \sigma_{ij})_{\rho_k = \text{const}} < 0$ if the wealth effects $\partial x_j/\partial E(w)$ and $\partial x_i/\partial E(w)$ can be neglected. Discuss how this result lends support to the statement that if the returns on two assets are positively (negatively) correlated, they are likely to be substitutes (complements).
(p) Discuss the possibilities of weakening some assumptions of the analysis.

17. Let a set of portfolios be characterized by random returns R which are log-normally distributed, i.e., $y = \log R \sim N(m, s)$, where $m = Ey$ and $s^2 = \operatorname{var} y$. Consider an investor exhibiting decreasing absolute and constant relative risk aversion with $-wu''(w)/u'(w) = c$, where $c > 0$.

(a) Show, by means of a linear transformation of the utility index, that all such $u(w)$ may be represented by

$$u(w) = w^{1-c} \quad \text{if} \quad 0 < c < 1,$$

$$u(w) = \log w \quad \text{if} \quad c = 1,$$

and

$$u(w) = -w^{1-c} \quad \text{if} \quad 1 < c < \infty.$$

(b) With $\pm w^{1-c} = \pm \exp[(1-c)\log w]$, show that expected utility becomes

$$\bar{u} = \pm(2\pi)^{-1/2}s^{-1}\int \exp[(1-c)y]\exp[-(y-m)^2/2s^2]\,dy.$$

(c) Show that this can be written as

$$\bar{u} = \pm\exp[(1-c)m + (1-c)^2 s^2/2]\left\{(2\pi)^{-1/2}s^{-1}\int \exp -\{y-[m+(1-c)s^2]\}^2/2s^2)\,dy\right\}.$$

Show that the value of the terms in the large braces is 1.

(d) Show that the mean μ and variance σ^2 of R are given by

$$\mu = \exp(m+s^2/2) \qquad \text{and} \qquad \exp(s^2) = 1 + (\sigma/\mu)^2.$$

Thus show that the indifference curves in the (μ, σ) plane are given by

$$\bar{u}(\mu, \sigma) = \pm\mu^{1-c}[1+(\sigma/\mu)^2]^{(c^2-c)/2}.$$

(e) Show that the slope of an indifference curve is

$$\frac{d\mu}{d\sigma} = \frac{c\sigma\mu}{[\mu^2 + \sigma^2(c+1)]} \geqq 0$$

for all $c > 0$, $c \neq 1$. Recall that this holds also for $c = 1$, logarithmic utility (Exercise CR-14).

(f) Show that the second derivative of an indifference curve is

$$\frac{d^2\mu}{d\sigma^2} = \frac{c - c(c+1)(\sigma/\mu)^4 - c^2(\sigma/\mu)^2}{\mu^{-5}[(c+1)\sigma^2 + \mu^2]^3}$$

and show that

$$\frac{d^2\sigma}{d\mu^2} \gtrless 0 \qquad \text{for} \quad \frac{\sigma}{\mu} \lesssim (1+c)^{-1/2}.$$

Thus $\mu(\sigma)$ is convex for small σ and concave for large σ.

(g) Show whether there is a discrepancy between this result and the result of Exercise CR-17, part (e).

18. In Lintner's paper it is shown that one may compute the efficient frontier by solving a certain fractional program or by solving a linear complementary problem of the form

$$w = Mz + q, \qquad w'z = 0, \qquad w \geqq 0, \quad z \geqq 0 \tag{1}$$

for the n-vectors w and z, where M is a given positive-definite matrix of order n and q is a given n-vector.

(a) Show that all positive-definite matrices, i.e., all M such that $x'Mx > 0$ if $x \neq 0$, have positive principal minors.

(b) Give an example to show that the converse of (a) is false.
(c) Show that if M has positive principal minors, then there exists $w \neq 0$ such that $w = Mz$ and for $i = 1, \ldots, n$, w_i and z_i have opposite signs (i.e., $w_i z_i \leqq 0$).
(d) Show that the system (1) always has at least one solution.
(e) Utilize (c) and (d) to show that the system (1) always has a unique solution.
(f) Suppose $z_i \geqq 0$ for $i = 1, \ldots, s$ but that z_i is unconstrained if $i = s+1, \ldots, n$. Develop a procedure for computing the solution to (1) that involves inverting an $(n-s) \times (n-s)$ matrix and solving a linear complementary problem of order s.

19. Consider the utility function

$$u(w) = t^2 w + (1/t) e^{tw} + k$$

over wealth w, where t is a parameter and k is a scaling factor.
(a) Show that u can be made to satisfy any reasonable scale of utility such as $u(-1) = -1$, $u(0) = 0$.
(b) Show that marginal utility is always positive and that u is concave iff t is nonpositive.
(c) Show that u has decreasing absolute risk aversion if $t \neq 0$.

Suppose the portfolio choice is between a risk-free asset having a nonnegative return and consuls, i.e., a bond with no redeeming rate. Hence, suppose w has the distribution $f(w)$ with mean μ and variance $\sigma^2 > 0$ and $w \geqq -1$. Suppose $t \leqq 0$.
(d) Show that expected utility is finite and equal to

$$[k+(1/t)] + [1+t^2]\mu + (t/2)e^{\theta}[\mu^2+\sigma^2]$$

for some θ. [*Hint*: Utilize a Taylor series expansion of the integral in the expression for $Eu(w)$ and the first mean value theorem.]
(e) Interpret the expression in (d) in the light of mean-variance analysis.
(f) It is tempting to implicitly differentiate twice the equation of constant expected utility and combine expressions to show that

$$\frac{d^2\mu}{d\sigma^2} = \frac{1}{\sigma}\left[1 + \left(\frac{d\mu}{d\sigma}\right)^2\right]\frac{d\mu}{d\sigma}$$

and then to conclude that the investor is a diversifier, since he has convex indifference curves. Unfortunately, this derivation is not valid because θ is a function of f and in particular μ and σ.
(g) Determine the correct expression for $d^2\mu\, d\sigma^2$ when f is normal, uniform, and triangular.
(h) Investigate the plunging behavior of the investor for various values of μ, σ, and t. In particular, show that whether or not the investor plunges does indeed depend upon the form of f and these parameters.

20. (nth-order stochastic dominance) Consider the set U_n of utility functions such that

$$U_n = \{u \mid u^{(k)} \geqq 0,\ k \text{ odd and } u^{(k)} \leqq 0,\ k \text{ even},\ k = 1, 2, \ldots, n\},$$

where $u^{(k)}$ denotes the kth derivative of u. To avoid technical problems, consider random variables X and Y with finite range $[a, b]$.

(a) Determine necessary and sufficient conditions for stochastic dominance in U_n, that is, for $Eu(x) \geqq Eu(Y)$ for all $u \in U_n$. [*Hint*: Find the appropriate $G_n \subset U_n$ such that $Eu(x) \geqq Eu(Y)$ for all $u \in U_n$ iff $Eu(x) \geqq Eu(Y)$ for all $u \in G_n$.]

(b) Find $\lim_n G_n = G_\infty$, and give necessary and sufficient conditioning for stochastic dominance over the set $U_\infty = \{u | u^{(k)} \geqq 0,\ k \text{ odd and } u^{(k)} \leqq 0,\ k \text{ even},\ k = 1, 2, \ldots\}$.

21. This problem is concerned with the qualitative behavior of portfolios consisting of a blend of one safe and one risky asset. The portfolio may be characterized in the following three ways: (1) by the percentage of its value held in the safe asset; (2) by its statistical properties, e.g., variance, etc.; and (3) by certainty equivalents, i.e., the rate of return that would make the investor indifferent between his risky portfolio and a safe asset.

Suppose that the investor maximizes expected utility of terminal wealth w, where his utility function u is strictly increasing, strictly concave, and thrice differentiable. Initial wealth is w_0. This risky asset yields a return per dollar of ρ_θ if the state of the world is θ, while the safe asset returns ρ_1 for all θ. Assume that $E\rho_\theta > \rho_1$ and that the risky asset does not dominate the safe asset, i.e., $\min_\theta \rho_\theta < \rho_1$. Let a_1 and $1 - a_1$ denote the fractions of initial wealth invested in the safe and risky assets, respectively. The investor may maximize expected utility by solving

$$U^* \equiv \max_{a_1} \sum_\theta u[(a_1 \rho_1 + (1 - a_1)\rho_\theta) w_0] \pi_\theta$$

where π_θ is the probability that state θ will occur.

(a) Show that the necessary and sufficient condition for optimality is

$$Eu'[w_0 (a_1 \rho_1 + (1 - a_1)\rho_\theta)](\rho_1 - \rho_\theta) = 0.$$

Let R denote the index of relative risk aversion $-u''w/u'$.

(b) Show that

$$\frac{da_1}{dw_0} \begin{matrix} \geqq \\ = \\ \leqq \end{matrix} 0 \quad \text{as} \quad \frac{dR}{dw} \begin{matrix} \geqq \\ = \\ \leqq \end{matrix} 0.$$

That is, the wealth elasticity of the demand for the riskless asset is greater than, equal to, or less than unity as relative risk aversion is an increasing, constant, or decreasing function of wealth. [*Hint*: Write the expression for da_1/dw_0 in terms of dR/dw.]

There are many statistics that may be used to characterize a portfolio, such as the mean, variance, and range. Let the mean return per dollar invested be $r = a_1 \rho_1 + \sum_\theta (1 - a_1)\rho_\theta \pi_\theta$. Then the variance of return per dollar invested is $\sigma^2 = \sum_\theta (a_1 \rho_1 + (1 - a_1)\rho_\theta - r)^2 \pi_\theta$. The range is $[r_L, r_U] = [\min_\theta r_\theta, \max_\theta r_\theta]$ where $r_\theta \equiv a_1 \rho_1 + (1 - a_1)\rho_\theta$.

(c) Show that

$$\frac{dr}{dw_0} \begin{matrix} \geqq \\ = \\ \leqq \end{matrix} 0, \quad \frac{d\sigma^2}{dw_0} \begin{matrix} \geqq \\ = \\ \leqq \end{matrix} 0, \quad \frac{dr_L}{dw_0} \begin{matrix} \leqq \\ = \\ \geqq \end{matrix} 0, \quad \text{and} \quad \frac{dr_U}{dw_0} \begin{matrix} \geqq \\ = \\ \leqq \end{matrix} 0 \quad \text{as} \quad \frac{dR}{dw} \begin{matrix} \leqq \\ = \\ \geqq \end{matrix} 0.$$

That is, the mean, variance, and range of the rate of return on the portfolio as a whole are increasing, constant, or decreasing functions of wealth as relative risk aversion is a decreasing, constant, or increasing function of wealth. [*Hint*: Utilize (b).]

We say that a random variable X with distribution function F is "more variable" than another Y with distribution G if all risk averters prefer Y to X, i.e. $\int u(x)\, dF(x) < \int u(y)\, dG(y)$.

Let $v_\theta \equiv r_\theta/r$ and note that $Ev_\theta = 1$.

(d) Interpret v_θ.

(e) Show that v_θ becomes more or less variable with increasing wealth as $dR/dw \gtrless 0$.

Define the certainty equivalent r^* as that certain return that would make the investor indifferent between his optimally chosen portfolio and investing all his wealth at r^*. Hence $u[r^* w_0] = U^*$.

(f) Show that

$$\frac{dr^*}{dw_0} \begin{matrix} \geqq \\ = \\ \leqq \end{matrix} 0 \quad \text{as} \quad \frac{dR}{dw} \begin{matrix} \leqq \\ = \\ \geqq \end{matrix} 0.$$

That is, the certainty eqiuvalent return is an increasing, constant, or decreasing function of wealth as relative risk aversion is a decreasing, constant, or increasing function of wealth.

Let $A \equiv -u''/u'$ denote the absolute risk-aversion function. Let $z_R \equiv (1-a_1)w_0$ be the investment in the risky asset.

(g) Show that

$$\frac{dz_R}{dw_0} \begin{matrix} \geqq \\ = \\ \leqq \end{matrix} 0 \quad \text{as} \quad \frac{dA}{dw} \begin{matrix} \leqq \\ = \\ \geqq \end{matrix} 0.$$

That is, the demand for the risky asset increases, remains unchanged, or decreases with initial wealth as there is decreasing, constant, or increasing absolute risk aversion. [*Hint*: Implicitly differentiate the first-order conditions.]

Let $\bar{w} \equiv w_0 \rho_1 + z_R(E\rho_\theta - \rho_1)$, $\sigma_w^2 \equiv E[w-\bar{w}]^2$, and Range $\equiv \max_\theta w_\theta - \min_\theta w_\theta$.

(h) Assume that $z_R > 0$. Show that

$$\frac{d\sigma_w^2}{dw_0} \gtrless 0 \quad \text{as} \quad \frac{d\,\text{Range}}{dw_0} \gtrless 0 \quad \text{as} \quad \frac{dA}{dw} \lessgtr 0.$$

That is, the variance and range of terminal wealth increase or decrease with initial wealth as there is decreasing or increasing absolute risk aversion.

Let the certainty equivalent level of wealth w^* be that level of wealth which, if invested n the safe asset, would yield the same expected utility as the optimally chosen portfolio, i.e., $u[w^* \rho_1] = U^*$.

(i) Show that

$$\frac{dw^*}{dw_0} \begin{matrix} \geqq \\ = \\ \leqq \end{matrix} 1 \quad \text{as} \quad \frac{dA}{dw} \begin{matrix} \leqq \\ = \\ \geqq \end{matrix} 0.$$

That is, the certainty equivalent level of wealth increases more than, the same as, or less than initial wealth as there is decreasing, constant, or increasing absolute risk aversion. [*Hint*: Check the sign of dp/dw_0 where $p \equiv w_0 - w^*$.]

(j) Compare all the results. Are there any surprises?

22. (Optimal reinsurance) Typically, insurance companies lay off some of the risk investments that they have assumed in the primary insurance market. This problem compares the risk-reducing ability of three types of reinsurance schemes.

Let the random variables $0 \leqq x \leqq 1$ and $S \geqq 0$ represent the percentage loss to be incurred on an insurance contract and the size (liability) of this contract. The total dollar loss incurred on a contract selected randomly from the company's portfolio is xS.

Let us consider the following three reinsurance schemes. Using "pro rata quota share" a percentage $100(1-\alpha)$, where $0 < \alpha < 1$, of the ceding company's liabilities is given to a reinsurance company. Hence αxS is the retained loss incurred on a contract selected randomly from the insurance company's portfolio.

With "excess of loss," a part of all losses exceeding a cutoff point M is ceded to the reinsurance company. Hence the retained loss is $\min\{xS, M\} \equiv xS \wedge M$. When using "sliding quota share" no liability is ceded unless the contract size S exceeds the cutoff point C in which case $100(S-C)/S\%$ of all liability is ceded. Hence the retained loss is $x(S \wedge C)$.

Let the expected value of a nondecreasing convex function of the retained loss denote the "risk level." It is clear that the choice of the parameters α, M, and C largely determines the risk level" of these three schemes. To make a valid comparison let us fix α and choose $M = M_\alpha$ and $C = C_\alpha$ so that $E(\alpha xS) = E(xS \wedge M_\alpha) = E(x(S \wedge C_\alpha))$, that is, each scheme has equal mean retained loss.

(a) Show that $C_\alpha > M_\alpha$. [*Hint*: Suppose the contrary.]

(b) Show that $xS \wedge M_\alpha$ is less risky that either $x(S \wedge C_\alpha)$ or αxS. [*Hint*: Utilize Theorem 3 in the Hanoch–Levy paper.]

(c) Interpret the results.

(d) What can one show concerning the variances of the three schemes?

23. (An unorthodox utility function with desirable properties) If $z = \tan t$, then $t = \tan^{-1} z$ is called the arctangent, denoted $\arctan z$. For $z \in E^1$,

$$|\arctan z| \leqq 1$$

and

$$\frac{d}{dz} \arctan z = \frac{1}{1+z^2}.$$

Consider the utility function $u(w) = \arctan(1+w)$ over wealth w.

(a) Show that for $w > 0$, u is positive, bounded, strictly increasing, strictly concave, has positive strictly decreasing absolute risk aversion, and positive increasing relative risk aversion.

(b) Show that the nonstrict versions of the properties just given obtain when $w \geqq 0$.

(c) Show that the limiting value of the relative risk-aversion index is 2.

Exercise Source Notes

Portions of Exercise 2 were based on Klevorick (1969); Exercises 3 and 4 were adapted from Leland (1971); Exercise 5 was based in part on Bessler and Veinott (1966); portions of Exercise 6 were adapted from Whitmore (1970); Exercise 7 is due to Professor S. L.

Brumelle; Exercises 8 and 9 were adapted from Menezes and Hanson (1971); Exercise 10 was adapted from Arrow (1971); Exercise 12 was adapted from Rothschild and Stiglitz (1970); Exercise 14 was adapted from Hakansson (1969a, b); Exercise 16 was based on Royama and Hamada (1967); Exercise 17 was adapted from Adler (1969); Exercise 18 was adapted from Murty (1972) and Ziemba *et al.* (1974); Exercise 19 was based on Glustoff and Nigro (1972) and Bierwag (1973); Exercise 21 was based on Cass and Stiglitz (1972); Exercise 22 was based on Lippman (1972); and the utility function in Exercise 23 was devised by Professor R. C. Grinold.

Part III

Static Portfolio Selection Models

INTRODUCTION

I. Mean–Variance and Safety-First Approaches and Their Extensions

This part of the book is concerned with static models of portfolio selection. The basic choice problem for the investor is the best allocation of his initial wealth among several risky assets. The discussion is primarily concerned with models that utilize the expected utility approach. In addition, the mean-variance and safety-first approaches are discussed in detail as are their relations to the expected utility approach.

The basic ideas of mean–variance analysis date at least to Hicks in the mid-1930s; see Hicks (1962) for some historical perspective. However, the approach did not achieve its full potential until it was given a thorough formulation in Markowitz's classic 1952 article [see Markowitz (1952)], and later in his book (Markowitz, 1959). Since then portfolio analysis has been dominated by the mean–variance approach. The notion is that an investor should limit consideration to those portfolios that are mean–variance efficient; i.e., those portfolios for which there does not exist an alternative portfolio that has at least as high a mean and a lower variance. Such a procedure limits the choice to those portfolios that are in the mean–variance efficient set. To determine an optimal portfolio from this set requires in addition that the investor express his preferences between mean and variance tradeoffs. The mean–variance approach has the same set of optimal portfolios as the expected utility approach if the utility function is quadratic (Exercise CR-1) or if the utility function is concave and the returns have a multivariate normal distribution (as shown in Lintner's paper in Part II).

Samuelson's first paper provides an alternative way to relate the expected utility and mean–variance approaches. He considers distributions that have the property that the returns from all securities converge to a (common) sure outcome when a given variable, which may be interpreted as time, goes to zero. These "compact" distributions possess the requisite property so that Taylor series approximations of the utility function of order 2 become exact as the variable goes to zero. Hence in the limiting case the utility function is "essentially" quadratic and the mean–variance and expected utility approaches are mutually consistent. He further argues that if one wishes to have the path near time zero of the approximate solutions essentially the same as that in the expected utility problem, then the Taylor series approximating problem should include more terms. Specifically, to obtain agreement on the first m derivatives at time zero, requires use of a $(m+2)$nd-order approximation. In Samuelson's analysis and in many other financial problems it is of interest to know that the optimal decision vector is a continuous or a differentiable function of a given parameter. Exercise ME-30 develops such results in the special case when the investor's utility function is a power or logarithmic function and it is optimal to invest in every available asset. For further results and information on this delicate topic the reader is referred to the discussion of the maximum theorem by Berge (1963) and to the work of Danskin (1967) and Hogan (1973). Not all distributions are "compact" in Samuelson's sense and the reader is asked to investigate the compactness properties of some common univariate distributions in Exercise CR-25. Samuelson's presentation does not delve deeply into many fine and subtle points of the analysis. The reader is therefore asked to provide some of this material and to verify some statements in the paper in Exercise CR-26. One can, of course, utilize Taylor series approximations for very general utility functions and random variables. In Exercise CR-2 the reader is asked to calculate bounds in the maximum error in objective value terms using such approximations. Taylor series approximations of order n yield explicit deterministic polynominal functions of order n in terms of the wealth variable. However, in terms of asset allocation variables (Exercise CR-12) the approximation yields a sum of n signomial functions (i.e., signed products of positive variables raised to arbitrary powers) which reduce to polynomial functions only in very special cases.

Samuelson's paper develops the notion that the mean and variance of wealth are approximately sufficient parameters for the portfolio selection model when the probability distribution of wealth is "compact." In the "compact" case, moments of order 3 and higher are small in magnitude relative to the first two moments of the portfolio return; hence a limiting approximation indicates that only the first two moments are relevant for optimal portfolio selection. Samuelson's presentation is heuristic. The paper by Ohlson presents a rigorous approach to the study of such approximations. Additionally Ohlson's analysis generalizes and extends the range of applicability of such asymptotic quadratic utility approximations. He shows that (essentially) it is sufficient if the third absolute moment vanishes at a faster rate than the first two moments, even if moments of order 4 and higher are infinite. Rather general utility functions also suffice, including the negative exponential, power, and logarithmic cases. The paper also discusses the specific case when the random returns have log-normal distributions, assuming that the mean and variance of the growth rates are linear in time. Such distributions are not "compact," yet the quadratic approximation is still valid. Exercise ME-31 illustrates a case where not all moments are finite, yet the asymptotic utility is a quadratic function of the asset proportions (see also Ohlson, 1974). For an alternative Taylor series justification for the validity of the mean–variance approximation in portfolio theory, see Tsiang (1972). Some discussion of Tsiang's paper appears in Borch (1974), Bierwag (1974), Levy (1974), and Tsiang (1974). See Chipman (1973) and Klevorick (1973) for a rigorous discussion concerning the existence of expected utility functions dependent on mean and variance parameters when the returns have normal or two equally likely point distributions.

One may compute the mean–variance efficient surface in several ways; for example, by solving a parametric qudratic program as discussed in Exercise CR-13. Paine (1966) illustrates this calculation when the returns follow the Markowitz–Sharpe (see Sharpe, 1970) diagonal model. When nonnegativity constraints are not present, one can, as in Exercise CR-14, explicitly solve for the optimal portfolio allocations. The efficient surface is then an explicit quadratic function of the stipulated mean portfolio return. See also Hart and Jaffee (1974) for a novel application of mean–variance analysis to the problem of financial intermediation. Their paper also develops some properties of the mean–variance efficient set when positive and negative holdings are allowed. When there is a risk-free asset the efficient surface is a ray in mean–standard deviation space, and one may find the optimal proportions by solving the fractional program or linear complementary problem developed in Lintner's paper. Exercise CR-3 discusses some aspects of the fractional program (see also Exercises II-CR-19 and 20 and II-ME-18). The continuity properties of the efficient surface in this case are explored in Exercise ME-3. The number of efficient portfolios in a given portfolio problem is determined largely by the constraints on the investment allocations. Exercise CR-11 illustrates how a useful constraint relaxation increases the number of efficient portfolios. The analysis is carried out for a numerical problem with one safe and two risky assets.

As indicated above, if the investment returns have normal distributions the mean–variance approach and the expected utility approach imply the same optimal portfolio behavior. That is, all optimal solutions of the expected utility problem are mean–variance (or equivalently mean-standard deviation) efficient. In their paper, Pyle and Turnovsky investigate whether or not an investor who operates with a safety-first criterion also has the same implied behavior. The basic notion in a safety-first criterion is that the investor wishes to minimize the chance of obtaining a large loss. There are several ways to formulate such a notion and Pyle and Turnovsky consider the following three alternative versions. Choose the investment allocations to: (a) minimize the probability of obtaining a return below a certain stipulated (aspiration) level; (2) provide the largest (fractile) return level that is achieved with at least a given probability level; and (3) provide the largest mean return such that a stipulated aspiration level is achieved with at least a given probability level. All of these criteria lead to linear indifference surfaces in mean–standard deviation space: hence if the mean–standard deviation efficient surface is convex but not linear there is a unique indifference surface that is tangent to the efficient surface. Hence there is a unique safety-first investor who will choose the same portfolio as that chosen by an investor with a given concave utility function. However, many concave utility functions may lead to the same optimal decision as that obtained by a given safety-first investor. When there is a risk-free asset the efficient surface is linear and the situation is largely indeterminate because the indifference surfaces are also linear. In Exercise CR-24 the reader is invited to show that these results also hold for a reformulated aspiration objective. The analysis also applies (Exercise CR-23) when the random returns have appropriate symmetric stable distributions instead of normal distributions and one utilizes a mean-dispersion analysis as discussed in Ziemba's paper. See also Exercise CR-21 for a related normal distribution case. Exercise ME-18 shows that the deterministic equivalent for the aspiration model that one obtains in the normal distribution case generally does not generate necessary nor sufficient deterministic equivalent sets when the returns have nonnormal distributions. Some properties of safety first like optimization problems are considered in Exercises ME-26 and CR-20. Exercise ME-2 considers a chance-constrained programming problem and the utility function that it induces. Agnew *et al.* (1969), Ahsan (1973), Bergthaller (1971), Dragomirescu (1972), Levy and Sarnat (1972b), Pyle and Turnovsky (1971), and Telser (1955–1956) consider additional applications of chance-constrained programming to the portfolio selection problem.

The paper by Ziemba is concerned with the expected utility portfolio model when the returns have Pareto–Levy distributions. These distributions are also termed stable because many of their members are closed under addition; they are of interest for the empirical explanation of asset price changes and other financial and economic phenomena. An important reason for this is that all limiting sums (that exist) of independent identically distributed random variables are stable. Thus it is reasonable to expect that empirical variables which are sums of random variables conform to stable laws. Stable distributions represent generalizations of the normal distribution and appropriate stable families are closed under addition. These distributions have four parameters. If two of these parameters (namely the skewness and characteristic exponent coefficients) are constant across investments, then a mean-dispersion analysis is consistent with the expected utility of wealth approach when preferences over alternative wealth levels are concave. The mean–dispersion analysis *per se* is considered in Exercise CR-7. In particular, if all means exist and are equal and investment alternatives are independent, then all optimal investment allocations are positive and are proportional to weighted ratios of the inverses of the dispersion parameters (as in the normal distribution case). A generalized risk measure and its properties are explored in Exercise ME-17. In order for expected utility to be finite it is required that for the large absolute values of wealth the utility function is not steeper than a power function whose order is not exceeded by the value of the characteristic exponent. Hence many common utility functions need to be appropriately modified since their limiting slopes are too steep. In Exercise CR-17 the reader is asked to investigate such boundedness conditions for some numerical functions and associated distributions. When the characteristic exponent is 1, in the Cauchy distribution case, there is no concave utility function that has finite expected utility (Exercise CR-8). It also develops that investors with utility functions that have finite expected utility are indifferent between all possible allocations of their funds between independent identically distributed Cauchy investments.

When there is a risk-free asset then appropriate stable investments induce a generalization of Tobin's separation theorem and the two-step procedure available for normally distributed investments as described in Lintner's paper (in Part II). In particular one may solve a fractional program to obtain the optimal relative proportions of the risky assets independent of the concave utility function. The fractional program generally has a unique solution and has appropriate generalized concavity properties so that the Kuhn–Tucker conditions are necessary and sufficient for optimality, and standard nonlinear programming algorithms may be used to find the optimal proportions. Since the probability density function of stable variates is not known except in very special cases, there is no apparent algorithm that can be used to find the optimal ratios of risk-free and composite assets. However, using tables compiled by Fama and Roll one may obtain a good approximate solution by solving a univariate search problem.

The two-stage procedure provides an efficient computational approach to solve portfolio problems when there are many assets that have stable distributions. The reason for the efficiency is that the expected utility problem, which involves n random variables and n decision variables, is solved by combining the solutions of two much easier problems. One is deterministic and involves n decision variables, while the other is stochastic but involves only a single random variable and a single decision variable. The procedure applies, in particular, for many classes of symmetric stable investments. A particularly interesting class of multivariate symmetric stable investments was developed by Press. His class allows for the decomposition of the investments into independent partitions in a way consistent with and motivated by the way such a decomposition might be utilized for joint normally distributed random variables. In addition to the discussion in Ziemba's paper, Press' class is considered in several problems. In Exercise CR-9 a mean-dispersion analysis is considered and the optimal asset proportions can be obtained explicitly when short sales are allowed. A similar problem is considered in Exercise ME-10 when there are several independent partitions of the random return vectors that have different characteristic exponents. An explicit solution is also available in this case. Exercise ME-9 considers the case when the utility function is a negative exponential. In this case there is an explicit deterministic concave program whose solution provides optimal asset proportions. When the characteristic exponent is 2 (in the multivariate normal distribution case) the deterministic program reduces to the quadratic program developed in Exercise CR-1. The reader is asked to verify some results used and stated in Ziemba's paper in Exercise CR-16. See Ohlson (1972a) for an analysis of the portfolio problem when the asset returns have log-stable distributions.

The mean–variance and related approaches are discussed in many papers and books. For additional discussion and results the reader may consult Archer and D'Ambrosio (1967), Borch (1968), Fama and Miller (1972), Levy and Sarnat (1972a), Markowitz (1959), Sharpe (1970), Szegö and Shell (1972) Smith (1971), and Tobin (1965).

II. Existence and Diversification of Optimal Portfolio Policies

In financial optimization problems it is important to know if an optimal decision or policy exists. If the objective function is continuous and the constraint set is compact, then, by Weierstrass' theorem, a maximizing point exists. However, it is not always possible or desirable to assume that the constraint set is compact. In general, of course, an investor may be led to arbitrarily large investment positions in some investments, and an optimal allocation does not exist. In his paper Leland considers conditions on the feasible region, the investment returns, and the utility function that guarantee that a maximizing point exists. He assumes that the investment choice is among a finite set of investment alternatives and that there is no arbitrarily large investment position in any asset or assets which offers a nonnegative net return with probability 1. It is supposed that the constraint set is closed and convex and contains the origin. The utility function is strictly increasing and concave in wealth. Existence of an optimal portfolio is then guaranteed if the utility function is bounded from above. If the expected returns in all investments are finite, then it is only necessary to assume that marginal utility converges to zero as wealth increases without limit. In Exercise CR-5 the reader is asked to determine whether or not an optimal allocation exists for some numerical examples. Exercise CR-6 relates the results to boundedness of the utility function and possible St. Petersburg paradoxes. The reader is asked to investigate some extensions of Leland's results in Exercise ME-5. Recently, Bertsekas (1974) has generalized Leland's results. In particular, he develops necessary and sufficient conditions for the existence of a maximizing point that are slightly weaker than Leland's sufficiency conditions.

In addition to the existence of optimal portfolio policies one is interested in how such policies are found numerically. The expected utility function generally has useful concavity properties and may be equivalently thought of as a deterministic nonlinear objective function whose variables are the asset proportions. The deterministic equivalent is generally implicit and one approach to the solution of such problems is to combine a numerical integration scheme with a standard nonlinear programming algorithm. This approach is considered in Exercise ME-13 for the general case (and in Exercise ME-14 for the multivariate normal case). In particular results are developed for the validity of the interchange of the differentiation and integration operators and for the differentiability of the deterministic equivalent. Such results are of great use when one wishes to utilize the Kuhn–Tucker conditions, as well as in algorithmic solution methods. If the utility function is concave (and not necessarily differentiable) and the random returns have an absolutely continuous density, then, under mild regularity assumptions the partial derivatives exist and are continuous. The reader is asked to investigate the continuity and differentiability properties of some numerical functions in Exercise CR-17. In some cases, such as those discussed in Exercise ME-15, one may determine an explicit deterministic equivalent; that is, the integral of the expected utility function exists in closed analytic form. One may then determine the optimal portfolio by applying a standard nonlinear programming algorithm such as the Frank–Wolfe algorithm which is discussed in Exercise ME-14. A question related to that of finding an explicit deterministic equivalent is when, as discussed in Exercise ME-4, the expectation and maximization operators may be reversed. Exercise ME-32 develops conditions under which expected utility of wealth is finite for unbounded nondecreasing concave utility functions if wealth is nonnegative or bounded from below. In particular expected utility is finite if the mean and utility of zero wealth are finite. In Exercise ME-16 the reader is invited to attempt to devise an efficient algorithm to solve the large-scale nonlinear program that results when the investment returns have discrete distributions. Exercise ME-12 is concerned with a portfolio problem involving bets on the occurrence of states of nature (Arrow–Debreu securities). When the utility function is logarithmic the optimal policy is to allocate funds in proportion to the probability that each state occurs independent of the payoffs. Unorthodox results also apply for power utility functions (Exercise ME-29). When the exponent is negative one is led to the startling conclusion that it is optimal to invest heavily in those securities that have the lowest returns.

Some critics of the mean-variance approach have suggested that for many investors the variance may not be an appropriate measure of risk because positive and negative deviations about the mean are equally weighted. An alternative is to consider the semivariance which gives weight only to negative deviations about some target point such as the mean. Such an approach induces a monotonic concave utility function that has quadratic and linear segments. Exercise ME-8 is concerned with the mean-semivariance approach and the computation of associated efficient surfaces. The calculation of the efficient boundary for a given parameter value is sufficient to determine asset proportions to maximize expected utility. A modification of the Frank–Wolfe method is developed in Exercise ME-25 to solve the problem. The objective has the required concavity and differentiability properties to guarantee convergence, and the algorithm should be quite efficient because the direction-finding problem may be solved in a trivial way in each iteration.

It is important to know when investors will diversify their holdings. As is shown in Exercise CR-10, it is sometimes optimal in mean-variance models to invest entirely in one asset even if all assets have equal mean returns. Samuelson considers quite general diversification problems in his second paper. He utilizes an expected utility framework and assumes that the utility function is strictly increasing, strictly concave and differentiable. The random investments are assumed to possess equal means and positive but finite variances. A natural question to pose is when is it desirable to invest equally in all securities? Samuelson shows that such an allocation is optimal if the investments have identical independent distributions or, more generally, if they have nontrivial symmetric distribution functions. He also shows that equal diversification will generally not be optimal unless some type of symmetry is present. The finiteness assumption on the variances is crucial for equal allocation to be the unique optimal solution. Indeed, as is shown in Exercise CR-8, all possible allocations between independent identically distributed Cauchy investments are equally good when expected utility is infinite. In fact for symmetric distributions, an equal allocation policy is optimal but generally not unique regardless of the finiteness of the means and variances.

Perhaps it is more basic to know when a particular investment will form a part of the optimal portfolio. Utilizing the same framework Samuelson shows that an investment whose return is distributed independently of the returns of the remaining investments will always be purchased if its mean is not exceeded by any of the remaining investments. Indeed if all investments have a common mean and have independent distributions with finite positive variances, they will enter positively in the optimal portfolio. When a mean-variance or mean–dispersion analysis is optimal, the investment proportions are inversely related to the variances or dispersions, respectively (Exercise CR-7). However, unless the strict independence assumption holds, it is not true that every investment in a group with equal means must enter the optimal portfolio.

Although it is natural to suppose that negative interdependence improves the case for diversification, the appropriate measure of interdependence is not clear. For example, if all investments have a common mean and have negative correlations, then all enter into the optimal portfolio if the utility function is quadratic or if the investments have a normal distribution. However, negative correlation is neither necessary nor sufficient for complete or even any diversification for general strictly concave functions (Brumelle, 1974) (Exercise ME-27). In the two-asset case a concept of negative dependence that leads to diversification is that both conditional means are strictly decreasing (Exercise ME-28). Diversification is also optimal if the conditional cumulative distribution functions are strictly increasing, as shown in Exercise CR-28. This is the opposite of the condition in Samuelson's Theorem IV (so that the sign in the expression above Theorem IV should be reversed). The reader is asked to investigate the diversification effects of a three-asset problem in Exercise CR-27. Hadar and Russell (1971, 1974) show how to derive the Samuelson results and some extensions using stochastic dominance concepts.

An important diversification problem arises in the study of optimal financial intermediation. Financial intermediaries issue claims on themselves and then use the proceeds to purchase other assets. Hence it is of interest to determine when the firm will sell deposits (short) and purchase loans (long). Exercise ME-6 considers this problem when there is also a risk-free asset. If the random investments are independent, there is intermediation if the mean loan return exceeds the risk-free rate, which in turn exceeds the mean deposit return rate. Such an ordering is sufficient but not necessary if the investments have increasing conditional means. If a mean-variance approach is optimal, one can determine the optimal allocations explicitly in terms of means and variances. Exercise ME-7 develops these results and additional sharper conditions for intermediation.

As an alternative to the maximization of the expected utility criterion, an investor may wish to minimize the expected disutility of his regret. Given a particular realization of the random returns, the regret of a particular decision is the maximum return that could have been obtained given that realization minus the return obtained with the given decision. Exercise ME-11 is concerned with diversification effects involved in such a formulation when the disutility is a strictly monotic and strictly increasing function of regret. If the distribution of returns is symmetric, then equal investment in each security is the unique minimax strategy and it also is the unique minimizer of expected disutility of regret. In the two-asset case it is always optimal to diversify if one utilizes the disutility approach and if the mean returns are equal regardless of the joint distribution of returns. In Pye's paper in Part V it is shown how this criterion can be used to explain the phenomenon of dollar cost averaging.

Specification requirements escalate extremely rapidly as one attempts to determine the joint distribution function when there are more and more securities in an investor's "universe." Hence an important concern of many investors relates to the choice of a proper size of the "universe" in order to provide a satisfactory level of diversification and expected utility. Exercise CR-15 develops a mean–variance-inspired method to experiment with different-sized security universes in a way that is computationally feasible.

Investment returns are often made under limited liability conditions in the sense that the investor can lose no more than his initial outlay. Hence gross investment returns are always nonnegative. Several common distributions might possibly be used to explain such security price changes. One such distribution that has some attractive analytic and methodological properties is the log-normal distribution. A variable has a log-normal distribution if its logarithm has a normal distribution. Many of the important properties of univariate and multivariate log-normal distributions are discussed in Exercises ME-20 and 21, respectively. Since log-normal variates are not closed under addition there does not appear to be an efficient method to solve portfolio problems that applies for general concave utility functions. However, for particular utility functions, one can develop approximations which have very attractive computational and qualitative properties. Such is the case when the utility function is logarithmic. Exercise ME-22 studies the qualitative properties of the optimal solution vector for this problem. Exercise ME-23 develops an approximating problem based on the assumption that the final wealth variate is log-normal. The approximating problem can be chosen from a surrogate family so that it has the remarkable property of possessing essentially all of the qualitative properties of the original problem. Moreover, the approximating problem is a single quadratic program, which may, as outlined in Exercise ME-24, be solved in an extremely simple way utilizing the Frank–Wolfe algorithm. Ohlson and Ziemba (1974) have developed similar results for the case when the investor's utility is a power function of wealth. Dexter *et al.* (1975) present computational results that support the approximation.

III. Effects of Taxes on Risk-Taking

The papers of Stiglitz and Naslund examine the effects of various taxation policies on an individual's risk-taking behavior. Stiglitz assumes that the investor is an expected utility maximizer, and analyzes the effects of taxation on the investor's allocation of wealth among a sure asset with nonnegative net return and a single risky asset with nonnegative gross return. There are assumed to be no borrowing or short-sale constraints. Stiglitz emphasizes two somewhat different measures of risk-taking: (1) the demand for risky assets, measured by the fraction of wealth devoted to the risky asset, and (2) "private risk-taking" (PRT), measured by the standard deviation of final wealth. He shows that a proportional wealth tax increases or decreases the demand for risky assets according as the Arrow–Pratt relative risk-aversion index increases or decreases with wealth, and that it increases or decreases PRT accordingly as the absolute risk-aversion index increases or decreases with wealth. Of more interest are proportional income taxes with various loss-offset provisions. A full loss-offset provision implies that the government shares the risks of loss as well as the potential gains. This reduces the variability of risky asset after-tax returns. Since the tax reduces expected return from riskless and risky assets in the same proportion, and reduces the variance of risky asset returns, it is plausible that such a tax may lead to increased demand for the risky asset. Stiglitz shows that this is indeed the case if either (i) the return on the safe asset is zero, (ii) absolute risk aversion is nondecreasing, or (iii) absolute risk aversion is decreasing and relative risk aversion is nondecreasing. He shows also that PRT remains unchanged if the risk-free return is zero, and that otherwise PRT increases or decreases according as absolute risk aversion is an increasing or decreasing function. The case of an income tax with no loss offset or with only partial loss offset is also examined. Not surprisingly, it is found that the demand for the risky asset is always less with no loss offset or with partial loss offset than with full loss offset. In Exercise CR-19 the reader is asked to examine the portfolio allocation effects of a 20% income tax with and without loss offset. The Stiglitz paper also considers in part the question of special provisions for capital gains. In the extreme case of a tax only on the safe asset, it is shown that the demand for the risky asset is increased if absolute risk aversion is nondecreasing, or if the relative risk aversion is not greater than unity. Exercise CR-4 examines the effects on a portfolio of two securities of a tax on one of the securities, assuming quadratic utility.

The Naslund paper examines the effects on portfolio allocation of a proportional income tax with full loss offset. The paper deals mainly with formulations of the portfolio problem *not* based on expected utility. In particular, the main emphasis is on safety-first (see the Pyle and Turnovsky paper) and chance-constrained programming models. The basic safety-first model adopted is that of minimizing the probability of returns below a preassigned "disaster" level d. As in the Pyle and Turnovsky paper, Naslund assumes that the problem of minimizing the Tchebychev upper bound on disaster probability will serve as an equivalent surrogate problem. The relation between these problems is examined in Exercises CR-20 and 21, where it is shown that the problems are equivalent for joint-normally distributed assets. For several risky assets which are not joint-normally distributed, the resulting portfolios may be quite different in the "true" and "surrogate" problems. This is illustrated in Exercise ME-18 for a two-asset case. In any case, assuming the validity of the surrogate problem, Naslund shows (for two assets) that an increase in taxes leads to increased risk-taking if the asset with higher yield is riskier, and to decreased risk-taking otherwise. The paper also discusses portfolio allocation in the so-called E-model of chance-constrained programming, which maximizes expected portfolio return subject to the constraint of attaining at least a given level of return B with at least a preassigned probability level α. For joint-normally distributed asset returns, the corresponding deterministic equivalent problem has a linear objective function and quadratic constraints. In this case, the Tobin–Lintner separation theorem is true, thus reducing the many-asset case to a case of one riskless and one risky asset. It is then easily shown that a proportional income tax leads to increased demand for the risky assets. When there are two risky assets and no riskless asset, it is shown that an increase in taxation will (essentially) lead risk averters to increase their demand for the risky asset with higher mean.

It is apparent from the Stiglitz and Naslund papers that the effects of taxation on risk-taking will depend on the investor's formulation of the portfolio problem (e.g., expected utility versus safety first), on his attitudes toward risk (e.g., decreasing absolute risk aversion), and on the nature of the tax itself (e.g., offset provisions). As an extension to the Stiglitz paper, the reader is invited in Exercise ME-19 to ponder the effects of a progressive (convex) income tax on the portfolio allocation of an expected utility maximizer. For additional material regarding taxation effects, the Lepper (1967) article is recommended. This article examines effects on portfolios of various loss-offset provisions, progressive rate structures, and capital gains provisions through the numerical solution of representative problems. See also Russell and Smith (1970) for a stochastic dominance approach to the taxation problem.

1. MEAN-VARIANCE AND SAFETY-FIRST APPROACHES AND THEIR EXTENSIONS

Reprinted from The Review of Economic Studies, Vol. XXXVII (4), October, 1970, pp. 537-542.
Paul A. Samuelson

The Fundamental Approximation Theorem of Portfolio Analysis in terms of Means, Variances and Higher Moments[1]

I

James Tobin [7, 8], Harry Markowitz [3, 4], and many other writers have made valuable contributions to the problem of optimal risk decisions by emphasizing analyses of means and variances. These writers have realized that the results can be only approximate, but have also realized that approximate and computable results are better than none.

Recently, Karl Borch [1] and Martin Feldstein [2] have re-emphasized the lack of generality of mean-variance analysis and evoked a reply from Tobin [9]. None of the writers in this symposium refer to a paper of mine (Samuelson, [6]) which suggested that most of the interesting propositions of risk theory can be proved for the general case with no approximations being involved. This same paper pointed out all the realms of applicability of mean-variance analysis and also its realms of non-applicability.

There is no need here to redescribe these arguments. But I think it important to re-emphasize an aspect of the mean-variance model that seems not to have received sufficient attention in the recent controversy, namely the usefulness of mean and variance in situations involving less and less risk—what I call " compact " probabilities. The present paper states and proves the two general theorems involved. In a sense, therefore, it provides a defence of mean-variance analysis—in my judgement the most weighty defence yet given. (In economics, the relevant probability distributions are *not* nearly Gaussian, and quadratic utility in the large leads to well-known absurdities). But since I improve on mean-variance analysis and show its exact limitations—along with those for any r-moment model—the paper can also be regarded as a critique of the mean-variance approach. In any case, the theorems here provide valuable insight into the properties of the general case. I should add that their general content has long been sensed as true by most experts in this field, even though I am unable to cite publications that quite cover this ground.

II

The Tobin-Markowitz analysis of risk-taking in terms of mean and variance alone is rigorously applicable only in the restrictive cases where the statistical distributions are normally Gaussian or where the utility-function to be maximized is quadratic. In only a limited number of cases will the central limit law be applicable so that an approximation to normality of distribution becomes tenable; and it is well known that quadratic utility has anomalous properties in the large—such as reduced absolute and relative risk-taking as wealth increases, to say nothing of ultimate satiation.

[1] Aid from the National Science Foundation is gratefully acknowledged, and from my M.I.T. students and co-researchers: Robert C. Merton, from whose conversations I have again benefited, and Dr. Stanley Fischer (now of the University of Chicago) whose 1969 M.I.T. doctoral dissertation, *Essays on Assets and Contingent Commodities* contains independently-derived results on compact distributions.

However, a defence for mean-variance analysis can be given (Samuelson, [6; p. 8]) along other lines—namely, when riskiness is " limited ", the quadratic solution " approximates " the true general solution. The present note states the underlying approximation theorems, and shows the exact limit of their accuracy.

III

Let the return from investing $1 in each of " securities " 1, 2, ..., n be respectively the random variable $(X_1, \ldots, X_n)$, subject to the joint probability distribution

$$\text{prob}\{X_1 \leqq x_1 \text{ and } X_2 \leqq x_2 \text{ and } \ldots X_n \leqq x_n\} = F(x_1, \ldots, x_n). \qquad \ldots(1)$$

Let initial wealth W be set (by dimensional-unit choice) at unity and $(w_1, w_2, \ldots, w_n)$ be the fractions of wealth invested in each security, where $\sum_1^n w_j = 1$. Then the investment outcome is the random variable $\sum_1^n w_j X_j$, and if $U[W]$ is the decision maker's concave utility, he is postulated as acting to choose $[w_j]$ to maximize expected utility, namely

$$\max_{\{w_i\}} \bar{U}[w_1, \ldots, w_n] = \int_0^\infty \ldots \int_0^\infty U\left[\sum_1^n w_j X_j\right] dF(X_1, \ldots, X_n). \qquad \ldots(2)$$

In general, the solution to this problem will *not* be the same as the solution to the quadratic case

$$\max_{\{w_i\}} \int_0^\infty \ldots \int_0^\infty \left\{U[1]+U'[1]\left[\sum_1^n w_j X_j - 1\right] + \tfrac{1}{2}U''[1]\left[\sum_1^n w_j X_j - 1\right]^2\right\} dF(X_1, \ldots, X_n). \qquad \ldots(3)$$

But now let us suppose that $F(\cdot, \ldots, \cdot)$ belongs to a family of " compact " or " small-risk " distributions, defined so that as some specified parameter goes to zero, all our distributions converge to a sure outcome. An appropriate family would be

$$F(x_1, \ldots, x_n) = P\left(\frac{x_1-\mu-\sigma^2 a_1}{\sigma\sigma_1}, \frac{x_2-\mu-\sigma^2 a_2}{\sigma\sigma_2}, \ldots, \frac{x_n-\mu-\sigma^2 a_n}{\sigma\sigma_n}\right). \qquad \ldots(4)$$

Here, the variables have been defined so that, as the parameter $\sigma \to 0$, it becomes ever more certain that the outcome for $(X_1, \ldots, X_n) = (\mu, \mu, \ldots, \mu)$, where μ might be $1+$the " safe " rate of interest. Fig. 1 illustrates for the one-dimensional case, $n = 1$, the meaning of such a " compact family ". As $\sigma \to 0$ all the probability piles up at μ. Note that no normality of distributions is involved—originally or (in any non-trivial sense) asymptotically.

By convention, $P(y_1, \ldots, y_n)$ is defined to have the properties

$$E[Y_i] = \int_0^\infty \ldots \int_0^\infty Y_i dP(Y_1, \ldots, Y_n) = 0,$$

$$E[Y_i^2] = \int_0^\infty \ldots \int_0^\infty Y_i^2 dP(Y_1, \ldots, Y_n) = 1, \qquad \ldots(5)$$

$$E[Y_i Y_j] = \int_0^\infty \ldots \int_0^\infty Y_i Y_j dP(Y_1, \ldots, Y_n) = r_{ij},$$

where $[r_{ij}]$ is a symmetric positive definite correlation matrix. Similarly higher moments $E[Y_i^{k_i} Y_j^{k_j} \ldots]$, for k's integers, can be defined. It follows then that

$$E[X_i] = \mu + \sigma^2 a_i, \quad E[X_i - E[X_i]]^2 = \sigma_i^2 \sigma^2, \text{ etc.} \qquad \ldots(6)$$

An explanation may be needed to motivate our putting the σ parameter in the numerator of P's arguments. If $[\mu+\sigma^2 a_i]$ had been replaced by the simple constants $[\mu_i]$, and if they were not all equal, then the securities with the largest μ_i would dominate the rest as $\sigma \to 0$. All other w's would either go to zero, or if borrowing and selling short were freely permitted—so that the w's need not be non-negative—infinitely profitable arbitrage would be possible. This bizarre, infinite case is of trivial interest. Hence, $\mu_i(\sigma)$ must as $\sigma \to 0$ approach a common limit.

Anyone familiar with Wiener's Brownian motion will identify σ with the square root of time, $\sqrt{t}$: in the numerator σ^2 appears because means grow linearly with time in Brownian motion; in the denominator σ appears because standard deviations grow like $\sqrt{t}$, with variance growing linearly. Dimensionally $a_i\sigma^2$ has the same dimension, namely dollars, as does X_i and $\sigma\sigma_i$; in Brownian terms a_i would be " dollars/time ". The inequality of instantaneous mean gains in Brownian motion is indicated by inequalities among the a_i, not among the μ_i. Although I have couched this heuristic explanation in terms of Brownian motion in time, the concept of a compact family is an independent and completely general one. Fortunately, it gains in importance because it does throw light on the reasons why enormous " quadratic " simplicities occur in continuous-time models, as the cited Fischer thesis make clear and as is evident in Merton [5].

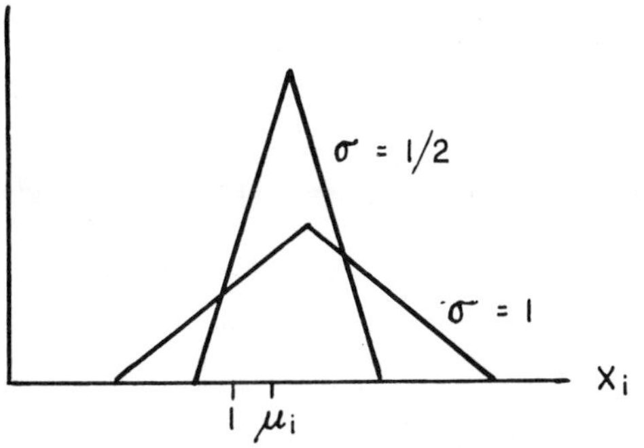

FIGURE 1
Example of family of compact probability densities.

Actually our family can be defined more generally than in terms of $[x_i - \mu - \sigma^2 a_i]/(\sigma\sigma_i)$. All that is required—and it *is* required if $[w_i(\sigma)]$, the optimal portfolio proportion as a function of the parameter σ, is to approach a unique and smooth limit $[w_i(0)]$—is that for the standardized variables $Z_i = X_i - \mu$

$$\lim_{\sigma \to 0} \frac{E[Z_i]}{E[Z_i^2]} = \frac{A}{B}, \quad \lim_{\sigma \to 0} \frac{E[Z_i^r]}{E[Z_i^2]} = \sigma^{r-2} C_r \quad (r = 3, 4, \ldots). \qquad \ldots(7)$$

The P family defined above does have this property, as can now be verified for the one-dimensional case, $X_1 = X$,

$$\begin{aligned} E[X] &= \int_0^\infty X dP\left(\frac{X-\mu-\sigma^2 a}{\sigma\sigma_1}\right) = \mu + \sigma^2 a, \\ E[(X-\mu-\sigma^2 a)^2] &= \sigma^2\sigma_1^2, \\ E[Z] &= \int_0^\infty (X-\mu) dP\left(\frac{X-\mu-\sigma^2 a}{\sigma\sigma_1}\right) = \sigma^2 a, \\ E[Z^2] &= E[(X-\mu-\sigma^2 a)^2] + (\sigma^2 a)^2, \\ &= \sigma_1^2\sigma^2 + \sigma^4 a^2 = \sigma^2(\sigma_1^2 + \sigma^2 a^2), \\ E[Z^3] &= E[\{(X-\mu-\sigma^2 a) + \sigma^2 a\}^3], \\ &= \sigma^3\sigma_1^3\mu_3 + \sigma\psi_3, \end{aligned} \qquad \ldots(8)$$

where μ_3 is the third moment around the mean of $P(y)$ and $\sigma\psi_3$ involves powers of σ higher than 3. Similarly, it can be shown that

$$E[Z^4] = \sigma^4\sigma_1^4\mu_4+\sigma\psi_4, \ldots, E[Z^r] = \sigma^r\sigma_1^r\mu_r+\sigma\psi_r, \quad r>2. \qquad \ldots(9)$$

An example of an admissible compact family that cannot quite be written in the $P(\cdot)$ form is

$$\text{prob}\{X_1 = 1+\sigma\} = \text{prob}\{X_1 = (1+\sigma)^{-1}\} = \tfrac{1}{2}, \qquad \ldots(10)$$

Here, $$E[X-1] = \frac{\sigma^2}{2} + \text{higher powers of } \sigma,$$

$$E[(X-1)^2] = 2\sigma^2 + \text{higher powers of } \sigma.$$

But this does satisfy our needed asymptotic conditions as defined in (7).

We can now state our fundamental approximation theorem

Theorem 1. *The solution to the general problem, $[w_i(\sigma)]$, does, as $\sigma\to 0$ have the property that $[w_i(0)]$ is the exact solution to the quadratic problem*

$$\max_{\{w_i^*\}} \int_0^\infty \ldots \int_0^\infty \left\{U[\mu]+U'[\mu]\left[\sum_1^n w_j^*X_j-\mu\right]+\frac{U''(\mu)}{2} \cdot \left[\sum_1^n w_j^*X_j-\mu\right]^2\right\} dP\left(\frac{X_1-\mu-\sigma^2a_1}{\sigma\sigma_1}, \ldots, \frac{X_n-\mu-\sigma^2a_n}{\sigma\sigma_n}\right),$$

i.e., $$\lim_{\sigma\to 0} w_i^*(\sigma) = w_i^*(0) = w_i(0) = \lim_{\sigma\to 0} w_i(\sigma).$$

But it is definitely *not* the case that the higher approximation implied by $w_i'(0) = w_i^{*\prime}(0)$ will hold. Theorem 2 will show that one must use cubic-utility 3-moment theory to achieve this higher degree of approximation, and that in general (r-moment, rth degree) utility theory must be used in order to get agreement between $[w_i(0), w_i'(0), \ldots, w_i^{[r-2]}(0)]$ and $[w_i^*(0), w_i^{*\prime}(0), \ldots, w_i^{*[r-2]}(0)]$.

Theorem 2. *The solution to the general problem above is related asymptotically to that of the r-moment problem*

$$\max_{\{w_i^{**}\}} \int_0^\infty \ldots \int_0^\infty \left\{\sum_0^r U^{[j]}(\mu)\frac{\left[\sum_1^n w_iX_i-\mu\right]^j}{j!}\right\} dP\left[\frac{X_1-\mu-\sigma^2a_1}{\sigma\sigma_1}, \ldots, \frac{X_n-\mu-\sigma^2a_n}{\sigma\sigma_n}\right]$$

by the high-contact equivalences

$$w_i(0) = w_i^{**}(0), w_i'(0) = w_i^{**\prime}(0), \ldots, w_i^{[r-2]}(0) = w_i^{**[r-2]}(0).$$

To prove the theorems most rapidly, note that if U possesses an exact Taylor's expansion, $[w_i(\sigma)] \equiv [w_i^{**}(\sigma)]$ is for $r = \infty$ trivially identical. Hence, $w_i(\sigma)$ is formally defined from the power series

$$\bar{U}(\sigma) = \max_{\{w_i\}} \sum_0^\infty U^{[j]}(\mu)\frac{E\left[\left(\sum_1^n w_iZ_i\right)^j\right]}{j!}. \qquad \ldots(11)$$

The first-order conditions for a regular optimum are

$$\frac{\partial \bar{U}}{\partial w_1} = \frac{\partial \bar{U}}{\partial w_2} = \ldots = \frac{\partial \bar{U}}{\partial w_n}, \qquad \ldots(12)$$

where each of these is a power series in all the w's, with coefficients that depend on the moments $E[Z_1^{k_1}Z_2^{k_2}...]$. However, the moments in the infinite series that are involved if we truncate the series at $r-1<\infty$ will be seen to be only the " lower " moments, in which $k_1+k_2+... \leqq r$. Also straightforward but tedious formal algebra shows that in the infinite power series for $w_i(\sigma)$, the coefficients of the terms of order σ^{r-2} or less do depend only on the coefficients of the $\bar{U}(\sigma)$ power series only up to the rth degree terms and rth moments. Hence, the indicated agreement of derivatives in Theorem 2 does hold. Theorem 1 is of course only a special, but important, sub-case of Theorem 2.

There are a number of obvious corollaries of the theorems. Thus, the well-known Tobin separation theorem, which is valid for quadratic utilities, will necessarily be asymptotically valid in the sense of becoming true as $\sigma \to 0$. I.e. $w_i(0)/w_j(0)$ for assets held along with cash will be independent of the form of $U[\cdot]$.

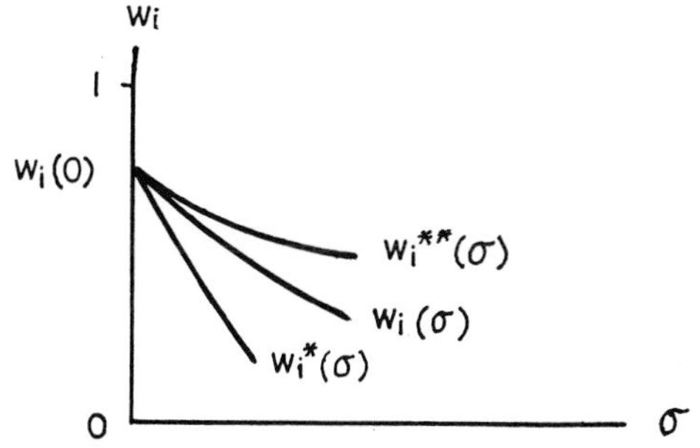

FIGURE 2

Although the quadratic mean-variance solution, $w_i^*(0)$, approaches the same intercept as the true solution, $w_i(0)$ the 3-moment cubic solution, $w_i^{**}(0)$, has higher contact with the true solution.

The exposition here has been heuristic, but can, subject to some hypotheses on the utility U and probability P functions, be made quite rigorous. The kind of formal algebra involved can be illustrated by the important case of cash versus one risky-asset (say a stock). Call cash X_0 and call the stock $X_1 = X$. Then the outcome for initial unit wealth is $(1-w)+wX = w(X-1)+1 = wZ+1$, where $w = w_1 = 1-w_0$, and our general problem becomes

$$\bar{U}(w) = \max_{\{w\}} \int_0^\infty U[wZ+1]dP\left(\frac{Z-\sigma^2 a}{\sigma}\right),$$

$$\bar{U}'(w) = 0 = \int_0^\infty ZU'[wZ+1]dP \qquad \text{...(13)}$$

$$= U'[1]\int_0^\infty ZdP + U''[1]\frac{w}{1}\int_0^\infty Z^2dP + U'''[1]\frac{w^2}{2!}\int_0^\infty Z^3dP + w^3R_3,$$

where dP is short for $dP[\sigma^{-1}(Z-\sigma^2 a)]$ and w^3R_3 involves higher powers of w than w^2.

If we truncate the series before the $U'''[1]$ term, we have the Tobin-Markowitz mean-variance approximation, with solution

$$w^*(0) = \lim_{\sigma\to 0} -\frac{a\sigma^2 U'[1]}{\sigma^2(\sigma_1^2+a^2\sigma^2)U''[1]} = \frac{a}{\sigma_1^2}\left(\frac{U'[1]}{-U''[1]}\right), \qquad \text{...(14)}$$

a not-surprising result. If from the above expression for $w^*(\sigma)$ we calculate $w^{*\prime}(0)$, we will not get the same result as if we had carried one more term in our truncated infinite expansion. Call this last value $w^{**}(\sigma)$ and define it as the root of the equation

$$0 = U'[1](a\sigma^2)+U''[1]w\sigma^2(\sigma_1^2+a^2\sigma^2)+U'''(1)\frac{w^2}{2}\sigma^3(\sigma_1^3\mu_3+\sigma\psi_3), \qquad \ldots(15)$$

where $\sigma^3\sigma\psi_3$ involves powers of σ beyond σ^3 and where the third moment of $P(y)$ has been earlier defined in (8). Then ignoring $\sigma^4\psi_3$, as we may, we find that $w^{**\prime}(0)$ most definitely *is* affected by the value of $U'''[1]\sigma_1^3\mu_3$. Hence, the quadratic approximation is not locally of " high " contact. Fig. 2 illustrates the phenomenon. Notice how higher-than second moments do improve the solution. But it also needs emphasizing that near to $\sigma = 0$, when " risk is quite limited ", the mean-variance result is a very good approximation. When the heat of the controversy dissipates, that I think will be generally agreed on.[1]

Massachusetts Institute of Technology PAUL A. SAMUELSON

First version received January 1970; *final version received March* 1970

REFERENCES

[1] Borch, K. " A Note on Uncertainty and Indifference Curves ", *Review of Economic Studies*, **36** (1969).

[2] Feldstein, M. S. " Mean-Variance Analysis in the Theory of Liquidity Preference and Portfolio Selection ", *Review of Economic Studies*, **36** (1969).

[3] Markowitz, H. " Portfolio Selection ", *The Journal of Finance*, **7** (1952).

[4] Markowitz, H. *Portfolio Selection: Efficient Diversification of Investments* (New York, John Wiley & Sons, 1959).

[5] Merton, R. C. " Lifetime Portfolio Selection under Uncertainty: the Continuous-Time Case ", *Review of Economics and Statistics*, **51** (1969).

[6] Samuelson, P. A. " General Proof that Diversification Pays ", *Journal of Financial and Quantitative Analysis*, **2** (1967).

[7] Tobin, J. " Liquidity Preference as Behaviour Towards Risk ", *Review of Economic Studies*, **25** (1958).

[8] Tobin, J. " The Theory of Portfolio Selection ", in F. H. Hahn and F. P. R. Brechling (eds.), *The Theory of Interest Rates* (London, Macmillan, 1965).

[9] Tobin, J. " Comment on Borch and Feldstein ", *Review of Economic Studies*, **36** (1969).

[1] See Markowitz [4] p. 121 for an argument that can be related closely to that here.

The Asymptotic Validity of Quadratic Utility as the Trading Interval Approaches Zero

*James A. Ohlson**

STANFORD UNIVERSITY
STANFORD, CALIFORNIA

I. Introduction and Summary

Samuelson [7] has developed the notion that the mean and variance of wealth are approximately sufficient parameters for the standard portfolio selection model when the probability distribution of wealth is "compact." The concept of a "compact" distribution refers to a decision context involving a family of distributions where the limiting moments of order 3 and higher are of a smaller magnitude relative to the first two moments of the portfolio return; in a limiting and approximating sense this would indicate that only the first two moments are relevant when selecting an optimal portfolio. One place where this mode of analysis arises is in models in which the portfolio revision horizon is small, and the limiting situation is one where the period spacing approaches zero. Models of these types are of increasing importance in portfolio theory, as is evident in the significant work by Merton [3–5].

The Samuelson [7] exposition is heuristic, and no rigorous weak conditions are presented as to when quadratic utility is valid in the asymptotic sense indicated above.[1] The formal analysis set forth by Samuelson requires unnecessarily sharp constraints on the utility of wealth function and the probability distribution. In fact, the important case of log-normally distributed returns will not involve convergent infinite Taylor expansions, and the same applies when utility is power or logarithmic.

In this paper a set of weak conditions is developed with respect to utility and probability distributions such that expected utility converges toward quadratic utility in the limit. It is shown that few restrictions are required on the probability distribution; essentially, it is sufficient if the third (absolute) moment vanishes at a faster rate than the first two moments, and moments of order 4 and higher need not be finite. Few assumptions are made about utility, aside from differentiability conditions. A not surprising result, therefore, is

* *Present address*: University of California, Berkely, California.

[1] Another heuristic discussion of the problem at hand is provided in a paper by Merton and Samuelson [6].

that quadratic utility is in general valid asymptotically whenever utility is negative exponential, power, logarithmic, and some obvious extensions of such functions.

In the second part of the paper exact compactness properties of the log-normal distribution are derived, assuming that expected growth rates and variance of growth rates of the individual assets are linear in time. This model will not imply a distribution which is compact in the strong Samuelson sense; however, it is sufficiently well behaved to satisfy the sufficient conditions evolved in this article.

II. The General Problem

Let $X = (X_1, X_2, \ldots, X_m)'$ denote the vector of random asset returns with a distribution function $G(x_1, \ldots, x_m; \theta(t)) \equiv Pr\{X_1 \leqq x_1, \ldots, X_m \leqq x_m; \theta(t)\}$. $\theta(t)$ expresses the parameters of the distribution G as a function of t, where t denotes the time span of the decision horizon. G, therefore, is determined by the value of t, and $t > 0$. Let $\lambda = (\lambda_1, \ldots, \lambda_m)'$ denote the vector of fractions of initial wealth invested in each asset, where $\sum_i \lambda_i = 1$ by convention. Further, let $U(W) \equiv U(\sum \lambda_i X_i)$ denote the investor's utility of wealth function, wealth being denoted by W, and suppose the investor wishes to maximize his expected utility. The optimal portfolio for any t, $\lambda^*(t)$, is then found by solving

$$\max_{\lambda \in D} \int_0^\infty \cdots \int_0^\infty U(\textstyle\sum \lambda_i x_i)\, dG(x_1, \ldots, x_m; \theta(t)), \tag{1}$$

where $D \equiv \{\lambda \mid \sum \lambda_i = 1;\, \lambda_i \geqq 0\}$, and no short sales are permitted. Also, it is assumed that $G(0, \ldots, 0; \theta(t)) = 0$, all $t > 0$. For notational simplicity, expected utility can be expressed as

$$EU(W) = \int_0^\infty U(W)\, dF(W), \tag{2}$$

keeping firmly in mind that the distribution of wealth, $F(W)$, and expected utility depend on both λ and t. Without loss of generality, suppose $\Pr\{|W-1| < \varepsilon\} \to 1$ as $t \to 0$, i.e., the portfolio return is approximately equal to unity for the short-period investment horizon.

The concern is to obtain a set of weak assumptions on $\lim_{t\to 0} F$ and U, so that the general problem (2) is equivalent to the quadratic utility problem

$$EU(W) = U(1) + U^{(1)}(1)\, E(W-1) + \tfrac{1}{2} U^{(2)}(1)\, E(W-1)^2 + EQ_3, \tag{3}$$

and where

$$\lim_{t\to 0} EQ_3/E(W-1)^n = 0 \qquad \text{for} \quad n = 1, 2, \quad \text{and all} \quad \lambda \in D.$$

That is, if $\lambda^{**}(t)$ is the solution to the surrogate problem

$$\max_{\lambda \in D} \{EU(W) - EQ_3\}, \tag{4}$$

then $\|\lambda^{**}(t) - \lambda^{*}(t)\| \to 0$ as $t \to 0$. One set of assumptions about F and U is implied in the cited Samuelson paper [7]. The restrictions on F are given by certain compactness characteristics:

*A*1:

$$\lim_{t\to 0} \frac{E(W-1)}{E(W-1)^2} = \frac{A}{B}; \qquad \lim_{t\to 0} \frac{E(W-1)^n}{E(W-1)^2} = t^{(n-2)/2} C_n; \qquad n = 3, 4, \ldots, \tag{5}$$

where $C_n \geqslant 0$ remains bounded as $t \to 0$.[2]

Alternatively, if $O(\cdot)$ denotes the usual asymptotic order symbol meaning "the same order as," and $o(\cdot)$ denotes "smaller order than," then assumption A1 entails $C_n = O(1)$, $E(W-1)^n = O(t)$, $n = 1, 2$, and $E(W-1)^n = O(t^{n/2})$ $[=o(t)]$, $n = 3, 4, \ldots$.[3]

As hypotheses on $U(W)$, it is first assumed that U is endowed with an exact Taylor's infinite expansion, i.e.,

*A*2:

$$EU(W) = E \sum_{j=0}^{\infty} U^{(j)}(1)(W-1)^j/j!. \tag{6}$$

Further, it is assumed that the expectation operator can be interchanged with the summation operator:

*A*3:

$$EU(W) = \sum_{j=0}^{\infty} U^{(j)}(1)\, E(W-1)^j/j! \equiv \sum_{j=0}^{\infty} h_j(t). \tag{7}$$

As a final requirement, if it is assumed (or proved) that

*A*4:

$$\lim_{t\to 0} \lim_{n\to\infty} \sum_{j=3}^{n} h_j(t) = \lim_{1/t,\, n\to\infty} \sum_{j=3}^{n} h_j(t), \tag{8}$$

then the convergence is uniform and $\sum_{j=3}^{\infty} h_j(t) = o(t)$. It is now obviously true that assumptions A1–A4 imply that (3) holds since

$$EU(W) = U(1) + U^{(1)}(1)\, E(W-1) + \tfrac{1}{2} U^{(2)}(1)\, E(W-1)^2 + o(t), \tag{9}$$

and the first two moments are of order $O(t)$.

[2] See Samuelson [7]. Note that Samuelson's symbol "σ" corresponds to $\sqrt{t}$ in this paper.

[3] Define $O(t^n) = h(t)$ iff $\lim_{t\to 0} |t^{-n} h(t)| = K > 0$, and $h(t) = o(t)$ iff $\lim_{t\to 0} |t^{-1} h(t)| = 0$. Further, note that $h(t) = O(t^n)$ implies $h(t) = o(t)$ iff $n > 1$.

Assumptions A1–A3 are extremely stringent; A1 requires that all moments are finite, and, as will be shown in subsequent analysis, this is quite unnecessary. Perhaps more important, even if all moments are finite and satisfy A1, it does not follow that A2 and A3 are always satisfied for a utility function which is everywhere differentiable an infinite number of times. In fact, not even A3 is implied by A2. As a first illustration, suppose $U(W) = \log W$, and $Pr\{K > W > 2\} > 0$ for every t, no matter how small. Although A1 may well be satisfied, it is still true that

$$E \sum_{j=0}^{\infty} U^{(j)}(1)(W-1)^j/j!$$

is not defined for any $t > 0$. Clearly, A2 is a nontrivial restriction. An example where A2 is true but A3 is not can also be easily provided. Consider the case in which $U(W) = -\exp\{-cW\}$, $c > 0$, and where $F(W)$ is a distribution with finite moments. Obviously, if $F(0) = 0$, then $E\,|\,U(W)\,| < \infty$ and A2 is satisfied. However, there are distributions which yield quadratic utility as $t \to 0$, but $\lim_{n\to\infty} |U^{(n)}(1)\, E(W-1)^n/n!| = \infty$ so the infinite series in (5) will not converge.[4] From both of these examples it is apparent that assumptions A1–A4 are not particularly useful even for the cases in which $U(W)$ or W or both are bounded.

In motivating the development of a set of weaker assumptions than those given so far (A1–A4), it is instructive to analyze the remainder term in expression (3), EQ_3. Now, for any $n \geqq 3$,

$$EQ_3 = E\left[\sum_{i=3}^{n} U^{(i)}(1)((W-1)^i/i!) + Q_{n+1}\right].$$

Assuming the first two moments are of order $O(t)$, one weak necessary condition for asymptotic quadratic utility is that $E(W-1)^n = o(t)$ for some $n \geqq 3$; if $E(W-1)^4 = O(t)$, say, it is then hardly plausible that $EQ_3 = o(t)$. More generally, it is desirable to state assumptions about F, such that these assumptions imply that no "information" is contained beyond the first two moments as $t \to 0$. The characteristic function of a random variable determines the distribution uniquely; it is therefore appealing to develop conditions on the moments such that the characteristic function is asymptotically determined by the first two moments alone. Preferably, the latter should hold even though higher order moments may not be finite. The following result provides some intuitively obvious and useful conditions.

Theorem I Suppose $E(W-1)$ and $E(W-1)^2$ are of order $O(t)$, and

[4] See Section II in this paper; there it is assumed the returns are log normally distributed.

$E|W-1|^3 = o(t)$. The characteristic function of $W-1$ is then given by

$$Ee^{iu(W-1)} = 1 + iuE(W-1) - \tfrac{1}{2}u^2E(W-1)^2 + o(t).$$

Proof The characteristic function of $(W-1) \equiv Z$ is defined as $Ee^{iuZ} = E\cos(uZ) + iE\sin(uZ)$. Applying Taylor's theorem for $\cos(\cdot)$ and $\sin(\cdot)$, the characteristic function is also equal to

$$E\left[\sum_{j=0}^{n-1} \frac{(iuZ)^j}{j!} + R_n(u)\right],$$

where

$$R_n(u) = \frac{(iuZ)^n}{n!}\left[\cos(u\varphi_1 Z) + i\sin(u\varphi_2 Z) - 1\right],$$

and φ_1, φ_2 are random variables depending on Z and n. Noting that, always,

$$|\cos(u\varphi_1 Z) + i\sin(u\varphi_2 Z) - 1| \leqq |\cos(u\varphi_1 Z)| + |\sin(u\varphi_2 Z)| + 1 \leqq 3$$

and

$$E|Z|^3 = o(t),$$

it follows that $E|R_3(u)| = o(t)$; hence,

$$Ee^{iuZ} = 1 + iuEZ - \tfrac{1}{2}u^2EZ^2 + o(t)$$

for all u.

The result just obtained submits that moments of order 4 and higher are of no particular significance and therefore need not be analyzed. Returning to the examination of the remainder EQ_3, additional evidence about this matter can be developed. The exact expression of Q_3 is

$$Q_3 = U^{(3)}(\varphi(W))(W-1)^3/3!,$$

where $\varphi(W)$ is a continuous function implied by Taylor's theorem. Now, if $U^{(3)}(\varphi(W))| \leqq 0(1)$ almost surely, as is the case if $|U^{(3)}(\varphi(W))| < K$ for every t with probability 1, then

$$|EQ_3| \leqq E|Q_3| \leqq 0(1)E|W-1|^3 = o(t).$$

The precise behavior of the function $\varphi(W)$ is rarely known, and bounds are not always easy to establish. However, applying appropriate truncations of the set of outcomes permits the enunciation of quite weak—and practical—sufficient conditions for time-asymptotic quadratic utility.

Theorem II Let $F(W)$ be a distribution function of W such that $F(0) = 0$ $E(W-1)^n = O(t)$, $n = 1, 2$, and $E|W-1|^3 = o(t)$. Suppose there exist two sets $S_2 = [0, \varepsilon)$ and $S_1 = [\varepsilon, \infty)$ with $0 \leqq \varepsilon < 1$ such that

(i) $\sup_{\varphi \varepsilon S_1} |U^{(3)}(\varphi)| < \infty$, and $U^{(3)}(\varphi)$ is either a nonnegative function or a nonpositive function for all $\varphi \varepsilon S_2$;

(ii) there exists some $t_0 > 0$ such that for all $t \varepsilon (0, t_0)$ the function $\partial F_t(W)/\partial W \equiv p_t(W)$ is continuous on S_2, and

$$0 = \lim_{t \to 0} \frac{1}{t} p(W) \leqq \frac{1}{t} p_t(W) \leqq \frac{1}{t_0} p_{t_0}(W); \qquad W \varepsilon S_2.$$

Then

$$EU(W) = U(1) + U^{(1)}(1)\, E(W-1) + \tfrac{1}{2} U^{(2)}(1)\, E(W-1)^2 + o(t).$$

Proof Taylor's theorem implies there is a function $\varphi(W)$ such that

$$U(W) = \sum_{i=0}^{2} U^{(i)}(1)(W-1)^i/i! + U^{(3)}(\varphi(W))(W-1)^3/3!$$

where

$$\sup\{W, 1\} \geqq \varphi(W) \geqq \inf\{W, 1\}.$$

The sets S_1 and S_2 are disjoint and $\Pr\{W \in S_1 \cup S_2\} = 1$; hence

$$EU(W) = I_1(t)/3! + I_2(t)/3! + \sum_{i=0}^{2} U^{(i)}(1)\, E(W-1)^i/i!,$$

where

$$I_1(t) \equiv \int_{S_1} U^{(3)}(\varphi(W))(W-1)^3\, dF_t(W),$$

and

$$I_2(t) \equiv \int_{S_2} U^{(3)}(\varphi(W))(W-1)^3\, dF_t(W).$$

The last two integrals are finite; this follows in that $U^{(3)}(\varphi(W))(W-1)^3$ is equal to a function which is integrable (finite) when $U(W)$ and the first two moments are finite.

Consider first the order of $I_1(t)$. Note that for this integral $\varphi(W) \geqq \min\{W, 1\} \geqq \varepsilon$ for all $W \in S_1$. Hence,

$$\sup_{W \in S_1} |U^{(3)}(\varphi(W))| \leqq \sup_{\varphi \in S_1} |U^{(3)}(\varphi)| = K < \infty,$$

where the finiteness follows by assumption (i). This further implies

$$|I_1(t)| \leqq \int_{S_2} |U^{(3)}(\varphi(W))||W-1|^3\, dF_t(W)$$

$$\leqq K \int_{S_1} |W-1|^3\, dF_t(W) \leqq KE|W-1|^3 = o(t).$$

To show that $I_2(t) = o(t)$, observe first that assumptions (i) and (ii) imply

$$\left|\frac{1}{t} I_2(t)\right| = \int_{S_2} |U^{(3)}(\varphi(W))| |W-1|^3 \frac{1}{t} p_t(W)\, dW = \int_{S_2} f_m(W)\, dW,$$

where dW denotes the usual Lebesgue measure and $t = 1/m$. Without any loss of generality, assume $t_0 = 1$ and it also follows directly from assumption (ii) that

$$0 = \lim_{m \to \infty} f_m(W) \leqq f_m(W) \leqq f_1(W); \qquad W \in S_2, \quad m = 1, 2, \ldots.$$

Since $|I_2(1)| < \infty$, apply the Lebesgue dominated convergence theorem to the limiting integral; this yields.

$$\lim_{m \to \infty} \int_{S_2} f_m(W)\, dW = \int_{S_2} \lim_{m \to \infty} f_m(W)\, dW = 0,$$

since $\lim_{m \to \infty} f_m(W) = 0$ for all $W \in S_2$. Therefore, $I_2(t)$ is of order $o(t)$.

Collecting all of the above yields the desired relation

$$EU(W) = U(1) + U^{(1)}(1)\, E(W-1) + \tfrac{1}{2} U^{(2)}(1)\, E(W-1)^2 + o(t).$$

The theorem is not the sharpest possible, but this is of little import since a number of generalizations and extensions are quite simple. For example, it is not required that $U(W)$ is everywhere differentiable. Specifically,

Corollary I Suppose $U(W)$ is three times differentiable, except at a finite number of points $\varepsilon_1 \ldots, \varepsilon_n$; $\varepsilon_i \neq 1$, with zero measure for all t. Then the theorem is satisfied if $\varepsilon < \varepsilon_1$; $S_1 \equiv [\varepsilon, \infty) - \sum_i [\varepsilon_i]$, and ε is the same as in the theorem.

It is also obvious that condition (ii) can be relaxed, since it is only a condition such that two limit operators can be interchanged.

Corollary II In condition (ii), the continuous density $1/t_0 P_{t_0}(W)$ can be replaced by any Lesbesgue measurable function $0 \leqq p'(W)$, provided

$$\int_{S_2} |U^{(3)}(\varphi(W))(W-1)^3|\, p'(W)\, dW$$

is finite.

In order to obtain more specific results with respect to different $U(W)$, it must be assumed that condition (ii) in the theorem is satisfied. However, one important exception emerges.

Corollary III Suppose $\sup_{\varphi \geqq 0} |U^{(3)}(\varphi)| < \infty$; then condition (ii) is redundant.

A number of interesting cases satisfy the last corollary. When $U(W)$ is negative exponential ($-\exp\{-cW\}$, $c>0$), then $U^{(3)}(W)$ is bounded for all $W \geqq 0$ and $EU^{(3)}(\varphi(W))(W-1)^3 = o(t)$. Also, if $U(W) = (c+W)^{\gamma}/\gamma$, $\gamma < 1$, $c > 0$, then $\sup_{\varphi \geqq 0} |U^{(3)}(\varphi)| = |(\gamma-2)(\gamma-1)| c^{\gamma-3}$ and the corollary is satisfied. The case when $U(W) = W^{\gamma}/\gamma$, or $\log W$, requires more care, and condition (ii) remains in force. However, condition (i) is satisfied for any fixed $\varepsilon > 0$; thus, it is easily verified that $\sup_{\varphi \geqq \varepsilon} |U(\varphi)| = K(\varepsilon) < \infty$ and that $U^{(3)}(\cdot)$ is nonnegative.

The second condition in the theorem is not as strong as it might appear at first glance. The requirement $\lim_{t\to 0}(1/t)p_t(W) = 0$ is in fact almost implied by the general assumption that $E|W-1|^3 = o(t)$. To see this, one only need to note that if $(1/t)p_t(W)$ exists as $t \to 0$ and $W < 1$, then

$$0 = \lim_{t\to 0} \frac{1}{t} E|W-1|^3 \geqq \lim_{t\to 0} \int_{W<1} |W-1|^3 \frac{1}{t} p_t(W)\, dW$$

$$\geqq \int_{W<1} |W-1|^3 \lim_{t\to 0} \frac{1}{t} p_t(W)\, dW = 0,$$

The second inequality follows from Fatou's lemma, which may be applied since $|W-1|^3 p_t(W)(1/t) \geqq 0$. Hence, one can in general expect that $p_t(W) = o(t)$ if $W < 1$. Of course, this is closely related to purely continuous Markovian processes, which also require that $p_t(W) = o(t)$. However, in Theorem II it is assumed that $t^{-1}p_t(W) \leqq t_0^{-1}p_{t_0}(W)$ for all $0 < t \leqq t_0$, $0 < W \leqq \varepsilon$. This appears to be a stronger statement than the Markovian requirement $p_t(W) = o(t)$. (The mode of convergence is uniform in the former case.) From a practical point of view, this does not seem to be of any particular significance, especially in view of Corollary II. In fact, as can be seen in the next section, a simple and direct proof demonstrates that log-normal returns satisfy the uniform convergence conditions of Theorem II.

For most theoretical and practical purposes, the proposition of time-asymptotic quadratic utility must be regarded robust. This appears to be particularly true with respect to the utility function, as long as the absolute third moment is finite and vanishes more rapidly than the first two moments, and some other moderate requirements on the probabilities are satisfied.

It should perhaps also be mentioned that there are both necessary and sufficient conditions such that $E|t^{-1}Q_3| \to 0$ as $t \to 0$. A slight reparametrization of $E|t^{-1}Q_3|$ produces an expression for which the probability measure is independent of t. The restated convergence problem is now one of convergence in mean, and necessary and sufficient conditions are well known in measure theory.[5] For purposes of the present paper, this result is not in a particularly useful form, and there is no reason to restate these conditions here.

[5] See, for example, Halmos [2, p. 108].

III. The Log-Normal Case

In this section the "compactness" characteristic of the log-normal distribution is derived. First the dominated convergence of the continuous density $(1/t)p_t(W)$ as $t, W \to 0$ is established; subsequently, exact order properties of $E(W-1)^n$ for $n = 1, 2, \ldots$, are proved. It is assumed that $(\log X_1, \ldots, \log X_m)'$ is distributed in the multivariate normal form with parameters

$$(E \log X_1, \ldots, E \log X_m)' \equiv (t\mu_1, \ldots, t\mu_m) \equiv t\mu$$

and

$$[\operatorname{cov}(\log X_i, \log X_j)] \equiv [t\sigma_{ij}] \equiv t \textstyle\sum.$$

In other words, the expected growth and variance of growth are linear in time, and all increments are independently distributed. For obvious reasons, this process is referred to as a stationary "geometric" Wiener process (see Merton [4]). As it turns out, the X_i's, and $W = \sum \lambda_i X_i$, are not quite "compact" in the sense of Samuelson; i.e., assumption A1 is not satisfied. However, the moments are sufficiently well behaved to assure quadratic utility as $t \to 0$; i.e., they will satisfy Theorem 1 and Corollary II for all $\lambda \in D$. For simplicity, we first analyze the univariate case and write $\mathscr{L}(X) \sim ML_1(t\mu, t\sigma^2)$ to denote that the probability law of X is log-normal with parameters $t\mu$ and $t\sigma^2$.

Theorem III Let $p_t(x)$ and $p_0(x)$ be the densities of two (independent) log-normal random variables specified by $ML_1(t\mu, t\sigma^2)$ and $ML_1(0, 1)$, respectively. Then, there exist a t_0 and ε such that

$$(1/t)\, p_t(x) \leqq p_0(x)$$

for all $t \in (0, t_0)$ and $x \in [0, \varepsilon) \equiv S_1$.

Proof The density $p_t(x) \propto x^{-1}(t\sigma^2)^{-1/2} \exp\{-\frac{1}{2}(\log x - t\mu)^2 (t\sigma^2)^{-1}\}$; hence if the function $g(y, t)$ is defined by

$$g(y,t) \equiv p_t(x)(t\sigma^2)^{1/2}[p_0(x)]^{-1} = \exp\{-\tfrac{1}{2}[(y-t\mu)^2(t\sigma^2)^{-1} - y^2]\}$$

where $y = \log x$, then it suffices to show that

$$\sigma^{-1} \lim_{t\to 0} t^{-3/2} \sup_{y \in S_1'} g(y,t) \equiv \lim_{t \to 0} \sup_{y \in S_1'} \frac{t^{-1} p_t(x)}{p_0(x)} = 0$$

where $S_1' \equiv (-\infty, \log \varepsilon)$, $t \in (0, t_0)$, and $g(y, t)$ is defined on $S_1' \mathscr{X} (0, t_0)$.

Let $t_0 = \inf\{\frac{1}{2}, |\mu|, |1/\mu|, \sigma^{-2}\}$ and let $\log \varepsilon \equiv \delta$ be any fixed number less than -1. It is then straightforward to verify that

$$\sup_{y \in S_1'} g(y,t) \leqq \exp\{-\tfrac{1}{2}(\delta^2 + 2\delta t)(t^{-1} - 1)\}$$

for all $y, t \in S_1'\mathscr{X}(0, t_0)$. Also, note that $(\delta^2 + 2\delta t)(t^{-1} - 1) > 0$ for all $t \in (0, t_0)$. Therefore, it suffices now to show that

$$t^{-3/2} \exp\{-\tfrac{1}{2}(\delta^2 + 2\delta t)(t^{-1} - 1)\} = \exp\{-\tfrac{3}{2}\log t - \tfrac{1}{2}(\delta^2 + 2\delta t)(t^{-1} - 1)\} \to 0$$

as $t \to 0$, or, alternatively,

$$\lim_{t \to 0} \frac{-\tfrac{1}{2}(\delta^2 + 2\delta t)(t^{-1} - 1)}{-\tfrac{3}{2}\log t} = -\infty.$$

That the last statement is true is easily shown by an application of L'Hôpital's rule.

The result just obtained can be extended to the general case where $F_t(W) \equiv \int_0^W p_t(s)\, ds$ is not a log-normal distribution function; i.e. $0 < \lambda_i < 1$ for at least one i. The calculations to show this are lengthy, although straightforward. Hence, only the argument is developed here; the formal verification of the omitted step requires no more than ordinary differentiation, and the specification of t_0 is similar to that of the previous theorem. Assume, without any loss of generality, that $\lambda_i > 0$ for all i, and let $p_{tm}(X_1, \ldots, X_m)$ denote the density of the vector $(X_1, \ldots, X_m)'$, where $\mathscr{L}(X) \sim ML_m(t\mu, t\sum)$, and let $\max \lambda_i = \lambda_1$. The density of $p_t(W)$ can be constructed from the (closed form) density $p_{tm}(\cdot)$ and a change of variables:

$$\begin{aligned} W &= \textstyle\sum \lambda_i X_i, & & X_1 = \lambda_1^{-1}\left[W - \left(\sum_{i=2}^{m} Y_i\right)\right], \\ Y_2 &= \lambda_2 X_2, & \text{or, equivalently} \quad & X_2 = \lambda_2^{-1} Y_2, \\ &\vdots & & \vdots \\ Y_m &= \lambda_m X_m & & X_m = \lambda_m^{-1} Y_m \end{aligned}$$

then

$$\begin{aligned} 0 &\leqq |J|\, p_t(W) \\ &= \int_{R'(W)} p_{tm}(\lambda_1^{-1}[W - Y_2 - \cdots - Y_m], \lambda_2^{-1} Y_2, \ldots, \lambda_m^{-1} Y_m)\, dY_2 \cdots dY_m, \end{aligned}$$

where $|J|$ is the determinant of the Jacobian, which depends solely on λ, and $R'(W)$ is the region of integration, which depends on W. Now if $W \in [0, \min \lambda_i) \equiv S_1 \equiv [0, \varepsilon)$, it can then be shown that there exists a $t_0 = t_0(\mu, \sum)$ such that

$$\log(1/t) + \log p_{tm}(\cdot) \leqq \log(1/t_0) + \log p_{t_0 m}(\cdot)$$

for all $W \in S_1, Y_2, \ldots, Y_m \in R'(W)$, and all $0 < t < t_0$. Thus, for any $\lambda \in D$, there exists a t_0 such that

$$0 \leqq (1/t)p_{tm}(\cdot) \leqq (1/t_0)p_{t_0 m}(\cdot)$$

for all $Y_2, \dots, Y_m \in R'(W)$. The integration operation over $R'(W)$ changes nothing; it follows immediately that $(1/t)p_t(W) \leqq (1/t_0)p_{t_0}(W)$; $W \in S_1$, $0 < t < t_0$. Subsequently, it will be shown that $E|W-1| = o(t)$ and, using Fatou's lemma, this implies $p_t(W) = o(t)$ except when $W = 1$. (This is also simple to verify directly.) Consequently, the condition (ii) in Theorem II is satisfied since $\varepsilon < 1$ and

$$0 = \lim_{t \to 0} (1/t)p_t(W) \leqq (1/t)p_{t_0}(W),$$

where $0 < t < t_0$ and $W \in S_1 \equiv [0, \varepsilon)$.

Next, the behavior of the moments around unity are analyzed as $t \to 0$; as previously, the case when $\lambda_i = 1$ is considered first, and generalizations are developed immediately thereafter.

Theorem IV Let $\mathscr{L}(X) \sim ML_1(t\mu, t\sigma^2)$ with $t, \mu, \sigma^2 > 0$. Then

$$\lim_{t \to 0} \frac{E(X-1)}{E(X-1)^2} = \frac{\mu + 1/2\sigma^2}{\sigma^2}, \qquad \text{(i)}$$

and

$$\lim_{t \to 0} \frac{E(X-1)^n}{E(X-1)^2} = \begin{cases} t^{(n-1)/2} c_n & \text{if} \quad n = 3, 5, 7, \dots, \\ t^{(n-2)/2} c_n & \text{if} \quad n = 4, 6, 8, \dots. \end{cases} \qquad \text{(ii)}$$

Alternatively, $E(X-1)^n = E(X-1)^{n-1} = O(t^{n/2})$, $n = 2, 4, \dots$.

Proof Note first that $EX^n = \exp\{t(n\mu + \frac{1}{2}n^2\sigma^2)\}$. Equation (i) is easy to verify and is given by Merton [3]. Thus, only (ii) will be formally proved here, and this by showing that $E(X-1)^n$ is of order $t^{n/2}$ if n is an even integer, and of order $t^{(n+1)/2}$ if n is an odd integer. Let $f(n,i) \equiv (n-i)\mu + (n-i)^2 \frac{1}{2}\sigma^2$ and $b(n,i) \equiv (-1)^i \binom{n}{i}$. Applying the binomial expansion and the closed-form expression of EX^n yields

$$\begin{aligned} E(X-1)^n &= E \sum_{i=0}^{n} (-1)^i \binom{n}{i} X^{n-i} \\ &= \sum_{i=0}^{n} b(n,i) \exp\{t[(n-i)\mu + (n-i)^2 \tfrac{1}{2}\sigma^2]\} \\ &= \sum_{i=0}^{n} b(n,i) \left(1 + \sum_{k=1}^{\infty} t^k f(n,i)^k / k!\right). \end{aligned}$$

Since $\sum_{i=0}^{n} b(n,i) = 0$ and $f(n,n) = 0$ the last expression is equal to

$$\sum_{i=0}^{n-1} b(n,i) \sum_{k=1}^{\infty} t^k f(n,i)^k/k! = \sum_{k=1}^{\infty} \sum_{i=0}^{n-1} t^k b(n,i) f(n,i)^k/k!$$
$$= \sum_{k=1}^{\infty} 1/k! \sum_{i=0}^{n-1} t^k b(n,i) f(n,i)^k.$$

Consider now $\sum_{i=0}^{n-1} t^k b(n,i) f(n,i)^k \equiv S(n,k)$. Applying the binomial expansion for $f(n,i)^k$ gives

$$S(n,k) = \sum_{i=0}^{n-1} t^k b(n,i) \sum_{j=0}^{k} \binom{k}{j} \mu^{k-j} (\tfrac{1}{2}\sigma^2)^j (n-i)^{k+j}$$
$$= t^k \sum_{j=0}^{k} \binom{k}{j} \mu^{k-j} (\tfrac{1}{2}\sigma^2)^j \sum_{i=0}^{n-1} b(n,i)(n-i)^{k+j}.$$

Lengthy but straightforward induction will show that[6]

$$\sum_{i=0}^{n-1} b(n,i)(n-i)^{k+j} \begin{cases} = 0 & \text{if} \quad k+j < n, \\ > 0 & \text{if} \quad k+j \geqq n, \end{cases}$$

and since

$$\sum_{j=0}^{k} \binom{k}{j} \mu^{k-j} (\tfrac{1}{2}\sigma^2)^j > 0 \qquad \text{and} \qquad j = 0, 1, \ldots, k,$$

we have that $S(n,k) = 0$ if and only if $2k \geqq n$. Consequently,

$$E(X-1)^n = t^{n/2}\eta_n + \text{higher powers of } t = O(t^{n/2}), \qquad n = 4, 6, \ldots,$$

and

$$E(X-1)^n = t^{(n+1)/2}\eta_n + \text{higher powers of } t = O(t^{(n+1)/2}),$$

$n = 3, 5, \ldots$, with $\eta_n > 0$.

The condition $\mu > 0$ is not crucial; for $\mu \leqq 0$ the theorem still holds if we read $O(t^n)$ as "at least of order t^n" for all odd $n \geqq 3$. (For $\mu = -\tfrac{3}{2}\sigma^2$ it may be shown that $E(X-1)^3$ is of the order t^3.) The relevant moments are of course $E(W-1)^n$, rather than $E(X_i - 1)^n$. This creates little difficulty, however; it is tedious but straightforward to show that $E(W-1)^n$ has moments of the same order as those stated in the last theorem. In any case, it suffices to demonstrate that $E|W-1|^3 = o(t)$. To see this, note first that $E(W-1)^n \leqq$

[6] This result is also implied, after a slight reparametrization, by a problem stated by Feller [1, p. 63].

$\sum_i E(X_i - 1)^n = O(t^{n/2})$ for $n = 2, 4$. Second, it can be shown that the inequality

$$E|W-1|^3 \leqq (E(W-1)^2 E(W-1)^4)^{1/2}$$

always holds.[7] Combining these results implies

$$E|W-1|^3 \leqq [O(t)O(t^2)]^{1/2} = O(t^{3/2}) = o(t).$$

Likewise, direct evaluation of the first two moments will show that they are of order $O(t)$ for all $\lambda \in D$, so the log-normal distribution satisfies the properties required for Theorem II and its corollaries.[8]

Some further, quite interesting, comments can be made about the log-normal distribution and time-asymptotic quadratic utility. The moments $E(W-1)^n$ are increasing very rapidly as n increases. In fact, for any t no matter how small, it may be shown that $E(W-1)^n$ increases faster than $n!S^{-n}$ $(0 < S < 1)$.[9] Consequently, for many utility functions one cannot expect that assumption A3 holds. Furthermore, although

$$\lim_{n\to\infty} |U^{(n)}(1)\, E(W-1)^n/n!| = \infty$$

it is still true that as $t \to 0$ quadratic utility prevails, the reason being no information is contained beyond the first two moments, as asserted by Theorem I.

It is tempting to conjecture that sums of log-normal random variables converge to a log-normal variable as $t \to 0$. However, this is not true in any *nontrivial* sense, as is also evident from Theorem I. The theorem asserts that the limiting characteristic function is identical for all distributions satisfying the hypotheses. Only the behavior of the first three moments is important as $t \to 0$, and the particular "type" of distribution is irrelevant in the limit. In view of this, it is clear that Merton's [4, Theorem II] result that wealth, for a *finite* period of time, is log-normally distributed, is conceptually distinct from the analysis in this paper. Essentially, the difference occurs because the

[7] Let s_1 and s_2 be any real numbers; then

$$0 \leqq E(s_1|Z| + s_2|Z|^2)^2 = s_1{}^2E|Z|^2 + 2s_1s_2E|Z|^3 + s_2{}^2E|Z|^4 \qquad \text{for all} \quad s_1, s_2,$$

and any distribution of Z; hence, the positive semidefiniteness of the quadratic form implies that the inequality is true.

[8] More precisely,

$$\lim \frac{1}{t} E(W-1) = \sum \lambda_i(\mu_i + \tfrac{1}{2}\sigma_{ii}) \qquad \text{and} \qquad \lim \frac{1}{t} E(W-1)^2 = \sum\sum \lambda_i\lambda_j\sigma_{ij}.$$

[9] This is not a surprising result, since the moment-generating function is nonconvergent for log normal random variables.

analysis here is "one short period," while the Merton analysis is "many short periods" (continuous portfolio revision is the limiting case). The many-short-periods context will, under appropriate assumptions, generate the product of a "large" number of independently distributed wealth relatives. The limiting distribution of the latter process will then be a log-normal distribution.

Finally, it is interesting to note that the third and fourth moments are of the same order. Referring to Theorem II in the cited Samuelson paper, this implies that a higher order approximation is not attained by adding a third moment and maximizing a cubic utility function.

Acknowledgment

Without implicating him, the author wishes to thank Markku Kallio for valuable discussions.

References

1. Feller, W., *An Introduction to Probability Theory and Its Applications*, Second edition, Volume 1. New York, 1957.
2. Halmos, P. R., *Measure Theory*. Van Nostrand-Reinhold, Princeton, New Jersey, 1950.
3. Merton, R. C., "Lifetime Portfolio Selection under Uncertainty: The Continuous-Time Case." *Review of Economics and Statistics* **51** (1969), 247–257.
4. Merton, R. C., "Optimum Consumption and Portfolio Rules in a Continuous-Time Model." *Journal of Economic Theory* **3** (1971).
5. Merton, R. C., "An Intertemporal Capital Asset Pricing Model." *Econometrica* **40** (1972).
6. Merton, R. C., and Samuelson, P. A., "Fallacy of the LogNormal Approximation to Optimal Portfolio Decision-Making over many Periods." *Journal of Financial Economics* **1** (1974).
7. Samuelson, P. A., "The Fundamental Approximation Theorem of Portfolio Analysis in Terms of Means, Variances, and Higher Moments." *Review of Economic Studies* **37** (1970).

SAFETY-FIRST AND EXPECTED UTILITY MAXIMIZATION IN MEAN-STANDARD DEVIATION PORTFOLIO ANALYSIS

David H. Pyle and Stephen J. Turnovsky *

I Introduction

DATING back to the original work by Markowitz and Tobin [7, 11], portfolio theory has usually followed the mean-standard deviation approach in which the investor is assumed to choose among alternative portfolios on the basis of a utility function defined in terms of the mean and standard deviation of the portfolio return. Because of the arbitrary nature of utility functions, more or less parallel with these developments there have been attempts to depart from the utility framework altogether and to invoke criteria based on more objective concepts. As the first of such objective criteria, Roy [8] suggested that investors have in mind some disaster level of returns and that they behave so as to minimize the probability of disaster. This criterion, along with some variants of it, since developed by others (Telser [10], Kataoka [5]), has become known as the safety-first criterion.

As we shall see, the various safety-first criteria also lead to optimization of expressions involving the mean and standard deviation. The objective of this paper is to compare this justification of mean-standard deviation analysis with the more conventional approach based on expected utility maximization. This relationship, although hinted at, has not been systematically studied in the literature (Lintner [6], Farrar [3]).

In particular we shall show that:

1) In the absence of a riskless asset, a correspondence can be established between the safety-first criterion and expected utility maximization when that maximization results in concave indifference curves in the mean-standard deviation space.

2) If a riskless asset is available then except in one special case the safety-first criterion does not lead to the traditional liquidity preference behavior.

* We thank Gordon Pye and Paul Cootner for helpful comments on an earlier draft of this paper. Needless to say they are not responsible for any errors which may remain.

II Expected Utility Maximization and the Safety-First Criteria

According to the expected utility approach to mean-standard deviation portfolio analysis, the investor's problem is to select a portfolio of the available assets so as to maximize some expected utility function of the form

$$E(u) = V(\mu, \sigma) \tag{1}$$

where μ is the expected value and σ is the standard deviation of z, the total one-period return on the portfolio. Furthermore, this utility function has the properties

$$\frac{\partial V}{\partial \mu} > 0, \quad \frac{\partial V}{\partial \sigma} < 0.$$

Given positive but diminishing marginal utility of wealth and a multivariate normal distribution of returns to the available assets, V is a concave function in the (μ, σ) plane.[1]

Turning to the safety-first principle, we shall state the following forms where $\bar{z}$ is the subsistence or disaster level of returns and α is the probability of disaster.[2]

(i) $\min Pr(z \leq \bar{z})$

(ii) $\max \bar{z}$ subject to $Pr(z \leq \bar{z}) \leq \alpha$

(iii) $\max \mu$ subject to $Pr(z \leq \bar{z}) \leq \alpha$

where Pr denotes probability.

Form (i) is the objective as originally stated by Roy, while (ii) is a later version proposed in a somewhat different context by Kataoka. These alternative forms both lead to indifference curves in the (μ, σ) plane whose slopes

[1] The original proof that V is a concave function under these conditions was given by Tobin [11]. As Fama [2] and others have pointed out, Tobin's proof based on any two parameter distribution is only valid for the normal. Feldstein [4] has presented a counter-example of a two-parameter distribution for which V is not concave. It should also be pointed out that only in special cases will $E(u)$ depend only on μ and σ. In general it will depend on higher moments as well.

[2] Recently Baumol [1] has suggested a related criterion in which the investor compares portfolios on the basis of $\mu, \bar{z}$ rather than μ, σ. Since this approach involves specifying a probability level of disaster for the individual, it effectively incorporates to some extent his preferences and therefore succeeds in narrowing down the Markowitz efficiency frontier.

depend on α. Moreover, for form (i), which shall be referred to as the "minimum α" safety-first criterion, the ordering of these indifference curves also depends on α, the probability of disaster, while in (ii), to be called the "maximum $\bar{z}$" criterion, the ordering depends on $\bar{z}$ the disaster level.[3] Telser proposed (iii) as an alternative to the "minimum α" criterion on the grounds that the existence of a riskless asset, cash, eliminated the need for minimizing the probability of disaster. We shall refer to this version as the "maximum μ" safety-first criterion.[4]

Both Roy and Telser obtain their criterion by making use of the Chebychev inequality. We will not follow them in this respect, but instead will proceed along the following lines. We shall assume that the distribution of the variable z can be fully described by two parameters so that it can be transformed into the standardized variable $\left(\frac{z-\mu}{\sigma}\right)$ which has a distribution function F such that $Pr(z \leq \bar{z}) \equiv F\left(\frac{\bar{z}-\mu}{\sigma}\right)$. Since F is monotonic, the "minimum α" version, which involves minimizing $Pr(z \leq \bar{z})$, is equivalent to

$$\min\left(\frac{\bar{z}-\mu}{\sigma}\right). \tag{2}$$

Similarly the constraint $Pr(z \leq \bar{z}) \leq \alpha$ is equivalent to $F\left(\frac{\bar{z}-\mu}{\sigma}\right) \leq \alpha$ and can be written in the form

$$\bar{z} \leq \mu + F^{-1}(\alpha)\sigma \tag{3}$$

where $F^{-1}(\alpha)$ is the inverse of F and is a constant, depending on the probability level α.[5] For any distribution we can find a critical probability level α^*, such that $F^{-1}(\alpha^*) = 0$ and $F^{-1}(\alpha) < 0$ for $\alpha < \alpha^*$ while $F^{-1}(\alpha) > 0$ for $\alpha > \alpha^*$. The magnitude of α^* reflects the skewness of the distribution, with $\alpha^* = 0.5$ indicating a distribution symmetrical about its mean. Normally one would expect the investor to choose a small value for α, so that as long as the distribution is not too skew, $\bar{z}$ is less than μ.

Using (2) and (3), the alternative versions of the safety-first criteria can be restated as follows:

$$\text{(i)}\quad \min\left(\frac{\bar{z}-\mu}{\sigma}\right) \tag{4}$$

$$\text{(ii)}\quad \max\ \mu + F^{-1}(\alpha)\sigma \tag{5}$$

$$\text{(iii)}\quad \max\ \mu \text{ subject to } \mu + F^{-1}(\alpha)\sigma \geq \bar{z} \tag{6}$$

III Relationship Between Approaches — No Riskless Asset

In this section a graphical demonstration of the relationship between mean-standard deviation analysis based on expected utility maximization and mean-standard deviation analysis based on the safety-first principle is given for the no riskless asset case. Following Markowitz and Sharpe [7, 9], we can derive the investor's opportunity locus AB in the (μ, σ) plane. This is done by obtaining the portfolio which minimizes the variance for given values of expected return and the locus AB can be shown to have the curvature indicated in figure 1. The investor's opportunity set of portfolios consists of those which are bounded below by APB.

Corresponding to the expected utility function $V(\mu, \sigma)$, we can draw a set of indifference

FIGURE 1. — EXPECTED UTILITY MAXIMIZATION: NO RISKLESS ASSET

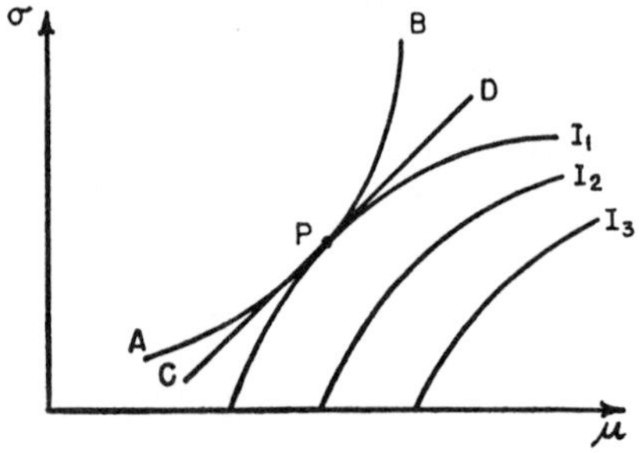

[3] By ordering, we mean the ranking of preferences as measured by the indifference curves.

[4] It is worth noting here that the "maximum μ" criterion requires investors to choose two parameters $\bar{z}$ and α. For the two other safety-first criteria only one parameter must be selected.

[5] Roy's use of the Chebychev inequality was an attempt to avoid the two parameter restriction. Whatever the merits of this approach in the context of our paper it is easy to show that the results we have obtained for the "minimum α" and "maximum μ" safety-first criteria are equally applicable to the Roy and Telser versions of these criteria which are derived using the Chebychev inequality. This follows from the fact that Roy's objective function is $\mu - \bar{z}/\sigma$ while Telser's version is max μ subject to $\mu - \alpha^{-\frac{1}{2}}\sigma \leq \bar{z}$ ([8, 10]).

curves which will have the indicated curvature.[6] The equilibrium portfolio P, is then found where a tangency occurs between an indifference curve and the efficiency frontier, AB. If the indifference curve passing through P is given by

$$V(\mu, \sigma) = \bar{V}$$

the slope of the indifference curve at P is found by substituting the coordinates of P into the equation

$$\frac{d\sigma}{d\mu} = -\frac{V_1}{V_2}. \qquad (7)$$

Since any monotonic transformation of the expected utility function V will also have a slope given by (7), it follows that there is an infinity of indifference curves which are tangential to AB at P.

Turning now to the safety-first principle, we shall commence with the "minimum α" criterion.

The indifference curves corresponding to (4) are given by

$$\frac{\bar{z} - \mu}{\sigma} = k, \qquad (8)$$

where k is a constant. We have already argued that one would expect α to be small, in which case $\bar{z} < \mu$ and $k < 0$. As k varies (4) defines a series of curves in the (μ, σ) plane having positive slopes equal to

$$\frac{d\sigma}{d\mu} = \frac{-1}{k} = \frac{-\sigma}{\mu - \bar{z}}. \qquad (9)$$

It follows from (8) that all these indifference curves pass through the point $(\bar{z}, 0)$ on the μ axis. This, of course, is very strange and among other things violates any law of transitive preferences. To avoid this difficulty we shall not define the indifference curves at the point $(\bar{z}, 0)$ but restrict ourselves to portfolios with strictly positive σ. Since we are really interested in choices among risky alternatives the exclusion of this point is not really a restriction. For $\sigma > 0$, these indifference curves imply an unambiguous ordering and further, since $d^2\sigma/d\mu^2 = 0$, they are linear. Noting that the quantity being minimized is α, the probability of disaster, it is clear that the ordering of these indifference curves depends on α. Minimizing α is equivalent to minimizing the slope of the indifference curves given by (8) and therefore in figure 2, the indifference curve I_2' will be preferred to I_1'. Thus, both the slope and the ordering of the indifference curves corresponding to "minimum α" safety-first criterion depend on α. As drawn in figure 2, the indifference curve I_1' has a tangency with AB at Q.

FIGURE 2. — "MINIMUM α" SAFETY-FIRST CRITERION: NO RISKLESS ASSET

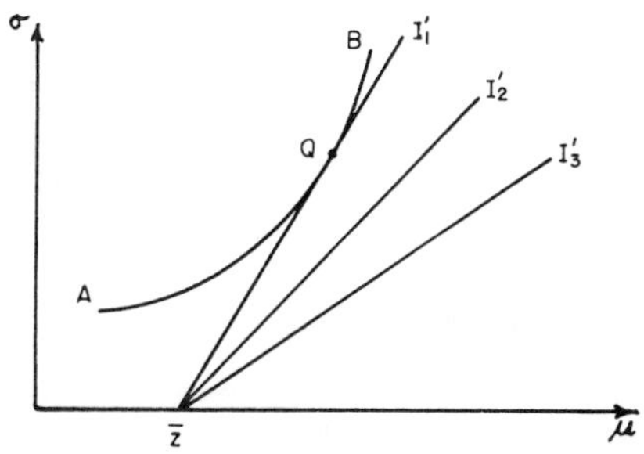

By a similar analysis it can be shown that (5) defines another set of indifference curves in the (μ, σ) plane such that these curves are a series of parallel straight lines with a positive slope dependent on α (since $d\sigma/d\mu = -1/F^{-1}(\alpha)$) and intercept dependent on $\bar{z}$. Since $\bar{z}$ is being maximized, the ordering of these curves depends on this parameter and the optimal $\bar{z}$ is determined by the tangency between a "maximum $\bar{z}$" indifference curve and the efficient frontier AB.

In general, there is no need for these two versions of the safety-first principle to lead to the same portfolio. Only if the $\bar{z}$ selected by the "minimum α" investor is the same as that which turns out to be optimal for the "maximum $\bar{z}$" investor will the two investors choose the same portfolio.[7]

The "maximum μ" version is somewhat different in that the optimum portfolio is not determined by any tangency conditions.[8] Since

[6] Because of the signs of $\partial V/\partial\mu$ and $\partial V/\partial\sigma$, south-easterly movements imply an increase in expected utility.

[7] Equivalently this could be stated as follows: If the α selected by the "maximum $\bar{z}$" investor is the same as turns out to be optimal for the "minimum α" investor then the two investors will behave the same. These two equivalent statements bring out the dual nature of these two versions of the safety-first principle.

[8] By the same token, it is not very sensible to define

the investor's portfolio must simultaneously lie within the opportunity set and satisfy the probability constraint (6), it must lie above the line AB and below the line RT. The feasible set of portfolios is therefore given by the set RST and is illustrated in figure 3.

FIGURE 3. — "MAXIMUM μ" SAFETY-FIRST CRITERION: NO RISKLESS ASSET

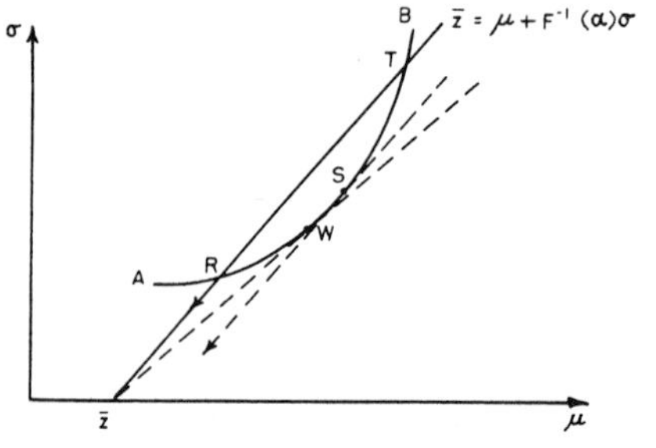

It is easy to see that the portfolio which maximizes expected returns among those included in this set is the one represented by T, which is therefore the one selected by the Telser safety-firster. It also can be seen that this portfolio is less conservative (in the sense that it leads the investor to select a portfolio with a larger mean and larger standard deviation) than that selected by a "minimum α" investor who selects the same disaster level $\bar{z}$, or by a "maximum $\bar{z}$" investor who selects the same probability level α. They will select the points W,S respectively. On the other hand, the "maximum μ" approach does create problems of feasibility that the other two variants do not have. By selecting both α and $\bar{z}$, the investor fixes the position of the line RT and it is quite possible that the values of these parameters he selects are such that this line fails to intersect the opportunity locus. In this case the "maximum μ" principle breaks down for there is no portfolio which will permit the investor to meet his disaster level with the required probability.

indifference curves in the mean-standard deviation plane for the "maximum μ" approach as could be done for the other approaches. The reason is that in this approach the optimality criterion, which is to maximize the expected return, does not involve σ. Consequently, one cannot get a very sensible "balancing off" between μ and σ such as portrayed by normal mean-standard deviation indifference curves.

We are now in a position to study the relationship between portfolio analysis based on expected utility maximization and that based on the safety-first principle. Referring to figure 1, imagine that an investor who maximizes expected utility chooses the portfolio P. Since the set of attainable portfolios bounded by APB is a convex set and since the set of points preferable to those lying along I_1 also forms a convex set, by the separating hyperplane theorem there exists a straight line CD passing through P which is tangential to both APB and I_1. If we use the "minimum α" safety-first criterion, we can equate the slope of this line to the slope given by (9), $\sigma/(\mu-\bar{z})$. For the given coordinates (μ_P, σ_P) of P, this determines a unique $\bar{z}$, say $\hat{\bar{z}}$. Thus, to the portfolio P chosen by an investor who maximizes expected utility, there corresponds a unique disaster level $\hat{\bar{z}}$ such that the same portfolio would also be chosen by a safety-first investor who minimizes the probability of his total portfolio return falling below this disaster level.

Similarly, if the "maximum $\bar{z}$" safety-first criterion is adopted, the slope of CD can be equated to $-(1/F^{-1}(\alpha))$. This determines a unique probability level, say $\hat{\alpha}$, such that an investor who maximizes $\bar{z}$ subject to the constraint that the probability of total return falling below $\bar{z}$ is equal to $\hat{\alpha}$ will select the same portfolio as an expected utility maximizing investor.

However, if we work back from these two safety-first criteria to expected utility maximization, the unique correspondence does not hold. In figure 2, the indifference curve I_1' which corresponds to the "minimum α" safety-first criterion has a tangency with AB at Q which determines a unique slope $\sigma_Q/(\mu_Q - \bar{z})$, where (μ_Q, σ_Q) are the coordinates of Q. We have already seen that there is an infinity of indifference curves derived from the expected utility maximization principle passing through a given point on AB with a prescribed slope. Hence, there is an infinity of utility functions that will lead the expected utility maximizing investor to select the same portfolio Q as does the "minimum α" safety-first investor. A similar conclusion holds for the "maximum $\bar{z}$" safety-first investor.

The "maximum μ" formulation requires the investor to select two parameters and as a result the correspondence to utility maximization is in general not unique. Referring to figure 3, we can imagine an expected utility maximizer selecting the portfolio T. There is an infinity of straight lines passing through T and each one of these is consistent with the behavior of a safety-first investor who selects T on the "maximum μ" principle. If, however, we specify either one of the two parameters, for example, by requiring the investor to choose a specific disaster level, then the correspondence is indeed unique. The reverse correspondence is as for the other two cases and need not be repeated.

In summary then, as long as there is no riskless asset, for any portfolio chosen by an expected utility maximizing investor with concave (μ, σ) indifference curves, we can always find a safety-first investor who will choose the same portfolio. For any portfolio chosen by any variant of the safety-first principle, there is in general an infinity of expected utility functions that will cause an investor who maximizes expected utility to behave in the same way.

IV Relationship Between Approaches — Riskless Asset

As we shall now demonstrate, the correspondence between the expected utility and the safety-first approaches to mean-standard deviation analysis breaks down seriously when the investor has the opportunity of holding a riskless asset. We shall assume that the riskless asset has a rate of return r and that the investor can either lend or borrow at this rate. Then, following Sharpe, the investor's efficiency locus is as drawn in figure 4.

The curved line AB is obtained as in section III, while REF is the tangent from the point (R,O) to AB where $R = rW$ and W is the investor's initial wealth. This line represents the opportunity locus for lending (segment RE) or borrowing (segment EF) at the riskless rate and holding the remainder of the investment balance (wealth less lending or plus borrowing) in portfolio E. Since the straight line REF dominates all parts of the curved line AB it now becomes the efficiency locus and the equilibrium portfolio is found where an indifference curve is tangent to REF.[9]

Assuming concave (μ, σ) indifference curves, we can always find a unique point of tangency between the efficiency locus REF and an indifference curve of an investor who maximizes expected utility. His optimum will be at a point such as P in figure 4 where he holds that portfolio of risky assets specified by E and borrows or lends a finite amount of the riskless asset.

FIGURE 4. — EXPECTED UTILITY MAXIMIZATION WITH RISKLESS ASSET

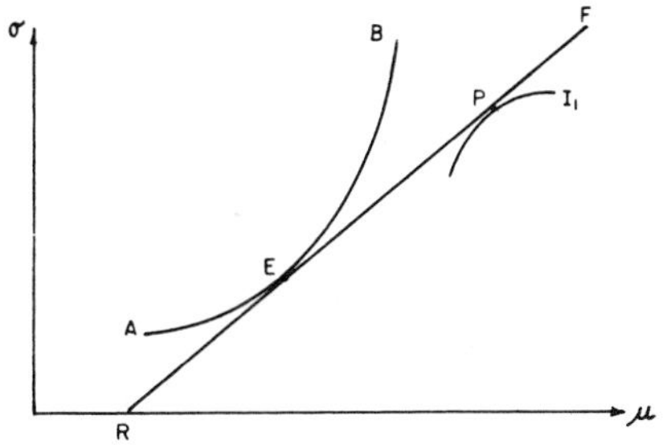

On the other hand, a "minimum α" or a "maximum $\bar{z}$" safety-first investor with straight line indifference curves can never achieve a tangency with REF except in the special case that one of his indifference curves coincides with REF, in which case an indeterminate solution results. If the slope of his indifference curves exceed that of REF, he will borrow an infinite amount (assuming that there is no limit) and move out in the direction of F.[10] Notice that although such an indifference curve implies unlimited borrowing, the investor is still holding a diversified portfolio of risky assets, namely that specified by E. If the slope is less than that of REF, the investor will move

[9] If we do not allow borrowing at the riskless rate, the efficiency locus becomes REB. When the tangency of an investor's expected utility indifference curve occurs along the segment EB of the efficiency locus, the results are the same as in the no riskless asset case. When the tangency occurs along the segment RE, the results are the same as for the case we are about to discuss in which the investor can obtain a riskless asset.

[10] This can be readily seen by superimposing the straight line indifference curves of figure 2 on figure 4 and bearing in mind that south easterly movements imply increasing expected utility.

towards R and hold all his wealth in the riskless asset.

In the case of the "minimum α" safety-first criterion, the investor chooses $\bar{z}$ and if, and only if, he selects R as his disaster or subsistence level will one of his indifference curves coincide with REF. For $\bar{z} > R$, this criterion leads to unlimited borrowing and for $\bar{z} < R$, it leads to an undiversified holding of the riskless asset.

For the "maximum $\bar{z}$" criterion, the investor selects α and the condition for the optimizing indifference curve to coincide with REF is

$$-\frac{1}{F^{-1}(\alpha)} = \frac{\sigma_E}{\mu_E - R} \tag{10}$$

where (μ_E, σ_E) are the coordinates of the point E.

This condition can be rewritten in the form

$$\alpha = F\left(\frac{R - \mu_E}{\sigma_E}\right) \tag{11}$$

which determines a value for α, which we shall denote α_E. Hence, the required α must be chosen equal to the probability that the rate of return on the only efficient purely risky portfolio is less than or equal to r the rate of return on the riskless asset.[11] For any $\alpha > \alpha_E$, an investor using this criterion would borrow an unlimited amount of the riskless asset; for $\alpha < \alpha_E$ he would hold only the riskless asset.

Without constraints on the parameters, these two versions of the safety-first criterion lead the investor to extreme strategies. Where his indifference curve coincides with REF or where it leads him to borrow an infinite amount, he will hold the same portfolio of risky assets as a risk-averse investor maximizing expected utility, but in general, he will not hold the same overall portfolio. Consequently it is not possible to establish any correspondence between these two approaches to mean-standard deviation analysis.

The implications of the "maximum μ" principle when a riskless asset is present can be studied by superimposing the relevant portions of figures 3 and 4 as is shown in figure 5.

The feasible region of portfolios must lie "above" the line REF and "below" the line $\bar{z}$Q.

[11] Equation (11) may be expressed in terms of rates of return by dividing the numerator and the denominator of the term in brackets by W.

FIGURE 5. — "MAXIMUM μ" SAFETY-FIRST CRITERION WITH RISKLESS ASSET

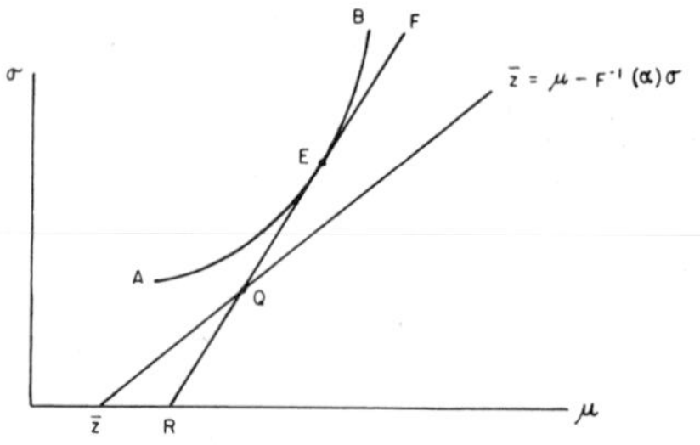

As it has been drawn, the feasible area is bounded by $\bar{z}$QR and consequently the investor will select the point Q. However, the feasible area does depend crucially on the relative positions of the two lines $\bar{z}$Q and REF. This in turn means that the behavioral implications depend on the relationship of $\bar{z}$ to R and of α to α_E, and a number of cases must be considered. These are summarized in the following table, which encompasses all possibilities.

Case		*Implied Behavior*
I	$\bar{z} < R, \alpha < \alpha_E$	Equilibrium at Q. Diversified portfolio with either finite borrowing or finite lending.
II	All z, $\alpha > \alpha_E$	Diversified portfolio, infinite borrowing.
III	$\bar{z} \leq R, \alpha = \alpha_E$	Diversified portfolio, infinite borrowing.
IV	$\bar{z} = R, \alpha < \alpha_E$	Equilibrium at R. Undiversified portfolio, consisting of all wealth invested in the riskless asset.
V	$\bar{z} > R, \alpha \leq \alpha_E$	Infeasible.

The portfolio chosen in Case I (illustrated in figure 5), is consistent with the behavior of an investor with concave (μ, σ) indifference curves, since it implies a diversified portfolio with a finite position in the riskless asset. Whether this point implies borrowing or lending of this asset depends on whether the point Q lies above or below E. The other three feasible cases imply extreme strategies, just as do the other two versions of the safety-first criterion. Notice, however, that unlike the other two versions an indeterminacy can never arise and in the case that $\bar{z}Q$ and REF coincide, infinite borrowing is implied. Hence, in all but Case I, the "maximum μ" criterion leads to an overall portfolio, which, in general, is not consistent with the behavior of an investor with concave (μ, σ) indifference curves.

The results obtained for the riskless asset case need not imply that the safety-first principle is without merit in the analysis of asset holding under uncertainty. It is conceivable that the safety-first principle might have an important role if this analysis was broadened to include the transactions demand as well as the investment demand for assets.[12] Furthermore, the value of the safety-first principle as an alternative justification for traditional mean-standard deviation analysis where distributions contain more than two parameters remains an open question.

REFERENCES

[1] Baumol, W. J., "An Expected Gain-Confidence Limit Criterion for Portfolio Selection," *Management Science*, 10 (Oct. 1963).

[12] This point was suggested to us by a referee.

[2] Fama, E. F., "Risk, Return and Equilibrium: Some Clarifying Comments," *Journal of Finance*, 23 (Mar. 1968).

[3] Farrar, D. E., *The Investment Decision Under Uncertainty* (Englewood Cliffs, N.J.: Prentice-Hall, 1962).

[4] Feldstein, M. S., "Mean-Variance Analysis in the Theory of Liquidity Preference and Portfolio Selection," *Review of Economic Studies*, 36 (Jan. 1969).

[5] Kataoka, S., "A Stochastic Programming Model," *Econometrica*, 31 (Jan.-Apr. 1963).

[6] Lintner, J., "The Valuation of Risk Assets and the Selection of Risky Investments in Stock Portfolio and Capital Budgets," this REVIEW, 47 (Feb. 1965).

[7] Markowitz, H., *Portfolio Selection* (New York: Wiley and Sons, 1959).

[8] Roy, A. D., "Safety-First and the Holding of Assets," *Econometrica*, 20 (July 1952).

[9] Sharpe, W. F., "Capital Asset Prices: A Theory of Market Equilibrium Under Conditions of Risk," *The Journal of Finance*, 19 (Sept. 1964).

[10] Telser, L., "Safety-First and Hedging," *Review of Economic Studies*, 23 (1955–6).

[11] Tobin, J., "Liquidity Preference as Behavior Towards Risk," *Review of Economic Studies*, 26 (Feb. 1958).

Reprinted from
THE REVIEW OF ECONOMICS AND STATISTICS
Published by Harvard University

Vol. LII, No. 1, February 1970

*Choosing Investment Portfolios When the Returns Have Stable Distributions**†

W. T. Ziemba

THE UNIVERSITY OF BRITISH COLUMBIA

This paper presents an efficient method for computing approximately optimal portfolios when the returns have symmetric stable distributions and there are many alternative investments. The procedure is valid, in particular, for independent investments and for multivariate investments of the classes introduced by Press and Samuelson. The algorithm is based on a two-stage decomposition of the problem and is analogous to the procedure developed by the author that is available for normally distributed investments utilizing Lintner's reformulation of Tobin's separation theorem.

A tradeoff analysis between mean (μ) and dispersion (d) is valid since an investment choice maximizes expected utility if and only if it lies on a μ–d efficient curve. When a risk-free asset is available one may find the efficient curve, which is a ray in μ–d space, by solving a fractional program. The fractional program always has a pseudo-concave objective function and hence may be solved by standard nonlinear programming algorithms. Its solution, which is generally unique, provides optimal proportions for the risky assets that are independent of the unspecified concave utility function. One must then choose optimal proportions between the risk-free asset and a risky composite asset utilizing a given utility function. The composite asset is stable and consists of a sum of the random investments weighted by the optimal proportions. This problem is a stochastic program having one random variable and one decision variable. Symmetric stable distributions have known closed-form densities only when α, the characteristic exponent, is $\frac{1}{2}$ (the arc sine), 1 (the Cauchy), or 2 (the normal). Hence, there is no apparent algorithm that will solve the stochastic program for general $1 < \alpha < 2$. However, one may obtain a reasonably accurate approximate solution to this program utilizing tables recently compiled by Fama and Roll. Standard nonlinear programming

* Presented by invitation at the NATO Advanced Study Institute on Mathematical Programming in Theory and Practice, Figueira da Foz, Portugal, June 12–23, 1972. This research was partially supported by the National Research Council of Canada Grant NRC-A7-7147, the Samuel Bronfman Foundation, The Graduate School of Business, Stanford University, and Atomic Energy Commission, Grant AT 04-3-326-PA #18.

† Reprinted from *Mathematical Programming in Theory and Practice*, P. L. Hammer and G. Zoutendijk, eds. North-Holland Publishing Company, 1974.

algorithms may be adapted to solve the portfolio problem approximately when the risk-free asset assumption is not made. Such a direct approach is also available when the risk-free asset assumption is made; however, the computations in the two-stage approach would generally be much less formidable.

I. Introduction

Stable distributions have increasing interest for the empirical explanation of asset price changes and other economic phenomena. An important reason for this is that all limiting sums (that exist) of independent identically distributed random variables are stable. Thus it is reasonable to suspect that empirical variables which are sums of random variables conform to stable laws. Such an observation has led to a substantial body of literature concerned with the estimation of the distributions of stock price changes[1] and other economic variables[2] utilizing stable distributions. Much of this literature focuses on the fact that the empirical distributions have more "outliers" and hence "fatter" tails than one would expect to be generated by a normal distribution. This leads to the conclusion that variance does not exist and that a normal distribution will not adequately fit the data.[3] The normal distribution has an important place in the theory of portfolio selection because the Markowitz theory [20] is then consistent with expected utility maximization. In this case it is well known that a portfolio maximizes expected utility if and only if it is mean-variance efficient. Since the normal is only one member of the stable family it is of interest to generalize the Markowitz theory to apply for more general classes of stable distributions. Samuelson [29] and Fama [5] have shown how this may be accomplished utilizing a mean-dispersion (μ–d) analysis when the random variables are independent or when they follow a Sharpe–Markowitz diagonal model (see Sharpe [30]), respectively. Utilizing the fact that linear combinations of multivariate stable distributions are

[1] See, for example, Blume [1], Fama [4], Roll [27], and Fama and Roll [7, 8]. A survey of this and other related research appears in the work of Fama [6].

[2] See, for example, Mandelbrot [17, 18]. The estimates in these papers and those mentioned in footnote 1 were generally performed on the logarithms of the changes in the economic variables. The theory in this paper is applicable only to data regarding absolute changes. However, if the changes are small, say of the order of 15% or less, then these variables and their logarithms are approximately equal and the results here are approximately valid for such data.

[3] There are other distributions besides the stable class that can be used to explain such data (see Press [24–26] and Mandelbrot and Taylor [19]). However, the distributions in these papers are not particularly convenient for the analysis of portfolio problems because they are not closed on addition.

univariate stable it is shown here that the μ–d analysis is valid as long as the mean vector exists.

The calculation of the μ–d curve is generally quite difficult because one must solve a parametric concave program. However, when a risk-free asset exists the efficient surface is a ray in μ–d space and a generalization of Tobin's separation theorem [31] obtains. One may then calculate the optimal proportions of the risky assets by solving a fractional program. The character of the fractional program depends, of course, on the assumptions made about the joint distribution of the stable random variables. However, in fairly general circumstances the fractional program has a pseudo-concave objective function and hence may be solved via a standard nonlinear programming algorithm. Typically the optimal solution is unique. This calculation comprises stage 1 of a two-stage procedure that will efficiently solve the portfolio problem when there are many random investments. In the second stage one introduces the investor's utility function and an optimal ratio between a stable composite asset and the risk-free asset must be chosen. The composite asset is a sum of the random investments weighted by the optimal proportions found in stage 1. The problem to solve is a stochastic program having a single random variable and a single decision variable. Such problems are generally easy to solve if the density of the random variable is known (as it is in the normal distribution case). Unfortunately, the density of the stable composite asset is known only in a few special cases. However, Fama and Roll [7], using series approximations due to Bergstrom, have tabulated, at discrete points, the density and cumulative distributions of a standardized symmetric stable distribution. These tables may be used to obtain a very good nonlinear programming approximation to the stochastic program. The solution of the nonlinear program generally provides a good approximation to the optimal solution of the stochastic program and hence of the portfolio problem.

Section II discusses the case when the random returns are independent. Some sufficient conditions for the expected utility and expected marginal utility to be bounded are given. The fractional program in this case is shown to have a strictly pseudo-concave objective, and it has a unique solution. The nonlinear programming approximation of the stochastic program is also described. Section III considers a class of multivariate stable distributions introduced by Press [22, 23]. This class generalizes the independent case and decomposes the dependence of the random variables into several independent subsets. Thus partial and full dependence may be handled in a convenient fashion. The optimization procedure described in Section II may be utilized for this class and a class of multivariate stable random variables introduced by Samuelson [29] as well. Section IV shows how one may find a portfolio that approximately maximizes expected utility when the risk-free asset assumption is not made, by utilizing standard nonlinear programming algorithms. The calculations in this direct approach are generally much more formidable than in the two-stage approach; hence it is generally preferable to use the latter approach if it is valid.

II. The Independent Case

We consider an investor having one dollar[4] to invest in assets $i = 0, 1, \ldots, n$. Assets $1, \ldots, n$ are random and they exhibit constant returns to scale so that if x_i is invested in i, then $\xi_i x_i$ is returned at the end of the investment period. The ξ_i are assumed to have independent stable distribution functions $F_i(\xi_i; \bar{\xi}_i, S_i, \beta, \alpha)$, $i = 1, \ldots, n$. A distribution function $F(y)$ is said to be stable if and only if for all positive numbers a_1 and a_2 and all real numbers b_1 and b_2 there exist a positive number a and a real number b such that

$$F\left(\frac{y-b_1}{a_1}\right) * F\left(\frac{y-b_2}{a_2}\right) = F\left(\frac{y-b}{a}\right), \tag{1}$$

where the * indicates the convolution operation. Equation (1) formalizes the statement that the stable family is precisely that class of distributions that is closed under addition of independent and identically distributed random variables. The F_i are unimodal, absolutely continuous, and have continuous densities f_i. The parameter $-1 \leqq \beta \leqq 1$ is related to the skewness of the distribution. When $\beta > 0$ (< 0) the distribution is skewed to the right (left). It is convenient for our purposes to assume that the distribution F_i is symmetric, in which case $\beta = 0$.[5] The parameter α is termed the characteristic exponent and absolute moments of order $< \alpha$ exist, where $0 < \alpha \leqq 2$. When $\alpha = 2$, F is the normal distribution and all moments exist. It will be convenient to assume that $1 < \alpha \leqq 2$ so that absolute first moments always exist and the value of α is the same for each F_i. The parameter $\bar{\xi}_i$ corresponds to the central tendency of the distribution which is the mean if $\alpha > 1$. The parameter S_i, assumed to be positive, refers to the dispersion of the distribution. It will be convenient to differentiate between S_i and $S_i^{1/\alpha}$. We will follow a suggestion of Beale and call this latter quantity the α-dispersion. When F_i is normal S_i is one-half the variance, in other cases it is approximately equal to the semiinterquartile range. In general $S_i^{1/\alpha}$ is proportional to the mean absolute deviation $E|\xi_i - \bar{\xi}_i|$.

[4] Without loss of generality, we may normalize the investment returns so that initial wealth is one dollar.

[5] The analysis in this section is valid though for $\beta \neq 0$ as long as β is the same for each F_i.

The f_i are known in closed form in only very special cases; hence the most convenient way to study the stable family utilizes the characteristic function

$$\psi_y(t) \equiv Ee^{ity} = \int_{-\infty}^{\infty} e^{ity} f(y)\, dy \qquad \text{where} \quad i = \sqrt{-1}.$$

The log characteristic function for a symmetric stable distribution $F_i(\xi_i;\ \bar{\xi}_i, S_i, 0, \alpha)$ is

$$\ln \psi_{\xi_i}(t_i) = i\bar{\xi}_i t_i - S_i |t_i|^{\alpha}.$$

Asset 0 is riskless and returns $\bar{\xi}_0$ with certainty and exhibits constant returns to scale so that if x_0 is invested, the return is $\bar{\xi}_0 x_0$. $x_0 < 0$ corresponds to borrowing at the risk-free rate. It is assumed that the investor may borrow or lend any amount at the constant rate of $\bar{\xi}_0$. This asset may be considered to have the degenerate stable distribution $F_0(\xi_0;\ \bar{\xi}_0, 0, 0, \alpha)$.

It is supposed that the investor wishes to choose the x_i to maximize the expectation of a utility function u of wealth $w = \xi' x$, where $\xi' \equiv (\xi_0, \xi_1, \ldots, \xi_n)$, $x' \equiv (x_0, x_1, \ldots, x_n)$, and primes denote transposition. Suppose that u is nondecreasing, continuously differentiable, and concave. (The continuously differentiable assumption is not crucial for the theory although it is useful for the algorithmic development.)

The investor's problem is

(1) Maximize

$$Z(x) \equiv E_{\xi} u(\xi' x),$$

$$\text{s.t.} \quad e'x = 1, \quad x_i \geqq 0, \quad i = 1, \ldots, n, \quad x_0 \text{ unconstrained},$$

where E_{ξ} represents expectation with respect to ξ, and $e \equiv (1, \ldots, 1)'$.

The portfolio $\xi' x$ is known (see, e.g., Fama [5]) to have the stable distribution $F(\xi' x) = F(\xi' x;\ \bar{\xi}' x, \sum_{i=0}^{n} S_i |x_i|^{\alpha}, 0, \alpha)$. It will be convenient to begin with some conditions on u which will guarantee that expected utility and expected marginal utility are bounded.

Theorem 1 Suppose w has the stable distribution $F(w; \bar{w}, S_w, 0, \alpha)$, where $|\bar{w}| < \infty$, $0 < S_w < \infty$, and $1 < \alpha \leqq 2$.

(a) If $|u(w)| \leqq L_1 |w|^{\beta_1}$ for some $\beta_1 < \alpha$ and some $0 < L_1 < \infty$, then $|Eu(w)| < \infty$.

(b) If $|u'(w)| \leqq L_2 |w|^{\beta_2}$ for some $\beta_2 < \alpha$ and some $0 < L_2 < \infty$, then $|Eu'(w)| < \infty$.

(c) If $|u'(w)| \leqq L_3 |w|^{\beta_3}$ for some $\beta_3 < \alpha - 1$ and some $0 < L_3 < \infty$ and[6]

[6] For convenience we will suppose, in (c), that $x_i \geqq 0$ for all i. If any x_i such as x_0 is unconstrained, then the modification that $(\partial/\partial x_i) S_i |x_i|^{\alpha} = \operatorname{sgn}(x_i)\, \alpha S_i |x_i|^{\alpha-1}$ may be used. Note that this partial derivative is continuous at $x_i = 0$, since $\alpha > 1$.

$w \equiv \xi' x \sim F(\xi' x;\ \bar{\xi}' x,\ \sum_{i=0}^{n} S_i x_i^{\alpha}, 0, \alpha)$, then

$$\left|\frac{\partial E_w u(\xi' x)}{\partial x_i}\right| = \left| E_w\left(\frac{\partial u(\xi' x)}{\partial x_i}\right)\right| < \infty, \qquad i = 0, 1, \ldots, n,$$

and the computation of the partials involves only a univariate integration utilizing a normalized variable $\tilde{w} \sim F(\tilde{w};\ 0, 1, 0, \alpha)$.

(d) The results in (a)–(c) remain valid if the hypotheses are modified to read $\lim_{|w|\to\infty} |u(w)|/|w|^{\beta_1} = v_1$ for some $0 \leqq v_1 < \infty$ and some $\beta_1 < \alpha$, $\lim_{|w|\to\infty} |u'(w)|/|w|^{\beta_2} = v_2$ for some $0 \leqq v_2 < \infty$ and some $\beta_2 < \alpha$, and $\lim_{|w|\to\infty} |u'(w)|/|w|^{\beta_3} = v_3$ for some $0 \leqq v_3 < \infty$ and some $\beta_3 < \alpha - 1$, respectively.

Proof See the Appendix.

Remark Most common utility functions have either unbounded expected utility or they are undefined over portions of the range of w, which is R. However, one may modify many of these utility functions by adding appropriate linear segments so that the utility functions are concave, nondecreasing, defined over the entire range of w, and have bounded expected utility. For the logarithmic, power, and exponential forms such modified utility functions are

$$\text{(a)}\quad u(w) = \begin{cases} \log \tau w & \text{if } w \geqq w_0, \\ (\log \tau w_0 - 1) + \dfrac{w}{w_0} & \text{if } w < w_0, \end{cases} \qquad w_0 > 0, \quad \tau > 0;$$

$$\text{(b)}\quad u(w) = \begin{cases} w^{\delta} & \text{if } w \geqq w_0, \\ (1-\delta)\, w_0^{\delta} + (\delta w_0^{\delta-1})\, w & \text{if } w < w_0, \end{cases} \qquad w_0 > 0, \quad 0 < \delta < 1;$$

$$\text{(c)}\quad u(w) = \begin{cases} -\exp(-\tau w) \\ \qquad \text{if } w \geqq w_0, \\ -(1+\tau w_0)\exp(-\tau w_0) + \tau \exp(-\tau w_0))\, w \\ \qquad \text{if } w < w_0, \end{cases} \qquad 0 > w_0 > -\infty, \quad \tau > 0.$$

Note that all polynomial utility functions of order 2 or more have unbounded expected utility unless the distribution of w is truncated from above and below.

It is our purpose to solve (1) using the following two-step procedure: (i) find an efficient surface independent of the utility function u; and (ii) given a particular u, find a maximizing point on this surface. Such a procedure for the case when the random returns are normally distributed was suggested by

Tobin's separation theorem [31] and implemented by Ziemba *et al.* [35] utilizing Lintner's [15] reformulation of the separation theorem. The analysis here is analogous and utilizes a mean–dispersion efficient surface.

The following theorem indicates that a point cannot solve (1) unless it lies on a mean–dispersion efficiency curve.

Theorem 2 Let G and H be two distinct distributions with finite means μ_1 and μ_2 and finite positive dispersions $D_1{}^{\alpha}$ and $D_2{}^{\alpha}$ $(1 < \alpha \leqq 2)$, respectively, such that $G(x) = H(y)$ whenever $(x-\mu_1)/D_1 = (y-\mu_2)/D_2$. Let $\mu_1 \geqq \mu_2$ and $G(z) > H(z)$ for some z. Then $\int u(w)\,dG(w) \geqq \int u(w)\,dH(w)$ for all concave nondecreasing u if and only if $D_1 \leqq D_2$.

Proof Sufficiency: Case (i): $\mu_1 = \mu_2 = 0$. Let $k \equiv D_2/D_1 \geqq 1$. Then

$$\int_{-\infty}^{\infty} u(w)\,dH(w) - \int_{-\infty}^{\infty} u(w)\,dG(w)$$

$$= \int_{-\infty}^{\infty} [u(kw)-u(w)]\,dG(w)$$

(since G is symmetric)

$$= \int_0^{\infty} \underbrace{\{(u(kw)-u(w))-(u(-w)-u(-kw))\}}_{\leqq\, 0 \text{ by concavity}}\,dG(w) \leqq 0.$$

Case (ii): $\mu_1 \geqq \mu_2$. Let $\tilde{G}$ and $\tilde{H}$ be the cumulative distribution functions for x and y, respectively, when their means are translated to zero. Let $\varepsilon \equiv \mu_1 - \mu_2 \geqq 0$.

$$\int u(w)\,dH(w) = \int u(w+\mu_2)\,d\tilde{H}(w) \leqq \int u(w+\mu_2)\,d\tilde{G}(w) \qquad \text{[by } (i)\text{]}$$

(since $\varepsilon \geqq 0$ and u is nondecreasing)

$$\leqq \int u(w+\mu+\varepsilon)\,d\tilde{G}(w) = \int u(w)\,dG(w).$$

Necessity: Let $\mu_1 = \mu_2 = 0$, $D_1 = 2 > 1 = D_2$, $\alpha = 2$, and $u(w) = -e^{-w}$; then

$$-\int_{-\infty}^{\infty} (e^{-w/2} - e^{-w})\,dG(w) > 0.$$

The proof is a generalization of the proof given by Hillier [13a] for the mean–variance case ($\alpha = 2$). The monotonicity assumption is not crucial, however; dropping the monotonicity assumption does not seem to add any apparent generality as free disposal of wealth always seems possible. A proof of Theorem 2 without the monotonicity assumption may be obtained by letting D_1^{α} replace σ_1^{2} in Hanoch and Levy's proof for the mean–variance case [12].

The theorem indicates that it is sufficient to limit consideration to only those points that lie on the mean–α-dispersion (μ–d) efficient surface. Let the risky assets $(x_1, \ldots, x_n) \equiv \hat{x}$, and let $\bar{\xi} \equiv (\bar{\xi}_1, \ldots, \bar{\xi}_n)$. Then the ($\mu$–$d$) surface corresponding to assets $(\xi_1, \ldots, \xi_n)$ may be obtained by solving[7]

$$(2) \quad \varphi(\beta) \equiv \max \bar{\xi}'\hat{x},$$

$$\text{s.t.} \quad e'\hat{x} = 1, \quad \hat{x} \geqq 0, \qquad \left\{\sum_{i=1}^{n} S_i x_i^{\alpha}\right\}^{1/\alpha} \leqq \beta,$$

for all $\beta > 0$.

Now (2) is a parametric concave program which would generally be difficult to solve, for all $\beta > 0$. It is easier to bypass the calculation of the μ–d curve if the risk-free asset is not available (see Section IV). However, when the risk-free asset does exist, as we are assuming, it is convenient to consider points that are linear combinations of $x_0 = 1$ (total investment in the risk-free asset) and points that lie on $\varphi(\beta)$. These combinations correspond to straight lines in μ–d space as well because the mean is linear and the α-dispersion measure is positively homogeneous [i.e., $f(\lambda\hat{x}) = \lambda f(\hat{x})$ for all $\lambda \geqq 0$] in the x_i. Clearly the best points will lie on the line L that is a support to the concave function $\varphi(\beta)$ (proof below); see Fig. 1. Such heuristic arguments indicate that the μ–d efficient surface is now L, and one gets the analog of the Tobin separation theorem for normal distributions which has the important implication that $x_i^*/\sum_{j=1}^{n} x_j^*$, $i \neq 0$, is independent of u and of initial wealth. The result may be stated as the

Separation Theorem Let the efficiency problem be

$$(2') \quad \max \bar{\xi}'x, \qquad \text{s.t.} \quad f(\hat{x}) \leqq \beta, \qquad e'x = w, \quad \hat{x} \geqq 0, \quad x_0 \text{ unconstrained.}$$

In (2') f is the α-dispersion measure, w is the initial wealth, and explicit consideration of the risk-free asset ($i = 0$) is allowed. Suppose that asset $i = 0$ has no dispersion and that borrowing or lending any amount is possible at the fixed rate $\bar{\xi}_0$. Assume that f is convex and homogeneous of degree 1 and that u is concave and nondecreasing.

[7] Samuelson [29] analyzes a problem similar to (2) in which one minimizes dispersion subject to the mean equalizing a given parameter. He notes that his problem is a convex program and he analyzes it using the Kuhn–Tucker conditions. His paper also contains some illustrative graphical results for some special cases. See also Fama [5] for a similar analysis, presented in the context of the Sharpe–Markowitz diagonal model, that is particularly concerned with diversification questions.

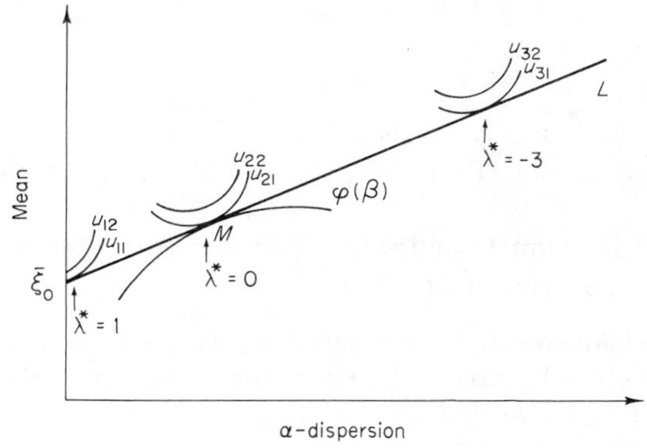

Fig. 1

(a) Total separation: If $x_0^* \neq 0$, the relative proportions invested in the risky assets, namely $x_i^*/\sum_{j=1}^n x_j^*$, $i \neq 0$, are independent of u and initial wealth w.

(b) Partial separation: If $x_0^* = 0$, all investment is in the risky assets and $\sum_{j=1}^n x_j^* = w$ and the x_j^* are independent of u.

Proof (a) Suppose $x_0^* \neq 0$. Since x^* solves (2′) it must satisfy the Kuhn–Tucker conditions:

(i) $f(\hat{x}) \leqq \beta, \quad \hat{x} \geqq 0, \quad e'x = w,$
(ii) (a) $\bar{\xi}_i - \lambda(\partial f/\partial x_i) - \mu \leqq 0, \quad i = 1, \ldots, n.$
(b) $\bar{\xi}_0 - \mu = 0, \quad \lambda \geqq 0,$
(iii) $x_i(\bar{\xi}_i - \lambda(\partial f/\partial x_i) - \mu) = 0, \quad i = 1, \ldots, n.$

It will suffice to show that for all $\sigma > 0$ there exists a $\gamma \neq 0$ such that $x^{**} \equiv (\gamma x_0^*, \sigma x_1^*, \ldots, \sigma x_n^*)$ also solves the Kuhn–Tucker conditions (which are necessary and sufficient) for all $\beta^{**} \geqq \beta\sigma$. Condition (i) is satisfied because

$$f(\hat{x}^*) \leqq \beta \quad \Rightarrow \quad f(\sigma\hat{x}^*) = \sigma f(\hat{x}^*) \leqq \sigma\beta \leqq \beta^{**},$$

$$\hat{x}^* \geqq 0 \quad \Rightarrow \quad \sigma\hat{x}^* \geqq 0,$$

$$x_0^* + \sum_{i=1}^n x_i^* = 1 \quad \Rightarrow \quad \gamma x_0^* + \sigma \sum_{i=1}^n x_i^* = 1$$

$$\text{for} \quad \gamma \equiv \frac{1 - \sigma \sum_{i=1}^n x_i^*}{x_0^*} \neq 0.$$

Now $\partial f/\partial x_i$ is homogeneous of degree 0, hence

$$\xi_i - \lambda \frac{\partial f(\sigma \hat{x}^*)}{\partial x_i} \leqq \xi_0 \qquad \text{iff} \quad \xi_i - \lambda \frac{\partial f(\hat{x}^*)}{\partial x_i} \leqq \xi_0;$$

and (ii) is satisfied.

If $x_i^* = 0$, (iii) is trivially satisfied.

If $x_i^* > 0$, (iii) is equivalent to (ii(a)) which is satisfied for $\sigma\hat{x}^*$ iff $\hat{x}^*$ satisfies (ii(a)).

By Theorem 2 an optimal solution to (2′) must solve (1); hence $\sum_{i=1}^n x_i^* = w$ and the x_i^* are independent of the u.

The proof and statement of the separation theorem given here are similar in spirit to that given by Breen [2]. Breen considered an efficiency problem in which one minimizes α-dispersion given that expected return is a stipulated level as well as some alternative assumptions regarding the risk-free asset.

For the analysis indicated in Fig. 1 to be valid it is necessary that the α-dispersion measure $f(\hat{x}) \equiv \{\sum_{i=1}^n S_i x_i^\alpha\}^{1/\alpha}$ be convex and that φ be a concave function of β. The function f is actually strictly convex as we now establish using

Theorem 3 (Minkowski's inequality) Suppose

$$M_r(y) \equiv \left\{(1/n) \sum_{i=1}^n y_i^r\right\}^{1/r}, \qquad n \geqq r > 1, \quad n < \infty,$$

and that a and b are not proportional (i.e., constants q_1 and q_2 not both zero do not exist such that $q_1 a = q_2 b$); then

$$M_r(a) + M_r(b) > M_r(a+b).$$

Proof See, e.g., Hardy *et al.* [13, p. 30].

Let $P \equiv \{\hat{x} \mid \hat{x} \geqq 0,\ e'\hat{x} = 1\}$.

Lemma 1 $f(\hat{x}) \equiv \{\sum_{i=1}^n S_i x_i^\alpha\}^{1/\alpha}$ is a strictly convex function of $\hat{x}$ on P if $2 \geqq \alpha > 1$ and $S_i > 0$, $i = 1, \ldots, n$.

Proof Let $a_i = \lambda S_i x_i^1$ and $b_i = (1-\lambda) S_i x_i^2$, $i = 1, \ldots, n$, where $\hat{x}^1 \neq \hat{x}^2$, and $0 < \lambda < 1$. Then by Minkowski's inequality

$$\left\{(1/n) \sum_{i=1}^n a_i^\alpha\right\}^{1/\alpha} + \left\{(1/n) \sum_{i=1}^n b_i^\alpha\right\}^{1/\alpha} > \left\{(1/n) \sum_{i=1}^n (a_i+b_i)^\alpha\right\}^{1/\alpha}$$

$$\Rightarrow$$

$$\left\{\sum_{i=1}^n [\lambda S_i x_i^1]^\alpha\right\}^{1/\alpha} + \left\{\sum_{i=1}^n [(1-\lambda) S_i x_i^2]^\alpha\right\}^{1/\alpha} > \left\{\sum_{i=1}^n [\lambda S_i x_i^1 + (1-\lambda) S_i x_i^2]^\alpha\right\}^{1/\alpha}.$$

Hence $\lambda f(\hat{x}^1) + (1-\lambda) f(\hat{x}^2) > f\{\lambda \hat{x}^1 + (1-\lambda)\hat{x}^2\}$.

Thus f is strictly convex unless a and b are proportional. But this requires that there exist constants q_1 and q_2 not both zero such that $q_1 a = q_2 b$ or that

$$q_1 \lambda S_i x_i^{\,1} = q_2 (1-\lambda) S_i x_i^{\,2} \quad \Rightarrow \quad q_1 \lambda x_i^{\,1} = q_2 (1-\lambda) x_i^{\,2}$$
$$(\text{since} \quad S_i > 0)$$

or that $x_i^{\,1}$ and $x_i^{\,2}$ are proportional since $q_1' \equiv q_1 \lambda$ and $q_2' \equiv q_2(1-\lambda)$ are both not zero. However, the condition that $\sum_{i=1}^n x_i^{\,1}$ and $\sum_{i=1}^n x_i^{\,2} = 1$ means that x^1 and x^2 cannot be proportional unless $x_i^{\,1} = x_i^{\,2}$ for all i, which is a contradiction.

Remark f is convex but not strictly convex on $M \equiv \{\hat{x} \mid \hat{x} \geqq 0\}$ because f is linear on every ray that contains the origin.

Lemma 2 φ is a concave function of $\beta > \beta_L$, where $\beta_L > 0$ is defined below, if $2 \geqq \alpha > 1$ and $S_i \geqq 0$, $i = 1, \ldots, n$.

Proof Let $K_\beta \equiv \{\hat{x} \mid e'\hat{x} = 1,\ \hat{x} \geqq 0,\ f(\hat{x}) \leqq \beta\}$. Let $\beta_L > 0$ be the smallest β for which $K_\beta \neq \varnothing$. Clearly $\beta_L \leqq \delta$, where $\delta^\alpha \equiv \min_i S_i$, and $K_\beta \neq \varnothing$ if and only if $\beta \geqq \beta_L$.

Choose $\beta_1 \geqq \beta_L$ and $\beta_2 \geqq \beta_L$. Since (2) has a linear objective function and a nonempty compact convex feasible region there is an optimal solution for all $\beta \geqq \beta_L$. Let optimal solutions when β equals β_1 and β_2 be $\hat{x}_1$ and $\hat{x}_2$, respectively. Consider

$$\beta_\lambda = (1-\lambda)\beta_1 + \lambda\beta_2, \qquad 0 \leqq \lambda \leqq 1,$$
$$\hat{x}_\lambda = (1-\lambda)\hat{x}_1 + \lambda\hat{x}_2.$$

Now $\hat{x}_\lambda$ is feasible when $\beta = \beta_\lambda$ because the constraint set of (2) is convex. By the concavity of $h(\hat{x}) \equiv \hat{\xi}'\hat{x}$,

$$h(\hat{x}_\lambda) \geqq (1-\lambda) h(\hat{x}_1) + \lambda h(\hat{x}_2) \tag{i}$$

But

$$h(\hat{x}_i) = \varphi(\beta_i) \tag{ii}$$

since $\hat{x}_i$ is optimal, $i = 1, 2$, and

$$\varphi(\beta_\lambda) \geqq h(\hat{x}_\lambda) \tag{iii}$$

since $\varphi(\beta_\lambda)$ denotes the maximum when $\beta = \beta_\lambda$. Combining (i)–(iii) gives

$$\varphi(\beta_\lambda) \geqq (1-\lambda)\varphi(\beta_1) + \lambda\varphi(\beta_2).$$

One may find the slope of L and the point M (see Fig. 1) by maximizing

$$(3)\quad g(\hat{x}) \equiv \frac{\sum_{i=1}^{n} \bar{\xi}_i x_i - \bar{\xi}_0}{\{\sum_{i=1}^{n} S_i x_i^{\alpha}\}^{1/\alpha}}, \qquad \text{s.t.} \quad \hat{x} \geqq 0, \quad e'\hat{x} = 1.$$

By letting $\tilde{\xi}_i = \bar{\xi}_i - \bar{\xi}_0$, $i = 1, \ldots, n$, and utilizing $e'\hat{x} = 1$,

$$g(\hat{x}) = \left\{\sum_{i=1}^{n} \tilde{\xi}_i x_i\right\} \Big/ \left\{\sum_{i=1}^{n} S_i x_i^{\alpha}\right\}^{1/\alpha}.$$

We will assume that $\tilde{\xi}'\hat{x} > 0$ for all "interesting" feasible $\hat{x}$. This is a very minor assumption since the fact that the $\hat{x} \geqq 0$ and $e'\hat{x} = 1$ implies that there always exists an $\hat{x}$ such that $\tilde{\xi}'\hat{x} > 0$ unless all $\tilde{\xi}_i \leqq 0$ in which case it is optimal to invest entirely in the risk-free asset. It will now be shown that (3) has a unique solution.

A differentiable function $\theta(x)\colon \Lambda \to R$ is said to be strictly pseudo-concave on $\Lambda \subset R^n$ if for all $x, \bar{x} \in \Lambda$, $x \neq \bar{x}$

$$(x - \bar{x})' \nabla\theta(\bar{x}) \leqq 0 \quad \Rightarrow \quad \theta(x) < \theta(\bar{x})$$

[∇ denotes the gradient operator so $\nabla\theta = (\partial\theta/\partial x_1, \ldots, \partial\theta/\partial x_n)'$]. Geometrically functions are strictly pseudo-concave if they strictly decrease in all directions that have a downward- or horizontal-pointing directional derivative. A normal distribution is such a function in R.

Theorem 4 Suppose the differentiable functions Ψ and ψ are defined on Λ, a convex subset of R^n, and that $\Psi > 0$ is concave and $\psi > 0$ is strictly convex. Then $\theta = \Psi/\psi$ is strictly pseudo-concave on Λ.

Proof $\nabla\theta = \{(\psi\nabla\Psi - \Psi\nabla\psi)/\psi^2\}$. Let $\bar{x} \in \Lambda$. Thus

$$\nabla\theta(\bar{x})'(x - \bar{x}) = \frac{\{\psi(\bar{x})\nabla\Psi(\bar{x}) - \Psi(\bar{x})\nabla\psi(\bar{x})\}'}{[\psi(\bar{x})]^2}(x - \bar{x}) \leqq 0, \qquad x \in \Lambda, \quad x \neq \bar{x}$$

$$\Rightarrow$$

$$\{\psi(\bar{x})\nabla\Psi(\bar{x}) - \Psi(\bar{x})\nabla\psi(\bar{x})\}'(x - \bar{x}) \leqq 0 \qquad (\text{since } \psi(\bar{x}) \neq 0)$$

$$\psi(\bar{x})[\Psi(x) - \Psi(\bar{x})] - [\Psi(\bar{x})\nabla\psi(\bar{x})]'(x - \bar{x}) \leqq \qquad \begin{pmatrix}\text{since } \Psi \text{ is concave} \\ \text{and } \psi(\bar{x}) > 0\end{pmatrix}$$

$$\psi(\bar{x})[\Psi(x) - \Psi(\bar{x})] + \Psi(\bar{x})[\psi(\bar{x}) - \psi(x)] < \qquad \begin{pmatrix}\text{since } \psi \text{ is strictly} \\ \text{convex and } \Psi(\bar{x}) > 0\end{pmatrix}$$

$$\psi(\bar{x})\Psi(x) - \Psi(\bar{x})\psi(x) \qquad \text{which implies that}$$

$$\theta(x) - \theta(\bar{x}) < 0 \qquad \begin{pmatrix}\text{since } \psi(\bar{x}) > 0 \\ \text{and } \psi(x) > 0\end{pmatrix}.$$

Lemma 3 Suppose $f: \Lambda \to R$, where Λ is a convex subset of R^n, is differentiable at x, and there is a direction d such that $\nabla f(x)'d > 0$. Then a $\sigma > 0$ exists such that for all $\tau, \sigma \geqq \tau > 0, f(x+\tau d) > f(x)$.

Proof See Zangwill ([32, p. 24].

Theorem 5 Suppose $\theta(x): \Lambda \to R$, where Λ is a convex subset of R^n, is strictly pseudo-concave. Then the maximum of θ, if it exists, is attained at most at one point $x \in \Lambda$.

Proof Case 1: $\exists x \in \Lambda$ such that $\nabla\theta(x) = 0$. Then by the strict pseudo-concavity of θ, i.e., $\nabla\theta(x)'(y-x) \leqq 0 \Rightarrow \theta(y) < \theta(x)$ for all $y \in \Lambda$, x is the unique maximizer.

Case 2: $\nexists x \in \Lambda$ such that $\nabla\theta(x) = 0$. Suppose x maximizes θ over Λ. Then $\forall y \neq x, \nabla\theta(x)'(y-x) \leqq 0 \Rightarrow \theta(y) < \theta(x)$ and x is clearly the unique maximum unless $\exists y \in \Lambda$ such that $\nabla\theta(x)'(y-x) > 0$. But by Lemma 3 that would imply that there exists a $\tau > 0$ such that $\theta(x+\tau(y-x)) > \theta(x)$, which contradicts the assumed optimality of x. (Note that the point $[x+\tau(y-x)] = [\tau y+(1-\tau)x] \in \Lambda$, by the convexity of Λ.)

Theorem 6 (3) has a unique solution.

Proof The function g may be seen to be strictly pseudo-concave on P by letting $\Psi(\hat{x}) = \bar{\xi}'\hat{x}$, which is positive and concave, and $\psi(\hat{x}) = \{\sum_{i=1}^n S_i x_i^\alpha\}^{1/\alpha}$, which is positive and strictly convex. Hence by Theorem 5, it has at most one maximizer. But P is compact and g is continuous; hence g is maximized at a unique point $x^* \in P$.

Suppose that the fractional program (3) has been solved to obtain $\hat{x}^*$. The problem then is to determine the optimal ratios of risky to nonrisky assets. The risky asset is

$$R = \xi'\hat{x}^* \sim F\left(\xi'\hat{x}^*;\ \bar{\xi}'\hat{x}^*,\ \sum_{i=1}^{n} S_i(\hat{x}_i^*)^\alpha, 0, \alpha\right) \equiv F(R;\ \bar{R}, S_R, 0, \alpha),$$

where $\xi \equiv \xi_1, \ldots, \xi_r$, and the best combination of R and the risk-free asset may be found by solving

$$\text{(4)} \qquad \max_{-\infty < \lambda \leqq 1} \Psi(\lambda) \equiv E_R u[\lambda\xi_0 + (1-\lambda)R].$$

Problem (4) is a stochastic program with one random variable (R) and one decision variable (λ). Since u is concave in w it follows that Ψ is concave in λ. Under the assumptions of Theorem 1, $\Psi(\lambda)$ and $d\Psi(\lambda)/d\lambda$ will be bounded.

Now

$$(5)\quad \Psi(\lambda) = \int_{-\infty}^{\infty} u[\lambda\xi_0 + (1-\lambda) R] f(R)\, dR$$

$$= \int_{-\infty}^{\infty} u[\lambda\xi_0 + (1-\lambda)\{\bar{R} + (S_R)^{1/\alpha}\tilde{R}\}] f(\tilde{R})\, d\tilde{R},$$

where the standardized stable variate.,

$$\tilde{R} \equiv \frac{R-\bar{R}}{(S_R)^{1/\alpha}} \sim F(\tilde{R}; 0, 1, 0, \alpha).$$

The continuous derivative of Ψ is

$$(6)\quad \frac{d\Psi(\lambda)}{d\lambda} = \int_{-\infty}^{\infty} \{\xi_0 - \bar{R} - (S_R)^{1/\alpha}\tilde{R}\}$$

$$\times \frac{du[\lambda\xi_0 + (1-\lambda)\{\bar{R} + (S_R)^{1/\alpha}\tilde{R}\}]}{dw} f(\tilde{R})\, d\tilde{R}.$$

Problems having the general form of (4) are generally easy to solve by combining a univariate numerical integration scheme with a search procedure that uses function or derivative evaluations (see Ziemba [35]). The difficulty here is that the density $f(\tilde{R})$ is not known in closed form except for very special cases such as the normal ($\alpha = 2$) and Cauchy ($\alpha = 1$) distributions. However, Fama and Roll [7] have utilized series expansions of $f(\tilde{R})$, due to Bergstrom, to tabulate approximate values of $f(\tilde{R})$ and $F(\tilde{R})$. The tables consider $\alpha = 1.0,\ 1.1, \ldots, 2.0$ and $\tilde{R} = \pm 0,\ 0.05, \ldots, 1.0,\ 1.1, \ldots, 2.0,\ 2.2, \ldots, 4.0,$ $4.4, \ldots, 6.0, 7.0, 8.0, 10.0, 15.0,$ and 20.0 (a grid of 50 points). One may then utilize the tables to get a good approximation to (5) of the form

$$(7)\quad \Psi(\lambda) \cong \sum_{j=1}^{m} P_j u[\lambda\xi_0 + (1-\lambda)\{\bar{R} + (S_R)^{1/\alpha}\tilde{R}_j\}],$$

where $P_j \equiv \Pr\{\tilde{R} = \tilde{R}_j\} > 0$, $\sum_{j=1}^{m} P_j = 1$, and $m = 100$ (recall that $\tilde{R}$ is symmetric). The approximation to (6) is

$$(8)\quad \frac{d\Psi(\lambda)}{d\lambda} \cong \sum_{j=1}^{m} P_j \{\xi_0 - \bar{R} - (S_R)^{1/\alpha}\tilde{R}_j\} \frac{du[\lambda\xi_0 + (1-\lambda)\{\bar{R} + (S_R)^{1/\alpha}\tilde{R}_j\}]}{dw}.$$

One may then obtain an approximate solution λ^a to (5) via a golden section search using (7) or a bisecting search using (8) (see Zangwill [32] for details on these search methods.) The approximate optimal portfolio is then [8]

$$x_0^a = \lambda^a \qquad \text{and} \qquad x_i^a = (1-\lambda^a)\hat{x}_i, \qquad i = 1, \ldots, n.$$

[8] Since the ξ_i are independent it is known (see Samuelson [28]) that $x_i^* > 0$ for each i whose $\bar{\xi}_i \geqq \min_{1 \leq j \leq n} \bar{\xi}_j$.

III. The Dependent Case

A set of random variables $r \equiv (r_1, \ldots, r_n)$ is said to be multivariate stable if every linear combination of r is univariate stable. The characteristic function of r is

$$\psi_r(t) \equiv \psi_r(t_1, \ldots, t_n) \equiv E^{it'r} = \int_{-\infty}^{\infty} \cdots \int_{-\infty}^{\infty} e^{it'r} g(r)\, dr_1 \ldots dr_n,$$

where g is the joint density of $(r_1, \ldots, r_n)$. If $g(r)$ is symmetric, then the log characteristic function (see Ferguson [10]) is

$$\ln \psi_r(t) = i\delta(t) - \gamma(t),$$

where the dispersion measure $\gamma(\lambda t_1, \ldots, \lambda t_n) = |\lambda|^\alpha \gamma(t_1, \ldots, t_n)$, i.e., positive homogeneous of degree α, $\gamma > 0$, and the central tendency measure δ is linear homogeneous. The characteristic exponent α is assumed to satisfy $0 < \alpha \leqq 2$. With specific choices of δ and γ one may generate many classes of (symmetric) multivariate stable laws.

Press [22] has developed a statistical theory along with associated estimation procedures [23] for particularly convenient choices of δ and γ. Let $\bar{r} \equiv (\bar{r}_1, \ldots, \bar{r}_n)$ be the mean of r, which will exist if $\alpha > 1$. Press sets $\delta(t) \equiv \bar{r}'t$ and $\gamma(t) = \frac{1}{2} \sum_{j=1}^{m} (t'\Omega_j t)^{\alpha/2}$ where $n \geqq m \geqq 1$ and each Ω_j is a positive-semidefinite matrix of order $n \times n$. The log characteristic function is then

$$\text{(9)} \quad \ln \psi_r(t) = i\bar{r}'t - \tfrac{1}{2} \sum_{j=1}^{m} (t'\Omega_j t)^{\alpha/2}.$$

The m in (9) indicates the number of independent partitions of the random variables $(r_1, \ldots, r_n)$. Hence if $m = 1$, all the random variables are dependent (as long as Ω_1 is positive definite). Then if $\alpha = 2$, the case of normal distributions,

$$\gamma(t) = \tfrac{1}{2} t'\Omega_1 t,$$

which is precisely half the variance of the linear sum $t'r$. If $m = n$, the random variables are independent. In this case with $\alpha = 2$ one may interpret $\gamma(t)$ as half the variance by setting the jjth coefficient of Ω_j equal to $\omega_j > 0$ (and all other coefficients equal to zero). Then

$$\gamma(t) = \tfrac{1}{2} \sum_{j=1}^{m} (t'\Omega_j t) = \tfrac{1}{2} \sum_{j=1}^{n} \omega_j t_j^2.$$

Thus the class of distributions defined by (9) allows for the decomposition of the $(r_1, \ldots, r_n)$ into independent parts in a way consistent with and motivated by the way such a decomposition might be utilized for joint-normally distributed random variables.

If $y \equiv (y_1, \ldots, y_n)$, then the linear combination $r'y$ has the symmetric univariate stable distribution[9] $F(r'y;\, \bar{r}'y, S_{r'y}, 0, \alpha)$, where

$$S_{r'y} \equiv \tfrac{1}{2} \sum_{j=1}^{m} (y'\Omega_j y)^{\alpha/2}.$$

The standardized variate

$$r_0 \equiv \frac{r'y - \bar{r}'y}{\{\frac{1}{2}\sum_{j=1}^{m} (y'\Omega_j y)^{\alpha/2}\}^{1/\alpha}} \sim F(r_0;\, 0, 1, 0, \alpha).$$

We will now suppose that the random returns in our portfolio problem $\hat{\xi} \equiv (\xi_1, \ldots, \xi_n)$ have the log characteristic function

$$\text{(10)} \quad \ln \psi_{\hat{\xi}}(t) = i\bar{\hat{\xi}}'t - \tfrac{1}{2} \sum_{j=1}^{m} (t'\Omega_j t)^{\alpha/2},$$

where $1 < \alpha \leqq 2$, the Ω_j are positive-semidefinite matrices of order $n \times n$, and $\bar{\hat{\xi}}$ denotes the mean vector of $\hat{\xi}$. We wish to show that a two-stage optimization procedure analogous to that in Section II is valid when the random variables have the log characteristic function (10) and the risk-free asset exists. Now Theorem 2 is valid in this case so that a mean–dispersion analysis may be used. The μ–d efficient surface may be found by solving.

$$\text{(11)} \quad \varphi_1(\beta) \equiv \max \bar{\hat{\xi}}'\hat{x}, \qquad \text{s.t.} \quad e'\hat{x} = 1, \quad \hat{x} \geqq 0,$$

$$\left\{\tfrac{1}{2} \sum_{j=1}^{m} (\hat{x}'\Omega_j \hat{x})^{\alpha/2}\right\}^{1/\alpha} \leqq \beta$$

for all $\beta > 0$. Now it is clear from Lemma 2 that $\phi_1(\beta)$ will be a concave function of $\beta > 0$ if $\Psi(\hat{x}) \equiv \{\frac{1}{2}\sum_{j=1}^{m} (\hat{x}'\Omega_j \hat{x})^{\alpha/2}\}^{1/\alpha}$ is a convex function of $\hat{x}$.

The pertinent facts relating to the convexity of Ψ are summarized in

Theorem 7 (a) (Independent case) Suppose $m = n$ and the jjth element of Ω_j is $\omega_j > 0$ (and all other elements are zero). Then Ψ is strictly convex on P.

(b) (Totally dependent case) Suppose $m = 1$ and Ω_1 is positive definite. Then f is strictly convex on any convex subset of R^n.

(c) (General case) Suppose $n \geqq m \geqq 1$, each $n \times n$ matrix is positive semidefinite, and for at least one j, $\hat{x}'\Omega_j \hat{x} > 0$ if $x \neq 0$. Then f is convex on M.

[9] Press [23] also proves that $Ar + b$ is multivariate symmetric stable of form (9) if A is any $m \times n$ constant matrix and b is any n vector of constants.

Proof (a) Let $S_j > 0$ be defined by $\omega_j = (1/\sqrt{2})\, S_j^{\alpha/2}$. Then

$$\Psi(\hat{x}) = \left\{\tfrac{1}{2} \sum_{j=1}^{m} (\hat{x}'\Omega_j \hat{x})^{\alpha/2}\right\}^{1/\alpha} = \sum_{j=1}^{m} S_j x_j^{\alpha},$$

which was shown to be strictly convex on *P* in Lemma 1.

(b)

$$\Psi(\hat{x}) = \left\{\sum_{j=1}^{m} (\hat{x}'\Omega_j \hat{x})^{\alpha/2}\right\}^{1/\alpha} = \{\tfrac{1}{2}(\hat{x}'\Omega_1 \hat{x})^{\alpha/2}\}^{1/\alpha}$$
$$= (\tfrac{1}{2})^{1/\alpha} (\hat{x}'\Omega_1 \hat{x})^{1/2} \equiv (\tfrac{1}{2})^{1/\alpha} \varphi_2(\hat{x}).$$

Hence Ψ will be strictly convex if φ_2 is strictly convex. Choose $\hat{x}^1 \neq \hat{x}^2$, and $0 < \lambda < 1$. Let $\hat{x}^\lambda = \lambda \hat{x}^1 + (1-\lambda)\hat{x}^2$ and suppose $z \equiv a'\hat{x}$ has variance $\hat{x}'\Omega_1 \hat{x}$. Then

$$\begin{aligned} \operatorname{var}(z^\lambda) &= \lambda^2 \operatorname{var}(z^1) + (1-\lambda)^2 \operatorname{var}(z^2) + 2\lambda(1-\lambda)\operatorname{cov}(z^1, z^2), \\ &< \lambda^2 \operatorname{var}(z^1) + (1-\lambda)^2 \operatorname{var}(z^2) + 2\lambda(1-\lambda)\{\operatorname{var}(z^1)\operatorname{var}(z^2)\}^{1/2}, \end{aligned}$$

since the fact that Ω_1 is positive definite implies that the correlation coefficient between z^1 and z^2 is less than unity. Taking the square root of both sides of this expression gives

$$\{\operatorname{var}(z^\lambda)\}^{1/2} < \lambda\{\operatorname{var}(z^1)\}^{1/2} + (1-\lambda)\{\operatorname{var}(z^2)\}^{1/2}.$$

(c) The proof requires some preliminary definitions that lead to a theorem whose application will yield the result.

Definitions A set $C \subset R^n$ is a convex cone if for any points c_1 and c_2 in C and any nonnegative numbers Δ_1 and Δ_2, the point $\Delta_1 c_1 + \Delta_2 c_2 \in C$.

A function ψ defined on a convex cone C is essentially positive if for all $c \in C$ such that $c \neq 0$, $\psi(c) > 0$.

Theorem 8 If a non-negative function ψ is defined on a convex cone c and is positively homogeneous and essentially positive, then it is convex if and only if it is quasi-convex.

Proof See Newman [21].

Now Ψ is essentially positive and positively homogeneous on the convex cone *M*. Now Ψ is quasi-convex if and only if the set $\{\hat{x} \mid \{\tfrac{1}{2}\sum_{j=1}^{m} (\hat{x}'\Omega_j \hat{x})^{\alpha/2}\}^{1/\alpha} \leqq q\}$ is convex for all $q \geqq 0$. Raising both sides of the inequality to the αth power gives $\{\hat{x} \mid \sum_{j=1}^{m} (\hat{x}'\Omega_j \hat{x})^{\alpha/2} \leqq 2q^\alpha\}$ and this set is convex for all $q' \equiv 2q^\alpha \geqq 0$ since Press [22] has shown that $\sum_{j=1}^{m} (\hat{x}'\Omega_j \hat{x})^{\alpha/2}$ is convex if $2 \geqq \alpha \geqq 1$.

The fractional program to be solved in stage 1 is then

(12) $$\max g_1(\hat{x}) \equiv \frac{\tilde{\xi}'\hat{x}}{\Psi(\hat{x})}, \qquad \text{s.t.} \quad \hat{x} \geqq 0, \quad e'\hat{x} = 1,$$

where each $\tilde{\xi}_i = \bar{\xi}_i - \bar{\xi}_0$.

A differentiable function $\theta(x)\colon \Lambda \to R$ is said to be pseudo-concave [32] on $\Lambda \subset R^n$ if for all $x, \bar{x} \in \Lambda$, $(x-\bar{x})'\nabla\theta(\bar{x}) \leqq 0 \Rightarrow \theta(x) \leqq \theta(\bar{x})$. Geometrically, functions are pseudo-concave if they are nonincreasing in all directions that have downward- or horizontal-pointing directional derivatives. Local maxima of pseudo-concave functions constrained over convex sets satisfying a constraint qualification are global maxima. Now g_1 is the ratio of an (assumed) nonnegative differentiable concave function to a strictly positive convex function and hence is pseudo-concave (the proof is analogous to that of Theorem 4). An optimal solution to (12) always exists, because the constraint region is compact, and may be found using a standard algorithm [32]. For cases (a) and (b) of Theorem 7 the solution will be unique since g_1 will be strictly pseudo-concave. It has not been possible to prove that $\Psi(\hat{x})$ is generally strictly convex (although one suspects that Ψ will usually have this property) hence (12) may have multiple solutions.

Once (12) has been solved to obtain $\hat{x}^*$, the problem is then to obtain the optimal ratios of risky to nonrisky assets. The risky asset is

$$R = \xi'\hat{x}^* \sim F(\xi'\hat{x}^*;\ \bar{\xi}'\hat{x}^*, \{\Psi(\hat{x}^*)\}^\alpha, 0, \alpha) \equiv F(R;\ \bar{R}, S_R, 0, \alpha).$$

The analysis is then precisely analogous to that in Section II: See Eqs. (4)–(8) and the accompanying discussion.

The essential features that one needs to utilize the two-stage decomposition are the risk-free asset assumption and the assumption that the random returns are multivariate symmetric stable variates having the same characteristic exponent α $(1 < \alpha < 2)$ and that the dispersion measure is convex and positively homogeneous. The assumption that each x_i is nonnegative is not crucial.

An additional class of multivariate stable distributions which has these features was suggested by Samuelson [29]. Suppose each $\xi_i = \sum_{j=1}^{k} a_{ij} Y_j$, where each Y_j has the independent stable distribution $F(Y_j;\ \bar{Y}_j, S_{Y_j}, 0, \alpha)$ and the a_{ij} are known constants. Then the portfolio $\hat{x}'\xi$ has the univariate stable distribution $F(\hat{x}'\xi;\ \hat{x}'\bar{\xi}, S_{\hat{x}'\xi}, 0, \alpha)$, where

$$S_{\hat{x}'\xi} \equiv \sum_{j=1}^{k} \left| \sum_{i=1}^{n} a_{ij} x_i \right|^\alpha (S_{Y_j})^\alpha.$$

It must be shown that

$$\Psi_1(\hat{x}) \equiv \left\{ \sum_{j=1}^{k} \left| \sum_{i=1}^{n} a_{ij} x_i \right|^\alpha (S_{Y_j})^\alpha \right\}^{1/\alpha}$$

is convex. Let $c_j \equiv (S_{Y_j})^\alpha > 0$ and $b_j' \equiv (a_{j1}, ..., a_{jn})$; then

$$\Psi_1(x) = \left\{\sum_{j=1}^{k} c_j \left|\hat{x}'b_j\right|^{\alpha}\right\}^{1/\alpha}.$$

Suppose that Ψ_1 is defined on R^n and that for some j, $|\hat{x}'b_j \neq 0|$ if $\hat{x} \neq 0$. Hence Ψ_1 is essentially positive and positively homogeneous on the convex cone R^n. Thus by Theorem 8, Ψ_1 is convex if and only if it is quasi-convex. Now Ψ_1 is quasi-convex if and only if $N \equiv \{\hat{x} \mid \{\sum_{j=1}^k c_j |\hat{x}'b_j|^\alpha\}^{1/\alpha} \leqq q\}$ is convex for all $q \geqq 0$, or equivalently, if $\{\hat{x} | \sum_{j=1}^k c_j |\hat{x}'b_j| \leqq q'\}$ is convex for all $q' \equiv q^\alpha \geqq 0$. Since each $|\hat{x}'b_j|^\alpha$ is convex and the $c_j > 0$, the sum $\sum_{j=1}^k c_j |\hat{x}'b_j|^\alpha$ is convex, hence the set N is convex.

IV. Dropping the Risk-Free Asset Assumption

Let us reconsider problem

(1) Maximize

$$Z(x) \equiv E_\xi u(\xi'x), \qquad \text{s.t.} \quad e'x = 1, \quad x \geqq 0.$$

Suppose that $\xi = (\xi_0, ..., \xi_n)'$ has a symmetric multivariate stable distribution G with characteristic exponent α $(1 < \alpha \leqq 2)$, having the log characteristic function

$$\ln \psi_\xi(t) = i\bar{\xi}'t - \gamma(t),$$

where $\gamma(t)$ is positive homgeneous of degree α. Suppose that a risk-free asset does not exist. Then one may (approximately) solve (1) as follows.[10]

For any feasible x, say x^k, $\xi'x$ has the univariate symmetric stable distribution $F(\xi'x^k;\ \bar{\xi}'x^k, \gamma(x^k), 0, \alpha)$. Now

$$\text{(13)} \quad Z(x^k) = \int \cdots \int u(\xi'x^k)\, dG(\xi)$$

$$= \int u[\bar{\xi}'x^k + \{\gamma(x^k)\}^{1/\alpha} \tilde{R}]\, f(\tilde{R})\, d\tilde{R},$$

where the standardized variate

$$\tilde{R} \equiv \frac{\xi'x^k - \bar{\xi}'x^k}{[\gamma(x^k)]^{1/\alpha}} \sim F(\tilde{R};\ 0, 1, 0, \alpha).$$

[10] It may be noted that the following procedure is also applicable when the risk free asset does exist but the lending and borrowing rates differ. Let $x_0 \geqq 0$ and $x_1 \geqq 0$ be the levels of borrowing and lending activities at constant rates $\bar{\xi}_0$ and $\bar{\xi}_1$, respectively. Then in (1) $\xi'x$ becomes $-\xi_0 x_0 + \bar{\xi}_1 x_1 + \bar{\xi}_2 x_2 + \cdots + \xi_n x_n$ and $e'x$ becomes $-x_0 + x_1 + \cdots + x_n$.

The continuous partial derivatives of Z are

$$(14)\quad \frac{\partial Z(x^k)}{\partial x_i} = \int \left\{ \bar{\xi}_i + \frac{1}{\alpha}[\gamma(x^k)]^{(1-\alpha)/\alpha} \frac{\partial \gamma(x^k)}{\partial x_i} \tilde{R} \right\} \times \frac{du[\bar{\xi}' x^k + \{\gamma(x^k)\}^{1/\alpha} \tilde{R}]}{dw} f(\tilde{R})\, d\tilde{R}, \qquad i = 0, \ldots, n.$$

Now using the Fama–Roll tables, as in Section II, one may obtain a good approximation to (13) and (14) via

$$(15)\quad Z(x^k) \cong \sum_{j=1}^{m} P_j u[\bar{\xi}' x^k + \{\gamma(x^k)\}^{1/\alpha} \tilde{R}_j],$$

and

$$(16)\quad \frac{\partial Z(x^k)}{\partial x_i} \cong \sum_{j=1}^{m} P_j \left\{ \bar{\xi}_i + \frac{1}{\alpha}[\gamma(x^k)]^{(1-\alpha)/\alpha} \frac{\partial \gamma(x^k)}{\partial x_i} \tilde{R}_j \right\} \times \frac{du[\bar{\xi}' x^k + \{\gamma(x^k)\}^{1/\alpha} \tilde{R}_j\}}{dw}, \qquad i = 0, \ldots, n,$$

respectively, where $P_j \equiv \Pr\{\tilde{R} = \tilde{R}_j\} > 0$, $\sum_{j=1}^{m} P_j = 1$, and $m = 100$.

One may then apply any standard nonlinear programming algorithm that uses function values and/or partial derivatives to solve (1) approximately, utilizing (15) and/or (16). If one uses an algorithm that utilizes only function values, such as the generalized programming algorithm (see Ziemba [33]), then for each evaluation of (15) one merely performs m function evaluations of the form $u(\cdot)$ and adds them up with weights P_j. The evaluation of the portfolio is more complicated since for each i ($i = 1, \ldots, n$) one must perform m function evaluations of the form $\{\cdot\}\, du(\cdot)/dw$ and add these up with the weights P_j.

One would suspect that it would be economically feasible to solve such approximate problems when there are say 40–60 investments, the grid m is say 20–50 points, and u, γ, and k are reasonably convenient. It is possible, of course, to apply this direct solution approach even when the risk-free asset exists. However, the two-stage decomposition approach appears simpler because one must solve a fractional program in n variables plus a nonlinear program in one variable that has m terms. In the direct approach one must solve one nonlinear program in n variables having m terms that may fail to be concave or pseudo-concave (because u is concave and γ is convex). Some numerical results are given by Ziemba [34].

Appendix

Proof of theorem 1

(a) $$\left|\int u(w)\,dF(w)\right| \overset{(i)}{\leqq} \int |u(w)|\,dF(w) \overset{(ii)}{\leqq} \int L_1\,|w|^{\beta_1}\,dF \overset{(iii)}{<} \infty.$$

(iii) is a well-known property of stable distributions (see, e.g., Feller [9]). (ii) is by assumption, while (i) follows because the middle term is finite (see Loève [16]).

(b) $$\left|\int du(w)/dw\;dF(w)\right| \overset{(i)}{\leqq} \int |du(w)/dw|\,dF(w) \overset{(ii)}{\leqq} \int L_2\,|w|^{\beta_2} \overset{(iii)}{<} \infty.$$

(*i*)–(iii) follow for the same reasons as in (a) since (b) is a special case of (a).

(c) Now $w \equiv \xi' x \sim F(\xi' x;\, \bar{\xi}' x,\, \sum_{i=0}^{n} S_i x_i^{\alpha}, 0, \alpha)$ if and only if

$$\tilde{w} \equiv \frac{w - \bar{\xi}' x}{\{\sum_{i=0}^{n} S_i x_i^{\alpha}\}^{1/\alpha}} \sim F(\tilde{w};\, 0, 1, 0, \alpha).$$

Notice that F corresponding to the standardized variable w does not depend on x. Now

$$E_w u(w) = \int u(w)\,dF(w) = \int u(\bar{\xi}' x + s(x)\tilde{w})\,dF(\tilde{w};\, 0, 1, 0, \alpha)$$

where $s(x) \equiv \{\sum_{i=0}^{n} S_i x_i^{\alpha}\}^{1/\alpha}$. Now if it is legitimate to differentiate under the integral sign

$$\begin{aligned}\frac{\partial Eu(w)}{\partial x_i} &= \int \frac{du[\bar{\xi}' x + s(x)\tilde{w}]}{dw}(\bar{\xi}_i + s_i'(x)\tilde{w})\,dF(\tilde{w};\, 0,1,0,\alpha)\\ &= \bar{\xi}_i \int \frac{du[\bar{\xi}' x + s(x)\tilde{w}]}{dw}\,dF(\tilde{w};\, 0, 1, 0, \alpha)\\ &\quad + s_i'(x)\int \tilde{w}\,\frac{du[\bar{\xi}' x + s(x)\tilde{w}]}{dw}\,dF(\tilde{w};\, 0, 1, 0, \alpha)\end{aligned}$$

where

$$s_i'(x) \equiv \frac{1}{\alpha}\left\{\sum_{i=0}^{n} S_i x_i^{\alpha}\right\}^{(1-\alpha)/\alpha} x_i^{\alpha-1}.$$

Now under the assumptions $\bar{\xi}_i$ and s_i' are finite; hence $\partial Eu(w)/\partial x_i$ is finite if the latter two integrals are finite, which they are using an argument as in (a) and the assumption on u'. Note that $|u'(w)\,w| \leqq |w|^{\beta}$ if and only if $|u'(w)| \leqq |w|^{\beta-1}$ and that $|u'(w)| \leqq |w|^{\beta-1}$ implies that $|u'(w)| \leqq |w|^{\beta}$ for $\beta > 1$. Also one may differentiate under the integral sign because u is continuously differentiable and the absolute expected marginal utility is bounded (see Loève [16]).

(d) The proofs remain valid as long as the appropriate absolute integrals remain finite. We show here for case (a) that $\int |u(w)|\, dF(w) < \infty$; similar arguments may be used to establish (b) and (c).

Suppose $v_1 > 0$. Then without loss of generality we may take $v_1 = 1$ (and $L_1 = 1$). Now $\lim_{|w| \to \infty} (|u(w)|/|w|^{\beta}) = 1$ if and only if for all ε such that $0 < \varepsilon < 1$, $\exists N_1 > 0$ such that $\forall |w| \geqq N_1, \big| |u(w)|/|w|^{\beta_1} - 1 \big| < \varepsilon$, or $\big| |u(w)| - |w|^{\beta_1} \big| < \varepsilon |w|^{\beta_1}$ hence $|u(w)| < (1+\varepsilon)|w|^{\beta_1}$ (i).

By the Cauchy convergence criterion (see Loève [16]), $\int_{-\infty}^{\infty} |w|^{\beta_1}\, dF(w) < \infty$ if and only if $\exists N_2$ such that

$$\forall n \geqq N_2, \qquad \int_n^{\infty} |w|^{\beta_1}\, dF(w) < \varepsilon/2 \qquad \text{and}$$

$$\forall n \leqq -N_2, \qquad \int_{-\infty}^{n} |w|^{\beta_1}\, dF(w) < \varepsilon/2.$$

Let $N \geqq \max(N_1, N_2)$. Then

$$\int_N^{\infty} |u(w)|\, dF(w) \overset{(\text{by i})}{<} (1+\varepsilon) \int_N^{\infty} |w|^{\beta_1}\, dF(w) \overset{(\text{since } \varepsilon < 1)}{<} 2 \int_N^{\infty} |w|^{\beta_1}\, dF(w) < \varepsilon$$

and

$$\int_{-\infty}^{-N} |u(w)|\, dF(w) < (1+\varepsilon) \int_{-\infty}^{-N} |w|^{\beta_1}\, dF(w) < 2 \int_{-\infty}^{-N} |w|^{\beta_1}\, dF(w) < \varepsilon.$$

Thus $\int_{-\infty}^{\infty} |u(w)|\, dF(w) < \infty$.

When $v = 0$ the same proof applies by letting ε and 1 replace $(1+\varepsilon)$ and (2), respectively.

Acknowledgment

Without implicating them, I would like to thank S. L. Brumelle, W. E. Diewert, J. L. Evans, J. Ohlson, C. E. Sarndal, C. Swoveland, and R. Vickson for some useful information and helpful discussions related to this work.

References

1. Blume, M., "Portfolio Theory: A Step Toward its Practical Application." *Journal of Business* **43** (1970), 152–173.
2. Breen, W., "Homogeneous Risk Measures and the Construction of Composite Assets." *Journal of Financial and Quantitative Analysis* **3** (1968), 405–413.

3. CHIPMAN, J., "On the Ordering of Portfolios in Terms of Mean and Variance." *Review of Economic Studies* **40** (1973), 167–190.
4. FAMA, E. F., "The Behavior of Stock Market Prices." *Journal of Business* **38** (1965), 34–105.
5. FAMA, E. F., "Portfolio Analysis in a Stable Paretian Market." *Management Science* **11** (1965), 404–419.
6. FAMA, E. F., "Efficient Capital Markets: A review of Theory and Empirical Work." *Journal of Finance* **25** (1970), 383–417.
7. FAMA, E. F., and ROLL, R., "Some Properties of Symmetric Stable Distributions." *Journal of the American Statistical Association* **63** (1968), 817–836.
8. FAMA, E. F., and ROLL, R., "Parameter Estimates for Symmetric Stable Distributions." *Journal of the American Statistical Association* **66** (1971), 331–338.
9. FELLER, W., *An Introduction to Probability Theory and Its Applications*, Vol. II, Wiley, New York, 1966.
10. FERGUSON, T., "On the Existence of Linear Regression in Linear Structural Relations." *University of California Publications in Statistics* **2** (1955), 143–166.
11. GNEDENKO, B. V., and KOLMOGOROV, A. N., *Limit Distributions for Sums of Independent Random Variables*. Addison-Wesley, Reading, Massachusetts, 1954.
12. HANOCH, G., and LEVY, H., "The Efficiency Analysis of Choices Involving Risk." *Review of Economic Studies* **36** (1969), 335–346.
13. HARDY, G. H., LITTLEWOOD, J. E., and POYLA, G., *Inequalities*, 2nd ed. Cambridge Univ. Press, London and New York, 1964.
13a. HILLIER, F. S., *The Evaluation of Risky Interrelated Investments*. North-Holland Publ., Amsterdam, 1969.
14. LEVY, P., *Theorie de l'Addition des Variables Aleataires*, 2nd ed. Gauthier-Villars, Paris, 1954.
15. LINTNER, J., "The Valuation of Risk Assets and the Selection of Risky Investments in Stock Portfolios and Capital Budgets." *Review of Economics and Statistics* **47** (1965), 13–37.
16. LOÈVE, M., *Probability Theory*, 3rd ed. Van Nostrand-Reinhold, Princeton, New Jersey, 1963.
17. MANDELBROT, B., "The Variation of Certain Speculative Prices." *Journal of Business* **36** (1963), 394–419.
18. MANDELBROT, B., "New Methods of Statistical Economics." *Journal of Political Economy* **61** (1963), 421–440.
19. MANDELBROT, B., and TAYLOR, H. M., "On the Distribution of Stock Price Differences." *Operations Research* **15** (1967), 1057–1062.
20. MARKOWITZ, H. M., *Portfolio Selection: Efficient Diversification of Investments*. Wiley, New York, 1959.
21. NEWMAN, P., "Some Properties of Concave Functions." *Journal of Economic Theory* **1** (1969), 291–314.
22. PRESS, S. J., "Multivariate Stable Distributions." *Journal of Multivariate Analysis* **2** (1972), 444–463.
23. PRESS, S. J., "Estimation in Univariate and Multivariate Stable Distributions." *Journal of the American Statistical Association* **67** (1972), 842–846.
24. PRESS, S. J., "A Compound Events Model for Security Prices." *Journal of Business* **40** (1967), 317–335.

25. Press, S. J., "A Modified Compound Poisson Process with Normal Compounding." *Journal of the American Statistical Association* **60** (1968), 607–613.
26. Press, S. J., "A Compound Poisson Process for Multiple Security Analysis." In *Random Counts in Scientific Work* (G. Patil, ed.). Pennsylvania State Univ. Press, University Park, Pennsylvania, 1970.
27. Roll, R., *The Behavior of Interest Rates.* Basic Books, New York, 1970.
28. Samuelson, P. A., "General Proof that Diversification Pays." *Journal of Financial and Quantitative Analysis* **2** (1967), 1–13.
29. Samuelson, P. A., "Efficient Portfolio Selection for Pareto-Levy Investments." *Journal of Financial and Quantitative Analysis* **2** (1967), 107–122.
30. Sharpe, W. F., "A Simplified Model for Portfolio Analysis." *Management Science* **9** (1963), 277–293.
31. Tobin, J., "Liquidity Preference as Behavior Towards Risk." *Review of Economic Studies* **25** (1958), 65–86.
32. Zangwill, W. I., *Nonlinear Programming: A Unified Approach.* Prentice-Hall, Englewood Cliffs, New Jersey, 1969.
33. Ziemba, W. T., "Solving Nonlinear Programming Problems with Stochastic Objective Functions." *Journal of Financial and Quantitative Analysis* **7** (1972), 1809–1827.
34. Ziemba, W. T., "Computational Aspects of Optimal Portfolio Construction: Relationship and Comparison of Mean-Dispersion and SD Approaches." In *Stochastic Dominance: An Approach to Decision Making Under Risk* (G. A. Whitmore and M. C. Findlay, eds.). To be published.
35. Ziemba, W. T., Parkan, C., and Brooks-Hill, F. J., "Calculation of Investment Portfolios with Risk Free Borrowing and Lending." *Management Science* **21** (1974), 209–222.

2. EXISTENCE AND DIVERSIFICATION OF OPTIMAL PORTFOLIO POLICIES

JOURNAL OF ECONOMIC THEORY 4, 35–44 (1972)

On the Existence of Optimal Policies under Uncertainty

HAYNE E. LELAND*

Stanford University, Stanford, California 94305

Received August 10, 1970

I. INTRODUCTION

Many economic theories presume that a feasible policy set contains an optimal policy. In portfolio theory, for example, theorists often presume that first order maximizing conditions are satisfied by a feasible portfolio, without consideration of whether the existence of an optimal policy is implied by, or even consistent with, the basic set of assumptions.

Theorists who have taken care to ensure the existence of solutions commonly have resorted to introducing explicit bounds on a closed policy or action set (in R^n), thereby rendering it compact. Given a continuous objective function, Weierstrass' Theorem can then be used to prove the existence of an optimal action. Hakansson [2], for example, bounds the feasible set of his portfolio model by requiring that terminal wealth be nonnegative with probability one.

But costs may be incurred by bounding the action set. Bounds often are difficult to justify on economic grounds.[1] Or they are so loose (e.g., "use of resources must not exceed world supply") that the question of existence is transformed to the question of whether the bounds are tight or slack at the optimum. But the answer to the latter question conditions the predictions of the theory as critically as the answer to the former. To escape Scylla and Charybdis, some theorists introduce constraints to ensure existence, and then assume these constraints are slack at the optimum! Clearly, this is little better than to have ignored the original existence questions.

* This work was supported by National Science Foundation Grant GS-2530 at the Institute for Mathematical Studies in the Social Sciences at Stanford University.

[1] Hakansson's requirement that final wealth be nonnegative, for example, seems rather restrictive. If an asset had normally distributed returns, no investment would occur no matter how large the expected return and how small the variance, as there always is a nonzero probability of an arbitrarily large loss.

An alternative approach to proving the existence of optimal policies is to assign further properties to the preference ranking over actions, as reflected by the objective function. If these properties alone ensure existence, there is no need to incur the costs associated with bounding the action set.

This paper considers a class of decision problems encompassing several of relevance to economists, including the choice of optimal portfolios and the choice of optimal inputs and outputs by a perfectly competitive firm (with constant or decreasing returns to scale) facing random prices. Weak, and indeed commonly assumed, properties of the utility function are shown to ensure the existence of optimal policies. Perhaps the most interesting implication of the results is that the perfectly competitive firm with constant returns to scale will have a determinate output when there is uncertainty regarding prices, and it has a risk-averse utility function over profits which is bounded above.

II. A Class of Decision Problems

The structure of a typical decision problem includes a set S of states of nature, a σ-algebra $\mathscr{S}$ of events, a set C of outcomes (hereafter assumed to be in R^1), and a set D of decisions, each element of which is characterized by a measurable function mapping S into C. In all decision problems, it is assumed that the decision-maker possesses a preference ranking over the elements of D. An *optimal policy* or action is defined as $d^* \in D$ such that $d^* \succsim d$ for all $d \in D$. Of course, such an action may not exist.

The problems examined in subsequent sections are a special class of the general decision problem. As above, elements include a set of states of nature and associated σ-algebra of events, and a set $C \subset R^1$ of outcomes. But the actions belong to a set X of feasible gambling positions in a set G of gambling opportunities.

Each element g of the *gambling opportunity set* G is characterized by a measurable net return function mapping S into a set of monetary returns $C^g \subset R^1$. $g(s)$, therefore, is the monetary return per unit of the gamble g, depending on the state of nature. If $g(s) = k$ for all $s \in S$, the "gamble" provides a sure return k. We assume the following:

G.1 G has a finite number of elements, $g^1, \ldots, g^n$.

G.2 G is independent of the actions of the decision-maker.

G.1 seems warranted, at least in the static case, by the finite number of stocks, bonds, products, etc., in which one can invest. *G*.2 is a straightforward "perfect market" assumption. Note that each gamble g^i maps S into C^i, and the set of gambles G maps S into $C^n \in R^n$, where C^n is the Cartesian product of the sets C^i, $i = i,..., n$.

An element x of the feasible action or *feasible gambling position set* X is characterized by a vector $(x^1,..., x^n)$ representing the number of units of gambles $(g^1,..., g^n)$ held. Each element of X, therefore, maps a point C^n, depending upon the state of nature, into a set of outcomes $C \subset R^1$, where

$$C = \{c = G(s)' x \mid G(s) \in C^n, x \in X\},$$

and $G(s)$ is the column vector of net return functions $g^1(s),..., g^n(s)$. As in the general decision problem, the choice of an action $x \in X$ is a choice of mapping from S into C, but via the gambling opportunity set G.

We assume actions can be ranked by expected utility.[2] That is, the ranking is consistent with a utility function over C, and a subjective probability measure over S, such that

$$x_1 \gtrsim x_2 \quad \text{iff} \quad E\{U[G(s)' x_1]\} \geqslant E\{U[G(s)' x_2]\}.$$

We impose the following restrictions on the feasible action set X and on the preference relation over X, as reflected by the utility function over C:

$X.1$ X is closed and (perhaps weakly) convex.

$X.2$ $0 \in X$, where 0 is the origin.

$X.3$ There exists an $H < \infty$ such that there is no $x_k \in X$ with the properties

(1) $G(s)' x_k \geqslant 0$, except on a set with probability measure zero, *and*

(2) $\| x_k \| \geqslant H$, where $\| x_k \|$ is the Euclidean norm of x_k.

$U.1$ $U'(c) > 0$ for all $c \in C$;

$U.2$ $U''(c) < 0$ for all $c \in C$.

Assumption $X.1$ necessarily limits the class of problems considered, but we shall indicate several important economic problems which satisfy this requirement. Typically, the feasible action set may be affected by prices of gambles, initial wealth, technological relationships, etc. Assumption $X.2$ states that no gambling at all is a feasible action. Assumption $X.3$ states that there is no arbitrarily large gambling position which offers a positive or zero return with probability one. It is equivalent to Hakansson's

[2] Axioms on a preference ranking leading to the conjoint imputation of a subjective probability distribution over states of nature and a utility function over outcomes are discussed by Savage [7]. His analysis requires S to have an infinite number of elements.

"no easy money" condition [2]: Any infinitely reproducible vector of gambling positions must have a nonzero probability of loss.

Assumption $U.1$ states that a larger outcome (money return) is always preferred, when that outcome occurs with certainty. If the utility function is presumed to be derived from the ranking over X alone, clearly C must be an interval or set of intervals. Assumption $U.2$ is equivalent to risk aversion: an amount with certainty will always be preferred to a gamble with equal expected value (see, e.g., Pratt [5]).

In the analysis which follows, we shall find it useful to define a function

$$V(x) = E\{U[G(s)'\, x]\}.$$

Assumptions $U.1$ and $U.2$ imply $V(x)$ is a continuous concave function with continuous partial derivatives.[3]

Economic problems which fall within the special class of problems introduced above include the portfolio problem and the selection of inputs and outputs by a perfectly competitive risk averse firm with constant returns to scale, when prices of inputs and/or outputs are uncertain.[4] Indeed, if the production set can be imbedded in a weakly convex set, the existence results derived in subsequent sections will hold.

III. The Existence Problem

Without further assumptions, there is no assurance that the class of problems considered in Section II will possess a maximal element. To prove this contention, consider the following example of portfolio choice. Let asset 1 return r with certainty, asset 2 return a or b with probability $\frac{1}{2}$ each, with $a > r > b$.

The investor seeks to

$$\max E[U(B_1)] \quad \text{subject to} \quad B_1 = (1 + r)\, x_1 + (1 + e_2)\, x_2\,, \qquad \text{where} \quad e_2 = a \text{ w.p. } \tfrac{1}{2} \quad = b \text{ w.p. } \tfrac{1}{2} \tag{1}$$

$$x_1 + x_2 = B_0\,. \tag{2}$$

[3] Note these statements are conditional on the existence of $V(x)$. If $E[g^i(s)] < \infty$, $i = 1,..., n$, there are no problems. If $V(x)$ is to exist for *any* gamble with infinite expected value, then U must be bounded; see [1]. For the relation between the existence of $V(x)$ and the existence of optimal policies, see the Conclusion.

[4] Risk aversion of utility over profit is justified on the basis that the firm is owned by risk averse investors; see Sandmo [6] and Leland [4]. In the portfolio (firm) case, X.3 implies there is no combination of assets (inputs and outputs) which yields a positive return (profit) with probability one, and is infinitely reproducible.

Substituting for x_1 from (2) gives an unconstrained maximization problem

$$\max_{x_2} E\{U[(1 + r) B_0 + (e_2 - r) x_2]\}. \tag{3}$$

Differentiating w.r.t. x_2 and writing out the expected value gives

$$dE[U(B_1)]/dx_2 = (\tfrac{1}{2})(a - r) U'(B_1{}^a) + (\tfrac{1}{2})(b - r) U'(B_1{}^b), \tag{4}$$

where $B_1{}^a = B_1$ when $e_2 = a$, $B_1{}^b = B_1$ when $e_2 = b$.

Suppose now $U(B_1)$ satisfies $U.1$ and $U.2$, and, furthermore,

$$\lim_{B_1 \to \infty} U'(B_1) = k_1 ; \qquad \lim_{B_1 \to -\infty} U'(B_1) = k_2 ,$$

where $k_2 > k_1 > 0$. Note that the first R.H.S. term of (4) is positive for all x_2, while the second term is negative for all x_2. It follows from $U.2$ that

$$dE[U(B_1)]/dx_2 \geqslant (\tfrac{1}{2})[(a - r) k_1 + (b - r) k_2] \qquad \text{for all } x_2 . \tag{5}$$

But if

$$k_1/k_2 > [-(b - r)]/(a - r), \qquad \text{implying} \quad [(a - r) k_1 + (b - r) k_2] > 0,$$

it follows from (5) that $dE[U(B_1)]/dx_2 > 0$ for any x_2. Therefore, X in this case *does not possess a maximal element*, for larger positions in x_2 are always preferred. As none of the assumptions on G, X, and U are violated in this example, we conclude that these assumptions are not sufficient for X to possess a maximal element.

In the Introduction, it is suggested that the approach of bounding the feasible set X has certain intuitive and operational drawbacks. A second approach, which we shall develop in subsequent sections, is to introduce further assumptions on the nature of the preference ranking over X. These assumptions, which seem to have considerable attraction in other areas of economic theory, will be shown to guarantee the existence of optimal policies.

IV. Sufficient Conditions for the Existence of Optimal Policies

We shall now show that, if the expected returns of all gambles are finite, the further condition

$U.3$a $$\lim_{c \to \infty} U'(c) = 0$$

is sufficient for X to have a maximal element. If we do not assume finite expected returns, the slightly stronger condition

$U.3b$ $\qquad U(c)$ bounded above

is a sufficient condition. To prove these results, several preliminary concepts and propositions are required.

DEFINITION 1. $Z = \{x \in X \mid V(x) \geqslant V(0)\}$.

Comment. Clearly, X will have a maximal element if and only if Z has a maximal element. Because V is continuous, Z is closed. If it is bounded, it will have a maximal element, as a continuous function reaches a maximum over a compact set. The object of the following analysis is to show that Z is bounded.

PROPOSITION I. *Z is convex.*

Proof. As X is convex, it remains only to show that for $x^1, x^2 \in X$ with $V(x^1), V(x^2) \geqslant V(0)$, then $V[\alpha x^1 + (1 - \alpha) x^2] \geqslant V(0)$ for all $\alpha \in (0, 1)$. But this follows immediately from the concavity of $V(x)$.

DEFINITION 2. $Z^* = \{x \in Z \mid \| x \| \geqslant H\}$.

Comment. $Z = Z^* \cup Z - Z^*$, where $Z - Z^* = \{x \in Z \mid \| x \| < H\}$. Clearly, $Z - Z^*$ is bounded. If Z^* is bounded, Z will be bounded, as the union of bounded sets is bounded.

DEFINITION 3. $Y = \{y \in R^n \mid \| y \| = 1, Hy \in Z\}$.

Comment. Y is the (compact) set of directions from the origin such that, when $x = Hy$, $x \in X$, and $V(Hy) \geqslant V(0)$.

PROPOSITION II. *$V(\lambda y)$ is a strictly concave function of λ for $y \in Y$.*

Proof. $V_{\lambda\lambda}(\lambda y) = \partial^2[E[U(\lambda e'y)]]/\partial\lambda^2 = E[(e'y)^2 U''(\lambda e'y)] < 0$, as $U''(c) < 0$ for all c by $U.2$, and $pr[(e'y)^2 = 0] < 1$ by $X.3$, since, by assumption, $\lambda y \in X$, when $\| \lambda y \| = \lambda = H$.

DEFINITION 4. $Y^* = \{x = \lambda y \mid y \in Y, \lambda \geqslant H, V(\lambda y) \geqslant V(0)\}$.

PROPOSITION III. $Z^* \subset Y^*$.

Proof. Choose any $z \in Z^*$. Define $y(z) = z/\| z \|$, and $\lambda(z) = \| z \|$. Therefore, $z = \lambda(z)\, y(z)$, where $\| y(z)\| = 1$. To show $z \in Y^*$, we must show $\lambda(z) \geqslant H$ and $y(z) \in Y$. The first follows directly, as $\lambda(z) = \| z \| \geqslant H$.

To show $y(z) \in Y$, we must show $Hy(z) \in Z$. As both O (the origin), $\lambda(z)\, y(z) \in Z$, the convexity of Z implies $Hy(z) \in Z$, because $H \in [0, \lambda(z)]$.

THEOREM I. *Let G, X, and U satisfy G.1, G.2; X.1, X.2, X.3; U.1, U.2, and U.3b: U(c) bounded above. Then, for every* $y \in Y$, *there exists a* $\lambda^*(y)$ *such that* $V[\lambda^*(y)\, y] = V(0)$, *and* $V(\lambda y) < V(0)$ *for* $\lambda > \lambda^*(y)$.

Proof. Let $\beta(s, y) = G(s)'\, y$. Note for $y \in Y$, $pr(\beta < 0) > 0$, by $X.3$. First, we wish to show for any $y \in Y$, we may find a $\lambda^{**}(y)$ such that $V[\lambda^{**}(y)\, y] < V(0)$. From the definition of V, we have

$$V(\lambda y) - V(0) = Pr(\beta < 0)\, E[U(\lambda\beta) - U(0) \mid \beta < 0] + Pr(\beta \geqslant 0)\, E[U(\lambda\beta) - U(0) \mid \beta \geqslant 0]. \tag{1}$$

From $U.2$,

$$\begin{aligned} Pr(\beta < 0)\, E[U(\lambda\beta) - U(0) \mid \beta < 0] &\leqslant Pr(\beta < 0)\, E[\lambda\beta U'(0) \mid \beta < 0] \\ &= \lambda U'(0)\, E[\beta \mid \beta < 0]\, Pr(\beta < 0) \\ &= \lambda K_1, \text{ say, where } K_1 < 0. \end{aligned}$$

The second RHS term of (1), which is positive, clearly is less than $C - U(0) = K_2$, where C is the upper bound of $U(c)$. Therefore, by choosing $\lambda^{**}(y) > -K_2/K_1$ (itself a function of y), $V[\lambda^{**}(y)\, y] < V(0)$. By the continuity of V in λ, and the fact that $V(Hy) \geqslant V(0)$ and $V[\lambda^{**}(y)\, y] < V(0)$, there exists a $\lambda^*(y) \in [H, \lambda^{**}(y)]$ such that $V[\lambda^*(y)\, y] = V(0)$.

Now assume that for some $\lambda^0 > \lambda^*(y)$, $V(\lambda^0 y) \geqslant V(0)$. By Proposition II, $V(\lambda y)$ is strictly concave in λ. Therefore $V[\lambda^*(y)\, y] > V(0)$, as $\lambda^*(y) \in (0, \lambda^0)$. But this is a contradiction, and we conclude for $\lambda > \lambda^*(y)$, $V(\lambda y) < V(0)$.

THEOREM II. *Let G satisfy G.1, G.2; and G.3:* $E[g^i(s)] < \infty$, $i = 1, \ldots, n$.

Let X satisfy X.1, X.2, and X.3;
Let U satisfy U.1, U.2, and U.3a: $\lim_{c \to \infty} U'(c) = 0$.

Then, as in Theorem I, there exists for every $y \in Y$ *a* $\lambda^*(y)$ *such that* $V[\lambda^*(y)\, y] = V(0)$, *and* $V(\lambda y) < V(0)$ *for* $\lambda > \lambda^*(y)$.

Proof. As in Theorem I, (1) holds and the first RHS term will be less than or equal to λK_1. The second RHS term can be written, for any $\delta > 0$, as

$$Pr(0 \leqslant \beta \leqslant \delta)\, E[U(\lambda\beta) - U(0) \mid 0 \leqslant \beta \leqslant \delta] \\ + Pr(\beta \geqslant \delta)\, E[U(\lambda\beta) - U(0) \mid \beta \geqslant \delta]. \tag{2}$$

Consider the first term of (2). From $U.2$, we have

$$\begin{aligned} &Pr(0 \leqslant \beta \leqslant \delta)\, E[U(\lambda\beta) - U(0) \mid 0 \leqslant \beta \leqslant \delta] \\ &\quad \leqslant \lambda Pr(0 \leqslant \beta \leqslant \delta)\, E[\beta \mid 0 \leqslant \beta \leqslant \delta]\, U'(0) \\ &\quad = \lambda K(\delta), \text{ say, where } K(\delta) > 0 \text{ and } K(\delta) \to 0 \text{ as } \delta \to 0. \end{aligned}$$

Choose δ, so that $K(\delta) < -(\frac{1}{3})\, K_1$. To consider the second term of (2), we need the following

LEMMA. *If $U(c)$ satisfies $U.1$, $U.2$, and $U.3a$: $\lim_{c\to\infty} U'(c) = 0$, then*

$$\lim_{c\to\infty} [U(c) - U(0)]/c = 0.$$

Proof. Note $U.2$ implies $[U(c) - U(0)]/c$ is monotonically decreasing in c; $U.1$ implies it is always nonnegative. Therefore, it must approach a limit. Assume it approaches a limit $\eta > 0$. Then $[U(c) - U(0)] \geqslant \eta c$, or, since $U'(c)$ approaches a limit by $U.1$ and $U.2$, $\lim_{c\to\infty} U'(c) \geqslant \eta$, a contradiction. Therefore, $\lim_{c\to\infty}[U(c) - U(0)]/c = 0$.

The lemma implies that, given any $\gamma > 0$, there exists a $\lambda(\gamma)$ such that, for $c > \lambda(\gamma)\, \delta$, then $U(c) - U(0) < \gamma c$. Therefore,

$$Pr(\beta \geqslant \delta)\, E[U(\lambda\beta) - U(0) \mid \beta \geqslant \delta] \leqslant \lambda\gamma E[\beta \mid \beta \geqslant \delta]\, Pr(\beta \geqslant \delta) \tag{3}$$

for all $\lambda > \lambda(\gamma)$. Note that the assumption $E[\beta] < \infty$ implies $E[\beta \mid \beta \geqslant \delta] < \infty$.

From (3), we may choose γ so that $\lambda\gamma E[\beta \mid \beta \geqslant \delta]\, Pr(\beta \geqslant \delta) \leqslant -(\frac{1}{3})K_1\lambda$, for all $\lambda > \lambda(\gamma)$. Then, for all $\lambda > \lambda(\gamma)$,

$$V(\lambda y) - V(0) \leqslant [K_1 - (\tfrac{1}{3})\, K_1 - (\tfrac{1}{3})\, K_1] = (\tfrac{1}{3})\, K_1\lambda < 0,$$

as required. The rest of the proof continues as in Theorem I.

PROPOSITION IV. *$\lambda^*(y)$ is a continuous function of y, $y \in Y$.*

Proof. We have shown for each $y \in Y$, there exists a λ^* such that $V(\lambda^* y) = k = V(0)$. Take any $y \in Y$, and notice V_{λ^*}, V_y exist and are continous in a neighborhood of λ^*, y. V_{λ^*} is negative, because V is strictly concave in λ, and $V(\lambda^* y) = V(0)$ implies $V(\lambda y)$ reaches a maximum at $\lambda_1 \in (0, \lambda^*)$, and $V_\lambda < 0$ for $\lambda > \lambda_1$. By the Implicit Function Theorem, λ^* is a continous function of y for all $y \in Y$.

PROPOSITION V. *$\lambda^*(y)$ reaches a maximum $\lambda^\dagger$ for $y \in Y$.*

Proof. The proposition follows immediately from Weierstrass' Theorem: $\lambda^*(y)$ is a continous function over a compact set Y, and therefore attains a maximal value $\lambda^\dagger$.

THEOREM III. *X possesses a maximal element, given the assumptions either of Theorem I or Theorem II.*

Proof. From Theorems I and II, it follows that for $\lambda > \lambda^\dagger$, $V(\lambda y) < V(0)$ for all $y \in Y$. Therefore the set Y^* contains elements which are at most distance $\lambda^\dagger$ from the origin: Y^* is bounded. But as $Z^* \subset Y^*$, Z^* is bounded. It was previously established that if Z^* is bounded, Z is bounded, and if Z is bounded, X has a maximal element.

V. CONCLUSION

We have derived conditions under which an important class of economic problems will possess a finite optimal policy even when the set of feasible policies is not bounded. Portfolio selection and input/output selections by a competitive firm facing random prices with constant returns to scale are included in the class of problems considered. The strongest requirement, that $U(c)$ is bounded above, is familiar in other areas of economic theory. When all gambles have finite expected values, the weaker condition $U'(c) \to 0$ as $c \to \infty$ is sufficient for the existence of optimal policies.

One word of warning. $U.1$ and $U.2$ are inconsistent with boundedness above *and* below, conditions necessary to resolve all possible St. Petersburg problems. But boundedness above and below alone do not imply the existence of optimal policies. The interested reader can verify that the utility function $U(Y) = Y/(Y + 1)$, $Y \geqslant 0$, $U(Y) = -Y/(Y - 1)$, $Y < 0$, will lead to an unboundedly large position in a gamble offering $(r + a)$ with probability p, $(r - a)$ with probability $(1 - p)$ for $p > \frac{1}{2}$, when initial wealth $B_0 = 0$, r is the riskless rate of interest, and $a > 0$.

ACKNOWLEDGMENTS

The author would like to thank Professors Arrow, Hildenbrand, Kurz, Majumdar, and Richter for their comments and suggestions. A referee provided suggestions which led to a considerably shorter proof of the main results.

References

1. K. Arrow, "Bernoulli Utility Indicators for Distributions Over Arbitrary Spaces." Technical Report No. 57, Stanford University, Stanford, Calif., 1958.
2. N. Hakansson, "Optimal Investment and Consumption Strategies for a Class of Utility Functions," Working Paper 101, WMSI, University of California at Los Angeles, Calif., 1966.
3. H. Leland, "Dynamic Portfolio Theory." Ph.D. Dissertation, Harvard University, Boston, Mass., 1968.
4. H. Leland, "The Theory of the Firm Facing Uncertain Demand," Technical Report No. 24, Department of Economics, Stanford University, Stanford, Calif., 1970.
5. J. Pratt, "Risk Aversion of the Small and in the Large," *Econometrica* **32** (1964), 122–136.
6. A. Sandmo, "On the Theory of the Competitive Firm Under Price Uncertainty," *American Economic Review* **51** (1971), 65–73.
7. L. J. Savage, "Foundations of Statistics," Wiley, New York, 1954.

Reprinted from *Journal of Financial and Quantitative Analysis* 2, 1-13 (1967).

GENERAL PROOF THAT DIVERSIFICATION PAYS**

Paul A. Samuelson*

"Don't put all your eggs in one basket," is a familiar adage. Economists, such as Marschak, Markowitz, and Tobin,[1] who work only with mean income and its variance, can give specific content to this rule--namely, putting a fixed total of wealth equally into independently, identically distributed investments will leave the mean gain unchanged and will minimize the variance.

However, there are many grounds for being dissatisfied with an analysis dependent upon but two moments, the mean and variance, of a statistical distribution. I have long used the following, almost obvious, theorem in lectures. When challenged to find it in the literature, I was unable to produce a reference--even though I should think it must have been stated more than once.

<u>Theorem I:</u> If U(X) is a strictly concave and smooth function that is monotonic for non-negative X, and $(X_1,\dots,X_n)$ are independently, identically distributed variates with joint frequency distribution

$$\text{Prob}\{X_1 \le x_1,\dots,X_n \le x_n\} = F(x_1)F(x_2)\dots F(x_n)$$

$$\text{with } E[x_i] = \int_{-\infty}^{\infty} X_i dF(X_i) = \mu_1$$

$$E[x_i - \mu_1]^2 = \int_{-\infty}^{\infty} (X_i - \mu_1)^2 dF(X_i) = \mu_2$$

$$\text{with } 0 < \mu_2 < \infty\ ,$$

* Massachusetts Institute of Technology.

** My thanks go to the Carnegie Corporation for providing me with a reflective year and to Mrs. F. Skidmore for research assistance.

then

$$E\left[U\left[\sum_{1}^{n}\lambda_j X_j\right]\right] = \int_{-\infty}^{\infty} \cdots \int_{-\infty}^{\infty} U\left[\sum_{1}^{n}\lambda_j X_j\right] dF(X_1) \ldots dF(X_n)$$

$$= \Psi(\lambda_1, \ldots, \lambda_n)$$

is a strictly concave symmetric function that attains its unique maximum, subject to

$$\lambda_1 + \lambda_2 + \ldots + \lambda_n = 1 \, , \quad \lambda_i \geq 0$$

at $\Psi\left(\frac{1}{n}, \ldots, \frac{1}{n}\right)$.

The proof is along the lines of a proof[2] used to show that equal distribution of income among identical Benthamites will maximize the sum of social utility.

$$\frac{\partial\Psi(\lambda_1,\ldots,\lambda_n)}{\partial\lambda_i} = \int_{-\infty}^{\infty} \cdots \int_{-\infty}^{\infty} X_i \, U'\left(\sum_{1}^{n}\lambda_j X_j\right) dF(X_1)\ldots dF(X_n)$$

is independent of i at $(\lambda_1,\ldots\lambda_n) = (1/n,\ldots 1/n)$ by symmetry.

The Hessian matrix with elements

$$\frac{\partial^2\Psi}{\partial\lambda_i\partial\lambda_j} = \int_{-\infty}^{\infty} \cdots \int_{-\infty}^{\infty} X_i X_j U''\left(\sum_{1}^{n}\lambda_k X_k\right) dF(X_1)\ldots dF(X_n)$$

is a Grammian negative definite matrix if $-U'' > 0$, $0 < \mu_2 < \infty$.

Hence, sufficient maximum conditions for a unique maximum are satisfied, namely

$$\frac{\partial\Psi\left(\frac{1}{n}, \ldots, \frac{1}{n}\right)}{\partial\lambda_1} = \cdots = \frac{\partial\Psi\left(\frac{1}{n}, \ldots, \frac{1}{n}\right)}{\partial\lambda_n}$$

$$\sum_{1}^{n}\sum_{1}^{n}\frac{\partial^2\Psi}{\partial\lambda_i\partial\lambda_j} y_i y_j < 0$$ for all non-negative λ's and not all y's vanishing.

2

Remarks: Differentiability assumptions could be lightened. It is not true, by the way, that $\frac{1}{2} X_1 + \frac{1}{2} X_2$ has a "uniformly more-bunched distribution" than X_1 or X_2 separately, as simple examples (even with finite μ_2) can show: still the risk averter will always benefit from diversification. The finiteness of μ_2 is important. Thus, for a Cauchy distribution $\frac{1}{2} X_1 + \frac{1}{2} X_2$ has the same distribution as either X_1 or X_2 separately; for the arc-sine Pareto-Lévy case, it has a worse distribution. The proof fails because the postulated E[U] cannot exist (be finite) for any concave U.

The General Case of Symmetric Interdependence

We can now drop the assumption of independence of distribution, replacing it by the less restrictive postulate of a symmetric joint distribution. I.e., we replace $F(x_1) \ldots F(x_n)$ by

$$\text{Prob}\{X_1 \leq x_1, X_2 \leq x_2, \ldots, X_n \leq x_n\} = P(x_1, x_2, \ldots, x_n)$$

where P is a symmetric function in its arguments. We can rule out, as trivial, the case where the x's are connected by an exact functional relation, which in view of symmetry would have to take the form

$$x_1 = x_2 = \ldots = x_n = x$$

$$\text{Prob}\{X \leq x\} = P(x) = P(x, \ldots, x) \quad .$$

We do stipulate finite means, variances, and covariances

$$E[x_i] = \int_{-\infty}^{\infty} \ldots \int_{-\infty}^{\infty} X_i dP(X_1, \ldots, X_n) = \mu$$

$$E[(x_i - \mu)(x_j - \mu)] = \int_{-\infty}^{\infty} \ldots \int_{-\infty}^{\infty} (X_i - \mu)(X_j - \mu) dP(X_1, \ldots, X_n)$$

$$= \sigma_{ij}$$

the elements of a positive definite Grammian matrix. A generalization of our earlier theory on diversification can now be stated.

Theorem II: For U(x) a smooth, strictly concave function, the maximum of the symmetric, concave function

$$E\left[U\left(\sum_1^n \lambda_j x_j\right)\right] = \int_{-\infty}^{\infty} \ldots \int_{-\infty}^{\infty} U\left(\sum_1^n \lambda_j X_j\right) dP(X_1, \ldots, X_n)$$

$$= \phi(\lambda_1, \ldots, \lambda_n) \quad ,$$

subject to

$$\lambda_1 + \lambda_2 + \cdots \lambda_n = 1, \quad \lambda_i \geq 0 ,$$

is given by $\phi\left(\frac{1}{n}, \ldots, \frac{1}{n}\right)$. Thus, diversification always pays.

The proof is exactly as before. By symmetry

$$\frac{\partial\phi\left(\frac{1}{n}, \ldots, \frac{1}{n}\right)}{\partial\lambda_1} = \cdots = \frac{\partial\phi\left(\frac{1}{n}, \ldots, \frac{1}{n}\right)}{\partial\lambda_n},$$

the necessary first-order conditions for the constrained maximum.

The Hessian matrix has elements

$$\frac{\partial^2 \phi}{\partial\lambda_i \, \partial\lambda_j} = \int_{-\infty}^{\infty} \cdots \int_{-\infty}^{\infty} X_i X_j U''\left(\sum_1^n \lambda_k X_k\right) dP(X_1, \ldots, X_n)$$

which, being the coefficients of a negative-definite Grammian matrix, do confirm the concavity of ϕ and therefore the maximum value at $\phi(1/n, \ldots, 1/n)$.

If equal diversification is to be mandatory, symmetry or some assumption like it is of course needed. To verify this obvious fact, suppose (x_1, x_2, x_3) to be independently distributed, with x_2 and x_3 having the same distribution $P(x_i)$, but with x_1 having a distribution that is identical with that of $\frac{1}{2} x_2 + \frac{1}{2} x_3$, namely

$$\text{Prob}\{X_1 \leq x_1\} = Q(x_1) = \int_{-\infty}^{\infty} P(2x_1 - 2s)\, dP(2s) .$$

In this case, symmetry tells us that wealth should be divided equally between the investments x_1 and $\frac{1}{2} x_2 + \frac{1}{2} x_3$, which is equivalent to investing in the (x_1, x_2, x_3) in the fractions $\left[\frac{1}{2}, \frac{1}{4}, \frac{1}{4}\right]$. Those who work with two moments, mean and covariance matrix, will find that minimum variance does not come at $(\lambda_1, \ldots, \lambda_n) = (1/n, \ldots, 1/n)$ when $\sigma_{ii} \neq \sigma_{jj}, \sigma_{ij} \neq \sigma_{rs}$.

It is possible, though, to prove that <u>some</u> positive diversification is mandatory under fairly general circumstances. Thus, in $(x_1, \ldots, x_n)$ let each have a common mean and each have finite but nonzero variance. Finally, suppose that one of the variables, say x_1, is independently

4

distributed from the rest. Then an optimal portfolio <u>must</u> involve $\lambda_1^* > 0$, with some positive investment in x_1, as shown in the following.

<u>Theorem III.</u> Let $(x_1, x_2, \ldots, x_n)$ be jointly distributed as $P(x_1)Q(x_2, \ldots, x_n)$ with common mean and finite positive variances

$$E[x_i] = \int_{-\infty}^{\infty} \cdots \int_{-\infty}^{\infty} X_i \, dP(X_1)\,dQ(X_2, \ldots, X_n) = \mu$$

$$0 < E[(x_i - \mu)^2] < \infty$$

and

$$E[U] = \int_{-\infty}^{\infty} \cdots \int_{-\infty}^{\infty} U\left[\sum_1^n \lambda_j X_j\right] dP(X_1)\,dQ(X_2, \ldots, X_n)$$

$$= \theta(\lambda_1, \lambda_2, \ldots, \lambda_n)$$

where, for $U'' < 0$, θ is a strictly concave function.

Then if

$$\theta(\lambda_1^*, \lambda_2^*, \ldots, \lambda_n^*) = \underset{\{\lambda_i\}}{\mathrm{Max}}\ \theta(\lambda_1, \ldots, \lambda_n) \text{ s.t. } \sum_1^n \lambda_j = 1,\ \lambda_i \geq 0,$$

necessarily $\lambda_1^* > 0$ and $\lambda_1^* < 1$.

This will first be proved for $n = 2$, since the general case can be reduced down to that case. Denoting $\partial\theta(\lambda_1, \lambda_2)/\partial\lambda_i$ by $\theta_i(\lambda_1, \lambda_2)$, we need only show the following to be positive

$$\theta_1(0,1) - \theta_2(0,1) = \int_{-\infty}^{\infty} X_1 \, dP(X_1) \int_{-\infty}^{\infty} U'(X_2)\,dP(X_2) - \int_{-\infty}^{\infty} X_2 U'(X_2)\,dP(X_2)$$

$$= E[x_2]E[U'(x_2)] - E[x_2 U'(x_2)]$$

$$= -E\left[\{x_2 - \mu\}\{U'(x_2) - E[U'(x_2)]\}\right] > 0$$

if $U''(x_2) < 0$, since the Pearsonian correlation coefficient between any monotone-decreasing function and its argument is negative.

We reduce $n > 2$ to the $n = 2$ case by defining

$$\lambda_1 x_1 + \sum_2^n \lambda_j x_j = \lambda_1 x_1 + \lambda_{II} x_{II}$$

where

$$x_{II} = \sum_2^n \frac{\lambda_j}{\lambda_{II}} x_j, \quad \sum_2^n \frac{\lambda_j}{\lambda_{II}} = 1, \text{ as definition of } \lambda_{II}.$$

5

To show that the optimal portfolio has the property

$$\theta(\lambda_1^*,\lambda_2^*,\ldots,\lambda_n^*) > \theta(0,\lambda_2,\ldots,\lambda_n) \text{ for } \sum_2^n \lambda_j = 1,$$

it suffices to show that

$$\theta(\lambda_1^*,\lambda_2^*,\ldots,\lambda_n^*) > \theta(0,\lambda_2^{**} \cdots \lambda_n^{**})$$

$$= \underset{\{\lambda_2,\ldots,\lambda_n\}}{\text{Max}} \theta(0,\lambda_2,\ldots,\lambda_n) \text{ for } \sum_2^n \lambda_j = 1$$

But now if we define

$$\theta(\lambda_1,\lambda_{II}) = \theta(\lambda_1,\lambda_{II}\lambda_2^{**},\ldots,\lambda_{II}\lambda_n^{**}),$$

we have an ordinary n=2 case, for which we have shown that $\lambda_1^* > 0$ and $\lambda_{II}^* > 0$.

Having completed the proof of Theorem III, we can enunciate two easy corollaries that apply to risky investments.

<u>Corollary I.</u> If any investment has a mean at least as good as any other investment, and is independently distributed from all other investments, it must enter positively in the optimal portfolio.

<u>Corollary II.</u> If all investments have a common mean and are independently distributed, all must enter positively in the optimum portfolio.

Can one drop the strict independence assumption and still show that every investment, in a group with identical mean, must enter positively in the optimum portfolio? The answer is, in general, no. Only if, so to speak, the component of an investment that is orthogonal to the rest has an attractive mean can we be sure of wanting it. Since a single counterexample suffices, consider joint normal-distributions (x_1,x_2), with common mean and where optimality requires merely the minimization of the variance of $\lambda_1 x_1 + \lambda_2 x_2, \Sigma\Sigma\sigma_{ij}, \lambda_i\lambda_j$, subject to $\Sigma\lambda_j = 1, \lambda_i \geq 0$. If one neglects the non-negativity constraints and minimizes the quadratic expression, one finds for the optimum

6

$$\lambda_1^* = \frac{\sigma_{22} - \sigma_{12}}{(\sigma_{22} - \sigma_{12}) + (\sigma_{11} - \sigma_{21})}$$

If $\sigma_{12} = \sigma_{21} < 0$, λ_1^* is definitely a positive fraction. But, if σ_{12} is sufficiently positive, as in the admissible case $(\sigma_{11}, \sigma_{12}, \sigma_{22}) = (2, 1.1, 1)$, λ_1^* would want to take on an absurd negative value and would, of course, in the <u>feasible</u> optimum be zero, even though x_1's mean is equal to x_2's. Naturally this is a case of positive intercorrelation.

If the assumption of independence is abandoned in favor of positive correlation, we have seen that positive diversification need not be mandatory. However, as Professor Solow pointed out to me, abandoning independence in favor of negative correlation ought to improve the case for diversification. We saw that this was true in the case of negative linear correlation between two investments. It is easy to prove for any number of investments with common mean, among which all the intercorrelations are negative, that total variance is at a minimum when each investment appears with positive weight in the portfolio. (Although there is no limit on the degree to which all investments can be positively intercorrelated, it is impossible for <u>all</u> to be strongly negatively correlated. If A and B are both strongly negatively correlated with C, how can A and B fail to be positively intercorrelated with each other? For 3 variables, the maximum common negative correlation coefficient is $-1/2$; for 4 variables, $-1/3$;....for n variables $-1/(n-1)$.)

The whole point of this paper is to free the analysis from dependence on means, variances, and covariances. What is now needed is the generalization of the concept of negative linear correlation of the Pearsonian type. The natural tool is found in the concept of conditional probability of each variable, say x_i, and the requirement that increasing all or any other variables x_j be postulated to reduce this conditional probability. Thus, define

$$\text{Prob}\{X_i \leq x_i \mid \text{each other } X_j = x_j\} = P(x_i \mid \underline{x}_i),$$

where $\underline{x}_i$ is the vector $(x_1, x_2, \ldots, x_{i-1}, x_{i+1}, \ldots, x_n)$

As always with conditional probabilities

$$P(x_i \mid \underline{x}_i) = P(x_1, x_2, \ldots, x_n) \div \int_{-\infty}^{\infty} dP(x_1, x_2, \ldots, X_i, \ldots, x_n)$$

where the last divisor $Q(\underline{x}_i) = Q(x_1, \ldots, x_{i-1}, x_{i+1}, \ldots, x_n)$ is assumed not to vanish.

The appropriate generalization of pair-wise negative correlation or negative interdependence is the requirement

$$\frac{\partial P(x_i\ \underline{x}\ )}{\partial x_j} < 0 \qquad j \neq i$$

Theorem IV, which I shall not prove, states that where the joint probability distribution has the property of negative interdependence as thus defined, and has a common mean expectation for every investment, $E[x_i] = \mu$, every investment must enter with positive weight in the optimal portfolio of a risk-averter with strictly concave U(x). Buying shares in a coal and in an ice company is a familiar example of such diversification strategy.

Having now shown that quite general conclusions can be rigorously proved for models that are free of the restrictive assumption that only two moments count, I ought to say a few words about how objectionably special the 2-moment theories are (except for textbook illustrations and simple proofs). To do this, I must review critically the conditions under which it is believed the mean-variance theories are valid.

1. If the utility to be maximized is a quadratic function of x, $U(x) = a_0 + a_1 x - a_2 x^2$,

$$\begin{aligned} E[U(x)] &= a_0 + a_1 E[x] + a_2 E[x]^2 - a_2 V(x) \\ &= a_0 + a_1 \mu + a_2 \mu^2 - a_2 \sigma_X^2, \\ &= f(\mu, \sigma) \end{aligned}$$

where μ = mean of x and σ^2 = variance of x

However, as Raiffa, Richter, Hicks, and other writers[3] have noted, the behavior resulting from quadratic utility contradicts familiar empirical patterns. [E.g., the more wealth I begin with the less will I pay for the chance of winning ($0,$K) with probabilities (1/2,1/2) if I have to maximize quadratic utility. Moreover, for large enough x, U begins

8

to decline -- as if having more money available begins to hurt a person.]

Anyone who uses quadratic utility should take care to ascertain which of his results depend critically upon its special (and empirically objectionable) features.

2. A quite different defense of 2-moment models can be given. Suppose we consider investment with less and less dispersion -- e.g., let

$$P(y_1,\ldots,y_n) \text{ have property } E[y_i] = 0 = \int_{-\infty}^{\infty}\cdots\int_{-\infty}^{\infty} Y_i\,dP(Y_1,\ldots,Y_n)$$

and

$$\text{Prob}\{X_1 \leq x_1,\ldots,X_n \leq x_n\} = P\left(\frac{x_1-\mu_1}{\alpha},\ldots,\frac{x_n-\mu_n}{\alpha}\right)_n$$

Then in the limit as $\alpha \to 0$, only the first 2 moments of $\Sigma_1 \lambda_j x_j$ will turn out to count in $E[U(x)] = f(\mu,\sigma,\ldots)$. In the extreme limit, even the second moment will count for less and less: for α small enough the mean money outcome will dominate in decision making. Similarly, when α is small, but not limitingly small, the third moment of skewness will still count along with the mean and variance; then the third-degree polynomial form of U(x) (its Taylor's expansion up to that point) will count. As Dr. M. Richter has shown in the cited paper, an nth degree polynomial for U(x) implies, and is implied, by the condition that only the first n statistical moments count.

3. If each of the constituent elements of $(x_1,\ldots,x_n)$ is normally distributed, then so will be $z=\Sigma\lambda_j x_j$ and then it will be the case that only the mean and variance of z matter for E[U(z)]. However, with limited liability, no x_j can become negative as is required by the normal distribution. So some element of approximation would seem to be involved. Is the element of approximation, or rather of lack of approximation, ignorable? No, would seem to be the answer if there is some minimum of subsistence of z or x at which marginal utility becomes infinite. Thus, consider U = logx, the Bernoulli form of logarithmic utility

$$E[U(x)] = \int_{-\infty}^{\infty} \log X\,dN\left(\frac{X-\mu}{\sigma}\right) = \lim_{a\to b}\int_{a}^{\infty} \log X\,dN\left(\frac{X-\mu}{\sigma}\right)$$

$$= -\infty \text{ for } b \leq 0,$$

9

where N(t) stands for the normal distribution with zero mean and unit variance.

Suppose that each constituent x_i takes on only non-negative values with variances all bounded by $M < \infty$. The central-limit theorem will still apply, so that $\Sigma_1^n x_j$ or $z_n = \Sigma\lambda_j x_j$, with certain weak restrictions on the spread of the λ's around 1/n, will approach a Gaussian distribution. Thus, let z_n have the distribution $P_n(z_n)$, with

$$\lim_{n\to\infty} P_n(z_n) = N\left(\frac{z_n - a_n}{b_n}\right) .$$

Knowing how treacherous are double limits, we dare not infer

$$\lim_{n\to\infty} E[\log z_n] = \int_{-\infty}^{\infty} \log Z \lim_{n\to\infty} P_n(Z_n) .$$

Actually, as $n\to\infty$ and each investment has its $\lambda_j > a/n$, each $\lambda_j x_j$ does have a smaller and smaller dispersion so that we might switch from reliance on the central limit theorem of normality to the 2-moment Taylor-expansion justification given in paragraph 2 above. The law of large numbers, which is even more basic than the central limit theorem involving normality, assures us that z_n becomes more and more tightly bunched around some positive value and this fact will make the quadratic approximation applicable in the limit.

4. A final defense of the mean-variance formulation, in which $E[U(x)]$ is replaced by $f(\mu,\sigma)$, comes when x belongs to a 2-parameter probability distribution $P(x; \theta_1, \theta_2)$.

Then

$$E[U(x)] = \int_{-\infty}^{\infty} U(X)dP(X;\theta_1,\theta_2) = g(\theta_1,\theta_2) ,$$

$$\mu = \int_{-\infty}^{\infty} XdP(X;\theta_1,\theta_2) = h_1(\theta_1,\theta_2) ,$$

$$\sigma = \left[\int_{-\infty}^{\infty} (X-\mu)^2 dP(X;\theta_1,\theta_2)\right]^{1/2} = h_2(\theta_1,\theta_2) .$$

Then, provided the Jacobian $\partial(\mu,\sigma)/\partial(\theta_1,\theta_2) \neq 0$, each θ_i can be

10

solved for as a function $\theta_i(\mu,\sigma)$, with

$$f(\mu,\sigma) = g[\theta_1(\mu,\sigma),\ \theta_2(\mu,\sigma)]\ .$$

So far so good, although even here one has to take care to verify that $P(x;\theta_1,\theta_2)$ has the properties needed to give $f(\mu,\sigma)$ the quasi-concavity properties used by the practitioners of the mean-variance techniques. And furthermore one cannot draw up the $f(\mu,\sigma)$ indifference contours once and for all from knowledge of the decision-makers risk preferences, but instead must redraw them for each new probability distribution $P(x;\theta_1,\theta_2)$ upon which the f functional depends.

But waiving these last matters, we must point out that the Markowitz efficient-portfolio frontier need not work to screen out (or rather in!) optimal portfolios. For even when each constituent x_i belongs to a common 2-parameter family, the resulting $z = \Sigma\lambda_j x_j$ will not belong to that family or to any 2-parameter family that is independent of the λ weightings. It suffices to show this in the case of statistical independencies. Thus, define the rectangular distribution

$$\begin{aligned} R(x;a,b) &= \frac{x-a}{b-a}\ , && a \leq x \leq b \\ &= 0\ , && x \leq a \\ &= 1\ , && x \geq b \end{aligned}$$

and

$$P(x_1,\ldots,x_n) = \prod_{i=1}^{n} R(x_i;a_i,b_i)\ .$$

Then, of course, $\Sigma\lambda_j x_j = z$ does not satisfy an R distribution but rather a 3n parameter distribution $P(z;\lambda_1,\ldots,\lambda_n;a_1,\ldots,a_n,b_1,\ldots,b_n)$. Let $(\lambda_1^{\dagger},\ldots,\lambda_n^{\dagger})$ be <u>a</u> point on the Markowitz efficiency frontier, with minimum variance $\Sigma\lambda_j^2\sigma_j^2$ subject to $\Sigma\lambda_j\mu_j = \mu$, $\Sigma\lambda_j = 1$. Then it need not be the case that the optimum $(\lambda_1^*,\ldots,\lambda_n^*)$ that maximizes $E[U(z)]$ will belong to the efficiency set $(\lambda_1^{\dagger},\ldots,\lambda_n^{\dagger})$! I do not recall this fact's being mentioned by those who speak of 2-parameter-family justifications of mean-variance analysis. Some quite different argument, such as that $n \to \infty$ and quadratic approximation to U then becomes increasingly good, will be needed to bring back the Markowitz frontier into more general applicability.

I do not wish to end on a nihilistic note. My objections are those of a purist, and my demonstrations in this paper have shown that even a purist can develop diversification theorems of great generality. But in practice, where crude approximations may be better than none, the 2-moment models may be found to have pragmatic usefulness.

FOOTNOTES

1. H. Makower and J. Marschak, "Assets, Prices, and Monetary Theory," Economica N.S. (Vol. V, 1938), pp. 261-88. Harry M. Markowitz, "Portfolio Selection," The Journal of Finance (Vol. VII, 1952), pp. 77-91 and Portfolio Selection: Efficient Diversification of Investments (New York: John Wiley & Sons, 1959); James Tobin, "Liquidity Preference as Behavior Towards Risk," Review of Economic Studies (Vol. XXV, 1958), pp. 65-86. The path-breaking article, E. D. Domar and R. A. Musgrave, "Proportional Income Tax and Risk-taking," Quarterly Journal of Economics (Vol. LVII, 1944), pp. 389-422 replaces variance by risk of loss (mean absolute loss) as dispersion parameters to be avoided. (This paper is reprinted in A.E.A. Selected Readings in Fiscal Policy and Taxation (Homewood, Ill.: Irwin, 1959).

2. P. A. Samuelson, "A Fallacy in the Interpretation of Pareto's Law of Alleged Constancy of Income Distribution," Essays in Honor of Marco Fanno, ed., Tullio Bagiotti, (Padua, Cedam-Casa Editrice Dott. Antonio Milani, 1966), pp. 580-584.

3. Howard Raiffa, unpublished Harvard Business School memos; Marcel K. Richter, "Cardinal Utility, Portfolio Selection and Taxation," Review of Economic Studies (Vol. XXVII, 1959), pp. 152-66; E. C. Brown, "Mr. Kaldor on Taxation and Risk Bearing," Review of Economic Studies (Vol. XXV, 1957), pp. 49-52; J. R. Hicks, "Liquidity," Economic Journal (Vol. LXXII, 1962), pp. 787-802, depicts a rediscovery of some of the Markowitz theory. Also see John Lintner, "Valuation of Risk Assets," Review of Economics and Statistics (Vol. XLVII, 1965), pp. 13-37, and "Optimum Dividends and Uncertainty," Quarterly Journal of Economics (Vol. LXXVIII, 1964), pp. 49-95 and unpublished appendix.

3. EFFECTS OF TAXES ON RISK TAKING

Reprinted from *Quarterly Journal of Economics* 83, 263-283 (1967).

THE EFFECTS OF INCOME, WEALTH, AND CAPITAL GAINS TAXATION ON RISK-TAKING *

J. E. Stiglitz

I. Introduction

In their pioneering article on the effects of taxation on risk-taking, Domar and Musgrave [1] showed that although the imposition of an income tax with full loss-offset might lead to less private risk-taking, total risk-taking would in fact increase.[2] If there were no loss-offset, they noted that the amount of risk-taking could either increase or decrease, although the presumption was for the latter. Their analysis rested on individual indifferences curves between risk and mean. The limitations of this kind of analysis are well known.[3]

The purpose of this note is to investigate the effects on the demand for risky assets of income, capital gains, and wealth taxation, with and without loss-offsets, using a general expected utility maximization model, and to examine the welfare implications of these alternative taxes.

* The research described in this paper was carried out under a grant from the National Science Foundation. After this paper was completed, a paper by J. Mossin, "Taxation and Risk-Taking: An Expected Utility Approach." *Economica,* XXXV (Feb. 1968), containing some of the results of Sections IV and VI was published. I am indebted to D. Cass and A. Klevorick for their helpful comments.

1. E. Domar and R. Musgrave, "Proportional Income Taxation and Risk-Taking," *this Journal,* LVI (May 1944), 388–422.

2. For further discussions, see C. Hall, Jr., *Fiscal Policy for Stable Growth* (New York: Holt, Rinehart and Winston, 1965); R. Musgrave, *The Theory of Public Finance* (New York: McGraw Hill, 1959); and M. Richter, "Cardinal Utility, Portfolio Selection, and Taxation," *Review of Economic Studies,* XXVII (June 1960), 152–66.

3. If the measure of risk is variance, then it requires a quadratic utility function, or that the returns from the asset be described by a two-parameter probability distribution. The quadratic utility function has some very peculiar properties, e.g., marginal utility becomes negative at finite incomes and the demand for risky assets decreases with wealth. See J. R. Hicks, "Liquidity," *Economic Journal,* LXXII (Dec. 1962), 787–802, and K. J. Arrow, *Some Aspects of the Theory of Risk Bearing* (Helsinki: Yrjö Johnssonin Säätiö, 1965.)

II. The Basic Model and Some Behavioral Hypotheses

An individual has initial wealth W_o. There are two assets in which he can invest his wealth. The risky asset yields a random return per dollar invested of $e(\theta)$ where θ has a probability distribution $F(\theta)$. The safe asset yields a sure rate of return per dollar invested of r. If we adopt the convention that $e(\theta)$ is nondecreasing with θ, then we can depict the pattern of returns for the two assets as in Figure Ia.[4] It is assumed that $e(\theta)$ does not depend on the amount invested in the risky asset, and that r is nonnegative. The individual wishes to maximize the expected utility of his wealth at the end of the period. If he invests $(1-a)$ of his wealth in the safe asset and a in the risky asset, then his wealth at the end of the period is

(1) $W=W_o(1+ae+(1-a)r)$.[5]

If we denote by E the expectations operator, then he wishes to maximize

(2) $E\{U(W)\}=\int\{U(W_o(1+ae(\theta)+(1-a)r))\}dF(\theta)$.

If $U''<0$, in the absence of taxes a necessary and sufficient condition for utility maximization is[6]

(3) $EU'(e-r)=0$.

In the literature on uncertainty, two measures of risk aversion have been extensively used: absolute risk aversion $-U''(W)/U'(W)=A(W)$ and relative risk aversion $-U''(W)W/U'(W)=R(W)$. Pratt[7] has shown that an individual is indifferent between an uncertain wealth with mean $E(W)$ and arbitrarily small variance σ^2; and a certain wealth of $EW-A(E(W))\frac{\sigma^2}{2}$ (which equals

4. The "safe" asset may also yield a random return and the analysis is unaffected, provided only that the safe asset is unambiguously safer. This means that if under some allocation of his wealth, the individual's income is greater in state θ than in state θ', then under any alternative allocation his income is not less in state θ than in state θ'. The pattern of returns can be depicted as in Figure Ib.

5. Throughout, we assume that his only source of wealth in the next period is from purchases of assets in this period.

6. This assumes that the individual can borrow as well as lend at r, and sell short as well as buy securities, i.e., a is not constrained. If a is constrained between $0\leqslant a\leqslant 1$, then (3) holds only for interior solutions; otherwise $a(1-a)EU'(e-r)=0$; $(1-a)EU'(e-r)\leqslant 0$; and $aEU'(e-r)\geqslant 0$. For the remainder of the analysis we shall assume that an interior maximum exists.

7. J. W. Pratt, "Risk Aversion in the Small and in the Large," *Econometrica,* Vol. 32 (Jan.-April 1964), pp. 122–36.

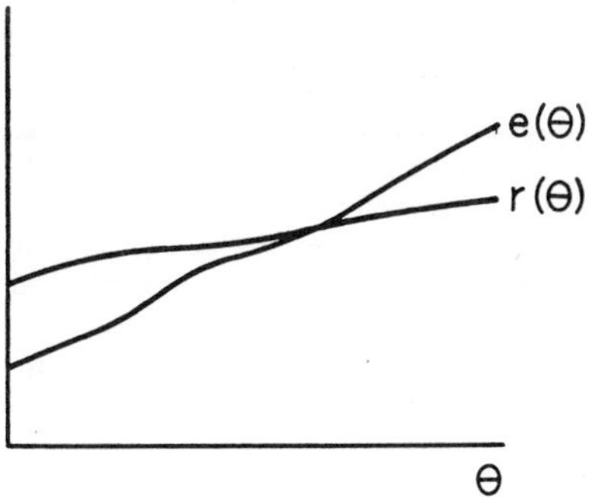

FIGURE IB

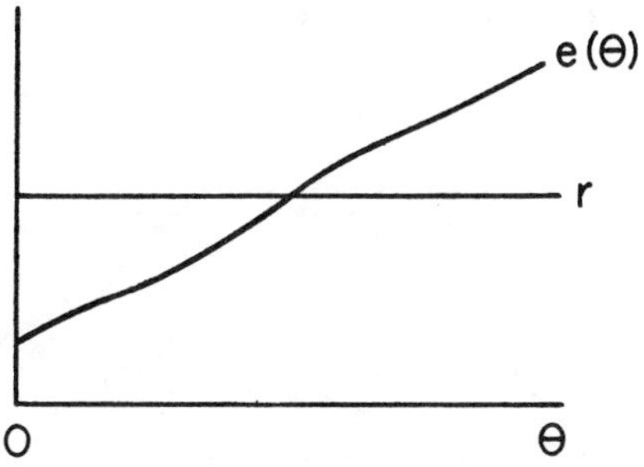

FIGURE IA
Patterns of Return of Safe and Risky Asset

$EW\left(1-R(E(W))\frac{\sigma'^2}{2}\right)$; where $\sigma'=\sigma/E(W)$, the coefficient of variation).

Arrow [8] has argued that:

A. Absolute risk aversion decreases as wealth increases.

B. Relative risk aversion increases as wealth increases.

These assumptions about the utility function are equivalent to the following assumptions about how the allocation to the risky asset changes as wealth increases:

A.′ As wealth increases, more of the risky asset is purchased, i.e., the risky asset is superior.

8. K. Arrow, *op. cit.*

B.′ As wealth increases, the *proportion* of one's wealth in the risky asset decreases.

It is easy to show that A and A′ and B and B′ are equivalent: (3) defines an implicit equation for a in terms of W_o. Using the implicit function theorem and integrating by parts, the result is immediate. This result can also be seen graphically as follows. We consider the special case where there are only two states of the world, θ_1 with probability p_1 and θ_2 with probability p_2. If the individual purchases only the safe security, his wealth at the end of the period is represented by the point S in Figure II, with

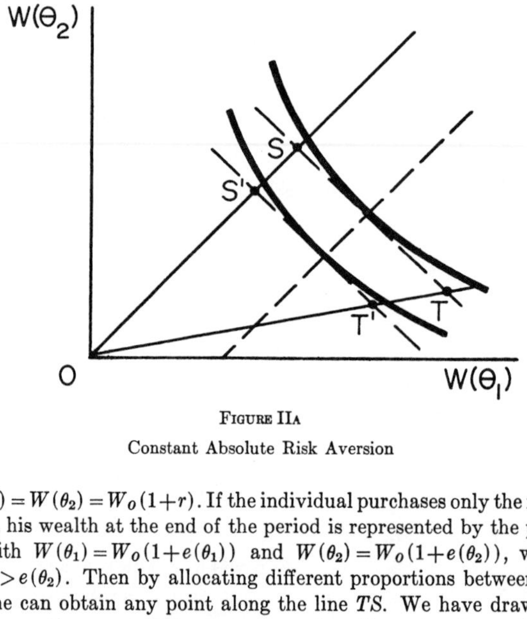

FIGURE IIA
Constant Absolute Risk Aversion

$W(\theta_1)=W(\theta_2)=W_O(1+r)$. If the individual purchases only the risky asset, his wealth at the end of the period is represented by the point T, with $W(\theta_1)=W_O(1+e(\theta_1))$ and $W(\theta_2)=W_O(1+e(\theta_2))$, where $e(\theta_1)>e(\theta_2)$. Then by allocating different proportions between the two he can obtain any point along the line TS. We have drawn in the same diagram indifference curves for different values of EU, where

(4) $$EU=p_1U(W(\theta_1))+p_2U(W(\theta_2)).$$

As W_O increases, the budget constraint giving different possible. values of $W(\theta_1)$ and $W(\theta_2)$ moves outward, but with unchanged slope. The individual maximizes expected utility at the point of tangency, i.e., where the marginal rate of substitution equals the slope of the budget constraint:

$$(5)\qquad \frac{p_1U'(W(\theta_1))}{p_2U'(W(\theta_2))}=\frac{e(\theta_2)-r}{e(\theta_1)-r}\,.$$

This means that the slope of an Engel curve is given by

$$(6)\qquad \frac{dW(\theta_2)}{dW(\theta_1)}=\frac{[-U''(W(\theta_1))/U'(W(\theta_1))]}{[-U''(W(\theta_2))/U'(W(\theta_2))]}=\frac{A(W(\theta_1))}{A(W(\theta_2))}\,.$$

The demand for the risky asset can be written as

$$(7)\qquad aW_O=\frac{W(\theta_1)-W_O(1+r)}{e(\theta_1)-\mathrm{r}}=\frac{W(\theta_2)-W_O(1+r)}{e(\theta_2)-\mathrm{r}}\,.$$

If the risky asset is neither superior nor inferior, then it is easy to see that the Engel curves must have a slope of 45°; since with increments in W_O, he purchases only the safe asset, the increments in $W(\theta_1)$ and $W(\theta_2)$ must be equal:

$$\frac{dW(\theta_1)}{dW_O}=(1+r)=\frac{dW(\theta_2)}{dW_O}\,.$$

From equation (5), this means that $\dfrac{A(W(\theta_1))}{A(W(\theta_2))}=1$ or absolute risk aversion, $-U''/U'$, is constant. If the risky asset is superior, the Engel curve must have a slope everywhere less than unity; since $W(\theta_2)<W(\theta_1)$, this means that absolute risk aversion must be declining.

If we divide all terms in equation (7) by W_O, we immediately see that if a is to remain constant, then the ratio of $W(\theta_1)$ to W_O must remain constant and the ratio of $W(\theta_2)$ to W_O must remain constant, i.e., $W(\theta_1)$ must be proportional to $W(\theta_2)$. All Engel curves must be straight lines through the origin, and have unitary elasticity (see Figure IIb), so from equation (6)

$$\frac{dln\,W(\theta_2)}{dln\,W(\theta_1)}=\frac{-[U''(W(\theta_1))W(\theta_1)/U'(W(\theta_1))]}{-[U''(W(\theta_2))W(\theta_2)/U'(W(\theta_2))]}=1\,,$$

which implies constant relative risk aversion. If a is to decline, as W_O increases, the Engel curves must bend upward (towards the "all safe" ray OS). The elasticity must be greater than unity, which implies increasing relative risk aversion. Conversely, if a is to increase as W_O increases, relative risk aversion must be decreasing.

The validity of these testable hypotheses can only be determined empirically. Certainly, the hypothesis that risky assets are not inferior seems reasonable. The second hypothesis is somewhat

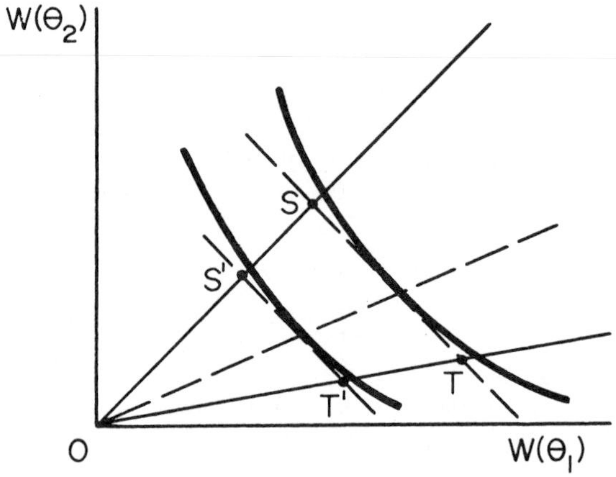

FIGURE IIB
Constant Relative Risk Aversion

more questionable. Arrow,[9] in addition to suggesting several theoretical reasons for its validity, has argued that the empirical evidence from studies on the demand for money also support it. Stiglitz [1] has raised some questions concerning these arguments, and the empirical evidence for it seems, at best, rather weak and contradictory. For instance, certain cross-section estate data leave some doubt whether individuals do allocate a larger percentage of the portfolio to safe assets as their wealth increases.[2]

In the next two sections we will show that if these hypotheses

9. K. Arrow, *op. cit.*
1. J. Stiglitz, "Review of *Some Aspects of the Theory of Risk Bearing,*" *Econometrica* (forthcoming).
2. R. Lampman, *Share of Top Wealth-Holders in National Wealth, 1922–56,* National Bureau of Economic Research (Princeton: Princeton University Press, 1962), provides the following data on average portfolio allocation to bonds and cash for different estate sizes (males):

	Age		
Size of Estate	30–40	55–60	75–80
70– 80,000	12%	20.2%	26.2%
100–120,000	11.5%	19.1%	23.5%
200–300,000	11.4%	15.3%	20.7%

There is some difficulty in interpreting the data, however, since the investment opportunities for rich people may be different from those for poor. Data of J. Spraos ("An Engel-Type Curve for Cash," *Manchester School,* XXV (May 1957), 183–89), for the United Kingdom, and D. S. Projector and G. S. Weiss (*Survey of Financial Characteristics of Consumers* (Washington: Federal Reserve Board, 1966)), show similar patterns.

are correct, then we can make some unambiguous statements about the effects of taxation on risk-taking, independent of the probability distribution of returns for the risky asset, but if these hypotheses are not correct, many of the conclusions of the original Musgrave-Domar analysis may no longer be valid. Our measure of risk-taking is the individual's demand for risky assets; this is simply measured by the fraction of his portfolio devoted to the risky asset. This corresponds to the Domar-Musgrave concept of total (or social) risk-taking. In contrast, there is no obvious corresponding measure of "private risk-taking." One natural measure, which we shall use, is the (subjectively perceived) standard deviation of wealth. In the absence of taxes, this is simply equal to $W_0 a\sigma$, where σ is the standard deviation of the risky asset.

III. Wealth Tax

We begin the discussion with an investigation of the effects of the wealth tax, since this is the simplest case to analyze. A proportional wealth tax at the rate t means that wealth at the end of the period is given by

$$W = W_0(1+(1-a)r+ae)(1-t). \tag{8}$$

It should be immediately apparent that changing the tax rate is just equivalent to changing W_0 in terms of the effect on risk-taking. Hence we immediately obtain:

Proposition 1(a). A proportional wealth tax increases, leaves unchanged, or decreases the demand for risky assets as the individual has increasing, constant, or decreasing relative risk aversion.

After tax private risk-taking, P, is given by

$$P = [E(W-E(W))^2]^{1/2} = [E(W_0 a(1-t))(e-Ee))^2]^{1/2} = W_0 a(1-t)\sigma$$

where $\sigma^2 = E(e-Ee)^2$. Whether private risk-taking increases or decreases from an increase in the tax rate depends on whether

$$-\frac{d\,ln\,P}{d\,ln(1-t)} = -\frac{d\,ln\,a}{d\,ln(1-t)} - 1 \gtreqless 0.$$

But implicit differentiation of the first order condition for expected utility maximization yields

$$\frac{da}{dt} = \frac{E\{-U''W(e-r)/(1-t)\}}{E\{-U''(e-r)^2 W_0\}}$$

$$=\frac{E\{-U''W_0a(e-r)^2\}-EU''W_0(1+r)(e-r)}{E\{-U''(e-r)^2W_0(1-t)\}}$$

so

$$-\frac{d\,ln\,a}{d\,ln(1-t)}=1-\frac{W_0(1+r)EU''(e-r)}{aE\{-U''(e-r)^2W_0\}}.$$

The denominator of the second term on the right-hand side is always positive. Hence whether $-\dfrac{d\,ln\,a}{d\,ln(1-t)}$ is greater or less than unity depends on the sign of

$$\begin{aligned}-EU''(e-r)&=E(-U''/U')U'(e-r)=E(A(W(\theta))\\&-A(W(\theta^*))U'(e-r)+A(W(\theta^*))EU'(e-r)\end{aligned}$$

where θ^* is defined by $e(\theta^*)=r$. The last term equals zero (since utility maximization requires $EU'(e-r)=0$). If absolute risk aversion increases with W, in those states of nature where

$$e(\theta)>e(\theta^*)=r,\ A(W(\theta))>A(W(\theta^*)),\ \text{so-}EU''(e-r)>0.$$

Similarly if absolute risk aversion decreases with W,-$EU''(e-r)<0$. Hence, we have shown:

Proposition 1(b). A proportional wealth tax increases, leaves unchanged, or decreases private risk-taking as the individual has increasing, constant, or decreasing absolute risk aversion.

IV. Income Taxation

The case of income taxation with full loss-offset is only slightly more difficult to analyze. We can write after-tax income, Y, as

(9) $$Y=W_0(1-t)\ [(1-a)r+ae],$$

and his wealth after tax is

$$W=W_0[1+(1-t)\ (ae+(1-a)r)].$$

We begin the analysis with the case of only two states of nature. In Figure IIIa we have drawn the before-tax budget constraint ST. Income is measured by the distance from, say, T to W_0 or S to W_0, so an income tax at the rate t reduces the returns from investing in only the safe asset or the risky asset to S' and T', respectively.

The after-tax budget constraint is the line joining T' to S'. It is clearly parallel to ST. Note, however, that a is not constant along a ray through the origin, but along a ray through the point W_0. Thus, it is immediately apparent that in this simple example if individuals have constant or increasing relative risk aversion, or increasing absolute risk aversion, risk-taking will increase. But if

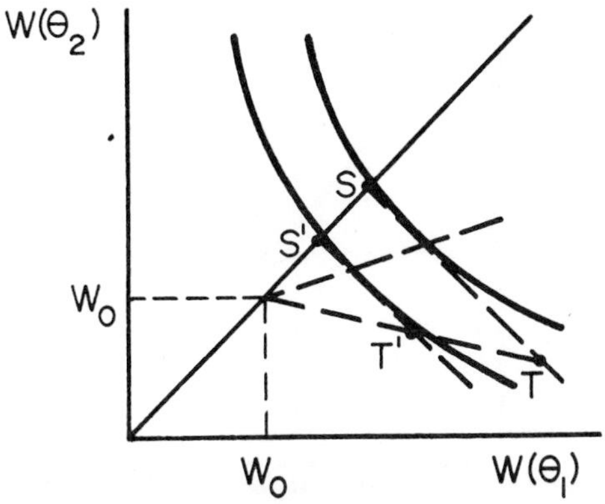

FIGURE IIIA

Income Tax: Demand for Risky Asset Unchanged

there is decreasing relative risk aversion, just the opposite may occur.

Before taking up the more general case, we should note the special case where $r=0$, i.e., the safe asset is money and has a zero rate of return. Then the after-tax budget constraint coincides with the before-tax budget constraint but $T'S$ is shorter than TS and the values of $W(\theta_1)$ and $W(\theta_2)$ are identical to their before-tax values: all the tax does is to induce individuals to hold more risky assets. (See Figure IIIb.)

In the more general case, only slightly stronger conditions are required to guarantee that a proportional tax will increase risk-taking. The condition for utility maximization is simply

$$EU'(e-r)=0.$$

It is of the same form as the no-tax condition, since both the risky and safe asset are taxed proportionately.

We wish to know, how does a change with t:

$$\frac{da}{dt}=\frac{-E\dfrac{U''Y}{1-t}(e-r)}{-EU''(e-r)^2(1-t)W_0}\,. \tag{10}$$

After-tax private risk-taking is again measured by

$$P=W_0a(1-t)\sigma$$

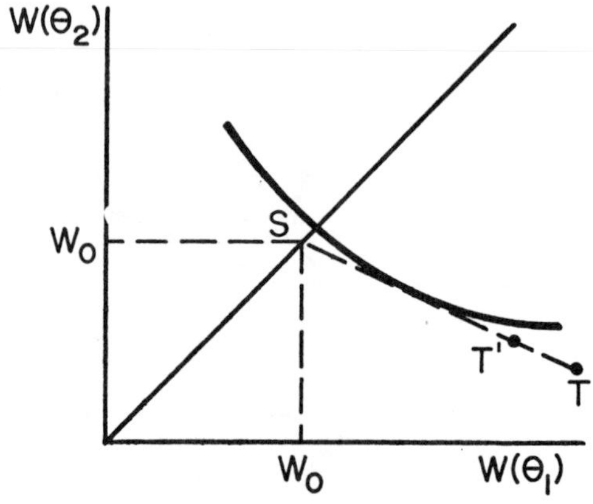

FIGURE IIIB
Income Tax: $r=0$.

so whether private risk-taking increases or decreases depends on whether

$$-\frac{d\,ln\,a}{d\,ln(1-t)} \gtrless 1.$$

But from (10), we immediately see that

$$(10)' \qquad -\frac{d\,ln\,a}{d\,ln(1-t)} = 1 - \frac{W_o r E U''(e-r)}{-aEU''(e-r)^2 W_o}.$$

The denominator of (10) is always positive, so whether increasing taxes leads to more or less investment in the risky assets depends on the numerator of equation (10). If we set $r=0$, from (10)′ it is immediately apparent that a increases, and in proportion to the change in $(1-t)$. It is also clear that if there is increasing absolute risk aversion, as in the quadratic utility function, the second term is unambiguously positive and the percentage increase in a from a percentage decrease in $(1-t)$ is greater than unity.[3]

If in equation (10), we recall that $Y=W-W_o$, we obtain the result that the numerator of (10) is equal to

$$E\left[\left(\frac{-U''W}{U'}\right)-\left(\frac{-U''W_o}{U'}\right)\right]\frac{U'(e-r)}{1-t}=E\{R(W(\theta))-R(W(\theta^*))\}$$

3. See above, pp. 269–70.

$$-W_O(A(W(\theta))-A(W(\theta^*)))\}\frac{U'(e-r)}{1-t}+\{R(W(\theta^*))$$
$$-W_OA(W(\theta^*))\}E\frac{U'(e-r)}{1-t}$$

where θ^* is defined as above. If there is constant or increasing relative risk aversion and constant or decreasing absolute risk aversion, the first term above is unambiguously positive, and by equation (9) the second term is identically zero. Thus, a is increased. But from (10)′, if there is decreasing absolute risk aversion, a is increased by a smaller percentage than the percentage decrease in $1-t$.

If there is decreasing relative and absolute risk aversion, investment in the risky asset may be unchanged or decreased as the result of the imposition of an income tax. To see this more clearly, consider the following utility function, and assume that the probability of $e<0$ is zero:

$$U(W)=\int_{W_O}^{W}A(W-W_O)^{\alpha}dW+U(W_O),\ A>0,\ \alpha<0.$$

Then, for $W>W_O$, marginal utility is just

$$U'=A(W-W_O)^{\alpha}>0$$

so absolute risk aversion is

$$-\frac{U''}{U'}=-\frac{d\ln U'}{dW}=\frac{-\alpha}{W-W_O}>0$$

and is decreasing, since

$$\frac{d\dfrac{-U''}{U'}}{dW}=\frac{\alpha}{(W-W_O)^2}<0$$

while relative risk aversion is

$$\frac{-U''W}{U'}=\frac{-\alpha W}{W-W_O}>0$$

and is decreasing,

$$\frac{d\dfrac{-U''W}{U'}}{dW}=\frac{\alpha W_O}{(W-W_O)^2}<0.$$

But since $\dfrac{-U''}{U'}(W-W_O)=\dfrac{-U''Y}{U'}=-\alpha$, a constant, it is clear that the numerator of equation (10) is zero. Thus, for a perfectly well-behaved utility function, with diminishing marginal utility and decreasing absolute risk aversion, the imposition of the income tax leaves the demand for risky assets unaffected. Similarly, we can construct examples where it decreases the demand for risky assets. We can summarize the results in the following proposition and table:

Proposition 2. Increased income taxes with full loss-offset lead to increased demand for risky assets if

(a) *the return to the safe asset is zero, or*

(b) *absolute risk aversion is constant or increasing, or*

(c) *absolute risk aversion is decreasing and relative risk aversion increasing or constant.*

If none of the above three conditions is satisfied, it is possible for increased taxes to reduce the demand for risky assets.

If $r>0$, *private risk-taking increases, is unchanged, or decreases as absolute risk aversion is increasing, constant, or decreasing. If* $r=0$, *private risk-taking is unaffected.*

TABLE I

EFFECTS OF INCOME TAX ON RISK-TAKING: $\frac{-d\,ln a}{d\,ln(1-t)}$

Relative Risk Aversion	Absolute Risk Aversion: Decreasing	Constant *	Increasing *
Decreasing	<1 but may be greater or less than zero.		
Constant	$0<\frac{-d\,ln a}{d\,ln(1-t)}<1$		
Increasing	$0<\frac{-d\,ln a}{d\,ln(1-t)}<1$	1	>1

* It is impossible to have constant or increasing absolute risk with nonincreasing relative risk aversion.

V. Special Treatment of Capital Gains

Our present tax laws however, do not treat all risks alike; indeed, one of the main justifications for the special capital gains provisions is that they encourage risk-taking. This, however, may not always be the case. Take, for instance, the extreme case of a tax only on the safe (or the relatively safe) asset, with no tax on the risky asset. It is easy to show that the demand for the risky asset increases or decreases as

$$(11) \qquad -W_0EU''r(1-a)(e-r(1-t))+EU'r \gtrless 0$$

and by arguments exactly analogous to those presented above, we can show:

Proposition 3(a). A tax on the safe asset alone will increase the demand for the risky asset if there is constant or increasing absolute risk aversion.

Rearranging terms in (11), we obtain the result that the sign of da/dt depends on the sign of

$$rE\left\{\frac{U''}{U'}W+1\right\}U'-W_0rE(U'')(1+e).$$

Under limited liability, $e>-1$, so the second term is unambiguously positive, while the first term is positive if relative risk aversion is always less than or equal to unity. Hence, we have shown

Proposition 3(b). A tax on the safe asset alone will increase the demand for risky assets if relative risk aversion is less than or equal to unity.

If there is decreasing absolute risk aversion and relative risk aversion is greater than unity it is surely possible for the tax on the safe asset to lead to less rather than more risk-taking. In Section VII, we shall compare this tax explicitly with a proportional income tax.

VI. No Loss Offset

We now examine the effects of no loss-offset provision in an otherwise proportional income tax. After-tax income is given by

$$Y = \begin{cases} W_0[1+ae+(1-a)r\,(1-t)] \text{ if } e<0. \\ W_0[1+(ae+(1-a)r)(1-t)] \text{ if } e\geqslant 0. \end{cases}$$

Diagrammatically, the after-tax budget constraint looks as depicted in Figure IVa, if r is greater than zero, or as in Figure IVb, if $r=0$. Observe that the value of $W(\theta_2)$ if the individual purchases only risky assets is unchanged (i.e., T and T' lie on the same horizontal line). We can see that there is an "income effect" and a "substitution effect," and in the cases discussed in the previous section, these will be of opposite signs, so that the net effect is ambiguous. It is easy to see, however:

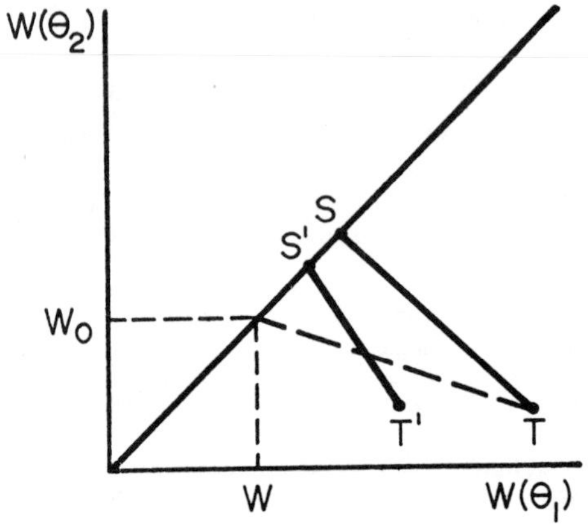

FIGURE IVA
Budget Constraints: No Loss-Offset

Proposition 4(a). For sufficiently large tax rates, the demand for risky assets is reduced.

To see this, all we have to observe is that for tax rates near 100 per cent, almost the entire portfolio is allocated to the safe asset since, as the tax rate approaches 100 per cent, the maximum return on the risky asset approaches zero and the expected return becomes negative. Since the indifference curves are convex, the demand curves for the different assets are continuous functions of the tax rate.

Moreover, it is easy to show that there will always be less risk-taking than with full loss-offset. Consider first the effects of partial offsetting, where we are allowed to deduct a portion of losses from the risky asset from other income. Income, when $e<0$, can be written

$$[ae(1-v)+(1-a)r(1-t)]W_o, v \leqslant t.$$

No loss-offset is the extreme case where $v=0$. Without loss of generality, we adopt the following convention about θ : θ is defined over the interval $[0,1]$, $e(0)=\min e$; $\frac{de(\theta)}{d\theta}>0$; $e(\theta')=0$.

Now, what happens as v is reduced? This will depend on the sign of

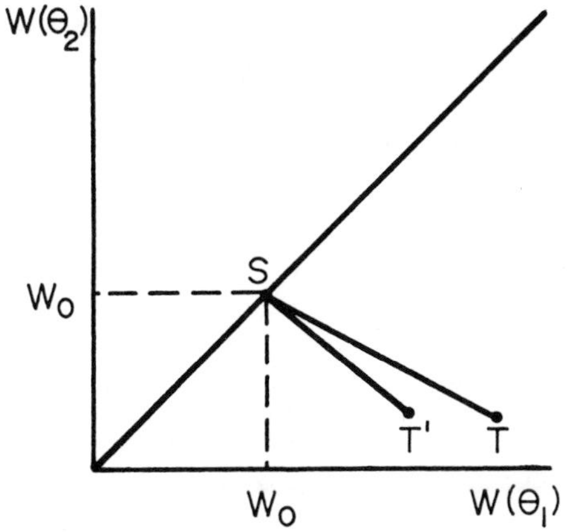

FIGURE IVB
Budget Constraints: No Loss-Offset with $r=0$.

$$\int_0^{\theta'}[-U''\{e(1-v)-r(1-t)\}eaW_0-U'e]dF(\theta)$$

which is unambiguously positive. Hence, we have:

Proposition 4(b). The demand for the risky asset is always less with no loss-offset or partial loss-offset than with full loss-offset.

We shall now attempt to find some more precise conditions under which risk-taking unambiguously increases or decreases. For simplicity we limit ourselves to the case where $r=0$. We can write the first order conditions for expected utility maximization as

$$\int_{\theta'}^{1}U'e(1-t)dF(\theta)+\int_0^{\theta'}U'edF(\theta)=0 \tag{12}$$

so the sign of da/dt is that of

$$\int_{\theta'}^{1}[-U''ae^2(1-t)W_0-U'e]dF(\theta) = \int_{\theta'}^{1}\left[\frac{-U''(W-W_0)}{U'}-1\right]U'edF(\theta): \tag{13}$$

We thus obtain:

Proposition 4(c). If $r=0$ the imposition of an income tax with no loss-offset decreases the demand for the risky asset if relative risk aversion is less than or equal to unity.[4]

4. The Bernoulli utility function, $U=lnW$, has constant relative risk

One more condition will now be derived. If we integrate equation (13) by parts, letting $H(\theta)=\int_{\theta'}^{\theta} U' e dF(\theta)$, and $a=\frac{Y}{W}$, we obtain

$$(14) \qquad -\int_{\theta'}^{1} H(\theta)\left\{\frac{d\frac{-U''W}{U'}}{dW}a+\frac{-U''W}{U'}\frac{da}{dW}\right\}\frac{dW}{d\theta}$$

$$+\left\{\left(\frac{-U''(W-W_0)}{U'}\right)_{\theta=1}-1\right\}H(1).$$

If there is increasing relative risk aversion the integral expression is negative. Assume that the maximum rate of return is m, so that $W \leqslant W_0(m+1)$; since $\left(-\frac{U''W}{U'}\right)\frac{W-W_0}{W} \leqslant \frac{-U''W}{U'} m/m+1$, for the second term to be negative, all that we require is that relative risk aversion be less than $m+1/m$ at its maximum. If m is 50 per cent, then relative risk aversion need only be less than 3. Thus we have:

Proposition 4(d). If $r=0$, the imposition of an income tax with no loss-offset decreases the demand for the risky asset if there is increasing relative risk aversion, and if the maximum value of relative risk aversion in the relevant region is less than $m+1/m$, where m is the maximum rate of return.

These results do tend to support the presumption that "social" risk-taking will be reduced by income taxes without loss-offset provisions.

Finally we shall show:

Proposition 4(e). If $r=0$ an income tax with no loss-offset reduces private risk-taking if there is decreasing absolute risk aversion.

If $\sigma^2 1=\int_{\theta'}^{1} e^2 dF(\theta)$ and $\sigma^2{}_2=\int_{0}^{\theta'} e^2 dF(\theta)$. It is easy to show that private risk-taking decreases if

$$-\frac{d\ln a}{d\ln(1-t)}<\frac{(1-t)^2\sigma^2{}_1}{(1-t)^2\sigma^2{}_1+\sigma^2{}_2}.$$

But from (12)

aversion of unity. For small values of W, if the utility function is bounded from below, relative risk aversion must be less than unity, while if the utility function is bounded from above, it must be greater than unity for large values of W. See K. Arrow, *op. cit.* (This is true provided for large and small W, $R(W)$ is monotonic.)

$$-\frac{d\,ln\,a}{d\,ln(1-t)} = \frac{{}_{\theta'}\int^{1}\{(-U'')e^{2}(1-t)^{2}W_{0} - U'e(1-t)/a\}dF(\theta)}{-{}_{\theta'}\int^{1}U''e^{2}(1-t)^{2}W_{0}dF(\theta) + {}_{0}\int^{\theta'}(-U'')e^{2}W_{0}dF(\theta)}$$
$$< \frac{{}_{\theta'}\int^{1} - U''e^{2}(1-t)^{2}W_{0}dF(\theta)}{-{}_{\theta'}\int^{1}U''e^{2}(1-t)^{2}W_{0}dF(\theta) - {}_{0}\int^{\theta'}U''e^{2}W_{0}dF(\theta)}$$
$$< \frac{(1-t)^{2}\sigma_{1}^{2}}{(1-t)^{2}\sigma^{2}{}_{1} + \sigma^{2}{}_{2}}$$

if $U''' > 0$. But

$$\frac{dA(W)}{dW} = -\frac{U'''}{U'} + \frac{U''^{2}}{U'^{2}} < 0$$

only if $U''' > 0$.

VII. Welfare Implications

Even if risk-taking is increased by a given type of tax, it is not clear that such a tax should be adopted: after all, risk-taking is not an end itself. Indeed, there are some who have argued that the stock market pools risk sufficiently effectively that there is no discrepancy between social and private risks, and hence no justification for governmental encouragement of risk-taking. It is important to observe, however, that some of the taxes considered may be more effective in obtaining a given end than others. Alternative taxes can be evaluated in terms of (a) losses in expected utility, (b) changes in demands for risky assets, and (c) revenues raised in *each* state of nature. Note that the last is much more stringent than comparisons simply between average revenues; two taxes may have the same expected revenue, but differ in the revenue they provide in different states of nature.

In this section we shall analyze the welfare implications of the preferential treatment of capital gains. To do this, we shall look at the polar case where risky assets are completely exempt from taxation, and compare its effects with those of alternative taxes.

A. Comparison of taxes with equal loss of expected utility.

First, we compare the tax on the safe asset only with a proportional income tax. From Figures Va and Vb it is clear that the *demand for risky assets may actually be larger with an income tax than with a tax on the safe asset only. ST* represents the before-tax budget constraint, *S'T'* the after-tax budget constraint for the income tax, *S''T* the after-tax budget constraint for the tax on the safe asset only. We have already noted that if the after-tax income

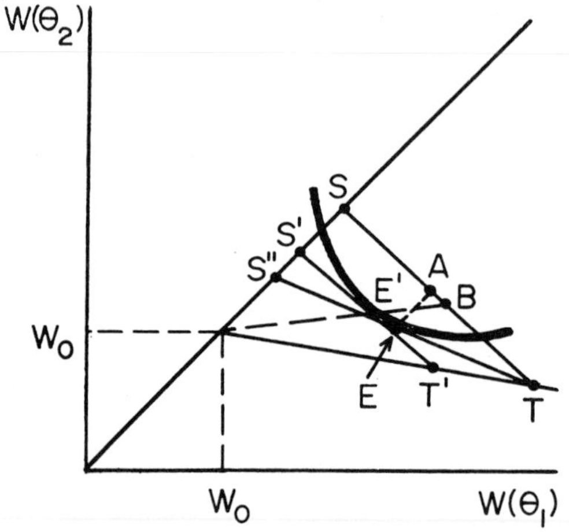

FIGURE VB

Demand for Risky Assets *Larger* for Income Tax than Preferential Treatment of Capital Gains

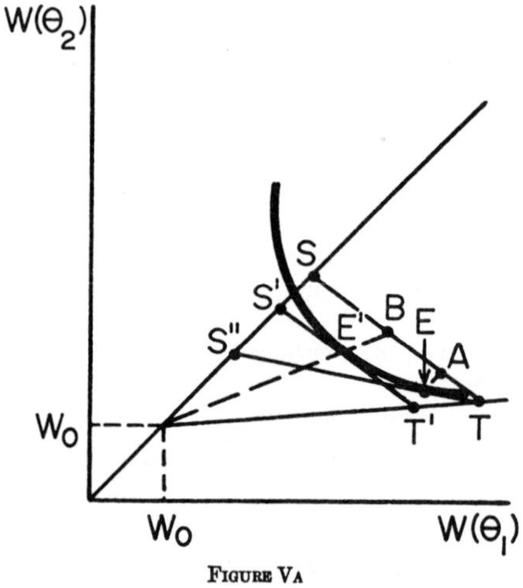

FIGURE VA

Demand for Risky Assets *Smaller* for Income Tax than Preferential Treatment of Capital Gains

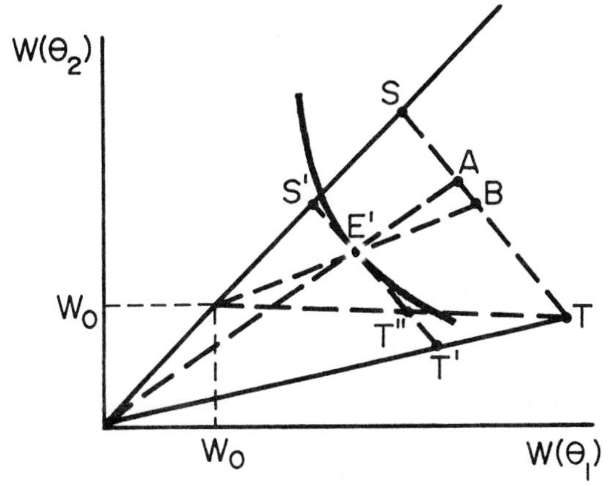

FIGURE VI

Comparison of Effects on Risk-taking of Income and Wealth Taxes

from the income tax is given by E', the before-tax income is given by B, where $E'B$ lies on a straight line through W_0. Since the tax on the safe asset only reduces income in both states of nature by the same amount, if the after-tax income from the tax on the safe asset only is E, the before-tax income is A, where EA is a 45° line. The fact that B may be closer or further from T than A implies that the demand for the risky asset may be larger or smaller with the income tax than with the tax on the safe asset only.

Revenue is measured by the vector EA for the tax on the safe asset only and $E'B$ for the income tax: one is larger in one state, the other in the other state. It can be seen from the diagram that, in general, there will not exist income taxes yielding the same revenue in each state of nature as yielded by the tax on the safe asset only.

Similar conclusions hold for a comparison between wealth taxes and the tax on the safe asset only. It is possible, however, to show that, for the same loss in expected utility, the income tax leads to more risk-taking than the wealth tax. See Figure VI. We have already noted that B is the before-tax wealth corresponding to the after-income-tax situation E', while A corresponds to the after-wealth-tax situation E', where EA lies on a straight line through the origin. But since A is closer to S than B, the result is immediate.

B. Comparison of taxes of equal revenue.

As noted above, we cannot make direct comparisons of different taxes with equal revenue in each state of nature. But we can compare the taxes indirectly by comparing each with a lump sum tax, i.e., a tax independent of the behavior of the individual although not of the state of nature. While the effect on risk-taking of a proportional income tax or a wealth tax is identical to that of the equal revenue lump sum tax, a tax on the safe asset only leads to more risk-taking but lower expected utility than the equal revenue lump sum tax. To see this, observe that in Figure VI the lump sum tax of equal revenue to the wealth tax is given by $E'A$ (for the income tax it is $E'B$). The after-lump-sum tax budget constraint is given by $S'T'$, so the equilibrium is still at E', and hence the taxes are equivalent in their effect on the demand for risky assets and on expected utility. In Figure VII, since the revenue from the tax on the safe asset only is EA, the after-lump-sum tax budget constraint is

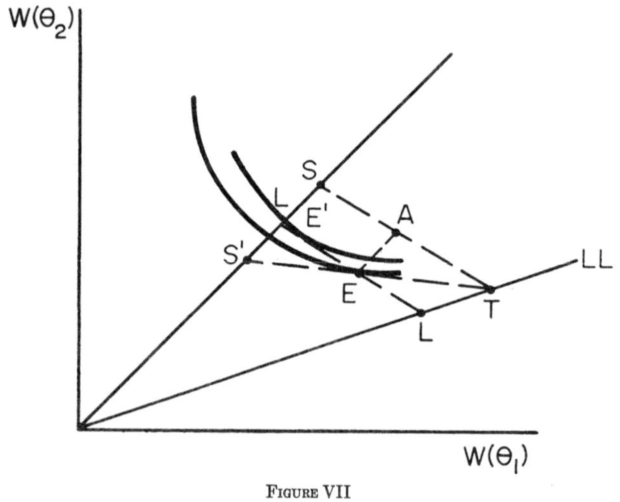

FIGURE VII

Comparison of Effects on Expected Utility of Tax on Safe Asset and Equal Revenue Lump Sum Tax

LL, and the equilibrium is given by E', which implies a higher demand for risky assets but at the cost of a lower level of utility for the tax on the safe asset than for the equal revenue lump sum tax.

The important point to observe is that even if one wished to encourage greater risk-taking, and even if preferential treatment of capital gains did this effectively, it is not clear that preferential treatment of capital gains is the most desirable way of encouraging risk-taking.

Cowles Foundation
Yale University

Reprinted from The Review of Economic Studies, Vol. XXXV (3), July 1968, B. Näslund, pp. 289-306.

Some Effects of Taxes on Risk-Taking[1]

INTRODUCTION

When the government alters the tax structure in society it will influence the direction of the flow of capital. Thus the tax might lead to a preference for investment in more risky projects or the opposite might be the case. Economists have for some time been interested in this problem and several suggestions exist in the literature as to the effect on the flow of capital of an alteration in taxes. A review of some of these suggestions is given below. There exists, however, no formulation of the investment decision under risk which is preferred by all economists, and it is therefore of some interest to include taxes in a few well-known formulations of decision-making under risk. If they all pointed in the same direction, one's belief in the validity of using economic theory as a guide to the policy aspects of this problem would be increased.

During the past two decades mathematical programming has been frequently used for applied decision-making. Recently there has been greater interest in including risk in these formulations. A discussion of the effect of taxes in one of these formulations is given below.

1. A BRIEF SURVEY OF PREVIOUS RESULTS

A well-known analysis of the effect of taxes on risk-taking is that of Domar and Musgrave [8].[2] They used the concepts of " yield ", which is the average return, and " risk " which the the sum of anticipated negative incomes weighted by their probabilities. The investor is assumed to have a utility function $U = U$ (yield, risk) where U increases with increasing yield and decreases with increasing risk. The investment possibilities between risk and yield are shown in Figure 1.

The proportional tax implies that the possibilities schedule moves to position P^1. The result is that the post-tax equilibrium is located at B, which implies higher pre-tax risk-taking at C.

Tobin [20] discusses the effect of tax on liquidity preference and also derives some results on the effect of taxes on risk-taking. He divides the assets into money and risky assets and the analysis is made under the assumption that there exists only one kind of risky asset. When there are several risky assets Tobin shows the important result that if an investor holds money then he holds the risky assets in certain fixed proportions which are independent of the size of his disposable wealth and of the form of the utility function.[3] Therefore the analysis can be made considering only one class of risky assets.

[1] This work has been supported by grant 220/65S from Statens Råd för Samhällsforskning. I am indebted to Professor M. Lovell of Carnegie Institute of Technology and to Civilekonom O. Ed and A. Linde at Stockholm University for interesting discussions, and to a referee for improvements in the organization of an earlier draft.

[2] For a critical evaluation of the Domar and Musgrave model, see Richter [16].

[3] This is called the separation theorem and is discussed further in Section 3a.

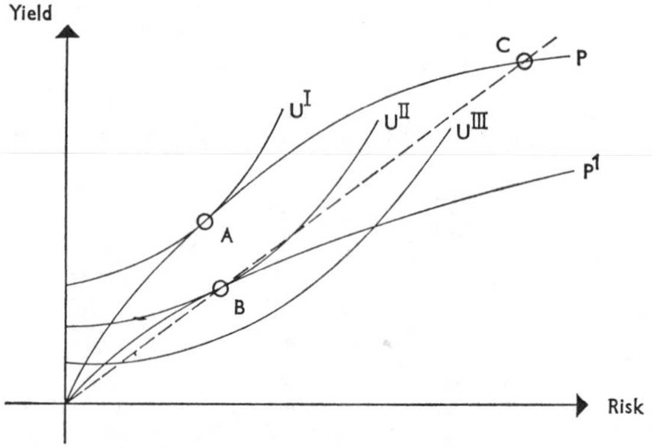

FIGURE 1

The fraction of wealth invested in risky assets is k. Assume that the mean and variance of the rate of return are μ_T and σ_T^2 respectively, if all the money is invested in risky assets. Then we obtain, for the portfolio selected

$$\mu = k\mu_T \qquad \text{...(1.1)}$$

and

$$\sigma = k\sigma_T. \qquad \text{...(1.2)}$$

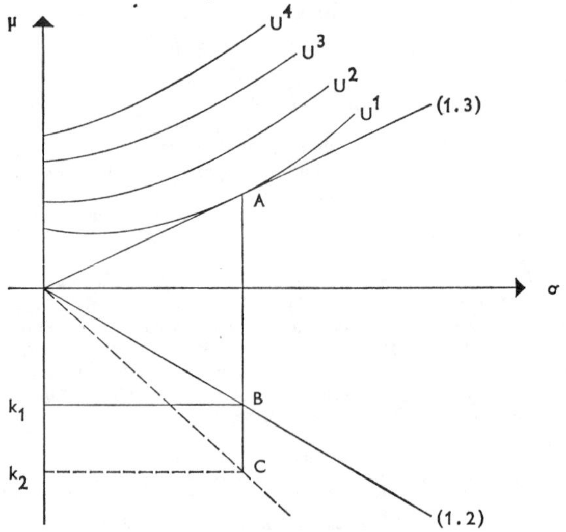

FIGURE 2

(1.1) and (1.2) give

$$\mu = \frac{\mu_T}{\sigma_T}\sigma. \qquad \text{...(1.3)}$$

Tobin uses a diagram of the type shown in Figure 2.

The curves in the first quadrant, U^1, U^2 etc. indicate expected utilities expressed in terms of mean and variance. The justification for expressing expected utility in terms of mean and variance only, is based either on the assumption that the utility-function is quadratic, or upon the assumption that the random variable is normally distributed (see Tobin [20]). The line in the first quadrant of Figure 3 is equation (1.3) and the line in the fourth quadrant represents equation (1.2).

If a linear tax is imposed (1.3) will remain unaltered since both μ_T and σ_T are multiplied by the same constant $(1-t)$ where t is the tax rate. Equation (1.2) will now take the form

$$\sigma = (1-t)k\sigma_T, \qquad \text{...(1.2a)}$$

and line (1.2) will move to the dotted position. Since the new position C corresponds to a higher value of $k(k_2 > k_1)$ the tax implies higher risk-taking.

Richter [16] deals with the problem along much the same lines as Tobin. He also uses von Neumann-Morgenstern utilities and he deals with two parameter distributions. He gives a fuller treatment of the lump sum tax case and above all he does not base his analysis on ratios of assets. This is very important in the case of borrowing since the effect of the tax is shown to be an increase both in money holdings and in risky assets which will not be seen if the analysis is based on asset ratios.

The analysis is based on quadratic utility functions.[1] The main result is the same as Tobin's, namely that increased taxation produces greater risk-taking in the society.

Lintner [9] gives an extension of Tobin's work by allowing for borrowing and lending at a specified rate of interest. As was shown above, Tobin considered the special case with no return on the riskless asset. Furthermore, Lintner studies the case with explicit constraints on the non-negativity of assets, and constraints indicating that the holdings of risk assets must not be greater than the total amount of assets available.

He utilizes the following notation:

r^x = interest rate on riskless assets,

$\bar{r}$ = mean return on portfolio,

w = ratio of gross investment in stocks to total net investment (stock plus riskless assets minus borrowing).[2]

The investor's net return per dollar of total net investment will be

$$\bar{y} = (1-w)r^x + w\bar{r} = r^x + w(\bar{r} - r^x), \qquad \text{...(1.4)}$$

where

$w < 1$ indicates that the investor holds riskless assets and

$w > 1$ indicates that the investor borrows.

[1] The assumption of quadratic utilities has been criticized by several authors as follows. The range over which the quadratic utility function is valid is limited by the condition that more return must produce higher utility. Attempts to widen this range lead to the consequence that the risk aversion represented by the function is not as strong as the one observed in experiments. Another consequence is that higher wealth leads to more risk aversion which does not seem to hold empirically. This has motivated the use of higher-ordered utilities. (See Puu [15].)

[2] $w = \dfrac{\text{stock}}{\text{stock} - \text{borrowing}}$ or

$w = \dfrac{\text{stock}}{\text{stock} + \text{lending}}.$

In addition to (1.4) we have

$$\sigma_y^2 = w^2\sigma_r^2. \qquad \text{...(1.5)}$$

Combining (1.4) and (1.5) we obtain

$$\bar{y} = r^x + \frac{\bar{r} - r^x}{\sigma_r}\sigma_y = r^x + \theta\sigma_y, \qquad \text{...(1.6)}$$

where

$$\theta = \frac{\bar{r} - r^x}{\sigma_r}.$$

Equations (1.5) and (1.6) can be represented as in Figure 3.

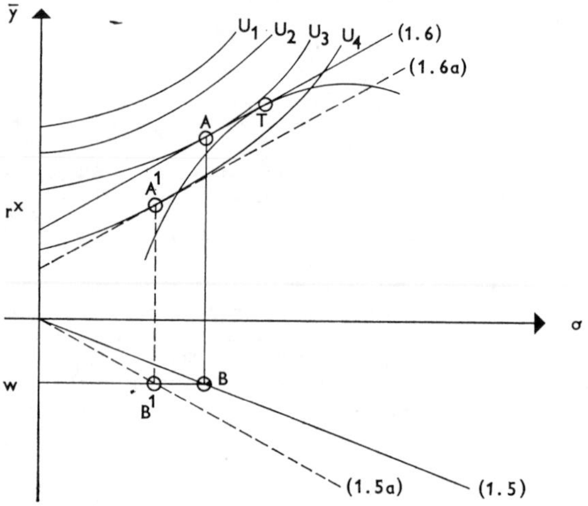

FIGURE 3

$U_1 > U_2 > U_3 > U_4$ indicate utility indifference curves. The point A is first determined and then the degree of lending/borrowing w is selected.

If we introduce a proportional tax, (1.5) will take the form

$$\sigma_y = w(1-t)\sigma_r, \qquad \text{...(1.5a)}$$

and (1.6) will take the form

$$\bar{y} = r^x(1-t) + \frac{\bar{r} - r^x}{\sigma_r}\sigma_y. \qquad \text{...(1.6a)}$$

These two curves are drawn with dotted lines in Figure 3. Thus we see that (1.6) moves down to the position (1.6a) and (1.5) swings to position (1.5a). The effects of the first factor is indicated by the difference between A and A^1 and the second effect by the difference $B - B^1$. Increased w means greater risk-taking. Whether w increases or not depends upon the size of these two effects. It seems that borrowing, i.e. operating to the right of T implies a move towards greater risk-taking since the effects there are much greater on the difference between (1.5a) and (1.5).

A somewhat different approach to the investment problem under risk is given in Roy [17]. He assumes that the investor makes investments in order to minimize the upper bound of the probability that the outcome will fall below a preassigned disaster level, d. The analysis is carried out by using the Bienamé-Tchebycheff inequality

$$P(|x-m| \geqq m-d) \leqq \frac{\sigma^2}{(m-d)^2}, \qquad \ldots(1.7)$$

where P stands for probability and $|\,.\,|$ indicates absolute value. Thus we have

$$P(|x-m| \geqq m-d) = P(x-m \geqq m-d) + P(m-x \geqq m-d) \leqq \frac{\sigma^2}{(m-d)^2},$$

and in particular,

$$P(m-x \geqq m-d) \leqq \frac{\sigma^2}{(m-d)^2}$$

which gives

$$P(x \leqq d) \leqq \frac{\sigma^2}{(m-d)^2}. \qquad \ldots(1.8)$$

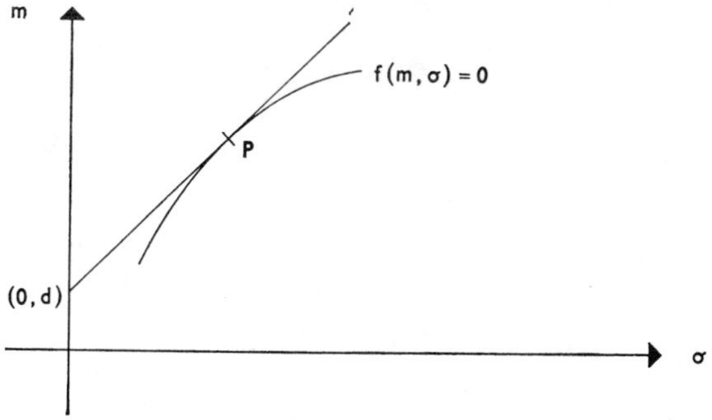

FIGURE 4

From (1.8) we have, in order to minimize $P(x \leq d)$, that it is sufficient (but certainly not necessary) that we minimize $\frac{\sigma^2}{(m-d)^2}$, which is equivalent to maximizing $\frac{m-d}{\sigma}$.[1]

The functional relationship between m and σ is denoted by $f(\sigma, m) = 0$. The function f is shown in Figure 4.

As was shown above, our aim is to maximize $\frac{m-d}{\sigma}$. Assume that we have found the maximum h. Then we have

$$m-d = \sigma \, . \, h$$

which is represented by a line in the (m, σ) plane with intercept d on the m-axis and slope h. Thus, graphically, our maximization problem is to find a line through $(0, d)$ with as high slope as possible but having a point on $f(m, \sigma) = 0$. This point is obtained at the point of tangency at P.

[1] We can observe here a similarity between Lintner's approach and that of Roy (see Lintner [9]) in that the θ function in (1.6) is identical to $\frac{m-d}{\sigma}$ if the disaster level is equal to the return on riskless assets.

Using this line of reasoning it is possible to derive (see Roy [17]) the solution for the two asset case,

$$x_1 = \frac{\lambda}{\sigma_1(1-r^2)}\left[\frac{m_1 - \dfrac{d}{M}}{\sigma_1} - \frac{r\left(m_2 - \dfrac{d}{M}\right)}{\sigma_2}\right], \qquad \text{...(1.9)}$$

$$x_2 = \frac{\lambda}{\sigma_2(1-r^2)}\left[\frac{m_2 - \dfrac{d}{M}}{\sigma_2} - \frac{r\left(m_1 - \dfrac{d}{M}\right)}{\sigma_1}\right], \qquad \text{...(1.10)}$$

$$x_1 + x_2 = M, \qquad \text{...(1.11)}$$

where

x_1 = amount invested in asset type 1,

x_2 = amount invested in asset type 2,

m_1 = estimated price of asset 1 with standard deviation σ_1,

m_2 = estimated price of asset 2 with standard deviation σ_2,

r^2 = correlation coefficient,

λ is determined by (1.11),

and M is the total amount of capital available for investment.

A proportional tax will affect both m_1 and σ_1 linearly. Therefore we can write (1.9) and (1.10) after tax as follows

$$x_1 = \frac{\lambda}{(1-t)\sigma_1(1-r^2)}\left[\frac{m_1(1-t) - \dfrac{d}{M}}{\sigma_1(1-t)} - \frac{r\left(m_2(1-t) - \dfrac{d}{M}\right)}{\sigma_2(1-t)}\right], \qquad \text{...(1.12)}$$

$$x_2 = \frac{\lambda}{(1-t)\sigma_2(1-r^2)}\left[\frac{m_2(1-t) - \dfrac{d}{M}}{\sigma_2(1-t)} - \frac{r\left(m_1(1-t) - \dfrac{d}{M}\right)}{\sigma_1(1-t)}\right]. \qquad \text{...(1.13)}$$

Dividing (1.12) by (1.13) gives

$$\frac{x_1}{x_2} = \left\{\frac{\dfrac{m_1}{\sigma_1} - \dfrac{d}{M\sigma_1(1-t)} - \dfrac{rm_2}{\sigma_2} + \dfrac{rd}{M\sigma_2(1-t)}}{\dfrac{m_2}{\sigma_2} - \dfrac{d}{M\sigma_2(1-t)} - \dfrac{rm_1}{\sigma_1} + \dfrac{rd}{M\sigma_1(1-t)}}\right\}\frac{\sigma_2}{\sigma_1},$$

which can be written

$$\frac{x_1}{x_2} = \left\{\frac{\left(\dfrac{m_1}{\sigma_1} - \dfrac{rm_2}{\sigma_2}\right)(1-t) + \dfrac{d}{M}\left(\dfrac{r}{\sigma_2} - \dfrac{1}{\sigma_2}\right)}{\left(\dfrac{m_2}{\sigma_2} - \dfrac{rm_1}{\sigma_1}\right)(1-t) + \dfrac{d}{M}\left(\dfrac{r}{\sigma_1} - \dfrac{1}{\sigma_2}\right)}\right\}\frac{\sigma_2}{\sigma_1}. \qquad \text{...(1.14)}$$

Writing (1.14) as

$$\frac{x_1}{x_2} = \frac{a(1-t)+b}{c(1-t)+d},$$

$$\frac{d\left(\dfrac{x_1}{x_2}\right)}{dt} = \frac{bc-ad}{(c(1-t)+d)^2} \qquad \text{...(1.15)}$$

and therefore the sign of (1.15) is determined by

$$bc-ad = \frac{m_1-m_2}{\sigma_1-\sigma_2}(1-r^2)\frac{d}{M}.$$

Thus if the asset with the higher yield is riskier (e.g. $m_1 > m_2$ and $\sigma_1 > \sigma_2$), an increase in taxes would produce higher risk-taking.

2. CHANCE—CONSTRAINED PROGRAMMING [1]

Work on the extension of the linear programming formulation to allow for randomness in the problems studied has been going on for over a decade. Many papers have been published on the subject, but among these contributions one can distinguish three main approaches, namely:

(*a*) stochastic linear programming,

(*b*) linear programming under uncertainty, and

(*c*) chance-constrained programming.

Method (*a*) was first developed by Tintner [19]. It enables the decision-maker to affect the probability distribution of the optimum by allocating the scarce resources to various activities prior to the realization of the random variables.

Method (*b*) partitions the decision problem into two or more stages. First a decision is made, then the random variables are observed and finally it is necessary to undertake corrective action to restore the proper form of those constraints that are violated due to the randomness in the problem.

The authors working in this field have been concerned with a decision-maker who minimizes the expected value of the functional. It is assumed that there always exist second stage variables that restore the problem to its proper form. Furthermore, it is assumed that the costs associated with these second stage variables can be specified.

The class of problems dealt with in chance-constrained programming can be described by the following formulation [2]

$$\max f(cx) \qquad \ldots(2.1)$$

subject to

$$P(Ax \leqq b) \geqq \alpha, \qquad \ldots(2.2)$$

where f is a concave function of x,

A is a random $m \times n$ matrix,

b is a random column vector with m elements,

c is a random row vector with n elements,

x is a column vector with n elements,

and P means probability and α is a column vector with m elements

$$0 \leqq \alpha_i \leqq 1.$$

Instead of seeking a maximum of (2.1) over *all* x subject to (2.2), we often require a maximum over only such vectors generated by some decision rule. In general, such a rule will involve a set of x values that depend upon the random variables A, b and c. That is,

$$x = D(A, b, c), \qquad \ldots(2.3)$$

[1] For a more complete discussion of mathematical programming under risk see Näslund [14].
[2] For a description of this class of problems, see Charnes and Cooper [3].

where the rules D are to be selected in accordance with certain prescribed properties, e.g. those that depend linearly on the random variables.[1]

A wider class of objectives has been studied or used for chance-constrained programming problems than for linear programming under uncertainty. The principal ones have been (*a*) the so-called E-type, where the functional is in the form of an expected value, (*b*) the V-type, where the functional has a quadratic form, for example, in connection with minimization of the variance, (*c*) the P-type, where the functional expresses the goal of maximizing the probability that at least a certain level of the functional is obtained.[2]

The most common way of solving chance-constrained programming problems is to transform the probabilistic constraints to their so-called deterministic equivalents. To illustrate, assume that we have a single probabilistic constraint of the form

$$P\left(\sum_{i=1}^{n} a_i x_i \geqq b_1\right) \geqq \alpha. \qquad \text{...(2.4)}$$

Assume further that b_1 is normally distributed with mean μ and variance σ^2. We can then write (2.4) as follows

$$\tfrac{1}{2} + F\left(\frac{\sum_{i=1}^{n} a_i x_i - \mu}{\sigma}\right) \geqq \alpha, \qquad \text{...(2.5)}$$

where

$$F(Z) = \frac{1}{(2\pi)^{\frac{1}{2}}} \int_0^Z e^{-\frac{1}{2}t^2}\, dt,$$

or

$$\sum_{i=1}^{n} a_i x_i - \mu \geqq F^{-1}(\alpha - \tfrac{1}{2})\sigma. \qquad \text{...(2.6)}$$

If we compare (2.4) and (2.6) we can see that the effect is to transform the probabilistic expression to a deterministic one. Moreover, in this special case the constraint is linear, and with a linear functional we have obtained a linear programming problem. This method of transforming more complicated probabilistic constraints will often give non-linear constraints and one has then to use some technique in the field of non-linear programming.[3]

Another method, described in Näslund and Whinston [13], makes use of the fact that under certain conditions constraints similar to (2.4) can be transformed to an integral whereupon variational methods can be used to solve the chance-constrained programming problem.

In the remainder of the paper several transformations of probabilistic constraints to their so-called deterministic equivalents will be made. We shall therefore conclude this section with a discussion of the method in detail.

The constraints that we are dealing with in this paper are of the form

$$P\,(x \geqq B) \geqq \alpha. \qquad \text{...(2.7)}$$

This means that the probability that the random variable x is greater than B must be greater than α.

In order to obtain a meaningful problem we need $\alpha > 0{\cdot}5$ (see Charnes, Cooper and Thompson [7]).

[1] For an example of an explicit form of (2.3) see Näslund and Whinston [12].
[2] For a discussion of these see A. Charnes and W. W. Cooper [4].
[3] For detailed description of this method for solving chance-constrained programming problems see [5] and [12].

In Figure 5 we have drawn certain parts of F.

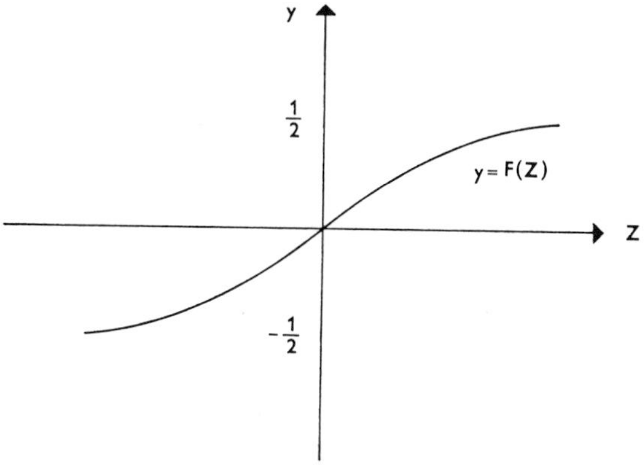

FIGURE 5

Assume that x is normally distributed with mean μ and variance σ^2. This means that (2.7) can be written as

$$\frac{1}{(2\pi\sigma^2)^{\frac{1}{2}}}\int_B^\infty e^{-\frac{(x-\mu)^2}{2\sigma^2}}\,dx \geqq \alpha \qquad \text{...(2.8)}$$

setting $\frac{x-\mu}{\sigma} = t$, we can write (2.8)

$$\frac{1}{(2\pi)^{\frac{1}{2}}}\int_{(B-\mu)/\sigma}^\infty e^{-\frac{1}{2}t^2}\,dt \geqq \alpha. \qquad \text{...(2.9)}$$

We shall suppose that $B<\mu$.[1] Then, since from the definition of F it follows that

$$F(-Z) = -F(Z),$$

we can write (2.9) as

$$\tfrac{1}{2}+F\left(\frac{\mu-B}{\sigma}\right) \geqq \alpha. \qquad \text{...(2.10)}$$

Taking the inverse of (2.10) we obtain

$$\frac{\mu-B}{\sigma} \geqq F^{-1}(\alpha-\tfrac{1}{2})\sigma. \qquad \text{...(2.11)}$$

In Figure 6 we show the form of F^{-1}

The inequality (2.11) can be written as follows,

$$\mu-\sigma F^{-1}(\alpha-\tfrac{1}{2}) \geqq B. \qquad (2.12)$$

This is the form used in most of the applications in the paper, and sometimes $F^{-1}(\alpha)$ will be denoted by F_α^{-1} which will be used as a shorthand for $F^{-1}(\alpha-\frac{1}{2})$.

[1] This assumption is natural in order to make the constraint (2.7) meaningful.

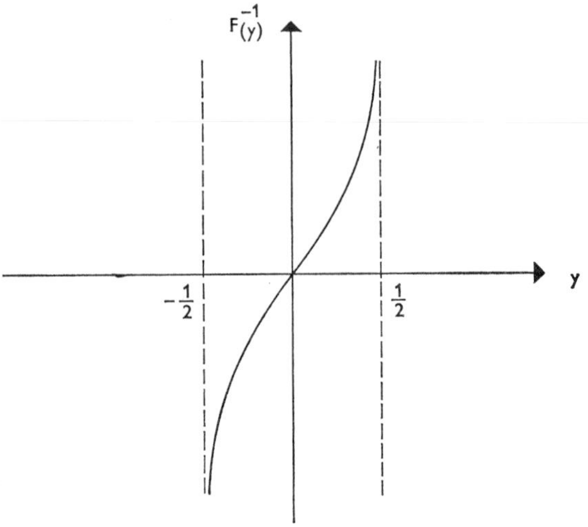

FIGURE 6

3. THE EFFECT OF TAXES IN CHANCE-CONSTRAINED MODELS

(a) *One riskless and several risky assets*

In this section the separation theorem of Tobin [20] plays a central role. This theorem states, as was mentioned above, that (for risk averters in purely competitive markets) the proportionate composition of the non-cash assets is independent of their share of the total asset holdings when utility functions are quadratic, or the rates of return are normally distributed.

This means that in the event that some of the riskless asset is held, the proportions in which the risky assets are held in an optimal portfolio are independent of wealth, the minimum return level and the risk level. If there existed a known distribution of stock prices, then all investors would hold the same proportions of risky assets and the theorem could be tested empirically.[1]

Proof of the separation theorem using the E-formulation of chance-constrained programming and assuming normally distributed asset returns with two constraints.

Assume that the investor can invest in one riskless asset, with return r_1, and in risky assets whose rate of return r_i are distributed with mean μ_i and variance σ_i^2, $(i = 2, \ldots, N)$. The amounts invested in the risky assets are denoted by x_i, $(i = 1, \ldots, N)$. We would then try to solve the problem

$$\max\left(r_1 x_1 + \sum_{i=2}^{N} \mu_i x_i\right) \qquad \ldots(3.1)$$

subject to

$$P\left(\sum_{i=1}^{N} r_i x_i \geqq B\right) \geqq \alpha, \qquad \ldots(3.2)$$

[1] It would be interesting to study the holdings of investment corporations from this point of view.

$$\sum_{i=1}^{N} x_i \leqq M, \quad x_i \geqq 0 \; (i = 1, \ldots, N).$$

Taking the deterministic equivalent of this programme yields

$$\max\left(r_1 x_1 + \sum_{i=2}^{N} \mu_i x_i\right)$$

subject to

$$r_1 x_1 + \sum_{i=2}^{N} \mu_i x_i - F^{-1}(\alpha)\left(\sum_{i=2}^{N} x_i^2 \sigma_i^2 + \sum_{\substack{i,j=2 \\ i \neq j}}^{N} x_i x_j \operatorname{cov}_{ij}\right)^{\frac{1}{2}} \geqq B, \qquad \ldots(3.3)$$

$$\sum_{i=1}^{N} x_i \leqq M, \quad x_i \geqq 0 \; (i = 1, \ldots, N),$$

where cov_{ij} is the covariance of x_i and x_j.

Using the Kuhn-Tucker conditions [1] we obtain (u_i indicates the dual variable associated with constraint i)

$$r_1 + r_1 u_1 - u_2 \leqq 0 \qquad \ldots(3.4)$$

$$\mu_2 + u_1\left(\mu_2 - F_\alpha^{-1} \frac{x_2 \sigma_2^2 + \sum_{j=3}^{N} x_j \operatorname{cov}_{2j}}{\left(\sum_{i=2}^{N} x_i^2 \sigma_i^2 + \sum_{\substack{i,j=2 \\ i \neq j}}^{N} x_i x_j \operatorname{cov}_{ij}\right)^{\frac{1}{2}}}\right) - u_2 \leqq 0 \qquad \ldots(3.5)$$

$$\mu_N + u_1\left(\mu_N - F_\alpha^{-1} \frac{x_N \sigma_N^2 + \sum_{j=2}^{N-1} x_j \operatorname{cov}_{Nj}}{\left(\sum_{i=2}^{N} x_i^2 \sigma_i^2 + \sum_{\substack{i,j=2 \\ i \neq j}}^{N} x_i x_j \operatorname{cov}_{ij}\right)^{\frac{1}{2}}}\right) - u_2 \leqq 0 \qquad \ldots(3.6)$$

If the investment in asset i is zero then the inequality in the ith expression of (3.5) to (3.6) holds; otherwise the equality holds. We are only interested in those assets which are held. Assuming that the riskless asset is held in the optimal solution, i.e. $x_1 > 0$, then expression (3.4) also holds as an equation. Then, substituting (3.4) in (3.5) and (3.6) and dividing (3.5) and (3.6) gives, after some simplification,

$$\frac{\mu_2 - r_1}{\mu_N - r_1} = \frac{\sigma_2^2 + \sum_{j=3}^{N} \frac{x_j}{x_2} \operatorname{cov}_{2j}}{\sigma_N^2 \frac{x_N}{x_2} + \sum_{j=2}^{N-1} \frac{x_j}{x_2} \operatorname{cov}_{Nj}} \qquad \ldots(3.7)$$

We get at most $(N-2)$ equations of the form of (3.7) in the $(N-2)$ ratios $\frac{x_j}{x_2}$ ($j = 3, \ldots, N$). These ratios are, if they exist, determined by the means, variances and co-variances *but are independent of the investor's risk preference F_α^{-1} and of his funds available for investment M.*

It follows from the separation theorem that we need only concern ourselves with two assets, all risky assets treated as one and the riskless asset.

Assuming that the investor maximizes expected return while imposing a probabilistic constraint, we can formulate the problem as follows.

$$\max Ekr_T \qquad \ldots(3.8)$$

[1] For a presentation of the Kuhn-Tucker conditions see e.g. Charnes, Cooper [1].

subject to

$$P\,(kr_T \geqq B) \geqq \alpha, \qquad ...(3.9)$$

where

r_T = return from risky assets if (the whole) total fund is invested in risky assets (assumed normally distributed),

k = proportion of total fund that is invested in risky assets,

μ_T = mean of r_T,

σ_T = standard deviation of r_T,

and B = risk level which the risky return must exceed at least with probability α.

Taking the deterministic equivalent of (3.9) we obtain

$$\mu - F_\alpha^{-1}\sigma \geqq B, \qquad ...(3.10)$$

where

$$\sigma = k\sigma_T, \text{ and } \mu = k\mu_T, \qquad ...(3.11)$$

so that

$$\mu = \frac{\mu_T}{\sigma_T}\sigma. \qquad ...(3.12)$$

In Figure 7 we have shown (3.10), (3.11) and (3.12).

According to (3.10) the solution must lie above the line (3.10). Since the solution must lie on the line given by (3.11) and also on the boundary of the constraint set determined by (3.10), the solution will be at A which corresponds to investing the proportion k_1 of the funds in risky assets. We can now use this formulation to study the effect of some economic factors on the solution.

I. *The effect of a proportional tax*

A proportional tax with the possibility of offsetting losses will not affect (3.12), but (3.11) will change to $\sigma = k(1-t)\sigma_T$ and (3.10) will remain unaltered, since none of the

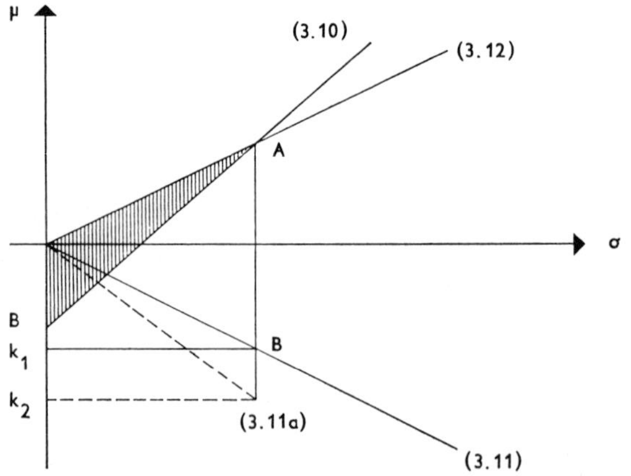

FIGURE 7

parameters is affected by the tax. The effect is therefore a shift of (3.11) to position (3.11a) as shown by the dotted line in Figure 7. This is exactly the same effect as was obtained by Tobin [20]. If point A agrees with the result obtained through expected utility maximization, then the effect of a proportional tax will also be the same.

II. *The effect of variations in the risk*

The risk may be altered by the government, since it can affect the environment in which the firm operates, or by the firm itself by improving its forecasting. If the standard deviation is decreased to half of its previous size this will lead to an alteration of (3.11) which takes the form

$$\sigma = k\frac{\sigma_T}{2}. \qquad \text{...(3.13)}$$

Furthermore (3.11) will alter and take the form

$$\mu = \frac{2\mu_T}{\sigma_T}\sigma. \qquad \text{...(3.14)}$$

Inequality (3.10) however remains unaltered since none of the parameters is affected. In Figure 8 we have represented the solution before the risk alteration by solid lines and after the alteration in dotted lines.

The effect of the decrease in standard deviation is increased risk-taking since $k_2 > k_1$. The same result is obtained by Tobin [20]. The point B would, in Tobin's case, be at a tangency between an indifference curve and (3.14). In our analysis we move along the constraint (3.10).

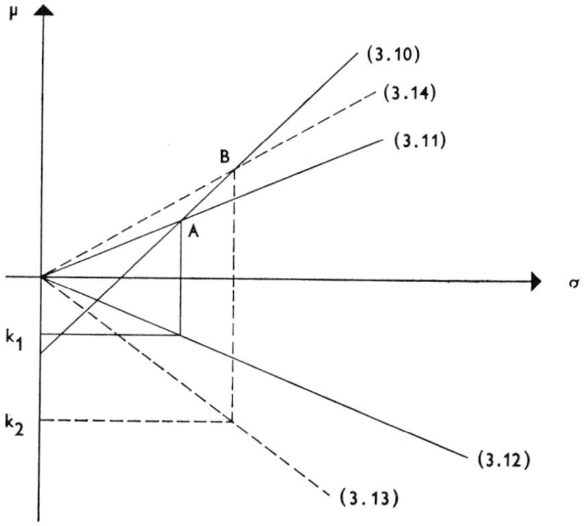

FIGURE 8

III. *Variations in risk aversion*

If the investor's attitude towards risk alters then this will alter the form of (3.10). A decrease in risk aversion can either be reflected by a decrease in α (dotted line (3.16) in Figure 9) or a decrease in the value of B (dotted line (3.15) in Figure 9).

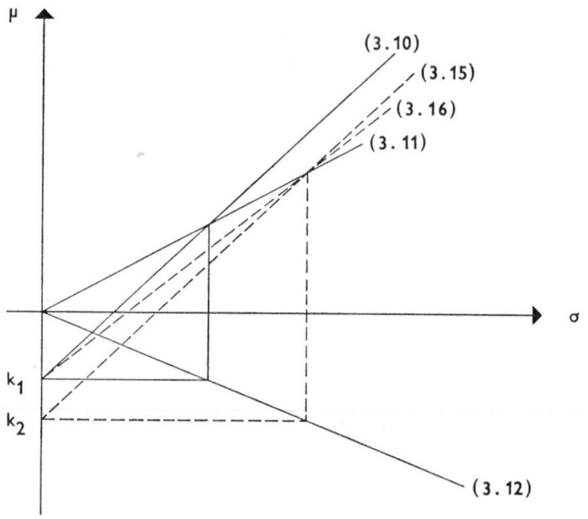

FIGURE 9

The discussion so far has assumed that there exists at least one asset which is riskless and which the investor holds. As a rule this riskless asset is assumed to be money. Considering the variations in purchasing power it may currently be a doubtful assumption to suppose that an investor considers money riskless. It might therefore be of some interest to study the case when all assets are risky.

(b) *All assets risky*

Assume that a choice is to be made between two investment projects. We assume that the amounts of money invested in these two projects are x_1 and x_2. The returns, arising in the next period are r_1 and r_2 per unit of x_1 and x_2 respectively. As in the previous discussion we assume that r_1 and r_2 are normally distributed with mean μ_1 and μ_2 and variance σ_1^2 and σ_2^2 respectively.

We maximize the expected return in the second period while requiring that the probability that the return is equal to or higher than a certain value B is at least α. The proportional tax is $t(0 \leqq t \leqq 1)$ and the lump sum tax is K. Thus we have the problem

$$\max E(x_1r_1(1-t)+x_2r_2(1-t)-K),$$

subject to

$$x_1+x_2 \leqq M,$$

$$P(x_1r_1(1-t)+x_2r_2(1-t)-K \geqq B) \geqq \alpha,$$

$$x_1, x_2 \geqq 0.$$

Taking the deterministic equivalents gives

$$\max \mu_1 x_1(1-t)+\mu_2 x_2(1-t)-K, \qquad \text{...(3.15)}$$

subject to

$$x_1+x_2 \leqq M, \qquad \text{...(3.16)}$$

$$x_1\mu_1(1-t)+x_2\mu_2(1-t)-F^{-1}(\alpha)(1-t)(\sigma_1^2x_1^2+2x_1x_2\,\mathrm{cov}_{12}+\sigma_2^2x_2^2)^{\frac{1}{2}}-K \geqq B,\; x_1,\, x_2 \geqq 0. \qquad \text{...(3.17)}$$

Since α is assumed to be $\geqq 0{\cdot}5$ (in order to make the constraint meaningful), we can interpret the third term of (3.17) as a reserve that will not, on the average, be used.

We now apply the Kuhn-Tucker conditions to the problem to yield

$$\mu_1(1-t)-u_1+u_2(1-t)\left[\mu_1-\frac{(\sigma_1^2x_1+x_2\,\mathrm{cov}_{12})F_\alpha^{-1}}{(\sigma_1^2x_1^2+2x_1x_2\,\mathrm{cov}_{12}+\sigma_2^2x_2^2)^{\frac{1}{2}}}\right]=0 \text{ and } x_1>0, \qquad \text{...(3.18)}$$

$$\mu_1(1-t)-u_1+u_2(1-t)\left[\mu_1-\frac{(\sigma_1^2x_1+x_2\,\mathrm{cov}_{12})F_\alpha^{-1}}{(\sigma_1^2x_1^2+2x_1x_2\,\mathrm{cov}_{12}+\sigma_2^2x_2^2)^{\frac{1}{2}}}\right]\leqq 0 \text{ and } x_1=0, \qquad \text{...(3.19)}$$

$$\mu_2(1-t)-u_1+u_2(1-t)\left[\mu_2-\frac{(\sigma_2^2x_2+x_1\,\mathrm{cov}_{12})F_\alpha^{-1}}{(\sigma_1^2x_1^2+2x_1x_2\,\mathrm{cov}_{12}+\sigma_2^2x_2^2)^{\frac{1}{2}}}\right]=0 \text{ and } x_2>0, \qquad \text{...(3.20)}$$

$$\mu_2(1-t)-u_1+u_2(1-t)\left[\mu_2-\frac{(\sigma_2^2x_2+x_1\,\mathrm{cov}_{12})F_\alpha^{-1}}{(\sigma_1^2x_1^2+2x_1x_2\,\mathrm{cov}_{12}+\sigma_2^2x_2^2)^{\frac{1}{2}}}\right]\leqq 0 \text{ and } x_2=0. \qquad \text{...(3.21)}$$

We are going to assume, as does Richter [16], that some investment takes place in the two projects x_1 and x_2. Thus conditions (3.18) and (3.20) hold. In order to evaluate the effect on x_1 and x_2 of a change in t, we assume (3.16) and (3.17) are both binding constraints and differentiate them with respect to t, as implicit functions of x_1, x_2 and t, to obtain

$$\frac{dx_1}{dt}+\frac{dx_2}{dt}=0, \qquad \text{...(3.22)}$$

and

$$\frac{dx_1}{dt}\left[\mu_1-\frac{F_\alpha^{-1}(\sigma_1^2x_1+x_2\,\mathrm{cov}_{12})}{(\sigma_1^2x_1^2 2x_1x_2\,\mathrm{cov}_{12}+\sigma_2^2x_2^2)^{\frac{1}{2}}}\right]+\frac{dx_2}{dt}\left[\mu_2-\frac{F_\alpha^{-1}(\sigma_2^2x_2+x_1\,\mathrm{cov}_{12})}{(\sigma_1^2x_1^2 2x_1x_2\,\mathrm{cov}_{12}+\sigma_2^2x_2^1)^{\frac{1}{2}}}\right]=\frac{B+K}{(1-t)^2}. \qquad \text{...(3.23)}$$

The factors multiplying $\frac{dx_1}{dt}$ and $\frac{dx_2}{dt}$ in (3.23) can be substituted for from (3.18) and (3.20) respectively. Expression (3.23) then takes the form

$$\frac{dx_1}{dt}\left[\frac{\frac{u_1}{1-t}-\mu_1}{u_2}\right]+\frac{dx_2}{dt}\left[\frac{\frac{u_1}{1-t}-\mu_2}{u_2}\right]=\frac{B+K}{(1-t)^2}. \qquad \text{...(3.24)}$$

Combining (3.22) and (3.24) gives

$$\frac{dx_1}{dt}=\frac{(B+K)\,.\,u_2}{(1-t)^2(\mu_2-\mu_1)}. \qquad \text{...(3.25)}$$

Noting the form of constraint (3.17), it follows from the Kuhn-Tucker theorem that under our assumptions $u_2<0$.

From (3.25) we can see that if $\mu_1>\mu_2$ then x_1 will be increased by an increase in t. If we assume that $\mu_1>\mu_2$ *and* $\sigma_1>\sigma_2$, we would get a movement toward the risky asset from the less risky one. The larger K, the greater would be this movement. If, however, $B+K<0$, the increase in the tax rate will decrease risk-taking.

In order to determine the effect of a change in K, we again differentiate (3.16) and (3.17) as equations with respect to K, and obtain

$$\frac{dx_1}{dK}+\frac{dx_2}{dK}=0 \qquad ...(3.26)$$

and

$$\frac{dx_1}{dK}\left[\mu_1-\frac{F_\alpha^{-1}(\sigma_1^2x_1+x_2\,\text{cov}_{12})}{(\sigma_1^2x_1^2+\sigma_2^2x_2^2)^{\frac{1}{2}}}\right]+\frac{dx_2}{dK}\left[\mu_2-\frac{F_\alpha^{-1}(\sigma_2^2x_2+x_1\,\text{cov}_{12})}{(\sigma_1^2x_1^2+\sigma_2^2x_2^2)^{\frac{1}{2}}}\right]=\frac{1}{1-t}. \qquad ...(3.27)$$

Using (3.18), (3.20) and (3.26), (3.27) can be written

$$\frac{dx_1}{dK}=\frac{u_2}{(1-t)(\mu_2-\mu_1)}. \qquad ...(3.28)$$

Assuming $\mu_1>\mu_2$ *and* $\sigma_1>\sigma_2$ we can draw the conclusion from (3.28) that increasing the lump sum part of the tax shifts investment to the risky one and the shift is greater the greater the proportional tax t.

For the case of perfect correlation, i.e. $\text{cov}_{12}=\sigma_1, \sigma_2$, the result may be obtained by using linear programming duality.

From (3.25) we see that the effect of a proportional tax depends upon the formulation of the constraint (3.17).

If $B+K<0$ the tax will decrease risk-taking,

$B+K=0$ the tax will have no effect on risk-taking,

$B+K>0$ the tax will increase risk-taking.

For a given α, B will be higher the more the decision-maker is a risk averter. He will tend to increase his risk-taking while the opposite would tend to be true for risk lovers. It is clearly brought out here that risk is increased when $B+K>0$, which means that if a decision-maker is concerned about the possibility of making profits then the tax will increase his willingness to take risks. This is due to the fact that the government is taking part of the risk by letting him offset losses against profits.

4. CONCLUSION

The effect of a proportional tax on the investment decision has been studied previously by several authors. These authors have used utility functions of various forms. The results indicate that increased taxation would increase risk-taking in the society.

Since many formulations of decision-making under risk exist in the literature, and since no one can be judged correct under all circumstances, it is important to see whether this conclusion is always true. Therefore some new models and one old but frequently discussed model are extended by including a proportional tax. Specifically, the chance-constrained programming formulation is discussed both for the case of one riskless asset and several risky ones, and for the case when all assets are risky. For the former case the important separation theorem due to Tobin is proved to hold.

The principal conclusion of the paper is that increased taxation will in general increase risk-taking in these models, but in some cases the opposite is true.

The analysis presented here could be carried out for the expected value of a quadratic function. Since we already have quadratic constraints, this would not complicate matters very much. Finally, it would be interesting to study the effect of a graduated tax.

Stockholm University B. NÄSLUND.

REFERENCES

[1] Charnes, A. and Cooper, W. W. *Management Models and Industrial Applications of Linear Programming* (John Wiley & Sons Inc., New York, 1961, I and II).

[2] Charnes, A., Cooper, W. W. and Thompson, G. L. " Chance-Constrained Programming and Related Approaches to Cost Effectiveness " O.N.R. Contract 760 (24) NR 047-048, Carnegie Institute of Technology, Pittsburgh, U.S.A.

[3] Charnes, A. and Cooper, W. W. " Chance-Constrained Programming ", *Management Science*, **6**, No. 1 (October 1959).

[4] Charnes, A. and Cooper, W. W. " Deterministic Equivalents for Optimizing and Satisficing under Chance-Constraints ", *Operations Research*, **11** (1963), 18-39.

[5] Charnes, A. and Cooper, W. W. " Normal Deviates and Chance Constraints ", *Journal of the American Statistical Association*, **57** (1962), 134-138.

[6] Charnes, A., Cooper, W. W. and Symonds, G. H. " Cost Horizons and Certainty Equivalents: An Approach to Stochastic Programming of Heating Oil Production ", *Management Science*, **4**, No. 3 (1958).

[7] Charnes, A., Cooper, W. W. and Thompson, G. L. " Characterizations by Chance-Constrained Programming " in R. L. Graves and P. Wolfe, eds., *Recent Advances in Mathematical Programming* (New York, 1963).

[8] Domar, E. and Musgrave, R. " Proportional Income Taxation and Risk-Taking ", *Quarterly Journal of Economics*, **57**.

[9] Lintner, J. " The Valuation of Risk Assets and the Selection of Risky Investments in Stock Portfolios and Capital Budgets ", *Review of Economics and Statistics* (June 1965).

[10] Markowitz, H. " Portfolio Selection ", *Journal of Finance*, **7**, No. 1 (1952), 77-91.

[11] Markowitz, H. M., *Portfolio Selection* (New York, 1959).

[12] Näslund, B. and Whinston, A. " A Model of Decision Making Under Uncertainty ", *Management Science* (January 1962).

[13] Näslund, B. and Whinston, A. " On some Variational Theorems ", *Indian Journal of Operations Research*, No. 1, 1965.

[14] Näslund, B. " Mathematical Programming Under Risk ", *The Swedish Journal of Economics* (October 1965).

[15] Puu, T. *Studier i Det Optimala Tillgångsvalets Teori* (Almqvist & Wicksell, Uppsala, 1964).

[16] Richter, M. K. " Cardinal Utility Portfolio Selection and Taxation ", *The Review of Economic Studies* (1960).

[17] Roy, A. D. " Safety First and the Holding of Assets ", *Econometrica* (July 1952).

[18] Samuelson, P. A. *Foundations of Economic Analysis* (Cambridge, 1958).

[19] Tintner, G. " Stochastic Linear Programming with Applications to Agricultural Economics ", in H. A. Antosiewics, ed. *Second Symposium on Linear Programming* (Washington: National Bureau of Standards, 1955).

[20] Tobin, J. " Liquidity Preference as Behaviour Towards Risk ", *The Review of Economic Studies* (1957).

[21] Van De Panne, C. and Popp, W. " Minimum-Cost Cattle Feed Under Probabilistic Protein Constraints ", *Management Science*, **9** (1963) No. 9, 405-430.

COMPUTATIONAL AND REVIEW EXERCISES

1. It is of interest to determine which quadratic functions of mean and variance are consistent with expected utility (of wealth) maximization.

(a) Show that the function $E(u) = a + b\bar{R} - c[(\bar{R})^2 + \sigma_R{}^2]$ is consistent, where $\bar{R}$ and $\sigma_R{}^2$ denote mean and variance, respectively.

(b) What restrictions on the parameters a, b, and c are needed?

(c) Show that the function $E(u) = \alpha\bar{R} - \beta\sigma_R{}^2$ is not consistent.

(d) Show that it is not possible to transform the function in (a) to that in (b). [*Hint*: Consider the shape of the indifference curves in mean–variance space.]

2. In Eqs. (2) and (3) of Samuelson's article in Chapter 1 the use of a Taylor series approximation is illustrated.

(a) Develop bounds on the maximum error than can occur with such an approximation when the series is truncated at two, three, and m terms.

(b) Calculate the exact amount of this error when the series is truncated at two terms for the case when $U(W) = W - 0.1W^2 + 0.01W^3$, $n = 2$, and X_1 and X_2 are uniformly distributed over the intervals $[0, 1]$ and $[-1, 2]$, respectively.

3. Refer to Lintner's paper in Part II. To determine the optimal proportions of risky assets associated with the Tobin separation theorem, one may maximize a function of the form $f(x) = a'x/(x'\Sigma x)^{1/2}$, over some convex set K, where Σ is positive definite and $0 \notin K$.

(a) Show that f is pseudoconcave on $K \cap \{x \mid a'x \geqq 0\}$. [*Hint*: Refer to Mangasarian's "Composition" paper in Part I.]

(b) Show that f is strictly pseudoconcave on $K \cap \{x \mid a'x > 0\}$. [*Hint*: Utilize Theorem 4 in Ziemba's paper.]

(c) Show that the optimal proportions are unique if $a'x > 0$ and K is compact. [*Hint*: Utilize Theorem 5 in Ziemba's paper.]

(d) Show that maximizing $f(x) = a'x/(x'\Sigma x)^{1/2}$ is equivalent to maximizing

$$h(x) \equiv \log f(x) = \log a'x - \tfrac{1}{2}\log x'\Sigma x.$$

(e) Show that f is not concave as Lintner states. [*Hint*: Let

$$\Sigma = \begin{pmatrix} 1 & -\frac{1}{2} \\ -\frac{1}{2} & 1 \end{pmatrix}, \qquad a = (1, 1)'.$$

Choose $x^1 = (20, 10)'$, $x^2 = (1, 1)'$, and $\lambda = 0.8$. Then show that $f(\lambda x^1 + (1-\lambda)x^2) < \lambda f(x^1) + (1-\lambda)f(x^2)$.]

(f) Show that h is concave. [*Hint*: Utilize the Cauchy–Schwarz inequality, i.e., $(u'v)^2 \leqq (u'u)(v'v)$ for arbitrary n vectors u and v, on the Hessian matrix of h.]

(g) When is h strictly concave?

4. The problem is to consider the effects of tax changes on portfolio selection when the return on only one type of security is taxable, such as when one wishes to analyze the effect of income changes between tax exempt and taxable securities. Assume the utility function over wealth is $u(w) = (1-b)w - bw^2$, where $0 < b < 1$.

(a) Show that u defines units of utility so that $u(-1) = -1$ and $u(0) = 0$, and that $Eu(w) = (1-b)\bar{w} - b(\bar{w}^2 + \sigma^2)$. Let the investor's initial wealth be M and suppose that x_1 and x_2 are the dollar amounts invested in securities 1 and 2, which provide random returns ξ_1 and ξ_2 per dollar invested, respectively. If t is the perecntage tax rate on asset 1, the decision problem is to maximize

$$Eu(w) = (1-b)[x_1(1-t)\bar{\xi}_1 + x_2\bar{\xi}_2] \\ -b[[x_1(1-t)\bar{\xi}_1 + x_2\bar{\xi}_2]^2 + [x_1^2(1-t)^2\sigma_1^2 + x_2^2\sigma_2^2]],$$

subject to $x_1+x_2 = M$, where it is assumed that ξ_1 and ξ_2 have independent distributions having means $\bar{\xi}_1$ and $\bar{\xi}_2$ and variances σ_1^2 and σ_2^2, respectively.

(b) Show that

$$\frac{dx_1}{dt} = \frac{u_{1t}-u_{2t}}{2u_{12}-u_{11}-u_{22}},$$

where u_{ij} is the second partial derivative of u with respect to i and j. [*Hint*: Totally differentiate the first order conditions with respect to t, then combine these two equations.]

(c) Show that

$$\frac{dx_1}{dt} = \frac{2bx_1(1-t)\bar{\xi}_1[(1-t)\bar{\xi}_1 - \bar{\xi}_2] + [(1-b) - 2b\bar{w}][(1-t)\bar{\xi}_1 - 2\bar{\xi}_2] + 4bx_2\sigma_2^2}{(1-t)[2u_{12}-u_{11}-u_{22}]},$$

where $\bar{w} = x_1(1-t)\bar{\xi}_1+x_2\bar{\xi}_2$.

(d) Show that dx_1/dt is nonnegative if $2\bar{\xi}_2 < (1-t)\bar{\xi}_1$.

(e) Show that the sign of dx_1/dt can be negative even if $(1-t)\bar{\xi}_1 > \bar{\xi}_2$.

(f) Interpret the results in (c) and (d).

5. Refer to Leland's paper.

(a) Verify that an optimal policy exists when

$$U = \log, \qquad X = \{(X_1, X_2) \mid X_1 \geqq 0,\ X_2 \geqq 0\}, \qquad S = \{1,2,3\},$$

where each state of nature is equally likely, $g^1(1) = 1$, $g^1(2) = -1$, $g^1(3) = \frac{1}{2}$, $g^2(1) = 2$, $g^2(2) = 1$, and $g^2(3) = -\frac{1}{2}$.

(b) Verify that an optimal policy does not exist if $g^1(1)$ is changed from 1 to ∞.

(c) Verify that an optimal policy does exist if $g^1(1)$ is changed from 1 to ∞ and $p(s=1)$ is changed from $\frac{1}{3}$ to 0.

(d) Verify that an optimal policy does not exist if $g^1(2)$ is changed from -1 to 0.

(e) Verify that an optimal policy does exist if $g^1(2)$ is changed from -1 to 0 and X is changed to $\{(X_1, X_2) \mid \infty > d \geqq X_1 \geqq 0,\ X_2 \geqq 0\}$.

6. Refer to Leland's paper. Find utility functions possessing the properties $U'(c) > 0$ and $U''(c) < 0$ for all $c \in C$ that are

(a) bounded from above and below;

(b) bounded from above but not from below; and

(c) bounded from below but not from above.

(d) Construct a St. Petersburg paradox for the utility function in (c), i.e., a gamble having infinite expected utility for which the probability that one makes a positive gain goes to zero as the price of the gamble increases.

(e) How might one construct a St. Petersburg paradox for the utility function in (b)?

(f) Does an optimal policy always exist for the utility function in (a)?

(g) Suppose that the utility function is

$$U(y) = \begin{cases} y/(y+1) & \text{if } y \geqq 0, \\ -y/(y+1) & \text{if } y < 0 \end{cases}$$

and a gamble is offered that returns

$$\begin{cases}(r+a) & \text{with probability} \quad p,\\ (r-a) & \text{with probability} \quad (1-p),\end{cases}$$

where $p > \frac{1}{2}$, $a > 0$, initial wealth $B_0 = 0$, and r is the riskless rate of interest. Show that there is no optimal policy.

7. Consider a portfolio problem associated with the choice of asset proportions $x_j \geqq 0$ of independent Pareto–Levy investments $\rho_1, \ldots, \rho_n$. Referring to the notation in Ziemba's paper, let ρ_j have the distribution $F(\rho_j; \bar{\rho}_j, S_j, 0, \alpha)$, where $S_j > 0$ and $1 < \alpha \leqq 2$.

(a) Show that the mean-α-dispersion efficiency frontier may be generated by solving

$$\min f(x) \equiv \sum S_j x_j^{\alpha}, \qquad \text{s.t.} \quad x_1 + \cdots + x_n = 1, \qquad \bar{\rho}_1 x_1 + \cdots + \bar{\rho}_n x_n = \mu,$$
$$x_1 \geqq 0, \cdots, x_n \geqq 0$$

for all μ.

(b) Show that the $=$ signs in (a) may be replaced by $\geqq$ signs as long as $\min S_j > 0$ and $\min \bar{\rho}_j < \mu < \max \bar{\rho}_j$.

(c) Let a Lagrangian expression for the problem in (a) be

$$L(x, v_1, v_2) = f(x) + v_1\left(1 - \sum x_j\right) + v_2\left(\mu - \sum \bar{\rho}_j x_j\right).$$

Develop the Kuhn–Tucker conditions for the problem in (a), and show that they have a unique solution under the assumptions in (b).

(d) Suppose investments $j = 1, \ldots, m \leqq n$ have equal means and $0 < S_1 \leqq \cdots \leqq S_m$; then these investments may be optimally blended to form a single composite stable asset ρ_c. Recall from Samuelson's paper in Part III, Chapter 3, that it is always optimal to allocate a positive amount of one's resources in each investment whose mean is at least $\min \bar{\rho}_j$. Show that such a result applies to this case even if $\alpha < 2$. Develop the efficiency problem to form the composite asset $\rho_c = \sum_{j=1}^{m} y_j^* \rho_j$. Show that $\alpha S_j (y_j^*)^{\alpha - 1} = v_1$.

(e) Show that the optimal asset proportions satisfy

$$\frac{y_j^*}{y_k^*} = \left(\frac{S_j}{S_K}\right)^{1/(1-\alpha)} \qquad \text{and that} \qquad y_j^* = \frac{S_j^{1/(1-\alpha)}}{\sum_{k=1}^{m} S_k^{1/(1-\alpha)}}.$$

Interpret these expressions. Notice that when $\alpha = 2$, the normal distribution case, the product of asset proportion and variance is equal for each j.

The results in (d)–(e) indicate that without loss of generality the assets may be assumed to have different means, say $\bar{\rho}_1 > \bar{\rho}_2 > \cdots > \bar{\rho}_n$. By Samuelson's results, $x_j^* > 0$ implies $x_{j-i}^* > 0$; hence the only pattern of zeros in $(x_1^*, \ldots, x_n^*)$ must have the form $(x_1^*, \ldots, x_r^*, 0, \ldots, 0)$.

(f) Develop the Kuhn–Tucker conditions under the assumption that r is known.

(g) Reduce the Kuhn–Tucker conditions to the system

$$S_j x_j^{\alpha-1} - a_{j1} S_1 \left(b_1 \mu + \sum_{j=3}^{r} b_{1j} x_j\right)^{\alpha} - a_{j2} S_2 \left(b_2 \mu + \sum_{j=3}^{r} b_{2j} x_j\right)^{\alpha} = 0, \qquad j = 3, \ldots, r,$$

where the a's and b's are rational functions of $\bar{\rho}_1$ and $\bar{\rho}_2$.

(h) Develop a procedure to find the optimal r using the system of nonlinear equations in (g). When is $r = 1$ or n?

8. Consider an investor with a utility function u. Suppose his investment alternatives $i = 1, \ldots, n$ have independent identically distributed Cauchy distributions $\rho_i \sim F(\rho; \bar{\rho}, S, 0, 1)$, $S > 0$ (notation follows Ziemba's paper). Assume that the investor wishes to allocate his initial wealth of one dollar among these investments in proportions x_i so as to maximize expected utility. Suppose $Z(x) \equiv Eu(\rho'x) < \infty$ and that the investment allocation x must belong to the compact convex set K.

(a) Show that $\rho'x$ has the same distribution for all $x \in K$. [*Hint*: Utilize characteristic functions.]
(b) Show that Z is a constant on K so that all $x \in K$ maximize expected utility.
(c) Can the "identically distributed" assumption be deleted in (a)?
(d) Show that there exists no nontrivial concave utility function for which $Eu(\rho'x) < \infty$ if the ρ_i have Cauchy distributions.

9. Consider Press' class of stable distributions as discussed in Ziemba's paper. Suppose $m = 1$ and Ω_1 is positive definite. Suppose one wishes to maximize return subject to a risk constraint.

(a) Show that the risk may be considered to be $x'\Omega_1 x$, where $x \equiv (x_1, \ldots, x_n)'$ is the investment allocation vector. Suppose short sales are allowed and consider the problem $\{\max \bar{\rho}'x - \lambda x'\Omega_1 x \mid e'x = 1\}$, where $\bar{\rho} \equiv (\bar{\rho}_1, \ldots, \bar{\rho}_n)'$ is the vector of mean returns, e is a vector of ones, and $\lambda \geqq 0$. Assume that $\bar{\rho}_i = \bar{\rho}_j$ for all i and j.
(b) Show that the optimal allocation vector is $x^* = \Omega_1^{-1}e/e'\Omega_1^{-1}e$.
(c) Suppose $\Omega_1 = \sigma^2 I$, where I is an identity matrix of order n and $\sigma^2 > 0$. Show that $x_i^* = 1/n$ for each i.
(d) Suppose

$$\Omega_1 = \begin{pmatrix} {\sigma_1}^2 & & 0 \\ & \ddots & \\ 0 & & {\sigma_n}^2 \end{pmatrix}.$$

Show that $x_i^* = (1/{\sigma_i}^2)/(\sum 1/{\sigma_j}^2)$ for each i.
(e) Relate the results in (c) and (d) to those in Samuelson's paper on diversification.
(f) Suppose $\bar{\rho}$ is general. Show that $x^* = \Sigma^{-1}\bar{\rho}/e'\Sigma^{-1}\bar{\rho}$.
*(g) Attempt to modify the results in (c) and (d) for the case when $\bar{\rho}$ is general.

Let $n = 3$ and $\bar{\rho} = (1, 1, 1)'$. Compute x^* when

$$\text{(h)}\quad \Omega_1 = \begin{pmatrix} 8 & 0 & 0 \\ 0 & 6 & 0 \\ 0 & 0 & 4 \end{pmatrix}; \qquad \text{and} \qquad \text{(i)}\quad \Omega_1 = \begin{pmatrix} 6 & 2 & 0 \\ 2 & 4 & 1 \\ 0 & 1 & 5 \end{pmatrix}.$$

(j) Compute x^* when $\bar{\rho} = (6, 7, 8)'$ when Ω_1 is as given in (h) and (i), respectively.

10. Let $\bar{\rho} \equiv (\bar{\rho}_1, \ldots, \bar{\rho}_n)'$ be the mean return vector and Σ, an $n \times n$ positive semidefinite matrix, be the variance–covariance matrix of the return vector ρ. Suppose $x = (x_1, \ldots, x_n)' \geqq 0$ represents the investment allocation vector. The mean–variance efficient points are those $x(\mu)$ that $\{\text{minimize } x'\Sigma x \mid e'x = 1, x \geqq 0, \bar{\rho}'x \geqq \mu\}$, where $e \equiv (1, \ldots, 1)'$ and μ is a stipulated mean portfolio return.

(a) Suppose $\bar{\rho}_1 = \cdots = \bar{\rho}_n \geqq \mu$ and that Σ has the property that $\sigma_{11} \leqq \sigma_{1j}$ for $j = 2, \ldots, n$. Show that $x^* = (1, 0, \ldots, 0)$ is always on the efficient frontier. [*Hint*: Utilize the Kuhn–Tucker conditions.] Show that x^* is the unique point (in x space) on the efficient frontier if $\sigma_{11} < \sigma_{1j}$ for $j = 2, \ldots, n$.
(b) Show that the results in (a) are unaltered if $\bar{\rho}_1 \geqq \mu$ and $\bar{\rho}_1 \geqq \bar{\rho}_j$ for $j = 2, \ldots, n$.
(c) Interpret the results in (a) and (b).

11. Relaxing the constraints imposed in portfolio problems generally admits more efficient portfolios. We say that a portfolio is efficient if for its expected rate of return, no portfolio exists with a lower variance of return. Suppose that the investor must allocate his wealth among two risky assets and a riskless asset having zero net return (cash). The two risky assets have mean returns $\bar{\rho}_1$ and $\bar{\rho}_2$ and variance–covariance matrix of returns

$$\Sigma = \begin{pmatrix} \sigma_{11}^2 & \sigma_{12} \\ \sigma_{12} & \sigma_{22}^2 \end{pmatrix}.$$

We allow short sales and margin loans and suppose that the investment allocations satisfy $2 \geqq x_i \geqq -1$, and $1 \geqq \sum x_i \geqq 0$. Assume that brokerage costs are negligible and that investors retain other assets which may be used to offset margin calls.

(a) Suppose there is perfect positive correlation, i.e., $\sigma_{12}^2 = \sigma_{11}\sigma_{22}$. Let

$$\Sigma = \begin{pmatrix} 0.0004 & 0.0006 \\ 0.0006 & 0.0009 \end{pmatrix} \quad \text{and} \quad (\bar{\rho}_1, \bar{\rho}_2) = (0.05, 0.10).$$

Show that the efficient surface is described as follows. A zero variance portfolio with expected return 0.0167 is $(\frac{1}{3}, -1, \frac{2}{3})$. That is, the investor holds one-third of his wealth in cash and invests two-thirds of his wealth in ρ_2, and he acquires a short position in ρ_1 equal to his net worth.

The next points are described by reducing cash to zero and purchasing ρ_2 until the portfolio is $(0, -1, 1)$, which has an expected return of 0.05 and standard deviation of 0.01. The next section of the frontier moves to $(0, 0, 1)$ if the rate of interest on margin loans is exactly equal to $r > 0.05$, or if not, to $(-1, 0, 2)$. If $0.05 < r < 0.10$, the frontier passes from $(0, 0, 1)$ to $(-1, 0, 2)$ and is linear between these points. If $r \geqq 0.10$, the efficient surface terminates at $(0, 0, 1)$. Plot this efficient surface, and also the efficient surface corresponding to the constraints $\sum x_i = 1$, $x_i \geqq 0$. Compare these plots paying particular attention to the number of admissible efficient portfolios.

(b) Suppose $\bar{\rho}_2 = 0.06$ and all other parameters are as in (a). Show that cash dominates the riskless portfolio $(\frac{1}{3}, -1, \frac{2}{3})$. Show that the portfolio $(0, 1, -0.67)$ is an efficient riskless portfolio having expected return 0.10. Show that the next set of efficient portfolios lies on a curve between $(0, 1, -0.67)$ and $(0, 1, 0)$, the following set along the straight line from $(0, 1, 0)$ to $(0, 0, 1)$, and the final set is again linear between $(0, 0, 1)$ and $(-1, 0, 2)$. Plot this efficient surface and compare it with the conventional efficient surface as in (a).

(c) Suppose there is zero correlation, i.e., $\sigma_{12} = 0$. Show that no portfolio with zero variance and positive expected returns exists. Show that the minimum variance portfolio is $x_1/x_2 = \sigma_{22}/\sigma_{11}$ *if* cash is excluded. Recall that this result also holds when short sales and margin loans are not allowed. When should short sales be made? Develop the efficient frontier when $\bar{\rho}_1 = 0.10$, $\bar{\rho}_2 = 0.12$, $\sigma_{11} = 0.05$, $\sigma_{22} = 0.08$, and $r = 0.10$.

(d) Consider the case in (c) with $\sigma_{12} = 0$ and other parameters general. Suppose $r \geqq x_1{}^*\bar{\rho}_1 + (1 - x_1{}^*)\bar{\rho}_2$, where $(x_1{}^*, 1 - x_1{}^*)$ is the minimum risk portfolio. Show that the investor will never acquire a margin loan.

(e) Investigate the case with perfect negative correlation, i.e., $\sigma_{12} = -\sigma_{11}\sigma_{22}$.

12. Consider an investor with utility function $u(w)$ over wealth w, and assume that the investor knows the distribution of w and that he is an expected utility maximizer.

(a) Suppose $u(w)$ is the polynomial $\sum_{i=0}^{n} a_i w^i$. Show that the expected utility may be expressed as a deterministic function of the first n moments of w.

(b) Suppose the expected utility depends only on the first n moments of w. Show that the utility function need not be a polynomial of order n.

(c) Under what conditions will a polynomial utility function be concave? Quasi-concave?

(d) Suppose that $w = \sum_{j=1}^{m} \rho_j x_j$, where the ρ_j have a known joint distribution, and that $u(w)$ is polynomial. Show that the expected utility is a polynomial function of the x_j, where the coefficients are certain moments of the ρ_j.

(e) Suppose u is general and has continuous derivatives of order n, and that $w = \sum_{j=1}^{m} \rho_j x_j$. A function $f(x)$ is called a signomial if $f(x) \equiv \sum a_i g_i(x)$, where each $g_i(x) = \prod_{j=1}^{m} x_j^{\alpha_j}$, with each α_j being an integer and each a_i being an arbitrary constant. Show how one may obtain an approximate signomial utility function using a Taylor series approximation.

(f) Suppose $u(\sum \rho_j x_j) = \sum u_j(\rho_j x_j)$. Show that the approximation in (e) reduces to a polynomial of degree n.

(g) Referring to (c), (e), and (f), is it true that the Taylor series approximation will result in a concave function of the x_j if u was a concave function of w?

13. Consider an investor who wishes to allocate B dollars among n investments. Suppose the return per dollar invested in i is ξ_i. Let $\xi = (\xi_1, \ldots, \xi_n)'$ and let the mean and variance–covariance matrix of ξ be $\bar{\xi}$ and Σ, respectively. One way to compute the frontier of the mean-variance efficient surface is to $\{\max \bar{\xi}'x \mid x'\Sigma x \leqq \alpha,\ e'x = B,\ x_i \geqq 0,\ i = 1, \ldots, s \leqq n\} \equiv \phi(\alpha)$, for all $\alpha \geqq 0$, where $e = (1, \ldots, 1)'$.

(a) Prove that ϕ is a concave function of α.

(b) Show that ϕ is still concave if the constraint $(x'\Sigma x)^{1/2} \leqq \alpha$ replaces the constraint $x'\Sigma x \leqq \alpha$.

(c) Illustrate the graphical and economic interpretation of (a) and (b).

(d) Show that the optimal x_i/x_j ratios are independent of B. [*Hint*: Utilize the Kuhn–Tucker conditions.]

(e) Utilize (d) to show that x is on the efficient surface if and only if $x(C/B)$ is on the efficient surface if the investor's initial wealth is C dollars. Hence, one may compute the efficient surface utilizing $B = 1$ and then scale the resulting optimal x_i so that they sum to B.

An alternative way to generate the efficient surface is to

$$\{\min x'\Sigma x \mid \bar{\xi}'x \geqq \beta,\ e'x = 1,\ x_i \geqq 0,\ i = 1, \ldots, s \leqq n\} \equiv \psi(\beta)$$

for all β.

(f) Prove that ψ is convex in β.

(g) When can the computations be limited to $\beta \geqq 0$?

Assume that the inequality constraints $x'\Sigma x \leqq \alpha$ and $\bar{\xi}'x \geqq \beta$ are binding for all $\alpha \geqq 0$ and β.

(h) Show that the two approaches give the same efficient surface. [*Hint*: Utilize the facts that x^* is optimal for $\{\max f(x) \mid x \in K\}$ if and only if it is optimal for $\{\max \alpha \mid f(x) \geqq \alpha,\ x \in K\}$ as long as $f(x) \geqq \alpha$ is binding at x^*, and that $\tilde{x}$ is optimal for $\{\min e(x) \mid x \in L\}$ if and only if it is optimal for $\{\min \beta \mid e(x) \leqq \beta,\ x \in L\}$ as long as $e(x) \leqq \beta$ is binding at $\tilde{x}$.]

14. Referring to Exercise 13, suppose $s = 0$, i.e., short sales are allowed in all securities. Consider the second formulation

$$f(\beta) = \{\min x'\Sigma x \mid \bar{\rho}'x \geqq \beta,\ e'x = 1\}. \tag{1}$$

(a) When is the constraint $\bar{\rho}'x \geqq \beta$ binding?

(b) Assuming that the constraint in (a) is binding show that (1) is equivalent to

$$\{\min x'\Sigma x - \lambda\bar{\rho}'x \mid e'x = 1\} \qquad \text{for} \quad \lambda \geqq 0. \tag{2}$$

(c) Interpret (2).

(d) Develop the Kuhn–Tucker conditions for (2) and show that the optimal x_i are linear functions of λ.

(e) Assuming that one wished to compute an efficient frontier, show how the result in (a) might simplify this calculation.

Consider problem (1) and suppose that Σ is positive definite and β, $\bar{\rho}$, and n are such that $\bar{\rho}'x \geqq \beta$ is always binding.

(f) Show that $x^* = a_1 + a_2\beta$ and $f(\beta) = b_1 + b_2\beta + b_2\beta^2$, where

$$a_1 \equiv \Sigma^{-1}(d\bar{\rho} - ec)/(c(adc-b)), \qquad a_2 \equiv \Sigma^{-1}(ac^2e - b\bar{\rho})/(c(adc-b)),$$

$$b_1 = a_1'\Sigma a_1, \qquad b_2 = 2a_1'a_2, \qquad b_3 = a_2'a_2,$$

$$a \equiv e'\Sigma^{-1}\bar{\rho}, \qquad b \equiv e'\Sigma^{-1}e, \qquad c \equiv \bar{\rho}'\Sigma^{-1}\bar{\rho}, \qquad \text{and} \qquad d \equiv \bar{\rho}'\Sigma^{-1}e.$$

[*Hint*: Develop the Kuhn–Tucker conditions for (1), then utilize the constraints to eliminate the Lagrange multipliers.]

(g) Consider the parametric problem $Z(t) \equiv \{\min g(x) \mid Ax = t,\ Bx = b\}$, where g is a convex function, A and B are constant matrices, b is a constant vector, and t is a vector parameter. Show that Z is a convex function. [*Hint*: Choose t^1 and t^2 from $\{t \mid \exists x\colon Ax = t,\ Bx = b\}$ and let x^1 and x^2 be optimal values when $t = t^1$ and $t = t^2$, respectively, and so on.]

(h) When is Z strictly convex? [*Hint*: Refer to Exercise I-ME-18.]

(i) Referring to (f), utilize (h) to show that f is strictly convex.

(j) Interpret the results in (f). What does it mean to have minimum variance equal to a quadratic function of mean return?

15. (A diversification experiment) Consider the model in Exercise 14. Estimate the mean and variance–covariance matrix for a group of m securities. Devise a preferential ordering scheme so that securities with higher numbers are less preferred, such as the scheme $i = 1$ for $\max_i \mu_i/\sigma_i^2$, etc.

(a) Calculate x_n^* and $f(\beta)_n$ for selected β values, where n $(1 \leqq n \leqq m)$ refers to the optimal portfolio when the n most preferred securities are considered to be the portfolio universe.

(b) Let

$$\psi_n = \frac{f(\beta)_n - f(\beta)_m}{f(\beta)_1 - f(\beta)_m}.$$

Show that $\psi_1 = 1$, $\psi_m = 0$, and that ψ is a decreasing function of n, where n takes on integer values from 1 to m.

(c) Interpret (b) as a measure of the diversification effect of a portfolio of size n.

(d) Compute the graph of ψ_n versus n for your data.

(e) Interpret the results. Is there an effect of the choice of β?

(f) Repeat the experiment with several other sets of data. Are the plots of ψ_n versus n substantially different?

*(g) Attempt to develop an efficient algorithm for choosing the best n out of m securities for the model in Exercise 14, when $m-n$ of the securities must have zero allocation weight.

16. Prove the following results that are utilized in Ziemba's paper:

(a) Suppose $\rho_1, \ldots, \rho_n$ have independent stable distributions $F(\rho; \bar{\rho}_i, S_i, 0, \alpha)$, where $S_i > 0$ and $1 < \alpha \leqq 2$. Show that $\Sigma \rho_i x_i$ has the stable distribution $F(\rho'x; \bar{\rho}'x, \sum S_i |x_i|^\alpha, 0, \alpha)$. [*Hint*: Utilize the multiplicative property of characteristic functions of sums of independent random variables.]

(b) Complete the proof of part (d) of Theorem 1 for cases (b) and (c).

(c) Show that all the α-dispersion measures used in the paper are positively homogeneous.

(d) Verify Minkowski's inequality: Suppose $M_r(y) \equiv \{(1/n)\sum_{i=1}^n y_i^r\}^{1/r}$, $\infty > n \geqq r > 1$, and that the n-vectors a and b are not proportional (i.e., constants q_1 and q_2 not both zero do not exist such that $q_1 a = q_2 b$). Then $M_r(a) + M_r(b) > M_r(a+b)$.

(e) Show that $\{\sum_{i=1}^{n} S_i x_i^{\alpha}\}^{1/\alpha}$ for $S_i > 0$ and $2 \geqq \alpha > 1$ is convex but not strictly convex on the nonnegative orthant.

(f) Prove Lemma 2. Attempt to show that ϕ is strictly concave if the $S_i > 0$.

(g) Show that a normal variate is strictly pseudo-concave on E.

(h) Illustrate a pseudo-concave function that is not strictly pseudo-concave.

(i) Prove Lemma 3. [*Hint*: Recall from calculus that

$$\lim_{\tau \to 0} \left[\frac{f(x+\tau d) - f(x)}{\tau} \right] = \nabla f(x)'d.]$$

(j) Verify the result in footnote 9.

*(k) Consider part (c) of Theorem 7. Attempt to prove under suitable assumptions that f is strictly convex on M.

(l) Prove Theorem 8.

(m) Show that $|\hat{x}'b_j|^{\alpha}$ is convex in $\hat{x} = (x_1, \ldots, x_n)$ for any given n-vector b_j.

*(n) Refer to the last paragraph of the paper. Attempt to find assumptions on u and γ so that the function Z in Eq. (13) is pseudo-concave.

17. Refer to Theorem 1 in Ziemba's paper and Exercise ME-13. Let $Z(x) \equiv E_{\rho}[f(x,\rho)]$, where $Z: K \to E, f: K \times \Xi \to E$, and K and Ξ are subsets of E^n and E^s, respectively. Investigate the continuity and differentiability properties of Z and $\partial Z/\partial x_j$ when

(a) $n = s = 1, f(x,\rho) = (x-\rho)^2$, and ρ has a normal distribution;

(b) as in (a), except that ρ has a symmetric stable distribution with characteristic exponent 1.5;

(c) as in (a), except that ρ has the stable distribution $F(\rho; 1, 1, 1, 1.5)$;

(d) as in (a), except that ρ has the stable distribution $F(\rho; 0, 1, -1, 1.5)$.

(e) $n = s = 1, f(x,\rho) = |x\rho|^{\alpha}$, and ρ is uniform on $[a, 1]$, for $\gamma = -\frac{1}{2}, \frac{1}{2}$, and 1 and $a = -1$ and 0.

(f) as in (e), except that γ has the symmetric stable distribution $F(\rho; 0, 1, 0, \alpha)$ for $\alpha = 1$, 1.5, and 2.

18. Verify the following statements made in the Stiglitz paper: Given a risk-free asset with net rate of return $r \geqq 0$ and a risky asset with net rate of return $\rho \geqq -1$, then:

(a) The amount invested in the risky asset is nondecreasing in wealth if $A(w) = -u''(w)/u'(w)$ is nonincreasing in w.

(b) The proportion of wealth held in the risky asset is nonincreasing in wealth if $R(w) = -wu''(w)/u'(w)$ is nondecreasing in w.

19. Suppose the investor in Exercise II-CR-22 has $w_0 = \$1000$ initially.

(a) Find the optimal policy if income is taxed at a rate of 20% with full loss offset.

(b) Formulate the investor's problem if positive income is taxed at a rate of 20% but negative income is not taxed.

20. If R is a random variable having mean $\bar{R}$ and variance σ_R^2, show;

(a) $P[|R-\bar{R}| \geqq \bar{R}-d] \leqq \sigma_R^2/(\bar{R}-d)^2$ (Tchebychev inequality).

(b) $P[R \leqq d] \leqq \sigma_R^2/(\bar{R}-d)^2$.

Suppose $R = w_0[1+r+a(\rho-r)]$, where $r \geqq 0$ and ρ is a random variable. Let $A \equiv \{a \mid \bar{R} \geqq d\}$. Let A^* be the solution set of the problem $\min\{P[R \leqq d] \mid a \in A\}$ and A^{**} be the solution set of the problem $\min\{\sigma_R^2/(\bar{R}-d)^2 \mid a \in A\}$.

(c) Show that $A^{**} \subset A^*$ and interpret this result. When does $A^{**} = A^*$?

(d) What is the corresponding result if a is not restricted to the set A?

21. Let $\rho_1, \ldots, \rho_n$ be joint-normally distributed with means $\mu_1, \ldots, \mu_n$ and variance–covariance matrix Σ. Let $R = w_0(1 + x_0 r + \sum_{i=1}^n x_i \rho_i)$, and let

$$K = \left\{ (x_0, x_1, \ldots, x_n) \mid x_0 + \sum_{i=1}^n x_i = 1,\ x_0 \geqq 0 \right\}.$$

(a) Show that $\min\{P[R \leqq d] \mid x \in K,\ \bar{R} \geqq d\}$ is equivalent to
$$\min\{\sigma_R^2/(\bar{R} - d)^2 \mid x \in K,\ \bar{R} \geqq d\}.$$
(b) Show that any solution x^* to the problem given above is mean-variance efficient.
(c) Show that $x_0^* = 0$ if $d > w_0(1+r)$, $x_0^* = 1$ if $d < w_0(1+r)$. Show that x_0^* is undetermined if $d = w_0(1+r)$.

22. Derive Eq. (1.9)–(1.15) and (3.15)–(3.21) in Näslund's paper.

23. Refer to the paper by Pyle and Turnovsky. Suppose the returns have a multivariate symmetric stable distribution as discussed in Section III of Ziemba's paper. Suppose that the α-dispersion measure $\{\gamma(t)\}^{1/\alpha}$ is convex and positively homogeneous and that $1 < \alpha \leqq 2$. Assume that the investor's objective is to maximize the expectation of a concave utility function u defined over wealth w and that $|Eu(w)| < \infty$.

(a) Consider the three versions of the safety-first principle (i)–(iii). Develop the deterministic equivalents that are analogous to (4)–(6), using the mean and α-dispersion in place of the mean and standard deviation, respectively.
(b) Show that the deterministic equivalents have sufficient generalized convexity properties so that the Kuhn–Tucker conditions will be necessary and sufficient for optimality.

Suppose there is no riskless asset.

(c) Show that the indifference curves between mean and α-dispersion are concave.
(d) Show that there exists a unique safety-first investor who will choose the same optimal portfolio as that chosen by the investor.
(e) Show that the converse of the statement in (d) is not true so that if $\hat{x}$ is an optimal safety-first decision, there may be many u's that lead to the choice $\hat{x}$.
(f) Interpret the result in (e) if u is strictly concave and hence for a given u there is a unique maximizing point.
(g) Consider the case when there is a risk-free asset. Show that the conclusions in Section IV also apply for the case at hand.

24. Refer to the paper by Pyle and Turnovsky. Consider the fractile objective

$$\{\min \alpha \mid \Pr\{z \leqq \bar{z}\} \leqq \alpha\}. \tag{1}$$

(a) Show that this problem may be restated as $\{\min \alpha \mid \bar{z} \leqq \mu + F^{-1}(\alpha)\sigma\}$.
(b) Assuming that F represents a standardized normal variate, show that $F^{-1}(\alpha)$ is quasi-concave so that the feasible region for the problem in (a) is convex. [*Hint*: F is strictly increasing, hence F^{-1} is strictly decreasing.]
(c) Suppose that $\bar{z}$ is fixed and is finite. Show that (1) has a unique solution.
(d) Show that the indifference curves in μ–σ space are linear.
(e) Show that the results in the Pyle–Turnovsky paper also apply to problem (1).

25. Refer to Samuelson's first article. Are the following univariate distributions $F(x)$ compact (assume that all relevant parameters are finite and nonzero.)

(a) $F(x) \sim$ normal (μ, σ^2)
(b) $F(x) \sim$ exponential (λ)
(c) $F(x) \sim$ uniform on $[a, b]$

(d) $F(x) \sim$ log-normal (μ, σ^2)
(e) $F(x) \sim$ stable $(x; \mu, S, 0, \alpha)$, $0 < \alpha \leqq 2$.

26. Refer to Samuelson's first article.
(a) Interpret the condition in (7) that

$$\lim_{\sigma \to 0} \frac{E[Z_i^r]}{E[Z_i^2]} = \sigma^{r-2} C_r \qquad \text{for} \quad r \geqq 3.$$

(b) Let $O(\alpha)$ have the property that $\lim_{\alpha \to 0}(O(\alpha)/\alpha) = M$, where $M < \infty$. Can the condition in (a) be weakened to

$$\lim_{\sigma \to 0} \frac{E[Z_i^r]}{E[Z_i^2]} = O(\sigma^{r-2})\, C_r?$$

If so, when will the $w_i(0)$ be continuous? Differentiable? [*Hint*: Consult Exercise ME-30.]
(c) Show that the P family defined in (4)–(5) has property (7) in the two-dimensional case.
(d) Verify the statements associated with the compact family defined in (10).
(e) Devise an example to show that cubic approximations, although generally better than quadratic approximations, are not always better.
(f) Verify and interpret Eq. (14).
(g) Suppose in Eq. (13) that $\sigma_1^2 = 1$, $a = \frac{1}{2}$, and $u(w) = -e^{-2w}$. Compute $w^*(0)$, $w^{**}(1)$, $w^{**}(\frac{1}{2})$, $w^{**}(\frac{1}{4})$, $w^{**}(\frac{1}{8})$, and $w^{**}(0)$.

27. Suppose the investor may choose among the following assets:

	Outcomes			
Assets	Net rate of return	Probability	Expectation of return	Standard deviation
A	0.05	1.00	0.05	0.00
B	0.10	0.33		
	0.05	0.33	0.05	0.041
	0.00	0.33		
C	0.20	0.33		
	0.10	0.33	0.10	0.082
	0.00	0.33		
D	0.30	0.083		
	0.10	0.833	0.10	0.082
	−0.10	0.083		
E	0.30	0.50		
	0.00	0.05	0.105	0.195
	−0.10	0.45		
F	0.0605	0.99		
	−1.00	0.01	0.05	0.105
G	1.10	0.01	0.05	0.105
	0.0395	0.99		

(a) Show that C dominates B.

Consider the following four utility functions:

r	$U_1(r)$	$U_2(r)$	$U_3(r)$	$U_4(r)$
−1.00	−1.00	−1.00	−1.00	−1.00
−0.10	−0.10	−0.082	−0.19	−0.82
0.00	0.00	0.00	0.00	0.00
0.0395	0.0395	0.0313	0.0806	0.0313
0.05	0.05	0.0395	0.1025	0.0395
0.0605	0.0605	0.0477	0.1246	0.0477
0.10	0.10	0.078	0.21	0.078
0.20	0.20	0.152	0.44	0.152
0.30	0.30	0.222	0.69	0.280
1.10	1.10	0.640	3.41	1.20

(b) Graph these utility functions. Discuss their properties.

(c) Compute the expected utility of each asset for the various utility functions and rank the corresponding assets. [*Hint*: Note that $U_1(r) = r$, $U_2(r) = 0.8r - 0.2r^2$, and $U_3(r) = 2r + r^2$.]

Exercise Source Notes

Portions of Exercise 3 were adapted from Ziemba *et al.* (1974); Exercise 4 was adapted from Penner (1964); portions of Exercises 7 and 8 were adapted from Samuelson (1967); portions of Exercise 9 were adapted from Press (1972); the idea for Exercise 10 was suggested by Professor J. Ohlson; Exercise 11 was adapted from Hester (1967); portions of Exercise 12 were adapted from Richter (1960) and Ziemba (1974); portions of Exercise 13 were adapted from Sharpe (1970); some results similar to those in Exercise 14 were independently derived in Merton (1972); and Exercise 27 was adapted from Tobin (1965).

MIND-EXPANDING EXERCISES

1. Suppose u is the exponential utility function $u(w) = 1 - e^{-aw}$, $a > 0$, and that wealth $w = \sum_{i=1}^{n} \xi_i x_i \equiv \xi' x$, where ξ_i is the return per dollar invested in i and x_i is the level of investment i. Suppose that the random vector ξ is joint-normally distributed with finite mean $\bar{\xi}$ and variance–covariance matrix Σ (all of whose elements are assumed finite). Show that the problem of maximizing the expected utility is equivalent to maximizing the concave quadratic function $\bar{\xi}'x - (a/2)\, x'\Sigma x$.

2. Consider an investor who wishes to allocate his resources B among investments $i = 1, \ldots, n$, which have returns per dollar invested of $(\xi_1, \ldots, \xi_n) \equiv \xi$, where the ξ_i have the joint-normal distribution $F(\xi_1, \ldots, \xi_n)$. Let the return $R \equiv \sum_{i=1}^{n} \xi_i x_i$. For any given $x = (x_1, \ldots, x_n)$, R has a normal distribution $f_x(R)$ which has mean $\bar{\xi}'x$ and variance $x'\Sigma x$, where $\bar{\xi}$ and Σ are the mean vector and variance–covariance matrix of ξ. The investor wishes to choose the x_i so that expected return is maximized subject to budget and nonnegativity constraints on the x_i and the constraint that the return is at least A with probability at least $1 - \alpha$, where $1 \geqq \alpha \geqq 0$.

(a) Formulate this chance-constrained programming problem.

(b) Obtain a deterministic equivalent for the problem.

(c) Consider the formulation in (a) for a general $f_x(R)$. Show that the prospect $p_r\{R = A+1\} = 1 + \varepsilon - \alpha$, and $p_r\{R = A-1\} = \alpha - \varepsilon$ is preferred to the prospect $p_r\{R = A + 10{,}000\} = 1 - \varepsilon - \alpha$, and $p_r\{R = A-1\} = \alpha + \varepsilon$ for any $\varepsilon > 0$.

(d) What does the result of (c) mean?

(e) Show that there is no utility function $u(R)$ that is consistent in an expected utility sense to the decision rule obtained from (b), if $\alpha \neq 0$.

(f) Show that the preference ordering can be represented by the utility function

$$u(R) = \begin{cases} -\infty & \text{if } \; R < A \; \text{ and } \; \alpha = 0 \\ R & \text{otherwise.} \end{cases}$$

(g) What does $\alpha = 0$ mean?

3. The Tobin separation theorem as described in Lintner's paper in Part II shows that the efficient surface is a ray if there exists a risk-free asset. It is of interest to know if the efficient surface converges to that ray if there is one asset that becomes less and less risky. Suppose there are n risky assets that have mean returns $\bar{\xi}_i$ and positive-definite variance–covariance matrix Σ, and a risky asset $i = 0$ that is independent of $i = 1, \ldots, n$ having mean $\bar{\xi}_0$ and variance β. One may calculate the mean–variance efficient surface by computing

$$l(\alpha, \beta) \equiv \left\{ \min \left[x' \begin{pmatrix} \beta & 0 & \cdots & 0 \\ 0 & \Sigma & & \\ \vdots & & & \\ 0 & & & \end{pmatrix} x \right]^{1/2} \,\middle|\, e'x = 1,\ \bar{\xi}'x \geqq \alpha,\ x \geqq 0 \right\} \qquad \text{(i)}$$

for all α, where $x \equiv (x_0, x_1, \ldots, x_n)'$ are the investment allocations in $i = 0, \ldots, n$, $e = (1, 1, \ldots, 1)$, and $\bar{\xi} = (\bar{\xi}_0, \bar{\xi}_1, \ldots, \bar{\xi}_n)$.

(a) Show that $l(\alpha, \beta)$ is strictly concave in α for all $\beta > 0$, assuming that $x^*(\alpha)$ is one-to-one. [*Hint*: Begin by showing that the matrix in the objective of (i) is positive definite.]

(b) Prove that $\lim_{\beta \to 0} l(\alpha, \beta)$ is linear in α.

4. In most stochastic optimizing models, the sequencing of the decision and random elements of the problem results in different model formulations, algorithms, and economic interpretations. If the utility of decision choice x and random event ξ is $u(\xi, x)$, then these two models are

$$Z_A \equiv \{\max [E_\xi u(\xi, x)] \mid x \in K\} \qquad \text{and} \qquad Z_B \equiv \{E_\xi [\max u(\xi, x)] \mid x \in K\},$$

where K is the domain of x, and E_ξ represents expectation with respect to ξ, where $\xi \in \Xi$.

(a) Show that $Z_B \geqq Z_A$.

(b) Suppose $u(\xi, x) = f(x) + g(x)\xi + h(\xi)$. Show that the optimal decisions under these approaches are identical.

(c) When is $Z_B = Z_A$ for the objective in (b)?

(d) Show that optimal decisions under these two aprpoaches are the same if K is the interval $I \equiv [a, b]$, $u(\xi, x) = n(\xi)\, m(x)$, where n is nonnegative and m is monotone nondecreasing on I.

(e) Illustrate other combinations of restrictions on u, K, and Ξ for which optimal decisions under these approaches are the same.

(f) Show that optimal decisions under these two approaches are not generally the same in the bilinear case. [*Hint*: Let $u(\xi, x) = \xi' x$, K be the interval $[0, b]$, and suppose ξ equals $+1$ and -1 with equal probabilities.]

5. Referring to Leland's paper, prove or disprove (via a counterexample) the following conjectures that there exist optimal policies under the stated assumptions.

(a) Suppose $U'(c) > 0$, $U''(c) > 0$ for all $c \leqq \alpha < \infty$, and $U''(c) < 0$ for all $c > \alpha$, $\lim_{c \to \infty} u'(c) = 0$, X.1–X.3 are satisfied, and $E[g^i(s)] < \infty$ for all i.

(b) Suppose $U'(c) > 0$, $U''(c) > 0$ for all $c \leqq \alpha < \infty$, and $U''(c) < 0$ for all $c > \alpha$, $U(c)$ is bounded from above, X.1–X.3 are satisfied and $E[g^i(s)] < \infty$ for all i.

(c) The same as (b) except that no restriction is made on $E[g^i(s)]$.

(d) Suppose $U'(c) > 0$, $U''(c) < 0$ for all $c \in C$, $\lim_{c \to \infty} U'(c) = 0$, and X.1–X.3 are satisfied.

(e)–(h) The same assumptions as in (a)–(d), except that $C = E$, $U''(c) > 0$ for all $c < 0$, and $U''(c) < 0$ for all $c > 0$ with $U''(0) = 0$.

6. Financial intermediaries issue claims on themselves and then use the proceeds to purchase other financial assets. Suppose that a firm can purchase or sell short three securities: $i = 0$ (a riskless asset having return r per dollar invested in it), $i = 1$ (loans), and $i = 2$ (deposits). Suppose ξ_j $(j = 1, 2)$ is the return per dollar invested in security j consisting of appreciation plus income yield, where the ξ_j have the joint density $F(\xi_1, \xi_2)$, having means $\bar{\xi}_1$ and $\bar{\xi}_2$, respectively. Let x_j represent the number of dollars invested in security j. The firm is then said to be a financial intermediary if the firm finds it optimal to engage in the sale of deposits $(x_2 < 0)$ and the purchase of loans $(x_1 > 0)$. Suppose that initial wealth is K and that the investor wishes to maximize the expected utility of terminal wealth w, where his utility function u is strictly concave, continuously differentiable, and strictly increasing.

(a) Show that $w = (\xi_1 - r)x_1 + (\xi_2 - r)x_2 + rK$.

(b) Let $F(x_1, x_2) \equiv E_\xi u(w)$. Show that F is a strictly concave function; make appropriate assumptions along the lines suggested in exercise I-ME-18.

(c) Show that the expected marginal utilities with respect to x_1 and x_2 are

$$F_1(x_1, x_2) = E_\xi [u'(w)(\xi_1 - r)]$$

and

$$F_2(x_1, x_2) = E_\xi [u'(w)(\xi_2 - r)],$$

respectively.

(d) Suppose $F_1(0, x_2) > 0$ for all $x_2 \leqq 0$ and $F_2(x_1, 0) < 0$ for all $x_1 \geqq 0$. Show that the optimum (if it exists) implies intermediation.

(e) Suppose ξ_1 and ξ_2 are independent. Show that a necessary and sufficient condition for intermediation is that $\bar{\xi}_1 > r > \bar{\xi}_2$.
(f) Interpret the result in (e), noting that $\bar{\xi}_j - r$ may be considered to be a risk premium for security j.

It is said that ξ_j and ξ_k are positively dependent if $\bar{\xi}_j(\xi_k)$ is strictly increasing in ξ_k.
(g) Show that $F_j(0, x_k)$ given positively dependent yields is greater than, equal to, or less than $F_j(0, x_k)$ given independently distributed yields as x_k is less than, equal to, or greater than zero.
(h) Utilizing the result of (g) show that a sufficient but not a necessary condition for intermediation is that $\bar{\xi}_1 > r > \bar{\xi}_2$.
(i) Develop restrictions on F and/or u that lead to a sufficient condition for intermediation when $\bar{\xi}_1 > r$ and $\bar{\xi}_2 > r$, and ξ_1 and ξ_2 are positively dependent.

7. Refer to the description of Exercise 6. Suppose that the firm has a utility function G defined over mean μ and variance σ^2, which has the property that $\partial G/\partial\mu > 0$ and $\partial G/\partial\sigma^2 < 0$. Let $\xi_j' \equiv \xi_j - r$, $j = 1, 2$; then $\mu = \bar{\xi}'x_1 + \bar{\xi}_2'x_2 + rK$ and $\sigma^2 = x_1\sigma_{11} + 2x_1x_2\sigma_{12} + x_2{}^2\sigma_{22}$.
(a) Show that the optimal investments are

$$x_1{}^* = \theta\frac{\bar{\xi}_1'\sigma_{22} - \bar{\xi}_2'\sigma_{12}}{\sigma_{11}\sigma_{22} - \sigma_{12}^2}, \qquad x_2{}^* = \theta\frac{\bar{\xi}_2'\sigma_{11} - \bar{\xi}_1'\sigma_{12}}{\sigma_{11}\sigma_{22} - \sigma_{12}^2}, \qquad \text{where} \quad \theta \equiv \frac{1}{2}\frac{\partial F/\partial\mu}{\partial F/\partial\mu^2}.$$

(b) Show that $\theta = \sigma^2/\mu$.
(c) Interpret (b) in the light of the definition of θ.
(d) Show that necessary and sufficient conditions for intermediation are $\bar{\xi}_i'\sigma_{22} > \bar{\xi}_2'\sigma_{12}$ and $\bar{\xi}_2'\sigma_{11} < \bar{\xi}_1'\sigma_{12}$.
(e) Show that given independence between the two yields the firm will choose its loan portfolio independent of the parameters of the deposit yield (and vice versa).
(f) Suppose there is a positive risk premium on loans and deposits, i.e., $\min(\bar{\xi}_1', \bar{\xi}_2') > 0$. Show that a necessary and sufficient condition for intermediation is $\rho\bar{\xi}_1'\sigma_2 > \bar{\xi}_2'\sigma_1$, where ρ is the correlation coefficient between ξ_1 and ξ_2.
(g) Interpret the result in (f).

8. Suppose the wealth that an investor has after making certain investments is w, which has a known cumulative distribution function F. The semivariance of w at a point h is defined to be

$$S_h = \int_{-\infty}^{h} (w-h)^2\, dF(w) > 0.$$

(a) Suppose that $h = \bar{w}$ and that F is symmetric. Show that $S_h = \frac{1}{2}\sigma^2$, where σ^2 is the variance of w.

Consider the utility function

$$u(w) = aw + b[\min\{w-h, 0\}^2].$$

(b) Graph u and note that it is quadratic for $w < h$ and linear for $w \geqq h$. When is u concave?
(c) Show that maximizing the expected utility is equivalent to maximizing $a\bar{w} + bS_h$.
(d) Show that the indifference curves $S_h = \phi(\mu)$ are strictly monotone increasing and strictly concave if $a > 0$ and $b < 0$.
(e) Devise a method for computing the μ–S_h efficient surface. [*Hint*: See Exercise 24.]
(f) Devise a method for computing the μ–s_h efficient surface, where s_h is the positive square root of S_h.

(g) Discuss the use of semivariance as a measure of a risky asset particularly in relation to variance. Illustrate some advantages and disadvantages.

(h) Suppose there are two projects A and B that have the following joint probability of outcome table per dollar invested in each investment:

		B	
	0	2	5
A 0	0.2	0.05	0.05
3	0.05	0.1	0
4	0.05	0.2	0.2

Perform μ–σ^2, μ–σ, μ–S_h, and μ–s_h analyses, where it is assumed that h and the investors' initial wealth is one dollar.

(i) Discuss the results of (h).

9. Suppose an investor's utility function is $u(w) = 1 - \lambda e^{-\gamma w} - (1-\lambda)e^{-\delta w}$, where $0 \leqq \lambda \leqq 1$, $\gamma > 0$, and $\delta > 0$.

(a) Show that u is concave, nondecreasing, bounded from above, and has decreasing absolute risk aversion.

Suppose that the investment returns have a multivariate stable distribution of Press' class as discussed in Ziemba's paper.

(b) Show that

$$Eu(w) = 1 - \lambda \exp[-\gamma\bar{\rho}'x + \gamma^{\alpha} r(x)] - (1-\lambda) \exp[-\delta\bar{\rho}'x + \delta^{\alpha} r(x)] \equiv \psi(x),$$

where $x \equiv (x_1, \ldots, x_n)' \geqq 0$ are the investment allocations, $1 < \alpha \leqq 2$, $r(x) \equiv \frac{1}{2}\sum_{j=1}^{m}(x'\Omega_j x)^{\alpha/2}$ is the dispersion of the portfolio $\rho'x$, and $\bar{\rho} \equiv (\bar{\rho}_1, \ldots, \bar{\rho}_n)'$ is the vector of mean returns. Investigate the boundedness properties of the expected utility integral.

(c) Show that r is convex.

(d) Show that ψ is concave.

Suppose u has constant absolute risk aversion.

(e) Show that $\lambda = 1$.

(f) Show that maximizing $Eu(w)$ is equivalent to maximizing the concave function

$$\gamma\bar{\rho}'x - (\gamma^{\alpha}/2) \sum_{j=1}^{m} (x'\Omega_j x)^{\alpha/2}.$$

(g) Show that the expression in (f) reduces to Freund's result (Exercise 1) when $\alpha = 2$.

10. Consider Press' class of multivariate stable distributions as discussed in Ziemba's paper. Suppose there are p_j assets whose prices follow a multivariate law with characteristic exponent α_j, $j = 1, \ldots, K$, and $\sum p_j = n$. Suppose each of these groupings is independent, has $m = 1$, and each $\Omega_1{}^j$ is positive definite. Let x^j denote the $p_j \times 1$ vector of asset allocations of type α_j, $\bar{\rho}$ the vector of mean returns, and suppose that short sales are allowed.

(a) Show that final wealth W has the log characteristic function

$$\log \phi_W(t) = itx'\bar{\rho} - \tfrac{1}{2} \sum_{j=1}^{K} |t|^{\alpha_j/2} (x^{j\prime}\Omega_1{}^j x^j)^{\alpha_j/2}.$$

(b) Show that W is not stable unless α_j is the same for all j.

(c) Note that the dispersion of w is

$$r(x) = \tfrac{1}{2} \sum_{j=1}^{K} (x^{j\prime}\Omega_1{}^j x^j)^{\alpha_j/2}.$$

Show that the x^j's that minimize $r(x)$ are the same as the minimizers of $x^{j\prime}\Omega_1{}^j x^j$ separately for each j.

(d) Utilize the result in (c) and that of Exercise 9 to show that

$$x^* = \frac{\Omega^{-1}\bar{\rho}}{e'\Omega^{-1}\bar{\rho}}, \qquad \text{where} \quad \Omega = \begin{pmatrix} \Omega_1{}^1 & & 0 \\ & \ddots & \\ 0 & & \Omega_1{}^K \end{pmatrix}$$

and that

$$x^{j*} = \frac{(\Omega_1{}^j)^{-1}\bar{\rho}_j}{e^1(\Omega_1{}^j)^{-1}\bar{\rho}},$$

where $\bar{\rho}_j$ is the portion of $\bar{\rho}$ corresponding to class α_j.

(e) Suppose $K = 2$, $n = 4$, $\Sigma_1 = \begin{pmatrix} 3 & 1 \\ 1 & 2 \end{pmatrix}$ and $\Sigma_2 = \begin{pmatrix} 4 & 2 \\ 2 & 3 \end{pmatrix}$, and $\bar{\rho}_i = \bar{\rho}_j$ for all ij. Show that $x^* = \frac{1}{39}(8, 16, 5, 10)$.

11. Consider an investor with a disutility function $-V$ over regret R. Suppose disutility increases with regret $(-V' > 0)$ and at an increasing rate reflecting risk aversion $(-V'' > 0)$. Let $x \equiv (x_1, \ldots, x_n)'$ denote the investment allocations where $x \geqq 0$ and $e'x = 1$, where $e \equiv (1, \ldots, 1)$ and initial wealth is 1. The investment returns are $\rho \equiv (\rho_1, \ldots, \rho_n)$, and have the joint distribution function $F(\rho)$. For a given x,

$$R(x, \rho) = \left\{ \max_{\rho \in \Xi} \rho' x \right\} - \rho' x = \max_i \{\rho_i\} - \rho' x$$

(assume that the ρ_i are nonnegative and Ξ is compact).

(a) Suppose that F is symmetric as defined in Samuelson's paper on diversification. Show that the solution $\hat{x} = (1/n, \ldots, 1/n)'$ is a minimax strategy, i.e., $S(\hat{x}) = \min_x \max_\rho R(x, \rho)$. [*Hint*: Consider perturbations away from $\hat{x}$.]

(b) Show that $\hat{x}$ is the unique minimax strategy.

To minimize expected regret, one must solve

$$\min_x E\{-V(R)\} = -\max_x E\{V(R)\}. \tag{1}$$

Let $W(x) = E\{V(R)\}$.

(c) Show that W is strictly concave, symmetric, $\hat{x}$ and increasing in each x_i, so that, as Samuelson shows in his paper on diversification, x is the unique optimal solution.

(d) Suppose F is general but $n = 2$ and $\bar{\rho}_1 = \bar{\rho}_2$. Show that $x_1{}^* > 0$, $x_2{}^* > 0$ if x^* solves (1). [*Hint*: Eliminate x_2 from (1) and investigate $dW(x_1)/dx_1$ at $x_1 = 0$.]

*(e) Attempt to generalize the result in (d) to the case $n = n$, $\bar{\rho}_1 = \cdots = \bar{\rho}_n$.

*(f) Attempt to prove that $x_i{}^* > 0$ if $\bar{\rho}_i \geqq \{\min \bar{\rho}_j \mid 1 \leqq j \leqq n,\ j \neq i\}$ when F is general and when the ρ_i are independent.

(g) Compare the results in (c)–(f) with those in Samuelson's paper.

12. Consider a portfolio problem where the only actions available are bets on the occurrence of states of nature. Let v_i be the return per dollar invested in i if state i occurs; investment in i produces no return if j occurs $(i \neq j)$. Assume that the number of states of nature is finite. Let x_i be the amount bet on the occurrence of i out of initial wealth one dollar. Suppose that the investor wishes to maximize expected utility of final wealth, and that the utility function is concave and differentiable.

(a) Suppose that the investor allocates all his resources and p_i is the probability that i occurs. Show that the Kuhn–Tucker conditions are

$$\begin{aligned} p_i v_i u'(v_i x_i) &= \lambda \qquad \text{if} \quad x_i > 0, \\ p_i v_i u'(0) \quad &\leqq \lambda \qquad \text{if} \quad x_i = 0, \end{aligned} \tag{1}$$

and $\sum x_i = 1$, $x_i \geqq 0$.

(b) Suppose $u = \log$. Show that $x_i^* > 0$ for all i and that the objective becomes

$$\sum p_i \log x_i + \sum p_i \log v_i.$$

Show that $x_i^* = p_i$ independently of the payoffs $\{v_i\}$. Interpret this optimal decision.

(c) Suppose the investor considers holding some of his wealth in reserve so that $\sum x_i \leqq 1$. Suppose $u'(0) = +\infty$. Show that the individual will invest all his resources if and only if there exists a system of bets such that the individual cannot lose; i.e., there exist $x_i \geqq 0$ such that $\sum x_i = 1$ and $a_i v_i \geqq 1$. [*Hint*: Utilize the Kuhn–Tucker conditions.]

(d) Show that the existence of the set of sure bets is sufficient for the investor to invest all his wealth even if $u'(0) < +\infty$. Show that this condition is not necessary, however, if $u'(0) < +\infty$. [*Hint*: Let u be linear, and suppose $p_j v_j > 1$ for some j.]

(e) Show that the optimal bets in (b) are not necessarily sure bets in the sense of (c) even if such a sure bet exists.

13. Under appropriate assumptions (see Exercise I-ME-17), functions defined by integrals have useful convexity properties. For example, $Z(x) \equiv E_\rho\{f(x,\rho)\}$ is convex in x if f is convex in x for each fixed ρ. It is generally difficult to find an explicit expression for Z; see, however, Exercise 15. Hence, it is of interest to devise methods to maximize functions defined by integrals that do not involve the explicit calculation of these integrals. Of particular interest are continuity and differentiability properties of Z, and verification of when it is legitimate to interchange the order of differentiation and integration. Suppose $Z\colon K \to E$, where K is a subset of E^n, and $\rho \in \Xi$ is a subset of E^s. A function $h(x,\rho)$ is said to be integrable with respect to ρ if $E_\rho |h(x,\rho)| < \infty$.

(a) Suppose that for all $x^0 \in K$ and $\varepsilon > 0$, $\exists \delta > 0$ such that $\|x - x^0\| < \delta$ for $x \in K$ implies that $|f(x,\rho) - f(x^0,\rho)| < \varepsilon$ for almost all $\rho \in \Xi$. Show that Z is continuous on K as long as f is integrable with respect to ρ.

(b) Show that the assumption in (a) is verified if f is continuous in x for each fixed ρ.

Under suitable regularity conditions, the monotone and dominated convergence theorems may be applied to obtain sufficient conditions for the differentiability of Z and the validity of the interchange. These theorems are proved by Loève (1963) and may be stated as follows.

Monotone Convergence Theorem Assume that e is integrable with respect to ξ. Let $g_n(x,\xi) \geqq e(\xi)$ and $g_n(x,\xi) \geqq g_{n-1}(x,\xi)$, and $\lim_{n\to\infty} g_n(x,\xi)$ exists; then

$$\lim_{n\to\infty} \left[\int g_n(x,\xi)\, dF(\xi)\right] = \int \left[\lim_{n\to\infty} g_n(x,\xi)\, dF(\xi)\right].$$

Dominated Convergence Theorem Suppose $|g_n(x,\xi)| \leqq h(\xi)$, with probability 1, where h is integrable. If $\lim_{n\to\infty} g_n(x,\xi) = g(x,\xi)$ with probability 1, then

$$\lim_{n\to\infty} \left[\int g_n(x,\xi)\, dF(\xi)\right] = \int \left[\lim_{n\to\infty} g_n(x,\xi)\right] dF(\xi) = \int g(x,\xi)\, dF(\xi).$$

(c) Suppose f is integrable with respect to ρ and continuously differentiable in x for almost all ρ. Assume that there exist integrable functions $e_1(\rho), \ldots, e_n(\rho)$ such that $|\partial f(x,\rho)/\partial x_j| \leqq e_j(\rho)$ for each j. Show that Z is continuously differentiable on the open set K and each

$\partial Z/\partial x_j = E_\rho[\partial f(x,\rho)/\partial x_j]$. [*Hint*: Utilize the dominated convergence theorem along with the mean-value theorem.]

A function $c(x): D \to R$ is said to be subdifferentiable on an open convex set $K \subset E^n$ if there exists a vector $s(x)$, termed a subgradient, such that

$$c(x^1) \geqq s(x)'(x^1 - x) + c(x) \qquad \text{for all} \quad x, x^1 \in K.$$

Let $s(x) = (s_1(x), \ldots, s_m(x))'$,. If c is convex, then c is differentiable except on a set of measure zero. Then $d_j^- c(x) \leqq s_j(x) \leqq d_j^+ c(x)$, where d_j^- and d_j^+ are, respectively, the left and right directional partial derivatives of c with respect to x_j at x. A function is differentiable at x if $s(x)$ is unique. In this case, $d_j^-(x) = d_j^+(x)$ for all j. Integration tends to smooth functions; hence Z will often be smoother than $f(x, \cdot)$. The smoothing is particularly effective when the density of ρ is absolutely continuous. In this case, under mild hypotheses, Z will be differentiable if $f(x, \cdot)$ is convex (or concave).

(d) Suppose K is open and that $f(x,\rho)$ is convex in x for all fixed $\rho \in \Xi$ and integrable with respect to ξ. Let $d_j^-(x,\xi)$ and $d_j^+(x,\xi)$ be the left and right directional derivatives of f with respect to x_j at x, given ξ. Suppose that the d_j^- and d_j^+ are bounded. Assume that for any given $x \in K$ and w_j^- (respectively w_j^+) the number of ξ such that $w_j^- = d_j^-(x,\xi)$ [respectively $w_j^+ = d_j^+(x,\xi)$] has measure zero. Assume that F is absolutely continuous. Show that Z is continuously differentiable on K and $\partial Z/\partial x_j = E_\xi[s_j(x,\xi)]$, where $s_j(x,\rho)$ is any number between $d_j^-(x,\rho)$ and $d_j^+(x,\rho)$, $j = 1, \ldots, m$. [*Hint*: Show that left and right partials of Z exist and appeal to the fact that partials of convex functions are always continuous if they exist.]

14. The results in Exercise 13 provide a strategy to solve problems of the form $\{\min Z(x) \equiv E_\rho\{f(x,\rho)\}$ without explicitly calculating Z. Instead, one evaluates Z and/or the $\partial Z/\partial x_j$ at various feasible points, using a numerical integration scheme. Such an approach might be used with nearly any nonlinear programming algorithm. A particularly simple algorithm that utilizes feasible directions is the Frank–Wolfe algorithm. The rationale for the algorithm is based on the fact that, under appropriate assumptions (see Exercise I-ME-5), a vector x^k solves $\{\min Z(x) \mid x \in K\}$ if and only if it solves $\{\min \nabla Z(x^k)'x \mid x \in K\}$. Suppose that K is the convex polytope $\{x \mid Ax \leqq b\}$, that Z is convex and continuously differentiable, and that the linear function $\nabla Z(\hat{x})'x$ is bounded from below on K for each $\hat{x} \in K$. The procedure is to linearize the objective about some feasible point x^k and then to solve a linear program to obtain a solution vector y^k. All points along the interval $[x^k, y^k]$ are feasible; hence, one may search along the interval for a minimum x^{k+1}. It develops that x^{k+1} will be a better solution than x_k. One then linearizes about x^{k+1} and repeats the process until $x^{k+1} = x^k$, i.e., $y^k = x^k$. The verification of the procedure may be made as follows.

(a) Show that if $\bar{x} \in K$ and $\nabla Z(\bar{x})'(x - \bar{x}) \geqq 0$, $\forall x \in K$, then $\bar{x}$ is a minimum for $Z(x)$ over K. [*Hint*: Utilize the convexity definition.]

(b) Suppose ϕ is convex on an interval $[a,b]$ and $\phi(a) < \phi(b)$. Show that for all $l \in [a,b)$, $\phi(l) < \phi(b)$.

(c) Suppose $x, \bar{x} \in K$, and $\nabla Z(x)'(x - \bar{x}) > 0$. Show that there exists a $\lambda^0 \geqq 0$ such that for all $\lambda \in [\lambda^0, 1)$, $Z\{\lambda x + (1-\lambda)\bar{x}\} < Z(x)$. [*Hint*: Let $\phi(\lambda) \equiv Z[\lambda x + (1-\lambda)\bar{x}]$ and utilize (b).]

(d) Suppose $x \in K$ and $Z(x) > Z(x^*) = \min\{Z(x) \mid x \in K\}$. Show that there exists an $\hat{x}$ such that $\nabla Z(x)'x > Z(x)'\hat{x}$.

(e) Show that

$$Z(x^*) = \min\{Z(x) \mid x \in K\}$$

if and only if

$$\nabla Z(x^*)'x^* = \min\{\nabla Z(x^*)'x \mid x \in K\}.$$

The linear programming subproblems have the form $\{\min \nabla Z(x^k)'x \mid x \in K\}$, which has a solution y^k. If $y^k \neq x^k$, then one must search for a $\lambda^k \in (0,1]$ that minimizes $Z\{\lambda y^k + (1-\lambda)x^k\}$.

(f) Draw a flow chart of the algorithm. Illustrate the number of numerical integrations that must be made in each iteration in the general case and in certain special cases.

(g) Show that the following bounds

$$Z(x^k) + \nabla Z(x^k)'(y^k - x^k) \leqq Z(x^*) \leqq Z(x^k),$$

obtain in each iteration. What is the best available lower bound at iteration k? Utilize this bound to develop an ε-optimal stopping rule for the algorithm.

(h) Illustrate some advantages and disadvantages of using such an algorithm when $f(x,\rho) = u(\rho'x)$.

(i) Show how one may evaluate Z and the $\partial Z/\partial x_j$ using only univariate integrations if the ρ_i are normal or symmetric stable, when the objective has the univariate form $u(\rho'x)$.

15. The algorithm in Exercise 14 illustrates some of the advantages and disadvantages of solving stochastic optimization problems by combining numerical integration schemes with standard nonlinear programming algorithms. There are a number of special cases in which the expression for $Z(x) \equiv E_\rho[f(x,\rho)]$ may be obtained explicitly, in which case the expensive numerical integrations may be dispensed with. Also, Exercises 1 and 9 illustrate cases in which explicit deterministic equivalents exist, whose solutions will provide optimal solutions for the stochastic problem. Some additional special cases are explored here.

(a) Suppose the vector ρ has only a finite number of possible realizations, say L. Find an explicit expression for $Z(x)$.

(b) Discuss the structure of the problem in (a) when L is very large or very small. When is L likely to be very small?

(c) Suppose f is a quadratic function of x and ρ, say $f(x,\rho) = a'x + b'\rho + (x',\rho')A\binom{x}{\rho}$. Show that $\bar{\rho}$ is a certainty equivalent for ρ, so that all vectors that solve $\{\max f(x,\bar{\rho}) \mid x \in K\}$ also solve $\{\max E_\rho[f(x,\rho)] \mid x \in K\}$.

(d) Attempt to generalize the result in (c).

(e) How do the results in (c) and (d) relate to the goal of maximizing expected return or expected return net of a constant times variance?

(f) Investigate the case when $f(x,\rho) = u(\rho'x)$ and u is a polynomial. What type of deterministic problem results?

Consider the special case when $f(x,\rho)$ is a penalty function, say $\sum g_j(x_j - \rho_j)$, where g_j represents the penalty attached to deviations at x_j away from ρ_j.

(g) Show that Z is a separable function of x, namely, $\sum Z_j(x_j)$, where each $Z_j(x_j) = E_{\rho_j} g_j(x_j - \rho_j)$ so that only marginal distributions are needed to calculate Z.

(h) Suppose each g_j is the piecewise quadratic function

$$\begin{array}{ll} a_j(x_j-\rho_j)^2 & \text{if} \quad x_j \geqq \rho_j, \\ b_j(\rho_j-x_j)^2 & \text{if} \quad x_j \leqq \rho_j. \end{array}$$

Compute expressions for Z when each ρ_i is uniform or normally distributed.

(i) Notice that the uniform case in (h) leads to a cubic function. Show that this result generalizes to piecewise polynomial functions so that if f is piecewise polynomial of degree n, then Z will be a polynomial function of degree $n+1$.

16. An investor has $\$B > 0$ and allocates $\$x_i$ to investment i. For each dollar invested in i, the random return of ρ_i is returned at the end of the investment horizon. Suppose that the utility function u over wealth $w = \rho' x = \sum \rho_i x_i$ is concave and continuously differentiable. If the investor is an expected utility maximizer, his decision problem is

$$\{\max E_\rho u(\rho' x) \mid \sum x_i = B,\ x_i \geqq 0\}. \tag{1}$$

Suppose the return vector ρ has the discrete distribution $p_r\{\rho = \rho^l\} = p_l > 0, l = 1, \dots, L < \infty$, and $\sum p_l = 1$. Then (1) is

$$\max Z(x) = p_1 u(\rho^{1'} x) + \cdots + p_L u(\rho^{L'} x), \qquad \text{s.t.}\ \sum x_i = B,\quad x_i \geqq 0. \tag{2}$$

(a) Show that Z is concave and continuously differentiable and

$$\frac{\partial Z(x)}{\partial x_i} = \sum_l p_l \left\{ \frac{\partial u[\rho^{l'} x]}{\partial w} \rho_i{}^l \right\}.$$

(b) Attempt to develop an efficient algorithm to solve (2) for large L that exploits the fact that each of the L terms in the objective is "similar."

(c) Suppose x_1 is unconstrained. Show that (2) becomes

$$\max v(x_2, \dots, x_n) = \{p_1 u[\sum \rho_i{}^1 x_i + \rho_1{}^1 (B - \sum x_i)] + \cdots + p_L u[\sum \rho_i{}^L x_i + \rho_1{}^L (B - \sum x_i)]\}$$

$$\text{s.t.}\quad x_i \geqq 0, \qquad \text{where}\quad i = 2, \dots, n.$$

(d) Show that v is concave and continuously differentiable.

(e) What is $\partial v / \partial x_i$? Note its similarity to the expression in (a).

(f) How does the algorithm in (b) simplify when applied to (3)?

(g) Investigate algorithmic simplifications when u is quadratic, $u = \log$, $\rho^l > 0$ for all l, and u is exponential.

17. Under certain assumptions, such as quadratic preferences or normally distributed random returns, the mean–variance approach is consistent with the expected utility approach. To provide generality to the risk–return approach, it is desirable to be able to consider a risk–return measure that is free of such restrictions on the utility function or the probability distributions. Let $\phi \equiv u(\bar{w}) - E[u(w)]$ be a generalized risk measure, where $\bar{w}$ is expected wealth and u is the utility function over wealth. Note that unlike the variance, ϕ depends on both u and the probability distribution of returns; and ϕ makes it possible to represent expected utility in terms of two parameters, $(\bar{w}, \phi)$, without imposing restrictions on the form of u or the distribution of w.

(a) Show that $\phi > 0$ (< 0) if u is strictly concave (strictly convex) and that $\phi = 0$ if u is linear.

(b) The risk premium π is defined by the equation $u(\bar{w} - \pi) = Eu(w)$, so that ϕ is related to π via $\phi = u(\bar{w}) - u(\bar{w} - \pi)$. Suppose u is strictly increasing; show that $\phi \gtreqqless 0$ if and only if $\pi \gtreqqless 0$.

(c) Suppose $u(w)$ is linearly transformed into $v(w) = au(w) + b$. Show that ϕ is transformed to $a\phi$.

(d) Let M_k, assumed finite, be the kth moment of w about $\bar{w}$. Suppose a Taylor's expansion of u about $\bar{w}$ exists in a neighborhood of $\bar{w}$ sufficiently large to include all w for which there is a nonzero probability of occurrence. Show that

$$\phi = -\sum_{k=2}^{\infty} \frac{u^{(k)}(\bar{w})}{k!} M_k,$$

where $u^{(k)}(\bar{w})$ is the kth derivative of u evaluated at $w = \bar{w}$.

(e) Suppose w is normally distributed. Show that ϕ is proportional to variance, i.e.,

$$\phi = a\sigma^2, \qquad \text{where}\quad a \equiv -\sum_{k=1}^{\infty} \frac{u^{(2k)}(\bar{w})}{2^k k!} \sigma^{2k-2}.$$

(f) Show that ϕ is proportional to variance if u is quadratic.

(g) At $\bar{w}$, Pratt's measure of local risk aversion is $r(\bar{w}) = -u^{(2)}(\bar{w})/u^{(1)}(\bar{w})$. Show that $\phi \approx \frac{1}{2}(u^{(1)}(\bar{w})r(\bar{w}))\sigma^2$ if terms of order $\geqq 3$ are dropped from the expression in (d).

(h) Show that the marginal rate of substitution between return and risk is equal to the marginal utility of wealth at $\bar{w}$, i.e., $d\phi/d\bar{w} = u^{(1)}(\bar{w})$.

18. Suppose that risky assets 1 and 2 have rates of return ρ_1 and ρ_2 that are nonnegative, independent, and exponentially distributed. Let $E\rho_i = \mu_i$ for $i = 1, 2$.

(a) Show that $\operatorname{var}\rho_i = \mu_i^2$.

(b) Show that, for $x_1, x_2 \geqq 0$, the distribution function of $x_1\rho_1 + x_2\rho_2$ is

$$P[x_1\rho_1 + x_2\rho_2 \leqq \xi] = \begin{cases} 1 + \dfrac{\mu_2 x_2 \exp(-\xi/\mu_2 x_2) - \mu_1 x_1 \exp(-\xi/\mu_1 x_1)}{\mu_1 x_1 - \mu_2 x_2}, & \xi \geqq 0, \\ 0, & \xi < 0. \end{cases}$$

Let $K = \{(x_0, x_1, x_2) \mid x_0, x_1, x_2 \geqq 0,\ x_0 + x_1 + x_2 = 1\}$. Let $R = w_0(1 + x_0 r + x_1\rho_1 + x_2\rho_2)$' and let $d > w_0(1+r)$.

(c) Show that the solution to the problem

$$\min\left\{\frac{\sigma_R^2}{(ER-d)^2}\ \middle|\ x \in K,\ ER \geqq d\right\}$$

is given by $x_0^* = 0$, $x_2^* = 1 - x_1^*$, and

$$\left[1 + \mu_2 - \frac{d}{w_0} + (\mu_1 - \mu_2)x_1^*\right]^2 [(\mu_1^2 + \mu_2^2)x_1^* - \mu_2^2]$$

$$= (\mu_1 - \mu_2)[\mu_2^2 + (\mu_1^2 + \mu_2^2){}_1^{2*}x - 2\mu_2^2 x_1^*] \qquad \text{if} \quad x_1^* \in (0, 1).$$

(d) For $x_0^* = 0$, show that the solution to $\min\{P[R \leqq d] \mid x \in K,\ ER \geqq d\}$ is given by $x_2^* = 1 - x_1^*$, $k = (d/w_0) - 1$, and

$$\mu_1\left(1 + \frac{k}{\mu_1 x_1^*}\right)\exp\left(\frac{-k}{\mu_1 x_1^*}\right) + \mu_2\left(1 + \frac{k}{\mu_2 x_2^*}\right)\exp\left(\frac{-k}{\mu_2 x_2^*}\right)$$

$$= (\mu_1 + \mu_2)\left[\mu_1 x_1^* \exp\left(\frac{-k}{\mu_1 x_1^*}\right) - \mu_2 x_2^* \exp\left(\frac{-k}{\mu_2 x_2^*}\right)\right] \Big/ [(\mu_1 + \mu_2)x_1^* - \mu_2],$$

if $x_1^* \in (0, 1)$.

(e) Show that solutions to (c) will generally not be solutions to (d), and vice versa. [*Note*: This demonstrates that, for problems involving many risky assets which are not normally distributed, the problem $\min \sigma_R^2/(ER-d)^2$ is generally neither necessary nor sufficient for the problem $\min P[R \leqq d]$.]

19. It is of interest to examine the possible effects on risk-taking of a more complex tax structure than a simple proportional tax. In particular, suppose that income Y is taxed: no tax for $Y \in (-\infty, B_0)$, tax $t_1(Y - B_0)$ for $Y \in [B_0, B_1)$, tax $t_1(B_1 - B_0) + t_2(Y - B_1)$ for $Y \in [B_1, B_2), \ldots$. Assume $t_i \in (0, 1)$ for all i, and $t_1 < t_2 < \ldots$. The intervals $(-\infty, B_0)$, $[B_0, B_1), \ldots$ are tax brackets.

(a) Show that after-tax income $T(Y)$ is a strictly increasing, concave function of before-tax income Y.

(b) Formulate the investor's problem of maximizing expected utility of terminal wealth.

Assume $B_0 = B > 0$ and $B_1 = \infty$, so that income below B is not taxed, but income in excess of B is taxed proportionately at a rate $t \in (0, 1)$.

(c) For a single risky asset having only two rates of return, as in the Stiglitz paper, sketch the before- and after-tax budget lines under various combinations of interest rate, initial wealth, and risky returns.
(d) Show by either graphical or analytical means that the effects of the tax on risk-taking are ambiguous, that is, show that risk-taking may increase or decrease depending on the problem details.
*(e) Try to find conditions under which unambiguous statements can be made regarding the effects of this tax on risk-taking.

Suppose that $B = \$2000$, $w_0 = \$1000$, and $t = 20\%$.
(f) For the investor of Exercise CR-19, formulate the optimality problem if short sales are not allowed, but borrowing is unlimited. Formulate an algorithm for solving this problem.

20. (Properties of univariate log-normal distributions) The notion that random investment returns are subject to limited liability in the sense that the investor can lose no more than his initial outlay leads to the study of distributions whose outcomes are never negative. The log-normal distribution is such a distribution and it arises naturally from modified central limit theorems. If X is an essentially positive variate ($0 < x < \infty$ for all outcomes x) and $Y \equiv \log X$ has a normal distibution with mean μ and variance σ^2, then X is said to have the log-normal distribution $\Lambda(x; \mu, \sigma^2)$. Hence

$$\Lambda(x) = \begin{cases} N(\log x) & \text{if} \quad x > 0, \\ 0 & \text{otherwise,} \end{cases}$$

and the density of X is

$$\frac{d\Lambda(x)}{dx} = \begin{cases} \dfrac{1}{x\sigma\sqrt{2\pi}} \exp\left[-\dfrac{1}{2\sigma^2}(\log x - \mu)^2\right] dx & \text{if} \quad x > 0, \\ 0 & \text{otherwise.} \end{cases}$$

Thus log-normal variates tend to arise in situations where the variate is a productof independent identically distributed (iid) elementary variates. It is possible to prove that if $X_1, X_2, \ldots$ is a sequence of iid variates, $E[\log X_i] = \mu < \infty$ and $\text{var}[\log X_i] = \sigma^2 < \infty$, then the geometric mean $[\prod_{j=1}^{n} X_j]^{1/n}$ is asymptotically distributed as $\Lambda(\mu, \sigma^2/n)$.

The study of moments of X is facilitated by the use of the moment-generating function (mgf) of Y; $\psi_Z(t) \equiv E[e^{tZ}]$ is the mgf of the random variable Z for any constant t (which is said to exist if ψ is finite in some neighborhood of $t = 0$).
(a) Show that the jth moment of Z about the origin equals

$$\left[\frac{d^j\psi_Z(t)}{dt^j}\right]_{t=0}.$$

(b) Show that the variance of Z equals

$$\left\{\frac{d^2\psi_Z(0)}{dt^2} - \left[\frac{d\psi_Z(0)}{dt}\right]^2\right\}.$$

(c) Show that $\psi_X(t)$ does not exist. (This means that the log-normal density cannot be uniquely derived in terms of its moment sequence. Hence a non-log-normal density can have the same moments as a given log-normal density.)
(d) Show that $\psi_Y(t) = \exp(\mu t + 1/2t^2\sigma^2)$ when $Y \sim N(\mu, \sigma)$. Hence, it follows that the jth moment of X about the origin is finite and equals $\{\exp(j\mu + 1/2j^2\sigma^2)\}$. [*Hint*: Complete the square.]

(e) Show that X has mean and variance equal to

$$\{\exp(\mu+1/2\sigma^2)\}$$

and

$$[\exp(2\mu+\sigma^2)(\exp(\sigma^2)-1)],$$

respectively.

(f) Illustrate the positive skewness of the log-normal distribution by graphing the distribution when $\mu = 1$ and $\sigma^2 = 4$. Indicate the relative positions of the mode $[\exp(\mu-\sigma^2)]$, median (e^μ), and mean $[\exp(\mu+1/2\sigma^2)]$.

The log-normal distribution is unfortunately not closed under addition so that $a_1 X_1 + a_2 X_2$ does not generally have a log-normal distribution if X_1 and X_2 have (independent) log-normal distributions and the a_i are constants. However, the log-normal distribution has a number of useful multiplicative properties since logs of multiplicative functions are additive.

(g) Suppose $X_i \sim \Lambda(\mu_i, \sigma_i^2)$ are independent, $i = 1, \ldots, n < \infty$. Show that $X_1 \cdots X_n \sim \Lambda(\sum \mu_i, \sum \sigma_i^2)$.

(h) Suppose $X \sim \Lambda(\mu, \sigma^2)$ and a, b, and $c > 0$ are constants $(c = e^a)$. Show that $cx^b \sim \Lambda(a+b\mu, b^2\sigma^2)$.

(i) Suppose $X_i \sim \Lambda(\mu_i, \sigma_i^2)$ are independent, $i = 1, \ldots, n < \infty$; a, $b_1, \ldots, b_n$, and $c > 0$ are constants $(c = e^a)$. Show that $c \prod_i X_i^{b_i} \sim \Lambda(a + \sum b_i \mu_i, \sum b_i^2 \sigma_i^2)$.

(j) Show that if $X \sim \Lambda(\mu, \sigma^2)$, its reciprocal $1/X \sim \Lambda(-\mu, \sigma^2)$.

(k) Suppose $X_i \sim \Lambda(\mu_i, \sigma_i^2)$ are independent, $i = 1, 2$. Show that

$$X_1/X_2 \sim \Lambda(\mu_1-\mu_2,\ \sigma_1^2+\sigma_2^2).$$

It is often of interest to consider moment distributions of log-normal variates.

(l) Suppose $X \sim \Lambda(\mu, \sigma^2)$, and λ_j is the jth moment of X about the origin. Show that $(\int_0^x X^j \, d\Lambda(X; \mu, \sigma^2))/\lambda_j \sim \Lambda(x: \mu + j\sigma^2, \sigma^2)$.

21. (Properties of multivariate log-normal distributions) Suppose $Y_1, \ldots, Y_m$ have the multivariate normal distribution $N(\mu, \Sigma)$, where μ is the mean vector and Σ is the $m \times n$ symmetric positive-definite variance–covariance matrix. Then the vector $X' = (X_1, \ldots, X_m) \equiv (\exp Y_1, \ldots, \exp Y_m)$ has the multivariate log-normal distribution $\Lambda(X; \mu, \Sigma)$.

(a) Show that the density of X is

$$f(X) = \frac{1}{(2\pi)^{m/2} |\Sigma|^{+1/2} (\prod X_i)} \exp\{-\tfrac{1}{2}(\log X - \mu)'\Sigma^{-1}(\log X - \mu)\},$$

where $\log X \equiv (\log X_1, \ldots, \log X_m)'$.

As in the univariate case, the study of moments of X is facilitated by using the moment-generating function of Y. For any real vector $t = (t_1, \ldots, t_m)$, the mgf of a random m-vector Z is $\psi_Z(t) = E[e^{t'Z}]$.

(b) Let $j = (j_1, \ldots, j_m)'$. Show that the jth cross moment of Z about the origin equals

$$\frac{\partial \sum_{j_i} \psi_Z(0)}{\partial t_1^{j_1}, \ldots, \partial t_m^{j_m}}.$$

(c) Show that the variance of w_i, the ith component of Z, equals

$$\left\{\frac{\partial^2 \psi_Z(0)}{\partial t_i^2} - \left[\frac{\partial^2 \psi_Z(0)}{\partial t_i}\right]^2\right\}.$$

(d) Show that $\psi_Y(t) = \exp(\mu' t + \frac{1}{2}t'\Sigma t)$ when $Y \sim N(\mu, \Sigma)$. So that the jth cross moment of X, namely $E(\pi X_i^{j_i})$, equals $\exp\{\mu' j + \frac{1}{2}j'\Sigma j\}$. Note that if $\mu = 0$, $(X_1^{j_1}, \ldots, X_m^{j_m})'$ has the same distribution as $(X_1^{-j_1}, \ldots, X_m^{-j_m})$.
(e) Use (d) to show that X has mean vector $\{\exp(\mu_1 + \frac{1}{2}\sigma_1^2), \ldots, \exp(\mu_m + \frac{1}{2}\sigma_m^2)\}'$.
(f) Use (d) to show that the covariance of X_i and X_j is

$$\exp\{\mu_i + \mu_j + \tfrac{1}{2}(\sigma_{ii} + \sigma_{jj} + 2\sigma_{ij})\} - \exp\{\mu_i + \mu_j + \tfrac{1}{2}(\sigma_{ii} + \sigma_{jj})\}.$$

(g) Show that marginal, joint marginal, and conditional distributions of X are multivariate log-normal.
(h) Suppose $b' = (b_1, \ldots, b_m)$ and $c > 0$ are constants $(c = e^a)$. Show that $c \prod_i X_i^{b_i}$ has the univariate log-normal distribution $\Lambda(a + b'\mu, b'\Sigma b)$.

22. The logarithmic function often provides a good representation as an investor's utility function as it has a number of attractive methodological properties and is analytically convenient. If investment returns have limited liability, i.e., gross returns are nonnegative, then the distribution of such returns may be well approximated by a multivariate log-normal distribution, particularly if it is believed that the investment returns are determined by the product of elementary random variates. This problem is concerned with properties of the optimal solution to the portfolio problem when one combines these two assumptions. Exercise 23 illustrates a solution approach for such problems.

Suppose initial wealth is normalized at 1 and x_i represents the investment in security i, $i = 1, \ldots, n$. Let ρ_i represent the random vector per unit investment in i. The vector $\rho \equiv (\rho_1, \ldots, \rho_n)'$ has the multivariate log-normal distribution $\Lambda(\mu, \Sigma)$, where each $\mu_i = E \log \rho_i$ and each $\sigma_{ij} = \text{cov}(\log \rho_i, \log \rho_j)$. Suppose that Σ is positive definite. The portfolio problem is

$$\{\max V(x, \mu, \Sigma) \equiv E_\rho \log(\rho' x) \,|\, e'x = 1,\ x \geqq 0\}, \qquad \text{where} \quad e = (1, \ldots, 1)'. \tag{P}$$

(a) Show that (P) has a unique optimal solution for all μ and Σ. [*Hint*: Show that V is finite, strictly concave, and continuous in x; then utilize part (g) of Exercise I-CR-15.]
(b) Show that V is strictly increasing in μ_i if $x_i > 0$ and that $V(\cdot, \mu + ke, \cdot) = V(\cdot, \mu, \cdot) + k$ for any scalar k. [*Hint*: Utilize a change of variables so that a new utility function can be determined that involves independent standardized normal variates.]
(c) Show that V is strictly increasing in σ_{ii} if $0 < x_i < 1$.
(d) Show that a small increase in all covariances of an asset decreases V. That is,

$$V(\cdot, \cdot, \Sigma) > V(\cdot, \cdot, A) \qquad \text{if} \quad A = \Sigma + \begin{pmatrix} & 0 & & k \\ & & & \vdots \\ & & & k \\ k & \cdots & k & 0 \end{pmatrix}$$

and $k > 0$ is sufficiently small. Make precise the notion that k is sufficiently small. [*Hint*: Consult Exercise 23, (c)–(f).]
*(e) Attempt to prove that V is a strictly decreasing function of σ_{ij} for $i \neq j$.

In addition to properties of V, it is of interest to examine the behavior of the optimal allocation vector as the parameters change and to examine some restated diversification properties. Let $x^*(\mu, \Sigma)$ denote the optimal allocation vector.

(f) Show that x_i^*/x_j^* is independent of the amount of capital to be invested. [*Hint*: Consult Vickson's paper on separation in Part II.]
(g) Show that $x^*(\mu + ke, \cdot) = x^*(\mu, \cdot)$ for an arbitrary constant k.
(h) Show that x_i^* is a strictly increasing function of μ_i if $0 < x_i^* < 1$.

(i) Show that for all $k_1, \ldots, k_m$, there exists a k^* such that $x^*(\cdot, \{\sigma_{ij}+k_i+k_j+k^*\}) = x^*(\cdot, \Sigma)$, where $\{\ \}$ means typical element. [*Hint*: Consult parts (e) and (f) of Exercise 23.]

A matrix A is said to be completely symmetric (cs) if

$$\{a_{ij}\} = \begin{cases} \alpha & \text{if } i = j, \\ \beta & \text{if } i \neq j, \end{cases} \qquad \text{where} \quad \alpha > \beta.$$

(j) Suppose that $\mu_1 = \cdots = \mu_m$ and Σ is cs (as well as positive definite). Show that each $x_i^* = 1/m$. [*Hint*: Show that the distribution function over returns is symmetric, then apply Theorem II in Samuelson's paper on diversification.]

(k) Suppose that $\mu_1 = \mu_2 = \cdots = \mu_m$ and that $\sigma_{ii}+\sigma_{jj}-2\sigma_{ij} = c$, where the constant c is independent of i and j when $i \neq j$. Show that each $x_i^* = 1/m$. [*Hint*: Utilize (i) and show that $k_1, \ldots, k_m$ exist so that $\{\sigma_{ij}+k_i+k_j+k^*\}$ is cs.]

(l) Utilize (k) so show that $x^* = (\frac{1}{2}, \frac{1}{2})'$ if $m = 2$ and $\mu_1 = \mu$ for all Σ. Explain why the result holds even when Σ is not cs. [*Hint*: Utilize (k).]

Let $\delta_{ij} \equiv \sigma_{ii}+\sigma_{jj}-2\sigma_{ij}$ so that

$$\delta_{ij} \begin{cases} > 0 & \text{if } i \neq j \\ = 0 & \text{if } i = j. \end{cases}$$

(m) Show that $\lambda_1^* = 1$ if $\mu_1 < \mu_i+\frac{1}{2}\delta_{1i}$, and that $\lambda_1^* = 0$ if $\mu_1 < \mu_i-\frac{1}{2}\Delta_{1i}$, where $i > 1$. [*Hint*: Utilize Jensen's inequality (Exercise I-CR-7) to show that $E\{\rho_1(\rho'x)^{-1}\} \geqq E\{\rho_1^{-1}(\rho'x)^{-1}\}$; then consider the Kuhn–Tucker conditions.]

(n) Utilize (m) to show that $0 < x_i^* < 1$, $i = 1, 2$, if and only if $|\mu_1-\mu_2| < \frac{1}{2}\delta_{12}$, where $m = 2$.

(o) Show that $x_i^* > 0$ if $E(x_1) \geqq E(x_i)$ for all i and $\sigma_{1i} = 0$ for each i. [*Hint*: Consult Theorem III in Samuelson's paper on diversification.]

23. This exercise illustrates an approximate solution approach to the problem in Exercise 22. The idea is to develop an approximate problem that is easy to solve and whose solution has essentially all the qualitative properties of problem (P). The basic approximation is to assume that the linear form $w = \sum \rho_i x_i$ is log-normal.

(a) Suppose w is log-normally distributed; show that

$$\theta_1 \equiv E[w] = \exp\{E(\log w)+\tfrac{1}{2}\operatorname{var}(\log w)\},$$

and

$$\theta_2 \equiv E[w^2] = \exp\{2E(\log w)+2\operatorname{var}(\log w)\}.$$

[*Hint*: Consult part (d) of Exercise 20.]

(b) Show that

$$\begin{aligned} E(\log w) &= 2\log\theta_1 - \tfrac{1}{2}\log\theta_2 = 2\log\textstyle\sum x_i \exp\{\mu_i+\tfrac{1}{2}\sigma_{ii}\} \\ &\quad -\tfrac{1}{2}\log\textstyle\sum\sum x_i x_j \exp\{\mu_i+\mu_j+\tfrac{1}{2}(\sigma_{ii}+\sigma_{jj}+2\sigma_{ij}) \\ &\equiv g(x, \mu, \Sigma). \end{aligned}$$

[*Hint*: Consult parts (e) and (f) of Exercise 21.]

(c) Show that $\{\max E\log[\sum x_j(\rho_j \pi \rho_i^{q_i})] \mid e'x = 1,\ x \geqq 0\}$ has the same optimal solution as $\max\{V(x, \lambda, \Sigma) \equiv E\log\sum x_j\rho_j \mid e'x = 1,\ x \geqq 0\}$, for any vector $q \equiv (q_1, \ldots, q_m)'$. [*Hint*: Show that the objective of the newly defined problem differs from that in the original problem by a constant.]

For the $m \times m$ symmetric positive-definite matrix $\Sigma = \{\sigma_{ij}\}$ and any integer $n \geqq m$, the set

$$Q_n(\Sigma) \equiv [\{(v_i+t)'(v_j+t)\} \mid v_1, \ldots, v_m \in R^n,\ t \in R^n, \quad \text{and} \quad v_i'v_j = \sigma_{ij}].$$

(d) Show that Q_n is nonempty and consists of positive-semidefinite and positive-definite matrices of order $m \times m$.
(e) Suppose $\Omega \in \bigcup_{n \geqq m} Q_n(\Sigma)$. Show that $V(\cdot,\cdot,\Sigma) = V(\cdot,\cdot,\Omega)$. [*Hint*: Utilize (c) and work with independent standardized normal variates in n space.]
(f) Show that for all $k_1, \ldots, k_m$ there exists a k^* such that

$$V(\cdot,\cdot,\Sigma) = V(\cdot,\cdot,\{\sigma_{ij}+k_i+k_j+k^*\}).$$

The result in (f) indicates a framework for the definition of a class of surrogate functions that are analytically convenient. Since $V(\cdot,\cdot,\Sigma)$ is essentially identical to $V(\cdot,\cdot,\Omega)$ if $\Omega \in Q_m(\Sigma)$, a natural surrogate family is defined as

$$\begin{aligned} g(x,\lambda,\Omega) &= g(x,\mu,\{\sigma_{ij}+k_i+k_j+k^*\}) \\ &= 2\log \sum x_i \exp[\mu_i + \tfrac{1}{2}(\sigma_{ii}+2k_i+k^*)] \\ &\quad -\tfrac{1}{2}\log \sum\sum x_i x_j \exp[[\mu_i+\mu_j+\tfrac{1}{2}(\sigma_{ii}+\sigma_{jj}+2\sigma_{ij}+4[k_i+k_j+k^*)]]. \end{aligned}$$

(g) Show that g is independent of k^*, hence k^* may be deleted from the expression for g.

It is a remarkable fact that there exist values of $k_1, \ldots, k_m$, say $\tilde{k}_1, \ldots, \tilde{k}_m$, such that g has essentially all the qualitative properties of V and the vector x that maximizes g may be found by solving a simple quadratic program.
(h) Let each $\tilde{k}_i \equiv \mu_i - \frac{1}{2}\sigma_{ii}$. Show that

$$g = \tilde{g} \equiv -\tfrac{1}{2}\log \sum\sum x_i x_j \exp\{-\mu_i-\mu_j-\tfrac{1}{2}\delta_{ij}\},$$

where $\delta_{ij} \equiv \sigma_{ii}+\sigma_{jj}-2\sigma_{ij}$.
(i) Let $B = \{b_{ij}\}$, where each $b_{ij} \equiv \exp\{-\mu_i-\mu_j-\frac{1}{2}\delta_{ij}\}$. Show that the vector $\tilde{x}$ that minimizes $\{x'Bx \mid e'x = 1,\ x \geqq 0\}$ also maximizes g.
(j) Prove that B is positive definite. [*Hint*: It suffices to show that $\{\exp(\sigma_{ij})\}$ is positive definite.]
(k) Show that g has essentially all of the qualitative properties that V has (as developed in Exercise 22). [*Hint*: Utilize the Kuhn–Tucker conditions to verify the diversification results.]
(l) Show that $\tilde{x}$ is not generally mean–variance efficient unless each asset has the same mean return.

24. Consider the problem $\{\min x'Bx \mid e'x = 1,\ x \geqq 0\}$, that arose in Exercise 23, where $x \in E^n$, $e = (1, \ldots, 1)'$, and B is positive definite. Apply the Frank–Wolfe algorithm (Exercise 14) to solve this problem.
(a) Show that the direction-finding problem at iteration k, namely

$$\{\min (x^{k\prime}B)y \mid e'y = 1,\ y \geqq 0\},$$

is trivial since $y_j^k = 1$, and $y_l^k = 0$ if $l \neq j$, where $(x^{k\prime}B)_j = \min_i (x^{k\prime}B)_i$.
(b) Show that the step-size problem at iteration k, namely

$$\{\min [\alpha x^k + (1-\alpha)y^k]'B[\alpha x^k + (1-\alpha)y^k] \mid 0 \leqq \alpha \leqq 1\},$$

has the very simple solution $\alpha^\kappa = \min\{\max(0,\hat{\alpha}^k), 1\}$, where

$$\hat{\alpha}^k \equiv \frac{y^{k\prime}By^k - x^{k\prime}By^k}{x^{k\prime}Bx^k + y^{k\prime}By^k - 2x^{k\prime}By^k}.$$

25. As an alternative to the mean–variance criterion, one can consider tradeoffs between mean and semivariance. For a random variable z and a constant t, the semivariance is $E_z[\min(0, z-t)]^2$; that is, only those values of z below t are considered of importance.

Some properties of the expected value semivariance criterion are discussed in Exercise 8. This exercise discusses a solution approach for such problems.

Let the investment allocations be $x = (x_1, \dots, x_n)'$ and the random return vector be $\rho = (\rho_1, \dots, \rho_n)'$. Hence end-of-period wealth is $w \equiv \rho' x$. The semivariance of w is $S_h(x) \equiv \int \min\{0, \rho' x - h\}^2 \, dF(\rho)$.

(a) Show that S_h is a convex function of x.

(b) Show that S_h is continuously differentiable and

$$\frac{\partial S_h(x)}{\partial x_i} = 2 \int \min\{0, \rho' x - k\}^2 \rho_i \, dF(\rho)$$

as long as the variance–covariance matrix of ρ has finite components. [*Hint*: Utilize the dominated convergence theorem (Exercise 13) along with the mean-value theorem and the Cauchy–Schwarz and triangle inequalities to show the existence of the partials.]

The efficiency problem is

$$\psi(\alpha) \equiv \min S_h(x), \qquad \text{s.t.} \quad x \in K, \quad \bar{\rho}' x \geqq \alpha,$$

where K is assumed to be convex and compact.

(c) Show that ψ is a convex function of α. Graph this function and interpret the meaning of its shape.

(d) Show that the efficient boundary may be found by solving $\{\min S_h(x) - \lambda \bar{\rho}' x \mid x \in K\}$ for all $\lambda \geqq 0$.

Suppose the ρ_i have the joint discrete distribution $p_r[\rho = \rho^l > 0, \ l = 1, \dots, L < \infty]$.

(e) Develop the expressions for $\bar{\rho}' x$, $S_h(x)$, and $\partial S_t(x)/\partial x_i$.

(f) Assume that $K \equiv \{x \mid Ax \leqq b, \ x \geqq 0\}$. Illustrate the use of the Frank–Wolfe algorithm (Exercise 14) to solve the efficiency problem for a given α.

(g) Discuss the advantages and disadvantages of the algorithm in (f). How could the algorithm be efficiently modified to solve the problem when α takes on many monotonic discrete values.

(h) Suppose that K takes the simple form $K = \{x \mid 0 \leqq x_i \leqq \beta_i, \ \sum x_i = 1\}$, for fixed constants $\beta_i \geqq 0$. Show how the algorithm simplifies. Find a simple way to solve the direction-finding problem (dfp). [*Hint*: Note that the dfp is a knapsack problem in continuous variables.]

(i) Show that a lower bound on the optimal solution value is

$$\max_{k \leqq l} \{S_h(x^k) - \lambda \bar{\rho}' x^k + \nabla S_h(x^j)'(y^k - x^k)\},$$

where l is the present iteration number, x^k is the feasible solution at iteration k, and y^k solves the direction-finding problem in iteration k.

26. (Chance constraints and the aspiration and fractile objectives) Consider the linear program

$$\min c^t x, \qquad \text{s.t.} \quad Ax \geqq b, \quad x \geqq 0,$$

where A is $m \times n$, b is $m \times 1$, c is $n \times 1$, and the decision vector x is $n \times 1$. When some or all of the coefficients of b and A are random variables one way to extend the formulation is by introducing marginal chance constraints by requiring that each row of $Ax \geqq b$ be satisfied most of the time, that is,

$$\Pr[A_i x \geqq b_i] \geqq \beta_i, \qquad i = 1, \dots, m,$$

where the $\beta_i \in [0, 1]$ are given constants.

(a) Suppose that A is fixed and that each b_i has the marginal cumulative distribution function G_i. Show that $P[A_i x \geqq b_i] \geqq \beta_i$ if and only if

$$x \in \{x \mid A_i x \geqq K_{\beta_i}\}, \qquad \text{where} \quad K_{\beta_i} = \{\min y \mid G_i(y) \geqq \beta_i\}.$$

Hence the chance-constrained program

$$\min c^t x, \qquad \text{s.t.} \quad P[A_i x \geqq b_i] = i, \quad i = 1, \ldots, m,$$

has the linear program

$$\min c^t x, \qquad A_i x \geqq K_{\beta_i}, \quad i = 1, \ldots, m,$$

as its deterministic equivalent.

(b) Show that $K_{\beta_i} = G_i^{-1}(\beta_i)$ if G_i has an inverse.

Suppose now that A as well as b is random. Let $w_i = [A_i, b_i]$ and $z = \binom{x}{-1}$. Then $P\{A_i \geqq b_i\}$ can be written as $P[w_i{}^t z \geqq 0]$. Assume that w_i has the multivariate normal distribution $w_i \sim N(\bar{w}_i, V_i)$, where $\bar{w}_i$ is the mean vector and V_i is the positive-semidefinite variance–covariance matrix of the w_i. Let $\psi(y)$ denote the cumulative distribution of a normally distributed random variable with mean 0 and variance 1 and K_β the β-fractile of ψ, namely the K_β such that $\psi(K_\beta) = \beta$ for any $\beta \in (0, 1)$.

(c) Show that $P[w_i{}^t z \geqq 0] \geqq \beta_i$ if and only if $\bar{w}_i{}^t z - K_{\beta_i}(z^t V_i z)^{1/2} \geqq 0$.

(d) Show that $(z^t V_i z)^{1/2}$ is a convex function of z.

(e) Use the Cauchy–Schwarz inequality [i.e., $u, v \in E^n$, $u^t v \leqq (u^t u)(v^t v)$] to show that $z^{0t} V_i z \leqq (z^{0t} V_i z^0)^{1/2} (z^t V_i z)^{1/2}$.

A function g is said to be subdifferentiable at x^0 if there exists a vector $s(x^0) \in E^n$ such that $g(x) \geqq s(x^0)^t (x - x^0) + g(x^0)$ for all $x \in E^n$.

(f) Use (e) to show that $(z^t V_i z)^{1/2}$ is a subdifferentiable function of $z \in E^n$ having subgradient vector

$$s(z) = \begin{cases} z^t V_i / (z^t V_i z)^{1/2} & \text{if} \quad z^t V_i z > 0, \\ 0 & \text{otherwise.} \end{cases}$$

(g) The chance-constrained problem

$$\min c^t z, \qquad \text{s.t.} \quad \Pr[w_i{}^t z \geqq 0] \geqq \beta_i, \quad i = 1, \ldots, m,$$

has the deterministic equivalent

$$\min c^t z, \qquad \text{s.t.} \quad w_i{}^t z - K_{\beta_i}(z^t V_i z)^{1/2} \geqq 0, \quad i = 1, \ldots, m. \tag{1}$$

Suppose $\beta_i \geqq 0.5$ for all i. Show that the set of points that satisfy the constraint of (1) form a convex set.

In chance-constrained programming models one can generally consider separately the objective and the constraints. When the vector c is random one can consider numerous objectives such as the expected utility criterion. In addition one can consider, as in the Pyle–Turnovsky paper, the aspiration or fractile objectives. The aspiration model may be written as

$$\min P[c^t x > d_0], \qquad \text{s.t.} \quad Ax \geqq b, \quad x \geqq 0$$

for a given aspiration level d_0, while the fractile model may be written as

$$\min d, \qquad \text{s.t.} \quad P[c^t x \leqq d] \geqq \alpha, \qquad Ax \geqq b, \quad x \geqq 0$$

for a given fractile $\alpha \in [0, 1]$. Suppose that $c \sim N(\bar{c}, V_0)$.

(h) Show that the fractile model has as its deterministic equivalent

$$\min \bar{c}^t x + K_\alpha (x V_0 x)^{1/2}, \qquad \text{s.t.} \quad Ax \geqq b, \quad x \geqq 0$$

where $K_\alpha = \psi^{-1}(\alpha)$.

(i) Suppose that V_0 is positive definite. Show that the aspiration model has as its deterministic equivalent

$$\min (\bar{c}^t x - d_0)/(xV_0 x)^{1/2}, \qquad \text{s.t.} \quad Ax \geqq b, \quad x \geqq 0,$$

assuming $x = 0$ is not optimal.

(j) Show that $x = 0$ is optimal for the aspiration model if and only if $d_0 \geqq 0$.

27. Suppose investments X and Y have a common mean and finite positive variances. Suppose an investor has initial wealth of one dollar and has a strictly increasing and strictly concave utility function over wealth w. To maximize expected utility, one must maximize $\{\phi(\lambda) = Eu[\lambda X + (1-\lambda)Y] \mid 0 \leqq \lambda \leqq 1\}$. It is of interest to investigate how diversification is affected by possible negative dependence between X and Y.

(a) Suppose X has a uniform distribution on $[0,1]$ and that $Y = -12X^2 + 10X - \frac{1}{2}$. Show that $\operatorname{cov}(X,Y) = -\frac{1}{6} < 0$ and $EX = EY = \frac{1}{2}$.

(b) Suppose that

$$u(X) = \begin{cases} X & \text{if} \quad X \leqq \frac{1}{10}, \\ \frac{1}{10} + \varepsilon(\frac{1}{10} - X)^2 & \text{if} \quad X \geqq \frac{1}{10}, \end{cases}$$

where $\varepsilon > 0$ is sufficiently small. Show that u is strictly increasing, strictly concave, and differentiable.

(c) Show that $\lambda^* = 1$ so that it is not optimal to diversify. [*Hint*: Show that $d\phi(1)/d\lambda = E[u'(X)(X-Y)] > 0$.]

(d) Suppose X and u are as defined in (a) and (b), and that $Y = 12X^2 - 10X + \frac{3}{2}$. Show that $EX = EY = \frac{1}{2}$, $\operatorname{cov}(X,Y) > 0$, and that $0 < \lambda^* < 1$. Hence, negative correlation is neither necessary nor sufficient for diversification.

(e) Show that $\lambda^* < 1$ if and only if $E[(X-Y)u'(X)] < 0$.

(f) Show that the hypotheses with regard to the result in (e) may be weakened so that u is concave and not necessarily differentiable or u is pseudo-concave. Show that the monotonicity assumption is crucial, however.

(g) Show that $0 < \lambda^*$ if and only if $E[(Y-X)u'(Y)] < 0$.

(h) Show that there is diversification, that is, $0 < \lambda^* < 1$, if and only if

$$\int_{y=-\infty}^{\infty} \int_{x=-\infty}^{z} (x-y)\, dF(x,y) \leqq 0 \qquad \text{and} \qquad \int_{x=-\infty}^{\infty} \int_{y=-\infty}^{z} (y-x)\, dF(x,y) \leqq 0$$

for all z, provided that $E[(X-Y)u'(X)] \neq 0 \neq E[(Y-X)u'(Y)]$.

(i) Use (h) to show that if all risk-averse investors should diversify between X and Y, then $EX = EY$. [*Hint*: Let $z \to \infty$.]

28. Refer to Exercise 27. It is of interest to develop a concept of negative dependence that always leads to diversification in the two-asset case.

(a) Suppose that $E(Y \mid X = x)$ is a strictly decreasing function of x. Show that it is optimal to hold some Y that is $\lambda^* < 1$. [*Hint*: Show that

$$E[(X-Y)u'(X)] = \operatorname{cov}[X, u'(X)] - \operatorname{cov}[E(Y \mid X), u'(X)] < 0.$$

(b) Suppose that $E(X \mid Y = y)$ is a strictly decreasing function of y. Show that it is optimal to hold some X. Hence if both conditional means are decreasing functions, the investor will always diversify.

(c) Show that if $P[Y \leqq y \mid X = x]$ is strictly increasing in x for each y, then $E(Y \mid X = x)$ is strictly decreasing in x, and that the converse is false.

(d) Show that if $E(Y \mid X = x)$ is strictly decreasing in x, then $\operatorname{cov}(X,Y) < 0$ and the converse is false.

(e) Utilize (a)–(c) to show that $P(Y \leqq y \mid X = x)$ and $P(X \leqq x \mid Y = y)$ strictly increasing in x and y, respectively, are sufficient for diversification.

(f) Note that the conditions in (e) under the differentiability hypothesis are

$$\frac{\partial}{\partial y} P[X \leqq x \mid Y = y] > 0 \qquad \text{and} \qquad \frac{\partial}{\partial y} P[Y \leqq y \mid X = x] > 0.$$

Compare this with Theorem IV in Samuelson's paper on diversification.

29. Suppose an investor has the power utility function $(1/\gamma)\, w^{\gamma}$ over wealth w. Let $r_i > 0$ be the return per dollar invested in i if state i occurs; investment in i produces no return if j occurs $(i \neq j)$. Assume that the number of states of nature is finite. Let $x_i \geqq 0$ be the amount bet on the occurrence of i out of initial wealth of one dollar. Suppose the investor wishes to maximize expected utility.

(a) Show that if x^* is an expected utility maximizing allocation, then each x_i^* is a decreasing function of r_i if and only if $\gamma < 0$.

(b) Show that each x_i^* is an increasing function of the r_i if and only if $0 < \gamma < 1$.

(c) Discuss the results in (a) and (b) and their relation to the results in Exercise 12.

30. In many financial optimization problems it is of interest to know that the optimal decision vector is a continuous or differentiable function of a given parameter. Such results were assumed in Samuelson's paper in Part III, Chapter 1.

Consider the pair of portfolio problems

$$\left\{\max W[x(\varepsilon), \varepsilon] \equiv E_{\xi} \frac{(\sum \xi_i x_i)^{\varepsilon}}{\varepsilon} \,\middle|\, x = (x_1, \ldots, x_n) \in X\right\}, \tag{A}$$

and

$$\{\max Z(x) = E_{\xi} \log(\textstyle\sum \xi_i x_i) \mid x \in X\}, \tag{B}$$

where $X \equiv \{x \mid e^t x = 1,\ x \geqq 0\}$ and $e^t = (1, \ldots, 1)$. Suppose the random vector ξ is positive and bounded from above so that $\Pr\{0 < \beta < \xi < \alpha\} = 1$, where each $\alpha_i < \infty$, that $f(\xi)$, the probability density function of ξ, exists and is continuous, and that $\xi' x^1 = \xi' x^2$ for all ξ implies $x^1 = x^2$.

(a) Show that W and Z are finite, continuous, strictly concave functions on X. Thus (A) and (B) have unique optimal solutions $x(\varepsilon)$ and $x(0)$, respectively.

(b) Utilize Exercise 13 to show that one may differentiate under the integral sign to calculate the partial derivatives of W and Z.

(c) Show that $\lim_{\varepsilon \to 0} [(V^{\varepsilon} - 1)/\varepsilon] = \log V$.

Thus as $\varepsilon \to 0$ one would suspect that $x(\varepsilon) \to x(0)$, since $(V^{\varepsilon} - 1)/\varepsilon$ and $V^{\varepsilon}/\varepsilon$ are equivalent for maximization purposes. Indeed Samuelson and Merton (1973) have devised a computational approach to calculate $x(\varepsilon)$ given knowledge of $x(0)$. Their procedure utilizes the expansion

$$x_i(\varepsilon) = x_i(0) + \varepsilon \frac{\partial x_i(0)}{\partial \varepsilon} + O(\varepsilon^2),$$

where $\lim_{\varepsilon \to 0} [O(\varepsilon^2)/\varepsilon^2] = L$, a positive constant. Such an expansion is valid if $x_i(0)$ is twice differentiable.

(d) Show that the Kuhn–Tucker conditions are necessary and sufficient for (A) and (B), and that for (A) they may be written as

$$\nabla_x W[x(\varepsilon), \varepsilon] \leqq \lambda(\varepsilon) e, \qquad x(\varepsilon) \geqq 0, \qquad e^t x(\varepsilon) = 1,$$
$$x^t[\lambda(\varepsilon) e - \nabla_x W[x(\varepsilon), \varepsilon]] = 0 \qquad \text{and} \qquad \lambda(\varepsilon) \geqq 0,$$

where ∇ is the gradient operator and $\lambda(\varepsilon)$ is a Lagrange multiplier.

(e) Show that as $\varepsilon \to 0$ these conditions converge to those for problem (B); hence $x(\varepsilon) \to x(0)$.

Assuming that each $x_i(\varepsilon) > 0$, the conditions in (d) simplify to

$$\nabla_x W[x(\varepsilon), \varepsilon] - \lambda(\varepsilon) e = 0, \qquad e^t x(\varepsilon) = 1.$$

(f) Show that a total differentiation of these $n+1$ equations yields the matrix equations

$$\begin{bmatrix} \nabla_{xx} W[x(\varepsilon), \varepsilon] & -e \\ -e^t & 0 \end{bmatrix} \begin{bmatrix} \nabla_\varepsilon x(\varepsilon) \\ \dfrac{\partial \lambda(\varepsilon)}{\partial \varepsilon} \end{bmatrix} = \begin{bmatrix} -\nabla_{x\varepsilon} W[x(\varepsilon), \varepsilon] \\ 0 \end{bmatrix},$$

where ∇_{xy} denotes the Hessian operator. Hence $x(\varepsilon)$ is differentiable if the Hessian matrix on the left-hand side is nonsingular.

(g) Let the $(n+1) \times (n+1)$ matrix

$$B = \begin{bmatrix} A & -e \\ -e^t & 0 \end{bmatrix},$$

where A is an $n \times n$ negative-definite matrix. Show that B^{-1} exists and may be partitioned as

$$\begin{bmatrix} A^{-1} & +X\theta^{-1}Y & -\theta^{-1}Y \\ & -\theta^{-1}Y & \theta^{-1} \end{bmatrix},$$

where $X \equiv -A^{-1}e$, $Y \equiv -e^t A^{-1}$, and $\theta \equiv -e^t A^{-1} e$. [*Hint*: Utilize the fact that B^{-1} exists if and only if $\theta \neq 0$, which it is since A^{-1} is negative definite.]

(h) Let $A(\varepsilon) \equiv \nabla_{xx} W[x(\varepsilon), \varepsilon]$. Utilize (f) to show that $x(\varepsilon)$ is differentiable and that

$$\nabla_\varepsilon x(\varepsilon) = -\left[A(\varepsilon)^{-1} - \frac{A(\varepsilon)^{-1} e e^t A(\varepsilon^{-1})}{e^t A(\varepsilon)^{-1} e}\right] \nabla_{x\varepsilon} W[x(\varepsilon), \varepsilon].$$

(i) Show that $\nabla_\varepsilon x(0)$ exists. [*Hint*: Show that

$$\lim \nabla_{xx} W[x(\varepsilon), \varepsilon] = \nabla_{xx} Z(x) \qquad \text{and that} \qquad \lim \nabla_{x\varepsilon} W[x(\varepsilon), \varepsilon]$$

exists. Hence $\lim_{\varepsilon \to 0} \nabla_\varepsilon x(\varepsilon) \equiv \nabla_\varepsilon x(0)$ exists.]

The expression in (g) may be used to verify that $x(\varepsilon)$ is also twice differentiable utilizing the following result.

(j) Suppose $B(y)$ is an $m \times m$ matrix whose elements are functions of a single variable y, i.e.,

$$B(y) = \begin{bmatrix} b_{11}(y) & \cdots & b_{1m}(y) \\ \vdots & & \vdots \\ b_{m1}(y) & \cdots & b_{mm}(y) \end{bmatrix}.$$

Assume that each b_{ij} is a differentiable function of y in some neighborhood $N(y^0, \delta)$, for some y^0 and $\delta > 0$ and that $|B(y)| \neq 0$ for all $y \in N(y^0, \delta)$. Show that

$$\frac{d}{dy} B(y)^{-1} = -B(y)^{-1} \frac{\partial}{\partial y}[B(y)]\, B(y)^{-1} \qquad \text{for all} \quad y \in N(y^0, \delta).$$

[*Hint*: Differentiate $B(y) B(y)^{-1} = I$.]

(k) Let

$$H(\varepsilon) \equiv -A(\varepsilon)^{-1} \frac{\partial}{\partial \varepsilon}[A(\varepsilon)]\, A(\varepsilon)^{-1}.$$

Utilize the result in (i) to show that

$$\nabla_{\varepsilon\varepsilon} x(\varepsilon) = \begin{bmatrix} \partial^2 x_1(\varepsilon)/\partial\varepsilon^2 \\ \vdots \\ \partial^2 x_n(\varepsilon)/\partial\varepsilon^2 \end{bmatrix} = -\left[\frac{A(\varepsilon)^{-1} - A(\varepsilon)^{-1}ee^t A(\varepsilon)^{-1}}{e^t A(\varepsilon)^{-1} e}\right]$$

$$= \nabla_{x\varepsilon\varepsilon} W[x(\varepsilon), \varepsilon](\nabla_\varepsilon x(\varepsilon) + e) - \nabla_{x\varepsilon} W[x(\varepsilon), \varepsilon]^t$$

$$\cdot \left[H(\varepsilon) - \frac{2e^t A(\varepsilon)^{-1} e H(\varepsilon) ee^t A(\varepsilon)^{-1} - A(\varepsilon)^{-1} ee^t A(\varepsilon)^{-1} e^t H(\varepsilon) e}{[e^t A(\varepsilon)^{-1} e]^2}\right].$$

(l) Show that $\nabla_{\varepsilon\varepsilon\varepsilon} x(\varepsilon)$ exists and in fact partial derivatives of $x(\varepsilon)$ of all orders exist.
*(m) Show that the preceding analysis is not applicable if any $x_i(\varepsilon) = 0$. Attempt to prove the existence of $\partial x_i(\varepsilon)/\partial\varepsilon$ in this case.

31. Consider a random variable X having the density function

$$p(x) = \begin{cases} \dfrac{1}{2\sigma^2}\dfrac{1}{X}\exp\left\{-\dfrac{|\log X - \mu|}{\sigma^2}\right\} & \text{if } X > 0, \\ 0 & \text{otherwise.} \end{cases}$$

This distribution may be called the log–Laplace distribution, since the distribution of $Y = \log X$ is the Laplace distribution with density

$$p(y) = \frac{1}{2\sigma^2}\exp\left\{-\frac{|y-\mu|}{\sigma^2}\right\}, \qquad -\infty < y < \infty.$$

Note that $EY = \mu$, and $\sigma^2 > 0$ is the dispersion parameter.
(a) As is the case with the log-normal distribution, the moments EX^k can be expressed in elementary form whenever they exist. Show that for any $k > 0$

$$EX^k = \begin{cases} \exp\{k\mu\}(1 - k^2\sigma^2)^{-1}, & 0 < \sigma k < 1 \\ \text{divergent} \to +\infty, & \sigma k \geqq 1. \end{cases}$$

[*Hint*: Note that EX^k is the moment-generating function of the Laplace distribution; therefore, apply the transform $(\log X - \mu)/\sigma^2 = y$ and evaluate the integral for positive and negative y separately.]
(b) Consider any function U with $U(1) = 0$, bounded on $(0, \infty)$, and three times differentiable. Show that

$$\lim_{t\to 0}(1/t)\,EU(X) = (\mu + \sigma^2)\,U^{(1)}(1) + 2\sigma^2 U^{(2)}(1)/2!,$$

where $U^{(1)}$ and $U^{(2)}$ are the first and second derivatives, respectively, and the distribution of X is log–Laplace with parameters $t\mu$ and $t\sigma^2$. [*Hint*: Show first that

$$\lim_{t\to 0}\frac{E(X-1)}{E(X-1)^2} = \frac{\mu+\sigma^2}{2\sigma^2} \qquad \text{and} \qquad \lim_{t\to 0}\frac{E(X-1)^3}{E(X-1)} = 0.$$

Second, express U as a Taylor series expansion in three terms, and note that $U^{(3)}$ is necessarily bounded on $(0, \infty)$ when U is bounded on the same region. Finally, using the facts given above, show that the third term in the series development vanishes faster than the two first terms (consult Ohlson's paper).]
(c) Part (b) suggests a model useful for portfolio selection analysis. Suppose the investment returns X_i are generated by the ("market model") process

$$\log X_i = \sqrt{t}\,\alpha_i + \sqrt{t}\,\beta_i + \sqrt{t}\,\gamma_i Y_i, \qquad i = 1, \ldots, m,$$

and where $Z, Y_1, \ldots, Y_m$ are independently but identically distributed random variables with the standardized Laplace density

$$\tfrac{1}{2}\exp\{-|y|\}.$$

Let $\sum_{i=1}^m \lambda_i = 1$, $\lambda_i \geqq 0$, and let U be the investor's utility of wealth function. Investigate what restrictions are required on U in order that

$$\lim_{t\to 0}\frac{1}{t} EU\left(\sum_{i=1}^{m} \lambda_i X_i\right)$$

is quadratic in $(\lambda_1, \ldots, \lambda_m)$. [Assume, without loss of generality that $U(1) = 0$]. Find the limit when it exists. [*Hint*: Consider first the case when U is bounded on $(0, \infty)$, and show that extremely mild conditions then are required on U. For the case where U is unbounded, consult Ohlson's paper. Note that *all* moments of EW^k, $k = 1, 2, \ldots$, do not exist, regardless of how small t is; hence, $EU(W)$ can never be expressed as a linear function of *all* moments EW^k, $k = 1, 2, \ldots$. (Compare this with Samuelson's paper, in which it is required that all moments exist.)]

32. Suppose an investor has a monotone, nondecreasing, differentiable, concave utility function $u(w)$ over wealth w. Assume further that $w \geqq 0$. We are concerned with the question of when the expected utility will be finite, given that u may be unbounded.

(a) Suppose $u'(0)$ and $E(w)$ are finite. Show that expected utility is finite. [*Hint*: First establish that $u'(w)$ is uniformly bounded, then utilize the differential definition of a concave function.]

It is appropriate to consider weakening the assumptions on the finiteness of marginal utility at zero wealth and/or the finiteness of the mean wealth level.

(b) Suppose u is a polynomial of degree n. Show that expected utility is never finite if $E(w)$ is infinite. [*Hint*: Begin by showing that if $E(w)$ is infinite, then so is $E(w^n)$ for all $n > 1$).]

(c) Show that the result in (b) also holds for any nonconstant, monotone, nondecreasing, concave utility function. Hence it is not possible to relax the assumption concerning the finiteness of the mean vector.

(d) Suppose that $E(w)$ is finite and $u(0)$ is finite with no specification regarding the finiteness of $u'(0)$ and $u(w_0)$ is finite for some $w_0 > 0$. Show that expected utility is finite.

(e) Extend the result in (d) to the case $w \geqq A$ for $A > -\infty$.

Exercise Source Notes

Exercise 1 was adapted from Freund (1956); Exercise 2 was adapted from Borch (1968); Exercise 4 was adapted from Madansky (1962) and Ziemba (1971); Exercises 6 and 7 were adapted from Pyle (1971); portions of Exercise 8 were adapted from Mao (1970); Exercises 9 and 10 were adapted from Press (1972); Exercise 11 was based on Pye (1974); Exercise 12 was based on Arrow (1971); portions of Exercises 13–15 were adapted from Ziemba (1972a, c); Exercise 17 was adapted from Stone (1970) utilizing notes written by Professor K. Nagatani; Exercise 20 was adapted from Aitchison and Brown (1966); Exercise 21 was adapted from notes provided by Professor J. Ohlson and Exercises 22 and 23 were adapted from Ohlson (1972b); Exercise 25 was adapted from Hogan and Warren (1972); Exercise 26 was adapted from Parikh (1968); Exercises 27 and 28 were adapted from Brumelle (1974); Exercise 29 was developed by Professor J. Ohlson; Exercise 30 was developed in collaboration with Professor W. E. Diewert; and Exercise 31 was developed by Professor J. Ohlson, and Exercise 32 was adapted from Arrow (1974) and Ryan (1974).

Part IV
Dynamic Models Reducible to Static Models

INTRODUCTION

This part of the book is concerned with stochastic dynamic models of financial problems that are reducible to static models. That is, problems where there exists a static program whose optimal solution provides the optimal action to take in the first period and an optimal strategy for the remaining periods conditional on preceding decision choices and realizations of random variables. There are three natural classes of such models. First, the dynamic nature of the model may be fictitious: Even though the model appears to have a dynamic character there is only one decision period. The second category concerns models in which there exists a deterministic equivalent (i.e., an explicit deterministic program whose set of optimal solutions is the same as that of the stochastic dynamic problem), but it is *not* generally possible to find an explicit analytical characterization for the deterministic equivalent. Models may be considered to be in this category whenever an optimal policy exists and a backward induction type of dynamic programming argument is valid. The major concern regarding such models is the determination of qualitative properties of the optimal decision policy using the deterministic equivalent. Third, the dynamic model may have the property that it is possible to determine an explicit static deterministic equivalent whose solution provides the requisite decision policy. Such a deterministic equivalent may depend on decision variables from only one period or it may depend on decision variables from several periods in a way that yields a static program. The original dynamic program is said to have a myopic policy if the optimal decision in each period can be determined without consideration of future decisions and random events. The dynamic program is said to yield zero-order decision rules if the optimal decision policy in each period is independent of the realizations of random variable occurrences in all future periods.

I. Models that Have a Single Decision Point

The paper by Wilson illustrates a capital budgeting model in the first category. He presents a method to add investment projects one at a time to provide successively higher values of the expected utility of the returns obtained over the n periods. It is assumed that the available project selection list is known at the beginning of the time horizon. Generally, projects may be commenced in any of several periods and they may have stochastic as well as deterministic interconnections with projects beginning in any of the n periods. Since projects are added one at a time, a static analysis is possible. Projects are always accepted if they have the property that a financing plan paid for by project revenues exists that will cover the project's costs for all states of the world. This weak sufficiency test may be checked for any particular project by solving a linear program. Projects that do not meet this stringent criterion may still be acceptable if there is a funding program that leads to an increase in expected utility. One may verify this "stronger" sufficient condition for project inclusion by solving a concave program. A relatively simple calculation can be made to determine if a project can be excluded from consideration in the present diversified package of projects. In Exercise CR-4, the reader is asked to provide direct rules for the inclusion and exclusion of projects in terms of a simple expected utility evaluation. Exercise ME-1 illustrates a two-period consumption–investment allocation problem that can be analyzed via a static analysis. Of particular interest is the relationship between the optimal investment decision that results when the random return takes on its mean value and the optimal investment decision resulting from the stochastic problem.

II. Risk Aversion over Time Implies Static Risk Aversion

The paper by Fama considers a multiperiod consumption–investment problem. The consumer's objective is to make investment allocations between several random investments and consumption withdrawals in each period to maximize the expected utility of lifetime consumption. The consumer's utility function is assumed to be a strictly increasing, strictly concave function of his lifetime consumption. It is not necessary in the development to assume that the utility function is intertemporally additive. It is assumed that the markets for consumption goods are perfect. Fama then shows that the consumer's behavior is indistinguishable from that of a risk-averse expected utility maximizer who has a one-period horizon. That is, the process of backward induction may be employed to reduce the multiperiod problem into an equivalent static problem. Moreover, the maximization and expectation operations preserve the concavity of the derived utility function at each stage of the induction. Fama's presentation is in a quite general setting and the results hold if the utility function is only concave (Exercise CR-1) or if the consumer's lifetime is uncertain. The results also provide a partial multiperiod justification for the static two-parameter portfolio models described in the papers by Lintner and Ziemba in Parts II and III, respectively. The basic idea that concavity properties of dynamic models are often preserved under maximization and expectation operations is a very useful one and dates back at least to early results by Bellman and Dantzig in the 1950s concerned with stochastic dynamic programming and linear programming under uncertainty, respectively. [See Wets (1966, 1972a, b) and Olsen (1973a, b) for recent results and additional references in this area.] Such reductions allow one to develop some qualitative characteristics of the optimal first-period decision in many dynamic models. However, the equivalent static program cannot generally be determined in explicit analytic form; hence quantitative properties of the optimal first-period decision cannot generally be ascertained.

In certain special cases, one can determine an explicit functional form for the derived utility function. Exercise CR-2 illustrates such a calculation when the utility function is quadratic and there are two risky assets. Conditions are developed that lead to optimal investment entirely in one of the risky assets. Exercise ME-2 illustrates the classic Simon (1956)–Theil (1957) certainty equivalence results. Preferences for state and/or decision variables are assumed to be quadratic. The constraints that link the state and decision variables are linear functions whose uncertainty enters additively. It is also assumed that all relevant maxima and expectations exist and that the maxima occur at interior points of any additional constraints. The backward induction process then yields a sequence of induced quadratic programs. It also develops that one can replace uncertain random variables by their conditional means to obtain a certainty equivalent. The certainty equivalent is an explicit deterministic quadratic program involving only variables from period 1 and whose optimal solution is an optimal first-period decision for the multiperiod problem. Exercise ME-3 develops similar results when it is not assumed that the constraints on the decision variables are nonbinding. The discussion proceeds by considering the following questions: (1) When is the optimal decision in each period independent of all random vectors and all other decisions; (2) when is the optimal decision in each period independent of all other decisions; and (3) when is it possible to develop an explicit deterministic static program whose solution provides an optimal first-period decision?

Conditions are presented that provide answers to each of these questions. They generally involve the assumption that the state vectors appear linearly in the preference function and in the period-by-period linkage constraints. For the first two questions, it is also necessary to assume that certain intertemporal separability conditions are satisfied by the decision and random vectors. All of the conditions provide for the optimality of zero-order decision rules so that optimal decisions for each period can be determined before any random variables are observed. The results are used to construct a multiperiod stochastic capital budgeting model in Exercise ME-4.

Some dynamic investment problems have the property that they possess simple optimal policies that are stationary in time. Exercise CR-5 considers the dynamic portfolio problem when the utility function is logarithmic and the investment choice in each period is between a risk-free asset and a lognormally distributed asset. Conditions are developed such that it is optimal in each period to invest totally in the risky asset. A similar result obtains if the utility function is a power function (Exercise CR-6) or if the investor receives additions to his wealth from sources exogenous to his investment choice (Exercise ME-6).

III. Myopic Portfolio Policies

The paper by Hakansson and several of the exercises are concerned with the problem of optimal portfolio myopia; that is, the determination of cases when the investor can make optimal decisions by considering only a one-period horizon. Such a result obtains if the derived utility function in each period is equivalent to the investor's utility function over final wealth. The utility functions are equivalent if they are the same except for a positive linear transformation.

Exercise ME-5 considers this question for concave utility functions when there is one risky and one risk-free asset and it is assumed that the maximum in each period occurs at an interior point. Myopia results obtain whenever the reciprocal of the Arrow–Pratt absolute risk aversion index is linear in wealth. When the intercept of the linear function is zero (i.e., the utility function has constant relative risk aversion), or if the risk-free asset is money (i.e., its net return is zero), there is true myopia. In the general case, there is partial myopia in the sense that the derived utility function is equivalent to the investor's utility function if one "pretends" that all the investor's wealth is invested in the risk-free asset. Exercise CR-3 illustrates the myopia and partial myopia results for power utility functions. In this case, myopia obtains when the investor does not receive additions to his wealth from sources exogenous to his investment choice; otherwise, there is partial myopia. Hakansson's paper considers the more realistic case when it is not permissible to assume that there are always interior maxima. If the investment returns are intertemporally independent, then the logarithmic and power utility functions induce true myopia. However, if the returns are dependent, then only the logarithmic function is myopic. He also shows that the partial myopia results and myopic results for the case when the interest rate is zero, as developed in Exercise ME-5, are true only in a highly restricted sense when there are constraints on the allowable investment allocations. Brennan and Kraus (1974) investigate the close relationship between these conditions for myopia and the Cass–Stiglitz conditions needed for portfolio separation as discussed in Vickson's paper in Part II.

Exercise ME-7 considers the myopia problem when the investment horizon becomes more and more distant. Turnpike-type results that induce true myopia are obtained when the reciprocal of the risk-aversion index is linear in wealth. Additional results obtain in the more general case when the reciprocal of the risk-aversion index equals a term proportional to wealth plus a bounded function of wealth. Hence the optimal portfolio choice depends on the current distribution of the random assets, and optimal asset holdings are proportional to current wealth when the utility function has "almost" constant relative risk aversion for large wealth. However, see Ross (1974a) for an example that illustrates how the turnpike result can fail for utility functions that are close to inelastic. For further results on turnpike portfolios, see Ross (1974a, b) and Hakansson (1974). For additional information and results concerning myopic policies, the reader may consult Arrow (1964), Dirickx and Jennergren (1973), Ignall and Veinott (1969), and Marglin (1963).

1. MODELS THAT HAVE A SINGLE DECISION POINT

MANAGEMENT SCIENCE
Vol. 15, No. 12, August, 1969
Printed in U.S.A.

INVESTMENT ANALYSIS UNDER UNCERTAINTY*†

ROBERT WILSON

Graduate School of Business, Stanford University

Investment projects are described in this paper as a pattern of uncertain cash flows over time; i.e., as a cash flow pattern over an event tree. Whereas the portfolio problem of selecting projects generally requires simultaneous consideration of all available projects, here we seek sufficient conditions for accepting and rejecting individual projects. These conditions are formulated as mathematical programming problems which are amenable to routine application at subordinate levels of an organization.

"Investment is, in essence, *present* sacrifice for *future* benefit. But the present is relatively well known, whereas the future is always an enigma. Investment is also, therefore, *certain* sacrifice for *uncertain* benefit." [1]

1. The State Description of Uncertain Events

When one says that the return in the next year from an investment project is uncertain, it usually means that the return will depend upon prevailing conditions or circumstances in the interim, but precisely which conditions will obtain is not known at present. If the prevailing conditions, or "state of the world," were known, however, then also the return would be known. That is, one knows well enough the effect of the prevailing conditions on the return to be able to predict the return that will be attained in each state of the world. Of course, how well the effects must be known depends upon the precision demanded of the prediction, but in practice it often suffices to take account only of major sources of uncertainty. A *state description* is a specification of possible states of the world which is, for purposes of analysis, sufficiently fine to enable one to regard the predicted return in each state as certain if that state should obtain. We will not formalize the construction of a state description, but rather take it to be a datum.

The time dimension of a state description can be made explicit by constructing an *event tree* specifying the sequence of states that can obtain. Figure 1 depicts an example of an event tree over four time periods. Each node of the tree represents a point in time, and each arc represents a state of the world prevailing up to the point in time of its successor node at the right-hand end point. The label $S_{ijk\cdots}$ on an arc records that the state is compounded of an event indexed by i in the first period, followed by an event indexed by j in the second period, then an event indexed by k in the third period, and so on. The essential feature of an event tree is the delineation of the possible events that can follow after the state of the world in one period to determine the state of the world in the succeeding period. It is important to recognize that, as defined here, a state embodies the entire sequence of events up to and including the period it describes. For this reason, the set of events that can possibly occur in the next period may be different for different nodes representing that same point in time; cf. Figure 1, in which the sets of events that might obtain in the last period depend upon the node attained at time 3.

* Received January 1967; revised March 1968.

† This study was supported, in part, by funds made available by the Ford Foundation to the Graduate School of Business, Stanford University. However, the conclusions, opinions and other statements in this publication are those of the author and are not necessarily those of the Ford Foundation.

[1] Quotation from J. Hirshleifer [1].

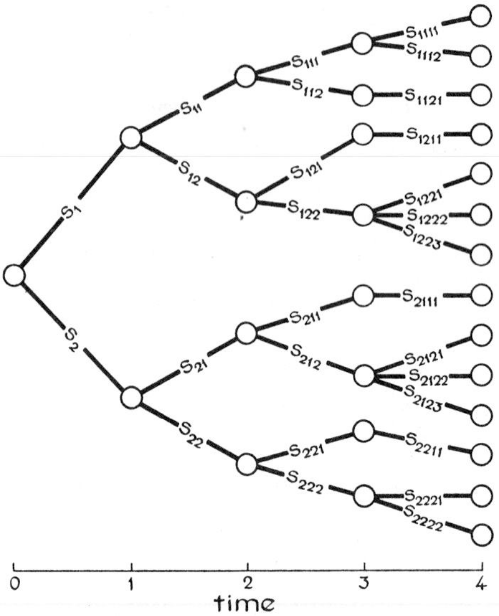

FIGURE 1. An Event Tree

We shall assume throughout that the number of possible states is finite, that each state (arc) has only one successor node, and that each node has only one predecessor arc.

2. The State-Relevant Description of Investment Projects

An investment project has many facets: besides the schedule of cash flows, it ordinarily includes allocations of material and human resources which depend upon the other activities of the firm as well as the sequence of states that will obtain. We shall capture the essential features in an abbreviated project description that will suffice for the purposes of analysis.

In an event tree, such as the one depicted in Figure 1, label each node by the indices of the state leading to it; specifically, label the initial node N_0 and then for each state $_{ijk\cdots}$ label the node following the arc with that label by $N_{ijk\cdots}$. Now for simplicity we shall assume that cash flows occur only at nodes of the event tree; that is, income and expenses for a project are accrued over each time period and then accounted for at the discrete points in time represented by the nodes. Consequently, given a *policy* (or program, or plan) specifying (1) for every state what the other activities of the firm will be, and (2) for that policy what the (feasible) allocation of resources to the project will be, one can specify at each node the predicted cash flow from the project.

To make this explicit, letting P denote the policy symbolically, a project is described by a *cash flow description*[2]

$$C(P) = [c_0 ; \{c_i\} ; \{c_{ij}\} ; \{c_{ijk}\} ; \cdots] \tag{1}$$

[2] The curly braces $\{\cdots\}$ in (1) are intended to imply an array of elements of the indicated form.

where $c_{ijk\cdots}$ denotes the net cash inflow at node $N_{ijk\cdots}$ assuming the policy P is pursued. Here, if $c_{ijk\cdots}$ is negative then it is interpreted as a net cash outflow.

In this format an investment project can also be represented as a cash flow description superimposed on an event tree. An example for a two-period project is depicted in Figure 2. For example, $c_0 = -\$100{,}000$ is the initial expenditure required to launch the project. All amounts depend, of course, on the policy assumed and might well be different for another policy.

The cash flow description is a flexible representational device. In particular, observe that deferred projects can be represented in this manner; for example, if a project were to be initiated at time 2 instead of time 0, then $c_0 = 0$, $c_i = 0$ for all nodes N_i at time 1, and the amounts c_{ij} for the nodes N_{ij} at time 2 would represent the initial expenditure required conditional on the state of the world over the interim two-year period.

Financing projects (stock issues, debentures, loans, etc.) can also be cast into this form. An issue of debentures, for example, would result in a positive cash inflow initially followed by net cash outflows at subsequent times for interest and sinking fund payments. In the case of a debenture the cash flows would be independent of the state of the world at each point in time, as well as of the policy, but this would not be true of a stock issue.

One might also construe a project as deletion of an activity already in the policy.

In practice it is important to keep in mind the dependence of the cash flow description upon the policy. Every firm is constrained by limited resources which bound the number and scope of the projects it can undertake. In what follows we shall allow for sequential revision of the policy as the investment analysis proceeds.

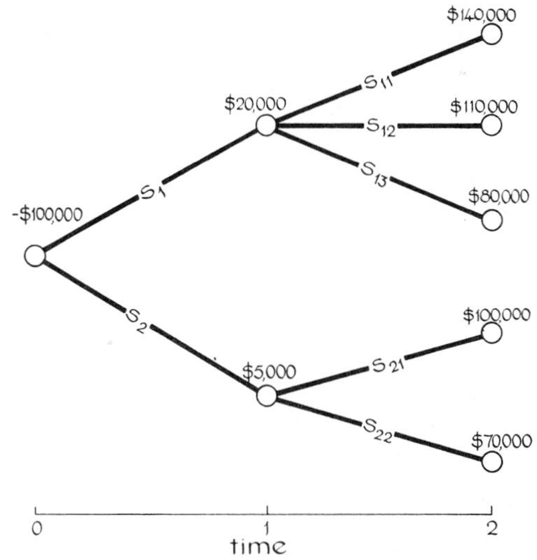

FIGURE 2. A Cash Flow Description for an Investment Project

3. The Structure of Investment Projects

So far we have introduced three basic concepts for investment analysis: a state description with its associated event tree, a policy, and a cash flow description for a project. In terms of these concepts, an investment project can be delineated abstractly as follows. A policy is given initially which will result in a known cash flow pattern over the event tree. To be analyzed is an amendment to that policy, called the investment project, which will result in a new policy and a new cash flow pattern over the event tree. The project is to be accepted if the new policy with its associated cash flow pattern is preferred to the old policy. Symbolically, denote the cash flow pattern for the initial policy, represented by P, by

$$D(P) = [d_0 ; \{d_i\} ; \{d_{ij}\} ; \cdots]. \tag{2}$$

In the usual case $D(P)$ is the *dividend stream* for the policy P. Letting P' represent the new policy,

$$D(P') = D(P) + C(P) \tag{3}$$

$$= [d_0 + c_0 ; \{d_i + c_i\} ; \{d_{ij} + c_{ij}\} ; \cdots], \tag{4}$$

where $C(P)$ is the cash flow description of the investment project, as displayed in (1). It should be noted that the project cash flow description is *defined* by the property $C(P) = D(P') - D(P)$; with this understanding, the cash flows $c_{ijk\cdots}$ can be interpreted as the net of all effects on cash flows for both the new and old activities, and thus take account of all dependencies among projects.[3]

For purposes of analysis, we shall place two restrictions on the set of investment projects to be considered for adoption. First, the set of available projects must be known; that is, all deferred projects, as well as the present projects, must be known. This says that we shall make no provision for investment opportunities that are unknown at present but which might arise at some future node of the event tree. On the other hand, since we allow deferred projects, this restriction does not rule out a sequential investment strategy; for example, one kind of deferred project provides for initiation of activities at a future point in time only if certain events are obtained in the interim. The second restriction, imposed in part by the definition of the state description, is that a project's cash flow description can be calculated for any policy. This restriction requires that one has available an optimization procedure, or at least some decision rule, for allocating resources among activities; therefore, for a given policy, each project has been already optimized with respect to resource allocation before it is brought up for consideration.

It is worth remarking that some sets of projects may not be feasible to adopt in total, due perhaps to logical restrictions requiring that one or another of two projects can be undertaken but not both. For example, one or two new factories can be built but not both. In this case, one or the other project can be proposed separately, but if one is already included in the policy then proposal of the second requires deletion of the first so that only the net effect of the second is considered. It is intended that the relation (3) embody such considerations, as is implied by the notational dependence of the cash flow description upon the policy.

[3] The notation $C(P, P')$ would be more accurate, but here we shall omit explicit reference to the dependence upon the new policy.

4. The Investment Problem Under Uncertainty

In principle, the object of the investment analysis is to select from the set of available projects a feasible package of projects to be undertaken which is preferred above all others. This package then constitutes an optimal sequential investment strategy. The strategy is sequential, of course, since it ordinarily includes deferred projects scheduled to be launched conditional upon, and in a way determined by, the interim states of the world. In practice, however, dependencies among projects usually preclude a complete analysis due to the combinatorial complexities involved in considering all possible packages. Instead, the more practical goal is established of successively seeking projects to adjoin to the existing policy. In this way, step by step, successively more preferred policies are attained. In part, of course, this approach is dictated by the administrative necessity of considering major projects one or two at a time. This goal, finding a desirable project to adjoin to an existing policy, is the one that will be considered henceforth. We shall, however, allow a project to have multiple aspects, such as both financing and expenditures, both present and deferred.

The object of choice in project selection is the dividend stream, as displayed in (3) and (4), for the resulting policy, or at least that is what we will assume. This means that each project is to be evaluated in terms of its effect on the dividend stream as measured by its cash flow description (1). It does not mean that retained earnings are prohibited, since the sequential investment strategy may well provide for reinvestment of returns in deferred projects; but, returns which are not reinvested are paid out as dividends, which surely must be mainly what matters to present (and potential) shareholders. The personal ambitions of professional managers, however, are prohibited as criteria for choices among projects.

For the methodology by which to evaluate alternative dividend streams we appeal to the axiomatic theory of consistent decision making under uncertainty (e.g., [5]). We suppose that each present and potential shareholder assesses a utility measure for dividend streams and a probability measure over the states of the world, and seeks to maximize the expected utility of his dividend stream (of course, if present and potential shareholders disagree appropriately than a mutually favorable exchange can be constructed). Although the dividends are shared jointly, nevertheless, the investment decision is made in common by the shareholders or their representative, the professional manager. There remains, therefore, the necessity of a transition from the utility and probability measures of the shareholders to comparable measures for the firm as an entity, as well as an accounting of the effects of security markets. The Appendix briefly discusses two approaches to this problem. Although the theory of this transition is incomplete, we shall in any case here presume that the firm possesses a utility measure for dividend streams and a probability measure over states of the world.

Let $U(d^0, d^1, d^2, d^3, \cdots)$ be the utility to the firm of a dividend stream for which d^0 is paid initially, d^1 is paid at time 1, and generally d^t is paid at time t. Also, let $p_{ijk\cdots}$ be the firm's assessed probability of state $S_{ijk\cdots}$. Of course, consistency requires that $\sum_{i,j,k,\cdots} p_{ijk\cdots} = 1$, and that the marginal probabilities of interim states satisfy, for example, $p_i = \sum_j p_{ij}$, $p_{ij} = \sum_k p_{ijk}$, *et cetera*. Consequently, if P is a policy with dividend stream $D(P)$ given by (2), then its expected utility is

$$(5) \qquad \sum_{i,j,k,\cdots} p_{ijk\cdots} U(d_0, d_i, d_{ij}, d_{ijk}, \cdots).$$

This imposes the following criterion for selecting investment projects: a project with the cash flow description $C(P)$ displayed in (1) is a desirable amendment to the policy

P if and only if

$$\sum_{i,j,k,\cdots} p_{ijk\cdots}[U(d_0 + c_0, d_i + c_i, \cdots) - U(d_0, d_i, \cdots)] \geqq 0. \tag{6}$$

The remaining discussion is devoted to procedures for implementing this criterion.

Since much of the remaining discussion is devoted to circumventing the calculations required by (6), it may be well to illustrate using the data from Figure 2. Suppose $U(d^0, d^1, d^2) = \sum_{t=0}^{2} \beta_t U_t(d^t)$, where $U_t(d) = d - a_t^2 d^2$. The calculations are given below for an example in which d_0, d_i, d_{ij}, etc are all zero under the present policy. Since the expected utility (1881.4) is positive

t	a_t	β_t	i	j	P_{ij}	U	$P_{ij}U$
0	10^{-6}	1.00	1	1	.30	33,217	9,965.1
1	$.5 \times 10^{-7}$	.95	1	2	.24	6,982	1,675.7
2	10^{-7}	.90	1	3	.06	−19,595	−1,175.7
			2	1	.32	−16,151	−5,168.3
			2	2	.08	−42,692	−3,415.4
							1,881.4

this project satisfies (6). The expected utility is the same as the utility of a stream consisting of \$1,884 at $t = 0$ and zero thereafter. By way of contrast, the expected present monetary value of the project, using the β_t's as discount factors, is \$3,020.

One restriction is imposed on the utility function U: in mathematical terms, we require that the utility function be concave. Essentially, this means (1) that the utility function must evidence risk aversion, or at least not risk preference; and (2) that the intertemporal indifference surfaces for dividends must be convex, so that there is always nonincreasing marginal utility from exchanging dividends at one time for another time. As a consequence of concavity, the utility function possesses gradients (the vector of partial derivatives) everywhere, although they need not be unique. In addition, of course, a utility function must always be nondecreasing in each of its arguments.

We shall not here develop the construction of utility and probability measures, but see [3] and [4].

5. The Methodology of Investment Analysis Under Uncertainty

In principle, the criterion (6) suffices entirely for the selection of investment projects: one simply computes the cash flow description for a project, and if (6) is satisfied, then the project is amended to the current policy. In practice, nevertheless, the situation is quite different. The main reason, in brief, is that a package of projects is considerably more than the sum of its components. To repeat, an investment is, in essence, a present certain sacrifice for future uncertain benefits. Conceptually, we can distinguish two characteristic rationale for investments. First, there is the *smoothing* effect. For example, a fairly even distribution of dividends over time is usually preferred to a lump-sum liquidation at any one time. Via investment, therefore, the firm can defer present dividends for future dividends so as to smooth the pattern of the dividend stream. Second, there is the *diversification* effect. For example, two projects each of which yields a large return in one state and a small return in another state, but the role of the states is opposite for the two projects, may be acceptable together when neither is separately; because, separately the large variation in returns is intolerable whereas together there is relatively little variation. Due to the importance of the smoothing and diversification effects, an important task in investment analysis is to construct projects and packages of projects which accomplish these desiderata. Yet, as baldly stated in (6) the expected utility criterion gives little help in this task: it completely embodies the valuation of smoothing and diversification effects, but only implicitly. In the subsequent discussion we shall take advantage of the special structure of investment problems to extend the expected utility criterion (6) in ways that are explicitly constructive in the design of projects.

For further discussion of investment analysis in a broader context one may consult Weingartner's dissertation [6] or his survey article [7], where many of the issues discussed here are also addressed.

6. Sufficient Conditions for Inclusion of Projects

There is, as mentioned before, an administrative impetus to consider projects singly, as well as an analytical simplification from doing so. Consider, then, a simple project that would deplete present dividends entirely but would certainly yield large returns in the future. As it stands, this project might be rejected because the initial drain on dividends is intolerable; that is, the cash flow description for the project violates the smoothing condition on the dividend stream. Common practice dictates the way out of this conundrum: finance the initial expenditure by borrowing funds. It is the systematic analysis of this solution that we now want to investigate.

The Weak Sufficiency Condition.

Suppose that in fact the initial expenditure could be financed by borrowing, and that repayment of the loan could be made entirely out of returns from the project; in this case the project plus the loan must surely be acceptable together since the dividend stream would not be decreased at any point in time. Although this is an extreme case, its occurrence is not infrequent in practice. For a formal analysis we shall depend upon a characteristic feature of financing projects: they usually evidence constant returns to scale, at least up to a point (e.g., borrowing twice as much requires paying back twice as much at each time).

Let $C(P)$ as displayed in (1) be the cash flow description of the project for the current policy P. Suppose that there are several kinds of financing projects (indexed by $l = 1, \cdots, L$) available with constant returns to scale. Let the cash flow description of financing project l be given by

$$a_l B^l(P) = [a_l b_0{}^l; \{a_l b_i{}^l\} ; \{a_l b_{ij}^l\} ; \cdots], \tag{7}$$

where a_l is the number of units of the funds source acquired. For example, a simple loan of \$1 might be of the form

$$B(P) = [\$1; \{-\$.40\} ; \{-\$.40\} ; \{-\$.40\} ; \{\$0\} ; \cdots] \tag{8}$$

whereupon (7) is

$$aB(P) = [\$a; \{-\$.40a\} ; \{-\$.40a\} ; \{-\$.40a\} ; \{\$0\} ; \cdots] \tag{9}$$

if a loan of $\$a$ is arranged. Now if the project is to cover the financing, then we require that

$$a_l b_{ijk\cdots}^l + c_{ijk\cdots} \geqq 0 \tag{10}$$

for every node $N_{ijk\cdots}$ of the event tree. A sure way to verify this condition is to solve the following linear programming problem:

$$\text{(11)}\qquad \begin{aligned} &\text{Maximize} && \textstyle\sum_{l=1}^{L} a_l b_0^{\,l} \\ &\text{Subject to} && a_l \geqq 0 \quad \text{for} \quad l = 1, \cdots, L, \quad \text{and to} \\ &&& \textstyle\sum_{l=1}^{L} a_l b_i^{\,l} + c_i \geqq 0 \quad \text{for all} \quad i, \\ &&& \textstyle\sum_{l=1}^{L} a^l b_{ij}^{\,l} + c_{ij} \geqq 0 \quad \text{for all} \quad (i, j), \\ &&& \textstyle\sum_{l=1}^{L} a_l b_{ijk}^{\,l} + c_{ijk} \geqq 0 \quad \text{for all} \quad (i, j, k), \\ &&& \text{et cetera.} \end{aligned}$$

If the objective value at a solution to this linear programming problem exceeds $-c_0$, then the project should be accepted. In concrete terms, if there exists a financing program that the project will cover, then it should be accepted.

As formulated in the linear program (11), the amount a_l of financing from source l must be nonnegative. If, however, one can either borrow or lend funds under the same circumstances, then the nonnegativity restriction can be omitted to obtain stronger results. In the extreme case, suppose that at each node of the event tree one can in addition either borrow or lend funds for one period. For each dollar borrowed (loaned) at time $t - 1$, let $\beta_{ijk\cdots}^{t}$ be the amount repaid (received) at node $N_{ijk\cdots}$ at time t. Denote the amount borrowed (or if negative, the amount loaned) under these circumstances at node $N_{ijk\cdots}$ at time t by $\alpha_{ijk\cdots}^{t}$. Then the linear programming problem can be revised to the following form:

$$\text{(12)}\qquad \begin{aligned} &\text{Maximize} && \alpha_0^{\,0} + \textstyle\sum_{l=1}^{L} a_l b_0^{\,l} \\ &\text{Subject to} && a^l \geqq 0 \quad \text{for} \quad l = 1, \cdots, L, \quad \text{and to} \\ &&& \alpha_i^{\,1} - \alpha_0^{\,0}\beta_i^{\,1} + \textstyle\sum_{l=1}^{L} a_l b_i^{\,l} + c_i \geqq 0 \quad \text{for all} \quad i, \\ &&& \alpha_{ij}^{2} - \alpha_i^{\,1}\beta_{ij}^{2} + \textstyle\sum_{l=1}^{L} a_l b_{ij}^{l} + c_{ij} \geqq 0 \quad \text{for all} \quad (i, j), \\ &&& \text{et cetera.} \end{aligned}$$

In this problem, the α's are unrestricted in sign. The dual of the linear program (12) is of special interest:

$$\text{(13)}\qquad \begin{aligned} &\text{Minimize} && \textstyle\sum_i \pi_i c_i + \sum_{i,j} \pi_{ij} c_{ij} + \sum_{i,j,k} \pi_{ijk} c_{ijk} + \cdots \\ &\text{Subject to} && \pi_i \geqq 0, \quad \pi_{ij} \geqq 0, \qquad \text{et cetera, and to} \\ &\text{(A)} && 1 = \textstyle\sum_i \pi_i \beta_i^{\,1}, \\ &&& \pi_i = \textstyle\sum_j \pi_{ij} \beta_{ij}^{2} \quad \text{for all} \quad i, \\ &&& \pi_{ij} = \textstyle\sum_k \pi_{ijk} \beta_{ijk}^{3} \quad \text{for all} \quad (i, j), \\ &&& \text{et cetera;} \end{aligned}$$

and to

$$\text{(B)} \quad b_0^{\,l} + \textstyle\sum_i \pi_i b_i^{\,l} + \sum_{i,j} \pi_{ij} b_{ij}^{l} + \cdots \leqq 0, \qquad l = 1, \cdots, L.$$

Here $\pi_{ijk\cdots}$ is the marginal valuation of funds at node $N_{ijk\cdots}$. The constraints (A) in (13) are the marginal efficiency conditions for the period-to-period borrowing and lending (the α's), and the constraints (B) are the marginal efficiency conditions for the regular funds sources (the a's). The constraints (B) in (13) amount to the following procedure: given a marginal valuation of funds, say

$$\Pi = [1; \{\pi_i\} ; \{\pi_{ij}\} ; \{\pi_{ijk}\} ; \cdots], \tag{14}$$

discount each funds source; then, if the discounted value is strictly negative then reject that source, and if the discounted value is positive then use that source: following the iterative procedures of the Simplex Method for linear programming, successive funds sources are utilized, and successive marginal valuations of funds are imputed until both the primal and dual constraints are satisfied and a solution is obtained. Note especially that one needs only to know the marginal valuation of funds at the terminals of the event tree to be able to compute all the remainder by working backwards through the constraints (A) in (13).

It should be remarked that sometimes the linear programs (11) and (12) will possess infinite solutions, in which case (13) will be infeasible; however, most linear programming routines are coded to report this circumstance to the user when it occurs. An infinite solution is, of course, sufficient for acceptance of the project.

In case there is some misunderstanding, it should be noted that acceptance of a project does not necessarily imply acceptance of the financing program used to justify it. Indeed, if several projects are accepted in this fashion, then surely one will want to reoptimize the financing program for the entire package of projects as a whole. This would be the case, for example, if minor projects were justified in this fashion at subordinate levels of the organization, after which top management would want to review the package of projects as well as the financing program.

A Stronger Sufficiency Condition.

The coverage condition (10) which was used to derive the linear program (11) is quite severe in that it requires coverage of the financing program by the project for *all* states of the world, no matter how improbable some might be. That is, the project must strictly dominate the financing program. By weakening this condition we can obtain a sufficient criterion for acceptance of a project which will allow a larger set of projects to be accepted. Correspondingly, however, the data inputs required are substantially greater.

As before, we take advantage of the characteristic constant-returns-to-scale of financing projects and seek to find an optimal financing program to complement the project. Precisely the same notation will be used as in the formulation of the linear program (11). In this case, however, we directly employ the utility and probability measures to evaluate variations in the dividend stream. Let $D(P)$ as displayed in (2) be the initial dividend stream, so that (5) gives the initial expected utility level. The maximal expected utility obtained by augmenting the policy P by a project with cash flow description $C(P)$, as displayed in (1), along with a complementary financing program (7) is, therefore, given by the solution to the following nonlinear programming problem:

Maximize

$$\sum_{i,j,k,\cdots} p_{ijk\cdots} U\Big(d_0 + c_0 + \sum_{l=1}^{L} a^l b_0^{\,l},\ d_i + c_i + \sum_{l=1}^{L} a^l b_i^{\,l},\ d_{ij} + c_{ij} + \sum_{l=1}^{L} a_l b_{ij}^l,\ \cdots\Big) \tag{15}$$

Subject to $a^l \geqq 0$ for $l = 1, \cdots, L$.

If the solution value of this nonlinear program exceeds (5), then the project should be accepted. Here again one might consider the circumstance in which funds can be either borrowed or loaned for some kind(s) of financing project(s), in which case the nonnegativity restriction imposed in (15) would be eliminated. Since the way to do this is obvious we shall here limit ourselves to the formulation (15). One word of caution, however: when one can either borrow or lend under favorable circumstances, and the firm has a strong preference for early dividends, an infinite solution to (15) is the rule rather than the exception. That is, the firm tends to borrow solely to pay dividends. Since this behavior trespasses on the assumption of constant returns to scale in financing, it is in practice necessary to bound the borrowing amounts by adding further constraints to (15).

The dual of the nonlinear program (15) is of some interest:

Minimize

$$\sum_{i,j,k,\cdots} p_{ijk\cdots} V(\lambda_0, \lambda_i, \lambda_{ij}, \lambda_{ijk}, \cdots) \tag{16}$$

Subject to

$$\lambda_0 b_0^l + \sum_i p_i \lambda_i b_i^l + \sum_{i,j} p_{ij} \lambda_{ij} b_{ij}^l + \cdots \leqq 0$$

For $l = 1, \cdots, L$.

Although it is not important for our present purposes, V is the conjugate utility function [2] defined by

$$V(\xi^0, \xi^1, \xi^2, \cdots) = \text{Supremum}_{d^0, d^1, d^2, \cdots} [U(d^0, d^1, \cdots) - \textstyle\sum_{t=0}^{T} \xi^t d^t], \tag{17}$$

and it is a convex function. In (16) we also use the marginal probabilities (p_i, p_{ij}, etc.) defined in §4. The important feature of the dual problem is the structure of the marginal efficiency conditions: a funds source is rejected if its expected discounted value is negative. The discount rate $\lambda_{ijk\cdots}$ can be interpreted as the marginal utility of funds at node $N_{ijk\cdots}$.

7. Sufficient Conditions for Exclusion of Projects

The sufficient conditions for *inclusion* of projects, developed above in Section 6, are valuable in large part because they provide automatic procedures which can be applied at subordinate levels of the organization to reduce the profusion of projects requiring top management consideration. Although these procedures can accept projects, nevertheless, they cannot reject projects. Their limitation is, essentially, that they address themselves only to the smoothing objective in investment analysis, while ignoring the diversification objective. The advantages of diversification derive from interrelationships among projects, whereas the mathematical programming problems (11) and (15) deal only with single projects (plus their complementary financing programs). Although rejected for inclusion by these procedures, a project may still be acceptable when combined with other projects to attain desired diversification. The discussion in this section is aimed at providing means for further analysis of projects not accepted by the inclusion criteria. At present, however, when projects are discrete "go or no-go" affairs there is no efficient procedure for designing packages of projects to meet diversification objectives: one must in an *ad hoc* fashion put together reasonable packages and then evaluate them using either the sufficiency criteria (11) and (15) or else the general criterion (6). Later we shall consider nondiscrete variable-intensity projects and obtain stronger results, but here attention is confined to criteria for *excluding* projects via automatic procedures so as to reduce the profusion of projects that must be considered in the design of packages to achieve diversification objectives. For projects, and packages of projects, in the no-man's-land between the inclusion and exclusion criteria, the general criterion (6) is the only remedy.

We are interested, therefore, in a criterion for excluding projects from further consideration in the design of diversified packages of projects. Here, we shall use the term project to label generic projects, projects with complementary financing, packages of projects, etc. Let the initial policy be represented by P and consider a project with cash flow description $C(P)$, as in (1), to be adjoined to the initial policy. The general criterion for acceptance is (6). It is a fundamental property of concave functions, such as the utility function U, that

$$(18)\quad U(d^0 + \Delta^0, d^1 + \Delta^1, d^2 + \Delta^2, \cdots) \leqq U(d^0, d^1, d^2, \cdots) + \textstyle\sum_{t=0}^{T} \Delta^t \partial U(d^0, d^1, d^2, \cdots)/\partial d^t.$$

Consequently,

$$(19)\quad \textstyle\sum_{i,j,k,\cdots} p_{ijk\cdots}[U(d_0 + c_0, d_i + c_i, d_{ij} + c_{ij}, \cdots) - U(d_0, d_i, d_{ij}, \cdots)] \leqq \textstyle\sum_{i,j,k,\cdots} p_{ijk\cdots}[\mu^0_{ijk\cdots}c_0 + \mu^1_{ijk\cdots}c_i + \mu^2_{ijk\cdots}c_{ij} + \cdots]$$

$$(20)\quad = \bar{\mu}_0 c_0 + \textstyle\sum_i p_i \bar{\mu}_i c_i + \textstyle\sum_{i,j} p_{ij}\bar{\mu}_{ij}c_{ij} + \cdots,$$

where in (19),

$$(21)\quad \mu^t_{ijk\cdots} = \partial U(d_0, d_i, d_{ij}, \cdots)/\partial d^t,$$

and in (20),

$$(22a)\quad \bar{\mu}_0 = \textstyle\sum_{i,j,k,\cdots} p_{ijk\cdots}\mu^0_{ijk\cdots}$$

$$(22b)\quad \bar{\mu}_i = \textstyle\sum_{j,k,\cdots} (p_{ijk\cdots}/p_i)\mu^1_{ijk\cdots},$$

$$(22c)\quad \bar{\mu}_{ij} = \textstyle\sum_{k,\cdots} (p_{ijk\cdots}/p_{ij})\mu^2_{ijk\cdots},$$

(22) et cetera.

Observe in (21) that $\mu^t_{ijk\cdots}$ is the marginal utility of funds at time t conditional on the terminal state $S_{ijk\cdots}$. Hence, in (22a) $\bar{\mu}_0$ is the expected marginal utility of funds at node N_0; in (22b) $\bar{\mu}_i$ is, conditional on the state S_i, the expected marginal utility of funds at node N_i; et cetera. In (22b), for example, $p_{ijk\cdots}/p_i$ is the probability of state $S_{ijk\cdots}$ conditional on the state S_i having obtained: indeed, often one will have originally assessed the probabilities in this form.

The relation (20) provides an exclusion criterion. Combining (20) with the general criterion (6), it follows that the project can be rejected as an amendment to the policy P if

$$(23)\quad c_0 + \textstyle\sum_i p_i\delta_i c_i + \textstyle\sum_{i,j} p_{ij}\delta_{ij}c_{ij} + \cdots < 0,$$

where

$$\delta_i = \bar{\mu}_i/\bar{\mu}_0, \ \delta_{ij} = \bar{\mu}_{ij}/\bar{\mu}_0, \qquad \text{etc., assuming } \bar{\mu}_0 > 0.$$

In part, the value of this exclusion criterion is its administrative simplicity when delegated to subordinates. For projects involving uniformly small cash flows, the difference between the left and right sides of (18) is small, and (23) is then nearly a strict criterion.

Of more fundamental importance, however, is that this exclusion criterion derives via (18) from a tangential linear approximation to the utility function U, which is the borderline case between risk aversion and risk preference. In concrete terms, therefore, the exclusion criterion says to reject any project of interest only to a gambler. Clearly, such a project will be of little value fo. diversification purposes (although under a different policy it might be, and if the investment analysis proceeds by successive amendments to the policy this must be kept in mind). On the other hand, any project which fails both the inclusion and exclusion criteria is a candidate for diversification purposes, since the inclusion criterion says that it cannot be justified singly or with complementary financing, and the exclusion criterion says that it is potentially valuable if, perhaps with other projects, it conforms to the firm's risk aversion. It is risk aversion, of course, that justifies diversification.

8. A Mathematical Programming Formulation

Up to now we have treated investment projects as discrete entities to be either accepted or rejected. More generally a project might be of variable intensity, although in the special case of a discrete project there would be only one feasible positive intensity. The formulation in this more general situation delineates the structure of the investment problem more clearly.

We pose the problem in terms of mathematical programming. Let the set of all available projects, including financing projects, be indexed by $n = 1, \cdots, N$. For an initial policy P let $D(P)$ as in (2) be the dividend stream, and for each project n let the cash flow description be

$$C_n(x \mid P) = [c_0^n(x); \{c_i^n(x)\}; \{c_{ij}^n(x)\}; \cdots], \tag{24}$$

where the vector of project intensities is $x = (x_1, \cdots, x_n, \cdots x_N)$. Often, of course, the cash flow description for project n will depend only on x_n. For the set of available projects define

$$\begin{aligned} \gamma_0(x) &= \textstyle\sum_{n=1}^N c_0^n(x), \\ \gamma_i(x) &= \textstyle\sum_{n=1}^N c_i^n(x) \quad \text{for all} \quad i, \\ \gamma_{ij}(x) &= \textstyle\sum_{n=1}^N c_{ij}^n(x) \quad \text{for all} \quad (i, j), \end{aligned} \tag{25}$$

et cetera.

We shall assume that each of the functions defined in (25) is concave; that is, that each evidences decreasing marginal returns; or, if the projects are independent, that there are nonincreasing returns to scale. The problem can then be formulated in terms of a mathematical program as follows:

$$\text{(26)} \quad \text{Maximize} \quad \textstyle\sum_{i,j,k,\cdots} p_{ijk\cdots} U(d_0 + \gamma_0(x), d_i + \gamma_i(x), d_{ij} + \gamma_{ij}(x), \cdots)$$

$$\text{Subject to} \quad x \in R,$$

where R is the set of feasible values of the intensity vector x. If any projects are discrete, then (26) is an integer concave programming problem. If the intensity of project n is restricted only to be nonnegative, then it is accepted at some positive intensity if and only if at a solution

$$\bar{\mu}_0 \gamma_0^n(x) + \textstyle\sum_i p_i \bar{\mu}_i \gamma_i^n(x) + \sum_{i,j} p_{ij} \bar{\mu}_{ij} \gamma_{ij}^n(x) + \cdots = 0 \tag{27}$$

and rejected if the left-hand side of (27) is negative, where as in (21) and (22) $\bar{\mu}_0$,

$\bar{\mu}_i$, $\bar{\mu}_{ij}$, etc. are the expected marginal utilities at the solution, and where $\gamma_0^n(x) = \partial\gamma_0(x)/\partial x_n$, $\gamma_i^n(x) = \partial\gamma_i(x)/\partial x_n$, etc. Observe that (27) corresponds closely to the exclusion criterion (23) developed in the case of discrete projects: the principle difference is that in the present case the intensity levels are varied continuously until the expected discounted valuation is zero, as in (27), since a positive value would imply that a further increase in the intensity level would be desirable.

In practice, discrete projects are the rule rather than the exception (except in financing), which limits the usefulness of the formulation (26). For this reason, specialization of the formulation to more structured situations (e.g., specialization of the event tree to the case of a Markov chain) appears to be of little practical value and is not dealt with here.

9. Summary

In summary, the formulation developed above deals directly with the pervasive effects of uncertainty in investment analysis. The basic concept underlying the formulation is the notion of a project's cash flow description superimposed on an event tree or state description. The formulation explicitly embodies measures of risk aversion and intertemporal income preferences. Projects of nearly any sort can be handled, and, in particular, heterogeneous sources of financing are considered.

Seeking the optimal design of a complementary financing program for a project yields inclusion criteria for its acceptance, and analysis of the borderline case between risk aversion and risk preference yields an exclusion criterion.

The practicality of the methods proposed is limited mainly by the difficulties of data collection, and for the next few years, by the size and speed of computing equipment.

Appendix. Construction of Preference Measures for a Firm

Two methods for constructing the preferences of a firm from the preferences of its owners have been developed recently. A direct extension of the economic theory of risk markets using a state description of uncertainty provides one approach [1], and the theory of cooperative games with sharing provides the other [8, 9, 10, 11]. Each approach offers certain advantages, so we will describe both very briefly.

The Economic Theory of Risk Markets.[4]

Assume a state description of uncertainty as in the text except that all of the states that might occur (for all time periods) are indexed by the single index $j = 1, \cdots$. Then the investment and financial policies of a firm describe a state distribution of returns to the various securities (common stock, bonds, etc.) issued by the firm; say, r_{jk} is the return in state j to securities of type k. Hence, the market value v of the firm is a function $v(\{r_{jk}\})$ of the state distribution of returns $\{r_{jk}\}$, and

$$\text{(A1)} \qquad v = \textstyle\sum_k p_k$$

where p_k is the market value of the securities of type k. Investors are assumed to optimize their consumption programs and investment portfolios, with the result that each investor assesses a marginal rate of substitution between present income and future income in state j from a security of type k, denoted by m_{ijk} (the dependence upon k

[4] I am indebted to Jacques Drèze and Jack Hirshleifer for discussions on this topic, although I bear sole responsibility for my conclusions.

is usually due to tax differentials).[5] Then any investor who holds the kth security will optimize his portfolio by equating the price and his valuation of the state distribution of returns,

$$p_k = \sum_j r_{jk} m_{ijk} . \tag{A2}$$

Let s_{ik} be the fraction of the firm's k^{th} security held by investor i: then combining (A1) and (A2) gives a formula for the market value of the firm:

$$v = \sum_i s_{ik} \sum_k p_k = \sum_k \sum_j r_{jk} \left(\sum_i s_{ik} m_{ikj} \right) ; \tag{A3}$$

or

$$v = \sum_{j,k} r_{jk} w_{jk} , \tag{A4}$$

where the shadow price w_{jk} for contingent returns in state j from security k is given by

$$w_{jk} = \sum_i s_{ik} m_{ijk} . \tag{A5}$$

Assuming its market value to be the firm's preference measure, the formula (A5) is particularly advantageous as an exclusion criterion, since in fact (A4) is the tangential approximation to $v(\{r_{jk}\})$ at $\{r_{jk}\}$.[6] For example, if a proposed state distribution of returns $\{r'_{jk}\}$ yields

$$v > \sum_{j,k} r'_{jk} w_{jk} \tag{A6}$$

then its adoption could only lower the firm's market value. Nevertheless, the problems of measuring the investors' marginal rates of substitution remain formidable, if only because of the incentive for an investor to report a distortion in order to bias favorably the computation of the shadow prices (A5).

The Theory of Cooperative Sharing.

The risk market approach has the deficiency of not indicating whether or how the firm's preference measure can be decomposed into a utility function and a probability assessment. The theory of cooperative sharing remedies this by casting the problem in a more general context. The securities of the firm can be construed as the means by which investors share the contingent returns. But, whereas a price-mediated market for contingent claims achieves a Pareto optimum constrained by the set of securities usually available, the theory of cooperative sharing assumes an unconstrained Pareto optimal sharing rule. This assumption allows a determination of the conditions under which a firm will possess risk preferences codified in a utility function and a consensus of probability assessments, as well as the method of compounding them from the investors' corresponding assessments. In practice, of course, the approximate validity of the assumption depends upon the variety of financial instruments issued by the firm or available in the market. For details one may consult [8, 9, 10, 11].

It should also be mentioned that in practice it is often desirable to assess a utility function, say $U(d^0, d^1, d^2, \cdots)$, for income streams as the composition of risk preferences and intertemporal preferences, each of which can be assessed independently. For example, let $u(d) = U(d, d, d, \cdots)$ be the utility of a constant stream, and let $f(d^0, d^1, d^2, \cdots)$ be a function which maps any variable stream into a constant stream to

[5] N. B. The marginal rate of substitution can ordinarily be factored into the product of the investor's assessed probability for the state and a marginal rate of utility substitution.

[6] The usual convexity restrictions on preferences and technologies must be assumed.

which the firm is indifferent; then $U(d^0, d^1, d^2, \cdots) = u(f(d^0, d^1, d^2, \cdots))$. Note especially that the function f embodies only intertemporal income preferences and is independent of risk preferences, which are reflected only in the utility function u measuring risk preferences among constant streams.

References

1. Hirshleifer, J., "Investment Decision under Uncertainty—Choice Theoretic Approaches," *The Quarterly Journal of Economics*, Vol. 74, No. 4 (November 1965), pp. 509–536.
2. Karlin, Samuel, *Mathematical Methods and Theory in Games, Programming, and Economics*, Vol. I, Addison-Wesley Publishing Company, Reading, Mass., 1959.
3. Pratt, John, "Risk Aversion in the Small and in the Large," *Econometrica*, Vol. 32, No. 1–2 (January-April 1964), pp. 122–136.
4. ——, Howard Raiffa, and Robert Schlaifer, "The Foundations of Decision under Uncertainty: An Elementary Exposition," *Journal of the American Statistical Association*, Vol. 59, No. 306 (June 1964), pp. 353–375.
5. Savage, Leonard J., *The Foundations of Statistics*, John Wiley & Sons, New York, 1954.
6. Weingartner, H. Martin, *Mathematical Programming and the Analysis of Capital Budgeting Problems*, Prentice-Hall, Inc., Englewood Cliffs, 1963.
7. ——, "Capital Budgeting of Interrelated Projects: Survey and Synthesis," *Management Science*, Vol. 12, No. 7 (March 1966), pp. 485–516.
8. Wilson, Robert, "On the Theory of Syndicates," *Econometrica*, Vol. 36, No. 1, (January 1968), pp. 119–132
9. ——, "A Pareto-Optimal Dividend Policy," *Management Science*, Vol. 13, No. 9 (May 1967) pp. 756–764.
10. ——, "The Structure of Incentives for Decentralization under Uncertainty," *La Decision*, M. Guilband (ed.), Centre National de la Recherche Scientifique, Paris, (to appear).
11. ——, "Decision Analysis in a Corporation," *IEEE Transactions on Systems Science and Cybernetics*, Vol. SSC-4, No. 3 (September 1968), 220–226.

2. RISK AVERSION OVER TIME IMPLIES STATIC RISK AVERSION

Reprinted from *The American Economic Review* **60**, 163-174 (1970).

Multiperiod Consumption-Investment Decisions

By EUGENE F. FAMA*

I. *The Problem*

The simplest version of the multiperiod consumption-investment problem considers a consumer with wealth w_1, defined as the market value of his assets at the beginning of period 1, which must be allocated to consumption c_1 and a portfolio investment $w_1 - c_1$. The portfolio will yield an uncertain wealth level w_2 at the beginning of period 2 which must be divided between consumption c_2 and investment $w_2 - c_2$. Consumption-investment decisions must be made at the beginning of each period, until the consumer dies and his wealth is distributed among his heirs. The consumer's objective is to maximize the expected utility of lifetime consumption.

Uncertainty models of the multiperiod consumption-investment problem have been considered by Edmund Phelps, Nils Hakansson, and Jan Mossin (1968). But their quite similar treatments place severe restrictions on both the form of the consumer's utility function and the process generating investment returns. For example, the most general model is Hakansson's. He assumes that the probability distributions of one-period portfolio wealth relatives[1] that will be available at any future period t are known for certain at period 1 and thus are independent of events that will occur between periods 1 and t. The consumer's utility function is assumed to be of the additive form

$$u(c_1, \ldots, c_t, \ldots) = \sum_{t=1}^{\infty} \alpha^{t-1} U(c_t), \qquad 0 < \alpha < 1,$$

so that the utility provided by consumption in period t cannot be affected by levels of consumption attained in other periods. Moreover, the one-period utility function $U(c_t)$ is assumed to be monotone increasing and strictly concave (i.e., marginal utility is positive and the consumer is a risk averter),[2] and $U(c_t)$ must imply either "constant risk aversion" or "constant proportional risk aversion," where these terms are as defined by John Pratt.

Hakansson shows that in this model the optimal consumption for any period t is a linear increasing function of wealth w_t. In addition, the portfolio opportunities that will be available in periods after t affect the optimal split of w_t between current consumption and investment, but the optimal *proportions* of portfolio funds invested in different assets at t depend only on the consumer's one-period utility function $U(c_t)$ and on the distributions of one-period wealth relatives associated with currently available portfolios. In essence, the choice of an optimal portfolio *mix* is "myopic" in the sense that it depends only on one-period utilities and returns.

But these appealingly simple results are direct consequences of the restrictions

* Graduate School of Business, University of Chicago. The author wishes to thank Jacques Drèze, Gary Eppen, Michael Jensen, a referee, and especially Merton H. Miller for their most helpful comments. Work on this project was supported by a grant from the National Science Foundation.

[1] The one-period wealth relative from t to t+1 is defined as total dollars of market value at t+1 per dollar nvested at t. It is thus one plus the one-period (percentage) return from t to t+1

[2] A general definition of concavity is provided in fn. 9.

imposed on utility functions and the process generating investment returns by the Phelps-Hakansson-Mossin models. The goal in this paper is to present a more general multiperiod consumption-investment model, but one which nevertheless leads to interesting hypotheses about observable aspects of consumer behavior. The main result is the proposition that if the consumer is risk averse (i.e., his utility function for *lifetime* consumption is strictly concave) and markets for consumption goods and portfolio assets are perfect,[3] then the consumer's observable behavior in the market in any period is indistinguishable from that of a risk averse expected utility maximizer who has a one-period horizon.

With this result it is then possible to provide a *multiperiod* setting for hypotheses about consumer behavior derived from *one-period* wealth allocation models, and these have been studied extensively.[4] One-period models assume, of course, that consumers have one-period horizons, but in most cases their behavioral propositions only require that consumers behave *as if* they are risk averse one-period expected utility maximizers, and this will be the case in the multiperiod model to be presented here. Thus perhaps the major contribution of this paper is in providing a means for bridging the gap between one-period and multiperiod models.

As a specific illustration we will later consider in detail the adjustments to the multiperiod model that are necessary to provide a multiperiod setting for the major propositions about consumer behavior associated with the one-period, two-parameter wealth allocation models of Harry Markowitz, James Tobin (1958, 1965), William Sharpe (1963, 1964), John Lintner (1965a, 1965b), and Fama (1965, 1968a). Indeed it will be shown that a multiperiod model in which the optimal portfolio for any period is "efficient" in terms of distributions of *one-period* portfolio wealth relatives requires few assumptions beyond those already made in the one-period models.[5]

II. *The Wealth Allocation Model*

First the multiperiod model must itself be developed. Let β_t, the "state of the world," signify the set of events (current and past prices, etc.) that constitutes history up to t. Thus β_{t-1} is a subset of β_t. If the state of the world at t is β_t, there will be $n(\beta_t)$ investment assets available to the consumer; the wealth relatives from t to t+1 for these assets are represented by the vector

$$(1)\qquad R = R(\beta_{t+1}) = (r_1(\beta_{t+1}), r_2(\beta_{t+1}), \ldots, r_{n(\beta_t)}(\beta_{t+1})),$$

so that a value of β_{t+1} implies a value of the vector of wealth relatives. If

$$H_{\beta_t} = (h_1, h_2, \cdots, h_{n(\beta_t)})$$

is the (nonnegative) vector of dollars invested in each asset at t in state β_t, the consumer's wealth at t+1 will be

$$(2)\qquad w_{t+1} = H_{\beta_t} R(\beta_{t+1})',$$

where w_{t+1} and $R(\beta_{t+1})$ are random vari-

[3] That is: (a) consumption goods and portfolio assets are infinitely divisible, (b) reallocations of consumption and investment expenditures are costless, and (c) the consumer's activities in any market have a negligible effect on prices. Such a "perfect markets" assumption is common to almost all wealth allocation models (one-period and multiperiod) and will be maintained throughout the discussion here.

[4] See, for example, Peter Diamond, Jacques Drèze and Franco Modigliani (1966), Fama (1965, 1968a, 1968b), Jack Hirshleifer (1965, 1966), Michael Jensen, John Lintner (1965a, 1965b).

[5] Measuring the dispersion of a distribution of wealth relatives with a single parameter such as the standard deviation or semi-interquartile range, a portfolio is efficient in the Markowitz sense if no other portfolio with the same expected one-period wealth relative has lower dispersion, and no other portfolio with higher expected wealth relative has the same or lower dispersion.

ables at t, and the primed variable denotes the transpose.

The consumer's behavior is assumed to conform to the von Neumann-Morgenstern expected utility model. Thus if for simplicity we initially assume he will die for certain[6] at the beginning of period $\tau+1$, and if the state of the world at $\tau+1$ is $\beta_{\tau+1}$, the consumer's utility for lifetime consumption is given by the "cardinal" function

$$U_{\tau+1}(C_{\tau+1} \mid \beta_{\tau+1}) = U_{\tau+1}(c_{1-k}, \ldots, c_1, \ldots, c_{\tau+1} \mid \beta_{\tau+1}),$$

where in general

$$C_t = (c_{1-k}, \ldots, c_1, \ldots, c_t)$$

is consumption from the beginning of his life, period $1-k$, through period t, and the consumption $c_{\tau+1}$ is in the form of a bequest.[7] The goal of the consumer in his consumption-investment decisions is to maximize the expected utility of *lifetime* consumption.

The consumer must make an optimal consumption-investment decision for period 1, taking into account that decisions must also be made at the beginning of each future period prior to $\tau+1$, and that these future decisions will depend on future events. Dynamic programming, with its "backward optimization," provides a natural approach. That is, to solve the decision problem for period 1, the consumer first determines optimal decisions for all contingencies for the decision problem to be faced at period τ. Then he determines optimal decisions for $\tau-1$, under the assumption that he will always make optimal decisions at τ. And so on, until he works his way back to the decision at period 1, which is then based on the assumption that optimal decisions will be made at each future period for any possible contingency.

Formally, optimal decisions for all w_τ and β_τ can be summarized by the function

$$U_\tau(C_{\tau-1}, w_\tau \mid \beta_\tau)$$

$$(3) \qquad = \max_{c_\tau, H_{\beta\tau}} \int_{\beta_{\tau+1}} U_{\tau+1}(C_\tau, HR' \mid \beta_{\tau+1}) \cdot dF_{\beta\tau}(\beta_{\tau+1}),$$

subject to the constraints

$$0 \leq c_\tau \leq w_\tau, \quad H_{\beta\tau} i' = w_\tau - c_\tau,$$
$$H_{\beta\tau} \geq 0_{n(\beta\tau)},$$

where $F_{\beta_\tau}(\beta_{\tau+1})$ is the distribution function of $\beta_{\tau+1}$ given state β_τ at τ; $O_n(\beta_\tau)$ is the null vector (i.e., a vector of zeros) with dimension $1 \times n(\beta_\tau)$; i is the sum vector (i.e., a vector of ones) which will always be assumed to have whatever dimension is needed for the purpose at hand.[8] The function $U_\tau(C_{\tau-1}, w_\tau | \beta_\tau)$ is the maximum of expected utility at τ as a function of realized past consumption $C_{\tau-1}$ and current wealth w_τ, given that the state of the world is β_τ.

More generally, for t=1, 2, . . ., τ, the process of backward optimization is summarized by the recursive relation

$$(4) \quad U_t(C_{t-1}, w_t \mid \beta_t) = \max_{c_t, H_{\beta t}} \int_{\beta_{t+1}} U_{t+1}(C_t, HR' \mid \beta_{t+1}) \cdot dF_{\beta_t}(\beta_{t+1}),$$

subject to

$$0 \leq c_t \leq w_t, \quad H_{\beta t} i' = w_t - c_t,$$
$$H_{\beta t} \geq 0_{n(\beta t)}.$$

[6] Later the model will be extended to allow for an uncertain period of death, (Section IV).

[7] Since the utility of a given $C_{\tau+1}$ can depend on the state of the world $\beta_{\tau+1}$, the model is consistent with the so-called "state preference" model introduced by Kenneth Arrow. As we shall see later, the more traditional framework in which utilities are not state dependent is just a special case.

For an axiomatic development of the expected utility model which implies the existence of subjective probabilities and allows for state dependent utilities (see Drèze). Drèze's analysis is in turn an extension of Savage's work.

[8] Keep in mind that, given (2), integrating over $\beta_{\tau+1}$ implies averaging over $R=R(\beta_{\tau+1})$ and thus over $w_{\tau+1}=H_{\beta_\tau}R'$.

The function $U_t(C_{t-1}, w_t|\beta_t)$ provides the maximum expected utility of *lifetime* consumption if the consumer is in state β_t at period t, his wealth is w_t, his past consumption was C_{t-1}, and optimal consumption-investment decisions are made at the beginning of period t and all future periods.

Expression (4) exemplifies a common feature of dynamic programming models. In general it is possible to *represent* the decision problem of any period t in terms of a derived objective function (in this case U_{t+1}) which is *explicitly* a function only of variables for t+1 and earlier periods, but which in fact summarizes the results of optimal decisions at t+1 and subsequent periods for all possible future events. Thus the recursive relation (4) *represents* the multiperiod problem as a sequence of "one-period" problems, though at any stage in the process the objective function used to solve the one-period problem summarizes optimal decisions for all future periods.

Representing the multiperiod consumption-investment problem as a sequence of one-period problems in itself says nothing about the characteristics of an optimal decision for any period. The main result of this paper is, however, the following.

Proposition 1. If the utility function for lifetime consumption $U_{\tau+1}(C_{\tau+1}|\beta_{\tau+1})$ has properties characteristic of risk aversion (specifically, if for all $\beta_{\tau+1}$, $U_{\tau+1}(C_{\tau+1}|\beta_{\tau+1})$ is monotone increasing and strictly concave in $C_{\tau+1}$), then for all t the derived functions $U_t(C_{t-1}, w_t|\beta_t)$ will also have these properties.[9]

[9] The monotonicity of $U_{\tau+1}$ says that the marginal utility of consumption in any period is positive, while strict concavity implies that for $0<\alpha<1$,

$$U_{\tau+1}(\alpha C_{\tau+1} + (1-\alpha)\hat{C}_{\tau+1} \mid \beta_{\tau+1}) > \alpha U_{\tau+1}(C_{\tau+1} \mid \beta_{\tau+1}) + (1-\alpha)U_{\tau+1}(\hat{C}_{\tau+1} \mid \beta_{\tau+1}),$$

where $C_{\tau+1}$ and $\hat{C}_{\tau+1}$ are any two consumption vectors that differ in at least one element. Geometrically, concavity says that a straight line between any two points on the function $U_{\tau+1}$ lies below the function. As in the case of the more familiar utility of money function, the concavity of $U_{\tau+1}$ implies risk aversion.

The proof of the proposition is presented in the Appendix.

III. *Implications*

Though at this point its importance is far from obvious, it is the concavity of the functions $U_t(C_{t-1}, w_t|\beta_t)$ for all t and β_t, as stated in Proposition 1, that will now allow us to bridge the gap between one-period and multiperiod wealth allocation models.

A. *The Utility of Money Function*

A foretaste of the discussion can be obtained by using the multiperiod model to derive the familiar utility of money function, most often discussed in the literature in connection with the expected utility model. If the state of the world at period 1 is β_1 and the consumer's past consumption has been $\hat{C}_0$, then for t=1 expression (4) yields

$$v_1(w_1 \mid \beta_1) = U_1(\hat{C}_0, w_1 \mid \beta_1)$$
$$= \max_{c_1, H_{\beta_1}} \int_{\beta_2} U_2(C_1, HR' \mid \beta_2) dF_{\beta_1}(\beta_2),$$

subject to

$$0 \leq c_1 \leq w_1, \quad H_{\beta_1}i' = w_1 - c_1,$$
$$H_{\beta_1} \geq 0_{n(\beta_1)}.$$

v_1 is the relevant utility function for timeless gambles taking place at period 1: that is, gambles where the outcome is known before the consumption-investment decision of period 1 is made. From Proposition 1, v_1 has the characteristics of a risk averter's utility of money function: that is, it is monotone increasing and strictly concave in w_1. Thus, though he obtains his utility of money function by a complicated process of backward optimization, and though his utility of money function in fact shows the expected utility of lifetime consumption associated with a given level of wealth at period 1, the consumer's behavior in choosing among timeless gambles is indistinguishable from that of a risk averter making a once-and-for-all decision. In other words, our analysis provides a multiperiod setting for the more traditional discussions of utility of money functions for risk averters, most of which abstract from the effects of future decisions.

B. *One-Period and Multiperiod Models: General Treatment*

More generally, when it comes time to make a decision at the beginning of any period t, t = 1, 2, . . . , τ, past consumption (equal, say, to $\hat{C}_{t-1}$) is known, so that the decision at t can be based on the function

$$v_{t+1}(c_t, w_{t+1} \mid \beta_{t+1}) = U_{t+1}(\hat{C}_{t-1}, c_t, w_{t+1} \mid \beta_{t+1})$$

Thus, for given wealth w_t and state of the world β_t, the consumer's problem at t can be expressed as

$$(5) \quad \max_{c_t, H_{\beta_t}} \int_{\beta_{t+1}} v_{t+1}(c_t, HR' \mid \beta_{t+1}) dF_{\beta t}(\beta_{t+1}),$$

subject to

$$0 \leq c_t \leq w_t, \quad H_{\beta t} i' = w_t - c_t,$$
$$H_{\beta t} \geq 0_{n(\beta_t)}.$$

Since, from Proposition 1, U_{t+1} is monotone increasing and strictly concave in (C_t, w_{t+1}), v_{t+1} is monotone increasing and strictly concave in (c_t, w_{t+1}). Thus, though the consumer faces a τ period decision problem, the function $v_{t+1}(c_t, w_{t+1} \mid \beta_{t+1})$, which is relevant for the consumption-investment decision of period t, has the properties of a risk averter's *one-period* utility of consumption-terminal wealth function. Though the consumer must solve a multiperiod problem, given v_{t+1} his observed behavior in the market is indistinguishable from that of a risk averse expected utility maximizer who has a one-period horizon.[10]

In itself, this result says little about consumer behavior. Its value derives from the fact that it can be used to provide a multiperiod setting for more detailed behavioral hypotheses usually obtained from specific one-period model. Since by design the multiperiod model is based on less restrictive assumptions than most one-period models,[11] adapting it to any specific one-period model will require additional assumptions. But as we shall now see, these are mostly restrictions already implicit or explicit in the one-period models. Little generality is lost in going from a one-period to a multiperiod framework.

C. *A Multiperiod Setting for One-Period, "Two-Parameter" Portfolio Models*

Given the concavity of v_{t+1}, (5) is formally equivalent to the consumption-investment problem of a risk averse consumer with state dependent utilities and a one-period horizon.[12] As such it can be used to provide a multiperiod setting for a wide

[10] But we must keep in mind that though $v_{t+1}(c_t, \mid w_{t+1} \mid \beta_{t+1})$ is only explicitly a function of variables for periods t and t+1, it shows the maximum expected utility of *lifetime* consumption, given optimal consumption-investment decisions in periods subsequent to t. Thus v_{t+1} depends both on tastes, as expressed by the function $U_{\tau+1}(C_{\tau+1} \mid \beta_{\tau+1})$, *and* on the consumption-investment opportunities that will be available in future periods.

The notion of summarizing market opportunities in a utility function should not cause concern. Indeed this is done when utility is written (as we have done throughout) as a function of consumption dollars; then we are implicitly summarizing the consumption opportunities (in terms of goods and services and their anticipated prices) that will be available in each period. We shall return to this point in the Appendix where the utility function $U_{\tau+1}(C_{\tau+1} \mid \beta_{\tau+1})$ for dollars of consumption will be derived from a more basic utility function for consumption goods.

[11] In particular, we have essentially assumed only that markets for consumption goods and portfolio assets are perfect, and that the consumer is a risk averter in the sense that his utility function for lifetime consumption is strictly concave.

[12] The term "state dependent utilities" refers to the fact that the function $v_{t+1}(c_t, w_{t+1} \mid \beta_{t+1})$ allows the utility of a given combination (c_t, w_{t+1}) to depend on β_{t+1}. Hirshleifer (1965, 1966) uses instead the term "state preference" to refer to this condition.

variety of one-period models such as, for example, the one-period model analyzed in detail by Hirschleifer.

But the theories of wealth allocation most thoroughly discussed in the literature are the one-period, two-parameter portfolio models of Markowitz, Tobin (1958, 1965), Sharpe (1963), and Fama (1965, 1968a). These have in turn been used by Sharpe (1964), Lintner (1965a, 1965b), Mossin (1966), and Fama (1968a, 1968b) as the basis of one-period theories of capital market equilibrium. The market equilibrium relationships between the one-period expected wealth relatives and risks of individual securities and portfolios derived from these models have in turn been given some empirical support by Marshall Blume and Michael Jensen. The remainder of this section will be concerned with using our model to provide a multiperiod setting for the apparently useful results of these one-period models.

The two-parameter portfolio models start with the assumption that one-period wealth relatives on asssets and portfolios conform to two-parameter distributions of the same general type. That is, the distribution for any asset or portfolio can be fully described once its expected value and a dispersion parameter, such as the standard deviation or the semi-interquartile range, are known.[13] It is then shown that if investors behave as if they try to maximize expected utility with respect to one-period utility functions $v_{t+1}(c_t, w_{t+1})$ that are strictly concave in (c_t, w_{t+1}), optimal portfolios will be efficient in terms of the two parameters of distributions of one-period wealth relatives.[14] The fact that optimal portfolios must be efficient then makes it possible to derive market equilibrium relations between expected wealth relatives and measures of risk for individual assets and portfolios.

But these models assume somewhat more about the utility function v_{t+1} than our multiperiod model. In particular, in the multiperiod model the function $v_{t+1}(c_t, w_{t+1}|\beta_{t+1})$, which is relevant for the consumption-investment decision of period t, is strictly concave in (c_t, w_{t+1}), but utility can be a function of the state β_{t+1} (i.e., utility can be state dependent). Thus to provide a multiperiod setting for the one-period two-parameter models it is sufficient to determine conditions under which v_{t+1} will be independent of β_{t+1}.

State dependent utilities in the derived functions v_{t+1} have three possible sources. First, tastes for given bundles of consumption goods can be state dependent. Second, as will be shown in Proposition 2 of the Appendix, utilities for given dollars of consumption depend on the available consumption goods and services and their prices, and these are elements of the state of the world. Finally, the investment opportunities available in any given future period may depend on events occurring in preceding periods, and such uncertainty about investment prospects induces state dependent utilities. Thus the most direct way to exclude state dependent utilities

[13] Since assets and portfolios must have distributions of the same two-parameter "type," the analysis is limited to the class of symmetric stable distributions, which includes the normal as a special case. Properties of these distributions are discussed, for example, in Fama (1965) and Benoit Mandelbrot, and a discussion of their role in portfolio theory can be found in Fama (1968a).

Alternatively, the results of the mean-standard deviation version of the two-parameter portfolio models can be obtained by assuming that one-period utility functions are quadratic in w_{t+1}. But strictly speaking, since the quadratic implies negative marginal utility at high levels of w_{t+1}, it is not a legitimate utility function. Moreover, the empirical evidence (see Marshall Blume, Fama (1965), Mandelbrot, and Richard Roll) that distributions of security and portfolio wealth relatives conform well to the infinite variance members of the symmetric stable class casts doubt on any model that relies on the existence of variances. Since the approach based on general two-parameter return distributions by-passes these problems, it seems simplest to lay the quadratic to rest, at least for the purpose of portfolio models.

[14] This concept of efficiency was defined in fn. 5.

is to asume that the consumer behaves as if the consumption opportunities (in terms of goods and services and their prices) and the investment opportunities (distributions of one-period portfolio wealth relatives) that will be available in any future period can be taken as known and fixed at the beginning of any previous period, and that the consumer's tastes for given bundles of consumption goods and services are independent of the state of the world.[15]

With these assumptions, the utility of a given (c_t, w_{t+1}) is independent of β_{t+1}, so that β_{t+1} can be dropped from $v_{t+1}(c_t, w_{t+1} | \beta_{t+1})$. Thus for given wealth, w_t, the decision problem facing the consumer at the beginning of any period t can be written as

$$\max_{c_t, H_{\beta_t}} \int_{R_{t+1}} v_{t+1}(c_t, HR')dF(R_{t+1})$$

Subject to

$$0 \leq c_t \leq w_t, \quad H_{\beta t}i' = w_t - c_t,$$
$$H_{\beta t} \geq 0_{n(\beta t)},$$

where $F(R_{t+1})$ is the distribution function for the vector of wealth relatives R_{t+1}.

Since Proposition 1 applies directly to this simplified version of the multiperiod model, at any period t the function $v_{t+1}(c_t, w_{t+1})$ is monotone increasing and strictly concave in (c_t, w_{t+1}) and is thus formally equivalent to the one-period utility function used in the standard treatments of the one-period, two-parameter portfolio models. If distributions of one-period security and portfolio wealth relatives are of the same two-parameter type, we have a multiperiod model in which the consumer's behavior each period is indistinguishable from that of the consumer in the traditional one-period, two-parameter portfolio models. From here it is a short step to develop a multiperiod setting for period-by-period application of the major results of the one-period, two-parameter models of market equilibrium.

IV. *Extensions: Uncertain Period of Death*

For simplicity, the development of Proposition 1 and its implications made use of the simplest version of the multiperiod consumption-investment model. In particular, it was assumed that (a) the consumer's resources at the beginning of any period t consist entirely of w_t, the value of the marketable assets carried into the period from previous periods; and that (b) the period of the consumer's death is known. But it is not difficult (indeed the major complications are notational) to extend the model to take account of the fact that the consumer has an asset, his "human capital," which will generate income in periods subsequent to t, but which cannot be sold outright in the market. The extended model would allow the ways that the consumer employs his human capital during t—his choice of occupation (s) and the division of his time between labor and leisure—to be at his discretion. It is also easy to extend the model to allow for opportunities the consumer may have to borrow against future labor income or against his portfolio.

But these extensions will not be pursued here.[16] We shall consider instead how the possibility of an uncertain period of death can be introduced into the simple wealth allocation model of Section II.

For simplicity, the analysis so far has assumed that the consumer dies for certain at the beginning of period $\tau+1$. But the model, exactly as stated in (4), is consistent with the probabilistic occurrence of

[15] It is important to note that some such assumptions are implicit in the one-period, two-parameter models themselves since they do not allow for the effects of state dependent utilities on the consumption-investment decision. Exactly these assumptions are quite explicit in the Phelps-Hakansson-Mossin models.

[16] They are discussed in detail in Fama (1969), an earlier version of this paper, which will be made available to readers on request.

death (and the distribution of the consumer's wealth among his heirs) in earlier periods. A subset of the events that comprise the state of the world, β_t, is the set of all events up to t that could affect the consumer's utility for any vector of lifetime consumption. A subset of these events could in turn be the life-death status of the consumer. Thus the state of the world at t might be defined as

$$\beta_t = (z_t, \hat{\beta}_t),$$

where the variable z_t represents the life-death status of the consumer and can take either the value a_t (indicating that the consumer is alive at t), or the value $d_{\bar{t}}$ (indicating that death occurred in some period $\bar{t} \leq t$), and where $\hat{\beta}_t$ is the set of all other elements of the state of the world. With this interpretation of β_t, it is easy to see that the model presented above (specifically, in (4)) is consistent with the possibility of probabilistic occurrence of death in periods prior to $\tau+1$.

Nevertheless some interesting insights into the role of the horizon period $\tau+1$ can be obtained by examining the effects of "probabilistic death" in a little more detail. When the period of death is uncertain, the consumer must make an optimal consumption-investment decision for period 1, taking into account that decisions must also be made at the beginning of any future period at which he is alive, but that the decision process will terminate as soon as he dies.

If the consumer is alive at τ, optimal decisions for all w_τ and β_τ can be summarized by the function

$$\begin{aligned} U_\tau(C_{\tau-1}, w_\tau \mid \beta_\tau) &= U_\tau(C_{\tau-1}, w_\tau \mid a_\tau, \hat{\beta}_\tau) \\ &= \max_{c_\tau, H_{\beta_\tau}} \int_{\beta_{\tau+1}} U_{\tau+1}(C_\tau, HR' \mid d_{\tau+1}, \hat{\beta}_{\tau+1}) \\ &\qquad \cdot dF_{\beta_\tau}(\beta_{\tau+1}), \end{aligned} \tag{6}$$

subject to

$$0 \leq c_\tau \leq w_\tau, \quad H_{\beta_\tau} i' = w_\tau - c_\tau,$$
$$H_{\beta_\tau} \geq 0_{n(\beta_\tau)},$$

where in this case $\beta_\tau = (a_\tau, \hat{\beta}_\tau)$ and $\beta_{\tau+1} = (d_{\tau+1}, \hat{\beta}_{\tau+1})$. Expression (6) is just (4) when $t=\tau$ and the consumer is alive at τ.

On the other hand, if the consumer dies at the beginning of any period t ($t=1, 2, \ldots, \tau+1$), his wealth is immediately distributed among his heirs, and expression (4) for the expected utility of his lifetime consumption becomes

$$\begin{aligned} U_t(C_{t-1}, w_t \mid \beta_t) &= U_t(C_{t-1}, w_t \mid d_t, \hat{\beta}_t) \\ &= \int_{\beta_{\tau+1}} U_{\tau+1}(C_t, 0_{\tau+1-t} \mid \beta_{\tau+1}) dF_{\beta_t}(\beta_{\tau+1}), \end{aligned} \tag{7}$$

where $w_t = c_t$ is his bequest, and in this case $\beta_t = (d_t, \hat{\beta}_t)$ and $\beta_{\tau+1} = (d_t, \hat{\beta}_{\tau+1})$.[17]

In his consumption-investment decision for any period prior to τ, the consumer must consider that he could be either alive or dead at the beginning of the following period. Assuming for simplicity that the occurrence of death is independent of other elements of the state of the world, let x_t be the conditional probability that the consumer will be alive at t, given that he is alive at $t-1$. Then, with (6) as a starting point, for $t=1, 2, \ldots, \tau-1$, the process of backward optimization summarized by (4) is now expressed by (7) and the recursive relations

$$\begin{aligned} U_t(C_{t-1}, w_t \mid z_t, \hat{\beta}_t) &= x_t U_t(C_{t-1}, w_t \mid a_t, \hat{\beta}_t) \\ &\quad + (1 - x_t) U_t(C_{t-1}, w_t \mid d_t, \hat{\beta}_t), \end{aligned} \tag{8}$$

$$\begin{aligned} U_t(C_{t-1}, w_t \mid a_t, \hat{\beta}_t) &= \max_{c_t, H_{\beta_t}} \int_{\beta_{t+1}} U_{t+1}(C_t, HR' \mid z_{t+1}, \hat{\beta}_{t+1}) \\ &\qquad \cdot dF_{\beta_t}(\beta_{t+1}), \end{aligned} \tag{9}$$

subject to

$$0 \leq c_t \leq w_t, \quad H_{\beta_t} i' = w_t - c_t,$$
$$H_{\beta_t} \geq 0_{n(\beta_t)}.$$

[17] If the consumer is not concerned with events subsequent to his death, then

$$U_t(C_{t-1}, w_t \mid d_t, \hat{\beta}_t) = U_{\tau+1}(C_t, O_{\tau+1-t} \mid d_t, \hat{\beta}_{\tau+1})$$

for $w_t = c_t$ and all $\hat{\beta}_{\tau+1}$ such that $\hat{\beta}_t$ is a subset of $\hat{\beta}_{\tau+1}$.

The function $U_t(C_{t-1}, w_t|a_t, \beta_t)$ in (9) provides the maximum expected utility of lifetime consumption and bequests if the consumer is alive in state β_t at period t, his wealth is w_t, his past consumption was C_{t-1}, and optimal consumption-investment decisions are made at the beginning of period t and all future periods at which he is alive.

Since (7)–(9) are just a special case of the model summarized by (4), Proposition 1 applies directly to the probabilistic death model. In this case the proposition implies that for all t and $\hat{\beta}_t$, $U_t(C_{t-1}, w_t|z_t, \hat{\beta}_t)$, $U_t(C_{t-1}, w_t|a_t, \hat{\beta}_t)$, and $U_t(C_{t-1}, w_t|d_t, \hat{\beta}_t)$ are monotone increasing and strictly concave in (C_{t-1}, w_t).

Finally, expressions (7)–(9) suggest an alternative to the "sure death" interpretation of the horizon $\tau+1$. For given wealth w_1, and state of the world β_1, the consumer's problem at period 1 is to choose c_1 and H_{β_1} which

$$\max_{c_1, H_{\beta_1}} \int_{\beta_2} U_2(C_1, HR' \mid z_2, \hat{\beta}_2) dF_{\beta_1}(\beta_2)$$

$$= \max_{c_1, H_{\beta_1}} \int_{\beta_2} [x_2 U_2(C_1, HR' \mid a_2, \hat{\beta}_2) + (1 - x_2) U_2(C_1, HR' \mid d_2, \beta_2)] dF_{\beta_1}(\beta_2).$$

Using (8) and (9) to expand this expression, it can be shown that the decision problem of period t has weight

$$x_t x_{t-1} \cdots x_2 = x(a_t \mid a_1)$$

in the expected utility for the decision of period 1. The probability $x(a_t|a_1)$ of being alive at t will decrease with t. Thus in general for some $t=\tau+1$, the decisions of periods $\tau+1$ and beyond will have negligible weight in the expected utility for the decision at period 1, so that in the decision at period 1 it is unnecessary to look beyond $\tau+1$.

More simply, since the consumer is likely to be dead, the effects of distant future decisions, which are unlikely to be made, can be ignored. And this result does not arise from discounting of future consumption, though the effect is the same. At period 1, consumption in period $\tau+1$ may be regarded as equivalent to consumption in period 1. But in the decision of period 1 the decision of $\tau+1$ is weighted by the probability that the consumer will be alive at that time, which reduces the importance of the future consumption in the current decision.

V. *Conclusion*

In sum, assuming only that markets for consumption goods and portfolio assets are perfect and that the consumer is risk averse in the sense that his utility function for lifetime consumption is strictly concave, it has been shown that though he faces a multiperiod problem, in his consumption-investment decision for any period the consumer's behavior is indistinguishable from that of a risk averter who has a one-period horizon. It was then shown how this result can be used to provide a multiperiod setting for the more detailed hypotheses about risk averse consumer behavior that are traditionally derived in a one-period framework.

Appendix

Proposition 1. If $U_{t+1}(C_t, w_{t+1}|\beta_{t+1})$ is monotone increasing and strictly concave (henceforth m.i.s.c.) in (C_t, w_{t+1}), then $U_t(C_{t-1}, w_t|\beta_t)$ is m.i.s.c. in (C_{t-1}, w_t).

Proof: The proof of the proposition relies primarily on straightforward applications of well-known properties of concave functions (cf. Iglehart). We first establish:

Lemma 1. If $U_{t+1}(C_t, w_{t+1}|\beta_{t+1})$ is m.i.s.c. in (C_t, w_{t+1}), the expected utility function

$$\int_{\beta_{t+1}} U_{t+1}(C_t, w_{t+1} \mid \beta_{t+1}) dF_{\beta t}(\beta_{t+1})$$

$$(10) \quad = \int_{\beta_{t+1}} U_{t+1}(C_t, H_{\beta t} R(\beta_{t+1})' \mid \beta_{t+1}) \cdot dF_{\beta t}(\beta_{t+1})$$

is strictly concave in $(C_t, H_{\beta t})$.

Proof: For any given value of β_{t+1}, and thus of $R=R(\beta_{t+1})$,

$$w_{t+1} = H_{\beta t}R(\beta_{t+1})'$$

is a linear and thus concave (though not strictly concave) function of H_{β_t}. Since by assumption $U_{t+1}(C_t, w_{t+1}\mid\beta_{t+1})$ is m.i.s.c. in (C_t, w_{t+1}), $U_{t+1}(C_t, H_{\beta_t}R(\beta_{t+1})'\mid\beta_{t+1})$ is strictly concave in (C_t, H_{β_t}).[18] Integrating over β_{t+1} in (10) preserves this concavity.

The remainder of the proof of Proposition 1 is then as follows. Let c_t^*, $H_{\beta t}^*$ and $\tilde{c}_t^*$, $\tilde{H}_{\beta t}^*$ be the optimal values of c_t and H_{β_t} in (4) for any two vectors (C_{t-1}, w_t) and $(\tilde{C}_{t-1}, \tilde{w}_t)$ that differ in at least one element. Let

$$\begin{aligned} C_{t-1} &= \alpha C_{t-1} + (1-\alpha)C_{t-1}, \\ \hat{w}_t &= \alpha w_t + (1-\alpha)\tilde{w}_t, \\ \hat{c}_t &= \alpha \overset{*}{c_t} + (1-\alpha)\overset{*}{\tilde{c}_t}, \\ \hat{H}_{\beta t} &= \alpha \overset{*}{H_{\beta t}} + (1-\alpha)\overset{*}{\tilde{H}_{\beta t}}, \quad 0<\alpha<1. \end{aligned}$$

To establish the concavity of $U_t(C_{t-1}, w_t\mid\beta_t)$, we must show that

$$\begin{aligned} &U_t(C_{t-1}, \hat{w}_t\mid\beta_t) \\ (11) \qquad &\quad > \alpha U_t(C_{t-1}, w_t\mid\beta_t) \\ &\quad + (1-\alpha)U_t(C_{t-1}, \tilde{w}_t\mid\beta_t). \end{aligned}$$

From Lemma 1, for $0<\alpha<1$,

$$\begin{aligned} &\int_{\beta_{t+1}} U_{t+1}(C_{t-1}, \hat{c}_t, \hat{H}_{\beta t}R(\beta_{t+1})'\mid\beta_{t+1})dF_{\beta t}(\beta_{t+1}) \\ &> \alpha\int_{\beta_{t+1}} U_{t+1}(C_{t-1}, \overset{*}{c_t}, \overset{*}{H_{\beta t}}R(\beta_{t+1})'\mid\beta_{t+1}) \\ (12) \qquad &\qquad\qquad \cdot dF_{\beta t}(\beta_{t+1}) \\ &+ (1-\alpha)\int_{\beta_{t+1}} U_{t+1}(C_{t-1}, \overset{*}{\tilde{c}_t}, \overset{*}{\tilde{H}_{\beta t}}R(\beta_{t+1})'\mid\beta_{t+1}) \\ &\qquad\qquad \cdot dF_{\beta t}(\beta_{t+1}) \\ &= \alpha U_t(C_{t-1}, w_t\mid\beta_t) \\ &\quad + (1-\alpha)U_t(C_{t-1}, \tilde{w}_t\mid\beta_t). \end{aligned}$$

Since the consumption-investment decision implied by $\hat{c}_t$, H_{β_t} is not necessarily optimal for the wealth level $\hat{w}_t$,

$$\begin{aligned} &U_t(C_{t-1}, \hat{w}_t\mid\beta_t) \\ &\geq \int_{\beta_{t+1}} U_{t+1}(\hat{C}_{t-1}, \hat{c}_t, \\ &\qquad \hat{H}_{\beta t}R(\beta_{t+1})'\mid\beta_{t+1})dF_{\beta t}(\beta_{t+1}), \end{aligned}$$

which, with (12) implies (11).[19]

The monotonicity of $U_t(C_{t-1}, w_t\mid\beta_t)$ in (C_{t-1}, w_t) follows straightforwardly from the monotonicity of $U_{t+1}(C_t, w_{t+1}\mid\beta_{t+1})$ in C_t. Thus the proposition is established.

Finally, as noted earlier (fn. 10), when utility is written (as we have done throughout) as a function of consumption dollars, we are implicitly summarizing the consumption opportunities (in terms of goods and services and their anticipated prices) that will be available in each period. We shall now conclude the paper by showing how a von Neumann-Morgenstern "cardinal" utility function for consumption dollars can be derived from a cardinal utility function for consumption commodities.

Let $q(\beta_t)=(q_1, q_2, \ldots, q_N(\beta_t))$ be the vector of quantities of $N(\beta_t)$ available commodities consumed during t in state β_t and let $p(\beta_t)=(p_1, p_2, \ldots, p_N(\beta_t))$ be the corresponding price vector. In *any* period or state one of the available consumption commodities is always "dollar gifts and bequests" which has price \$1 per unit. At the horizon $\tau+1$, dollar gifts and bequests, denoted $w_{\tau+1}$, is the only available consumption good. Let

$$Q_\tau = (q(\beta_{1-k}), \ldots, q(\beta_1), \ldots, q(\beta_\tau))$$

be the vector representing lifetime consumption of commodities, and let $V(Q_\tau, w_{\tau+1}\mid\beta_{\tau+1})$ be the consumer's utility of lifetime con-

[18] If $f(x_1, x_2, \ldots, x_N)=f(X)$ is m.i.s.c. in X, and if $x_i=g_i(y_1, y_2, \ldots, y_n)=g_i(Y)$, $i=1, 2, \ldots, N$, is concave (though not necessarily strictly concave) in Y, then $f(g_1(Y), g_2(Y), \ldots, g_N(Y))=f(G(Y))$ is strictly concave in Y. (See, e.g., H. G. Eggleston, p. 52.)

[19] It is assumed that $(\hat{c}_t, \hat{H}_{\beta t})$ is a feasible consumption-investment decision for the wealth level $\hat{w}_t$, or equivalently, that the set of feasible values of $(c_t, H_{\beta t})$ is convex. But this is a weak assumption that will be met, for example, when the constraints on c_t and $H_{\beta t}$ are equations such as $H_{\beta t}i'=w_t-c_t$ or linear inequalities such as $0\leq c_t\leq w_t$ or $\underline{H}\leq H\leq\overline{H}$, where $\underline{H}$ and $\overline{H}$ are vectors of lower and upper bounds on quantities invested in each asset.

sumption, given state $\beta_{\tau+1}$ at $\tau+1$, and where $\beta_{1-k} \subset \ldots \subset \beta_1 \subset \ldots \subset \beta_{\tau+1}$. The utility function for *dollars* of consumption can then be defined as

$$U_{\tau+1}(C_{\tau+1} \mid \beta_{\tau+1}) = \max_{Q_\tau} V(Q_\tau, w_{\tau+1} \mid \beta_{\tau+1}) \tag{13}$$

Subject to

$$C_{\tau+1} = (p(\beta_{1-k})q(\beta_{1-k})', \ldots, p(\beta_\tau)q(\beta_\tau)', w_{\tau+1}) = (c_{1-k}, \ldots, c_\tau, c_{\tau+1}).$$

The role of $\beta_{\tau+1}$ in $U_{\tau+1}$ is twofold. First, psychological attitudes towards current and past consumption (or "tastes") may depend on the state of the world. Second, even if tastes for consumption *commodities* are not state dependent (so that $\beta_{\tau+1}$ can be dropped from V), the utility of any stream of *dollar* consumption expenditures depends on the history of the set of available consumption commodities and their prices, both of which are subsumed in $\beta_{\tau+1}$.

A utility function $U_{\tau+1}(C_{\tau+1} | \beta_{\tau+1})$ which has the properties required by Proposition 1 can then be obtained from $V(Q_\tau, w_{\tau+1} | \beta_{\tau+1})$ as follows.

Proposition 2. If $V(Q_\tau, w_{\tau+1} | \beta_{\tau+1})$ is m.i.s.c. in $(Q_\tau, w_{\tau+1})$, then $U_{\tau+1}(C_\tau, w_{\tau+1} | \beta_{\tau+1})$ is m.i.s.c. in $(C_\tau, w_{\tau+1})$.

Proof: Let Q_τ^* be the optimal value of Q_τ in (13) for $(C_\tau, w_{\tau+1})$ and let $\tilde{Q}_\tau^*$ be optimal for $(\tilde{C}_\tau, \tilde{w}_{\tau+1})$, where the vectors $(C_\tau, w_{\tau+1})$ and $(\tilde{C}_\tau, \tilde{w}_{\tau+1})$ differ in at least one element. For $0<\alpha<1$, let

$$(\hat{Q}_\tau, \hat{w}_{\tau+1}) = \alpha(Q_\tau^*, w_{\tau+1}) + (1-\alpha)(\tilde{Q}_\tau^*, \tilde{w}_{\tau+1}),$$

$$(\hat{C}_\tau, \hat{w}_{\tau+1}) = \alpha(C_\tau, w_{\tau+1}) + (1-\alpha)(\tilde{C}_\tau, \tilde{w}_{\tau+1}).$$

Then the strict concavity of V implies

$$V(\hat{Q}_\tau, \hat{w}_{\tau+1} \mid \beta_{\tau+1}) > \alpha V(Q_\tau^*, w_{\tau+1} \mid \beta_{\tau+1}) + (1-\alpha)V(\tilde{Q}_\tau^*, \tilde{w}_{\tau+1} \mid \beta_{\tau+1}).$$

Or equivalently,

$$V(\hat{Q}_\tau, \hat{w}_{\tau+1} \mid \beta_{\tau+1}) > \alpha U_{\tau+1}(C_\tau, w_{\tau+1} \mid \beta_{\tau+1}) + (1-\alpha)U_{\tau+1}(C_\tau, \tilde{w}_{\tau+1} \mid \beta_{\tau+1}).$$

Since $\hat{Q}_\tau$ is a feasible but not necessarily an optimal allocation of $\hat{C}_\tau$, an optimal allocation must have utility at least as high as that implied by $\hat{Q}_\tau$, so that

$$U_{\tau+1}(\hat{C}_\tau, \hat{w}_{\tau+1} \mid \beta_{\tau+1}) > \alpha U_{\tau+1}(C_\tau, w_{\tau+1} \mid \beta_{\tau+1}) + (1-\alpha)U_{\tau+1}(C_\tau, \tilde{w}_{\tau+1} \mid \beta_{\tau+1}),$$

and the concavity of $U_{\tau+1}$ is established.

To establish the monotonicity of $U_{\tau+1}$ in C_τ, simply note that if the dollars available for consumption in any period are increased, consumption of at least one commodity can be increased without reducing consumption of any other commodity, so that utility must be increased. An optimal reallocation of consumption expenditures must do at least as well.

REFERENCES

K. J. **Arrow**, "The Role of Securities in the Optimal Allocation of Risk-Bearing," *Rev. Econ. Stud.*, Apr. 1964, *31*, 91–96.

M. E. **Blume**, "The Assessment of Portfolio Performance: An Application of Portfolio Theory," unpublished doctoral dissertation, Grad. School of Business, Univ. Chicago 1968.

P. A. **Diamond**, "The Role of A Stock Market in a General Equilibrium Model with Technological Uncertainty," *Amer. Econ. Rev.*, Sept. 1967, *57*, 759–76.

J. H. **Drèze**, "Fondements Logiques de la Probabilité Subjective et de l'Utilité," *La Decision*, Centre National de la Recherche Scientifique, Paris 1961, 73–87.

——— and F. **Modigliani**, "Epargne et Consommation en Avenir Aleatoire," *Cahiers du Seminaire D'Econometrie*, 1966, *9*, 7–33.

H. G. **Eggleston**, *Convexity*. Cambridge, Mass. 1958.

E. F. **Fama**, "The Behavior of Stock Market Prices," *J. Bus. Univ. Chicago*, Jan. 1965, *38*, 34–105.

———, "Multiperiod Consumption-Investment Decisions," Report No. 6830, Center for Mathematical Studies in Business and Economics, Univ. Chicago, rev. May 1969.

———, "Portfolio Analysis in a Stable Paretian Market," *Manage. Sci.*, Jan. 1965, *12*, 404–19.

———, (a) "Risk, Return, and Equilibrium," Report No. 6831, Center for Mathematical Studies in Business and Economics, Univ. Chicago, June 1968.

———, (b) "Risk, Return, and Equilibrium: Some Clarifying Comments," *J. Finance*, Mar. 1968, *23*, 29–40.

N. **Hakansson**, "Optimal Investment and Consumption Strategies for a Class of Utility Functions," Working Paper No. 101, Western Management Science Institute, Univ. California, Los Angeles, June 1966

J. **Hirshleifer**, "Investment Decision Under Uncertainty: Applications of the State-Preference Approach," *Quart. J. Econ.*, May 1966, *80*, 252–77.

———, "Investment Decision Under Uncertainty: Choice-Theoretic Approaches." *Quart. J. Econ.*, Nov. 1965, *79*, 509–536.

D. L. **Iglehart**, "Capital Accumulation and Production for the Firm: Optimal Dynamic Policy," *J. Manage. Sci.*, Nov. 1965, *12*, 193–205.

M. **Jensen**, "Risk, the Pricing of Capital Assets, and the Evaluation of Investment Portfolios," *J. Bus. Univ. Chicago*, Apr. 1969, *42*, 167–247.

J. **Lintner**, (a) "Security Prices, Risk, and Maximal Gains from Diversification," *J. Finance*, Dec. 1965, *20*, 587–615.

———, (b) "The Valuation of Risk Assets and the Selection of Risky Investments in Stock Portfolios and Capital Budgets," *Rev. Econ. Statist.*, Feb. 1965, *47*, 13–37.

N. **Liviatan**, "Multiperiod Future Consumption as an Aggregate," *Amer. Econ. Rev.*, Sept. 1966, *56*, 828–40.

B. **Mandelbrot**, "The Variation of Certain Speculative Prices," *J. Bus. Univ. Chicago*, Oct., 1963, *36*, 394–419.

H. **Markowitz**, *Portfolio Selection: Efficient Diversification of Investments.* New York 1959.

J. **Mossin**, "Equilibrium in a Capital Asset Market," *Econometrica*, Oct. 1966, *34*, 768–83.

———, "Optimal Multiperiod Portfolio Policies." *J. Bus. Univ. Chicago*, Apr. 1968, *41*, 215–29.

Edmund Phelps, "The Accumulation of Risky Capital: A Sequential Utility Analysis," *Econometrica*, Oct. 1962, *30*, 729–43; reprinted in *Risk Aversion and Portfolio Choice.* D. Hester and J. Tobin, eds., New York 1967, 139–53.

J. **Pratt**, "Risk Aversion in the Small and in the Large," *Econometrica*, Jan.-Apr. 1964, *32*, 122–136.

R. **Roll**, "Efficient Markets, Martingales, and the Market for U.S. Government Treasury Bills," unpublished doctoral dissertation, Grad. School of Business, Univ. Chicago 1968.

L. **Savage**, *The Foundations of Statistics.* New York 1954.

W. F. **Sharpe**, "Capital Assets Prices: A Theory of Market Equilibrium under Conditions of Risk," *J. Finance*, Sept. 1964, *19*, 425–42.

———, "A Simplified Model for Portfolio Analysis," *J. Manage. Sci.*, Jan. 1963, *10*, 277–93.

J. **Tobin**, "Liquidity Preference as Behavior Towards Risk," *Rev. Econ. Stud.*, Feb. 1958, *25*, 65–86.

———, "The Theory of Portfolio Selection." *The Theory of Interest Rates.* F. H. Hahn and F. P. R. Brechling, eds, London 1965. Ch. 1.

J. **von Neumann**, and O. **Morgenstern**, *Theory of Games and Economic Behavior*, 3rd ed, Princeton 1953.

3. MYOPIC PORTFOLIO POLICIES

Reprinted from THE JOURNAL OF BUSINESS OF THE UNIVERSITY OF CHICAGO
Vol. 44, No. 3, July 1971

ON OPTIMAL MYOPIC PORTFOLIO POLICIES, WITH AND WITHOUT SERIAL CORRELATION OF YIELDS

NILS H. HAKANSSON

I. INTRODUCTION

In a recent paper, Mossin[1] attempts to isolate the class of utility functions of terminal wealth, $f(x)$, which, in the sequential portfolio problem, induces myopic utility functions of intermediate wealth positions. Induced utility functions of short-run wealth are said to be myopic whenever they are independent of yields beyond the current period; that is, they are positive linear transformations of $f(x)$. Mossin concludes (1) that the logarithmic function and the power functions induce completely myopic utility functions; (2) that, when the interest rate in each period is zero, all terminal wealth functions such that the risk tolerance index $-f'(x)/f''(x)$ is linear in x induce completely myopic utility functions of short-run wealth; (3) that, when interest rates are not zero, the last class of terminal wealth functions induces partially myopic utility functions (only future interest rates need be known); and (4) that all of the preceding is true whether the yields in the various periods are serially correlated or not. With the exception of the last assertion, the same conclusions are reached by Leland.[2] The purpose of this note is to show that the second and third statements are true only in a highly restricted sense even when yields are serially independent, and that, when investment yields in the various periods are statistically dependent, only the logarithmic function induces utility functions of short-run wealth which are myopic.

II. PRELIMINARIES

In this and the next three sections, the following notation will be employed:

x_j: amount of investment capital at decision point j (the beginning of the jth period);
M_j: number of investment opportunities available in period j;
S_j: the subset of investment opportunities which it is possible to sell short in period j;
$r_j - 1$: rate of interest in period j;
β_{ij}: proceeds per unit of capital invested in opportunity i, where $i = 2, \ldots, M_j$, in the jth period (random variable); that is, if we invest an amount θ in i at the beginning of the period, we will obtain $\beta_{ij}\theta$ at the end of that period;
z_{1j}: amount lent in period j (negative z_{1j} indicates borrowing) (decision variable);
z_{ij}: amount invested in opportunity i, $i = 2, \ldots, M_j$, at the beginning of the jth period (decision variable);
$f_j(x_j)$: utility of money at decision point j;
$z^*_{1j}(x_j)$: an optimal lending strategy at decision point j;
$z^*_{ij}(x_j)$: an optimal investment strategy for opportunity i, $i = 2, \ldots, M_j$, at decision point j.

$$F_j(y_2, y_3, \ldots, y_{M_j}) \equiv \Pr\{\beta_{2j} \le y_2, \beta_{3j} \le y_3, \ldots, \beta_{M_j j} \le y_{M_j}\}\ ; \quad \bar{z}_j \equiv (z_{2j}, \ldots, z_{M_j j})\ .$$

[1] Jan Mossin, "Optimal Multiperiod Portfolio Policies," *Journal of Business* 41 (April 1968): 215–29.

[2] Hayne Leland, "Dynamic Portfolio Theory" (Ph.D. thesis, Harvard University, 1968).

As most portfolio models do, we assume, in addition to stochastically constant returns to scale, perfect liquidity and divisibility of the assets at each (fixed) decision point, absence of transaction costs, withdrawals, capital additions, and taxes, and the opportunity to make short sales. Furthermore, we assume, until Section VI, that the yields in the various periods are stochastically independent.

Since the end-of-period capital position is given by the proceeds from current savings, or the negative of the repayment of current debt plus interest, plus the proceeds from current risky investments, we have

$$x_{j+1} = r_j z_{1j} + \sum_{1=2}^{M_j} \beta_{ij} z_{ij} \qquad j = 1, 2, \ldots, \quad (1)$$

where

$$\sum_{i=1}^{M_j} z_{ij} = x_j \qquad j = 1, 2, \ldots. \quad (2)$$

Combining (1) and (2) we obtain

$$x_{j+1} = \sum_{i=2}^{M_j} (\beta_{ij} - r_j) z_{ij} + r_j x_j \qquad j = 1, 2, \ldots. \quad (3)$$

Let us now assume that $f_J(x_J)$ is given for some horizon J. Then, as Mossin shows, we may write, by the principle of optimality,[3]

$$f_j(x_j) = \max_{\bar{z}_j \in Z_j(x_j)} E[f_{j+1}(x_{j+1})] \qquad j = 1, \ldots, J-1 \quad (4)$$

where $Z_j(x_j)$ is the set of feasible investments at decision point j given that capital is x_j. When there are two assets in each period (i.e., $M_j = 2$ for all j) and $Z_j(x_j) = \{z_{2j}\colon 0 \leq z_{2j} \leq x_j\}$, that is, borrowing and short sales are ruled out, Mossin concludes that $f_j(x_j) = a_j f_J(x_j) + b_j$ (where $a_j > 0$ and b_j are constants), $j = 1, \ldots, J-1$, that is, that the induced short-run utility functions at decision points $1, \ldots, J-1$ are completely myopic, if and only if (1) $f_J(x)$ is either logarithmic or a power function, or (2) $r_1 = r_2 = \ldots = r_{J-1} = 1$ and $f_J(x)$ is one of

$$f_J(x) = -e^{-\mu x}\,; \quad (5)$$

$$f_J(x) = \log(x + \mu)\,; \quad (6)$$

$$f_J(x) = \frac{1}{\lambda - 1}(\lambda x + \mu)^{1-1/\lambda} \qquad \lambda \neq 0, \lambda \neq 1\,, \quad (7)$$

where $\mu \neq 0$ and λ are constants. Note that λ and μ cannot both be negative.

While the first conclusion is beyond dispute, the second is incorrect, as are the conclusions concerning partial myopia in general, except in a severely restricted sense. We shall first demonstrate the assertion in the preceding case and then show that it also holds when borrowing and short sales are not ruled out.

III. AN EXAMPLE

Assume that $r_1 = r_2 = \ldots = r_{J-1} = 1$ and that there is only one risky opportunity (i.e., $M_j = 2$) in each period. Moreover, assume that the proceeds β_{2j} of

[3] Richard Bellman, *Dynamic Programming* (Princeton, N.J.: Princeton University Press, 1957).

this opportunity are

$$\beta_{2j} = \begin{cases} 0 \text{ with probability } 1/2 \\ 3 \text{ with probability } 1/2 \end{cases} \qquad \text{all } j \quad (8)$$

and that

$$f_J(x_J) = (x_J + d)^{1/2} \qquad d > 0 . \quad (9)$$

(9) clearly belongs to the class (7). (4) and (3) now give

$$f_j(x_j) = \max_{0 \le z_{2j} \le x_j} E\{f_{j+1}[(\beta_{2j} - 1)z_{2j} + x_j]\} \qquad j = 1, \ldots, J-1 , \quad (10)$$

where $f_J(x_J)$ is given by (9).

It is easily verified that

$$z^*_{2,J-1}(x_{J-1}) = \begin{cases} \text{does not exist} & x_{J-1} < 0 \\ x_{J-1} & 0 \le x_{J-1} < d . \\ 1/2(x_{J-1} + d) & x_{J-1} \ge d \end{cases} \quad (11)$$

Thus,

$$f_{J-1}(x_{J-1}) = \begin{cases} 1/2(3x_{J-1} + d)^{1/2} + 1/2d^{1/2} & 0 \le x_{J-1} < d \\ a_{J-1}(x_{J-1} + d)^{1/2} & x_{J-1} \ge d \end{cases} , \quad (12)$$

where

$$a_{J-1} = 1/2[(1/2)^{1/2} + 2^{1/2}] . \quad (13)$$

We now observe that $f_{J-1}(x)$ is a positive linear transformation of $f_J(x)$ *only* for $x \ge d$; for $x < d$, $f_{J-1}(x) < a_{J-1}f_J(x)$.

Proceeding with the solution to (10), we obtain

$$f_j(x_j) = \begin{cases} g_j(x_j) & 0 \le x_j < b_j \\ a_j(x_j + d)^{1/2} & x_j \ge b_j \end{cases} \qquad j = 1, \ldots, J-1 , \quad (14)$$

where a_j is a positive constant, $g_j(x_j) < a_j(x_j + d)^{1/2}$ for $0 \le x_j < b_j$,

$$b_j = d + 2b_{j+1} \qquad (b_J = 0), j = 1, \ldots, J-1 , \quad (15)$$

and

$$z^*_{2j}(x_j) = \begin{cases} h_j(x_j) & 0 \le x_j < b_j \\ 1/2(x_j + d) & x_j \ge b_j \end{cases} \qquad j = 1, \ldots, J-1 . \quad (16)$$

It is easily determined that $h_j(x_j)$ is highly irregular except for $j = J - 1$.

When $J = 11$ and $d = 1{,}000$, we obtain from (15) that $b_1 = 1.023$ million. Thus, when the horizon is ten periods distant, the optimal amount to invest in opportunity 2 is, in this example, proportional to $x_j + d$ only if initial wealth x_1 exceeds \$1 million by a substantial margin. Furthermore, while $f_j(x)$ is a positive linear transformation of $f_J(x)$ for $x \ge b_j$, $j = 1, \ldots, J-1$, it is not for $x < b_j$, that is, for $x_1 < 1.023$ million, $x_2 < 511{,}000$, etc., in the above example. Since the constant $b_j > 0$ depends on the distribution functions $F_j, \ldots, F_{J-1}$, the short-run utility functions induced by the terminal utility function (9) are clearly not myopic. In other words, to make an optimal decision at decision point j, not only F_j but $F_{j+1}, \ldots, F_{J-1}$ must be known.

IV. BORROWING AND SOLVENCY

The nonmyopic nature of the induced utility functions $f_1(x_1), f_2(x_2), \ldots, f_{J-1}(x_{J-1})$ in the preceding example is clearly attributable to the constraint

$$0 \le z_{2j} \le x_j \qquad j = 1, \ldots, J-1 , \quad (17)$$

which precludes borrowing and short sales. We shall now relax this constraint.

Case I.—(17) will first be replaced with

$$0 \leq z_{2j} \leq mx_j \qquad m > 1 \qquad j = 1, \ldots, J-1, \quad (18)$$

that is, $100/m$ is assumed to be the percentage margin requirement. The solution to (10) for $J - 1$ now becomes

$$f_{J-1}(x_{J-1}) = \begin{cases} 1/2[(1-m)x_{J-1} + d]^{1/2} \\ \quad + 1/2[(2m+1)x_{J-1} + d]^{1/2} & 0 \leq x_{J-1} < \dfrac{d}{2m-1} \\ a_{J-1}f_J(x_{J-1}) & x_{J-1} \geq \dfrac{d}{2m-1} \end{cases}$$

$$z^*_{2,J-1}(x_{J-1}) = \begin{cases} mx_{J-1} & 0 \leq x_{J-1} < \dfrac{d}{2m-1} \\ 1/2(x_{J-1} + d) & x_{J-1} \geq \dfrac{d}{2m-1}, \end{cases} \quad (19)$$

and the total solution is represented by (14)–(16) with $b_j > 0$, $j = 1, \ldots, J-1$. Consequently, the optimal portfolio policy is nonmyopic in the case of constraint (18) also.

Case II.—Let us now introduce an absolute borrowing limit of m, that is, substitute

$$0 \leq z_{2j} \leq x_j + m \qquad m > 0 \qquad j = 1, \ldots, J-1 \quad (20)$$

for (17). When $m < d/2$ the solution to (9) is again given by (14)–(16) with $b_j > 0$, $j = 1, \ldots, J-1$. However, when $m \geq L = d/2$, the solution becomes

$$f_j(x_j) = a_j(x_j + d)^{1/2} \qquad j = 1, \ldots, J-1;$$

$$z^*_{1j}(x_j) = 1/2(x_j - d) \qquad j = 1, \ldots, J-1; \quad (21)$$

$$z^*_{2j}(x_j) = 1/2(x_j + d) \qquad j = 1, \ldots, J-1; \quad (22)$$

that is, the optimal investment policy would seem to be completely myopic on the basis of our assumptions. But L clearly depends on $F_{j+1}, \ldots, F_{J-1}$. Thus to know whether $m \geq L$, knowledge of future returns is necessary. Consequently, the optimal investment policy is not myopic in Case II either.

Let us now consider the realism of assumptions (18) and (20). With respect to (20), we observe from (21) that borrowing takes place, considering decision point $J - 1$, only when $x_{J-1} < d$. By (3), (8), (21), and (22), we obtain

$$x_J = \begin{cases} 1/2(x_{J-1} - d) \text{ with probability } 1/2 \\ 2x_{J-1} + d \text{ with probability } 1/2. \end{cases}$$

Thus, the terminal wealth position has a 1/2 chance of being negative if and only if $x_{J-1} < d$, that is, when borrowing takes place. If the first event ($\beta_{2,J-1} = 0$) takes place and the investor declares bankruptcy at time J, the lender will stand to lose the entire loan of $|1/2(x_{J-1} - d)|$.

The point here is that it would be unreasonable for anyone to lend money to his investor when his optimal strategy calls for it; that is, m should be zero in (20)—which converts (20) to (17). In fact, (18) and (20) may be said to be inconsistent with the portfolio model itself. This is because a wealthier investor (one whose wealth exceeds d) with the same preferences and probability beliefs as a poorer one is, by (21), a possible lender to the poorer one whose wealth is less than d. But the model assumes that lending is safe, that is, that all loans are repaid with probability 1 while, as we have seen, the poorer investor may not be able to repay.

Case III.—A borrowing arrangement that is consistent with the assumed risklessness of lending is one which permits borrowing to the extent that ability to repay, that is, solvency, is guaranteed. Thus, a reasonable constraint on borrowing and short sales, with considerable intuitive appeal as well, is given by

$$\Pr\{x_{j+1} \geq 0\} = 1 \qquad j = 1, \ldots, J-1 . \quad (23)$$

When (23) is substituted for (17), the solution to (10) is the same as when (17) is used; that is, it is given by (14), (15), and (16). Thus, myopia is not optimal in this case either.

V. THE GENERAL CASE

It is readily verified that the conclusions of Sections III and IV are not changed if the number of risky investment opportunities is arbitrary. Moreover, the conclusions hold for all of the functions (5), (6), and (7) whenever $\mu > 0$, both with no borrowing and in each of Cases I–III. Finally, when $r_j \neq 1, j = 1, \ldots, J-1$, partial myopia, as defined by Mossin, is not optimal either in any of the preceding cases. It should be noted that a solution need not exist in Case III unless the "no-easy-money condition" holds.[4] A generalization of this condition (for the case when yields are serially correlated) is given in Section VI. In the most general version of Case III, the set $Z_j(x_j)$ in (4) is given by those $\bar{z}_j$ which satisfy

$$z_{ij} \geq 0 \qquad i \notin S_j \quad (24)$$

and (23).

When $\mu = 0$ in (6) and (7), [(5) is of no interest when $\mu \leq 0$], complete myopia is optimal in both Cases I and III but not in Case II, as is easily shown. The Mossin-Leland conclusions concerning complete myopia when $r_1 = r_2 = \ldots = r_{J-1} = 1$ and partial myopia do not apply in (6) and (7) when $\mu < 0$ either, except in Case III, as we shall demonstrate below. In doing so, we shall also show that, when $\mu \neq 0$, (5), (6), and (7) imply that the optimal investment policies at decision points 1, $\ldots, J-1$ are never myopic in the presence of explicit borrowing limits of any kind, with one exception.

When a solution to the portfolio problem at decision point $J-1$ exists in the presence of constraints (24) only, the optimal lending strategy $\hat{z}_{1,J-1}(x_{J-1})$ has the form $\hat{z}_{1,J-1}(x_{J-1}) = (1 - \lambda A_{J-1})x_{J-1} - A_{J-1}\mu$ in the case of (6) ($\lambda = 1$) and (7) and the form $\hat{z}_{1,J-1}(x_{J-1}) = x_{J-1} - B_{J-1}$ in the case of (5), where A_{J-1} and B_{J-1} are constants, generally positive,[5] which depend on F_{J-1}.[6]

Let us consider (6) and (7) when $\lambda, \mu > 0$. Since A_{J-1}, and hence, $-\hat{z}_{1,J-1}(x_{J-1})$,

[4] Nils Hakansson, "Optimal Investment and Consumption Strategies under Risk for a Class of Utility Functions," *Econometrica* 38 (September 1970): 587–607.

[5] Nonpositive A_{J-1} and B_{J-1} imply that total short sales exceed or equal total long investments.

[6] Nils Hakansson, "Risk Disposition and the Separation Property in Portfolio Selection," *Journal of Financial and Quantitative Analysis* 4 (December 1969): 401–16.

may be arbitrarily large, any finite borrowing limit has a chance to be binding. Consequently, for $f_{J-1}(x)$ to be a positive linear transformation of $f_J(x)$ in the presence of a borrowing limit, it must be a positive linear transformation of $f_J(x)$ *whether the borrowing limit is binding or not.* From Section IV, it is apparent that, when $M_j = 2$ and (17) holds, a necessary and sufficient condition for $f_{J-1}(x)$ to be a positive linear transformation of $f_J(x)$ is that $z^*_{2,J-1}(x_{J-1})$ has the form $z^*_{2,J-1}(x_{J-1}) = a_{2,J-1}(\lambda x_{J-1} + \mu)$, where $a_{2,J-1}$ is a nonnegative constant. When there is more than one risky asset and (24) holds, this condition generalizes to

$$z^*_{i,J-1}(x_{J-1}) = a_{i,J-1}(\lambda x_{J-1} + \mu) \qquad i = 2, \ldots, M_{J-1}, \quad (25)$$

where the $a_{i,J-1}$ are constants, nonnegative only for $i \notin S_{J-1}$.[7] It is now clear that the optimal investment strategy $\bar{z}^*_{J-1}$ will have the form (25) if and only if (1) the borrowing limit is not binding or (2) the borrowing limit has the form

$$-z_{1,J-1}\left(= \sum_{i=2}^{M_{J-1}} z_{i,J-1} - x_{J-1}\right) \leq (\lambda C_{J-1} - 1)x_{J-1} + C_{J-1}\mu$$
$$\lambda C_{J-1} > 1, \qquad \lambda, \mu > 0\,. \quad (26)$$

The latter assertion follows from (2) and the fact that this form of the borrowing limit does give the solution (25) for any F_{J-1}, as is easily verified; moreover, only (26) is capable of giving a solution of form (25) when the borrowing limit is binding. Since knowledge of whether any given borrowing limit is binding or not requires knowledge of F_{J-1}, it follows that $f_{J-1}(x)$ is myopic in the presence of a borrowing limit only if this limit has the form (26). By induction, $f_1(x), \ldots, f_{J-1}(x)$ are then myopic in the case of (6) and (7) for $\lambda, \mu > 0$ if and only if the borrowing limit in period j is given by

$$(\lambda C_j - 1)x_j + C_j\mu \qquad \lambda C_j > 1, \qquad \lambda, \mu > 0, \qquad j = 1, \ldots, J-1\,. \quad (27)$$

When $\mu < 0$ or $\lambda < 0$ in (6) and (7), any borrowing limit would, to be consistent with myopia, again have to have the form (27). But when $\mu < 0$, we must have $\lambda > 0$ and vice versa so that (27) cannot be nonnegative for all $x_j > 0$ for which borrowing may be desired, a basic requirement of any "true" borrowing limit. The situation in the case of function (5) is analogous. As a result, $f_1(x_1), \ldots, f_{J-1}(x_{J-1})$ can never be myopic for (5), (6), and (7) when $\lambda < 0$ or $\mu < 0$ in the presence of a borrowing limit.

Turning now to the solvency constraint (23), we obtain whenever a solution exists for (6) and (7) that the greatest lower bound on b such that $\Pr\{x_J < b\} > 0$, for any decision at decision point $J-1$ which satisfies (24), is $K_{J-1}(\lambda x_{J-1} + \mu) + r\hat{z}_{1,J-1}(x_{J-1})$, where K_{J-1} is a constant which depends on F_{J-1}. Since $f_J'(-\mu/\lambda) = \infty$ for $\lambda > 0$, we obtain, letting $x_J \equiv K_{J-1}(\lambda x_{J-1} + \mu) + r\hat{z}_{1,J-1}(x_{J-1})$, that $\lambda x_J + \mu > 0$, which implies, since λ and μ cannot both be negative, $x_J > 0$ when $\mu \leq 0$. Thus the solvency constraint (23) is not binding when $\mu \leq 0$ but may be when $\mu > 0$. Consequently, the induced utility functions $f_1(x), \ldots, f_{J-1}(x)$ are myopic for the class (6) and (7) when $\mu < 0$ in the presence of (24) and the solvency constraint (23).

[7] Ibid.

In sum then, when $\mu \neq 0$ and interest rates are zero, the induced utility of wealth functions $f_1(x_1), \ldots, f_{J-1}(x_{J-1})$ are myopic in the presence of borrowing constraints if and only if the borrowing limits are of the form (27) and $f_J(x)$ is of the form (6) (7) with $\lambda,\mu > 0$; in the presence of the solvency constraint, $f_1(x_1), \ldots, f_{J-1}(x_{J-1})$ are myopic only if $f_J(x)$ has the form (6) or (7) with $\mu < 0$ (and hence $\lambda > 0$).

It should also be noted that, when $\mu < 0$, $f_J(x)$ is undefined for $x < |\mu/\lambda|$ and $f_j(x_j)$, $j = 1, \ldots, J-1$ is undefined for small x_j, a significant drawback. In addition, the relative risk aversion index $-xf_j''(x)/f_j'(x)$ is decreasing for these functions, whereas Arrow,[8] for example, suggests that plausible utility functions of money exhibit increasing relative risk aversion.

VI. SERIALLY CORRELATED YIELDS

We shall now consider the sequential investment problem when yields are serially correlated. In contrast to Mossin's assertion,[9] we shall find that the optimal investment policy is myopic in this case only for a small subset of the terminal utility functions which induce myopic short-run utility functions when returns are serially independent.

For simplicity, we assume that yields and the interest rate obey a Markov process. A distinction between risk due to general market forces, called the economy, and risk due to individual assets and periods is made. As a result, the assumptions and notation of Section II are modified as follows:

x_j: amount of investment capital at decision point j;
N_j: number of states of the economy at decision point j;
M_{jm}: number of investment opportunities available at decision point j, given that the economy is at state m at that time;
S_{jm}: the subset of investment opportunities which it is possible to sell short at decision point j, given that the economy is in state m at that time;
$r_{jm} - 1$: interest rate in period j, given that the economy is in state m at decision point j ($r_{jm} > 1$);
β_{ijmn}: proceeds at the end of period j, given that the economy is in state n at that time, per unit of investment in opportunity i, $i = 2, \ldots, M_{jm}$, at decision point j, given that the economy was in state m at that time;
p_{jmn}: probability that the economy makes a transition from state m to state n in period j

$$\left(p_{jmn} \geq 0, \sum_{n=1}^{N_{j+1}} p_{jmn} = 1\right);$$

z_{1jm}: amount lent in period j, given that the economy is in state m at decision point j (negative z_{1jm} indicate borrowing) (decision variable);
z_{ijm}: amount invested in opportunity i, $i = 2, \ldots, M_{jm}$, at decision point j, given that the economy is in state m at that point;
$f_{jm}(x_j)$: utility of money at decision point j, given that the economy is in state m at that time;
$\overset{*}{z}_{1jm}$: an optimal lending policy for state m at decision point j;
$\overset{*}{z}_{ijm}$: an optimal investment policy for state m at decision point j, $i = 2, \ldots, M_{jm}$.

$$\bar{z}_{jm} \equiv (z_{2jm}, \ldots, z_{M_{jm}jm}) ; \quad v_{ijm} \equiv \frac{z_{ijm}}{x_j} ; \qquad i = 1, \ldots, M_{jm} .$$

$$\bar{v}_{jm} \equiv (v_{2jm}, \ldots, v_{M_{jm}jm}) ; \quad F_{jmn}(y_2, \ldots, y_{M_{jm}}) \equiv \Pr\{\beta_{2jmn} \leq y_2, \ldots, \beta_{M_{jm}jmn} \leq y_{M_{jm}}\} .$$

[8] Kenneth Arrow, *Aspects of the Theory of Risk-bearing* (Helsinki: Yrjö Jahnssonin Säätiö, 1965).
[9] Mossin, p. 222.

Clearly, v_{ijm} represents the proportion of capital x_j invested in opportunity i at decision point j, given that the economy is in state m at that point; thus

$$v_{1jm} = 1 - \sum_{i=2}^{M_{jm}} v_{ijm} .$$

It will be assumed that the joint distribution functions F_{jmn} are independent with respect to j. In addition, we postulate that the $\{\beta_{ijmn}\}$ satisfy the following conditions:

$$\Pr\{0 \leq \beta_{ijmn} < \infty\} = 1 , \qquad i = 2, \ldots, M_{jm} , \quad (28)$$

all j, m, and n

$$\Pr\left\{\sum_{i=2}^{M_{jm}} (\beta_{ijmn} - r_{jm})\theta_i < 0\right\} > 0 \quad (29)$$

for all j, all m, some n for which $p_{jmn} > 0$, and all finite θ_i such that $\theta_i \geq 0$ for all $i \notin S_{jm}$ and $\theta_i \neq 0$ for at least one i. (29) is a modification of the "no-easy-money-condition" for the case when the lending rate equals the interest rate.[10] This condition states that no combination of risky investment opportunities exists in any period which provides, with probability 1, a return at least as high as the (borrowing) rate of interest; no combination of short sales is available for which the probability is zero that a loss will exceed the (lending) rate of interest; and no combination of risky investments made from the proceeds of any combination of short sales can guarantee against loss. (29) may be viewed as a condition which the prices of all assets must satisfy in equilibrium.

(3) is now replaced by the conditional difference equations

$$x_{j+1}|\,mn = \sum_{i=2}^{M_{jm}} (\beta_{ijmn} - r_{jm})z_{ijm} + r_{jm}x_j \quad j = 1, \ldots, J-1, \text{ all } m, n , \quad (30)$$

and (4) becomes, for $j = 1, \ldots, J-1$ and all m,

$$f_{jm}(x_j) = \max_{\bar{z}_{jm}} \sum_{n=1}^{N_{j+1}} p_{jmn}E[f_{j+1,n}(x_{j+1}|\,mn)] , \quad (31)$$

where $f_{Jm}(x_J)$ is given for all m, subject to

$$z_{ijm} \geq 0 \qquad i \notin S_{jm} , \quad (32)$$

and

$$\Pr\{x_{j+1}|\,mn \geq 0\} = 1 \qquad n = 1, \ldots, N_{j+1} . \quad (33)$$

VII. OPTIMAL MYOPIC POLICIES

On the basis of the finite yield assumption (28) and the "no-easy-money-condition" (29), we obtain the following:

Theorem.—Let r_{jm}, F_{jm}, and p_{jmn} be defined as in Section VI and let $u(x)$ be a monotone increasing and strictly concave function for all $x \geq 0$. Then the functions

$$h_{jm}(\bar{v}_{jm}) \equiv \sum_{n=1}^{N_{j+1}} p_{jmn}E\left\{u\left[\sum_{i=2}^{M_{jm}} (\beta_{ijm} - r_{jm})v_{ijm} + r_{jm}\right]\right\} , \quad (34)$$

[10] Hakansson, "Optimal Investment . . ." (see n. 4 above).

subject to

$$v_{ijm} \geq 0 \qquad i \notin S_{jm}\,; \qquad (35)$$

and

$$\Pr\left\{\sum_{i=2}^{M_{jm}} (\beta_{ijmn} - r_{jm})v_{ijm} + r_{jm} \geq 0\right\} = 1 \qquad n = 1, \ldots, N_{j+1} \qquad (36)$$

have maxima for all $j = 1, \ldots, J-1$ and all m. Moreover, the maximizing vectors, $\bar{v}^*_{jm}$, are finite and unique. The proof may be found in Hakansson (1968).[11]

Let us now assume that

$$f_{Jm}(x) = x^{1/2} \qquad \text{all } m\,; \qquad (37)$$

and let k_{jm} denote the maximum of (34) subject to (35) and (36) when $u(x) = x^{1/2}$; that is,

$$k_{jm} \equiv \sum_{n=1}^{N_{j+1}} p_{jmn} E\left\{\left[\sum_{i=2}^{M_{jm}} (\beta_{ijmn} - r_{jm})v^*_{ijm} + r_{jm}\right]^{1/2}\right\} \qquad j = 1, \ldots, J-1 \quad \text{all } m\,. \qquad (38)$$

By the theorem, we know that k_{jm} exists.

Let us now determine $f_{J-1,m}(x_{J-1})$. From (31) we obtain for all m

$$f_{J-1,m}(x_{J-1}) = \max_{\bar{z}_{J-1,m}} \sum_{n=1}^{N_J} p_{J-1,mn} E[(x_J \mid mn)^{1/2}]\,, \qquad (39)$$

subject to

$$z_{i,J-1,m} \geq 0 \qquad i \notin S_{J-1,m} \qquad (40)$$

and

$$\Pr\left\{\sum_{i=2}^{M_{J-1,m}} (\beta_{i,J-1,mn} - r_{J-1,m})z_{i,J-1,m} + r_{J-1,m}x_{J-1} \geq 0\right\} = 1 \qquad n = 1, \ldots, N_J\,. \qquad (41)$$

By (29) and (41), $f_{J-1}(x_{J-1})$ does not exist for $x_{J-1} < 0$. For $x_{J-1} \geq 0$, (39) may be written, since (32) and (33) are equivalent to (35) and (36) when $x_{J-1} > 0$,

$$f_{J-1,m}(x_{J-1}) = x_{J-1}^{1/2} \max_{\bar{v}_{J-1,m}} \sum_{n=1}^{N_J} p_{J-1,mn} E\left\{\left[\sum_{i=2}^{M_{J-1,m}} (\beta_{i,J-1,mn} - r_{J-1,m})v_{i,J-1,m} + r_{J-1,m}\right]^{1/2}\right\}, \qquad (42)$$

subject to

$$v_{i,J-1,m} \geq 0 \qquad i \notin S_{J-1,m} \qquad (43)$$

and

$$\Pr\left\{\sum_{i=2}^{M_{J-1,m}} (\beta_{i,J-1,mn} - r_{J-1,m})v_{i,J-1,m} + r_{J-1,m} \geq 0\right\} = 1 \qquad n = 1, \ldots, N_J\,. \qquad (44)$$

By the theorem, we now obtain that $f_{J-1,m}(x_{J-1})$ exists for all m and $x_{J-1} \geq 0$ and is given by, using (38),

$$f_{J-1,m}(x) = k_{J-1,m}x^{1/2} \qquad \text{all } m\,. \qquad (45)$$

[11] Nils Hakansson, "Optimal Entrepreneurial Decisions in a Completely Stochastic Environment," *Management Science: Theory* 17 (March 1971): 427–49.

By (31), the expression to be maximized at decision point $J-2$, given that the economy is in state m at that time, becomes

$$\sum_{n=1}^{M_{J-2,m}} p_{J-2,mn} k_{J-1\,n} E[(x_{J-1} | mn)^{1/2}] . \quad (46)$$

Since the constants $k_{J-1,n}$ will in general be different for different n, (46) and, therefore, the optimal portfolio $\bar{z}^*_{J-2,m}$, depend on the yields in period $J-1$. Thus the optimal investment policy is *not* myopic at decision point $J-2$; neither is it myopic at decision points $1, \ldots, J-3$, which is easily shown by induction.

The existence of positive constants $a_1, \ldots, a_{J-1}$ and of constants $b_{11}, \ldots, b_{J-1,N_{J-1}}$ such that

$$f_{jm}(x) = a_j f_{Jm}(x) + b_{jm} \quad \text{all } m , \quad j = 1, \ldots, J-1 \quad (47)$$

are clearly both necessary and sufficient for myopia to be optimal in this model. As noted, (45) violates (47) for $j = J-1$ whenever $N_{J-1} > 1$. However, when

$$N_j = 1 , \quad j = 1, \ldots, J , \quad (48)$$

(47) is satisfied; but (48) also implies that yields are statistically independent in the various periods. This confirms Mossin's result that the optimal investment policy is myopic when returns are stochastically independent over time and the terminal utility function is $x^{1/2}$.

Let us now assume that $f_{Jm}(x)$ has the form

$$f_{Jm}(x) = \log x \quad \text{all } m . \quad (49)$$

Letting $H_{J-1,m}$ denote the maximum of (34) subject to (35) and (36) when $u(x) = \log x$, that is,

$$H_{J-1,m} \equiv \sum_{n=1}^{N_J} p_{J-1,mn} E\left\{\log\left[\sum_{i=2}^{M_{J-1,m}} (\beta_{i,J-1,mn} - r_{J-1,m}) v^*_{i,J-1,m} + r_{J-1,m}\right]\right\} , \quad (50)$$

we obtain from (31)–(33), solving recursively,

$$f_{jm}(x) = \log x + b_{jm} \quad \text{all } m \quad j = 1, \ldots, J-1 , \quad (51)$$

(where $b_{J-1,m} = H_{J-1,m}$, all m), which is consistent with (47). As a result, the induced utility functions

$$f_{11}(x), \ldots, f_{1N_1}(x), \ldots, f_{J-1\,1}(x), \ldots, f_{J-1,N_{J-1}}(x)$$

are myopic when the terminal utility function is logarithmic, both when yields are serially correlated and when they are not.

Just as in the case of (37), which is a special case of (7), the optimal investment policy is not myopic for *any* function (7) (whether $\mu = 0$ or not), nor for any function (5), when yields are serially dependent, as is easily verified. Since interest rates are assumed to be positive and state-dependent, myopia is not optimal for log $(x + \mu)$, $\mu \neq 0$ either, or any other nonlogarithmic function, under serial dependence. Thus, when yields are serially correlated, only the logarithmic utility function of terminal wealth induces short-run utility of wealth functions which are myopic when yields in the various periods are nonindependent, contrary to Mossin's assertion.[12]

VIII. CONCLUDING REMARKS

In view of the difficulty of estimating future yields and their apparent serial correlation, the myopic property of the logarithmic utility functions is, of course, highly significant. However, this function also has other attractive properties.[13] Perhaps the most important of these is the property that maximization of the expected logarithm of end-of-period capital subject to (32) and (33) in each period also maximizes the expected growth rate of capital, whether returns are serially correlated[14] or not.[15]

As Mossin points out, the portfolio decision is in general not independent of the consumption decision. A realistic model of the investor's decision problem must, therefore, include consumption as a decision variable and a preference function for evaluating consumption programs. Consumption-investment models of this type have been developed by Hakansson, both when investment yields are serially correlated[16] and when they are not.[17]

[12] Mossin, p. 222.

[13] Some of the properties are reviewed in Nils Hakansson and Tien-Ching Liu, "Optimal Growth Portfolios When Yields Are Serially Correlated," *Review of Economics and Statistics* 52 (November 1970): 385–94.

[14] Ibid.

[15] Henry Latané, "Criteria for Choice among Risky Ventures," *Journal of Political Economy* 67 (April 1959): 144–55; and Leo Breiman, "Optimal Gambling Systems for Favorable Games," *Fourth Berkeley Symposium on Probability and Mathematical Statistics* (Berkeley: University of California Press, 1961).

[16] Hakansson, "Optimal Entrepreneurial Decisions . . ." (see n. 11 above).

[17] Hakansson, "Optimal Investment . . ." (n. 4 above); and "Optimal Investment and Consumption Strategies under Risk, an Uncertain Lifetime, and Insurance," *International Economic Review* 10 (October 1969): 443–66.

COMPUTATIONAL AND REVIEW EXERCISES

1. Modify the hypotheses of Proposition 1 in Fama's article by assuming that U is concave rather than strictly concave.

(a) Prove a similar proposition to that given in Proposition 1.

(b) Are any of the behavioral implications of the paper changed if one utilizes the result from this modified proposition?

*(c) Attempt to verify Proposition 1 using Fama's assumptions in light of the fact that footnote 18 is in error (see Exercise I-ME-18).

2. (Computing derived utility functions) Consider an investor having initial wealth $w \geqq 0$ and a quadratic utility function for wealth, $u(\xi) = \xi - k\xi^2$, $k > 0$. Suppose there are two risky assets whose rates of return ρ_1, ρ_2 are independent and have means and variances (μ_1, σ_1^2), (μ_2, σ_2^2), respectively. The investor wishes to determine the proportions x^1, x^2 to be invested in assets 1 and 2 in such a way as to maximize expected utility of terminal wealth:

$$\max_{x \in K} Eu(w\rho^T x), \qquad \text{where} \quad K = \{x = (x^1, x^2) \mid e^T x = 1,\, x \geqq 0\}, \quad e = (1,1)^T.$$

Let $x^1 = \frac{1}{2}(1+z)$, $x^2 = \frac{1}{2}(1-z)$.

(a) Show that $x \in K$ iff $z \in [-1, 1]$.

Let $U(w) = \max_{x \in K} Eu(w\rho^T x)$. Note that $U(\cdot)$ would be the "derived utility function" relevant to the first period of a two-period investment problem. Let $\phi(z, w) = Eu(w\rho^T x)$ for $x \in K$. Suppose that $\mu_1 > \mu_2$.

(b) Show that $\phi(z, w) = -A(w)z^2 + B(w)z + C(w)$, where

$$A(w) = \tfrac{1}{4}kw^2[\sigma_1^2 + \mu_1^2 + \sigma_2^2 + \mu_2^2 - 2\mu_1\mu_2],$$
$$B(w) = \tfrac{1}{2}w[\mu_1 - \mu_2 - kw(\sigma_1^2 + \mu_1^2 - \sigma_2^2 - \mu_2^2)],$$
$$C(w) = \tfrac{1}{2}(\mu_1 + \mu_2)w - \tfrac{1}{4}kw^2[\sigma_1^2 + \mu_1^2 + \sigma_2^2 + \mu_2^2 + 2\mu_1\mu_2],$$

and that ϕ is a concave function of z for fixed $w \geqq 0$.

(c) Show that

$$\frac{\partial}{\partial z}\phi(z, w)]_{z=-1} = \tfrac{1}{2}(\mu_1 - \mu_2)w + kw^2(\sigma_2^2 + \mu_2^2 - \mu_1\mu_2),$$

$$\frac{\partial}{\partial z}\phi(z, w)]_{z=+1} = \tfrac{1}{2}(\mu_1 - \mu_2)w - kw^2(\sigma_1^2 + \mu_1^2 - \mu_1\mu_2).$$

(d) Show that for sufficiently small w, it is optimal to invest completely in the asset ρ_1, and show that for such w,

$$U(w) = \mu_1[w - k'w^2] \qquad \text{where} \quad k' = k(\sigma_1^2 + \mu_1^2)/\mu_1. \tag{d1}$$

Let

$$w_1 = \frac{\frac{1}{2}(\mu_1 - \mu_2)}{k|\sigma_1^2 + \mu_1^2 - \mu_1\mu_2|}, \qquad w_2 = \frac{\frac{1}{2}(\mu_1 - \mu_2)}{k|\sigma_2^2 + \mu_2^2 - \mu_1\mu_2|}.$$

(e) If $\mu_1 \leqq (\sigma_2^2 + \mu_2^2)/\mu_2$, show that (d1) holds only for $w \in [0, w_1]$. Show that for $w > w_1$

$$U(w) = \alpha + \beta[w - k''w^2], \tag{e1}$$

where

$$\alpha = \frac{1}{4k}\left\{\frac{(\mu_1-\mu_2)^2}{\sigma_1^2+\mu_1^2+\sigma_2^2+\mu_2^2-2\mu_1\mu_2}\right\},$$

$$\beta = \frac{\mu_1\sigma_2^2+\mu_2\sigma_1^2}{\sigma_1^2+\mu_1^2+\sigma_2^2+\mu_2^2-2\mu_1\mu_2},$$

and

$$k'' = k\left\{\frac{\sigma_1^2\mu_2^2+\sigma_2^2\mu_1^2+\sigma_1^2\sigma_2^2}{\mu_1\sigma_2^2+\mu_2\sigma_1^2}\right\}.$$

Let $w_- = \min\{w_1, w_2\}$, $w_+ = \max\{w_1, w_2\}$.

(f) If $\mu_1 > (\mu_2^2+\sigma_2^2)/\mu_2$, show that (d1) holds if $w \in [0, w_-]$. For $w > w_+$, show that

$$U(w) = \mu_2[w-k'''w^2], \tag{f1}$$

where

$$k''' = k\frac{\sigma_2^2+\mu_2^2}{\mu_2}.$$

(g) In (f) show that (e1) holds for $w \in [w_-, w_+]$.

(h) Discuss the results above for the special cases $\mu_2 = r$, $\sigma_2 = 0$ (ρ_2 is money) and $\mu_1 = r$, $\sigma_1 = 0$ (ρ_1 is money.)

(i) Discuss the case $\mu_1 = \mu_2$.

3. Given a utility function u, a risky asset with rate of return $\rho \geqq 0$, and a risk-free asset with rate of return unity, the derived utility function $U(w)$ is

$$U(w) = \max_{x\in[0,w]} Eu[w+x(\rho-1)].$$

Let $h \equiv \rho-1$ be a random variable uniformly distributed over $[\alpha,\beta]$ with $\beta > \alpha > 0$, Consider the derived utility function $U(w)$ for the class u such that $-u'(\xi)/u''(\xi) = \mu+\lambda\xi$, for constants λ and μ. According to Exercise ME-5, this class of utility function propagates unchanged, up to a positive linear transformation, if the maximization problem has an *interior* solution.

(a) For $u(\xi) = -e^{-a\xi}$, show that $x^* = w$, i.e., the maximum does not occur at an interior point. Thus

$$U(w) = \frac{e^{-aw}}{\beta-\alpha}\left[\frac{e^{-a\beta w}-e^{-a\alpha w}}{aw}\right].$$

Note that the function $U(w)$ is not equivalent to $u(w)$.

(b) For $u(\xi) = (a+\xi)^c$, show that

$$U(w) = \frac{1}{(c+1)(\beta-\alpha)}\frac{1}{w}\left[(a+w(\beta+1))^{c+1} - (a+w(\alpha+1))^{c+1}\right].$$

Note that $U(w)$ is not equivalent to $u(w)$ except when $a = 0$.

(c) For $u(\xi) = \log(\xi+a)$, show that

$$U(w) = -1+\frac{1}{(\beta-\alpha)w}\left\{\begin{matrix}[a+(\beta+1)w]\log(a+(\beta+1)w)\\ -[a+(\alpha+1)w]\log(a+(\alpha+1)w)\end{matrix}\right\}.$$

Show that U is equivalent to u (in the sense of a positive linear transformation) if and only if $a = 0$.

4. Referring to the article by Wilson consider the cash flow pattern for the initial policy of $[d_0, \{d_i\}, \{d_{ij}\}, \ldots]$, and the cash flow pattern for an investment project of $[c_0, \{c_i\}, \{c_{ij}\}, \ldots]$.

(a) Show that if

$$u\left(c_0 + d_0, \sum_i p_i(c_i + d_i), \sum_{ij} p_{ij}(c_{ij} + d_{ij}), \ldots\right)$$

$$\leqq \sum_{i, j, k, \ldots} p_i\, p_{ij}\, p_{ijk} \cdots u(d_0, d_i, d_{ij}, \ldots),$$

or in a more compact notation, $u\{E(C+D)\} \leqq Eu(D)$, then exclude the project.

(b) Show that if

$$\sum_{i, j, k, \ldots} p_i\, p_{ij}\, p_{ijk} \cdots u(c_0 + d_0,\ c_i + d_i,\ c_{ij} + d_{ij}, \ldots)$$

$$\geqq u\left(d_0, \sum_i p_i d_i, \sum_{ij} p_{ij} d_{ij}, \sum_{ijk} p_{ijk} d_{ijk}, \ldots\right)$$

then include the project.

5. Consider an expected utility maximizer with logarithmic utility function for terminal wealth, $u(w) = \log w$. Assume that he can invest in a risk-free asset returning $r \geqq 1$ and a risky asset returning $\rho \geqq 0$ per dollar invested. Suppose that ρ is log-normally distributed (i.e., $\log \rho$ is normal), with $E\rho = \mu$, $\operatorname{var} \rho = \sigma^2$. Recall from Exercise II-CR-18 that if w_0 is initial wealth and $a \in [0, w_0]$ is the amount invested in the risky asset, then for a 1-period problem, $a = w_0$ is optimal if $\mu \geqq r(1 + \sigma^2/\mu^2)$. In an n-period problem, suppose that the return r on money can vary between periods but always satisfies the inequality above.

(a) Show that the optimal policy is to invest totally in the risky asset in each period.

(b) Calculate the probability distribution of terminal wealth under the optimal policy.

$$\left[\textit{Hint}: \quad \frac{w_n}{w_0} = \frac{w_n}{w_{n-1}} \cdot \frac{w_{n-1}}{w_{n-2}} \cdot \ldots \cdot \frac{w_1}{w_0}.\right]$$

(c) Calculate the expected value and most probable value of terminal wealth w_n.

(d) Let $\mu = 1.10$, $\sigma = 0.05$, and let $r = 1.065$ for all periods. Evaluate numerically the expected and most probable wealth ratios w_{10}/w_0 for a 10-period problem. Evaluate the probability that the optimal investment performs better than money, that is,

$$\Pr[w_{10} \geqq r^{10} w_0].$$

6. Referring to Exercise 5, suppose that the investor's utility function for terminal wealth is $u(w) = w^\alpha$, $\alpha \in (0, 1)$. Suppose that

$$\mu(1 + (\sigma^2/\mu^2))^{\alpha - 1} \geqq r.$$

(a) Show that the optimal 1-period policy is to invest totally in the risky asset.

(b) Show that the optimal n-period policy is to invest totally in the risky asset in each period.

Exercise Source Notes

Exercises 2 and 3 were adapted from Mossin (1968c), and Exercise 4 is due to Professor S. L. Brumelle.

MIND-EXPANDING EXERCISES

1. Consider a two-period investment problem where the investor has initial resource Y which is either consumed in period 1, C_1, or in period 2, C_2. If $I = Y - C$ is invested in period 1, then $C_2 = \xi f(I)$, where $f \geqq 0$, $f' > 0$, and $f'' < 0$ (primes denote differentiation) and ξ has a known nonnegative distribution. It is desired to maximize the expectation of the additive utility function, $u(\cdot) = u_1(C_1) + u_2(C_2)$, where $u_i' > 0$, $u_i'' < 0$. The problem is to

$$\{\max \phi(I) \equiv u(C_1) + E_\xi u_2[\xi f(I)] \,|\, 0 \leqq I \leqq Y\}.$$

(a) Show that ϕ is strictly concave on $[0, Y]$.
(b) Show that the optimal investment I_v may be found by solving the equation,

$$-u_1'(Y-I) + \frac{f'(I)}{f(I)} E_\xi\, g[\xi f(I)] = 0,$$

where $g(I) \equiv I u_2'(I)$ provided that there is an interior maximum and all expectations are finite.
(c) Under what conditions is I_v equal to 0 or Y?
(d) For the deterministic problem in which ξ takes on only its mean value $\bar{\xi}$ develop the optimizing conditions for I_d as in (b) and (c).
(e) Show that $\phi'(Y) \geqq 0$ implies $I_v \geqq I_d$ and that $\phi'(0) < 0$ implies $I_v \leqq I_d$. In other cases when $0 < I_v < Y$, $\phi'(I_v) = 0$. Show that $I_v \leqq I_d$ if and only if $\phi'(I_d) \leqq 0$.
(f) Let $\lambda = f(I_d)$ and suppose $0 < I_d < Y$. Show that a necessary and sufficient condition for $I_d \geqq I_v$ is that $g(\lambda\bar{\xi}) \geqq E_\xi\, g(\lambda\xi)$.
(g) Show that the condition in (f) means that g must be concave, and that $u''' \leqq 0$.
(h) Show that $I_d = I_v$ if $u_2(C_2) = a \log C_2 + bC_2 + d$, $a \geqq 0$, $b \geqq 0$.
(i) Interpret (g) and (h).

2. (Certainty equivalence results) In uncertain dynamic financial models, one seeks a first-period decision and a policy that indicates the decisions to be made in succeeding periods as a function of accumulated information. In theory, most problems of this nature may be solved via dynamic programming; however, the computational cost is often prohibitive. It is important in problems of this nature to know what information is required for the calculation of optimal decisions in each period and when certain minimal amounts of information are sufficient. Consider a situation where a decision maker chooses a sequence of decisions $x_1, \ldots, x_T$ $(T \geqq 2)$. After each decision vector x_t is chosen, a random vector ϕ_t is observed without error and a state vector y_t is uniquely determined. The decision-maker's objective is to choose the constrained decision vectors such that the expected sum of the period-by-period preferences is maximized.

$$\max_{x_1 \in K_1} E\phi_1 \left\{ F_1(x_1, y_1, \phi_1) + \max_{x_2 \in K_2} \left[E\phi_2 | \phi_1 \left\{ F_2(x_2, y_2, \phi_2) + \cdots + \max_{x_T \in K_T} [E\phi_T | \phi_{T-1} F_T(x_T, y_T, \phi_T)] \right\} \right] \right\},$$

$$\begin{aligned}
\text{s.t.} \quad H_1(x_1, y_1, \phi_1) &= 0,\\
H_2(x_1, x_2, y_1, y_2, \phi_1, \phi_2) &= 0,\\
&\;\;\vdots\\
H_T(x_1, \ldots, x_T, y_1, \ldots, y_T, \phi_1, \ldots, \phi_T) &= 0,
\end{aligned}$$

where $E\phi$ represents mathematical expectation with respect to ϕ, $\phi_t \mid \phi_{t-1}$ means ϕ_t given $\phi_{t-1}, \ldots, \phi_1$, the K_t are the constraints on the decision choices, the F_t are real-valued scalar functions that attach cardinal utility to the various values of the vectors (x_t, y_t, ϕ_t), and the H_t are real vector-valued functions.

Suppose that the K_t are never binding, the random variables are additive, the H_t are linear in the decision and state variables, i.e.,

$$H_t(\cdot) = \sum_{\tau=1}^{t} (A_{t\tau} x_\tau + B_{t\tau} y_\tau) + C_t(\phi_1, \ldots, \phi_t),$$

where B_{tt} has full column rank, and the F_t are quadratic in the decision and state variables, i.e.,

$$F_t(\cdot) = \begin{pmatrix} x_t \\ y_t \end{pmatrix}' D_t(x_t', y_t')' + d_t(\phi_t).$$

(a) Utilize a backward induction argument to show that the optimal decision in each period may be found by solving the deterministic quadratic program that results when all future random vectors are replaced by their conditional means. What assumption must be made about the D_t?

(b) Find an equivalent deterministic quadratic program depending only on x_1 whose solution will provide an optimal initial decision x_1^*.

(c) Show that x_1^* is a linear function of ϕ_1.

(d) Develop a multiperiod portfolio selection model based on the model in (a). What must be assumed about the asset value generation process?

3. (Optimality of zero-order decision rules) Referring to the formulation of Exercise 2, suppose that the constraints K_t are meaningful.

(a) Suppose

$$F_t(\cdot) = f_t(x_t) + g_t(y_t) + h_t(\phi_t),$$

$$H_t(\cdot) = \sum_{\tau=1}^{t} h_{t\tau}(x_\tau) + \sum_{\tau=1}^{t} l_{t\tau}(y_\tau) + c_t(\phi_1, \ldots, \phi_t)$$

(i.e., separability), and that

$$g_t = a_t' y_t + b_t, \qquad l_{t\tau}(\cdot) = D_{t\tau} y_\tau + d_{t\tau}$$

(i.e., linearity in state), where D_{tt} has full column rank, for $t = 1, \ldots, T$. Show that the $H_t(\cdot) = 0$ may be put into the form $\bar{h}_{t1}(x_1) + \cdots + \bar{h}_{tt}(x_t) - y_t + \rho_t = 0$, where ρ_t is a random vector, $t = 1, \ldots, T$.

(b) Utilize (a) to show that the x_t^* are independent of $\phi_1, \ldots, \phi_T$ and x_τ^*, $\tau \neq t$, $t, \tau = 1, \ldots, T$.

(c) Show that the x_t^* are the solution of

$$\max_{x_t \in K_t} \left\{ f_t(x_t) + \sum_{\tau=t}^{T} a_\tau' \bar{h}_{\tau t}(x_t) \right\}, \qquad t = 1, \ldots, T.$$

(d) When is the problem in (c) a concave program?

(e) Let ψ_t be a subvector of ϕ_t, i.e., the elements of ψ_t are a subset of the elements of ϕ_t, and suppose that the ψ_t have known intertemporally independent distributions. Suppose

$$F_t(\cdot) = f_t(x_t, \psi_t) + g_t(y_t) + h_t(\phi_t),$$

$$H_t(\cdot) = \sum_{\tau=1}^{t} h_{t\tau}(x_\tau, \psi_\tau, \ldots, \psi_t) + \sum_{\tau=1}^{t} l_{t\tau}(y_\tau, \psi_\tau) + c_t(\phi_1, \ldots, \phi_t)$$

(i.e., partial separability) and

$$g_t(\cdot) = a_t(\psi_t)'y_t + b_t \qquad \text{and} \qquad l_{t\tau}(\cdot) = D_{t\tau}(\psi_\tau)y_\tau + d_{t\tau},$$

where D_{tt} has full column rank for almost all ψ_t (i.e., linearity in state). Show that the $x_t{}^*$ are independent of the $x_\tau{}^*$, $t \neq \tau$, but may depend on the ψ_t.

(f) Show that the $x_t{}^*$ may be found by solving a problem of the form

$$\max_{x_t \in K_t} \{\bar{f}_t(x_t) + \bar{M}_t(x_t)\}, \qquad \text{where} \quad \bar{f}_t(x_t) = E_{\psi_t}\{f_t(x_t, \psi_t)\}.$$

(g) When is the problem in (f) a concave program?

(h) Suppose ψ_t is as in (e), and that

$$F_t(\cdot) = f_t(x_t, \psi_t) + a_t(\psi_t)'y_t + h_t(\phi_t),$$

$$H_t(\cdot) = U_t(x_1, \ldots, x_t, \psi_t, \ldots, \psi_\tau) + \sum_{\tau=1}^{t} D_{t\tau}(\psi_\tau)y_\tau + c_t(\phi_1, \ldots, \phi_t),$$

and D_{tt} has full rank for almost all ψ_t, $t = 1, \ldots, T$ (i.e., partial separability and linearity in the state vectors). Show that there exists an explicit deterministic program involving $x_1, \ldots, x_T$, whose solution provides an optimal first-period decision $x_1{}^*$. When is this problem a concave program?

(i) Discuss the limitations of the results in (b)–(h), noting the key assumptions: that the state vectors are unconstrained (except indirectly via the K_t) and that they display linear returns to scale, and discuss the way the uncertainty drops out of the problem.

4. The results in Exercise 3 may be used to construct an easily solvable dynamic model of capital asset choice under uncertainty.

Consider a business enterprise, whose aim is to maximize the discounted sum of expected dividend payments plus the firm's terminal valuation. In each period resources accrue from exogenous sources, previous investments, and from borrowing. The firm allocates these resources between investment projects, investment lending, and dividend payments (which, if negative, amount to additional financing for the firm by the present stockholders). However, a certain liquidity position, which depends on the level of borrowing, must be maintained.

The following notation is used:

M_t	Random exogenous cash flow in period t
ψ_t	A random vector that is observed in period t
W_t	Cash withdrawn in the form of dividends in period t
x_{jt}	Cash invested in project j in period t
K_t	Set of available decision possibilities in period t
m_t	Number of different projects available in period t
u_t	Marginal utility of dividends in period t
a	Marginal utility of terminal wealth
$h_{jt}(x_{jt}, \psi_\tau)$	Cash return in period τ of x_{jt} dollars invested in project j in period t if random event ψ_τ is observed in period τ
w_t	Cash borrowed in period t
$b_{t\tau}(w_t)$	Cash owed in period τ if w_t dollars are borrowed in period t, $\tau > t$, $t = 1, \ldots, T$, $\tau = 2, \ldots, L$
$c_{1t}(w_1) + \cdots + c_{tt}(w_t)$	Minimum cash that must be carried as liquidity in period t if w_τ is borrowed in periods $\tau = 1, \ldots, t$, $t = 1, \ldots, T$
v_t	Cash lent above and beyond liquidity requirement in period t
$l_{t\tau}(v_t)$	Cash received in period τ if v_t dollars are loaned in period t, $\tau > t$, $t = 1, \ldots, T$, $\tau = 2, \ldots, L$

G	Terminal wealth
d_t	Time T "present worth" factor for cash inflows in periods $t > T$
z_t	$(w_t, v_t, x_{1t}, \ldots, x_{m_t, t})$ = decision vector in period t
T	Time horizon
L	Last period in which returns from invested funds are received, and by definition, $c_{0\tau}(\cdot) = b_{0\tau}(\cdot) = c_{0\tau}(\cdot) = 0$

The decision procedure, in each period t, is to observe the random vector $\phi_t \equiv (M_t, \psi_t)$ and then to choose a decision vector z_t. The decision problem is

$$E_{\phi_1} \max_{z_1 \in K_1} \left[u_1 W_1 + E_{\phi_2 | \phi_1} \left\{ \max_{z_2 \in K_2} \left\{ u_2 W_2 + \cdots + E_{\phi_T | \phi_{T-1}} \left[\max_{z_T \in K_T} [u_T W_T + aG] \right] \right\} \right\} \right],$$

s.t.

$$\sum_{j=1}^{m_\tau} x_{j\tau} - \sum_{t=1}^{\tau-1} l_{t\tau}(v_t) + v_\tau + \sum_{t=1}^{\tau} c_{t\tau}(w_t) + \sum_{t=1}^{\tau-1} b_{t\tau}(w_t) - w_\tau + W_\tau$$
$$= M_\tau + \sum_{t=1}^{\tau-1} \sum_{j=1}^{m_t} \bar{h}_{jt}^{\tau}(x_{jt}, \psi_\tau), \qquad \tau = 1, \ldots, T,$$

where

$$G \equiv \sum_{\tau=T+1}^{L} d_\tau \left[\sum_{t=1}^{T} \left\{ \sum_{j=1}^{m_t} \bar{h}_{jt}^{\tau}(x_{jt}) + l_{t\tau}(v_t) - b_{t\tau}(w_t) \right\} \right] + \sum_{t=1}^{T} c_{tT}(w_t),$$

E_ϕ represents mathematical expectation with respect to ϕ, bars above functions refer to expected functional values, and $\phi_t | \phi_{t-1}$ means ϕ_t, given $\phi_1, \ldots, \phi_{t-1}$.

(a) Show how the following types of (separable) constraints may be accommodated via the K_t: scarce material, manpower and other resource restrictions, prohibition of multiple projects, all-or-nothing projects, mutually exclusive and contingent projects.

(b) Assume that

(1) the joint distribution of $M_1, \ldots, M_T, \psi_1, \ldots, \psi_L$ does not depend on the decision vectors $z_1, \ldots, z_T$;

(2) the random vectors $\psi_1, \ldots, \psi_L$ have known intertemporally independent distributions that do not depend on $M_1, \ldots, M_T$;

(3) all relevant expectations and maxima exist; and

(4) K_t does not depend on z_τ, if $t \neq \tau$, for all t and τ in $\{1, 2, \ldots, T\}$.

Show that optimal decisions in periods $t = 1, \ldots, T$ may be found by solving the static deterministic programs

$$\max_{z_t \in K_t} \left\{ a \left[\sum_{\tau=T+1}^{L} d_\tau r_{t\tau}(z_t) + c_{tT}(w_t) \right] + u_t\, g_t(z_t) + \sum_{\tau=T+1}^{T} u_\tau [r_{t\tau}(z_t) - c_{t\tau}(w_t)] \right.$$
$$\left. + \sum_{\tau=T+1}^{T} u_\tau [r_{t\tau}(z_t) - c_{t\tau}(w_t)] \right\},$$

where

$$g_t(z_t) \equiv w_t - v_t - \sum_{j=1}^{m_t} x_{jt} - c_{tt}(w_t),$$

$$r_{t\tau}(z_t) \equiv \left\{ \sum_{j=1}^{m_t} \bar{h}_{jt}^{\tau}(x_{jt}) + l_{t\tau}(v_t) - b_{t\tau}(w_t) \right\} \qquad \text{for} \quad t = 1, \ldots, T.$$

[*Note*: Since decisions in each period are made after the observation of random vectors, the results in Exercise 3 must be modified slightly to be used here.]

(c) Discuss the solution properties of the problem in (b), noting in particular when the objective is concave.

A key assumption leading to the result of (b) is that essentially all of the functions considered are period-by-period separable. In many cases, this assumption is not justified. For example, decision choices in the early periods may affect the feasible alternatives in later periods. If this happens, the result in (b) is not applicable. However, it is possible to modify the hypotheses of the model to consider nonseparable constraints. Again, the decision procedure in each period t is to observe the random vector ϕ_t and then to choose a decision vector z_t. The decision problem is

$$E_{\phi_1} \max_{z_1} \left[u_1 W_1 + E_{\phi_2|\phi_1} \left\{ \max_{z_2} \left[u_2 W_2 + \cdots + E_{\phi_T|\phi_{T-1}} \left\{ \max_{z_T} [u_T W_t + aG] \right\} \right] \right\} \right],$$

s.t.

$$\sum_{j=1}^{m_\tau} x_{j\tau} - l_\tau(v_1, \dots, v_{\tau-1}) + v_\tau + c_\tau(w_1, \dots, w_\tau) + b_\tau(w_1, \dots, w_{\tau-1}) - w_\tau + W_\tau$$
$$= M_\tau + s_\tau(x_{11}, \dots, x_{m_{\tau-1}, \tau-1}, \psi_\tau), \qquad \tau = 1, \dots, T. \qquad \text{(i)}$$
$$z \equiv (z_1, \dots, z_T) \in K,$$

where

$$G \equiv \sum_{\tau=T+1}^{L} d_\tau [\bar{s}_\tau(x_{11}, \dots, x_{m_T, T}) + l_\tau(v_1, \dots, v_T) - b_\tau(w_1, \dots, w_T)] + c_T(w_1, \dots, w_T); \qquad \text{(ii)}$$

$s_\tau(x_{11}, \dots, x_{m_t, t}, \psi_\tau)$	Cash return in period τ of previous investments; $x_{11}, \dots, x_{m_t, t}$ if ψ_τ is observed in period τ, $t < \tau$, $t = 1, \dots, T$, $\tau = 2, \dots, L$;
$l_\tau(v_1, \dots, v_t)$	Cash due in period τ if $v_1, \dots, v_t$ is loaned in periods $1, \dots, t$, $t < \tau$, $t = 1, \dots, T$, $\tau = 2, \dots, L$;
$b_\tau(w_1, \dots, w_t)$	Cash owed in period τ if $w_1, \dots, w_t$ is borrowed in periods $1, \dots, t$, $t < \tau$, $t = 1, \dots, T$, $\tau = 2, \dots, L$;
$c_t(w_1, \dots, w_t)$	Minimum cash that must be carried as liquidity in period t if $w_1, \dots, w_t$ are borrowed in periods $\tau = 1, \dots, t$, $t = 1, \dots, T$;
K	Set of available decision possibilities and by definition, $l_0(\cdot) = b_0(\cdot) = c_0(\cdot) = 0$.

(d) Interpret Eqs. (i) and (ii).

(e) Show how the following types of nonseparable constraints may be accommodated in K:

(i) expected group payback restriction,

$$\sum_{t=1}^{\tau-1} \sum_{j=1}^{mt} x_{jt} \leqq \sum_{t=1}^{\tau} \bar{s}_t(x_{11}, \dots, x_{m_{t-1}, t-1})$$

for, say, all $\tau \geqq t'$; and

(ii) net debt repayments in each period are not greater than some fraction of (expected) earnings in that period,

$$b_t(w_1, \dots, w_{t-1}) - w_t \leqq \alpha_{1t} \bar{s}_t(x_{11}, \dots, x_{m_{t-1}, t-1}),$$

where α_{1t} is a given constant between 0 and 1, $t = 2, \dots, T$.

Develop a constraint that states that net debt in each period is not greater than some fraction of the net worth of the firm.

(f) Show that under assumptions (1)–(3) of (b) optimal decisions for periods $t = 1, \dots, T$

may be found by solving the static deterministic program:

$$\max_{z \in K} a\left[\sum_{\tau=T+1}^{L} d_\tau\{\bar{s}_\tau(x_{11}, \ldots, x_{m_T, T}) + j_\tau(z_1, \ldots, z_T)\} + c_T(w_1, \ldots, w_T)\right]$$
$$+ \sum_{t=2}^{T} u_t\{\bar{s}_t(x_{11}, \ldots, x_{m_{t-1}, t-1}) + j_t(z_1, \ldots, z_{t-1}) + h_t(z_1, \ldots, z_t)\}$$
$$+ u_1 h_1(z_1),$$

where

$$h_t(z_1, \ldots, z_t) \equiv w_t - v_t - \sum_{j=1}^{m_t} x_{jt} - c_t(w_1, \ldots, w_t),$$
$$j_\tau(z_1, \ldots, z_t) \equiv l_\tau(v_1, \ldots, v_t) - b_\tau(w_1, \ldots, w_t), \qquad t < \tau.$$

(g) Discuss the solution properties of the problem in (f).

(h) Develop similar models to the two preceding models that apply for situations in which the allocation of resources to projects in each period occurs before the random amount of available resources is known.

5. Consider an investor having initial wealth $w \geqq 0$ and utility function u ($u' > 0$, $u'' < 0$). The investor wishes to determine the optimal amount $x \in [0, w]$ to be invested in a risky asset having rate of return $\tilde{\rho} \geqq 0$, with the remainder $(w-x)$ to be invested in a risk-free asset having rate of return $r \geqq 1$. The investor's "derived utility function" for wealth

$$U(w) = \max_{x \in [0, w]} E[u(w-x)r + x\tilde{\rho}]$$

will generally differ from his original utility function u. It is of interest to examine conditions under which U and u are equivalent, that is, conditions under which $U(w) = bu(w) + c$ (where $b > 0$ and c may, however, depend on the market, i.e., on r and $\tilde{\rho}$). We wish to prove the following.

Theorem If the problem given above has an *interior* maximum, a necessary and sufficient condition for the equivalence of U and u is $-(u'(\xi))/(u''(\xi)) = \alpha\xi + \beta$, with $\beta \neq 0$ iff $r = 1$. Necessity is proved first: The return R on the investment is $R = (w-x)r + \tilde{\rho}x = wr + x\tilde{h}$, where $\tilde{h} = \tilde{\rho} - r$.

(a) Let

$$\tilde{h} = \begin{cases} h > 0 & \text{with probability } p_1, \\ -h & \text{with probability } p_2 = 1 - p_1. \end{cases}$$

Show that the optimal x^* satisfies the equation

$$p_1 u'(wr + hx^*) - p_2 u'(wr - hx^*) = 0.$$

(b) If $\rho \equiv p_1/p_2$ and $k(w)$ is the solution of the equation

$$u'(wr - k) = \rho u'(wr + k), \tag{b1}$$

show that $x^* = k/h$.

By assumption, $U(w) = p_1 u(wr + k) + p_2 u(wr - k) = bu(w) + c$.

(c) Show that

$$u'(wr + k) = \frac{b}{2rp_1} u'(w) \tag{c1}$$

and

$$u'(wr - k) = \frac{b}{2rp_2} u'(w). \tag{c2}$$

(d) By differentiating (c1) and (c2) with respect to p_1, show that

$$1 = \frac{\partial k/\partial p_1}{\partial k/\partial p_1} = \frac{(b'(p_1)/b(p_1))-(1/p_1)}{-b'((p_1)/b(p_1))-(1/p_2)} \cdot \frac{u'(wr+k)/u''(wr+k)}{u'(wr-k)/u''(wr-k)};$$

differentiating (b1) and (b2) with respect to w, show that

$$\frac{u'(wr+k)/u''(wr+k)}{u'(wr-k)/u''(wr-k)} = \frac{1}{p}\frac{r+(\partial k/\partial w)}{r-(\partial k/\partial w)}.$$

Thus, show that $\partial k/\partial w = c(p_1, r)$ (some constant); hence $k = cw+f$.

(e) From (c1), show that

$$u''(wr+cw+f)\,[c'(p_1)\,w+f'(p_1)] = \left(-\frac{b}{2rp_1{}^2} + \frac{b'(p_1)}{2rp_1}\right) u'(w);$$

hence,

$$-\frac{u'(x)}{u''(x)}\frac{1}{c'w+f'} = \frac{1}{(1/p_1)-(b'(p_1)/b(p_1))}, \qquad \text{where} \quad x \equiv wr + wc + f.$$

(f) Show that

$$-\frac{u'(w)}{u''(w)} = p_1 \frac{r+c(p_1)}{1-p_1(b'(p_1)/b(p_1))} \cdot [c'(p_1)\,w+f'(p_1)],$$

so

$$-\frac{u'(w)}{u''(w)} = \alpha w + \beta, \qquad \alpha, \beta \text{ constant.} \tag{f1}$$

(g) Show that the solution to (f1) is of the form $u'(w) = A(w+a)^l$, where $l = 1/\alpha$ and $a = \beta/\alpha$. For f and c, as in (d), show that we must have $a = (a-f)/(r-c) = (a+f)/(r+c)$; hence $r = 1$ or $a = f = 0$.

This proves the necessity part of the theorem. Now prove sufficiency.

Sufficiency: In the general case, $Eu(R) \equiv E[u(wr+x\tilde{h})]$ is maximized for $x = x^*$, satisfying $E[u'(R)\tilde{h}] = 0$.

(h) Assuming $-u'(R^*)/u''(R^*) = \beta+\alpha R^*$, show that $x^* = c(\beta+\alpha rw)$, where c is a constant. [*Hint*: $-u'(R^*) = (\beta+\alpha R^*)\,u''(R^*)$. Multiply both sides by $\tilde{h}$ and take expectations.]

(i) Show that $Eu'(R^*) = bu'(rw)$, where b is a constant. [*Hint*: Show that

$$\beta + \alpha R^* = \frac{1}{r}(\beta+\alpha rw)\frac{dR^*}{dw} = -\frac{1}{r}\frac{dR^*}{dw}\frac{u'(rw)}{u''(rw)}.\,]$$

(j) Show that if $r = 1$ or $\beta = 0$, U is equivalent to u.

(k) Prove the following:

Theorem If the expected utility maximization problem has an interior solution $x^* \in (0, w)$, a necessary and sufficient condition for $U(w)$ to be equivalent to $u(rw)$ is that

$$-\frac{u'(\xi)}{u''(\xi)} = \beta + \alpha\xi.$$

In the context of a multiperiod investment problem, the equivalence of $U(w)$ and $u(rw)$ is known as *partial myopia*, and it states that the "induced utility function" $U(w)$ can be computed by pretending that all of the investor's wealth is invested in a risk-free asset with rate of return r. The equivalence of $U(w)$ and $u(w)$ is known as *complete myopia*, and states that in a multiperiod investment problem, the "induced utility function" remains unchanged. In both cases, the multiperiod investment problem is simplified to a sequence of one-period problems. See, however, the Hakansson paper for difficulties arising from endpoint maxima.

6. Referring to Exercise 5, suppose that the investor's utility for terminal wealth is $u(w) = \log(w+k)$, where $k > 0$ is constant. Assume that $\mu \geqq r(1+(\sigma^2/\mu^2))$, and assume that borrowing and short sales are not allowed.

(a) Show that the optimal policy for a 1-period problem is to invest totally in the risky asset. [*Hint*: If w_0 is initial wealth and $\phi(a)$ is the expected utility for an investment of a in the risky asset, then $\phi'(w_0) \geqq \phi'(w_0+k/r)$.]

We now demonstrate that the optimal policy for an n-period problem is to invest totally in the risky asset in each period. If X is a normal random variate with mean m and variance s^2, let $N(x; m, s^2) = \Pr[X \leqq x]$. Let

$$U_1(w) = \max_{a \in [0,\, w]} Eu[(w-a)r+a\rho].$$

(b) Show that $U_1'(w) = (1/w)F(w/k)$, where

$$F(x) = 1 - \int [1+xe^y]^{-1}\, dN(y; m, s^2)$$

for appropriate m and s^2.

For notational simplicity, let $N(x) = N(x; m, s^2)$ in the following. Let $\Phi_2(a) = EU_1[(w-a)r+a\rho]$, $a \in [0, w]$.

(c) Show that

$$\Phi_2'(w) = \frac{1}{w}\int (1-re^{-x})\, F\left(\frac{w}{k}e^x\right) dN(x).$$

(d) Show that $F((w/k)e^x)$ is positive and strictly increasing in x.

(e) Using (d), show that $\Phi_2'(w) \geqq 0$. This establishes the result for $n = 2$.

(f) Show by induction that the result is true for all n. [*Hint*: If $\Phi_{n+1}(a) = EU_n[(w-a)r+a\rho]$, show that

$$\Phi_{n+1}'(w) = \frac{1}{w}\int (1-re^{-x})\, F_{n+1}\left(\frac{w}{k}e^x\right) dN(x),$$

where

$$F_{n+1}(y) = \int \cdots \int F(y \exp(x_1+\cdots+x_n))\, dN(x_1)\cdots dN(x_n).]$$

7. Suppose that for each period $t = 1, 2, \ldots, T$ an investor can dispose of his wealth in a risk-free asset with net rate of return $r_t \geqq 0$ and n risky assets with net rate of return vector $e_t = (e_t^1, \ldots, e_t^n)'$, where $e_t^i \geqq -1$ $(1 \leqq i \leqq n)$. Assume that the e_t have distribution functions $F_t(e_t)$ and that they are intertemporally independent. Suppose that unlimited borrowing and short sales are allowed. For $t = 1, 2, \ldots, T$, let w_t be the wealth at the *end* of period t, x_t^0 the amount held in the risk-free asset, and x_t the vector of risky asset holdings during period t. Let w_0 be initial wealth. Note that $w_t = (1+r_t)x_t^0 + (I+e_t)'x_t$ and $x_t^0 + I'x_t = w_{t-1}$, where $I = (1, 1, \ldots, 1)'$ is a vector of ones.

(a) Show that $w_t = (1+r_t)+(e_t - r_t I)'x_t$.

Let U be the utility function for terminal wealth w_T. Define the indirect or induced utility functions recursively as

$$V_{t-1}(w_{t-1}) = \max_{x_t} E[V_t(w_t)],$$
$$V_T(w_T) = U(w_T).$$

Assume that U satisfies the equation $-U'(w_T)/U''(w_T) = a + bw_T$ $(b \geqq 0)$.

(b) Show that V_t satisfies $-V_t'(w_t)/V_t''(w_t) = (a/c_t) + bw_t$, where $c_t = \prod_{s=t}^{T}(1+r_s)$. [*Hint*: $V_{T-1}(w_{T-1}) = \max_{x_T} \Phi_T(w_{T-1}; x_T)$, where

$$\Phi_T(w_{T-1}; x_T) = EU[(1+r_T)w_{T-1} + (e_T - r_T I)'x_T].$$

If $b \neq 0, 1$,

$$\Phi_T = k(1+r_T)^{\gamma}((bw_T + a/(1+r_T))^{\gamma} E[1 + b(e_T - r_T I)'z_T]^{\gamma},$$

where $\gamma = 1 - 1/b$ and $z_T = x_T/[a + bw_T(1+r_T)]$. Establish similar results for $b = 0$ and 1. Prove the result for $t = T-1$, and use backward induction.]

(c) Show that the optimal solution x_t^* is of the form $x_t^* = z_t^*[(a/c_t) + b(1+r_t)w_{t-1}]$ where z_t^* depends only on b, r_t, and $F_t(e_t)$.

(d) If $a = 0$, show that x_t^* is independent of future rates of return on the riskless and risky assets.

(e) If $a \neq 0$, show that the result in (d) is true in the limit $T \to \infty$ for any fixed t.

Now assume that the utility function U satisfies the more general equation $-U'(w)/U''(w) = bw + f(w)$, where $|f(w)| \leqq k < \infty$ for $w \in R$. Assume that $c_t = \prod_{s=t}^{T}(1+r_s) \to \infty$ as $(T-t) \to \infty$, and assume that $dw_t^*/dw_{t-1} \geqq 0$ with probability 1 when optimal x_t^* is chosen.

(f) Show that $-V_t'(w_t)/V_t''(w_t) = bw_t + f_t(w_t)$, where

$$|f_t(w_t)| \leqq k/c_t \qquad \text{for} \quad w_t \in R.$$

[*Hint*: Use backward induction: $-V_{t-1}'(w_{t-1})/V_{t-1}''(w_{t-1}) = bw_{t-1} + f_{t-1}(w_{t-1})$, where

$$f_{t-1} = \frac{E_t[(dw_t^*/dw_{t-1})V_t''(w_t^*)f_t(w_t^*)]}{(1+r_t)E_t[(dw_t^*/dw_{t-1})V_t''(w_t^*)]}.\Bigg]$$

(g) Show that $\sup_{w_t \geqq 0} |bw_t + (V_t'(w_t)/V_t''(w_t))| \to 0$ as $T \to \infty$ for any fixed t.

(h) Show that $x_t^*(w_{t-1}) \to z_t^* b(1+r_t)w_{t-1}$ as $T \to \infty$ for any fixed t.

[*Note*: (h) states that the portfolio depends only on the current distribution of assets ("myopia"), and asset holdings are proportional to current wealth, provided that the utility function for terminal wealth has "almost" constant relative risk aversion for large wealth. For difficulties arising from borrowing/short sales constraints, and from intertemporal dependence of the e_t (see the Hakansson paper in Part IV, Chapter 3 and Exercise CR-3).]

Exercise Source Notes

Exercise 1 was adapted from Mirman (1971); Exercise 2 illustrates the Simon (1956)–Theil (1957) theorem; Exercise 3 was adapted from Ziemba (1971); Exercise 4 was adapted from Ziemba (1969); Exercise 5 was adapted from Mossin (1968c); and Exercise 7 was adapted from Leland (1972).

Part V
Dynamic Models

INTRODUCTION

I. Two-Period Consumption Models and Portfolio Revision

The last part of this book is concerned with dynamic financial models. Two-period models are the simplest dynamic models; they are considered in Chapter 1. The Drèze–Modigliani article presents an extension of Irving Fisher's classic theory of savings to the case where there is uncertainty. In such a model the consumer must allocate his wealth between current and future consumption so as to maximize expected utility. Typically the consumer's actual wealth is uncertain because his future labor and other income are not known with certainty. The rate of return on his savings is also generally random. In advanced versions of the model the consumer may engage himself in a selection of an optimal portfolio utilizing the initial period's funds that were not consumed.

The Drèze–Modigliani article focuses on the following three questions:

(1) What are the determinants of risk aversion for future consumption?
(2) What is the impact of uncertainty about future resources on current consumption?
(3) When is it possible to separate consumption decisions and portfolio choices?

They consider these questions based on the extended Fisher model of savings. Such a model is somewhat primitive but does lend itself to the development of a number of key qualitative economic propositions. They utilize standard assumptions in the analysis, such as concavity and smoothness of the cardinal utility function, knowledge of the distribution functions of the random variables involved, and that the marginal propensity to consume is positive but less than unity. Exercise CR-1 concerns itself with some utility functions that satisfy the latter requirement. If all uncertainty is removed before the first period's consumption decision must be made, then the prospects are said to be timeless; otherwise they are temporally uncertain. As is well known from the theory of economic information, temporally uncertain prospects are never preferred to timeless uncertain prospects for general utility functions and probability distributions. Or in other words, the expected value of perfect information concerning the random variables is nonnegative. Exercise ME-4 concerns itself with the development of upper as well as lower bounds on the value of information.

The inferiority of temporal over timeless uncertain prospects has implications for the willingness of consumers to bear future risks. Drèze and Modigliani develop a (local) measure of risk aversion for delayed risk which, like the

Arrow–Pratt measure in the timeless context, represents "twice the risk premium per unit of variance for infinitesimal risks." It develops that the risk premium for a delayed risk equals the expected value of perfect information plus the risk premium for the same risk when timeless. Also not surprisingly, it can be seen that the aversion for delayed risks increases with the curvature of the indifference loci in consumption space. It also develops that there is less inferiority of temporal as opposed to timeless uncertainty in the case of uncertainty regarding the rates of return. Exercise ME-2 is also concerned with the different effects on consumption and savings decisions that occur because of uncertainty concerning either future income or yield on capital investments. Finally Exercise ME-5 is concerned with the effect of "increased" uncertainty with respect to an investor's future income and/or his rate of return on invested capital, on the consumption–savings–portfolio decisions of an expected utility maximizing investor faced with a two-period horizon. Exercise ME-29 considers similar results when the investor has an infinite planning horizon.

In regard to their second question Drèze and Modigliani consider a portfolio selection model in which the investor has access to perfect insurance and asset markets and can allocate his funds between a safe asset and several risky assets. They show that (at optimality): (1) the ratio of the expected marginal utilities of present and future consumption equals the gross rate of return on the safe asset; (2) the expected marginal utility per unit investment is the same for all assets; and (3) the expected marginal utility per unit's worth of insurance on each source of income is the same. Under the perfect market assumptions, they show that the optimal consumption level in period 1 is independent of the insurance and asset policy chosen by the consumer. Hence the portfolio and consumption–savings decisions separate and may be optimally made in a sequential manner.

The situation is, of course, more complicated when markets are not perfect. They develop some results concerned with the relative magnitudes of the optimal consumption level under uncertainty and the optimal consumption level when the uncertain elements are replaced by certainty equivalents. The certainty equivalents provide deterministic problems whose optimal solutions provide the same utility as the maximum expected utility in the uncertain problem. Similar results were also developed in Exercise IV-ME-1. Finally Drèze and Modigliani relate the perfect and imperfect market results. Exercise ME-3 shows that in the extended Fisher model of savings it is not generally true that a risky asset is not an inferior good if the utility function is not additive. Exercise CR-5 is concerned with the class of utility functions defined over present and future consumption that have the property that the preferences for gambles concerned with present (future) consumption are independent of the level of future (present) consumption. In Exercise CR-6 the reader is asked to develop explicit solutions and risk-aversion functions in a two-period consumption–savings model when the utility function has a modified Cobb–Douglas form. Sandmo (1972) surveys much of the research pertaining to two-period models of consumption decisions under uncertainty. Exercise CR-2 describes the formulation of a simple two-period portfolio problem when the random returns have discrete distributions. Exercise ME-1 discusses how one may determine upper and lower bounds on the optimal solution value in a general two-period decision model. The reader is asked in Exercise CR-17 to formulate and solve a three-period portfolio problem when there is one safe and two risky assets.

The Drèze–Modigliani paper and its related exercises provide a number of important qualitative results concerning the consumption, savings, and portfolio decisions of an investor having a two-period horizon. To obtain these results a number of strong assumptions were needed such as the absence of transaction costs, taxes, and liquidity problems, and the existence of interior maxima. Naturally in the real world one is faced with such complexities and it is thus important to include these complexities in dynamic models. Models that include some or all of these features are often termed portfolio revision models. Unfortunately such considerations generally provide optimization problems that are extremely difficult to solve and to analyze qualitatively.

The Bradley–Crane paper provides a simple model of portfolio revision that includes most of the desired features yet is computationally feasible. They are concerned with a bond portfolio; however, their model may be applied to other types of investment as well. In each period the manager of the portfolio has an inventory of bonds and cash on hand. Based on current market conditions and an assessment of future interest rates and cash flows, the manager must decide which bonds to buy, sell, and hold given his available budget. Their model is completely linear and they utilize a clever adaption of the Dantzig–Wolfe decomposition method to provide an efficient solution scheme. The key to the efficiency of this iterative scheme is the fact that the subproblems may be easily solved using a simple recursion. Despite the strong linearity assumptions, the model is quite rich because it considers the effect of taxes, transaction costs, liquidity problems, and risk limitations. The model is designed in a managerial spirit that requires and utilizes a number of significant data and judgmental inputs. Although the objective is the maximization of final wealth, they attempt to capture the essence of a risk-averse nonlinear preference function over final wealth with the introduction of constraints which limit each period's realized net capital loss. Under the assumption that taxes are levied at constant rates, tax considerations are introduced into the model by defining certain coefficients as after tax. Similarly, transaction costs are taken into account by supposing that bonds are purchased at the "asked" price and sold at the "bid" price. For an empirical application of the model, see Bradley and Crane (1973). The reader is asked to clarify some aspects of the model in Exercise CR-7. Exercises CR-8 and 9 illustrate a two-period linear programming under uncertainty approach to the bond portfolio problem. Exercises CR-3 and 4 describe two simple models of portfolio revision for portfolios that consist of a single asset. Taxes, transaction costs, and liquidity considerations are shown to affect the advisability of a switch from one asset to another.

For further work on the portfolio revision problem, the reader may consult Chen *et al.* (1971, 1972), Pogue (1970), Smith (1967, 1971), and Tobin (1965). Unfortunately none of these sources provides a very satisfying model from a theoretical standpoint; rather their thrust is largely concerned with *ad hoc* extensions of static portfolio models. See Zabel (1973) for an interesting extension of the Phelps–Hakansson model discussed in the next section, to include proportional transaction costs.

II. Models of Optimal Capital Accumulation and Portfolio Selection

The seminal work of Phelps (1962) extended the Ramsey (1928) model of lifetime saving to include uncertainty. In Phelps' problem, the choice in each period was between consumption or investment in a single risky asset, with the objective being the maximization of expected utility of lifetime consumption. Phelps assumed the utility function was additive in each period's utility for consumption, and he obtained explicit solutions when each utility function was the same and a member of the isoelastic marginal utility family. The papers reprinted in Chapter 2 generalize and extend the work of Phelps in several directions: Portfolio choice is included in the Samuelson and Hakansson papers, and more general utility functions are treated in the Neave paper. These papers also constitute generalizations of the multiperiod consumption–investment papers of Part IV.

Neave treats the multiperiod consumption–investment problem for a preference structure given by a sum of utility functions of consumption in different periods, including a utility for terminal bequests. Future utilities are discounted by an "impatience" factor, compounded in time. In this paper, the portfolio problem is neglected, with the only decisions at any time being the amount of wealth to be consumed, the nonnegative remainder being invested in a single risky asset. In each period, terminal wealth consists of a certain, fixed income plus gross return on investment, and this terminal wealth becomes initial wealth for the following period, from which consumption and investment decisions are again to be made, and so forth. Neave is concerned with conditions under which risk-aversion properties are preserved in the induced utility functions for wealth. He demonstrates the important result that the property of nonincreasing Arrow–Pratt absolute risk aversion is preserved under rather general conditions. For intertemporally independent risky returns, if the single-period utilities for consumption and the utility for terminal bequests exhibit nonincreasing absolute risk aversion, then all the induced utility functions for wealth also have this property. Mathematically, this result arises from the fact that convex combinations of functions having decreasing absolute risk aversion also have this property (see the Pratt paper, Part II). Thus, the decreasing absolute risk-aversion property is preserved under the expected-value operation, and is further preserved under maximization, as Neave shows. The paper specifically includes the possibility of endpoint as well as interior solutions in the presentation.

Neave also considers the behavior of the Arrow–Pratt relative risk-aversion index of induced utility for wealth. He presents some sufficient conditions under which the property of nondecreasing relative risk aversion is preserved. Since this property is not, in general, preserved under convex combinations, results can be obtained only under special conditions. Unfortunately, these conditions are not easily interpreted, and involve the specific values of the optimal decision variables as well as the properties of the utility functions. In general, however, nondecreasing relative risk aversion obtains for sufficiently large values of wealth. In Exercise ME-6 the reader is asked to attempt to widen Neave's classes of utility functions that generate desirable risk-aversion properties for the induced utility functions. Neave also relates properties of the relative risk-aversion measures to wealth elasticity of demand for risky investment. Arrow (1971) performed a similar study in the context of choice between a risk-free and a risky asset, with no consumption. In the present case, the choice is between consumption and risky investment, with no risk-free asset. Neave shows that if the utilities for consumption and terminal wealth in any period exhibit constant relative risk aversion equal to unity, then the wealth elasticity for risky investment in that period is also equal to unity. He further shows that the wealth elasticity of risky investment exceeds unity if and only if the relative risk-aversion index of expected utility for wealth is less than or equal to that of utility for consumption in the relevant period.

The Samuelson paper studies the optimal consumption–investment problem for an investor whose utility for consumption over time is a discounted sum of single-period utilities, with the latter being constant over time and exhibiting constant relative risk aversion (power-law functions or logarithmic functions). Samuelson assumes that the investor possesses in period t an initial wealth w_t which can be consumed or invested in two assets. One of the assets is safe, with constant known rate of return r per period, while the other asset is risky, with known probability distribution of return z_t in period t. The z_t are assumed to be intertemporally independent and identically distributed. Samuelson thus generalizes Phelps' model to include portfolio choice as well as consumption.

Terminal wealth w_{t+1} in period t consists of gross return on investment, and becomes initial wealth governing the consumption–investment decisions in period $t+1$. The problem is to determine the fraction of wealth to be consumed and the optimal mix between riskless and risky investment in each period (given an initial wealth w_0), so as to maximize expected utility of lifetime consumption. The usual backward induction procedure of dynamic programming yields a sequence of induced single-period utility functions U_t for terminal wealth in period t, with optimal consumption and investment in period t being determined by maximizing expected utility of consumption plus terminal wealth. Assuming interior maxima in each period, a recursive set of first-order conditions is derived for the optimal consumption–investment program.

The explicit form of the optimal solution is derived for the special case of utility functions having constant relative risk aversion. Samuelson shows that the optimal portfolio decision is independent of time, of wealth, and of the consumption decision at each stage. This optimal portfolio is identical with the optimal portfolio resulting from the investment of one dollar in the two assets so as to maximize expected utility of gross return, using the instantaneous utility for consumption as the "utility for wealth" function in the optimization procedure. Furthermore, in each period the optimal amount to consume is a known fraction of initial wealth in the period, with the optimal fraction being dependent only on the subjective discount factor for consumption over time, the return on the risk-free asset, the probability distribution of risky return, and the relative risk-aversion index. For logarithmic utilities, the optimal consumption fraction further simplifies to a function of the subjective discount factor alone, and is independent of the asset returns.

Samuelson utilizes his solutions to analyze optimal consumption and investment behavior as a function of time of life. As has already been mentioned, the optimal mix of riskless versus risky investment is (for constant relative risk aversion) completely independent of time of life. Samuelson relates the optimal consumption over time to the asset returns and the subjective discount factor, and discusses conditions under which the consumer–investor will save during early life and dissave in old age.

Throughout the paper, Samuelson assumes the validity of interior maxima in the optimization problems. The validity of such an assumption in the multiperiod portfolio problem was seriously questioned in the Hakansson paper of Part IV, and in Exercises IV-CR-3, 5, and 6. When consumption decisions are included, as they are in the present case, the question of boundary solutions is presumably even more important than it was in the pure investment case. The reader may wish to ponder whether these considerations can materially affect Samuelson's conclusions. Exercise CR-16 shows that a stationary policy is not optimal in a multiperiod mean–variance model even under strong independence assumptions.

The Hakansson paper studies a generalization of the problem treated by Samuelson. Hakansson's assumptions regarding utility for consumption are identical to Samuelson's, except for an assumed infinite lifetime. The investment possibilities facing Hakansson's investor are, however, more general. The individual is assumed to possess a (possibly negative) initial capital position plus a noncapital steady income stream which is known with certainty. In each period, borrowing and lending at a known, time-independent rate of interest, and investment in a number of risky assets with known joint distribution of returns, are possible. As usual, the risky asset returns are assumed to be independent identically distributed (iid) over time. It is further assumed that a subset of the risky assets may be sold short. In each period, initial wealth to be consumed and invested consists of the certain income plus gross return on the previous period's investment. The objective is maximization of the expected utility of consumption over an infinite lifetime.

Hakansson introduces two important and apparently realistic constraints on the risky returns and the investor's decision variables. The capital market is assumed to impose a so-called no-easy-money restriction on the risky asset returns. This restriction ensures that no mix of risky assets exists which provides with probability 1 a return exceeding the risk-free rate, and that no mix of long and short sales exists which guarantees against loss in excess of the risk-free lending rate. The consumption–investment decision variables are assumed to satisfy the so-called solvency constraint. This requires that the investor always remain solvent with probability 1, i.e., that in any period t, the investor's initial capital position plus the capitalized value of the future income stream be nonnegative with certainty.

Hakansson uses the familiar backward induction procedure of dynamic programming to set up a functional equation for the optimal return function (i.e., the maximum expected utility of present and future consumption, given any value of initial wealth). Since the risky returns are iid and the lifetime infinite, the functional equation obtained is stationary over time: The optimal decision variables in period t depend only on initial wealth (in period t), on the single-period asset returns, and on the relative risk-aversion index of consumption. Hakansson's treatment of the infinite horizon dynamic programming problem is intuitively plausible, but not wholly satisfactory from a rigorous standpoint. (The results are correct, however.) First, he assumes that the optimal return function exists and satisfies the functional equation; in a totally rigorous development, these facts need proof. Next, he verifies the form of the optimal solution by direct substitution, that is, by showing that the assumed form of solution actually solves the functional equation. Here the question of uniqueness arises, and is not treated satisfactorily in the development. One way of treating the problem rigorously would be to deal first with the finite horizon problem (as Samuelson does) and subsequently to go to the infinite horizon limit. At each stage in the finite horizon backward induction procedure, one would have a functional equation similar to Hakansson's (but rigorously justified). As shown in the paper, the decision variable feasibility region defined by the solvency constraint is compact and convex, and (as in Hakansson's proof of his lemma and its corollaries) the functional equation has a finite, unique solution. By considering the limiting behavior of the optimal solution and optimal return function, the given infinite horizon results could be justified. For the case of a power-law utility function unbounded from above, Hakansson needs a restriction on the utility function parameters, the optimal risky asset returns, and the subjective discount factor, in order to obtain a solution. In a limiting argument as outlined above, such a restriction would be required in order to obtain a finite return function in the limit of infinite horizon.

For constant relative risk-aversion utilities, the form of the optimal solution is (i) a fixed fraction of initial wealth plus capitalized future income is consumed in each period; and (ii) a fixed proportion of initial wealth plus capitalized future income is invested in each of the risky assets, with the optimal asset proportions being independent of time, initial wealth, capitalized income stream, and impatience to consume. The remainder of initial wealth is, of course, invested in the risk-free asset. For the special case of logarithmic utilities, the optimal consumption fraction is completely independent of all asset returns, and is a function only of the impatience factor. Hakansson also considers the interesting case of constant absolute risk aversion (exponential utilities), and states without proof the form of the optimal policy under special restrictions. The conditions required for the solution and the form of the optimal policy are more difficult to interpret than the corresponding results for the constant relative risk-aversion case. The reasons for this different behavior in the two classes of utilities may be understood as follows. For a power-law (or logarithmic) utility, expected utility of terminal wealth decomposes into a product (or sum) of utility of initial wealth and expected utility of rate of return on one dollar of investment (since terminal wealth equals initial wealth times rate of return). Such a factorization does not obtain for exponential utilities. See Exercise IV-CR-3 for a clarification of this point in a simple example.

A number of conclusions regarding consumption and investment behavior are derived as corollaries to the optimal solution. As stated previously, the optimal level of consumption is a fixed fraction of initial wealth plus capitalized future income. Furthermore, the optimal amount of consumption increases with increasing impatience to consume, as expected. The optimal solution is such that the individual borrows when poor and generally lends when rich. Additional behavioral implications of the solutions are discussed in the paper.

In the Samuelson–Hakansson additive utility models of this section, the individual's consumption versus saving and investment behavior may vary as a function of time of life, but attitudes toward risk do not. In these models, the same risky portfolios are chosen in youth and in old age. In Exercise ME-21 a multiplicative utility model, based on a paper by Pye (1972), is developed. In this model, risk aversion in portfolio selection increases or decreases with age according as risk tolerance is greater or less than that of the logarithmic utility function. (Note that the logarithmic utility is a member of both the additive and multiplicative families.) In the multiplicative family, risk aversion in portfolio selection depends on impatience to consume as well as on age. The multiplicative family thus allows greater richness in portfolio selection behavior over time. On the other hand, consumption behavior over time is simpler in the multiplicative model: Given the impatience factor and the length of remaining life, the propensity to consume is identical for all members of the multiplicative family, and is independent of present and future investment opportunities. Furthermore, in the infinite horizon limit, portfolio selection behavior changes over time, but consumption behavior does not: A constant fraction of wealth is consumed in every period.

To summarize, the papers of Chapter 2 show that an optimal lifetime consumption–investment program is simple to calculate for additive utilities belonging to the constant relative risk-aversion family. The optimal portfolio remains unchanged throughout life for asset returns which are iid over time. Consumption in any period is always proportional to wealth (including capitalized future income), the proportionality factor being dependent in general on impatience to consume, on the capital market, and possibly also on time of life. The optimal portfolio and the optimal consumption fraction are easily determined by solving a static, one-period, pure portfolio problem using utility for consumption as the appropriate utility for wealth function. Effects due to age-dependent risk aversion can be incorporated through a multiplicative form of utility function. In this case the optimal consumption program may be calculated (in the absence of exogenous income, at least) without solving any "optimization" problem and without reference to the capital market. Optimal portfolio choice depends on time, however, and must be continually recalculated in successive periods. In both the additive and multiplicative cases, a useful form of myopia obtains: In each period, past or future realized random outcomes need not be known in choosing the optimal portfolio, and past outcomes affect present consumption decisions only by fixing the available wealth.

After studying the papers it will likely be apparent to the reader that some of the conclusions would remain unchanged under a slight weakening of hypotheses. In particular, the assumption of identically distributed asset returns over time may be dropped (while retaining independence). For additive utilities having constant relative risk aversion, the optimal amount of consumption will still be proportional to wealth. The relative risk aversion of induced utility for wealth will remain constant over time. The optimal asset proportions at any time t will be independent of wealth, consumption, and past or future asset returns, and will depend only on the nature of the capital market at time t. Calculation of the optimal consumption fraction at time t will, however, generally require the solution of all "future" consumption and portfolio problems, and will thus be much more complex than in the iid case. In a recent paper, Hakansson (1971c) studied a significant generalization of his model. In Hakansson's newer paper, the individual's impatience is variable, his lifetime is stochastic, and the capital assets (both risk free and risky) change in time. The individual consumes, borrows (or lends), invests in risky assets, and purchases life insurance. For additive utilities having constant relative risk aversion, solutions are obtained for the optimal lifetime consumption, investment, and insurance-buying program. For logarithmic utilities, Miller (1974a) analyzed rigorously the optimal consumption strategies over an infinite lifetime with a stochastic (possible nonstationary) income stream and a single, safe asset. Rentz (1971–1973) has analyzed optimal consumption and portfolio policies with life insurance, for stochastic lifetimes and changing family size. Additional extensions are given by Hakansson (1969b), Long (1972), and Neave (1973).

III. Models of Option Strategy

In the first two chapters of Part V, the emphasis is more on questions of "preferences" than on "randomness": Probabilities play a role only as weighting factors of utility through the expected value operation. The papers in Chapters 3 and 4 seem to emphasize questions of a more "probabilistic" nature. In particular, considerations involving stochastic processes and the dynamical behavior of random occurrences over time assume an importance lacking in previous parts of the book. Chapter 3 deals with optimal stock or bond option strategies, and with optimal sequential strategies for divestiture of a risky asset. The first Pye paper and the Taylor paper treat bond and stock option strategies from a "stopping rule" viewpoint. In a stopping rule problem, the decision maker observes (for a cost) a stochastic process and must at any decision point choose either to continue observing for at least another period or to stop observing and take a terminal action. Upon stopping, a terminal reward is received which may include future as well as immediate payoffs. The problem is to determine when to stop so as to optimize some overall average reward function. The dynamic programming approach taken in the first two papers of Chapter 3 is common to most stopping problem situations, and many other similar applications can be treated using essentially the same techniques. The article by Breiman (1964) should be consulted for a clear account of general stopping problems. The Kalymon paper also discusses optimal bond option strategy. In Kalymon's paper, present option strategy is affected by the infinite chain of future option strategies. Because of this long-term view, Kalymon's model does not fall into the stopping rule category in any obvious or fruitful manner. Finally, the second Pye paper treats the problem of optimal selling strategy for an asset having randomly fluctuating prices. A unique feature of this paper is its use of the minimax regret principle in the decision process, as opposed to the expected utility principle which has dominated much of the work in this book.

The first paper by Pye introduces the idea of a call option on a bond issue, and discusses the optimal refunding strategy for callable issues in an environment of both deterministically and randomly fluctuating interest rates. A call option permits the borrower to recall the issue prior to maturity. If the issue is prematurely refunded, the borrower must pay the bond holder a contractually determined premium (call price) for the privilege. Although bonds may in some cases be recalled in order to remove contractual restrictions or to retire debt, the refunding decision is treated primarily as a means to increase profits in a world of fluctuating interest rates. When interest rates are falling, it is profitable to recall bonds and refinance the issue at the higher prevailing prices. The relationship of bond prices to short-term interest rates is reviewed in Exercise CR-10. For deterministic interest rates, Pye shows that the present value of profit from the call option is optimized by calling the issue at that point in time when the instantaneous rate of growth of profit from doing so equals the market rate of interest. For perfect capital markets, Pye shows that a callable issue will always sell for less than an identical noncallable issue, and will also never sell for more than the current call price. In the uncertainty case, Pye assumes that the market rate of interest is governed by a stationary Markov chain, i.e., that the probability distribution of interest rates at time $(t+1)$ depends only on the realized rate of interest at time t. He further assumes that the borrower is an expected present-value maximizer, with discounting taken at the randomly fluctuating market rate of interest. He then sets up a dynamic programming recursion relation for the optimal solution, and derives an explicit form for the optimal policy under the following reasonable assumptions: (1) that higher interest rates at time t lead to higher rates at time $(t+1)$ (higher in the sense of first-degree stochastic dominance), and (2) that the scheduled call prices are a nonincreasing, convex function of time t. In this case, the optimal policy is to call the bond at time t if and only if the instantaneous market rate of interest $\rho(t)$ is less than a critical value $\rho^*(t)$, where $\rho^*(t)$ is nondecreasing in t. The intuitive plausibility of this policy should be evident to the reader. An alternative derivation of the optimal policy is contained in Exercise ME-7, and a linear programming approach is outlined in ME-8. In Exercise CR-11 the reader is asked to calculate the optimal option strategy in a numerical example. Exercise CR-12 analyzes the problem in CR-11 from the points of view presented in Exercises ME-7 and 8. Exercise ME-12 considers the possibility of weakening Pye's Markovian assumption regarding behavior of interest rates.

The Taylor paper discusses the optimal call option strategy for purchases on the stock market. In this context, the purchase of a call option allows the holder to buy a block of shares either at a specified call price, or at the current market price, at any time during a specified interval. For an investor, the option serves to guard against a possible price rise in the near future. For a speculator, the option ensures a profit if the market price rises above the call price. In both cases, the option holder's problem is to determine when, if ever, to exercise the option (that is, whether to buy at the current market price, buy at the option price, or wait another period). Taylor treats these problems using two models of daily stock-price changes: the absolute and the geometric random walk models. In the absolute random walk model, successive daily price differences are a sequence of independent identically distributed (iid) random variables, while in the geometric model, successive daily price ratios are iid. For the investor's problem of minimizing expected purchase price, Taylor shows that the optimal policy in the case of "falling" prices is to do nothing until the last day the call is effective, then purchase at the lower of the call or market prices. For "rising" prices, the optimal policy is shown to be of the form: Purchase on the open market if the current price with n periods to go is less than a critical value a_n; otherwise wait another day before deciding. The critical numbers a_n are less than or equal to the call price, and are nonincreasing in n. For the absolute random walk model with normally distributed daily price changes, the minimum expected price paid for the stock is easily calculated, allowing the decision maker to assess whether the purchase of a call option would be profitable in the expected-value sense. Simplified results are also obtained for this case in the limit of large n. For the speculator, maximization of expected profit leads to optimal policies which are essentially opposite to those of the investor. For "rising" prices the optimal strategy is to do nothing until the last possible day, while for "falling" prices the call should be exercised with n days to go if and only if the market price exceeds a critical value b_n. The numbers b_n are greater than or equal to the call price, and are nondecreasing in n. Here too, for the absolute random walk model, with normally distributed daily price changes, the expected profit from ownership of a call option is easily calculated.

The Kalymon paper contains a very general treatment of the bond refunding problem. Because of its generality, the article is one of the technically most difficult papers in this book. Kalymon assumes that the firm possesses an accurate forecast of future net financing requirements (determined exogenously). These requirements are to be met through issuance of a bond. At any decision point, an outstanding bond issue may be recalled at a refunding cost having a fixed component as well as a component proportional to the size of the issue. At the same time, a new bond may be issued by paying a fixed cost plus a cost proportional to the size of the new issue. Both the size and the term to maturity of the new issue are decision variables. The coupon rate payable on a bond issued at time t is assumed to be a function of the term to maturity and the long-term interest rate prevailing at time t. It is further assumed that the long-term interest rates are given by a (nonstationary) Markov process. The model includes concave penalty costs for failing to meet financing requirements, and a linear bonus for funds in excess of requirements. The various penalties, bonuses, and transaction costs are allowed to be nonstationary, that is, to vary from period to period. A bond refunding policy is sought which minimizes expected discounted costs, where the discount factor is generally nonstationary, and is chosen to reflect the firm's time preferences.

Kalymon shows that the optimal refunding policy has the following form: Keep the issue if the current coupon rate r is less than a certain critical level r^*; otherwise refund. The critical level r^* generally depends on the size of the current bond issue, the current term to maturity, and the current long-term interest rate. For the special case of level debt financing (i.e., constant future new financing requirements), it is optimal to keep the size of all bond issues constant, and r^* thus depends only on the term to maturity and current long-term interest rate. The arguments given in the paper supporting this conclusion seem weak, and the reader is invited in Exercise ME-13 to devise a rigorous proof. Additional simplifications in the optimal policy are proved for infinite length bonds in a nonstationary, additive random walk model of long-term interest rates. Kalymon also discusses the calculation of optimal debt level. Assuming that the term to maturity has been decided, it is shown that the size of the new issue can be restricted to the set consisting of zero and the period financing requirements during the term to maturity. The choice of bond size is thus reduced from a continuous, to a much simpler finite search problem. This reduction in problem size results from the piecewise concavity of the optimal expected cost as a function of bond size. Naturally, one would generally also need to perform a search over the term to maturity in order to arrive at the true optimal policy. A number of simplified tests are also presented which can guide the firm in managing its bond strategy. These tests are necessary but not sufficient conditions for optimality of the refund decision.

In the fourth paper of Chapter 3, Pye deals with optimal strategies for selling an asset. The asset's price over time is assumed to follow an arithmetic random walk, with successive price changes being a sequence of iid random variables. Pye characterizes the decision-maker's objective as being the minimization of maximum possible regret. For a given policy and a given random outcome, regret is defined to be the loss incurred by not following the best policy for the specific outcome. Thus, if the decision maker employed a reliable clairvoyant to inform him of the maximum price which would actually be realized, the optimal policy would be to sell the entire asset at the highest price. Failure to follow this policy would lead to regret, since profits would be lower than necessary. Not having access to a clairvoyant, the decision maker might choose instead to adopt a policy which leads to the smallest regret under worst possible conditions. By following such a policy the decision maker can be sure that even if events turn out badly, failure to adopt the policy would have exposed him to potentially worse outcomes. This is a conservative policy, but one which is clearly appropriate in cases where the decision maker is, for example, accountable for his actions after the fact. It is, in general, a much less conservative policy than always assuming the worst.

One apparently common policy recommendation for asset selling is that of "dollar averaging." In such a policy, the decision maker sells equal quantities of the asset in each period. (In the case of security buying, equal quantities of dollars are "sold" in each period.) Pye shows that dollar averaging is an optimal nonsequential policy when the maximum possible single-period price increases and decreases are equal. In general, nonsequential minimax regret policies are not unique, and dollar averaging is merely one of an infinite number of optimal policies. In Exercise CR-13, another simple optimal nonsequential policy is outlined. Pye also derives the optimal nonsequential selling strategy for a policy of maximizing expected utility of total revenue. For any strictly concave utility function, the optimal policy is to sell all of the asset on the last day. In general, none of the infinitely many nonsequential minimax policies can have this form. Thus, an expected utility of revenue principle cannot lead to dollar averaging.

Pye considers next the question of sequential minimax regret policies. Here the decision at each stage is conditioned on actual, realized past outcomes. Dynamic programming is used to derive the general form of an optimal policy. In the nomenclature of the Ziemba paper of Part I, the state variable at stage t consists of two dimensions: (1) the difference between the maximum price occurring up through time t and the price at time t, and (2) the quantity of asset still held at time t. The optimal return function at stage t is defined to be the regret at time t given that an optimal minimax sequential policy is followed at t and subsequently. A functional equation is derived for the optimal return function using backward induction. Pye shows that the return function is convex and decreasing in each state dimension, and for a fixed state, is decreasing in time. These properties are used to derive the form of the optimal policy, which is to sell at time t any portion of the asset in excess of a critical quantity. This critical value is a decreasing function of the first state dimension (i.e., the difference between maximum realized price up through t and the price at t). A numerical example is given to show that the critical value is not, in general, a monotonic function of time. In behavioral terms, the optimal sequential policy dictates that more of the asset is sold when the price drops than when it rises. A recent paper by Bawa (1973) discusses the optimal minimax regret policy for selling an indivisible asset (i.e., an asset which can only be sold as a whole). This model and its solution are presented in Exercise CR-14.

IV. Cash Balance Management

The applications of stochastic dynamic programming in finance constitute a large and growing literature. There are a number of important papers in this literature which could not be reprinted in this book because of space limitations. In particular, several papers dealing with the important problem of cash balance management have necessarily been excluded. We have attempted instead to cover the essential features of these works in the Exercises. The exercises are limited primarily to mathematical questions and readers who are especially interested in the cash balance problem are urged to consult the original articles.

The problem of cash balance management is similar in many respects to the problem of inventory management for a physical good. The firm retains an inventory of cash to meet its daily requirements for money. The actual level of cash on hand fluctuates due to deposits and withdrawals associated with the firm's operations. The fluctuations in cash balance are, to a certain extent, unpredictable. Thus uncertainty exists regarding future cash requirements. In the literature on cash balance management, it is universally assumed that deposits to and withdrawals from the cash balance are describable by a sequence of iid random variables with known distribution (or else are known with certainty).

Associated with the firm's cash balance are certain costs. First, there is an opportunity cost for holding cash in a nonproductive capacity. Second, there are penalty costs associated with delay in meeting requirements for cash. As in any "inventory" problem, the specification of penalty costs in actual situations might be far from simple, but it is surely true that the costs are real nevertheless, and may have important effects on optimal operating policies. Finally, there are costs associated with the deliberate altering of the cash balance level, such as the costs of "paperwork," and brokerage fees. The objective is assumed to be the minimization of discounted or average cost of the firm's cash balance.

Although the cash balance problem, as outlined above, is formally quite similar to the problem of inventory management, it also differs from the latter in a number of important ways. First, since both withdrawals and deposits occur, the instantaneous "demand" for cash can be negative as well as positive.

Second, operating policies can involve decreases as well as increases of cash inventory (for example, by transferring a quantity of cash to an earning account or productive asset). Because of these differences, the cash balance problem is mathematically more complex than the ordinary inventory problem, even if the cost structures are similar in the two cases.

The similarity between the cash balance problem and the standard inventory problem was originally exploited by Baumol (1952), who applied to cash holdings the classical "lot-size" model of inventory management. This model was extended by Miller and Orr (1966) to include uncertainty. Miller and Orr assumed the following costs: (1) a holding cost proportional to the average cash balance level and (2) a fixed cost for transfers to or from the cash balance. They assumed that changes in the cash level could be adequately described by a Bernoulli random walk, with known probability p of an increase by a fixed amount m, and probability $(1-p)$ of a decrease by m. The firm was assumed to employ the following simple policy for cash balance management. Let the cash balance wander freely within the control limits 0 and $h > 0$, and return it instantly to a fixed, intermediate value z upon reaching either control limit. The objective was to determine the upper control limit h and return point z so as to minimize the long-run average cost of the cash balance.

In the Miller and Orr theory the behavior of the cash balance level may be described as a random walk with reflecting boundaries. A number of useful methods for random walk problems are given by Feller (1962). To make the present treatment as self-contained as possible, several of the necessary mathematical tools have been developed in the exercises. In the cash balance problem the random walk process starts anew after each transfer from the control limits to the return point. The times between successive cash transfer form a sequence of nonnegative iid random variables. Significant economies can often be achieved by studying general properties of systems which "restart" themselves in this manner, and Exercise ME-14 presents a number of simple but useful results pertaining to such renewal processes. For a given (stationary) operating policy, the cash balance levels form a stationary, finite Markov chain. Exercise ME-15 introduces some definitions relating to Markov chains, and proves a number of important facts about recurrent events and the limiting behavior in time of chains having finitely many states. For applications to the cash balance problem, it is necessary to know the asymptotic behavior of average reward functions defined on a Markov chain. This topic is discussed in Exercise ME-16. It is shown that the long-run average cost is given by the expected value of the cost function, taken with respect to the asymptotic probability distribution. This is a typical example of a so-called *ergodic* theorem, and is true under more general circumstances. Although the general treatment of such problems can become highly technical, the results for finite Markov chains are reasonably simple to obtain. For a general introduction to ergodic theory, the book by Breiman (1968) should be consulted. Additional results on first passage times for finite Markov chains are obtained in Exercise ME-17, and applied to a numerical example in Exercise CR-15. The results in Exercises ME-14–ME-17 are applied to the problem of random walk with reflecting boundaries in Exercise ME-18. For additional material on Markov chains, the reader may consult Bharucha-Reid (1960), Breiman (1968), Feller (1962), Hillier and Lieberman (1967), Howard (1960), Kemeny and Snell (1960), Kushner (1971), and Ross (1970). For renewal theory and some of its applications, see Breiman (1968), Cox (1962), Feller (1962, 1966), and Ross (1970).

The basic results of Miller and Orr, as generalized by Weitzman (1968), are given in Exercise ME-19. In their original paper of 1966, Miller and Orr assumed equal costs for transfers to or from the cash balance. Weitzman generalized the results to unequal transfer costs, and studied the sensitivity of the optimal policy as a function of the transfer costs' ratio. As noted in part (e) of Exercise ME-19, the optimal policy is only slightly affected by this ratio, for realistic values. In a recent paper, Miller and Orr (1968) have studied in some detail the adequacy of their original assumptions. They demonstrate that the assumption of a Bernoulli random walk can be generalized without significant effect on the optimal cash balance policy. They show how more general cash balance dynamics can be well approximated by mixtures of Bernoulli random walks, and they present methods for the solution of such problems. They discuss the possibility of more complex cost structures, including both fixed and proportional components for cash transfers. They also treat the interesting case of a "three asset" model, where in addition to idle cash there exist two alternative earning accounts: a high yield, relatively illiquid account, and a lower yield, highly liquid "buffer" account. Unfortunately, space limitations prevent the inclusion of most of these interesting topics. However, the results for fixed plus proportional transfer costs are outlined in Exercise ME-20.

The Miller and Orr theory of cash balance management assumed the firm's operating procedure to be of the simple control-limit type. In many "inventory theoretic" cases, especially those involving complex cost structures (non-convexities, price breaks, etc.), such partial optimization over a set of simple policies is the most that can be achieved, since the actual form of truly optimal policies may be unknown. Even if optimal policies are known, they may be sufficiently complicated from an operating point of view that sub-optimization (over simple policies) is actually preferable in practical cases. This view, based on the paper by Karlin (1958), is the one adopted by Miller and Orr. Other work on the cash balance problem concerns the actual form of optimal policies. Eppen and Fama (1968) treated the cash balance problem using linear programming. They assumed linear holding and penalty costs, and transfer costs having both fixed and proportional components. The possible cash balance states were assumed to be discrete; and withdrawals or deposits were assumed to be a sequence of discrete iid random variables. The objective was minimization of discounted expected cost. For any given policy, changes in the cash balance thus become a Markov chain, and the cash balance problem becomes a standard Markovian decision problem. As noted by Manne (1960), such problems can be reformulated in a linear programming framework. See Exercise I-ME-20 for a linear programming approach to general Markovian decision problems, and Exercise ME-8 for applications to bond-option strategies. A common drawback of the standard linear programming formulation is the large size of the resulting problem. As noted by de Ghellinck and Eppen (1967), significant reductions in problem size obtain for the special but important class of so-called separable Markovian decision problems. In separable problems, the cost of making a decision k when in state i reduces to a sum of a function of k and a function of i; and the transition probabilities given decision k and state i are independent of i. The properties hold for the cash balance problem. First, costs are clearly additive in the states and decisions. Second, the firm decides to start the current period in a particular state, so the transition probabilities are actually independent of the previous period's ending state. Exercise ME-9 presents the de Ghellinck and Eppen formulation for linear programming in discounted Markovian decision problems, and gives another proof of the optimality of stationary policies. Exercise ME-10 outlines their theory of linear programming for separable Markovian decision problems, in the special case of the cash balance problem. It was this formulation of the problem that Eppen and Fama (1968) used in their study. They solved a large number of examples numerically, and found that, in all cases, the optimal policy had the following simple $(u, U; D, d)$ form. Move the cash balance down to D when it exceeds a control limit d, move it up to U when it becomes less than a control limit u, and do nothing otherwise.

In a recent paper, Eppen and Fama (1969) studied the cash balance problem using discounted stochastic dynamic programming. The basic problem formulation was that of their earlier 1968 paper, except that fixed components of transfer costs were assumed to be absent. By starting with a finite horizon problem and then going to the infinite horizon limit, they were able to show that the optimal policy was of the following simple (U, D) form: Move the cash balance down to D whenever it exceeds D, move it up to U when it becomes less than U, and do nothing otherwise. Their proof of this result is presented in Exercise ME-11. The form of the optimal policy with fixed transfer costs included appears to be somewhat uncertain at this time. Neave (1970) has studied this problem, and has found that the optimal policy is either of the simple $(u, U; D, d)$ form, or else is of the form $(u, U, u^+; d^-, D, d)$: Move the cash balance down to D or up to U whenever it is greater than d

or less than u; either do nothing or move the cash balance to U whenever it is between u and u^+; either do nothing or move the cash balance to D whenever it is between d^- and d; and move the cash balance to some (unknown) value between u^+ and d^- whenever it reaches this region. The simple $(u, U; D, d)$ policy is a special case of this more complex policy; whether or not it is optimal in the general case appears to be an unsolved problem. Part (j) of Exercise ME-11 presents a special case for which optimality of the $(u, U; D, d)$ policy is known to be true. Girgis (1969) proved the optimality of such a simple policy when at least one of the fixed transfer costs is zero. Recently, Vial (1972) established the optimality of the simple policy for the general case in a continuous-time model, with changes in the cash balance level being given by a diffusion process.

For additional information on the cash balance problem, the highly readable and informative book by Orr (1970) is recommended. It discusses all of the mathematical cash balance models up to 1970, and, in addition, presents empirical analysis, discussions of institutional constraints, and economic consequences of various cash balance models. Except for Miller and Orr (1968) and one chapter by Orr (1970), most existing cash balance models have treated the problem in a "two asset" setting, where in addition to idle cash, there exists a single alternative earning account. A more realistic model may require the existence of several alternative earning accounts, each having its own "risks" and transfer costs. The ultimate cash balance model would thus combine portfolio choice with inventory theoretic considerations. A tentative step in this direction has been taken by Eppen and Fama (1971), who treat the cash balance problem in a "three asset" setting, using stochastic dynamic programming. For an interesting treatment of sequential policies for bank money management in a minimax regret framework, see Pye (1973). For further results and information on the cash balance problem, the reader may consult Daellenback and Archer (1969), Frost (1970), Heyman (1973), Homonoff and Mullins (1972), Orgler (1969, 1970), Porteus (1972), and Porteus and Neave (1972).

V. The Capital Growth Criterion and Continuous-Time Models

The papers by Breiman and Thorp are concerned with Kelly's capital growth criterion for long-term portfolio growth. The criterion states that in each period one should allocate funds to investments so that the expected logarithm of wealth is maximized. Hence the investor behaves in a myopic fashion using the stationary logarithmic utility function and the current distribution of wealth.

Kelly (1956) supposed that the maximization of the exponential rate of growth of wealth was a very desirable investment criterion. Mathematically the criterion is the limit as time t goes to infinity of the logarithm of period t's wealth relative to initial wealth divided by t. If one considers Bernoulli investments in which a fixed fraction of each period's present wealth is invested in a specific favorable double-or-nothing gamble, then the criterion is easily shown to be equivalent to the maximization of the expected logarithm of wealth. Exercise CR-18 illustrates the calculations involved and related elementary properties of Kelly's criterion.

The desirability of Kelly's criterion was further enhanced when Breiman [Breiman (1961) and his 1960 article reprinted here], showed that the expected log strategy produced a sequence of decisions that had two additional desirable properties. First, if in each period two investors have the same investment opportunities and one uses the Kelly criterion while the other uses an essentially different strategy (i.e. the limiting difference in expected logs is infinite), then in the limit the former investor will have infinite times as much wealth as the latter investor with probability 1. Second, the expected time to reach a preassigned goal is asymptotically (as the goal increases) least with the expected log strategy. Latané (1959) provided an earlier intuitive justification of the first property.

Breiman (1961) established these results for discrete intertemporally independent identically distributed random variables. In this case, as in the Bernoulli case, a stationary policy is optimal. In Exercise ME-22 the reader is asked to develop these and related results for the Bernoulli case. The main technical tool involved in the proof is the Borel strong law of large numbers, which states that with probability 1 the limiting ratio of the number of successes to trials equals the Bernoulli probability of a success. Breiman's paper reprinted here proves the first result for positive bounded random variables. Results from the theory of martingales form the basis of proof of his conclusions. Exercise CR-19 provides the reader with some elementary background on martingales. The reader is asked in Exercise CR-20 to consider some questions concerning Breiman's analysis.

The proof of the results in Breiman's paper involves difficult and advanced mathematical tools. Exercise ME-23 shows how similar results may be proved in a simple fashion using the Chebychev inequality. The crucial assumption is that the variance of the relative one-period gain be finite in every investment period. Breiman's assumption that the random returns are finite and bounded away from zero is sufficient but not necessary for the satisfaction of this assumption. Exercise CR-22 presents an extension of Breiman's model to allow for more general constraint sets and value functions. The development extends his Theorem 1 and indicates that no strategy has higher expected return than the expected log strategy in any period. The model applies to common investment circumstances that include borrowing, transactions and brokerage costs, taxes, possibilities of short sales, and so on. In Exercise

ME-24 the reader is invited to consider the relationship between properties 1 and 2, to attempt to verify the validity of property 2 for Breiman's model reprinted here, and to consider the discrete asset allocation case.

Thorp's paper provides a lucid expository treatment of the Kelly criterion and Breiman's results. He also discusses some relationships between the max expected log approach and Markowitz's mean–variance approach. In addition he points out some of the misconceptions concerning the Kelly criterion, the most notable being the fact that decisions that maximize the expected log of wealth do not necessarily maximize expected utility of terminal wealth for arbitrarily large time horizons. The basic fallacy is that points that maximize expected log of wealth do not generally maximize the expected utility of wealth if an investor has nonlogarithmic utility function; see Exercise ME-25 for one such example and Thorp and Whitley (1973) for a general analysis. See Markowitz (1972) for a refutation, in a limited sense, of the fallacy if the investor's utility function is bounded. For some enlightening discussion of this and other fallacies in dynamic stochastic investment analysis see Merton and Samuelson (1974) and Exercise ME-26. Miller (1974b) shows how one can avoid the fallacy altogether by utilizing what is called the utility of an infinite capital sequence criterion. Under this criterion, the investor's utility is assumed to depend only on the wealth levels in one or more periods infinitely distant from the present; that is, capital is accumulated for its own sake, namely its prestige. In this formulation the time and expectation limit operators are reversed (from the conventional formulation) and hence the improper limit exchange that yields the fallacy does not need to be made. One disadvantage of this formulation is that the admissible utility functions generally are variants of the unconventional form: limit infimum of the utility of period t's wealth. The reader is invited in Exercise CR-21 to consider some questions concerning Thorp's paper. In Exercise ME-27 the reader is asked to determine whether or not good decisions obtained from other utility functions have a property "similar" to the expected log strategy when they produce infinitely more expected utility.

For additional discussion and results concerning the Kelly criterion the reader may consult Aucamp (1971), Breiman (1961), Dubins and Savage (1965), Goldman (1974), Hakansson (1971a, b, d), Hakansson and Miller (1972), Jen (1971, 1972), Latané (1959, 1972), Markowitz (1972), Roll (1972), Samuelson (1971), Samuelson and Merton (1974), Thorp (1969), Young and Trent (1969), and Ziemba (1972b).

For an elementary presentation of the theory of martingales the reader may consult Doob (1971). More advanced material may be found in the work of Breiman (1968), Burrill (1972), and Chow *et al.* (1971).

The highly technical Merton paper discusses the optimal consumption–investment problem in continuous time. Because of its heavy reliance on stochastic differential equations and stochastic optimal control theory, the paper may be quite intimidating upon first reading. To make the paper intelligible to readers unfamiliar with these mathematical concepts, we deviate from our previous policy of keeping formalism out of the introduction: Appendix A contains a brief, intuitive introduction to stochastic differential equations and stochastic control theory. Although not rigorous, the arguments are, hopefully, sufficiently plausible as to enable the reader to understand Merton's paper with relative ease.

Merton assumes the investor's utility for lifetime consumption to be an integral of utilities for instantaneous consumption over time. This is a natural, continuous-time version of the additive utility assumption in discrete time. The general formalism of the optimal control problem is outlined first for utilities that vary arbitrarily in time and is specialized in later sections to the pure "impatience" case, as in the Samuelson and Hakansson discrete time models. Most of Merton's explicit solutions pertain to utility functions of the hyperbolic absolute risk aversion (HARA) class, that is, to the class where the reciprocal of the Arrow–Pratt absolute risk-aversion index is linear in wealth. This class contains as special cases utilities having constant absolute or relative risk aversion. The individual must decide at each point in time how to allocate his existing wealth to between consumption and investment in a number of financial assets. The risky asset returns are governed by a known Markov process. The objective is maximization of expected utility of lifetime consumption, including terminal bequests.

In the spirit of dynamic programming, there exists at each time t a derived utility function for wealth w, namely, the maximum expected utility of future lifetime consumption given that wealth is w at time t. The instantaneous consumption–investment problem requires the maximization of utility of consumption and terminal wealth at time t, or rather, at an "infinitesimal" time later than t. For risky asset returns governed by a stationary log-normal process (see Appendix A), terminal wealth at the end of the "infinitesimal" time span is almost deterministic. The expected utility maximization thus involves only means and variances of risky returns, as in the Samuelson paper of Part III, Chapter 1. The portfolio selection aspect of the continuous-time consumption–investment problem is thus solved exactly using mean–variance analysis. As in the Lintner and Ziemba papers of Part III, this implies that the optimal portfolio possesses the mutual fund separation property: All investors will choose a linear combination of two composite assets which are, moreover, independent of individual preferences. If there exists a risk-free asset, it may be chosen as one of the mutual funds, and all portfolios reduce to a mix of this risk-free asset and a single composite risky asset which is the same for all investors. The returns on the mutual fund are also governed by a stationary log-normal process. Note that such a result is only true in continuous time:

In discrete time, a nonnegative linear combination of log-normally distributed random returns is not log-normally distributed. Through the use of a continuous-time formulation, the consumption–investment problem is thus reduced to a two-asset model with consumption. Using this reduction, Merton obtains explicit solutions for the optimal consumption–investment program for utility functions of the HARA class. As in the papers of Chapter 2, the optimal amount of consumption and the optimal amount of investment are linear functions of wealth. Furthermore, when a steady noncapital income stream is included, this remains true with "wealth" reinterpreted as present wealth plus future income (capitalized at the risk-free rate of return).

Merton discusses the effects on optimal consumption and investment due to the possibility of large but rare random events. Specifically, among such rare but significant events which he discusses are (1) a bond which is otherwise risk free, may suddenly become worthless, or (2) the investor dies. The modeling of such processes in continuous time is achieved through the use of Poisson differential equations (see Appendix A). Merton derives the optimal consumption–investment policy for choice between a log-normally distributed common stock and a "suddenly worthless" bond, for utilities having constant relative risk aversion. As before, the optimal amount of consumption is linear in wealth. The optimal investment in the common stock is an increasing function of the probability of bond default, as expected. Merton also discusses the effect of uncertain lifetime, using a stationary Poisson process to model the arrival of the consumer's death. He shows that the optimal consumption–investment problem in this case (with no bequests) is identical to that of an infinite-lifetime model, with the consumer's utility discounted by a subjective rate of time preference equal to the reciprocal of life expectancy.

There are a number of other topics covered in the paper which we mention only briefly. These pertain to the important and usually neglected question of nonstationary random asset returns. Merton presents solutions for the optimal consumption–investment problem in three such nonstationary models. In the first model, he assumes there exists an asymptotic, deterministic price curve $P(t)$, toward which the stochastic asset prices tend (in expected value) in the distant future. In the second model, he assumes that the risky return is given by a log-normal distribution whose mean is itself a random process of a special form. Finally, in the third model, he assumes that the risky return is given by a stationary log-normal process whose parameters are unknown, but must be estimated from past behavior of the random variable. Explicit solutions for these three models are obtained in a two-asset setting, for utility functions having constant absolute risk aversion. Exercise ME-28 considers a deterministic continuous-time consumption–investment model. If utility is intertemporally additive, then it is possible to develop explicit optimality criteria that yield an algorithm for constructing an optimal policy.

Appendix A.
An Intuitive Outline of Stochastic Differential Equations and Stochastic Optimal Control

R. G. Vickson

UNIVERSITY OF WATERLOO

In many modeling situations, continuous-time formulations may be preferable to discrete time formulations, for precisely the same reasons that derivatives are simpler than finite differences and integrals are simpler than finite sums. However, when the models possess stochastic elements, the interpretation of continuous-time processes becomes somewhat delicate. In this outline we treat the continuous-time cases as limits of discrete time cases, when the length of the discrete intervals tends to zero. Conceptually there is no loss of rigor in this procedure: This is precisely the manner in which many mathematically impeccable presentations *define* the properties of continuous-time processes. [see, e.g., Breiman (1968)]. What is missing in this outline is any form of proof. We remain throughout at the level of plausibility arguments.

I. Itô Processes

A Brownian motion is a stochastic process $\{z(t)\}$ having stationary, independent increments and satisfying a continuity property:

$$\lim_{\delta \searrow 0} (1/\delta) \cdot P[|z(t+\delta)-z(t)| \geqq k] = 0 \qquad \text{for all} \quad k > 0.$$

It can be shown (Breiman, 1968) that for any such process $\{z(t)\}$, $t \geqq 0$, with $z(0) = 0$, $z(t)$ is normally distributed, with $Ez(t) = \mu t$ and $\operatorname{var} z(t) = \sigma^2 t$ for some real μ and σ.

Consider now a stochastic process $\{X(t)\}$ whose dynamics is that of a deterministic, memoryless law (first-order ordinary differential equation), perturbed by a random disturbance at each point in time. In particular, suppose that for sufficiently small $\delta > 0$ we have

$$\delta X(t) = X(t+\delta) - X(t) = f(X(t), t) \cdot \delta + g(X(t), t) \cdot z(\delta) + o(\delta), \qquad \text{(i)}$$

where $\{z(t)\}$ is a Brownian motion having $\mu = 0$, $\sigma^2 = 1$. Recall that $o(\delta)$ stands for terms which tend to zero faster than δ, as $\delta \searrow 0$. Since $Ez(\delta) = 0$ and $Ez^2(\delta) = \delta$, the "size" of the random variable $z(\delta)$ is of the order of $\sqrt{\delta}$. By writing $\delta = dt$, $z(\delta) = dz$, and $\delta X(t) = dX$, Eq. (i) becomes, as $\delta \to 0$,

$$dX = f(X, t)\, dt + g(X, t)\, dz. \qquad \text{(ii)}$$

This must *not* be interpreted as a differential equation in any ordinary sense; in fact, the derivative dz/dt in the ordinary sense exists almost *nowhere*. It is best to think of (ii) as a shorthand notation for (i).

An Itô process is a stochastic process $\{X(t)\}$ which satisfies Eq. (ii).

Consider now a smooth function $F(\cdot, t)$ of the process $\{X(t)\}$. The random variable $Y(t) = F(X(t), t)$ also satisfies a stochastic differential equation. Consider $\delta Y(t) = Y(t+\delta) - Y(t)$. Using Taylor's expansion, we have

$$\delta Y(t) = \frac{\partial F}{\partial X}\delta X(t) + \frac{\partial F}{\partial t}\cdot\delta + \frac{1}{2}\frac{\partial^2 F}{\partial X^2}[\delta X(t)]^2 + o(\delta).$$

Apply Eq. (i) and retain terms of order δ only. This gives $[\delta X(t)]^2 = g^2(X,t)[z(\delta)]^2$, the terms in δ^2 and $\delta \cdot z(\delta)$ having dropped out. Furthermore, to first order in δ, $[z(\delta)]^2$ can be equated to its expected value, since fluctuations around this value are of higher order. Thus, we can write $[z(\delta)]^2 = E[z(\delta)]^2 = \delta$ to first order, so that

$$\delta Y(t) = \frac{\partial F}{\partial X}\delta X + \frac{\partial F}{\partial t}\cdot\delta + \frac{1}{2}\frac{\partial^2 F}{\partial X^2}g^2\cdot\delta + o(\delta). \tag{iii}$$

Formally, we could write this as

$$dY = \frac{\partial F}{\partial X}dX + \frac{\partial F}{\partial t}dt + \frac{1}{2}\frac{\partial^2 F}{\partial X^2}(dX)^2, \tag{iv}$$

with the convention that $dz\,dt = 0$ and $(dz)^2 = dt$.

Although the stochastic process $\{Y(t)\}$ does not have derivatives in the ordinary sense, the conditional expected value $E_t\,\delta Y(t)/\delta = E[\delta Y(t)/\delta \mid X(t)]$ will generally possess a limit as $\delta \searrow 0$. In fact, using (i) and recalling that $Ez(\delta) = 0$, we have from (iii):

$$\frac{E_t\,\delta Y(t)}{\delta} = f(X,t)\frac{\partial F}{\partial X} + \frac{\partial F}{\partial t} + \frac{1}{2}g^2(X,t)\frac{\partial^2 F}{\partial X^2} + o(1). \tag{v}$$

Defining $\dot{Y}(t) = \lim_{\delta \searrow 0} E_t\,\delta Y(t)/\delta$, we find

$$\dot{Y}(t) = \mathscr{L}_{X,t}[Y(X,t)], \tag{vi}$$

where $\mathscr{L}_{X,t}$ is the differential operator

$$\mathscr{L}_{X,t} = f(X,t)\frac{\partial}{\partial X} + \frac{\partial}{\partial t} + \frac{1}{2}g^2(X,t)\frac{\partial^2}{\partial X^2}\,. \tag{vii}$$

For multidimensional processes $\{X(t)\}$, these results generalize in an analogous fashion.

EXAMPLE

Geometric Brownian motion $\{X(t)\}$ is defined as the Itô process which solves

$$dX = aX\,dt + bX\,dz \qquad \text{for constants } a \text{ and } b.$$

To solve for $X(t)$, let $Y = \log X$ and use Eq. (iv). Thus

$$dY = (a - \tfrac{1}{2}b^2)\,dt + b\,dz.$$

We have $\log X(t)/X(0) = Y(t) - Y(0) = (a - \frac{1}{2}b^2)\,t + bz(t)$; thus $\log X(t)/X(0)$ is normally distributed, with mean $(a - \frac{1}{2}b^2)\,t$ and variance bt. We may therefore also refer to $\{X(t)\}$ as a (stationary) log-normal process.

II. Poisson Differential Equations

Consider a stochastic process $\{q(t)\}$, $t \geqq 0$, such that the value of $q(t)$ jumps by an amount y (a random variable) at times T which are arrival times of a Poisson process. Thus $\lambda\delta + o(\delta)$ is the probability of a jump occurring during $[t, t+\delta)$; and given that a jump occurs, it has amplitude y. The random variables y and T are assumed to be independent and stationary. Let $\delta q(t) = q(t+\delta) - q(t)$ be the jump in q during $[t, t+\delta)$. Then

$$\delta q(t) = \begin{cases} 0 & \text{w.p.} \quad 1 - \lambda\delta \\ y & \text{w.p.} \quad \lambda\delta \end{cases}$$

to lowest order in δ.

A more general model for a stochastic process $\{X(t)\}$ having possibly nonstationary random jumps at random times is

$$\delta X(t) = X(t+\delta) - X(t) = f(X,t)\cdot\delta + g(X,t)\cdot\delta q(t) + o(\delta), \qquad \text{(viii)}$$

which can be rewritten formally as

$$dX = f(X,t)\,dt + g(X,t)\,dq. \qquad \text{(ix)}$$

Let $Y(t) = F(X(t),t)$, where F is a smooth function. Define

$$\dot{Y}(t) = \lim_{\delta \searrow 0} E_t\,\delta Y(t)/\delta.$$

As in the case of Itô processes, we can obtain a compact expression for $\dot{Y}(t)$ which involves only "ordinary" mathematical quantities. We have

$$E_t\,\delta Y(t)/\delta = (1 - \delta\lambda)\,[F(X+f\delta, t+\delta) - F(X,t)]/\delta$$
$$+ (\delta\lambda/\delta)\cdot E_y[F(X+f\delta+gy, t+\delta) - F(X,t)].$$

This gives

$$\dot{Y}(t) = f \cdot \frac{\partial F}{\partial X} + \frac{\partial F}{\partial t} + \lambda E_y[F(X+gy, t) - F(X, t)]. \tag{x}$$

III. Stochastic Optimal Control

A typical stochastic optimal control problem for an Itô process is

$$\max_v E \int_0^T f(X(t), v, t)\, dt, \tag{xi}$$

with

$$dX = f(X, v, t)\, dt + g(X, v, t)\, dz, \tag{xii}$$

where v is a "control" function, constrained to be in some infinite-dimensional set V. To obtain optimality conditions for this problem, it is convenient to replace it by an approximating discrete time problem:

$$\max_{\{v_i\}} E \sum_{i=0}^{N} \delta \cdot f(X_i, v_i, t_i) \tag{xiii}$$

s.t.

$$\delta X_i = f(X_i, v_i, t_i) \cdot \delta + g(X_i, v_i, t_i) \cdot z_i(\delta), \tag{xiv}$$

and $v \in V$. Using the principle of optimality, the optimal control v_i at stage i, given $X_i = x_i$, is the solution to

$$J_i(x_i, t_i) = \max_{v_i} [\delta \cdot f(x_i, v_i, t_i) + E_i J_{i+1}(x_i + \delta X_i, t_i + \delta)], \tag{xv}$$

subject to (xiv) above and the constraints on v_i. To lowest order in δ, Eq. (xv) can be written as

$$J_i(x_i, t_i) = \max_{v_i} \{\delta \cdot [f(x_i, v_i, t) + \dot{J}_i(x_i, t_i)] + J_i(x_i, t_i)\}, \tag{xvi}$$

where

$$\dot{J}_i(x, t) = \lim_{\delta \searrow 0} E_i[J_{i+1}(x_i + \delta X_i, t_i + \delta) - J_i(x_i, t_i)]/\delta.$$

Defining

$$\phi_i(x, v, t) = f(x, v, t) + \dot{J}_i(x, t),$$

we have from (xvi):

$$\phi_i(x_i, v_i, t_i) \leqq 0 \qquad \text{for all} \quad v_i,$$

with equality holding for an optimal v_i. In the limit of continuous time this condition becomes

$$0 = \max_v \{f(x, v, t) + \mathscr{L}_{x,v}[J(x, t)]\}, \qquad \text{(xvii)}$$

where $\mathscr{L}_{x,v}$ is given by an equation like (vii).

Note that the stochastic aspects have dropped out of the problem in this formulation. Rather than being a random variable X governed by a stochastic differential equation, the state variable x in Eq. (xvii) is now simply a real parameter in the control problem. Of course, the dynamics of the original Itô process $\{X(t)\}$ are reflected in the differential operator $\mathscr{L}$. The optimality condition (xvii) is similar to the Hamilton–Jacobi equation of deterministic control theory.

There are a number of references which interested readers may find helpful. For continuous-time stochastic processes, see Breiman (1968), the classic work of Doob (1953), Itô and McKean (1964), Jazwinski (1970), Kushner (1967, 1971), Loève (1963), and Wong (1971). Except for Breiman and Loève, these references all contain extensive material on stochastic differential equations, with Jazwinski (1970) and Kushner (1971) being the most lucid. For an elementary treatment of deterministic optimal control, see Bellman (1957, 1961), Kopp (1962), Mangasarian (1966), and Nemhauser (1966). For a rigorous discussion of the extremely important Pontryagin maximum principle, see the classic work by Pontryagin *et al.* (1962), or the equally rigorous but more lucid new treatment by Boltyanskii (1971). For a discussion of stochastic optimal control, see Kushner (1967) or the somewhat more lucid Kushner (1971). In Kushner's treatment, as well as that of the Merton paper and this Appendix, the class of admissible control functions is limited to "smooth" cases (differentiable functions, or continuous functions satisfying a Lipschitz condition). For optimal control problems, it is undesirable to be limited in this way, since "bang-bang" controls, which are known to be globally optimal in numerous deterministic problems [e.g., Boltyanskii (1971)], are artificially eliminated. The theory of stochastic optimal control with a broader class of admissible controls is quite technical. See Davis and Varaiya (1973) for a general treatment and references to previous work. The paper of Vial (1972) previously alluded to describes, heuristically, the application of discontinuous controls to the optimal cash balance problem.

1. TWO-PERIOD CONSUMPTION MODELS AND PORTFOLIO REVISION

JOURNAL OF ECONOMIC THEORY 5, 308–335 (1972)

Consumption Decisions under Uncertainty*

JACQUES H. DRÈZE

Center for Operations Research and Econometrics, Université Catholique de Louvain, Louvain, Belgium

AND

FRANCO MODIGLIANI

Massachussets Institute of Technology, Cambridge, Massachusetts 02139

Received May 15, 1970

INTRODUCTION

This paper deals with three issues related to consumption decisions under uncertainty, namely, (i) the determinants of risk aversion for future consumption; (ii) the impact of uncertainty about future resources on current consumption and (iii) the separability of consumption decisions and portfolio choices. These issues are discussed in the context of a simple model introduced, together with our assumptions, in Section 1. The first issue is motivated and treated in Section 2, the conclusions of which are summarized in Proposition 2.5. The other two issues are treated in Section 3 under the assumption that there exist perfect markets for risks, and in Section 4 under the converse assumption.

Some technical results needed in the text are collected in Appendices A, B and C; a simple graphical illustration of our major result, Theorem 3.3, is given in Appendix D.[1]

* The research underlying this paper was initiated while the authors were both affiliated with the Graduate School of Industrial Administration, Carnegie-Mellon University; the support of that institution, and at a later stage of the Sloan School of Management, Massachussets Institute of Technology, is gratefully acknowledged. The authors also wish to thank Albert Ando and Ralph Beals for their helpful assistance at an early stage of this work, as well as Louis Gevers, Agnar Sandmo and Joseph Stiglitz for their critical reading of the final manuscript.

[1] An earlier summary version of this paper, written in French, has appeared in the *Cahiers du Séminaire d'Econométrie* [5].

1. The Model and the Assumptions

1.1. Following Fisher [6], we study the problem faced by a consumer who must allocate his total wealth y between a flow of current (or "initial") consumption c_1 and a residual stock $(y - c_1)$ out of which future consumption c_2 (including bequests) will be financed. We restrict our attention to the aggregate values of present and future consumption, or equivalently to a single-commodity, two-period world.

We conceive of the consumer's wealth y as being the sum of two terms:

1. The (net) market value of his assets, plus his labor income during the initial period, to be denoted altogether by y_1;

2. The present value of his future labor income, plus additional receipts from sources other than his current assets.

Denote by y_2 the value of the second term, discounted back to the *end* of the initial period; and by r the real rate of interest prevailing over that initial period[2]; y and c_2 are then defined by

$$y = y_1 + y_2(1 + r)^{-1}; \quad c_2 = (y - c_1)(1 + r) = (y_1 - c_1)(1 + r) + y_2 . \tag{1.1}$$

Usually, when a decision about current consumption is made, y_1 may be taken as known with certainty, but y_2 and r may not: future labor income and real rates of return on assets are, in most cases, imperfectly known *ex ante*. In our simple two-period model, we conceive of the uncertainty about y_2 and r as being removed only at the end of the initial period—hence, *after* c_1 has been chosen. Accordingly, we refer to uncertain prospects for y_2 and/or *are* as *temporal uncertain prospects* (time will elapse before the uncertainty is removed), and we refer to this type of uncertainty as being "temporal" or "delayed." By contrast, if the uncertainty is to be entirely removed before the choice of c_1, we speak of *timeless uncertain prospects.*

1.2. Relying upon the theory of decision under uncertainty, as developed by von Neumann and Morgenstern [16], Savage [15], etc..., we start from assumptions about probability and utility, instead of the more natural axioms about choice.

For analytical convenience, their results are strengthened into

Assumption I. Every uncertain prospect is described by a (subjective) mass or density function $\phi(y_2, r)$ with finite moments of at least first and

[2] In a single-commodity world, real rates are well defined; a multiplicity of assets, with different rates of return, is introduced in Section 3.

second order. The distribution function corresponding to $\phi(y_2, r)$ will be denoted by $\Phi(y_2, r)$.[3]

ASSUMPTION II. There exists a cardinal utility function $U(c_1, c_2)$, real valued, continuous and continuously differentiable at least three times.[4]

In addition, we introduce two assumptions that go beyond consistency requirements but reflect behavior patterns that we regard as generally encountered in reality.

In the first place, we assume that neither present nor future consumption is an inferior commodity, so that both c_1 and c_2 increase when y (or y_1 with y_2 and r constant) increases. An alternative statement is that the "marginal propensity to consume," as defined in appendix formula (A.4), is everywhere positive but less than one, i.e.,

ASSUMPTION III. $1 > dc_1/dy_1 > 0$.

In the second place, we assume that the consumer's preferences among consumption vectors are convex, and that his choices among uncertain prospects reflect risk aversion, or possibly risk neutrality; that is,[5]

ASSUMPTION IV. U is concave.

Various properties of U, derived from assumptions II, III and quasi-concavity of U, are collected in Appendix A.

1.3. The (cardinal) indirect utility function corresponding to $U(c_1, c_2)$ may be written $V(y, r)$, where

$$V(y, r) =_{\text{def}} \max_{c_1} U(c_1, (y - c_1)(1 + r)).$$

Let $r = r^0$ be the sure and only rate of interest at which a consumer may lend and borrow; then $V(y, r^0)$, a function of y alone, is the *cardinal utility function for wealth* relevant to the analysis of choices among *timeless* uncertain prospects. In other words, if a cardinal utility function for wealth were derived from observations about choices between *timeless* uncertain

[3] For notational convenience, we use the integral symbol without introducing parallel statements in the notation of discrete random variables; the standard symbol E is used for the expectation operator when there is no ambiguity about the underlying mass or density function.

[4] Thus, U is defined up to a linear increasing transformation; if the consumer were to choose between the certainty of consuming (c_1, c_2) and the prospect of consuming either (c_1', c_2') or (c_1'', c_2'') with respective probabilities π and $1 - \pi$, he would never prefer the former alternative if $U(c_1, c_2) \leqslant \pi U(c_1', c_2') + (1 - \pi)U(c_1'', c_2'')$.

[5] As argued elsewhere by one of us [4], risk preference may be excluded without loss of generality, if one assumes the availability on the market of fair gambling opportunities.

prospects, when the market rate of interest for safe loans is r^0 and y_2 is known, then such a function would coïncide with $V(y, r^0)$ up to an increasing linear transformation.

In the language of demand theory, $V(y, r^0)$ measures utility cardinally for movements along the Engel curve corresponding to r^0, by assigning utility levels to the successive indifference curves crossed by that Engel curve. Provided dc_1/dy is continuous (as implied by assumption II) and satisfies assumption III, the Engel curve will have a point in common with every indifference curve and the assignment of utility levels to these curves will be exhaustive. One may then *construct* the cardinal utility function $U(c_1, c_2)$ by relying simultaneously on two independent and familiar tools, namely,

1. Indifference curves, as revealed by choices among sure vectors of present and future consumption;

2. A cardinal utility function for wealth, as revealed by choices among timeless uncertain prospects.

2. Temporal Prospects, the Value of Information and Risk Preference

2.1. If a consumer owns a temporal uncertain prospect $\phi(y_2, r)$ that he cannot or does not wish to exchange for some other prospect, his expected utility is given by

$$\max_{c_1} \int U\{c_1, (y_1 - c_1)(1 + r) + y_2\}\, d\Phi(y_2, r). \tag{2.1}$$

The solution to this maximization problem determines the optimal current consumption $\hat{c}_1$. Future consumption is a random variable defined by (1.1), with $c_1 = \hat{c}_1$. We shall assume that c_2 so defined is nonnegative, identically in y_2 and r, so that the density of c_2 is defined by

$$\psi(c_2) = \int \phi\{c_2 - (y_1 - \hat{c}_1)(1 + r), r\}\, dr \tag{2.2}$$

with first and second moments

$$\bar{c}_2 = (y_1 - \hat{c}_1)(1 + \bar{r}) + \bar{y}_2,$$

$$\sigma_{c_2}^2 = (y_1 - \hat{c}_1)^2\, \sigma_r^{\,2} + \sigma_{y_2}^2 + 2(y_1 - \hat{c}_1)\, \sigma_{ry_2}. \tag{2.3}$$

Had the *same* uncertain prospect been timeless, so that the value of y_2 and r were known to our consumer before his choice of c_1, then his expected utility would have been

$$\int \max_{c_1} U\{c_1, (y_1 - c_1)(1 + r) + y_2\} \, d\Phi(y_2, r)$$
$$= \int V\{y_1 + y_2(1 + r)^{-1}, r\} \, d\Phi(y_2, r). \tag{2.4}$$

It is immediately verified, by application of a well-known theorem,[6] that

$$\max_{c_1} \int U\{c_1, (y_1 - c_1)(1 + r) + y_2\} \, d\Phi(y_2, r)$$
$$\leqslant \int \max_{c_1} U\{c_1, (y_1 - c_1)(1 + r) + y_2\} \, d\Phi(y_2, r). \tag{2.5}$$

The difference between the right and left hand sides of (2.5) is "the expected value of perfect information" (EVPI), well known to the statisticians.[7]

The meaning of (2.5) may be conveyed somewhat informally, as follows:

PROPOSITION 2.1. *A temporal uncertain prospect is never preferred to the timeless uncertain prospect described by the same mass or density function, no matter what the consumer's utility function may be.*[8]

2.2. The general inferiority of temporal over timeless uncertain prospects has implications for the willingness to bear risk in a temporal context. One convenient way of capturing these implications rests upon the "risk aversion function" introduced by Pratt [12] for timeless uncertainty about total resources (wealth).

Let $r = r^0$ be given; Pratt's (absolute) risk aversion function is then given by $(-V_{yy}/V_y)_{r^0}$. This quantity, which is equal to "twice the risk

[6] See, e.g., Marschak [9, p. 201]. The theorem may be stated as follows: "Let g be a function of the decision variable d and of the random variable x with density $f(x)$; then: $\int_x \max_d g(d, x)f(x)\, dx \geqslant \max_d \int_x g(d, x)f(x)\, dx$."

[7] In order to get a measure that does not depend upon the choice of units for the utility function, one should divide both sides of (2.5) by some appropriate index of marginal utility — like $U_1 = V_y$, or U_2 — so as to measure the EVPI in the same units as consumption, either current or future.

[8] The mass or density functions must, of course, be kept identical, not only "theoretically" but also "practically," if spurious contradictions are to be avoided; thus, a consumer with strong risk aversion may prefer a temporal prospect that is marketable to a similar timeless one that is not; the appropriate density for the temporal prospect is then given by the certainty of its market value and our proposition is not applicable.

premium per unit of variance for infinitesimal risks" when the consumer's wealth is y and $r = r^0$, is a local measure of risk preference. It is, however, related to risk aversion in the large: if one consumer has a greater local risk aversion than another at all wealth levels y, then (and only then) he has greater risk aversion in the large—in the sense that he would exchange *any* timeless uncertain prospect $\chi(y)$ against the certainty of an amount which would be unacceptable to the other consumer.[9]

We shall now derive a (local) measure of risk aversion for *delayed* risks which, like the Pratt measure in the timeless context, represents "twice the risk premium per unit of variance for infinitesimal risks."

We begin with a given r (say $r \equiv r^0$) and income prospects $\phi(y_2)$. If such prospects are *timeless*, the random outcome will be known at time 1 but *paid* at time 2: the choice of c_1 still occurs under certainty. It is readily verified that the risk aversion function relevant for such prospects is

$$\frac{-1}{1+r^0}\left(\frac{V_{yy}}{V_y}\right)_{r^0} = \left(\frac{-V_{y_2y_2}}{V_{y_2}}\right)_{r^0}.$$

When such prospects become temporal, on the other hand, the appropriate risk aversion function is $(-U_{22}/U_2)_{\hat{c}_1}$. This can be verified as follows. Let $\hat{c}_1$ be the first period consumption that is optimal for a given temporal uncertain prospect $\phi(y_2, r)$, and let $\psi(c_2)$ be defined as in (2.2). The expected utility of the prospect is then $\int U(\hat{c}_1, c_2)\, d\Psi(c_2)$.

Clearly, if U_{22} does not change sign over the range of $\psi(c_2)$, then

$$\int U(\hat{c}_1, c_2)\, d\Psi(c_2) \gtreqless U(\hat{c}_1, \bar{c}_2) \text{ according as } U_{22} \gtreqless 0. \quad (2.6)$$

Confining attention to infinitesimal risks, we have

$$\int U(\hat{c}_1, c_2)\, d\Psi(c_2) \simeq U(\hat{c}_1, \bar{c}_2) + (\sigma_{c_2}^2/2)\, U_{22}(\hat{c}_1, \bar{c}_2) =_{\text{def}} U(\hat{c}_1, c_2'), \quad (2.7)$$

thereby defining c_2' implicitly.

Furthermore, $U(\hat{c}_1, c_2') \simeq U(\hat{c}_1, \bar{c}_2) + (c_2' - \bar{c}_2)\, U_2(\hat{c}_1, \bar{c}_2)$, so that

$$c_2' \simeq \bar{c}_2 + \frac{\sigma_{c_2}^2}{2}\frac{U_{22}(\hat{c}_1, \bar{c}_2)}{U_2(\hat{c}_1, \bar{c}_2)}, \quad \text{implying} \left(-\frac{U_{22}}{U_2}\right)_{\hat{c}_1,\bar{c}_2} \simeq \frac{2(\bar{c}_2 - c_2')}{\sigma_{c_2}^2}. \quad (2.8)$$

Thus, $(-U_{22}/U_2)_{\hat{c}_1,\bar{c}_2}$ is equal to "twice the risk premium per unit of variance y_2 for infinitesimal *delayed* risks." For given $\hat{c}_1$, the function

[9] See Pratt [12, p. 122] and Sections 3–5; Arrow has independently introduced the same concept in [2].

$(-U_{22}/U_2)_{\hat{c}_1}$ is a local measure of risk aversion at all levels of c_2 in the same sense as $-V_{y_2y_2}/V_{y_2}$ provides such a measure at all levels of y_2 for timeless uncertain prospects. One must, however, be careful to realize that the value of $(-U_{22}/U_2)_{\hat{c}_1}$ is in general not independent of $\hat{c}_1$[10]: It measures risk aversion along a particular cut of the utility function orthogonal to the c_1 axis, but the measure may not be the same, at a given level of c_2, for different choices of c_1 (different cuts of the utility function by parallel planes).

2.3. It follows from Appendix formula (A.15) that, *at any point in* (c_1, c_2) *space*,

$$\frac{-U_{22}}{U_2} = \frac{-V_{y_2y_2}}{V_{y_2}} + \left(\frac{dc_1}{dy_2}\right)^2 \left(\frac{d^2c_2}{dc_1^2}\bigg|_U\right) \geqslant -\frac{V_{y_2y_2}}{V_{y_2}}, \tag{2.9}$$

where $V_{y_2y_2}/V_{y_2}$ and dc_1/dy_2 are computed along the Engel curve going through *that point*, and where the inequality follows from assumptions III and IV.

The risk premium for a delayed risk must be equal to the sum of (i) the expected value of perfect information, and (ii) the risk premium for the same risk when timeless.[11] Thus the second term on the right hand side of (2.9) measures "twice the expected value of perfect information per unit of variance y_2 for infinitesimal risks." That second term is invariant under monotonic transformations of the utility function;[12] it is the product of two factors, of which the second one is most easily interpreted. $d^2c_2/dc_1^2\,|_U$ is a (local) measure of curvature of the indifference loci. As shown in the appendix formulas (A.5, A.6), it also measures the reciprocal of the substitution effect on c_1, of a rise in r.[13] That curvature of the indifference loci should be relevant to assess the superiority of timeless

[10] A necessary and sufficient condition for $-U_{22}/U_2$ to be everywhere independent of c_1 is that $U = f(c_1) + g(c_1)\,h(c_2)$.

[11] Indeed, the total premium paid to dispose of a given delayed risk should be the same, whether the uncertain prospect be exchanged outright for a sure amount, or whether it be exchanged first (at some premium) for an identical but timeless prospect, to be converted next into a sure amount.

[12] It is thus observed that, for *infinitesimal* risks, the expected value of perfect information depends only upon ordinal properties of U; of course, this strong and somewhat surprising result does not hold more generally.

[13] This effect is usually referred to as the "substitution term of the Slutsky equation;" it measures the response of c_1 to a *compensated* change in the rate of interest; the Slutsky equation, however, is typically expressed in terms of the "price" $(1 + r)^{-1}$ rather than in terms of the interest rate r.

over temporal uncertain prospects is readily seen if one contrasts extreme situations. At one extreme, suppose that the indifference curves are nearly linear in the vicinity of the equilibrium point: the consumer is almost indifferent about the allocation of his total resources between c_1 and c_2, which are almost perfect substitutes, the curvature is close to nil, and the response of c_1 to a compensated change in the rate of interest would be very large. Obviously, for such a consumer, delayed uncertainty is not appreciably different from timeless uncertainty, since the opportunity to gear c_1 exactly to total resources matters little to him. At the other extreme, suppose that the indifference curves are very close to right angles in the vicinity of the equilibrium point: the consumer has very exacting preferences for the allocation of his total resources between c_1 and c_2, which are strongly complementary, the curvature is very pronounced, and the response of c_1 to a compensated change in the rate of interest would be negligible. For such a consumer, delayed uncertainty is very costly, due to the imperfect allocation which it entails: the utility of a consumption plan with given present value $c_1 + c_2(1 + r)^{-1}$ decreases rapidly when the allocation departs from the preferred proportions. Thus, as formula (2.9) shows, *the aversion for delayed risks grows as curvature of the indifference loci increases*, or, to use more operational terms, consumers who would respond strongly to a (compensated) change in the rate of interest are *relatively* better suited to carry delayed risks.

The role of the other factor, the marginal propensity to consume, is again most easily understood by looking at limiting situations. If $dc_1/dy = 0$, then the optimum c_1 can be chosen without exact knowledge of total resources, so that perfect information is worthless. At the other extreme, a person who wants to consume all his resources now because he derives no satisfaction from later consumption is ill-suited to bear delayed risks: since he can only afford to consume now the resources he is sure to own, the uncertain prospect carries no more utility for him than the certainty of its worst outcome. In general, *the inferiority of temporal over timeless uncertain prospects will be the more severe, the larger the marginal propensity to consume (other things being equal).*

2.4. We now turn briefly to the case where y_2 is given (say $y_2 = \bar{y}_2$) and r is a random variable with density $\phi(r)$. Our problem is to compare timeless with temporal uncertain prospects about r. One can readily verify that the Pratt "risk-aversion function" for timeless gambles about r is $-(y_1 - c_1)(V_{rr}/V_r)_{\bar{y}}$.[14] Similarly, when $\sigma^2_{c_2} = (y_1 - \hat{c}_1)^2 \sigma_r^2$, we see from (2.7) that $-(y_1 - \hat{c}_1)^2(U_{22}/U_2)_{\hat{c}_1}$ is the appropriate corresponding meas-

[14] Note from (A.11) that $V_r = V_{y_2}(y_1 - c_1)$ has the same sign as $(y_1 - c_1)$: an increase in r affects utility positively for a lender, negatively for a borrower.

ure for temporal gambles. Furthermore, it follows from formula (A.15) that, *at any point in* (c_1, c_2) *space,*

$$-(y_1 - c_1)^2(U_{22}/U_2) = -(y_1 - c_1)(V_{rr}/V_r) + (dc_1/dr)^2(d^2c_2/dc_1^2 \,|_U)$$
$$\geqslant -(y_1 - c_1)(V_{rr}/V_r), \qquad (2.10)$$

where V_{rr}/V_r and dc_1/dr are computed along the offer-curve going through that point. Formula (2.10) admits of the same interpretation as (2.9), so that the second term on the right hand side measures "twice the expected value of perfect information per unit variance r for infinitesimal risks"—a nonnegative quantity that is again invariant under *monotonic* transformations of the utility function. We notice that this quantity vanishes when $dc_1/dr = 0$: if current consumption is insensitive to r, it is also insensitive to σ_r^2 (at least locally), and uncertainty about r is of no concern in choosing c_1. For people with positive asset holdings $(y_1 - c_1 > 0)$, dc_1/dr is unrestricted as to sign on *a priori* grounds; the absence of empirical evidence pointing strongly to either a positive or a negative sign is perhaps an indication that dc_1/dr, whatever its sign, may not be appreciably different from zero, thus pointing towards a small value for the expected value of perfect information about r, and *a less pronounced inferiority of temporal over timeless uncertainty in the case of rates of return than in the case of income.*[15]

2.5. Summarizing our discussion of (2.9)–(2.10), we have:

PROPOSITION 2.5. *A consumer's willingness to bear delayed risks, as measured by his risk aversion function for temporal prospects* $(-U_{22}/U_2)_{\hat{c}_1}$, *will be the lower:*

(i) *the lower his willigness to bear immediate risks, as measured by his risk aversion function for timeless prospects* $(-V_{yy}/V_y)_r$;

(ii) *the larger his marginal propensity to consume* dc_1/dy, *and/or the responsiveness of his current consumption to the rate of interest* $|\, dc_1/dr \,|$;

(iii) *the lower, in absolute value, the substitution effect of a change in the rate of interest on his current consumption* $|\, S \,|$.

[15] The general case of joint uncertainty about y_2 and r is a straightforward extension of the foregoing analysis; (2.9) and (2.10) combine to

$$(-U_{22}/U_2)\,\sigma_{c_2}^2 = (-U_{22}/U_2)(\sigma_{y_2}^2 + (y_1 - c_1)^2\sigma_r^2 + 2(y_1 - c_1)\sigma_{ry_2})$$
$$= -\left(\frac{V_{y_2y_2}}{V_{y_2}}\sigma_{y_2}^2 + \frac{V_{rr}}{V_r}(y_1 - c_1)\sigma_r^2 + 2\frac{V_{ry_2}}{V_r}(y_1 - c_1)\sigma_{ry_2}\right)$$
$$+ \left(\left(\frac{dc_1}{dy_2}\right)^2\sigma_{y_2}^2 + \left(\frac{dc_1}{dr}\right)^2\sigma_r^2 + 2\frac{dc_1}{dr}\frac{dc_1}{dy_2}\sigma_{ry_2}\right)\left(\frac{d^2c_2}{dc_1^2}\,\Big|_U\right).$$

3. Consumption and Portfolio Decisions with Perfect Markets

3.1. We now turn to the following questions: (i) How does uncertainty about future resources affect current consumption? (ii) What is the relationship between consumption decisions and portfolio choices? The first question may be raised irrespective of the nature and source of uncertainty, but cannot be answered until some reference criterion is chosen; the second question is appropriate only when savings may be invested in a variety of assets, and the consumer is free to *choose* his portfolio mix.

The portfolio problem traditionally considered in the literature involves a perfectly safe asset, yielding a rate of return r_0, and n risky assets yielding uncertain rates of return. The consumer is free to allocate his savings (wealth) among these $n + 1$ assets. If there is no uncertainty about future income, then any uncertainty affecting future resources is "chosen" or "endogenous," since it results entirely from portfolio choices (all the savings could have been invested in the safe asset). And a natural reference criterion, in assessing the impact of uncertainty on current consumption, is the value of c_1 that would be optimal if indeed all the savings were yielding the *sure* rate r_0. This is a more natural reference than the (more traditional) optimal c_1, given the *expected* rate of return on the chosen portfolio. Indeed, under assumption IV, a consumer would not choose a risky portfolio unless its expected return were higher than r_0; but no portfolio yielding the certainty of that expected return is available on the market; and knowing how $\hat{c}_1$ stands relative to the expected value criterion would not tell us how endogenous uncertainty actually affects c_1.

This argument can be extended to income uncertainty if one assumes the existence of insurance markets where an uncertain future income with density $\phi(y_2)$ can be exchanged against the certainty of some sure income y_2^0. In the presence of perfect markets for both income and assets, all uncertainty is "chosen" or "endogenous," and it is natural to compare $\hat{c}_1$ with the consumption that would be optimal if c_2 were equal to $y_2^0 + (y_1 - c_1)(1 + r_0)$ with certainty.

Such is the case treated in this section. It turns out that with perfect markets, a particular *ordinal* property of the utility function determines unambiguously how uncertainty affects consumption, *and* whether consumption and portfolio decisions are separable.

First-order conditions for optimal decisions are given in 3.2. We then prove a certainty equivalence theorem in 3.3 and interpret it in 3.4. The property of the utility function mentioned above is discussed in 3.5. In Section 4, we then turn to the case where the uncertain prospect faced by a consumer is not chosen, but given "exogenously" (at least on the income side). And we conclude that section with some remarks on the response of current consumption to availability of market opportunities for sharing risks.

3.2. We now introduce a general model designed to analyze simultaneous decisions about $\phi(y_2, r)$ and $\hat{c}_1$ under perfect insurance and asset markets.[16] These decisions are assumed to maximize expected utility over the class of all prospects, the market value of which does not exceed that of $\phi(y_2, r)$.

Let there be one perfectly safe asset, yielding a rate of return r_0, and n risky assets yielding the uncertain rates of return r_j, $j = 1 \cdots n$. The amounts invested in these $n+1$ assets will be denoted by $(x_0, x_1 \cdots x_n)$. Let furthermore future earnings y_2 be the sum of m components y_{i2}, $i = 1 \cdots m$; and let z_{i1} be the *present* value of y_{i2} on the insurance market.[17] Denote by $(1-\alpha_i)$, $i = 1 \cdots n$, the *fraction* of y_{i2} that a consumer chooses to *sell* on the insurance market; his current wealth and future (net) earnings then become $y_1 + \sum_{i=1}^m (1-\alpha_i) z_{i1}$ and $\sum_{i=1}^m \alpha_i y_{i2}$, respectively. Given a current consumption c_1, his portfolio of assets must satisfy the constraint

$$y_1 + \sum_{i=1}^{m} (1-\alpha_i) z_{i1} - c_1 = x_0 + \sum_{j=1}^{n} x_j,$$

or (3.1)

$$x_0 = y_1 - c_1 + \sum_{i=1}^{m} (1-\alpha_i) z_{i1} - \sum_{j=1}^{n} x_j,$$

[16] In [5, Section 6], we have used a slightly different formulation, based upon the notion that labor income (current and future) results from activities among which the consumer divides his *time*, of which a fixed quantity is available; it was also assumed that earnings from a given activity were proportional to the amount of time devoted to it, and that one of the activities entailed a perfectly safe income; the activities themselves did not appear as arguments of the utility function. Under that formulation, earnings per unit of time from the safe activity provide an implicit "insurance value" for the earnings per unit of time from any of the risky activities.

[17] This "insurance value" may be defined in a number of ways: one of them is straightforward insurance of professional income (including unemployment and medical insurance); another is suggested in footnote 16; another still is provided by the purchase (or short sale) of a portfolio of assets perfectly negatively (or positively) correlated with y_{i2}. One might also consider a "states of the world" model [1, 3, 7] with m states and define:

$$y_{i2} = \begin{cases} \text{future earnings, if state } i \text{ obtains,} \\ 0, \text{ otherwise;} \end{cases}$$

$z_{i1} = y_{i2}$ times the current price of a unit claim contingent on state i.

Our formal analysis is consistent with any of these interpretations, or combinations thereof, so long as z_{i1} is well-defined, independently of the amount of "coverage" that our consumer buys on y_{i2}.

and his future consumption is defined by

$$c_2 = \sum_{i=1}^{m} \alpha_i y_{i2} + x_0(1 + r_0) + \sum_{j=1}^{n} x_j(1 + r_j)$$

$$= \sum_{i=1}^{m} \alpha_i y_{i2} + \left(y_1 - c_1 + \sum_{i=1}^{m} (1 - \alpha_i) z_{i1}\right)(1 + r_0) + \sum_{j=1}^{n} x_j(r_j - r_0)$$

$$= z_2 + (y_1 - c_1)(1 + r_0) + \sum_{i=1}^{m} \alpha_i(y_{i2} - z_{i2}) + \sum_{j=1}^{n} x_j(r_j - r_0), \qquad (3.2)$$

where

$$z_{i2} =_{\text{def}} z_{i1}(1 + r_0) \qquad \text{and} \qquad z_2 =_{\text{def}} \sum_{i=1}^{m} z_{i2}\,.$$

Given the joint density $\phi(y_{12} \cdots y_{m2}, r_1 \cdots r_n)$, the simultaneous choice of an asset portfolio $(x_1 \cdots x_n)$, an insurance portfolio $(1 - \alpha_1 \cdots 1 - \alpha_m)$ and a consumption level c_1, is then arrived at by solving the following problem:

$$\max_{c_1, \alpha_1 \cdots \alpha_m, x_1 \cdots x_n} \int U \left\{ c_1, z_2 + (y_1 - c_1)(1 + r_0) + \sum_{i=1}^{m} \alpha_i(y_{i2} - z_{i2}) + \sum_{j=1}^{n} x_j(r_j - r_0) \right\} d\Phi(y_{12} \cdots y_{m2}, r_1 \cdots r_n), \qquad (3.3)$$

subject to whatever constraints prevail on the maximizing variables. We shall assume that such constraints, if any, are never binding, and that the solutions to (3.3) are given by the first-order conditions (3.4)–(3.6):[18]

$$\partial EU/\partial c_1 = E(U_1 - U_2(1 + r_0)) = 0, \quad \text{or} \quad EU_1/EU_2 = 1 + r_0\,; \qquad (3.4)$$

$$\partial EU/\partial x_j = E(U_2(r_j - r_0)) = 0 \quad \text{or} \quad EU_2 r_j/EU_2 = r_0\,; \qquad (3.5)$$

$$\partial EU/\partial \alpha_i = E(U_2(y_{i2} - z_{i2})) = 0, \quad \text{or} \quad EU_2 y_{i2}/EU_2 = z_{i2}\,. \qquad (3.6)$$

These results admit of the following economic interpretation:

PROPOSITION 3.2. *Under perfect insurance and asset markets, any solution to problem* (3.3) *has the following properties*:

(3.4) *The ratio of the expected marginal utilities of present and future consumption is equal to one plus the rate of return on the* safe *asset*;

[18] Thus, we assume that all solutions to (3.4)–(3.6) satisfy $c_1 \geqslant 0$ and $c_2 \geqslant 0$ identically in $(y_{12} \cdots y_{m2}, r_1 \cdots r_n)$, plus whatever conditions might be imposed on the α_i's and x_j's; clearly, the model lends itself to a more courageous formulation with inequality constraints. The second-order conditions follow naturally from Assumption IV.

(3.5) *The expected marginal utility of a unit investment in every asset is the same*; *the expected value of the rate of return on every asset*, weighted by the marginal utility of future consumption, *is equal to the rate of return on the safe asset*;

(3.6) *The expected marginal utility of a unit worth of insurance on every source of earnings is the same*; *the expected value of the earnings from any source*, weighted by the marginal utility of future consumption, *is equal to the insurance value of these earnings.*

Clearly, if there exist perfect asset markets, but no insurance markets, the solution to (3.3) with all α_i's equated to one's is still given by (3.4)–(3.5), and if there exist neither asset nor insurance markets, the solution is given by (3.4).

Furthermore, if the rate of return on the entire portfolio, namely,

$$r_0 + \left(\sum_{j=1}^{n} (r_j - r_0)\, x_j \Big/ \left(x_0 + \sum_{j=1}^{n} x_j\right)\right),$$

is still denoted by r, and since $y_2 =_{\text{def}} \sum_{i=1}^{m} y_{i2}$, (3.5)–(3.6) imply

$$EU_2 r/EU_2 = r_0, \quad EU_2 y_2/EU_2 = z_2 = EU_2 \sum_{i=1}^{m} (\alpha_i y_{i2} + (1 - \alpha_i)\, z_{i2}) \Big/ EU_2 . \tag{3.7}$$

3.3 We now state and prove a theorem that has an immediate bearing on consumption and portfolio decisions with perfect markets. It does, however, admit of a somewhat broader interpretation, which justifies the notation "y_2^*, r^*, c_1^*" introduced in the statement of the theorem.

THEOREM 3.3. *Let* $y_2^* =_{\text{def}} Ey_2 U_2(\hat{c}_1, c_2)/EU_2(\hat{c}_1, c_2)$, $r^* =_{\text{def}} ErU_2(\hat{c}_1, c_2)/EU_2(\hat{c}_1, c_2)$ *and define* $c_1^* = c_1^*(r^*, y_2^*)$ *by*

$$U_1(c_1^*, (y_1 - c_1^*)(1 + r^*) + y_2^*) - (1 + r^*)\, U_2(c_1^*, (y_1 - c_1^*)(1 + r^*) + y_2) = 0.$$[19]

Then

$$\partial^2 \frac{U_1}{U_2} \Big/ \partial c_2^2 \gtreqless 0 \text{ (identically in } c_2 \text{, given } \hat{c}_1) \text{ implies } \hat{c}_1 \gtreqless c_1^*.$$

Proof. The proof is based upon appendix Lemma C.2. Let $U_1(\hat{c}_1, c_2)/U_2(\hat{c}_1, c_2) - (1 + r^*) =_{\text{def}} f(c_2)$; we may rewrite (3.4) as

$$0 = \int U_2 f(c_2)\, d\Psi(c_2) =_{\text{def}} \int h(c_2)\, d\Psi(c_2), \tag{3.8}$$

[19] That is, c_1^* is the level of current consumption that would be optimal given $y_2 \equiv y_2^*$, $r \equiv r^*$; $\hat{c}$, is still the optimal level given $\phi(y_2, r)$.

where $U_2 = U_2(\hat{c}_1, c_2)$ is a function of c_2. Let then

$$c_2^* =_{\text{def}} \frac{Ec_2U_2}{EU_2} = \frac{E((y_1 - \hat{c}_1)(1 + r) + y_2)\, U_2}{EU_2}$$
$$= (y_1 - \hat{c}_1)(1 + r^*) + y_2^* \text{ by (3.7).}$$

Lemma C.2 then implies

$$f(c_2) \begin{array}{c}\text{concave}\\ \text{linear}\\ \text{convex}\end{array} \Rightarrow \int h(c_2)\, d\Psi(c_2) \begin{array}{c}\leqslant\\ =\\ \geqslant\end{array} f(c_2^*) \int U_2\, d\Psi(c_2). \tag{3.9}$$

Now, $\int U_2\, d\Psi(c_2) > 0$, and $f(c_2)$ is a linear function of U_1/U_2, whose concavity properties (in c_2) are determined by the sign (assumed constant) of $\partial^2(U_1/U_2)/\partial c_2^2$; consequently, (3.8) and (3.9) together imply

$$\frac{\partial^2(U_1/U_2)}{\partial c_2^2} \begin{array}{c}\leqslant\\ =\\ \geqslant\end{array} 0 \Rightarrow U_1(\hat{c}_1, c_2^*)/U_2(\hat{c}_1, c_2^*) - (1 + r^*) \begin{array}{c}\geqslant\\ =\\ \leqslant\end{array} 0$$
$$\Rightarrow U_1(\hat{c}_1, c_2^*) - (1 + r^*)\, U_2(\hat{c}_1, c_2^*) \begin{array}{c}\geqslant\\ =\\ \leqslant\end{array} 0. \tag{3.10}$$

Assumption III and the definition of c^* imply that

$$U_{11}(c_1^*, c_2^*) - (1 + r^*)\, U_{21}(c_1^*, c_2^*) < 0,$$

so that

$$U_1(\hat{c}_1, c_2^*) - (1 + r^*)\, U_2(\hat{c}_1, c_2^*) \gtreqless 0 \Leftrightarrow \hat{c}_1 \lesseqgtr c_1^*.$$

The theorem then follows from (3.10). Q.E.D.

3.4. When there exist perfect markets for income insurance and for assets, then (3.4)–(3.7) imply that y_2^* is equal to the insurance value of future income (z_2) and r^* is equal to the market sure rate of return (r_0).[20] Suppose that $\partial^2(U_1/U_2)/\partial c_2^2 = 0$; Theorem 3.3 then implies that current consumption $\hat{c}_1$ is equal to the level (c_1^*) that would be optimal if all income were insured ($y_2 \equiv y_2^*$) and all savings were held in the safe asset ($r \equiv r^*$). *This result holds independently of the actual insurance policy and asset portfolio chosen by the consumer.* Hence, endogenous uncertainty has no impact on consumption. Furthermore, y_2^* ($= z_2$) and r^* ($= r_0$) being directly observable market values, $\hat{c}_1$ ($= c_1^*$) may be chosen first (as a function of z_2 and r_0), the optimal insurance policy and asset portfolio being determined thereafter (jointly, for this given $\hat{c}_1$).

[20] Some readers may find it more convenient to transpose this interpretation to the situation where there is no uncertainty about future income, so that $y_2 \equiv y_2^*$.

642/5/3-2

Consumption and portfolio decisions may be taken sequentially and are "separable", in that sense.[21]

When $\partial^2(U_1/U_2)/\partial c^2_2 \neq 0$, then the sign of that quantity is also the sign of the impact of endogenous uncertainty on current consumption. It is noteworthy that U_1/U_2, hence its second derivative, is invariant under monotonic transformations of the utility function, and thus independent of risk aversion. There thus exist ordinal preferences, consistent with our assumptions, such that endogenous uncertainty results in increased consumption, and alternative preferences such that the opposite result holds. In the latter case, the consumer chooses an uncertain prospect which yeilds a higher expected utility than the sure prospect of identical market value, but he simultaneously chooses to consume less in the first period-postponing the (uncertain) benefit to the second period. Such behavior is consistent with risk aversion, in spite of the saying that "a bird in hand is worth two in the bush."

3.5. It is appropriate at this point to inquire about the meaning of the rather unfamiliar quantity $\partial^2(U_1/U_2)/\partial c_2{}^2$ which controls the response of consumption to endogenous risk, and to inquire whether there is ground for supposing that some sign is more plausible than another.

First we recall that U_1/U_2 is a familiar quantity, the slope of the indifference curve; hence $\partial(U_1/U_2)/\partial c_2$ is the rate of change of the slope of the indifference curves as we increase c_2 for fixed c_1. That derivative must have a positive sign by Assumption III (c_1 is not an inferior good). The function $(\partial^2(U_1/U_2)/\partial c_2{}^2)_{\bar{c}_1}$ measures the curvature of U_1/U_2 as a function of c_2.

An intuitive explanation of the relevance of $\partial^2(U_1/U_2)/\partial c_2{}^2$ for consumption decisions under uncertainty is provided in Appendix D, by means of a simple graphical illustration.

More generally, the following can be said:

(i) $\partial^2(U_1/U_2)/\partial c_2{}^2 = 0$ identically in c_1 and c_2 if and only if $U(c_1, c_2) = F(g(c_1) + h(c_1) \cdot c_2)$, $F' > 0$, $h > 0$ (see Appendix B). That is, $\partial^2(U_1/U_2)/\partial c_2{}^2 \equiv 0$ is the *ordinal* property of $U(c_1, c_2)$ that is *necessary* for risk neutrality in terms of c_2, and sufficient for such neutrality to obtain under a monotonic transformation of U.

(ii) given any $r > -1$, $y > 0$ and u, there exist y_1 and $y_2 = (y - y_1)(1 + r)$ such that $dc_1/dr = 0$; when $dc_1/dr = 0$, then d^2c_1/dr^2 has the sign of $\partial^2(U_1/U_2)/\partial c_2{}^2$ (see formulas A.7–A.8). That is, $\partial^2(U_1/U_2)/\partial c_2{}^2 = 0$ is the *ordinal* property of $U(c_1, c_2)$ that is *necessary* for

[21] An extension of these propositions to an n-period model, $n > 2$, has been provided by Pestieau [11], under the additional assumption of homothetic indifference surfaces.

a zero interest-elasticity of consumption at all r and *sufficient* for this situation to obtain under an appropriate time-distribution of income.

Concluding heuristically about the case of perfect insurance and asset markets, we would like to suggest as *a rough first approximation* that uncertainty has little impact on current consumption, and that consumption decisions are for practical purposes separable from portfolio decisions. The lack of empirical evidence pointing towards a substantial interest–elasticity of consumption and the intuitive appeal of the separability proposition lend support to this conclusion.

4. Consumption and Portfolio Decisions without Perfect Markets

4.1. When there do not exist perfect markets, with prices at which an arbitrary uncertain prospect can be evaluated and exchanged, then uncertainty is no longer endogenous, and a new reference criterion must be introduced to replace market value. Expected value then seems to be a natural criterion; it calls for comparing $\hat{c}_1$, that maximizes EU given $\phi(y_2, r)$, with $\bar{c}_1$ that would maximize U given $y_2 \equiv \bar{y}_2$ and $r \equiv \bar{r}$.

Theorem 3.3 has some implications for the relation of $\hat{c}_1$ to $\bar{c}_1$, but these implications are limited in scope. Specifically, it follows from the definition in Theorem 3.3 that

$$y_2{}^* = EU_2 y_2/EU_2 = \bar{y}_2 + [\text{cov}(U_2, y_2)/EU_2]. \tag{4.1}$$

When r is nonstochastic ($r \equiv r^*$), then it follows from $U_{22} < 0$ and $c_2 = y_2 + (y_1 - \hat{c}_1)(1 + r)$ that $\text{cov}(U_2, y_2) < 0$ and $y_2{}^* < \bar{y}_2$. In view of Assumption III, this entails $c_1{}^* < \bar{c}_1$. Consequently, $\partial^2(U_1/U_2)/\partial c_2{}^2 \leqslant 0$ implies $\hat{c}_1 \leqslant c_1{}^* < \bar{c}_1$; the relationship of $\hat{c}_1$ to $\bar{c}_1$ is indeterminate only when $\partial^2(U_1/U_2)/\partial c_2{}^2 > 0$.

Unfortunately, when r is stochastic, this line of reasoning is no longer valid. Indeed, $\text{cov}(U_2, y_2) = \text{cov}(U_2, c_2) - (y_1 - \hat{c}_1)\,\text{cov}(U_2, r)$. Whereas $\text{cov}(U_2, c_2) < 0$ still follows from $U_{22} < 0$, the sign of the second term is indeterminate: both $y_1 - \hat{c}_1$ and $\text{cov}(U_2, r)$ are arbitrary as to sign (r could be negatively correlated with c_2, if the returns on the *chosen* portfolio were negatively correlated with labor income).

4.2. A different line of analysis has been pursued, still for the case where r is nonstochastic, by Leland [8] and Sandmo [14]. Broadly speaking, their results point to diminishing absolute risk aversion as a sufficient condition for $\hat{c}_1 < \bar{c}_1$, where r is nonstochastic. Remember that $-U_{22}(c_2, \hat{c}_1)/U_2(c_2, \hat{c}_1)$ has been defined in Section 2 as the "absolute risk aversion" function relevant for temporal risks. Starting from any point in (c_1, c_2)-space, one may wonder whether $-(U_{22}/U_2)$ increases, decreases or remains constant when the starting point is displaced in some particular direction. Leland [8] considers a move along the (tangent to the) indifference curve through (c_1, c_2): c_2 increases and c_1 is simultaneously decreased to keep utility constant. Leland assumes that such a move *decreases* absolute risk aversion, and derives as an implication that current consumption diminishes if the variance of y_2 increases, the expectation of y_2 being kept constant. In other words, such an "increase in risk" reduces current consumption.

Sandmo [14] assumes that $-U_{22}/U_2$ decreases with c_2 and increases with c_1; he then defines an "increase in risk" as a multiplicative shift in the distribution of y_2 combined with an additive shift that keeps the mean constant. His assumptions imply that such an increase in risk reduces current consumption.[22]

We will now state and prove (Section 4.3) a theorem and a corollary that generalize the analysis of Leland and Sandmo. An interpretation of our results is given in 4.4, where it is also explained how Theorem 4.3 generalizes these related results. Finally, we come back in Section 4.5 to the relevance of market opportunities for consumption decisions under uncertainty.

4.3. The condition appearing in Theorem 4.3 refers to the behavior of the absolute risk aversion function along budget lines with slope $dc_2/dc_1 = -(1 + r^*)$. Define indeed

$$R(c_1, c_2, r^*) = \frac{\partial - (U_{22}/U_2)}{\partial c_1} - (1 + r^*) \frac{\partial - (U_{22}/U_2)}{\partial c_2}.$$

The sign of R determines whether absolute risk aversion increases (> 0), decreases (< 0) or remains constant ($= 0$) when c_1 increases and c_2 decreases along the budget line $c_2 = (y_1 - c_1)(1 + r^*) + y_2$. In this definition r^* is still given by (3.7) and satisfies $EU_1/EU_2 = 1 + r^*$.

THEOREM 4.3. *Let $y_2^\dagger$ be such that*

$$\max_{c_1} U(c_1, (y_1 - c_1)(1 + r^*) + y_2^\dagger) = EU(\hat{c}_1, (y_1 - \hat{c}_1)(1 + r) + y_2) \tag{4.2}$$

and let $c_1^\dagger$ be the value of c_1 maximizing the left hand side of (4.2). *Then* $R \gtreqless 0$ *(identically in c_2 given $\hat{c}_1$) implies* $\hat{c}_1 \lesseqgtr c_1^\dagger$.

[22] Related results have been established under the additional assumption of additive (cardinal) utility, e.g., by Mirman [10] or by Rothschild and Stiglitz [13]. The latter paper clarifies in a basic way the concept of "increase in risk."

Proof. The proof is based upon appendix Lemma C.1. For convenience, it is broken into three easy steps.

(i) We first notice that

$$U_1(\hat{c}_1, c_2) - (1 + r^*)\, U_2(\hat{c}_1, c_2) =_{\text{def}} h(c_2) = f(U(\hat{c}_1, c_2)), \quad (4.3)$$

with

$$f'(U) = [U_{12} - (1 + r^*)\, U_{22}]/U_2 . \quad (4.4)$$

Indeed, differentiating both sides of (4.3) with respect to c_2, we verify: $dh/dc_2 = U_{12} - (1 + r^*)\, U_{22} = f'(U) \cdot U_2$, which satisfies (4.4);

$$\begin{aligned} d^2h/dc_2^2 &= U_{122} - (1 + r^*)\, U_{222} = f''(U) \cdot U_2^2 + f'(U) \cdot U_{22} \\ &= df'(U)/dc_2 \cdot (dc_2/dU) \cdot U_2^2 + f'(U) \cdot U_{22} \\ &= \frac{(U_{122} - (1 + r^*)\, U_{222})\, U_2 - U_{22}(U_{12} - (1 + r^*)\, U_{22})}{U_2^2} \\ &\quad \times \frac{U_2^2}{U_2} + U_{22} \cdot \frac{U_{12} - (1 + r^*)\, U_{22}}{U_2}, \end{aligned}$$

and so on for higher derivatives. We notice in the process that

$$\begin{aligned} U_2 f''(U) &= \frac{(U_{122} - (1 + r^*)\, U_{222})\, U_2 - U_{22}(U_{12} - (1 + r^*)\, U_{22})}{U_2^2} \\ &= \frac{\partial(U_{22}/U_2)}{\partial c_1} - (1 + r^*) \frac{\partial(U_{22}/U_2)}{\partial c_2} = -R. \end{aligned} \quad (4.5)$$

(ii) In view of (4.5) and $U_2 > 0$, $R \gtreqless 0$ implies that f is a $\begin{matrix}\text{concave}\\ \text{linear}\\ \text{convex}\end{matrix}$ function of U. Let c_2' be such that $U(\hat{c}_1, c_2') = \int U(\hat{c}_1, c_2)\, d\Psi(c_2)$; Lemma C.1 then implies

$$R \gtreqless 0 \Rightarrow \int h(c_2)\, d\Psi(c_2) \lesseqgtr h(c_2'). \quad (4.6)$$

By (3.4) and the definition of h, $\int h(c_2)\, d\Psi(c_2) = 0$; therefore,

$$R \gtreqless 0 \Rightarrow h(c_2') = U_1(\hat{c}_1, c_2') - (1 + r^*)\, U_2(\hat{c}_1, c_2') \gtreqless 0. \quad (4.7)$$

(iii) By definition,

$$\begin{aligned} U(\hat{c}_1, c_2') &= U(c_1^\dagger, (y_1 - c_1^\dagger)(1 + r^*) + y_2^\dagger) \\ &= \max_{c_1} U(c_1, (y_1 - c_1)(1 + r^*) + y_2^\dagger). \end{aligned}$$

Since $d^2U/dc_1^2 < 0$, this implies

$$U_1(\hat{c}_1, c_2') - (1 + r^*) U_2(\hat{c}_1, c_2') \gtreqless 0 \Leftrightarrow \hat{c}_1 \lesseqgtr c_1^\dagger.$$

Combining this with (4.7), we conclude

$$R \gtreqless 0 \Rightarrow \hat{c}_1 \lesseqgtr c_1^\dagger. \qquad \text{Q.E.D.}$$

COROLLARY. $r \equiv r^*$ implies $c_1^\dagger \leqslant \bar{c}_1$.

Proof. $U_{22} \leqslant 0$ implies

$$\begin{aligned} U(c_1^\dagger, (y_1 - c_1^\dagger)(1 + r^*) + y_2^\dagger) &= EU(\hat{c}_1, (y_1 - \hat{c}_1)(1 + r) + y_2) \\ &\leqslant U(\hat{c}_1, (y_1 - \hat{c}_1)(1 + \bar{r}) + \bar{y}_2) \\ &\leqslant U(\bar{c}_1, (y_1 - c_1)(1 + r^*) + \bar{y}_2). \end{aligned}$$

It then follows from Assumption III that $c_1^\dagger \leqslant \bar{c}_1$. Q.E.D.

4.4. When $r \equiv r^*$, $R < 0$ means that absolute risk aversion $(-U_{22}/U_2)_{\hat{c}_1}$ diminishes when c_2 increases *thanks to* the additional savings implied in a decrease of c_1. We shall refer to this situation as "endogenously diminishing absolute risk aversion." Combining Theorem 4.3 and its corollary, we have the result that $R \leqslant 0$ implies $\hat{c}_1 \leqslant \bar{c}_1$.

This conclusion is consistent with those reached by Leland and Sandmo. In the case of infinitesimal risks, the three conclusions are identical, although the assumptions are not quite identical—indicating that the assumptions used are sufficient, but not necessary, for the conclusion. Theorem 4.3 clarifies that issue, by showing that $R \leqslant 0$ is *necessary and sufficient* for $\hat{c}_1 \leqslant c_1^\dagger$; when $U_{22} < 0$ and $\sigma^2_{y_2} > 0$, then $c_1^\dagger < \bar{c}_1$. Thus, $R < 0$ is not necessary for $\hat{c}_1 < \bar{c}_1$, but it is necessary for $\hat{c}_1 < c_1^\dagger$. The three-way implication in Theorem 4.3 is thus a generalization of the other results.

Our sharper result may be interpreted as follows. The impact of the uncertainty about future income on current consumption may be decomposed into an income effect and a substitution effect. The income effect corresponds to the fact that the expected utility of $\phi(y_2)$ is less than the utility of $\bar{y}_2$—it is only equal to the utility of $y_2^\dagger \leqslant \bar{y}_2$. This income effect alone would call for setting $\hat{c}_1 = c_1^\dagger < \bar{c}_1$: the income effect is always negative under risk aversion. But in addition there is room for a substitution effect: keeping expected utility constant, uncertainty about y_2 may still affect current consumption. Theorem 4.3 states that the sign of the substitution effect is the sign of R: risk aversion alone does not imply that the substitution effect is negative, but endogenously diminishing absolute risk aversion does. The implications of risk aversion are thus unambiguously defined.

A strong case may be made for regarding endogenous risk aversion as a meaningful, operational concept. Arrow [2] argues as follows that absolute risk aversion for total wealth may reasonably be expected to decrease with wealth: "If absolute risk aversion increased with wealth, it would follow that as an individual became wealthier, he would actually decrease the amount of risky assets held" (p. 35). In that argument, wealth is used as a primitive concept, and the increase in wealth is treated as exogenous. The argument may, however, be reformulated for the case where assets are acquired with savings and used to finance future consumption. One would then say: "If absolute risk aversion for c_2 increased with c_2 along a budget line, it would follow that as an individual accumulated more wealth, he would actually decrease the amount of risky assets held." One may thus consider that standard arguments invoked to discuss increasing versus decreasing absolute risk aversion for "wealth" apply almost verbatim to "risk aversion for c_2 along a budget line," that is, to endogenous risk aversion.

The arguments for decreasing absolute risk aversion are perhaps not compelling (the argument quoted above lacks generality when there are more than two assets), but it is a general conclusion that $\hat{c}_1$ is less than, equal to or greater than its "expected utility" certainty equivalent according to whether absolute risk aversion for c_2 decreases, remains constant or increases with c_2 along budget lines defined by r^*—with some plausibility arguments in favor of the "decreasing" case.

4.5. There remains now to relate the results of Sections 3 and 4. This will be done in three steps.

(i) *When there exist perfect insurance and asset markets*, then y_2^* (as defined in Theorem 3.3) $\leqslant y_2^\dagger$ (as defined in Theorem 4.3). Indeed, define y_2^{00} by

$$\max_{c_1} \underset{r}{E}\, U(c_1, (y_1 - c_1)(1 + r) + y_2^{00}) = U(c_1^\dagger, (y_1 - c_1^\dagger)(1 + r^*) + y_2^\dagger)$$
$$= \underset{r, y_2}{E}\, U(\hat{c}_1, (y_1 - \hat{c}_1)(1 + r) + y_2).$$

Because r is the rate of return on the *chosen* portfolio, whereas the sure rate r^* was available, we must have $y_2^\dagger \geqslant y_2^{00}$. Similarly, because the chosen future income could have been exchanged against the certainty of y_2^*, $y_2^{00} \geqslant y_2^*$. It follows that $c_1^* \leqslant c_1^\dagger$. Furthermore, if $\hat{c}_1 \leqslant c_1^*$, then $\hat{c}_1 \leqslant c_1^\dagger$, revealing that $\partial^2(U_1/U_2)/\partial c_2^2 \leqslant 0$ implies $R \leqslant 0$. This may be verified through formula (A.16) which may be rewritten as

$$\frac{\partial^2(U_1/U_2)}{\partial c_2^2} = R - \frac{U_{22}}{U_2}\frac{\partial(U_1/U_2)}{\partial c_2} + \left(\frac{U_1}{U^2} - (1 + r^*)\right)\frac{\partial - (U_{22}/U_2)}{\partial c_2}. \tag{4.8}$$

On the right hand side of (4.8), the second term is positive, and the third vanishes when $U_1/U_2 = 1 + r^*$. Hence, at the value of c_2 for which $U_1/U_2 = 1 + r^*$, $\partial^2(U_1/U_2)/\partial c_2{}^2 \leqslant 0$ implies $R < 0$; if R does not change sign over the range of $\psi(c_2)$, that sign must be negative. We may then conclude that $\partial^2(U_1/U_2)/\partial c_2{}^2 \leqslant 0$ is consistent with endogenously diminishing absolute risk aversion.

(ii) *When perfect insurance markets do not exist*, then: if $\hat{c}_1 = c_1{}^{\dagger}$, the availability of insurance would definitely increase $\hat{c}_1$; if $\hat{c}_1 = c_1{}^*$, the availability of insurance would increase (decrease) $\hat{c}_1$ if the insurance prices were such that the consumer would buy (sell) some insurance on his whole future income.

The first proposition is immediate: new insurance opportunities could only raise expected utility, irrespective of the insurance prices; this would also raise $y_2{}^{\dagger}$, hence $c_1{}^{\dagger}$, hence $\hat{c}_1 = c_1{}^{\dagger}$. The second proposition can be verified as follows: To say that insurance becomes available at a price such that the consumer would buy some on his whole future income means that $EU_2(y_2 - z_2) < 0$, or $EU_2 y_2/EU_2 < z_2$ (where z_2 is still the insurance value of future income). Hence, $y_2{}^*$ would increase from its present level to the level z_2, through insurance purchase, and $\hat{c}_1 = c_1{}^*$ would similarly rise. A similar reasoning applies to the selling case.[23]

(iii) In the absence of perfect markets for assets *and* insurance, separability of consumption and portfolio decision is rather implausible; actually, we do not know of any reasonable conditions under which that situation obtains.

Appendix A

Under certainty, the maximum of $U(c_1, c_2)$, under the budget constraint

$$c_2 = (y_1 - c_1)(1 + r) + y_2 \tag{A.1}$$

[23] The existence of markets with prices at which the consumer would *sell* insurance might seem remote, under generalized risk aversion. This remark is well-taken when the risks of different consumers or groups of consumers are sufficiently independent, or even negatively correlated, so that insurance can reduce everybody's risks simultaneously. On the other hand, when the risks of most consumers are strongly positively correlated, equality of supply and demand in the insurance markets calls for prices at which there will be sellers as well as buyers; the less risk-averse consumers will then be sellers: they will find it profitable to accept a greater variability of c_2 but will offset partly this added variability by reducing c_1 (for reasons indicated in Appendix D). In such cases, the organization of the insurance market need not stimulate total consumption.

is defined by the first- and second-order conditions

$$U_1 - U_2(1 + r) = 0, \qquad U_{11} - 2U_{12}(1 + r) + U_{22}(1 + r)^2 < 0. \tag{A.2}$$

Let

$$S =_{\text{def}} U_2(U_{11} - 2U_{12}(1 + r) + U_{22}(1 + r)^2)^{-1} < 0. \tag{A.3}$$

Through total differentiation of the first-order condition in (A.2), we find

$$\frac{dc_1}{dy} = \frac{dc_1}{dy_1} = -S(1 + r)\,\frac{\partial(U_1/U_2)}{\partial c_2},$$
$$\frac{dc_1}{dy_2} = S\,\frac{\partial(U_1/U_2)}{\partial c_2}, \qquad \text{where} \quad \frac{\partial(U_1/U_2)}{\partial c_2} = \frac{U_{12} - (U_1/U_2)\,U_{22}}{U_2}; \tag{A.4}$$

$$\begin{aligned} dc_1/dr &= ((y_1 - c_1)/(1 + r))(dc_1/dy_1) + S \\ &= -S\left((y_1 - c_1)\,\frac{\partial(U_1/U_2)}{\partial c_2} - 1\right). \end{aligned} \tag{A.5}$$

In (A.5), $((y_1 - c_1)/(1 + r))(dc_1/dy_1)$ measures the income effect, and $S\,(< 0)$ the substitution effect, of a change in r on c_1. The absolute value of S is also a measure of curvature of the indifference surfaces:

$$-1/S = (d^2c_2/dc_1^2)_U = (-1/U_2)(d^2U/dc_1^2)_{U_1=U_2(1+r)}\,. \tag{A.6}$$

For any given values of r and $y = y_1 + y_2(1 + r)^{-1}$, dc_1/dr is equal to 0 provided $y_1 = \hat{y}_1$ and $y_2 = (y - \hat{y}_1)(1 + r)$, where

$$\hat{y}_1 = c_1 + \left(\frac{\partial(U_1/U_2)}{\partial c_2}\right)^{-1} > c_1\,. \tag{A.7}$$

One then finds that $(dc_2/dr)_{y_1=\hat{y}_1} = \hat{y}_1 - c_1 = (\partial(U_1/U_2)/\partial c_2)^{-1}$ and

$$\begin{aligned} \left.\frac{d^2c_1}{dr^2}\right|_{y_1=\hat{y}_1} &= \frac{\partial(dc_1/dr)}{\partial c_2}\,\frac{dc_2}{dr} = -S(\hat{y}_1 - c_1)\,\frac{\partial^2(U_1/U_2)}{\partial c_2^2} \\ &= -S\left(\frac{\partial(U_1/U_2)}{\partial c_2}\right)^{-1}\frac{\partial^2(U_1/U_2)}{\partial c_2^2}\,. \end{aligned} \tag{A.8}$$

This expression has the sign of $\partial^2(U_1/U_2)/\partial c_2^2$, where

$$\frac{\partial^2(U_1/U_2)}{\partial c_2^2} = \frac{U_{122} - (U_1/U_2)\,U_{222}}{U_2} - 2\,\frac{U_{22}}{U_2}\,\frac{\partial(U_1/U_2)}{\partial c_2}\,. \tag{A.9}$$

The indirect utility function $V(y, r)$ is defined by

$$V(y, r) = U(c_1(y, r), c_2(y, r)|\ U_1 = U_2(1 + r)), \tag{A.10}$$

and its partial derivatives are evaluated by

$$
\begin{aligned}
V_y &= U_1(dc_1/dy) + U_2(dc_2/dy) \\
&= U_1(dc_1/dy) + U_2(1+r)(1 - (dc_1/dy)) = U_2(1+r) = U_1, \\
V_r &= U_1(dc_1/dr) + U_2(dc_2/dr) \qquad \text{(A.11)} \\
&= U_1(dc_1/dr) + U_2(y_1 - c_1 - (1+r)(dc_1/dr)) = U_2(y_1 - c_1).
\end{aligned}
$$

Proceeding further in this manner, one finds:

$$V_{yy} = U_{22}(1+r)^2 - (U_2/S)(dc_1/dy)^2 \geqslant U_{22}(1+r)^2; \qquad \text{(A.12)}$$

$$V_{rr} = U_{22}(y_1 - c_1)^2 - (U_2/S)(dc_1/dr)^2 \geqslant U_{22}(y_1 - c_1)^2; \qquad \text{(A.13)}$$

$$V_{ry} = U_{22}(1+r)(y_1 - c_1) - (U_2/S)(dc_1/dy)(dc_1/dr). \qquad \text{(A.14)}$$

Since $dy/dy_2 = (1+r)^{-1}$, one may define $V_{y_2} = (1+r)^{-1} V_y = U_2$, $V_{y_2y_2} = (1+r)^{-2} V_{yy}$, and write in view of (A.12)–(A.14):

$$
\begin{aligned}
-\frac{U_{22}}{U_2} &= -\frac{V_{y_2y_2}}{V_{y_2}} + \left(\frac{dc_1}{dy_2}\right)^2 \left(\frac{-1}{S}\right) \\
&= -\frac{V_{y_2y_2}}{V_{y_2}} + \left(\frac{dc_1}{dy_2}\right)^2 \left(\frac{d^2c_2}{dc_1^2}\Big|_U\right) \geqslant -\frac{V_{y_2y_2}}{V_{y_2}}, \\
-\frac{U_{22}}{U_2} &= -\frac{V_{rr}}{V_r(y_1 - c_1)} + \frac{1}{(y_1 - c_1)^2}\left(\frac{dc_1}{dr}\right)^2 \left(\frac{-1}{S}\right) \qquad \text{(A.15)} \\
&= -\frac{V_{ry_2}}{V_r} + \frac{1}{y_1 - c_1}\frac{dc_1}{dy_2}\frac{dc_1}{dr}\left(\frac{-1}{S}\right).
\end{aligned}
$$

$-(U_{22}/U_2)$ is defined in the text as the absolute risk aversion function for future consumption. Its partial derivatives satisfy:

$$
\begin{aligned}
&\frac{\partial - (U_{22}/U_2)}{\partial c_1} - (1+r)\frac{\partial - (U_{22}/U_2)}{\partial c_2} \\
&= -\frac{U_{122} - (1+r)\,U_{222}}{U_2} + \frac{U_{22}}{U_2}\frac{U_{12} - (1+r)\,U_{22}}{U_2} \\
&= -\frac{\partial^2(U_1/U_2)}{\partial c_2^2} - \frac{U_{22}}{U_2}\frac{\partial(U_1/U_2)}{\partial c_2} + \left(\frac{U_1}{U_2} - (1+r)\right)\frac{\partial - (U_{22}/U_2)}{\partial c_2}, \qquad \text{(A.16)} \\
&\frac{\partial - (U_{22}/U_2)}{\partial c_1} - \frac{U_1}{U_2}\frac{\partial - (U_{22}/U_2)}{\partial c_2} \\
&= -\frac{U_{122} - (U_1/U_2)\,U_{222}}{U_2} + \frac{U_{22}}{U_2}\frac{U_{12} - (U_1/U_2)\,U_{22}}{U_2} \\
&= -\frac{\partial^2(U_1/U_2)}{\partial c_2^2} - \frac{U_{22}}{U_2}\frac{\partial(U_1/U_2)}{\partial c_2} \geqslant -\frac{\partial^2(U_1/U_2)}{\partial c_2^2}, \qquad \text{(A.17)}
\end{aligned}
$$

as can be readily verified, starting from (A.9).

Appendix B[24]

Let $f(x, y)$ have the property that $f_x/f_y = a(x) + b(x)y$, $f_y \neq 0$. We wish to show that $f(x, y) = F(g(x) + h(x) \cdot y)$.

For $f(x, y) =$ constant, we have $f_x + f_y y' = 0$, with $y' = dy/dx\,|_{f\ \text{constant}}$, or $y' = -(f_x/f_y)$, so that:

$$y' = -a(x) - b(x)y. \tag{B.1}$$

The solution of this ordinary differential equation is readily verified to be

$$y = e^{-B(x)}\left(-\int a(x)\, e^{B(x)} dx + C\right), \tag{B.2}$$

where $B(x) = \int b(x)\, dx$; we may write (B.2) as

$$g(x) + h(x) \cdot y = c, \tag{B.3}$$

with $h(x) = e^{B(x)}$, $g(x) = \int a(x)\, e^{B(x)}\, dx$; since (B.3) is equivalent to "$f(x, y) =$ constant," our hypothesis is verified.

Appendix C

Lemma C.1. *Let $h(x) = f\{g(x)\}$, where f is differentiable in g and g is continuous in x; let furthermore $\phi(x)$ be any density such that $\int h(x)\, d\Phi(x)$ and $\int g(x)\, d\Phi(x)$ exist and are finite. Define x^0 (not necessarily unique) implicitly by $\int g(x)\, d\Phi(x) = g(x^0)$. Then*

$$f\{g(x)\}\ \begin{matrix}\text{concave}\\ \text{linear}\\ \text{convex}\end{matrix}\ \text{ in } g \text{ over the range of } \phi \text{ implies } \int h(x)\, d\Phi(x) \begin{matrix}\leqslant\\ =\\ \geqslant\end{matrix} h(x^0).$$

Proof.

$$f\{g(x)\}\ \begin{matrix}\text{concave}\\ \text{linear}\\ \text{convex}\end{matrix}\ \text{ in } g \Rightarrow h(x) = f\{g(x)\}$$

$$\begin{matrix}\leqslant\\ =\\ \geqslant\end{matrix} f\{g(x^0)\} + \{g(x) - g(x^0)\} \cdot f'(g)|_{g(x^0)}\,.$$

This in turn implies

$$\int h(x)\, d\Phi(x) \begin{matrix}\leqslant\\ =\\ \geqslant\end{matrix} f\{g(x^0)\} + f'(g)|_{g(x^0)} \int \{g(x) - g(x^0)\}\, d\Phi(x)$$
$$= f\{g(x^0)\} = h(x^0),$$

since $\int \{g(x) - g(x^0)\}\, d\Phi(x)$ vanishes by definition of x^0. Q.E.D.

[24] We are grateful to Wlodzimierc Szwarc for this result.

LEMMA C.2. *Let $h(x) = g(x)f(x)$, where f is differentiable and g is continuous in x, let furthermore $\phi(x)$ be any density such that $\int h(x)\, d\Phi(x)$, $\int g(x)\, d\Phi(x)$ and $\int xg(x)\, d\Phi(x)$ exist and are finite. Define*

$$x^0 = \int xg(x)\, d\Phi(x) \Big/ \int g(x)\, d\Phi(x).$$

Then $f(x)$ $\begin{matrix}\text{concave}\\ \text{linear}\\ \text{convex}\end{matrix}$ *in x over the range of ϕ implies*

$$\int h(x)\, d\Phi(x) \substack{\leq\\=\\\geq} f(x^0) \int g(x) d\Phi(x).$$

Proof. $f(x)$ $\begin{matrix}\text{concave}\\ \text{linear}\\ \text{convex}\end{matrix}$ in $x \Rightarrow f(x) \substack{\leq\\=\\\geq} f(x^0) + (x - x^0) f'(x)|_{x^0}$.

This in turn implies:

$$\int h(x)\, d\Phi(x) \substack{\leq\\=\\\geq} f(x^0) \int g(x)\, d\Phi(x) + f'(x)|_{x^0} \cdot \int (x - x^0)\, g(x)\, d\Phi(x)$$

$$= f(x^0) \int g(x)\, d\Phi(x)$$

since $\int (x - x^0)\, g(x)\, d\Phi(x)$ vanishes by definition of x^0. Q.E.D.

APPENDIX D

Figure 1 may be helpful to illustrate the role of $\partial^2(U_1/U_2)/\partial c_2{}^2$ in our problem, as well as the way in which a reduction of c_1 is equivalent to a reduction in the risk about c_2.

In the figure, line I is the sure budget equation $c_2 = (y_1 - c_1)(1 + \bar{r}) + \bar{y}_2$ with slope $-(1 + \bar{r})$. The point $\bar{c}$ with coordinates $(\bar{c}_1, \bar{c}_2)$ represents the chosen point on this budget equation, and the rising line EE depicts the Engel curve through $\bar{c}$, drawn linear for graphical convenience. Now suppose $\bar{y}_2$ is replaced by a very simple uncertain prospect $\phi(y_2)$ in terms of which y_2 will assume the value $(\bar{y}_2 + \delta)$ or the value $(\bar{y}_2 - \delta)$ with equal probability. In the figure the lines labelled IIA and IIB represent the (mutually exclusive) budget equations corresponding to each of these alternatives. Suppose further that, when confronted with $\phi(y_2)$, the consumer wonders whether he should consume $\bar{c}_1$ or alternatively reduce his first period consumption from $\bar{c}_1$ to $c_1' = \bar{c}_1 - \epsilon$.

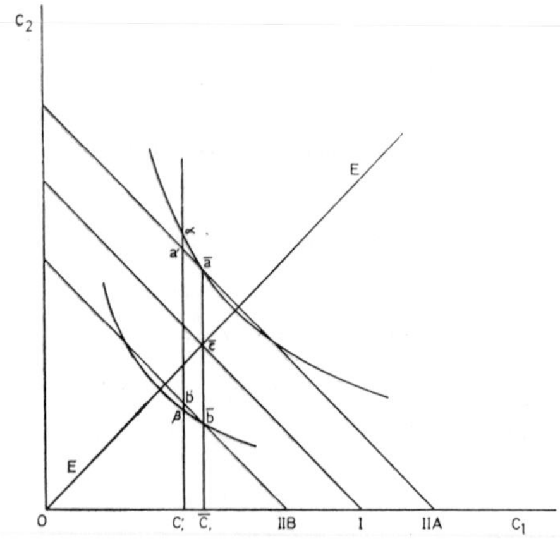

FIGURE 1

If the consumer stays at $\bar{c}_1$, we see from the figure that he will end up at the consumption points $\bar{a}$ or $\bar{b}$, with equal probability. Letting $\bar{a}$ and $\bar{b}$ denote also the c_2 coordinate of the corresponding points in the figure, the mutually exclusive and equally likely outcomes can be described by the two consumption vectors $(\bar{c}_1, \bar{a})$, $(\bar{c}_1, \bar{b})$ having expected utility: $\bar{U} = (1/2)(U(\bar{c}_1, \bar{a}) + U(\bar{c}_1, \bar{b}))$. Similarly, if he moves to c_1', we see that his expected utility can be expressed as $U' = (1/2)(U(c_1', a') + U(c_1', b'))$. To compare the two consumption decisions, first draw the indifference curve through $\bar{a}$ and let it intersect the line $c_1 = c_1'$ at the point α. Hence, by construction, $U(\bar{c}_1, \bar{a}) = U(c_1', \alpha)$. By drawing similarly the indifference curve through $\bar{b}$, we locate the point β such that $U(\bar{c}_1, \bar{b}) = U(c_1', \beta)$. Hence $\bar{U} = \frac{1}{2}(U(c_1', \alpha) + U(c_1', \beta))$. In other words, moving back from c_1' to $\bar{c}_1$ is entirely equivalent, in terms of $E(U)$, to remaining at c_1' but exchanging the random variable c_2' taking the values a' and b' with equal probability against a new random variable, call it γ_2, which takes the values α and β with equal probability. Thus our problem can be reduced to the question: Given c_1', is the prospect $\{c_2'\}$ better or worse than the prospect $\{\gamma^2\}$? One point is immediately apparent from Fig. 1: since $\beta < b' < a' < \alpha$, it follows that $\mathrm{var}(\gamma_2) > \mathrm{var}(c_2')$. In other words *increasing consumption* from c_1' to $\bar{c}_1$ has the effect of *increasing the variance* of the outcome. The reason is not far to seek. Increasing c_1 is equivalent to gearing current consumption more nearly to the higher income; this will produce an improvement *if* the larger y_2 obtains, but a deterioration if the smaller y_2 obtains. It will thus make things even brighter when they would be bright anyway and even more dismal if the worse turns out, hence it increases the variance of the outcome.[25] On this account then, if there is risk aversion, c_2' will be preferred to γ_2, and hence c_1' will be preferred to $\bar{c}_1$. Or, to put it differently, uncertainty tends to increase savings because the increase in savings has the effect of trimming the uncertainty of the outcome.

But this "variance effect" is not the end of the story: In choosing between the uncertain prospects $\{c_2'\}$ and $\{\gamma_2\}$, one must also compare their mean values.[26] If $\bar{\gamma}_2 = \frac{1}{2}(\alpha + \beta) < c_2' = \frac{1}{2}(a' + b')$, then c_2' dominates γ_2 in both respects and hence c_1' will be unequivocally preferred to $\bar{c}_1$ (even under risk indifference). But if the above inequality is reversed, then no definite conclusion can be reached. We will now show that the relation between $\bar{\gamma}_2$ and $\bar{c}_2'$ is precisely controlled by the sign of $\partial^2(U_1/U_2)/\partial c_2^2$.

To this end we first observe from Fig. 1 that the two mutually exclusive values of c_2', namely a' and b', can be expressed as follows:

$$a' = \bar{a} + \epsilon(1 + \bar{r}), \qquad b' = \bar{b} + \epsilon(1 + \bar{r}). \tag{D.1}$$

We further note that, for ϵ small, the slope of the chord joining the points $\bar{a}$ and α in the figure can be approximated by that of the indifference curve through $\bar{a}$, namely, $U_1(\bar{c}_1, \bar{a})/U_2(\bar{c}_1, \bar{a}) =_{\text{def}} (U_1/U_2)_{\bar{a}}$. Similarly, the slope of the chord joining $\bar{b}$ and β can be approximated by the slope of the indifference curve through $\bar{b}$, $(U_1/U_2)_{\bar{b}}$. We therefore have

$$\alpha \simeq \bar{a} + \epsilon(U_1/U_2)_{\bar{a}}, \qquad \beta \simeq \bar{b} + \epsilon(U_1/U_2)_{\bar{b}}. \tag{D.2}$$

Hence:

$$\begin{aligned}\bar{c}_2' - \bar{\gamma}_2 &\simeq \epsilon\left((1 + \bar{r}) - \frac{1}{2}\left(\frac{U_1}{U_2}\bigg|_{\bar{a}} + \frac{U_1}{U_2}\bigg|_{\bar{b}}\right)\right) \\ &= \epsilon\left(\frac{U_1}{U_2}\bigg|_{\bar{c}} - \frac{1}{2}\left(\frac{U_1}{U_2}\bigg|_{\bar{a}} + \frac{U_1}{U_2}\bigg|_{\bar{b}}\right)\right).^{27}\end{aligned} \tag{D.3}$$

Since $\bar{a}$ and $\bar{b}$ are symetrically located about $\bar{c}$, in the neighborhood of that point, it follows that

$$\frac{\partial^2(U_1/U_2)}{\partial c_2^2}\bigg|_{\bar{c}} \gtreqless 0 \Leftrightarrow \left\{\frac{1}{2}\left(\frac{U_1}{U_2}\bigg|_{\bar{a}} + \frac{U_1}{U_2}\bigg|_{\bar{b}}\right) \gtreqless \frac{U_1}{U_2}\bigg|_{\bar{c}=(a+b)/2}\right\} \Leftrightarrow \bar{c}_2' - \bar{\gamma}_2 \lesseqgtr 0. \tag{D.4}$$

[25] It can be verified from the figure that, for ϵ sufficiently small, the above conclusion must necessarily hold as long as the indifference curves are convex and the Engel curve has a finite positive slope. For an extension to more complicated densities, see Theorem 5.1 in [5].

[26] With infinitesimal risks, higher moments need not be considered.

[27] The last step follows from the fact that the point $\bar{c} = (c_1, c_2)$ is the point of tangency of the budget equation I with an indifference curve, hence at $\bar{c}$, the slope of the indifference curve $(U_1/U_2)_{\bar{c}}$ is precisely $(1 + \bar{r})$.

Accordingly, this "mean effect" will reinforce the "variance effect" whenever $\partial^2(U_1/U_2)/\partial c_2^2 \leqslant 0$, establishing unequivocally the superiority of c_1' over $\bar{c}_1$; but it will work in opposite direction when $\partial^2(U_1/U_2)/\partial c_2^2 > 0$. Theorem 3.3 may be interpreted as showing that the "variance effect" vanishes at c_1^*, through optimal choice of the prospect ϕ; the sign of $\hat{c}_1 - c_1^*$ is then determined by the "mean effect", i.e., by the sign of $\partial^2(U_1/U_2)/\partial c_2^2$.

REFERENCES

1. K. J. ARROW, Le rôle des valeurs boursières pour la répartition la meilleure des risques, *Econométrie* (1953), 41–48: English translation: The role of securities in the optimal allocation of risk bearing, *Rev. Econ. Studies* **31** (1964), 91–96.
2. K. J. ARROW, "Aspects of the Theory of Risk Bearing," Yrjö Jahnsson Foundation, Helsinki, 1965.
3. G. DEBREU, "Theory of Value," Wiley, New York, 1959.
4. J. DRÈZE, Market allocation under uncertainty, *European Econ. Rev.*, **2** (1971), 133–165.
5. J. DRÈZE AND F. MODIGLIANI, Epargne et consommation en avenir aléatoire, *Cah. Sém. Econ.* **9** (1966), 7–33.
6. I. FISHER, "Theory of Interest," Macmillan, New York, 1930.
7. J. HIRSHLEIFER, Investment decision under uncertainty: Choice theoretic approaches, *Quart. J. Econ.* **79** (1965), 509–536.
8. H. E. LELAND, Saving and uncertainty: The precautionary demand for saving, *Quart. J. Econ.* **82** (1968), 465–473.
9. J. MARSCHAK, Towards an economic theory of organization and information, in "Decision Processes" (Thrall, Coombs and Davis, Eds.), pp. 187–220, Wiley, New York, 1954.
10. L. J. MIRMAN, Uncertainty and optimal consumption decisions, *Econometrica* **39** (1971), 179–185.
11. P. PESTIEAU, Epargne et consommation dans l'incertitude: un modèle à trois périodes, *Recherches Economiques de Louvain* **2** (1969), 63–88.
12. J. W. PRATT, Risk aversion in the small and in the large, *Econometrica* **32** (1964), 122–136.
13. M. ROTHSCHILD AND J. E. STILGITZ, Increasing risk, I and II, *J. Econ. Theory* **2** (1970), 225–243; **3** (1971), 66–84.
14. A. SANDMO, The effect of uncertainty on saving decisions, *Rev. Econ. Studies* **37** (1970), 353–360.
15. L. J. SAVAGE, "The Foundations of Statistics," Wiley, New York, 1954.
16. J. VON NEUMANN AND O. MORGENSTERN, "Theory of Games and Economic Behavior," Princeton University Press, Princeton, 1944.

MANAGEMENT SCIENCE
Vol. 19, No. 2, October, 1972
Printed in U.S.A.

A DYNAMIC MODEL FOR BOND PORTFOLIO MANAGEMENT*†

STEPHEN P. BRADLEY AND DWIGHT B. CRANE‡

Harvard University

The bond portfolio problem is viewed as a multistage decision problem in which buy, sell, and hold decisions are made at successive (discrete) points in time. Normative models of this decision problem tend to become very large, particularly when its dynamic structure and the uncertainty of future interest rates and cash flows are incorporated in the model. In this paper we present a multiple period bond portfolio model and suggest a new approach for efficiently solving problems which are large enough to make use of as much information as portfolio managers can reasonably provide. The procedure utilizes the decomposition algorithm of mathematical programming and an efficient technique developed for solving subproblems of the overall portfolio model. The key to the procedure is the definition of subproblems which are easily solved via a simple recursive relationship.

1. Introduction

The bond portfolio problem can be viewed as a multistage decision problem in which portfolio actions are taken at successive (discrete) points in time. At each decision period, the portfolio manager has an inventory of bonds and/or cash on hand. (In the context of this paper, "bonds" refers to all fixed income and maturity securities which might be held in a portfolio.) Based upon present credit market conditions and his assessment of future interest rates and cash flows, the manager must decide which bonds to hold in the portfolio over the next time period, which bonds to sell, and which bonds to purchase from the marketplace. These decisions are made subject to a constraint on total portfolio size, which may be larger or smaller than the previous period's constraint depending upon whether a cash inflow or outflow occurred. At the next decision period, the portfolio manager faces a new set of interest rates and a new portfolio size constraint. He must then make another set of portfolio decisions which take into account the new information. This decision-making process, which is repeated over many time periods, is dynamic in the sense that the optimal first period decision depends upon the actions which will be taken in each future period for each uncertain event.

This description of the bond portfolio problem is potentially applicable to a number of portfolio management problems, such as the management of marketable securities held by nonfinancial corporations. The most important example, though, is the management of security portfolios in commercial banks. Thus, the motivating force and focus of the modeling approach reported here is the commercial bank problem.

Normative models of this decision problem, or variants of it, tend to become large and difficult to solve when many of its characteristics are taken into account. Hence, previous approaches to the problem have tried to limit its size by one or more of the

* Received May 1971; revised November 1971, January 1972.

† The research for this paper was funded by the Cambridge Project and by the Division of Research, Graduate School of Business Administration, Harvard University. An earlier version of this paper was presented to the Eleventh American Meeting of The Institute of Management Sciences at Los Angeles, California, October 1970.

‡ The authors wish to acknowledge the valuable assistance of William Giaque, including his development of the computer program for the model.

following techniques: ignoring the dynamic structure of the problem, excluding or restricting the uncertainty, limiting the number of assets considered, and including only one or two time periods.

Some approaches to the problem which included multiple assets and periods, such as the linear programming models of Chambers and Charnes [1] and Cohen and Hammer [5], do not consider either the dynamic structure or the uncertainty. Chance-constrained programming approaches suggested by Charnes and Littlechild [2] and Charnes and Thore [3], among others, do incorporate some uncertainty, but they are not dynamic. Cheng [4] also includes uncertainty in a mathematical programming framework, using a programming model to select a solution from prespecified portfolio strategies. His approach, however, is still not dynamic, since the strategies enumerated at the beginning of the model cannot be made conditional on future uncertain events. Eppen and Fama [11] have developed a dynamic programming formulation with uncertain cash flows, but they solve only a two-asset problem. In a later paper Eppen and Fama [12] specified a three-asset problem and illustrated the nature of its solution. Finally, Daellenbach and Archer [9] have presented a two-asset one-liability dynamic programming model with uncertain cash flows.

Two approaches, sequential decision theory and two-stage programming under uncertainty, have been used to solve multiple asset, two-period portfolio problems which explicitly incorporated uncertainty in both cash flows and interest rates. The problem is amenable to these approaches, but its size grows rapidly with either approach as the number of time periods is increased. The decision-theoretic approach of Wolf [14] provided an optimal solution for a one-period problem. In the two-period case, however, it is necessary to enumerate all possible portfolio strategies for the first decision period in order to guarantee an optimal solution. The linear programming under uncertainty approach, suggested by Cohen and Thore [6] and Crane [7], can optimally and efficiently solve a two-period problem, but the number of constraints is greatly increased by the addition of more periods.

In this paper we present a multiple period bond portfolio model and suggest a new approach for efficiently solving problems which are large enough to make use of as much information as portfolio managers can reasonably provide. The procedure utilizes the decomposition algorithm of mathematical programming and an efficient technique developed for solving subproblems of the overall portfolio model. The key to the procedure is the definition of subproblems which are easily solved via a simple recursive relationship. Optimal solutions for the whole portfolio are found by iterating between the subproblems and the decomposition master program.

2. Mathematical Formulation

The model primarily focusses upon selection of bank investment portfolio strategies, although it potentially could be expanded to include other bank assets. This concentration on the investment portfolio requires the assumption that portfolio decisions are residual to other bank decisions such as loan policy. The size of the portfolio is therefore determined outside the model framework by the flow of loans, deposits, and borrowed funds. As described below, an uncertain amount of funds is made available to the portfolio over time and the optimal investment strategy must take this liquidity need into account. A second external constraint is also imposed upon investment strategy by the bank's available capital. This affects the riskiness of the investment portfolio which the bank is willing to tolerate and is expressed through limits on maximum capital losses.

The structure of the portfolio model presented here assumes that decisions are made at the start of each period. At the beginning of the process, the manager starts with a known portfolio and he faces a known set of interest rates. For ease of exposition it is assumed that the initial portfolio is cash, but this assumption can be easily relaxed. He can invest in any of a finite number of asset categories which can represent maturity groups and/or types of bonds, such as U. S. Governments and municipals.

The results of this initial investment decision are subject to some uncertainty, represented by a random "event" which occurs during the first period. An event is defined by a set of interest rates and an exogenous cash flow. For example, one event might represent a tightening of credit conditions in which interest rates increase and the portfolio size has to be decreased to finance a rise in loans. It is assumed that there are a finite number of such events which have a discrete probability distribution. In addition, the portfolio manager knows the probability of each event, or it is appropriate for him to behave as if he did. Because of this assumption, the paper treats risk and uncertainty synonymously.

At the start of the second period, another set of investment decisions is made, taking into account the initial period decisions and the random event which occurred. A second random event determines the outcome of these decisions and is then followed by a third-period decision. This process in general continues for n decision periods with the nth random event determining the horizon value of the portfolio.

The mathematical formulation of the problem can be specified for a large number of time periods, but for expositional convenience the formulation below has been limited to three periods. These are a sufficient number to illustrate how the problem size can be extended.

In the three-period formulation, bonds can be purchased at the start of periods 1, 2, and 3 denoted by the subscript n. They can be sold or held at the start of any succeeding period m, where subscript $m = 2, 3$. Categories of bonds are denoted by superscript k. The random events which can occur in periods 1 and 2 are identified by parenthetical terms e_1 and e_2, respectively.

The decision variables of the model are buy (b_n^k), sell $(s_{n,m}^k)$, and hold $(h_{n,m}^k)$, where each variable is expressed in dollars of initial purchase price and is constrained to be greater than or equal to zero. Each of these decision variables is conditional upon events which precede the time of the decision. For example, a sequence of decisions might be: $b_1^k = \$1{,}000$, $h_{1,2}^k(e_1) = \$1{,}000$, and $s_{1,3}^k(e_1, e_2) = \$1{,}000$. In this sequence \$1,000 was invested in type k bonds at the start of period 1. After random event e_1 occurred, the bonds were held in period 2. Then after event e_2 the bonds were sold at the start of period 3. The cash flow from the sale might be more or less than \$1,000 depending upon whether the sale price was more or less than \$1,000. (The definition of the decision variables and parameters is summarized in Table I.)

In addition to the exogenous cash flows associated with random events, there are endogenous cash flows associated with the portfolio decisions. There is an income yield (y_n^k) stemming from the semiannual coupon interest on the bonds and a capital gain or loss $(g_{n,m}^k)$ which occurs when bonds are sold. Both are expressed as a percent of initial purchase price. It is assumed that taxes are paid when income and/or gains are received, so that the y and g coefficients are defined as after-tax. Transaction costs are taken into account by adjusting the gain coefficient for the broker's commission, i.e., bonds are purchased at the "asked" price and sold at the "bid" price.

The model is designed on a cash accounting rather than on the accrual basis normally used by banks for reporting purposes. On balance, this is a more accurate method be-

TABLE I
Definition of Variables

$b_n^k(\cdot)$	= *amount* of security class k *purchased* at beginning of period n; in dollars of initial purchase price
$s_{n,m}^k(\cdot)$	= *amount* of security class k, which was purchased at the beginning of period n and *sold* at the beginning of period m; in dollars of initial purchase price
$h_{n,m}^k(\cdot)$	= *amount* of security class k which was purchased at the beginning of period n and *held* (as opposed to sold) at the beginning of period m; in dollars of initial purchase price
$g_{n,m}^k(\cdot)$	= capital *gain* or loss on security class k purchased at the beginning of period n and sold at the beginning of period m; in percent of initial purchase price
$y_n^k(\cdot)$	= income *yield* from interest coupons on security class k, purchased at the beginning of period n; in percent of initial purchase price
$f_n(\cdot)$	= incremental amount of *funds* either made available to or withdrawn from the portfolio at the beginning of period n; in dollars
$L(\cdot)$	= upper bound on the realized capital loss (after taxes) from sales during a year; in dollars
$(\cdot)$	indicates that the variable is conditional on the sequence of uncertain events which precede it

cause it takes into account when cash is actually available for reinvestment. It is not completely accurate, however, because it ignores the tax effects of amortization of bond premiums and accretion of discounts of municipal bonds which might be contained in the starting portfolio. If this were judged to be a significant problem, it potentially could be solved by setting up separate bond categories for each combination of maturity and initial purchase price and then adjusting the cash flow and book value.

The decisions of the model are constrained by a set of cash flow constraints which express a limit on the funds available for the portfolio. We assume one security class is a risk free asset, and hence we are always fully invested. At the start of the first time period a flow of funds, f_1, is made available for the portfolio. Given our simplifying assumption that there are no bonds in the portfolio before the start of the model, the first-period cash flow constraint simply limits the sum of all purchases to be equal to f_1:

$$\sum_k b_1^k = f_1 . \tag{2.1}$$

Before the start of the second period an uncertain event occurs in which a new set of interest rates obtains and there is an exogenous cash flow to or from the portfolio. New investments in the portfolio at the start of the second period are limited by this cash flow, plus the income received from bonds held in the portfolio during the previous period and the cash generated from any sales at the start of the period[1]

$$-\sum_k y_1^k b_1^k - \sum_k (1 + g_{1,2}^k(e_1)) s_{1,2}^k(e_1) + \sum_k b_2^k(e_1) = f_2(e_1), \quad \forall e_1 \in E_1 . \tag{2.2}$$

Note that the second-period decision variables, $s_{1,2}^k(e_1)$ and $b_2^k(e_1)$, and the gain or loss resulting from security sales, $g_{1,2}^k(e_1)$, are conditional upon the event which occurs in the first period. Since there are a number of such events, there must be one constraint of type (2.2) for each event which can occur.

[1] For expositional convenience, this and subsequent equations assume that interest is paid each period. In the actual model, if the initial two periods were three months each and interest were paid in six-month intervals, there would be no cash flows until the end of the second period.

The cash flow constraint for the start of period 3, and any additional periods in a larger model, is analogous to that of period 2:

$$\begin{aligned} &-\textstyle\sum_k y_1^k h_{1,2}^k(e_1) - \sum_k (1 + g_{1,3}^k(e_1, e_2)) s_{1,3}^k(e_1, e_2) - \sum_k y_2^k(e_1) b_2^k(e_1) \\ &- \textstyle\sum_k (1 + g_{2,3}^k(e_1, e_2)) s_{2,3}^k(e_1, e_2) + \sum_k b_3^k(e_1, e_2) = f_3(e_1, e_2), \\ &\qquad \forall (e_1, e_2) \in E_1 \times E_2 . \end{aligned} \tag{2.3}$$

The number of terms in this constraint is much larger because it is necessary to account separately for bonds purchased at different points in time. For example, bonds sold at the start of period 3 could have been purchased at the start of either period 1 or 2 and the gain or loss which results depends upon this distinction. As with constraint (2.2), it is necessary to have one constraint of type (2.3) for each uncertain event sequence e_1, e_2.

The sale of securities is constrained by inventory balance equations which require that the amount of a security class sold at the start of period m must be less than or equal to the amount of that security held during the previous period. The difference between the amount previously held and that which is sold is the amount to be held during the next period:

$$-b_1^k + s_{1,2}^k(e_1) + h_{1,2}^k(e_1) = 0, \tag{2.4}$$

$$-h_{1,2}^k(e_1) + s_{1,3}^k(e_1, e_2) + h_{1,3}^k(e_1, e_2) = 0, \tag{2.5}$$

$$-b_2^k(e_1) + s_{2,3}^k(e_1, e_2) + h_{2,3}^k(e_1, e_2) = 0, \quad \forall k \in K \quad \text{and} \quad (e_1, e_2) \in E_1 \times E_2 . \tag{2.6}$$

It is important to point out that this formulation of the problem includes security classes that mature before the time horizon of the model. This is accomplished by setting the hold variable for a matured security to zero (actually dropping the variable from the model). This has the effect, through the inventory balance constraints, of forcing the "sale" of the security at the time the security matures. The value of the gain coefficient reflects the fact that the security matures at par with no transactions cost.

Theoretically, we would like to solve this problem with the objective of maximizing preference for terminal assets. This function would be difficult to specify though, since it involves preference for assets over time. In addition, problems of realistic size would be hard or even impossible to solve numerically because such objective functions are nonlinear. Therefore, in lieu of a nonlinear preference function, we have added a set of constraints which limit the realized net capital loss per year. This approach is similar to the "safety first principle" which some, e.g., Telser [13], have used in portfolio models. In our model, a limit on downside risk is expressed as an upper limit on capital losses. These constraints lead to some hedging behavior as the possibility of an increase in rates causes some short maturities to be held to avoid losses on sale.

Loss constraints are particularly appropriate for banks, in part because of a general aversion to capital losses, but also because of capital adequacy and tax considerations. Measures of adequate bank capital, such as that of the Federal Reserve Board of Governors, relate the amount of capital required to the amount of "risk" in the bank's assets. Thus, a bank's capital position affects its willingness to hold assets with capital loss potential. Tax regulations are also important, because capital losses can be offset against taxable income to reduce the size of the after-tax loss by roughly 50%. As a result, the amount of taxable income, which is sometimes relatively small in commercial banks, imposes an upper limit on losses a bank is willing to absorb.

The constraints which limit losses are of the following form:

$$\begin{aligned} -\textstyle\sum_k g_{1,2}^k(e_1)s_{1,2}^k(e_1) - \sum_k g_{1,3}^k(e_1, e_2)s_{1,3}^k(e_1, e_2) \\ - \textstyle\sum_k g_{2,3}^k(e_1, e_2)s_{2,3}^k(e_1, e_2) \leqq L(e_1, e_2), \qquad \forall(e_1, e_2) \in E_1 \times E_2. \end{aligned} \tag{2.7}$$

Since risk aversion is contained only in the loss constraints, the objective function of this paper is the maximization of the expected "value" of the portfolio at the end of the final period. This value consists of the cash received in the final period from interest plus the value of securities held in the portfolio. It is not obvious, though, how the value of the portfolio should be measured, since it is likely to contain some unrealized gains and losses. Should these gains or losses be calculated before or after taxes? Before or after transaction costs? At one extreme it would be possible to assume that the portfolio would be sold at the horizon so that its final value would be after taxes and transaction costs. This approach would tend to artificially encourage short maturities in the final portfolio, since they have low transaction costs. The alternative approach of valuing the portfolio before taxes and transaction costs is equally unrealistic. To solve this dilemma we have assumed it is possible for portfolio managers to express an equivalent cash value which can be substituted for the expected market value of the portfolio. This value represents the cash amount which the portfolio manager would be willing to receive in exchange for the final portfolio, i.e., he would be indifferent between receiving the cash and holding the portfolio.

Assuming the event space is finite and letting $p(e_1, e_2)$ be the probability of event sequence e_1, e_2 we have for the objective function:

$$\begin{aligned} \text{Max} \textstyle\sum_{e_1,e_2} p(e_1, e_2)[\sum_k (y_1^k + v_{1,3}^k(e_1, e_2))h_{1,3}^k(e_1, e_2) \\ + \textstyle\sum_k (y_2^k(e_1) + v_{2,3}^k(e_1, e_2))h_{2,3}^k(e_1, e_2) \\ + \textstyle\sum_k (y_3^k(e_1, e_2) + v_{3,3}^k(e_1, e_2))b_3^k(e_1, e_2)], \end{aligned} \tag{2.8}$$

where $v_{n,m}^k(e_1, e_2)$ is the expected cash value per dollar of book value for security class k, purchased at the beginning of period n and held at the end of the last period m. This coefficient is obtained from the actual equivalent cash values as follows:

$$v_{n,3}^k(e_1, e_2) = \textstyle\sum_{e_3} p(e_3 \mid e_1, e_2)v_{n,4}^k(e_1, e_2, e_3), \tag{2.9}$$

where $p(e_3 \mid e_1, e_2)$ is the probability of event e_3 given the events e_1 and e_2.

This mathematical formulation is completely general in the sense that it holds for arbitrary probability distributions on the uncertain events and for an arbitrary number of time periods. However, in such a general form it should be clear that no algorithm exists for its solution. We must at least restrict the event space to be finite in order that the formulation given will have a finite number of equations and a finite number of decision variables. Further, even assuming a linear objective function and a finite event space, the problem is in general much too large to solve by ordinary linear programming. Hence, we are left with the dilemma of deciding on the number of events in each probability space and the number of time periods as well as choosing a solution strategy.

Assume for the moment that the event space for each time period has exactly D points of nonzero probability. Further, let there be a total of n time periods with n_i periods in year i. Then, if k is the number of different security classes, the number of equations can be calculated as follows:

cash flow $\qquad 1 + D + D^2 + \cdots + D^{n-1}$,

net capital loss $D^{n_1} + D^{n_1+n_2} + \cdots,$
inventory balance $k[D + 2D^2 + \cdots + nD^{n-1}].$

By far the largest group of constraints are the inventory balance constraints. Consider the following table indicating the number of each type of constraint under a variety of assumptions.

Constraint	$k = 8$					
	$D = 3$			$D = 5$		
	$n_1 = 2, n_2 = 1$	$n_1 = 3, n_2 = 1$	$n_1 = 4, n_2 = 1$	$n_1 = 2, n_2 = 1$	$n_1 = 3, n_2 = 1$	$n_1 = 4, n_2 = 1$
Cash Flow	13	40	121	31	256	781
Net Capital Loss	9	27	81	25	125	625
Inventory Balance	168	816	3408	440	3440	20,000
Total Constraints	190	883	3601	496	3821	21,406

It is clear that even for a relatively small number of events and time periods the problem size rapidly becomes completely unmanageable. However, it is also clear that the main difficulty lies with the inventory balance constraints. Hence, an efficient solution procedure should treat these constraints implicitly instead of explicitly. Of the methods that can treat some constraints implicitly, decomposition seems to be the most appropriate.

3. The Decomposition Approach

In the decomposition procedure some of the constraints are treated implicitly as subproblems of the original mathematical formulation. Solutions to these subproblems generate information in the form of columns for a linear programming approximation to the original model, called the *restricted master program*. By iterating between the subproblems and the restricted master, which is always smaller than the original model, it is possible to find an optimal solution to the complete problem. In fact, a number of smaller problems are solved several times instead of solving a large complex problem directly. The effectiveness of the method depends crucially upon the ease with which the subproblems can be solved.

In the bond model we are able to define subproblems which have a unique recursive solution procedure that is very efficient. Consider a *commodity* to be a security class purchased at the start of a given period. For example, security class k purchased at the start of period 1 would be one commodity and the same security class purchased at the start of period 2 would be another. The inventory balance constraints associated with each commodity in the complete formulation define a subproblem.

To illustrate the subproblems, consider first the period 1 commodities. Security class k is available for purchase at the start of period 1. If a decision to purchase is made, this security class is then available for sale at the start of period 2 or it may be held during the second period. The amount that was held is then available for sale or holding at the start of the third period. This multistage problem involving a sequence of buy, sell, and hold decisions can be solved with a recursive procedure described in the next section. The problem's "state" in period t is defined by the amount of the commodity held in t. This amount is constrained by the inventory balance equations (2.4) and (2.5) which limit the amount sold to be less than or equal to the amount held before the decision.

Note that if security class k is purchased at the start of period 2 its purchase price and income yield are conditional upon the random event which occurred during period 1. Thus, a subproblem is defined for each commodity and each sequence of random events which precede the purchase date of that commodity.

The inventory balance equations of type (2.6) define the subproblems for each commodity purchased at the start of period 2 and each preceding event sequence. As would be expected, these equations have no variables in common in equations (2.4) and (2.5), since each set of inventory balance equations simply keeps track of the holdings of each commodity. The commodity definition adopted leads to a relatively large number of subproblems. The rationale, however, is that the state variable of each subproblem is then one dimensional, i.e., the amount of the commodity held.

We will in discussing the model present the results in terms of only three time periods; however, this is only for ease of exposition as the results hold for an arbitrary number of time periods. The constraints for the subproblems corresponding to the k security classes that may be purchased in period one are $\forall k$ as follows:

$$\begin{aligned} &-b_1^k + s_{1,2}^k(e_1) + h_{1,2}^k(e_1) = 0, \qquad \forall e_1 \in E_1, \\ (3.1)\quad &-h_{1,2}^k(e_1) + s_{1,3}^k(e_1, e_2) + h_{1,3}^k(e_1, e_2) = 0, \qquad \forall (e_1, e_2) \in E_1 \times E_2 \\ &b_1^k \geqq 0, \quad s_{1,2}^k(e_1) \geqq 0, \quad h_{1,2}^k(e_1) \geqq 0, \quad s_{1,3}^k(e_1, e_2) \geqq 0, \quad h_{1,3}^k(e_1, e_2) \geqq 0. \end{aligned}$$

The constraints for the subproblems corresponding to the k security classes that may be purchased in period two, conditional on events in period one, are $\forall k$ and $\forall e_1 \in E_1$ as follows:

$$(3.2)\quad \begin{aligned} &-b_2^k(e_1) + s_{2,3}^k(e_1, e_2) + h_{2,3}^k(e_1, e_2) = 0, \qquad \forall e_2 \in E_2, \\ &b_2^k(e_1) \geqq 0, \quad s_{2,3}^k(e_1, e_2) \geqq 0, \quad h_{2,3}^k(e_1, e_2) \geqq 0. \end{aligned}$$

For eight security classes, three time periods, and three uncertain events in each period there are thirty-two subproblems. There are eight subproblems of the form (3.1) corresponding to the eight commodities available for purchase at the start of period 1 and twenty-four of the form (3.2). The basic structure of the subproblems for an arbitrary number of time periods and uncertain events should be evident, since the subproblem constraints merely reflect the holdings of each commodity for each possible sequence of uncertain events. Note that each commodity is purchased only once and then sold or held during subsequent periods.

The interesting point to note about the subproblem constraints is that they are homogeneous systems of equations (i.e., zero right-hand sides). In decomposition the fundamental theorem employed is that a convex polyhedral set may be represented by a convex combination of its extreme points plus a nonnegative combination of its extreme rays [10]. The subproblems of (3.1) and (3.2) have only one extreme point, all decision variables equal to zero. For any nonzero point satisfying the subproblem constraints, a scalar times that point also satisfies the constraints and hence with a linear objective function there exists an associated unbounded solution. As a result we need only consider the extreme rays of the subproblems. It is critical for an effective procedure to be able to construct these extreme rays in an efficient manner. In the next section we show how to do this.

The restricted master for the decomposition scheme defining subproblems for each commodity and event sequence is given in Figure 1. Note that the usual convex combination constraints (e.g., $\sum \lambda_i = 1$) are not present, since we are only dealing with nonnegative combinations of extreme rays. Further, there are no subproblem con-

Maximize

$$\sum_k \{\sum_n [\sum_{e_1e_2} p(e_1, e_2)(y_1^k + v_{1.3}^k(e_1, e_2))\bar{h}_{1.3.n}^k(e_1, e_2)]\lambda_{1.n}^k$$
$$+ \sum_{e_1} \sum_n [\sum_{e_2} p(e_1, e_2)(y_2^k(e_1) + v_{2.3}^k(e_1, e_2))\bar{h}_{2.3.n}^k(e_1, e_2)]\lambda_{2.n}^k(e_1)$$
$$+ \sum_{e_1e_2} p(e_1, e_2)(y_3^k(e_1, e_2) + v_{3.3}^k(e_1, e_2))b_3^k(e_1, e_2)\}$$

subject to

$$\pi_1 : \sum_k \{\sum_n [\bar{b}_{1.n}^k]\lambda_{1.n}^k\} = f_1,$$
$$\pi_2(e_1): \sum_k \{\sum_n [-y_1^k\bar{b}_{1.n}^k - (1 + g_{1.2}^k(e_1))\bar{s}_{1.2.n}^k(e_1)]\lambda_{1.n}^k$$
$$+ \sum_{e_1}\sum_n [\bar{b}_{2.n}^k(e_1)]\lambda_{2.n}^k(e_1)\} = f_2(e_1), \qquad \forall e_1 \in E_1,$$
$$\pi_3(e_1, e_2): \sum_k \{\sum_n [-y_1^k\bar{h}_{1.2.n}^k(e_1) - (1 + g_{1.3}^k(e_1, e_2))\bar{s}_{1.3.n}^k(e_1, e_2)]\lambda_{1.n}^k$$
$$+ \sum_{e_1} \sum_n [-y_2^k(e_1)\bar{b}_{2.n}^k(e_1) - (1 + g_{2.3}^k(e_1, e_2))\bar{s}_{2.3}^k(e_1, e_2)]\lambda_{2.n}^k(e_1)$$
$$+ b_3^k(e_1, e_2)\} = f_3(e_1, e_2), \qquad \forall (e_1, e_2) \in E_1 \times E_2,$$
$$\pi_4(e_1, e_2): \sum_k \{\sum_n [-g_{1.2}^k(e_1)\bar{s}_{1.2.n}^k(e_1) - g_{1.3}^k(e_1, e_2)\bar{s}_{1.3.n}^k(e_1, e_2)]\lambda_{1.n}^k$$
$$+ \sum_{e_1} \sum_n [-g_{2.3}^k(e_1, e_2)\bar{s}_{2.3.n}^k(e_1, e_2)]\lambda_{2.n}^k(e_1)\} \leqq L(e_1, e_2), \quad \forall (e_1, e_2) \in E_1 \times E_2,$$
$$\lambda_{1.n}^k \geqq 0, \quad \lambda_{2.n}^k(e_1) \geqq 0, \quad b_3^k(e_1, e_2) \geqq 0.$$

FIGURE 1. Restricted master.

straints associated with buying a security in period three, since there are no subsequent sell or hold decisions. Hence these decision variables are not represented in terms of extreme points and extreme rays of a subproblem. Notationally, the λ_n's refer to the weighting in the restricted master of the nth solution of the appropriate subproblem, and the bars over the decision variables indicate subproblem solutions that have been generated. The notation is not precise here as the subscript n is different for each subproblem.

For our problem, we can initiate the decomposition procedure by constructing a feasible solution to our original problem. This solution is translated into the coefficients for an initial restricted master which is then solved. The optimal dual solution associated with the constraints of the restricted master, denoted by $\pi_n(\cdot)$, is indicated in Figure 1. These dual variables are used to modify the objective functions of the subproblems at each iteration, and these subproblems are then solved as indicated in the next section. The resulting extreme rays are then added as columns to the restricted master, and the procedure is repeated. The algorithm terminates when no new extreme rays are generated at some iteration. A proof of the convergence of the method is given in Dantzig [10].

4. Solving the Subproblems

The key to successfully using the decomposition procedure is an efficient solution procedure for the subproblems. In this section we present an extremely efficient procedure that may be easily implemented on a computer. The subproblems for each security class that may be purchased in period one are $\forall k$

Maximize

$$
\begin{aligned}
&-[\pi_1 - y_1^k \textstyle\sum_{e_1} \pi_2(e_1)]b_1^k + \sum_{e_1} [(1 + g_{1,2}^k(e_1))\pi_2(e_1) \\
&+ g_{1,2}^k(e_1) \textstyle\sum_{e_2} \pi_4(e_1, e_2)]s_{1,2}^k(e_1) + \sum_{e_1} [y_1^k \sum_{e_2} \pi_3(e_1, e_2)]h_{1,2}^k(e_1) \\
&+ \textstyle\sum_{e_1e_2} [(1 + g_{1,3}^k(e_1, e_2))\pi_3(e_1, e_2) + g_{1,3}^k(e_1, e_2)\pi_4(e_1, e_2)]s_{1,3}^k(e_1, e_2) \\
&\cdot \textstyle\sum_{e_1e_2} [p(e_1, e_2)(y_2^k(e_1, e_2) + v_{2,3}^k(e_1, e_2))]h_{2,3}^k(e_1, e_2)
\end{aligned}
\tag{4.1}
$$

subject to

$$-b_1^k + s_{1,2}^k(e_1) + h_{1,2}^k(e_1) = 0, \qquad \forall e_1 \in E_1,$$

$$-h_{1,2}^k(e_1) + s_{1,3}^k(e_1, e_2) + h_{1,3}^k(e_1, e_2) = 0, \qquad \forall (e_1, e_2) \in E_1 \times E_2,$$

$$b_1^k \geqq 0, \quad s_{1,2}^k(e_1) \geqq 0, \quad h_{1,2}^k(e_1) \geqq 0, \quad s_{1,3}^k(e_1, e_2) \geqq 0, \quad h_{1,3}^k(e_1, e_2) \geqq 0.$$

The subproblems for each security class that may be purchased in period two conditional on the events of period one are $\forall k$ and $\forall e_1 \in E_1$

Maximize

$$\begin{aligned}(4.2) \qquad & -[\pi_2(e_1) - y_2^k(e_1) \textstyle\sum_{e_2} \pi_3(e_1, e_2)]b_2^k(e_1) \\ & + \textstyle\sum_{e_2} [(1 + g_{2,3}^k(e_1, e_2))\pi_3(e_1, e_2) + g_{2,3}^k(e_1, e_2)\pi_4(e_1, e_2)]s_{2,3}^k(e_1, e_2) \\ & + \textstyle\sum_{e_2} [p(e_1, e_2)(y_2^k(e_1, e_2) + v_{2,3}^k(e_1, e_2))]h_{2,3}^k(e_1, e_2)\end{aligned}$$

subject to

$$-b_2^k(e_1) + s_{2,3}^k(e_1, e_2) + h_{2,3}^k(e_1, e_2) = 0, \qquad \forall e_2 \in E_2,$$

$$b_2^k(e_1) \geqq 0, \qquad s_{2,3}^k(e_1, e_2) \geqq 0, \qquad h_{2,3}^k(e_1, e_2) \geqq 0.$$

The basic structure of the subproblems in each case is identical. When we consider more than three time periods, we have the same structure only with necessary additional variables and constraints. The underlying structure is simply that the hold variables determine the amount of the commodity on hand at the start of the subsequent period. This amount on hand is then either sold or held and the process repeated.

These subproblems turn out to be very easy to solve due to their simple structure and homogeneity. First, every unbounded solution of subproblems of the form (4.1) has $b_1^k > 0$. If not, the subproblem constraints imply the null solution. Similarly, every unbounded solution of subproblems of the form (4.2) has $b_2^k(e_1) > 0$. Further, if in solving a subproblem it is profitable to buy one unit of a commodity it is profitable to buy as much as possible, since the objective function is linear and the constraints are homogeneous. Let us look at the problem of buying one unit of a particular commodity. If it is profitable to do so we have constructed a ray yielding arbitrarily large profit for an arbitrarily large multiple of that solution. Hence we have the following *ray finding problem* which determines the selling and holding strategy that maximizes the return from buying one unit of a security purchased in period one.

Maximize

$$\begin{aligned}& \textstyle\sum_{e_1} [(1 + g_{1,2}^k(e_1))\pi_2(e_1) + g_{1,2}^k(e_1) \sum_{e_2} \pi_4(e_1, e_2)]s_{1,2}^k(e_1) \\ (4.3) \qquad & + \textstyle\sum_{e_1} [y_2^k(e_1) \sum_{e_2} \pi_3(e_1, e_2)]h_{1,2}^k(e_1) \\ & + \textstyle\sum_{e_1 e_2} [(1 + g_{1,3}^k(e_1, e_2))\pi_3(e_1, e_2) + g_{1,3}^k(e_1, e_2)\pi_4(e_1, e_2)]s_{1,3}^k(e_1, e_2) \\ & + \textstyle\sum_{e_1 e_2} [p(e_1, e_2)(y_1^k + v_{1,3}^k(e_1, e_2))]h_{1,3}^k(e_1, e_2)\end{aligned}$$

subject to

$$s_{1,2}^k(e_1) + h_{1,2}^k(e_1) = 1, \qquad \forall e_1 \in E_1,$$

$$-h_{1,2}^k(e_1) + s_{1,3}^k(e_1, e_2) + h_{1,3}^k(e_1, e_2) = 0, \qquad \forall (e_1, e_2) \in E_1 \times E_2,$$

$$s_{1,2}^k(e_1) \geqq 0, \quad h_{1,2}^k(e_1) \geqq 0, \quad s_{1,3}^k(e_1, e_2) \geqq 0, \quad h_{1,3}^k(e_1, e_2) \geqq 0.$$

This ray-finding problem has an extremely simple solution which can be easily seen by ooking at its dual which is:

Minimize

$$\sum_{e_1} w_1(e_1)$$

subject to

$$w_1(e_1) \geqq (1 + g_{1,2}^k(e_1))\pi_2(e_1) + g_{1,2}^k(e_1) \sum_{e_2} \pi_4(e_1, e_2), \quad \forall e_1 \in E_1,$$

$$(4.4) \quad \begin{aligned} w_1(e_1) - \sum_{e_2} w_2(e_1, e_2) &\geqq y_2^k(e_1) \sum_{e_2} \pi_3(e_1, e_2), \\ w_2(e_1, e_2) &\geqq (1 + g_{1,3}^k(e_1, e_2))\pi_3(e_1, e_2) \\ &\quad + g_{1,3}^k(e_1, e_2)\pi_4(e_1, e_2), \qquad \forall (e_1, e_2) \in E_1 \times E_2. \\ w_2(e_1, e_2) &\geqq p(e_1, e_2)(y_1^k + v_{1,3}^k(e_1, e_2)), \end{aligned}$$

Since we wish to minimize $\sum_{e_1} w_1(e_1)$, and $w_1(e_1)$ and $w_2(e_1, e_2)$ are linked only by the constraints

$$(4.5) \qquad w_1(e_1) \geqq \sum_{e_2} w_2(e_1, e_2) + y_2^k(e_1) \sum_{e_2} \pi_3(e_1, e_2),$$

we need only choose $w_2(e_1, e_2)$ as small as possible and then choose $w_1(e_1)$ as small as possible. Hence:

$$(4.6) \quad \begin{aligned} w_2(e_1, e_2) &= \text{Max}\,[(1 + g_{1,3}^k(e_1, e_2))\pi_3(e_1, e_2) + g_{1,3}^k(e_1, e_2)\pi_4(e_1, e_2); \\ &\qquad p(e_1, e_2)(y_1^k + v_{1,3}^k(e_1, e_2))], \qquad \forall (e_1, e_2) \in E_1 \times E_2, \\ w_1(e_1) &= \text{Max}\,[(1 + g_{1,2}^k(e_1))\pi_2(e_1) + g_{1,2}^k(e_1) \textstyle\sum_{e_2} \pi_4(e_1, e_2); \\ &\qquad \textstyle\sum_{e_2} w_2(e_1, e_2) + y_2^k(e_1) \sum_{e_2} \pi_3(e_1, e_2)], \qquad \forall e_1 \in E_1, \end{aligned}$$

constitutes a recursive algorithm for finding the optimal solution to the dual of the ray-finding problem. If the optimal value of the objective function of the dual exceeds the purchase cost of the commodity, which is

$$(4.7) \qquad \pi_1 - y_1^k \sum_{e_1} \pi_2(e_1),$$

then we have found a *profitable* ray of the subproblem. The actual ray is given by $b_1^k = 1$ and a sequence of zeros and ones for the sell and hold variables, a one corresponding to a tight constraint in the dual of the ray-finding problem. If the set of tight constraints in this dual problem is not unique then this merely implies an indifference between selling and holding at some stage, and the choice may be made arbitrarily. This solution procedure is a general one and holds for an arbitrary number of events and time periods.

At each iteration, all profitable rays found are added as columns to the restricted master. The restricted master is then solved again yielding new dual variables that modify the objective functions of the subproblems, and the process is repeated. If no profitable ray is found for any subproblem, then the process terminates and we have the optimal solution. The value of the optimal solution is merely given by the nonnegative combination of the subproblem solutions as indicated by the final restricted master. In general we need *not* add any unprofitable ray to the restricted master. However, a ray that is unprofitable at one iteration may become profitable at a future iteration as the objective functions of the subproblems are modified by the dual variables. Hence, if one profitable ray is generated at an iteration all new rays generated, profitable or not, are in fact added to the restricted master as columns. As the restricted master is augmented by more and more columns, those columns not in the current basis are retained provided storage limitations permit. As storage limitations become binding, those columns that price out most negatively are dropped.

5. Conclusions

The three-period three-event problem used to illustrate the model structure is perhaps misleadingly small for illustrating the requirements of a realistic bond portfolio problem. Let us consider a four-period model covering two years where the periods are 3 months, 3 months, 6 months, and 1 year. Variable length periods are chosen to reflect our greater uncertainty about events more distant in the future. Let the security classes represent maturity categories of U. S. Government bills, notes, and bonds, e.g., 3 months, 6 months, 1 year, 2, 3, 5, 10, and 20 years. Buy, sell, and hold decisions must be made for each of these security classes at the start of each period conditional on the preceding sequence of uncertain events. After the decisions in each period a random event occurs which determines the set of interest rates (i.e., the Government yield curve) and the exogenous cash flow for the subsequent decision.

The number of random events in each time period depends on the degree of detail desired and the resulting problem size. For the four-period model under consideration we take a five-point approximation to the distribution of interest rates and exogenous cash flow in each of the three-month periods and a three-point approximation in the six-month and one-year periods. Our motivation is that over the first six months some reasonable forecasting of interest rates is currently possible; however, beyond that point merely assigning probabilities to the three events corresponding to rising, unchanging, and falling rates is difficult. Since a tightening of credit conditions is normally associated with an increase in rates, an exogenous cash outflow is assumed to occur with a rate increase, representing a need for funds in other parts of the institution. Similarly, no exogenous cash flow is assumed to occur with constant rates and a cash inflow is associated with falling rates. The resulting general formulation has 2,421 constraints and 5,328 variables which is clearly a large problem. By applying decomposition the resulting linear programming restricted master has 181 constraints and 248 subproblems each of which can be solved very quickly. The number of decision variables will of course remain the same; however, a large number of these will be zero.

With a model of this size the data requirements are rather extensive. This data estimation problem stems from the need for portfolio managers to specify for each event sequence both the interest rate structure and its associated probability. We have simplified this effort somewhat in our model by requiring only three future interest rate estimates for each future yield curve: the one-year rate, the twenty-year rate, and the highest rate of the forecast yield curve. Our research indicates the remainder of the yield curve can be satisfactorily approximated using the following equation:[2]

$$y = at^b e^{ct},$$

where

y = yield to maturity,

t = time to maturity,

a,b,c = parameters estimated from the three points on the yield curve.

[2] An equation of this type was suggested to the authors by Kenneth E. Gray.

Further research on the assessment problem is currently underway. Initial results are reported in Crane [8].

Preliminary computational experience with the model has been very promising. A model with eight security classes, three time periods, and three events per period has been solved on an IBM 360/65 with a running time of 68 seconds. It is difficult to draw strong conclusions from running times for different programs, but our model can also be solved as a linear programming under uncertainty model with 190 constraints. Running the model in this form using IBM's Mathematical Programming System on the 360/65 required 213 seconds. Furthermore, the MPS version was aided in this comparison because our program had to generate the data matrix while MPS did not.

Although our program seems to solve the problem faster than MPS, the more important advantage of our approach is its ability to handle large problems which would require an inordinate number of constraints if they were specified as linear programming under uncertainty models. The solution procedure proposed in this paper will hopefully permit models large enough to solve realistic descriptions of the portfolio management problem. Further research is needed, though, to improve the ability of portfolio managers to supply the needed assessments for larger models and to study the sensitivity of model solutions to increases in the number of time periods and the number of events per period.

References

1. Chambers, D. and Charnes, A., "Intertemporal Analysis and Optimization of Bank Portfolios," *Management Science*, Vol. 7, No. 4 (July 1961), pp. 393–410.
2. Charnes, A. and Littlechild, S. C., "Intertemporal Bank Asset Choice with Stochastic Dependence," Systems Research Memorandum No. 188, The Technological Institute, Northwestern University, April 1968.
3. —— and Thore, Sten, "Planning for Liquidity in Financial Institutions: The Chance-Constrained Method," *The Journal of Finance*, Vol. XXI, No. 4 (December 1966), pp. 649–674.
4. Cheng, Pao Lun, "Optimum Bond Portfolio Selection," *Management Science*, Vol. 8, No. 4 (July 1962), pp. 490–499.
5. Cohen, Kalman J. and Hammer, Frederick S., "Linear Programming and Optimal Bank Asset Management Decisions," *The Journal of Finance*, Vol. XXII, No. 2 (May 1967), pp. 147–165.
6. —— and Thore, Sten, "Programming Bank Portfolios Under Uncertainty," *Journal of Bank Research*, Vol. 1, No. 1 (Spring 1970), pp. 42–61.
7. Crane, Dwight B., "A Stochastic Programming Model for Commercial Bank Bond Portfolio Management," *Journal of Financial and Quantitative Analysis*, Vol. VI, No. 3 (June 1971), pp. 955–976.
8. ——, "Assessment of Yields on U.S. Government Securities," Graduate School of Business Administration, Harvard University, Working Paper No. 71-18, December 1971.
9. Daellenbach, Hans G. and Archer, Stephen H., "The Optimal Bank Liquidity: A Multi-Period Stochastic Model," *Journal of Financial and Quantitative Analysis*, Vol. IV, No. 3 (September 1969), pp. 329–343.
10. Dantzig, George B., *Linear Programming and Extensions*, Princeton University Press, Princeton, N.J., 1963.
11. Eppen, Gary D. and Fama, Eugene F. "Solutions for Cash Balance and Simple Dynamic Portfolio Problems with Proportional Costs," *Journal of Business*, Vol. 41 (January 1968), pp. 94–112.
12. —— and ——, "Three Asset Cash Balance and Dynamic Portfolio Problems," *Management Science*, Vol. 17, No. 5 (January 1971), pp. 311–319.
13. Telser, Lester, "Safety First and Hedging," *Review of Economic Studies*, Vol. 23 (1955–1956).
14. Wolf, Charles R., "A Model for Selecting Commercial Bank Government Security Portfolios," *The Review of Economics and Statistics*, Vol. LI, No. 1 (February 1969), pp. 40–52.

2. MODELS OF OPTIMAL CAPITAL ACCUMULATION AND PORTFOLIO SELECTION

JOURNAL OF ECONOMIC THEORY 3, 40–53 (1971)

Multiperiod Consumption–Investment Decisions and Risk Preference

EDWIN H. NEAVE

Graduate School of Management, Northwestern University, Evanston, Illinois 60201

Received February 17, 1970

The decision problem of a consumer who must make regular allocations of wealth to consumption expenditures and to risky investment is considered. Sufficient conditions for the consumer's multiperiod utility function to exhibit both decreasing absolute and increasing relative risk aversion are determined. Essentially, the multiperiod utility function exhibits decreasing absolute risk aversion if each period's utility of consumption exhibits decreasing absolute risk aversion; the multiperiod function exhibits increasing relative risk aversion for large levels of wealth if each period's utility exhibits increasing relative risk aversion and if consumption of wealth is approximately linear.

1. INTRODUCTION

This paper considers the risk preferences exhibited by a consumer who must allocate his available wealth at the beginning of each of a series of equally spaced time periods between consumption and risky investment. In each period the wealth placed in the risky investment stochastically increases or decreases in value, and is at the same time augmented by receipt of a certain[1] income. The consumer's behavior is assumed to be motivated by expected utility maximization, and utility is assumed to be additive between time periods. It is further assumed that each period's utility of consumption function exhibits both decreasing absolute and increasing relative risk aversion. Then with certain other specifications, it will be shown that the consumer's decisions when made with reference to multiperiod time horizons are always characterized by the property of decreasing absolute risk aversion, and generally also by increasing relative risk aversion.[2]

These results are of interest because in the consumption–investment

[1] As the ensuing discussion will suggest, this assumption is made primarily for convenience. Only minor formal changes are needed to consider a stochastically varying income.

[2] For discussions of risk preference measures, see Refs. [2, 12]. Measures of absolute and relative risk aversion are defined at the beginning of Section 2.

models which have recently appeared, preservation of decreasing absolute and increasing relative risk aversion under maximization and expectation operations is considered only for the special cases in which these measures are constant (see, e.g., Refs. [6, 8, 13]), or else in terms of examples (such as in Ref. [9]). On the other hand, when the measures are not assumed to be constant, their properties have been examined in the context of single-period decision models, as is the case in the classic papers [2,12], and in the extensions developed in Ref. [14]. This paper's results thus contribute to integration of the directions that investigations involving the Arrow–Pratt measures have taken.

The outline of the remaining sections of the paper is as follows. In Section 2 some new properties of risk aversion measures are developed, with emphasis being given to the behavior of the risk aversion measures under expectation operations. In the paper's third section, the preservation of both decreasing absolute and increasing relative risk aversion under maximization is considered in the context of the consumption–investment decision problem already outlined. The fourth section considers some extensions of the measures to the analysis of relations between consumption–investment decisions and wealth. One of the characteristics of the present approach is to consider measures of risk aversion somewhat more generally as concavity measures, and to use the measures for comparisons of the concavity properties of the additive utility functions.

2. Some Properties of Risk Aversion Measures

This section is concerned primarily with studying the behavior of absolute and relative risk aversion measures under expectation operations. For the most part, this section develops results which will be employed in Section 3 to study the consumer's multiperiod consumption–investment decisions. However, the first two lemmas are not employed in the sequel; they are included for the sake of their general interest.

Definition. If f is a thrice-differentiable function defined on the real line, then f is said to exhibit decreasing absolute risk aversion iff $r_f' \leqslant 0$, where $r_f = -f''/f'$.

Lemma 1. *If f is a function defined on the integers 1, 2,..., and if f can be written in the form*

$$f(n) = \sum_{t=0}^{n} k^{g(t)},$$

where $k \in (0, 1)$ and $g(t)$ is an increasing function for all $t \geqslant 0$, then $g(t)$ is concave for all $t \geqslant 0$ iff r_f is decreasing.

Proof. The function g is concave

$$\Leftrightarrow 2g(m) \geqslant g(m+1) + g(m-1)$$
$$\Leftrightarrow k^{2g(m)} \leqslant k^{g(m+1)} + k^{g(m-1)}$$
$$\Leftrightarrow k^{g(m)}/k^{g(m-1)} \leqslant k^{g(m+1)}/k^{g(m)}$$
$$\Leftrightarrow \frac{f(m) - f(m-1)}{f(m-1) - f(m-2)} \leqslant \frac{f(m+1) - f(m)}{f(m) - f(m-1)}$$
$$\Leftrightarrow r_f \text{ is decreasing.}$$

(See Ref. [12, p. 128] for justification of this last step.)

The next result relates the properties of decreasing absolute and increasing relative risk aversion for the case of strictly increasing functions.

DEFINITION. If f is a thrice-differentiable function defined on the real line, f is said to exhibit increasing relative risk aversion iff $(r_f^*)' \geqslant 0$, where $r_f^* = -xf''/f'$.

LEMMA 2. *Let g, h be strictly increasing functions such that r_g^* and r_h^* are increasing, and such that for every[3] x, $r_g(x) \geqslant r_h(x)$ and $r_g^*(x) \leqslant r_h^*(x)$. Let $f = g + h$. Then r_f^* is an increasing function.*[4]

Proof. The assumption $r_g \geqslant r_h$

$$\Rightarrow -g''/g' \geqslant -h''/h'$$
$$\Rightarrow -g''h' \geqslant -h''g' \text{[since both } g \text{ and } h \text{ are strictly increasing functions]}$$
$$\Rightarrow (r_h^* - r_g^*)(g'h'' - g''h') \geqslant 0$$
$$\Rightarrow (g')^2 (r_g^*)' + g'h'(r_h^*)' + g'h''(r_h^* - r_g^*) + (h')^2 (r_h^*)' + g'h'(r_g^*)' + g''h'(r_g^* - r_h^*) \geqslant 0$$
$$\Rightarrow [(g''r_g^* + g'(r_g^*)' + h''r_h^* + h'(r_h^*)')(g' + h') - (g'r_g^* + h'r_h^*)(g'' + h'')] = (g' + h')^2 (r_f^*)' \geqslant 0 \text{ [as may be seen by noting that } r_f^* = (r_g^*g' + r_h^*h')/(g' + h')]$$
$$\Leftrightarrow r_f^* \text{ is an increasing function.}$$

Q.E.D.

[3] For brevity in the subsequent development this type of statement will be written $r_g \geqslant r_h$.

[4] Note that if the functions g and h of the lemma are strictly increasing and strictly concave, $r_g(x) \geqslant r_h(x) > 0$. This implies that $r_g^*(x) = xr_g(x) \leqslant r_h^*(x) = xr_h(x)$ iff $x \leqslant 0$. Accordingly, in this case the hypotheses of the lemma are satisfied only by nonpositive values of x.

The next two lemmas are concerned with the behavior of the absolute risk aversion measure under expectation operations. In these lemmas, as well as in the remainder of the paper, only situations in which the expectations exist and are finite will be considered.

LEMMA 3. (Pratt). *Let g, h be functions such that r_g and r_h are decreasing, and let $f = t_1 g + t_2 h$; $t_1, t_2 \geqslant 0$. Then r_f is a decreasing function.*

Proof. The proof consists of showing that $r_f' \leqslant 0$; see Ref. [12, p. 132].

LEMMA 4. *Let $f(x) = E\{g(k + \zeta x)\}$, where ζ is a random variable defined on any interval $(\subseteq, \infty)$. Then if g exhibits decreasing absolute risk aversion, f exhibits decreasing absolute risk aversion.*

Proof. For any fixed value of ζ, say ζ_0, the function $g(k + \zeta_0 x)$ exhibits decreasing absolute risk aversion. Lemma 3 states that the nonnegative sum of decreasingly absolute risk averse functions is decreasingly absolute risk averse. Then since expectation can be regarded as a nonnegative summation operation whether ζ is discrete or continuous, it follows that f exhibits decreasing absolute risk aversion.

Some results concerning the behavior of the relative risk aversion measure under expectation operations will next be considered.

LEMMA 5. *If g is a strictly increasing, strictly concave function which exhibits constant relative risk aversion, and if $k > 0$, then r_h^* is increasing, where $h(x) = g(x + k)$.*

Proof. Note that $r_h^*(x) = -xh''(x)/h'(x) = -xg''(x + k)/g'(x + k)$. Since $r_g^*(x) = m > 0$, $r_g(x) = m/x$, and $r_g(x + k) = m/(x + k)$. But $r_h^*(x) = xr_g(x + k) = xm/(x + k)$, an increasing function of x.

COROLLARY. *If $k < 0$, $g(x + k)$ exhibits decreasing relative risk aversion.*

Lemma 5 does not imply that $E\{g(x + \rho)\}$ exhibits constant relative risk aversion when $E(\rho) = 0$, as is shown by considering

$$(\tfrac{2}{3}) \ln(x + 1) + (\tfrac{1}{3}) \ln(x - 2),$$

where $\ln(x)$ denotes the natural logarithm of x. The measure of relative risk aversion for this function is $(3x^3 - 6x^2 + 9x)/(3x^3 - 6x^2 - 3x + 6)$, a decreasing function.

EXAMPLE. The nonnegative sum of increasingly relative risk averse functions is not necessarily increasingly relative risk averse.

Let $g(x) = \ln x$; $h(x) = \ln(x + 5)$; $f = g + h$.

Then $r_f^*(x) = (x^{-1} + x(x + 5)^{-2})/(x^{-1} + (x + 5)^{-1})$, $r_f^*(1) = 55.5/63$, and $r_f^*(2) = 53/63$.

The next lemma can be used to show that increasing or decreasing relative risk aversion is preserved under expectation operations for sufficiently large values of x.

LEMMA 6. *Let ζ be a random variable defined on any finite closed interval $\subseteq (0, \infty)$, let k be a real number, and let $f(x) = E\{g(k + \zeta x)\}$, where g is a strictly increasing function. Then if $\lim_{x\to\infty} r_g^*(x)$ exists, for each $\epsilon > 0$ there is x_ϵ such that $|r_f^*(x) - r_g^*(x)| < \epsilon$ for $x \geqslant x_\epsilon$.*

Proof. Let $\epsilon > 0$. For each fixed value of ζ, define $x_{\epsilon,\zeta}$ to be a number such that

$$|(\zeta x/(k + \zeta x))\, r_g^*(k + \zeta x) - r_g^*(x)| < \epsilon \text{ for } x \geqslant x_{\epsilon,\zeta}. \tag{2.2}$$

Finite numbers $x_{\epsilon,\zeta}$ exist because $\lim_{x\to\infty} (\zeta x/(k + \zeta x)) = 1$ and because $\lim_{x\to\infty} r_g^*(x)$ exists. Then let $x_\epsilon = \text{lub}\,\{x_{\epsilon,\zeta}\}$. Now the function r_f^* can be written as

$$r_f^*(x) = E\{(\zeta^2 x/(k + \zeta x))\, g'(k + \zeta x)\, r_g^*(k + \zeta x)\}/E\{g'(k + \zeta x)\zeta\}. \tag{2.3}$$

The results of the lemma can now be established by multiplying both sides of inequality (2.2) by $g'(k + \zeta x)\zeta > 0$ and taking expectations to give

$$|E\{g'(k + \zeta x)\zeta\}(r_f^*(x) - r_g^*(x))| < E\{g'(k + \zeta x)\zeta\}\epsilon \quad \text{for} \quad x \geqslant x_\epsilon, \tag{2.4}$$

and finally, dividing both sides of inequality (2.4) by $E\{g'(k + \zeta x)\zeta\} > 0$.

Q.E.D.

The results of Lemma 6 can be employed to argue that if r_g^* is increasing, then r_f^* is increasing for sufficiently large values of x.

Some useful properties of the risk aversion measures now having been established, we turn to a consideration of the multiperiod decision problem sketched in the Introduction. The behavior of the risk aversion measures under maximization will be explored in the context of this consumption–investment allocation problem.

3. MULTIPERIOD CONSUMPTION–INVESTMENT DECISIONS AND RISK PREFERENCE

This section considers the multiperiod risk preferences of a consumer faced with recurring consumption–investment decisions. At the beginning of each period the consumer is required to allocate his current wealth between consumption and an income-generating asset. He is then assumed to receive an income from sale of labor services and to realize gains or losses on his risky investment, after which he is required to make another allocation decision.

It will be assumed that utilities are additive in time, and that the single-period functions exhibit decreasing absolute and increasing relative risk aversion; with these conditions and some additional qualifications it will be shown that the consumer's multiperiod decisions are made in a manner which exhibits both decreasing absolute and increasing relative risk aversion. Thus, sufficient conditions for single-period risk preferences to be reflected in the derived utility functions of the sequential maximization problem will be exhibited.

To proceed formally, define the problem

$$\max_{a,b} E\left\{\sum_{n=1}^{N} \beta^{n-1} u_n(a_n, b_n)\right\} \text{ subject to } a_n + b_n = A_n, \quad a_n \in [0, A_n],\ A_{n+1} = k + \zeta a_n, \tag{3.1}$$

where

$u_n(a_n, b_n) = v_n(b_n)$;

$n = 1, 2, \ldots, N-1$;

$u_N(a_N, b_N) = v_N(b_N) + \beta w_N(k + \zeta a_N)$;

$v_n(\ ) =$ utility of consumption in period[5] n;

$b_n =$ dollar value of consumption in period n;

$\beta w_N(\ ) =$ beginning-period utility of wealth[6] possessed at end of period N;

$k =$ nonnegative cash income during each period[7];

$a_n =$ amount of wealth invested in period n;

[5] Economists differ concerning the importance of assuming these functions to be bounded. Reference [2] discusses the types of paradox which can arise if functions are not bounded; Ref. [13] maintains that qualifications respecting boundedness are not of prime importance. For simplicity we omit the boundedness assumption.

[6] Wealth is assumed to have utility at the problem's horizon because the individual is supposed to regard it either as a bequest (which provides him satisfaction) or as a proxy for consumption at points in time beyond those included in his planning problem. The need for (and desirability of) including the bequest motive disappears in infinite-horizon formulations of the problem. For discussions of this point, see Refs. [3, 8, 13].

[7] It is assumed that transactions between cash and the risky asset are costless; hence, it becomes unnecessary to distinguish between the forms in which wealth may be held at the beginning of any period.

ζ = stochastic rate of return on wealth invested in period n; $\zeta \in [y_1, y_2] \subseteq (0, \infty)$;

A_n = total wealth available at beginning of period n.

Discount factors or time preference are assumed to be reflected in the transformation factor $\beta \in (0, 1)$.

To continue the analysis, redefine Eq. (3.1) recursively as

$$U_n(A) = \max_{a\in[0,A]} E\{v_n(A-a) + \beta U_{n+1}(k + \zeta a)\}, \qquad n = 1, 2,..., N,$$
$$U_{N+1}(A) \equiv w_N(A), \tag{3.2}$$

where subscripts discernible from the context of the recursive formulation have been dropped, and the equality constraints are incorporated in Eqs. (3.2). The functions U_n now represent the expected utility of following an optimal policy from the beginning of period n to the end of period N, given that the wealth at the beginning of period n was A.

A set of conditions sufficient for the functions U_n to exhibit decreasing absolute and increasing relative risk aversion for all $n = 1, 2,..., N$ will now be presented.[8] These characteristics of the multiperiod return functions are obtained inductively and proceed from a consideration of the one-period problem.

Let

$$U_N(A) = \max_{a\in[0,A]} E\{v_N(A-a) + \beta w_N(k + \zeta a)\}, \tag{3.3}$$

and write

$$U_N(A) = v_N(c) + z_N(s), \quad c(A) = A - a^*(A), \quad s(A) = a^*(A), \tag{3.4}$$

where $a^*(A)$ satisfies Eq. (3.3).

THEOREM 1. *If the functions v_N and w_N of Eq.* (3.3) *are strictly concave and if r_v and r_w are nonincreasing functions,*[9] *then U_N is a strictly concave decreasingly absolute risk averse function.*

Proof. It will first be shown that r_U is nonincreasing, thus establishing that U_N is decreasingly absolute risk averse. Three cases are considered in this part of the proof.

(i) $s = 0$. In this case $U_N(A) = v_N(A) +$ const; hence

$$r_U'(A) \leqslant 0 \Leftrightarrow r_v'(A) \leqslant 0.$$

[8] An extension of the above results to infinite-horizon problems using techniques developed in Ref. [4] can be effected.

[9] No ambiguity arises from our writing r_v and r_w rather than r_{v_N} and r_{w_N}, because the time subscripts can be inferred from the context of the discussion.

(ii) $s = A$. In this case $U_N(A) = \text{const} + z_N(A)$; hence r_U' is determined by r_z', and the sign of r_z' is nonpositive by Lemma 4.

(iii) $s \in (0, A)$. It is useful first to establish some preliminary results. By hypothesis and by definition of c and s,

$$-v_N'(c) + z_N'(s) \equiv 0. \tag{3.5}$$

Therefore,

$$U_N'(A) = v_N'(c)\, c' + z_N'(s)\, s' = v_N'(c)(1 - s') + z_N'(s)\, s',$$

so that by Eqs. (3.4) and (3.5),

$$U_N'(A) = v_N'(c) = z_N'(s). \tag{3.6}$$

Furthermore,

$$U_N''(A) = v_N''(c)\, c' = z_N''(s)\, s'. \tag{3.7}$$

Finally, total differentiation of (3.5) gives

$$s' = v_N''(c)/[v_N''(c) + z_N''(s)] \in (0, 1) \tag{3.8}$$

by the strict concavity of v_N and w_N and by the fact that strict concavity is preserved under expectation operations.[10]

Accordingly,

$$r_U(A) = r_v(c)\, c' = r_z(s)\, s', \tag{3.9}$$

and it is desired to show that $r_U'(A) \leqslant 0$. Now $s' \in (0, 1)$ by Eq. (3.8), and r_v, r_z are positive functions by strict concavity of v_N and z_N. Let $[B_1, B_2]$ be any interval on which s' is monotone. Then on $[B_1, B_2]$ either s', c', or both must be nonincreasing. Suppose s' is nonincreasing. Then $r_U(A) = r_z(s)\, s'$ must be nonincreasing since it is a product of nonnegative, nonincreasing functions. If c' is nonincreasing, the same result obtains by taking $r_U(A) = r_v(c)\, c'$ and by noting that Eq. (3.8) implies c to be a nondecreasing function of A.

The process may be continued for all intervals on which s' is monotone, thus completing the first part of the proof.

The fact that U_N will be strictly concave is proved in Ref. [7], where it is shown that if $g(x, y)$ is a (strictly) concave function and if C is a convex set, then $f(x) = \max_{y \in C}\{g(x, y)\}$ is a (strictly) concave function. Q.E.D.

THEOREM 2. *Suppose that the functions v_N and w_N of Eqs.* (3.3) *and* (3.4) *are strictly concave, and that the functions r_v^* and r_z^* are nondecreasing functions. Let $\epsilon > 0$. If $s''(s' - \sigma) \leqslant 0$, or if $s''(s' - \sigma) > 0$ and*

$$|s''| \leqslant \min\left\{\frac{(1 - s')|s' - \sigma|}{A + \epsilon}, \frac{s'|s' - \sigma|}{A + \epsilon}\right\},$$

[10] This last fact is quite well-known. See, e.g., Ref. [5].

where $s'' = s''(A)$, $s' = s'(A)$, *and* $\sigma = s/A$, *then* U_N *is increasingly relative risk averse.*

Proof. Several cases must be considered, and it will be noted that some hypotheses of the theorem can be relaxed in each case.

(i) If $s = 0$ or if $s = A$, the conclusion that $r_U{}^*$ is increasing follows immediately using the reasoning employed in Theorem 1.

(ii) Suppose $s \in (0, A)$. Then from Eq. (3.9),

$$r_U{}^*(A) = Ar_v(c)\, c' = Ar_z(s)\, s', \tag{3.10}$$

so that[11]

$$r_U{}^*(A) = \eta_c r_v{}^*(c) = \eta_s r_z{}^*(s), \text{ where } \eta_f = xf'(x)/f(x). \tag{3.11}$$

The reasoning of Theorem 1 can be applied whenever it can be shown that η_c and η_s are not both decreasing. Such a demonstration requires examining four cases, some preliminary discussion of which is useful. For given A and for $\epsilon > 0$, Taylor's theorem may be used to write

$$s(A + \epsilon) \approx s(A) + \epsilon s'(A) = \sigma A + \epsilon\varphi$$

and

$$s'(A + \epsilon) \approx s'(A) + \epsilon s''(A) = \varphi + \epsilon\delta;$$

$\sigma \in (0, 1)$ by hypothesis and $\varphi \in (0, 1)$ by Eq. (3.8). Consider now the functions $\eta_c(B)$ and $\eta_s(B)$, which may be written, respectively, as $(1 - \varphi)/(1 - \sigma)$ and φ/σ when $B = A$, and as

$$(1 - \varphi - \epsilon\delta)/\left[(1 - \sigma) - \frac{\epsilon\gamma}{A + \epsilon}\right] \text{ and } (\varphi + \epsilon\delta)/\left[\sigma + \frac{\epsilon\gamma}{A + \epsilon}\right]$$

when $B = A + \epsilon$, using the notation $\varphi = \sigma + \gamma$.
Noticing that

$$(1 - \varphi)/\left[(1 - \sigma) - \frac{\epsilon\gamma}{A + \epsilon}\right] < (1 - \varphi)/(1 - \sigma) \text{ and } \varphi/\left[\sigma + \frac{\epsilon\gamma}{A + \epsilon}\right] < \varphi/\sigma$$

cannot both obtain simultaneously, considerations of continuity suggest

[11] An economic interpretation of Eq. (3.11) is given in Section 4. Note that $(1 - \sigma)\eta_c + \sigma\eta_s \equiv 1$, but the identity does not permit the conclusion that η_c and η_s cannot decrease simultaneously.

that the magnitude of δ be investigated to obtain the conclusions sought. Four cases must be considered:

(a) $\gamma \geqslant 0, \delta \leqslant 0$. Then $(1-\varphi-\epsilon\delta)/\left[(1-\sigma)-\dfrac{\epsilon\gamma}{A+\epsilon}\right] \geqslant (1-\varphi)/[1-\sigma]$.

(b) $\gamma \leqslant 0, \delta \geqslant 0$. Then $(\varphi+\epsilon\delta)/\left[\sigma+\dfrac{\epsilon\gamma}{A+\epsilon}\right] \geqslant \varphi/\sigma$.

(c) $\gamma \geqslant 0, \delta \geqslant 0$.

In this case the conclusion sought is the same as in case (*a*). By hypothesis,

$$0 \leqslant \delta \leqslant \frac{(1-\varphi)\,\gamma}{A+\epsilon} \leqslant \frac{(1-\varphi)\,\gamma}{(1-\sigma)(A+\epsilon)}$$

$$\Rightarrow \epsilon\delta \leqslant \frac{(1-\varphi)\,\epsilon\gamma}{(1-\sigma)(A+\epsilon)}$$

$$\Rightarrow (1-\varphi)(1-\sigma)-(1-\sigma)\,\epsilon\delta \geqslant (1-\varphi)(1-\sigma)-\frac{(1-\varphi)\,\epsilon\gamma}{A+\epsilon}$$

$$\Rightarrow (1-\varphi-\epsilon\delta)/\left(1-\sigma-\frac{\epsilon\gamma}{A+\epsilon}\right) \geqslant \frac{1-\varphi}{1-\sigma}.$$

(d) $\gamma \leqslant 0, \delta \leqslant 0$. In this case the conclusion sought is the same as in case (b). By hypothesis,

$$0 \geqslant \delta \geqslant \frac{\varphi\gamma}{A+\epsilon} \geqslant \frac{\varphi\gamma}{\sigma(A+\epsilon)}$$

$$\Rightarrow \epsilon\delta \geqslant \frac{\varphi\epsilon\gamma}{\sigma(A+\epsilon)}$$

$$\Rightarrow \sigma\varphi+\sigma\epsilon\delta \geqslant \sigma\varphi+\frac{\varphi\epsilon\gamma}{A+\epsilon}$$

$$\Rightarrow (\varphi+\epsilon\delta)/\left(\sigma+\frac{\epsilon\gamma}{A+\epsilon}\right) \geqslant \frac{\varphi}{\sigma}.$$

This completes the proof of the theorem.

Remarks. In view of Lemma 6 and the example preceding it, only the weaker conclusion that $r_U{}^*(A)$ is increasing for sufficiently large A can be obtained if the hypotheses of the theorem are altered to have $r_w{}^*$ rather than $r_z{}^*$ nondecreasing. However, this conclusion could be rendered slightly sharper if boundedness of the utility functions were assumed, since as Ref. [2] has indicated, in the case of bounded functions relative risk aversion can be decreasing only on intervals the sum of whose lengths is finite.

642/3/1-4

Finally, although the result is implied by the statement of Theorem 2, it is useful to present the next lemma because of the alternative method of proof employed.

LEMMA 7. *Let the functions v_N and z_N be defined as in Eqs.* (3.3) *and* (3.4). *If the functions r_v^* and r_z^* are nondecreasing and if c is linear on some interval $[B_1, B_2]$, then $r_U^*(A)$ is nondecreasing on $[B_1, B_2]$.*

Proof. The linearity of c and Eq. (3.8) together imply that $c'(A) = 1 - t$, and $s'(A) = t$, $t \in (0, 1)$, for $A \in [B_1, B_2]$. Therefore, $(1 - \sigma)\,\eta_c = (1 - t)$ and $\sigma\eta_s = t$. As A increases, either σ, $1 - \sigma$, or both must be nonincreasing, so that either η_c, η_s, or both must be nondecreasing. But since from Eq. (3.11), $r_U^*(A) = \eta_c r_v^*(c) = \eta_s r_z^*(s)$, it follows that r_U^* is nondecreasing on $[B_1, B_2]$.

It is now time to consider the problem of extending Theorems 1 and 2 to any finite-horizon[12] problem. Analogous to the definitions of Eqs. (3.3) and (3.4) let[13]

$$U_n(A) = \max_{a \in [0, A]} E\{v_n(A - a) + \beta U_{n+1}(k + \zeta a)\}, \tag{3.12}$$

$$U_n(A) = v_n(c) + z_n(s). \tag{3.13}$$

Completion of an inductive step relating absolute risk aversion in periods $n + 1$ and n will first be undertaken.

THEOREM 3. *If the functions v_n and U_{n+1} of Eqs.* (3.12) *and* (3.13) *are strictly concave, and if r_v and $r_{U_{n+1}}$ are nonincreasing functions, then U_n is a strictly concave decreasingly absolute risk averse function.*

Proof. The proof is exactly the same as for Theorem 1.

We turn now to the task of relating relative risk aversion in periods $n + 1$ and n.

THEOREM 4. *Suppose that the functions v_n and U_{n+1} of Eqs.* (3.12) *are strictly concave, and that r_v^* and r_z^* are nondecreasing functions.*[14] *Let $\epsilon > 0$. If*

$$s''(s' - \sigma) \leqslant 0,$$

[12] As noted before, extensions to infinite-horizon problems will not be dealt with in the present investigation.

[13] As indicated in footnote 9, no ambiguity arises from our omitting to write the functions as c_n and s_n.

[14] To be precise, it should be stipulated that r_z^* is nondecreasing for $x \geqslant x_1$; see the remarks following Lemma 5. This complication is omitted to simplify the proof.

or if

$$s''(s' - \sigma) > 0 \text{ and } | s'' | \leqslant \min \left\{ \frac{(1 - s')| s' - \sigma |}{A + \epsilon}, \frac{s' | s' - \sigma |}{A + \epsilon} \right\},$$

where $s'' = s''(A)$, $s' = s'(A)$, *then* U_n *is an increasingly relative risk averse function.*

Proof. The proof is exactly the same as for Theorem 2.

Note that, as indicated in the remarks following Theorem 2, the hypotheses of Theorem 4 imply that $r^*_{U_n}$ will be nondecreasing whenever the functions c_n (and consequently the functions s_n) are linear.

Having developed sufficiency conditions for the functions U_n to exhibit decreasing absolute and increasing relative risk aversion, it is now appropriate to turn to a discussion of the uses of these measures in studying wealth elasticities. It is rather unfortunate that, since the measure of absolute risk aversion is preserved under more general assumptions than is the measure of relative risk aversion, the former's ability to characterize the sensitivity of the optimal policy precisely is less strong than that of the latter.

4. Wealth Elasticities and Risk Preference

This section discusses how the concavity measures studied in the context of the multiperiod decision model of Section 3 can be employed to develop properties of the policy functions c_n and s_n. The present work provides some results additional to those in Refs. [2, 14], thus further illustrating the usefulness of the Arrow–Pratt concavity measures.

The first of these extensions is concerned with developing benchmark measures of the elasticity of demand for risky investment.

Lemma 8. *If both* v_n *and* z_n *exhibit constant relative risk aversion equal to unity*[15] *the wealth elasticity of demand for risky investment is also equal to unity.*

Proof. By Eq. (3.8),

$$s' = v_n''(c)/[v_n''(c) + z_n''(s)].$$

By constant relative risk aversion,

$$v_n''(c) = -v_n'(c)/c$$

[15] Usually the measure of relative risk aversion for bounded functions will hover around unity; see Ref. [2].

and

$$z_n''(s) = -z_n'(s)/s = -v_n'(c)/s;$$

the last equality following by conditions analogous to those of Eq. (3.5). Therefore,

$$s' = \frac{v_n''(c)}{z_n''(s) + v_n''(c)} = \frac{-v_n''(c)}{v_n'(c)\left(\frac{1}{s} + \frac{1}{c}\right)} = \frac{-v_n''(c)\, cs}{v_n'(c)\, A} = \frac{s}{A} = \sigma,$$

again by constant relative risk aversion. Therefore, $\eta_s = s'/\sigma = 1$. Q.E.D.

The method of proof used in Lemma 8 may be employed in making useful benchmark statements about the elasticity of demand for risky investments in more general situations by writing $r_z^*(x) = 1 + \epsilon(x)$ (where, for example, $\epsilon(x)$ may be a decreasing positive function) and again making appropriate substitutions in the expression for s'. An alternative to this procedure is given in the following lemma.

LEMMA 9. *In any period n, the wealth elasticity of risky investment exceeds unity iff* $r_z^*(s) \leqslant r_v^*(c)$.

Proof. We have

$$\eta_s = \frac{s'}{\sigma} = \frac{Av_n''(c)}{s(v_n''(c) + z_n''(s))}$$

$$= 1 - \frac{sz_n''(s) - cv_n''(c)}{s(v_n''(c) + z_n''(s))} \geqslant 1$$

$$\Leftrightarrow sz_n''(s) - cv_n''(c) \geqslant 0$$

$$\Leftrightarrow \frac{sz_n''(s)}{z_n'(s)} - \frac{cv_n''(c)}{v_n'(c)} \geqslant 0 \qquad \text{[by conditions analogous to those of Eq. (3.5)]}$$

$$\Leftrightarrow -r_z^*(s) + r_v^*(c) \geqslant 0.$$

LEMMA 10. *If the functions* r_v^* *and* r_z^* *are nondecreasing, then* r_U^* *is nondecreasing whenever the wealth elasticity of either consumption or risky investment is nondecreasing.*

Proof. The result is an interpretation of Eq. (3.11):

$$r_U^*(A) = \eta_c r_v^*(c) = \eta_s r_z^*(s). \qquad \text{Q.E.D.}$$

5. Conclusions

Given that some relations between multiperiod decisions and risk preferences have been established, it is clearly desirable to explore further the nature of the conditions under which these relationships obtain. Is it possible, for example, to generalize the results of this paper to the two-asset decision problem considered in Ref. [14]? To what extent can the sufficiency conditions of Section 3 be weakened and yet yield the same results? What conditions are necessary for some of the results of Section 3? As these questions suggest, additional work remains to be done before the problem of extending risk preference measures to multiperiod models is well understood.

References

1. K. J. Arrow, Comment, *Rev. Econ. Statist.* **45** (1963), 24–26.
2. K. J. Arrow, "Aspects of the theory of risk-bearing," Yrjö Jahnssonin Säätiö, Helsinki, 1965.
3. K. J. Arrow and M. Kurz, Optimal consumer allocation over an infinite horizon, *J. Econ. Theory* **1** (1969), 68–91.
4. D. Blackwell, Discounted dynamic programming, *Ann. Math. Statist.* **36** (1965), 226–235.
5. G. D. Eppen and E. F. Fama, Cash balance and simple dynamic portfolio problems with proportional costs, *Internat. Econ. Rev.* **10** (1969). 119–133.
6. N. H. Hakansson, Optimal investment and consumption strategies under risk for a class of utility functions, *Econometrica*, **38** (1970), 587–607.
7. D. L. Iglehart, Capital accumulation and production: Optimal dynamic policies, *Management Sci.* **12** (1965), 193–205.
8. R. C. Merton, Lifetime portfolio selection under uncertainty: The continuous-time case, *Rev. Econ. Statist.* **50** (1969), 247–257.
9. J. Mossin, Optimal multiperiod portfolio policies, *J. Business* **41** (1968), 215–229.
10. E. H. Neave, "The stochastic cash balance problem: Further results," Northwestern University Graduate School of Management, Working Paper 19–69, 1969.
11. E. H. Neave, "Multiperiod money demand functions of stochastic variables: A formal approach," Northwestern University Graduate School of Management, Working Paper, in preparation.
12. J. W. Pratt, Risk aversion in the small and in the large, *Econometrica* **32** (1964), 122–136.
13. P. A. Samuelson, Lifetime portfolio selection by dynamic stochastic programming, *Rev. Econ. Statist.* **50** (1969), 239–246.
14. A. Sandmo, Portfolio choice in a theory of saving, *Swed. J. Econ.* **70** (1968), 106–122.
15. J. E. Stiglitz, The effects of income, wealth, and capital gains on risk-taking, *Quart. J. Econ.* **83** (1969), 263–283.

ERRATA

p. 46 (507). Before Theorem 1, read:

> In the succeeding development, we say that $s(A)$ is an interior solution if $s(A) \in (0, A)$ for all A, and that $s(A)$ is a boundary solution if either $s(A) = A$ or $s(A) = 0$ for all A.

p. 46 (507) ff. Statements of Theorems 1 through 4 should end with:

> ...whenever $s(A)$ is either a boundary or an interior solution.

Reprinted from *The Review of Economics and Statistics* **51**, 239-246 (1969).

LIFETIME PORTFOLIO SELECTION BY DYNAMIC STOCHASTIC PROGRAMMING

Paul A. Samuelson *

Introduction

MOST analyses of portfolio selection, whether they are of the Markowitz-Tobin mean-variance or of more general type, maximize over one period.[1] I shall here formulate and solve a many-period generalization, corresponding to lifetime planning of consumption and investment decisions. For simplicity of exposition I shall confine my explicit discussion to special and easy cases that suffice to illustrate the general principles involved.

As an example of topics that can be investigated within the framework of the present model, consider the question of a "businessman risk" kind of investment. In the literature of finance, one often reads; "Security A should be avoided by widows as too risky, but is highly suitable as a businessman's risk." What is involved in this distinction? Many things.

First, the "businessman" is more affluent than the widow; and being further removed from the threat of falling below some subsistence level, he has a high propensity to embrace variance for the sake of better yield.

Second, he can look forward to a high salary in the future; and with so high a present discounted value of wealth, it is only prudent for him to put more into common stocks compared to his present tangible wealth, borrowing if necessary for the purpose, or accomplishing the same thing by selecting volatile stocks that widows shun.

Third, being still in the prime of life, the businessman can "recoup" any present losses in the future. The widow or retired man nearing life's end has no such "second or n^{th} chance."

Fourth (and apparently related to the last point), since the businessman will be investing for so many periods, "the law of averages will even out for him," and he can afford to act almost as if he were not subject to diminishing marginal utility.

What are we to make of these arguments? It will be realized that the first could be purely a one-period argument. Arrow, Pratt, and others[2] have shown that any investor who faces a range of wealth in which the elasticity of his marginal utility schedule is great will have high risk tolerance; and most writers seem to believe that the elasticity is at its highest for rich — but not ultra-rich! — people. Since the present model has no new insight to offer in connection with statical risk tolerance, I shall ignore the first point here and confine almost all my attention to utility functions with the same relative risk aversion at all levels of wealth. Is it then still true that lifetime considerations justify the concept of a businessman's risk in his prime of life?

Point two above does justify leveraged investment financed by borrowing against future earnings. But it does not really involve any increase in relative risk-taking once we have related what is at risk to the proper larger base. (Admittedly, if market imperfections make loans difficult or costly, recourse to volatile, "leveraged" securities may be a rational procedure.)

The fourth point can easily involve the innumerable fallacies connected with the "law of large numbers." I have commented elsewhere[3] on the mistaken notion that multiplying the same kind of risk leads to cancellation rather

* Aid from the National Science Foundation is gratefully acknowledged. Robert C. Merton has provided me with much stimulus; and in a companion paper in this issue of the REVIEW he is tackling the much harder problem of optimal control in the presence of continuous-time stochastic variation. I owe thanks also to Stanley Fischer.

[1] See for example Harry Markowitz [5]; James Tobin [14], Paul A. Samuelson [10]; Paul A. Samuelson and Robert C. Merton [13]. See, however, James Tobin [15], for a pioneering treatment of the multi-period portfolio problem; and Jan Mossin [7] which overlaps with the present analysis in showing how to solve the basic dynamic stochastic program recursively by working backward from the end in the Bellman fashion, and which proves the theorem that portfolio proportions will be invariant only if the marginal utility function is iso-elastic.

[2] See K. Arrow [1]; J. Pratt [9]; P. A. Samuelson and R. C. Merton [13].

[3] P. A. Samuelson [11].

than augmentation of risk. I.e., insuring many ships adds to risk (but only as $\sqrt{n}$); hence, only by insuring more ships and by *also* subdividing those risks among more people is risk on each brought down (in ratio $1/\sqrt{n}$).

However, before writing this paper, I had thought that points three and four could be reformulated so as to give a valid demonstration of businessman's risk, my thought being that investing for each period is akin to agreeing to take a $1/n^{\text{th}}$ interest in insuring n independent ships.

The present lifetime model reveals that investing for many periods does not *itself* introduce extra tolerance for riskiness at early, or any, stages of life.

Basic Assumptions

The familiar Ramsey model may be used as a point of departure. Let an individual maximize

$$\int_0^T e^{-\rho t}\, U[C(t)]\,dt \tag{1}$$

subject to initial wealth W_0 that can always be invested for an exogeneously-given certain rate of yield r; or subject to the constraint

$$C(t) = rW(t) - \dot{W}(t) \tag{2}$$

If there is no bequest at death, terminal wealth is zero.

This leads to the standard calculus-of-variations problem

$$J = \underset{\{W(t)\}}{\text{Max}} \int_0^T e^{-\rho t}\, U[rW - \dot{W}]\,dt \tag{3}$$

This can be easily related [4] to a discrete-time formulation

$$\text{Max} \sum_{t=0}^{T} (1+\rho)^{-t}\, U[C_t] \tag{4}$$

subject to

$$C_t = W_t - \frac{W_{t+1}}{1+r} \tag{5}$$

or,

$$\underset{\{W_t\}}{\text{Max}} \sum_{t=0}^{T} (1+\rho)^{-t}\, U\left[W_t - \frac{W_{t+1}}{1+r}\right] \tag{6}$$

[4] See P. A. Samuelson [12], p. 273 for an exposition of discrete-time analogues to calculus-of-variations models. Note: here I assume that consumption, C_t, takes place at the beginning rather than at the end of the period. This change alters slightly the appearance of the equilibrium conditions, but not their substance.

for prescribed (W_0, W_{T+1}). Differentiating partially with respect to each W_t in turn, we derive recursion conditions for a regular interior maximum

$$\frac{(1+\rho)}{1+r} U'\left[W_{t-1} - \frac{W_t}{1+r}\right] = U'\left[W_t - \frac{W_{t+1}}{1+r}\right] \tag{7}$$

If U is concave, solving these second-order difference equations with boundary conditions (W_0, W_{T+1}) will suffice to give us an optimal lifetime consumption-investment program.

Since there has thus far been one asset, and that a safe one, the time has come to introduce a stochastically-risky alternative asset and to face up to a portfolio problem. Let us postulate the existence, alongside of the safe asset that makes \$1 invested in it at time t return to you at the end of the period $\$1(1 + r)$, a risk asset that makes \$1 invested in, at time t, return to you after one period $\$1Z_t$, where Z_t is a random variable subject to the probability distribution

$$\text{Prob}\,\{Z_t \leqq z\} = P(z). \qquad z \geqq 0 \tag{8}$$

Hence, $Z_{t+1} - 1$ is the percentage "yield" of each outcome. The most general probability distribution is admissible: i.e., a probability density over continuous z's, or finite positive probabilities at discrete values of z. Also I shall usually assume independence between yields at different times so that $P(z_0, z_1, \ldots, z_t, \ldots, z_T) = P(z_t)P(z_1)\ldots P(z_T)$.

For simplicity, the reader might care to deal with the easy case

$$\text{Prob}\,\{Z = \lambda\} = 1/2 = \text{Prob}\,\{Z = \lambda^{-1}\}, \qquad \lambda > 1 \tag{9}$$

In order that risk averters with concave utility should not shun this risk asset when maximizing the expected value of their portfolio, λ must be large enough so that the expected value of the risk asset exceeds that of the safe asset, i.e.,

$$\frac{1}{2}\lambda + \frac{1}{2}\lambda^{-1} > 1 + r, \text{ or}$$

$$\lambda > 1 + r + \sqrt{2r + r^2}.$$

Thus, for $\lambda = 1.4$, the risk asset has a mean yield of 0.057, which is greater than a safe asset's certain yield of $r = .04$.

At each instant of time, what will be the optimal fraction, w_t, that you should put in the risky asset, with $1 - w_t$ going into the safe asset? Once these optimal portfolio fractions are known, the constraint of (5) must be written

$$C_t = \left[W_t - \frac{W_{t+1}}{[(1-w_t)(1+r) + w_t Z_t]} \right]. \tag{10}$$

Now we use (10) instead of (4), and recognizing the stochastic nature of our problem, specify that we maximize the expected value of total utility over time. This gives us the stochastic generalizations of (4) and (5) or (6)

$$\underset{\{C_t, w_t\}}{\text{Max}}\ E \sum_{t=0}^{T} (1+\rho)^{-t} U[C_t] \tag{11}$$

subject to

$$C_t = \left[W_t - \frac{W_{t+1}}{(1+r)(1-w_t) + w_t Z_t} \right]$$

W_0 given, W_{T+1} prescribed.

If there is no bequeathing of wealth at death, presumably $W_{T+1} = 0$. Alternatively, we could replace a prescribed W_{T+1} by a final bequest function added to (11), of the form $B(W_{T+1})$, and with W_{T+1} a free decision variable to be chosen so as to maximize (11) + $B(W_{T+1})$. For the most part, I shall consider $C_T = W_T$ and $W_{T+1} = 0$.

In (11), E stands for the "expected value of," so that, for example,

$$E\,Z_t = \int_0^\infty z_t dP(z_t)\ .$$

In our simple case of (9),

$$EZ_t = \frac{1}{2}\lambda + \frac{1}{2}\lambda^{-1}.$$

Equation (11) is our basic stochastic programming problem that needs to be solved simultaneously for optimal saving-consumption and portfolio-selection decisions over time.

Before proceeding to solve this problem, reference may be made to similar problems that seem to have been dealt with explicitly in the economics literature. First, there is the valuable paper by Phelps on the Ramsey problem in which capital's yield is a prescribed random variable. This corresponds, in my notation, to the $\{w_t\}$ strategy being frozen at some fractional level, there being no portfolio selection problem. (My analysis could be amplified to consider Phelps'[5] wage income, and even in the stochastic form that he cites Martin Beckmann as having analyzed.) More recently, Levhari and Srinivasan [4] have also treated the Phelps problem for $T = \infty$ by means of the Bellman functional equations of dynamic programming, and have indicated a proof that concavity of U is sufficient for a maximum. Then, there is Professor Mirrlees' important work on the Ramsey problem with Harrod-neutral technological change as a random variable.[6] Our problems become equivalent if I replace $W_t - W_{t+1}[(1+r)(1-w_t) + w_t Z_t]^{-1}$ in (10) by $A_t f(W_t/A_t) - nW_t - (W_{t+1} - W_t)$ let technical change be governed by the probability distribution

$$\text{Prob}\,\{A_t \leq A_{t-1}Z\} = P(Z);$$

reinterpret my W_t to be Mirrlees' per capita capital, K_t/L_t, where L_t is growing at the natural rate of growth n; and posit that $A_t f(W_t/A_t)$ is a homogeneous first degree, concave, neoclassical production function in terms of capital and efficiency-units of labor.

It should be remarked that I am confirming myself here to regular interior maxima, and not going into the Kuhn-Tucker inequalities that easily handle boundary maxima.

Solution of the Problem

The meaning of our basic problem

$$J_T(W_0) = \underset{\{C_t, w_t\}}{\text{Max}}\ E \sum_{t=0}^{T} (1+\rho)^{-t} U[C_t] \tag{11}$$

subject to $C_t = W_t - W_{t+1}[(1-w_t)\ (1+r) + w_t Z_t]^{-1}$ is not easy to grasp. I act now at $t = 0$ to select C_0 and w_0, knowing W_0 but not yet knowing how Z_0 will turn out. I must act now, knowing that one period later, knowledge of Z_0's outcome will be known and that W_1 will then be known. Depending upon knowledge of W_1, a new decision will be made for C_1 and w_1. Now I can only guess what that decision will be.

As so often is the case in dynamic programming, it helps to begin at the end of the planning period. This brings us to the well-known

[5] E. S. Phelps [8].

[6] J. A. Mirrlees [6]. I have converted his treatment into a discrete-time version. Robert Merton's companion paper throws light on Mirrlees' Brownian-motion model for A_t.

one-period portfolio problem. In our terms, this becomes

$$J_1(W_{T-1}) = \underset{\{C_{T-1},w_{T-1}\}}{\text{Max}} U[C_{T-1}] + E(1+\rho)^{-1}U[(W_{T-1} - C_{T-1})\{(1-w_{T-1})(1+r) + w_{T-1}Z_{T-1}\}^{-1}]. \quad (12)$$

Here the expected value operator E operates only on the random variable of the next period since current consumption C_{T-1} is known once we have made our decision. Writing the second term as $EF(Z_T)$, this becomes

$$EF(Z_T) = \int_0^\infty F(Z_T)dP(Z_T|Z_{T-1},Z_{T-2},\ldots,Z_0)$$

$$= \int_0^\infty F(Z_T)dP(Z_T),$$ by our independence postulate.

In the general case, at a later stage of decision making, say $t = T-1$, knowledge will be available of the outcomes of earlier random variables, $Z_{t-2},\ldots$; since these might be relevant to the distribution of subsequent random variables, conditional probabilities of the form $P(Z_{T-1}|Z_{T-2},\ldots)$ are thus involved. However, in cases like the present one, where independence of distributions is posited, conditional probabilities can be dispensed within favor of simple distributions.

Note that in (12) we have substituted for C_T its value as given by the constraint in (11) or (10).

To determine this optimum (C_{T-1}, w_{T-1}), we differentiate with respect to each separately, to get

$$0 = U'[C_{T-1}] - (1+\rho)^{-1}EU'[C_T]\{(1-w_{T-1})(1+r) + w_{T-1}Z_{T-1}\} \quad (12')$$

$$0 = EU'[C_T](W_{T-1} - C_{T-1})(Z_{T-1}-1-r)$$
$$= \int_0^\infty U'[(W_{T-1} - C_{T-1})\{(1-w_{T-1}(1+r) - w_{T-1}Z_{T-1}\}](W_{T-1}-C_{T-1})(Z_{T-1}-1-r)dP(Z_{T-1}) \quad (12'')$$

Solving these simultaneously, we get our optimal decisions (C^*_{T-1}, w^*_{T-1}) as functions of initial wealth W_{T-1} alone. Note that if somehow C^*_{T-1} were known, (12″) would by itself be the familiar one-period portfolio optimality condition, and could trivially be rewritten to handle any number of alternative assets.

Substituting (C^*_{T-1}, w^*_{T-1}) into the expression to be maximized gives us $J_1(W_{T-1})$ explicitly. From the equations in (12), we can, by standard calculus methods, relate the derivatives of U to those of J, namely, by the envelope relation

$$J_1'(W_{T-1}) = U'[C_{T-1}]. \quad (13)$$

Now that we know $J_1[W_{T-1}]$, it is easy to determine optimal behavior one period earlier, namely by

$$J_2(W_{T-2}) = \underset{\{C_{T-2},w_{T-2}\}}{\text{Max}} U[C_{T-2}] + E(1+\rho)^{-1}J_1[(W_{T-2}-C_{T-2})\{(1-w_{T-2})(1+r) + w_{T-2}Z_{T-2}\}]. \quad (14)$$

Differentiating (14) just as we did (11) gives the following equations like those of (12)

$$0 = U'[C_{T-2}] - (1+\rho)^{-1}EJ_1'[W_{T-2}]\{(1-w_{T-2})(1+r) + w_{T-2}Z_{T-2}\} \quad (15')$$

$$0 = EJ_1'[W_{T-1}](W_{T-2} - C_{T-2})(Z_{T-2} - 1-r)$$
$$= \int_0^\infty J_1'[(W_{T-2} - C_{T-2})\{(1-w_{T-2})(1+r) + w_{T-2}Z_{T-2}\}](W_{T-2} - C_{T-2})(Z_{T-2} - 1-r)dP(Z_{T-2}). \quad (15'')$$

These equations, which could by (13) be related to $U'[C_{T-1}]$, can be solved simultaneously to determine optimal (C^*_{T-2}, w^*_{T-2}) and $J_2(W_{T-2})$.

Continuing recursively in this way for T−3, T−4, ..., 2, 1, 0, we finally have our problem solved. The general recursive optimality equations can be written as

$$\begin{cases} 0 = U'[C_0] - (1+\rho)^{-1}EJ'_{T-1}[W_0]\{(1-w_0)(1+r) + w_0Z_0\} \\ 0 = EJ'_{T-1}[W_1](W_0 - C_0)(Z_0 - 1-r) \end{cases}$$

$$\cdots\cdots\cdots$$

$$0 = U'[C_{T-1}] - (1+\rho)^{-1}EJ'_{T-t}[W_t]\{(1-w_{t-1})(1+r) + w_{t-1}Z_{t-1}\} \quad (16')$$

$$0 = EJ'_{T-t}[W_{t-1} - C_{t-1})(Z_{t-1} - 1-r), \quad (t = 1,\ldots,T-1). \quad (16'')$$

In (16′), of course, the proper substitutions must be made and the E operators must be over the proper probability distributions. Solving (16″) at any stage will give the optimal decision rules for consumption-saving and for portfolio selection, in the form

$$C^*_t = f[W_t; Z_{t-1},\ldots,Z_0]$$
$$= f_{T-t}[W_t]$$ if the Z's are independently distributed

$$w^*_t = g[W_t; Z_{t-1}, \ldots, Z_0]$$
$$= g_{T-t}[W_t] \text{ if the } Z\text{'s are independently distributed.}$$

Our problem is now solved for every case but the important case of infinite-time horizon. For well-behaved cases, one can simply let $T \to \infty$ in the above formulas. Or, as often happens, the infinite case may be the easiest of all to solve, since for it $C^*_t = f(W_t)$, $w^*_t = g(W_t)$, independently of time and both these unknown functions can be deduced as solutions to the following functional equations:

$$0 = U'[f(W)] - (1+\rho)^{-1} \int_0^\infty J'[(W - f(W))\{(1+r) - g(W)(Z-1-r)\}][(1+r) - g(W)(Z-1-r)]dP(Z) \tag{17'}$$

$$0 = \int_0^\infty U'[\{W - f(W)\}\{1 + r - g(W)(Z-1-r)\}][Z-1-r]. \tag{17''}$$

Equation (17′), by itself with $g(W)$ pretended to be known, would be equivalent to equation (13) of Levhari and Srinivasan [4, p. f]. In deriving (17′)–(17″), I have utilized the envelope relation of my (13), which is equivalent to Levhari and Srinivasan's equation (12) [4, p. 5].

Bernoulli and Isoelastic Cases

To apply our results, let us consider the interesting Bernoulli case where $U = \log C$. This does not have the bounded utility that Arrow [1] and many writers have convinced themselves is desirable for an axiom system. Since I do not believe that Karl Menger paradoxes of the generalized St. Petersburg type hold any terrors for the economist, I have no particular interest in boundedness of utility and consider $\log C$ to be interesting and admissible. For this case, we have, from (12),

$$J_1(W) = \underset{\{C,w\}}{\text{Max}} \log C + E(1+\rho)^{-1}\log[(W-C)\{(1-w)(1+r)+wZ\}]$$
$$= \underset{\{C\}}{\text{Max}} \log C + (1+\rho)^{-1}\log[W-C] + \underset{\{w\}}{\text{Max}} \int_0^\infty \log[(1-w)(1+r) + wZ]dP(Z). \tag{18}$$

Hence, equations (12) and (16′)–(16″) split into two independent parts and the Ramsey-Phelps saving problem becomes quite independent of the lifetime portfolio selection problem. Now we have

$$0 = (1/C) - (1+\rho)^{-1}(W-C)^{-1} \text{ or } C_{T-1} = (1+\rho)(2+\rho)^{-1}W_{T-1} \tag{19'}$$

$$0 = \int_0^\infty (Z-1-r)[(1-w)(1+r) + wZ]^{-1}dP(Z) \text{ or } w_{T-1} = w^* \text{ independently of } W_{T-1}. \tag{19''}$$

These independence results, of the C_{T-1} and w_{T-1} decisions and of the dependence of w_{T-1} on W_{T-1}, hold for all U functions with isoelastic marginal utility. I.e., (16′) and (16″) become decomposable conditions for all

$$U(C) = 1/\gamma\, C^\gamma, \qquad \gamma < 1 \tag{20}$$

as well as for $U(C) = \log C$, corresponding by L'Hôpital's rule to $\gamma = 0$.

To see this, write (12) or (18) as

$$J_1(W) = \underset{\{C,w\}}{\text{Max}} \frac{C^\gamma}{\gamma} + (1+\rho)^{-1}\frac{(W-C)^\gamma}{\gamma}\int_0^\infty [(1-w)(1+r) + wZ]^\gamma dP(Z)$$
$$= \underset{\{C\}}{\text{Max}} \frac{C^\gamma}{\gamma} + (1+\rho)^{-1}\frac{(W-C)^\gamma}{\gamma} \times \underset{w}{\text{Max}} \int_0^\infty [(1-w)(1+r) + wZ]^\gamma dp(Z). \tag{21}$$

Hence, (12″) or (15″) or (16″) becomes

$$\int_0^\infty [(1-w)(1+r) + wZ]^{\gamma-1}(Z-r-1)\,dP(Z) = 0, \tag{22''}$$

which defines optimal w^* and gives

$$\underset{\{w\}}{\text{Max}} \int_0^\infty [(1-w)(1+r) + wZ]^\gamma dP(Z) = \int_0^\infty [(1-w^*)(1+r) + w^*Z]^\gamma dP(Z) = [1 + r^*]^\gamma, \text{ for short.}$$

Here, r^* is the subjective or util-prob mean return of the portfolio, where diminishing marginal utility has been taken into account.[7] To get optimal consumption-saving, differentiate (21) to get the new form of (12′), (15′), or (16′)

[7] See Samuelson and Merton for the util-prob concept [13].

$$0 = C^{\gamma-1} - (1+\rho)^{-1}(1+r^*)^\gamma (W-C)^{\gamma-1}. \tag{22'}$$

Solving, we have the consumption decision rule

$$C^*_{T-1} = \frac{a_1}{1+a_1} W_{T-1} \tag{23}$$

where

$$a_1 = [(1+r^*)^\gamma/(1+\rho)]^{1/\gamma-1}. \tag{24}$$

Hence, by substitution, we find

$$J_1(W_{T-1}) = b_1 W^\gamma{}_{T-1}/\gamma \tag{25}$$

where

$$b_1 = a_1{}^\gamma(1+a_1)^{-\gamma} + (1+\rho)^{-1}(1+r^*)^\gamma(1+a_1)^{-\gamma}. \tag{26}$$

Thus, $J_1(\cdot)$ is of the same elasticity form as $U(\cdot)$ was. Evaluating indeterminate forms for $\gamma = 0$, we find J_1 to be of log form if U was.

Now, by mathematical induction, it is easy to show that this isoelastic property must also hold for $J_2(W_{T-2})$, $J_3(W_{T-3}), \ldots,$ since, whenever it holds for $J_n(W_{T-n})$ it is deducible that it holds for $J_{n+1}(W_{T-n-1})$. Hence, at every stage, solving the general equations (16′) and (16″), they decompose into two parts in the case of isoelastic utility. Hence,

Theorem:

For isoelastic marginal utility functions, $U'(C) = C^{\gamma-1}$, $\gamma < 1$, the optimal portfolio decision is independent of wealth at each stage and independent of all consumption-saving decisions, leading to a constant w^*, the solution to

$$0 = \int_0^\infty [(1-w)(1+r)+wZ]^{\gamma-1}(Z-1-r)\,dP(Z).$$

Then optimal consumption decisions at each stage are, for a no-bequest model, of the form

$$C^*_{T-i} = c_i W_{T-i}$$

where one can deduce the recursion relations

$$c_1 = \frac{a_1}{1+w_1},$$

$$a_1 = [(1+\rho)/(1+r^*)^\gamma]^{1/1-\gamma}$$

$$(1+r^*)^\gamma = \int_0^\infty [(1-w^*)(1+r) + w^* Z]^\gamma\, dP(Z)$$

$$c_i = \frac{a_1 c_{i-1}}{1+a_1 c_{i-1}} = \frac{a^i{}_1}{1+a_1+a^2{}_1+\ldots+a^i{}_1} < c_{i-1}$$

$$= \frac{a^i{}_1(a_1-1)}{a^{i+1}{}_1-1}, \quad a_1 \neq 1$$

$$= \frac{1}{1+i}, \quad a_1 = 1.$$

In the limiting case, as $\gamma \to 0$ and we have Bernoulli's logarithmic function, $a_1 = (1+\rho)$, independent of r^*, and all saving propensities depend on subjective time preference ρ only, being independent of technological investment opportunities (except to the degree that W_t will itself definitely depend on those opportunities).

We can interpret $1+r^*$ as kind of a "risk-corrected" mean yield; and behavior of a long-lived man depends critically on whether $(1+r^*)^\gamma \gtreqless (1+\rho)$, corresponding to $a_1 \lesseqgtr 1$.

(i) For $(1+r^*)^\gamma = (1+\rho)$, one plans always to consume at a uniform rate, dividing current W_{T-i} evenly by remaining life, $1/(1+i)$. If young enough, one saves on the average; in the familiar "hump saving" fashion, one dissaves later as the end comes sufficiently close into sight.

(ii) For $(1+r^*)^\gamma > (1+\rho)$, $a_1 < 1$, and investment opportunities are, so to speak, so tempting compared to psychological time preference that one consumes nothing at the beginning of a long-long life, i.e., rigorously

$$\lim_{i\to\infty} c_i = 0, \qquad a_1 < 1$$

and again hump saving must take place. For $(1+r^*)^\gamma > (1+\rho)$, the *perpetual* lifetime problem, with $T = \infty$, is divergent and ill-defined, i.e., $J_i(W) \to \infty$ as $i \to \infty$. For $\gamma \leqq 0$ and $\rho > 0$, this case cannot arise.

(iii) For $(1+r^*)^\gamma < (1+\rho)$, $a_1 > 1$, consumption at very early ages drops only to a limiting positive fraction (rather than zero), namely

$$\lim_{i\to\infty} c_i = 1 - 1/a_1 < 1, \quad a_1 > 1.$$

Now whether there will be, on the average, initial hump saving depends upon the size of $r^* - c_\infty$, or whether

$$r^* - 1 - \frac{(1+r^*)^{\gamma/1-\gamma}}{(1+\rho)^{1/1-\gamma}} > 0.$$

This ends the *Theorem*. Although many of the results depend upon the no-bequest assumption, $W_{T+1} = 0$, as Merton's companion paper shows (p. 247, this *Review*) we can easily generalize to the cases where a bequest function $B_T(W_{T+1})$ is added to $\Sigma^T{}_0\,(1+\rho)^{-t}U(C_t)$. If B_T is itself of isoelastic form,

$$B_T \equiv b_T(W_{T+1})^\gamma/\gamma,$$

the algebra is little changed. Also, the same comparative statics put forward in Merton's continuous-time case will be applicable here, e.g., the Bernoulli $\gamma = 0$ case is a watershed between cases where thrift is enhanced by riskiness rather than reduced; etc.

Since proof of the theorem is straightforward, I skip all details except to indicate how the recursion relations for c_i and b_i are derived, namely from the identities

$$\begin{aligned} b_{i+1}W^\gamma/\gamma &= J_{i+1}(W) \\ &= \operatorname*{Max}_C \{C^\gamma/\gamma \\ &\quad + b_i(1+r^*)^\gamma(1+\rho)^{-1}(W-C)^\gamma/\gamma\} \\ &= \{c^\gamma_{i+1} + b_i(1+r^*)^\gamma \\ &\quad (1+\rho)^{-1}(1-c_{i+1})^\gamma\}\, W^\gamma/\gamma \end{aligned}$$

and the optimality condition

$$\begin{aligned} 0 &= C^{\gamma-1} - b_i(1+r^*)^\gamma(1+\rho)^{-1}(W-C)^{\gamma-1} \\ &= (c_{i+1}W)^{\gamma-1} - b_i(1+r^*)^\gamma(1+\rho)^{-1} \\ &\quad (1-c_{i+1})^{\gamma-1}W^{\gamma-1}, \end{aligned}$$

which defines c_{i+1} in terms of b_i.

What if we relax the assumption of isoelastic marginal utility functions? Then w_{T-j} becomes a function of W_{T-j-1} (and, of course, of r, ρ, and a functional of the probability distribution P). Now the Phelps-Ramsey optimal stochastic saving decisions do interact with the optimal portfolio decisions, and these have to be arrived at by simultaneous solution of the nondecomposable equations (16′) and (16″).

What if we have more than one alternative asset to safe cash? Then merely interpret Z_t as a (column) vector of returns $(Z^2{}_t, Z^3{}_t, \ldots)$ on the respective risky assets; also interpret w_t as a (row) vector $(w^2{}_t, w^3{}_t, \ldots)$, interpret $P(Z)$ as vector notation for

$$\begin{aligned} &\text{Prob}\,\{Z^2{}_t \leqq Z^2, Z^3{}_t \leqq Z^3, \ldots\} \\ &= P(Z^2, Z^3, \ldots) = P(Z), \end{aligned}$$

interpret all integrals of the form $\int G(Z)dP(Z)$ as multiple integrals $\int G(Z^2,Z^3,\ldots)dP(Z^2,Z^3,\ldots)$. Then (16″) becomes a vector-set of equations, one for each component of the vector Z_t, and these can be solved simultaneously for the unknown w_t vector.

If there are many consumption items, we can handle the general problem by giving a similar vector interpretation to C_t.

Thus, the most general portfolio lifetime problem is handled by our equations or obvious extensions thereof.

Conclusion

We have now come full circle. Our model denies the validity of the concept of businessman's risk; for isoelastic marginal utilities, in your prime of life you have the same relative risk-tolerance as toward the end of life! The "chance to recoup" and tendency for the law of large numbers to operate in the case of repeated investments is not relevant. (Note: if the elasticity of marginal utility, $-U'(W)/WU''(W)$, rises empirically with wealth, and if the capital market is imperfect as far as lending and borrowing against future earnings is concerned, then it seems to me to be likely that a doctor of age 35–50 might rationally have his highest consumption then, and certainly show greatest risk tolerance then — in other words be open to a "businessman's risk." But not in the frictionless isoelastic model!)

As usual, one expects w^* and risk tolerance to be higher with algebraically large γ. One expects C_t to be higher late in life when r and r^* is high relative to ρ. As in a one-period model, one expects any increase in "riskiness" of Z_t, for the same mean, to decrease w^*. One expects a similar increase in riskiness to lower or raise consumption depending upon whether marginal utility is greater or less than unity in its elasticity.[8]

Our analysis enables us to dispel a fallacy that has been borrowed into portfolio theory from information theory of the Shannon type. Associated with independent discoveries by J. B. Williams [16], John Kelly [2], and H. A. Latané [3] is the notion that if one is investing for many periods, the proper behavior is to maximize the *geometric* mean of return rather than the arithmetic mean. I believe this to be incorrect (except in the Bernoulli logarithmic case where it happens [9] to be correct for reasons

[8] See Merton's cited companion paper in this issue, for explicit discussion of the comparative statical shifts of (16)'s C^*_t and w^*_t functions as the parameters (ρ, γ, r, r^*, and $P(Z)$ or $P(Z_1,\ldots)$ or $B(W_T)$ functions change. The same results hold in the discrete-and-continuous-time models.

[9] See Latané [3, p. 151] for explicit recognition of this point. I find somewhat mystifying his footnote there which says, "As pointed out to me by Professor L. J. Savage (in correspondence), not only is the maximization of G [the geometric mean] the rule for maximum expected utility in connection with Bernoulli's function but (in so far as certain approximations are permissible) this same rule is approximately valid for all utility functions." [Latané, p. 151, n.13.] The geometric mean criterion is definitely too conservative to maximize an isoelastic utility function corresponding to positive γ in my equation (20), and it is definitely too daring to maximize expected utility when $\gamma < 0$. Professor Savage has informed me recently that his 1969 position differs from the view attributed to him in 1959.

quite distinct from the Williams-Kelly-Latané reasoning).

These writers must have in mind reasoning that goes something like the following: If one maximizes for a distant goal, investing and reinvesting (all one's proceeds) many times on the way, then the probability becomes great that with a portfolio that maximizes the geometric mean at each stage you will end up with a larger terminal wealth than with any other decision strategy.

This is indeed a valid consequence of the central limit theorem as applied to the additive logarithms of portfolio outcomes. (I.e., maximizing the geometric mean is the same thing as maximizing the arithmetic mean of the logarithm of outcome at each stage; if at each stage, we get a mean log of $m^{**} > m^{*}$, then after a large number of stages we will have $m^{**}T >> m^{*}T$, and the properly normalized probabilities will cluster around a higher value.)

There is nothing wrong with the logical deduction from premise to theorem. But the implicit premise is faulty to begin with, as I have shown elsewhere in another connection [Samuelson, 10, p. 3]. It is a mistake to think that, just because a w^{**} decision ends up with almost-certain probability to be better than a w^{*} decision, this implies that w^{**} must yield a better expected value of utility. Our analysis for marginal utility with elasticity differing from that of Bernoulli provides an effective counter example, if indeed a counter example is needed to refute a gratuitous assertion. Moreover, as I showed elsewhere, the ordering principle of selecting between two actions in terms of which has the greater probability of producing a higher result does not even possess the property of being transitive.[10] By that principle, we could have w^{***} better than w^{**}, and w^{**} better than w^{*}, and also have w^{*} better than w^{***}.

[10] See Samuelson [11].

REFERENCES

[1] Arrow, K. J., "Aspects of the Theory of Risk-Bearing" (Helsinki, Finland: Yrjö Jahnssonin Säätiö, 1965).

[2] Kelly, J., "A New Interpretation of Information Rate," *Bell System Technical Journal* (Aug. 1956), 917–926.

[3] Latané, H. A., "Criteria for Choice Among Risky Ventures," *Journal of Political Economy* 67 (Apr. 1959), 144–155.

[4] Levhari, D. and T. N. Srinivasan, "Optimal Savings Under Uncertainty," Institute for Mathematical Studies in the Social Sciences, Technical Report No. 8, Stanford University, Dec. 1967.

[5] Markowitz, H., *Portfolio Selection: Efficient Diversification of Investment* (New York: John Wiley & Sons, 1959).

[6] Mirrlees, J. A., "Optimum Accumulation Under Uncertainty," Dec. 1965, unpublished.

[7] Mossin, J., "Optimal Multiperiod Portfolio Policies," *Journal of Business* 41, 2 (Apr. 1968), 215–229.

[8] Phelps, E. S., "The Accumulation of Risky Capital: A Sequential Utility Analysis," *Econometrica* 30, 4 (1962), 729–743.

[9] Pratt, J., "Risk Aversion in the Small and in the Large," *Econometrica* 32 (Jan. 1964).

[10] Samuelson, P. A., "General Proof that Diversification Pays," *Journal of Financial and Quantitative Analysis* II (Mar. 1967), 1–13.

[11] ———, "Risk and Uncertainty: A Fallacy of Large Numbers," *Scientia*, 6th Series, 57th year (April-May, 1963).

[12] ———, "A Turnpike Refutation of the Golden Rule in a Welfare Maximizing Many-Year Plan," Essay XIV *Essays on the Theory of Optimal Economic Growth*, Karl Shell (ed.) (Cambridge, Mass.: MIT Press, 1967).

[13] ———, and R. C. Merton, "A Complete Model of Warrant Pricing that Maximizes Utility," *Industrial Management Review* (in press).

[14] Tobin, J., "Liquidity Preference as Behavior Towards Risk," *Review of Economic Studies*, XXV, 67, Feb. 1958, 65–86.

[15] ———, "The Theory of Portfolio Selection," *The Theory of Interest Rates*, F. H. Hahn and F. P. R. Brechling (eds.) (London: Macmillan, 1965).

[16] Williams, J. B., "Speculation and the Carryover," *Quarterly Journal of Economics* 50 (May 1936), 436–455.

ECONOMETRICA

VOLUME 38 September, 1970 NUMBER 5

OPTIMAL INVESTMENT AND CONSUMPTION STRATEGIES UNDER RISK FOR A CLASS OF UTILITY FUNCTIONS[1]

BY NILS H. HAKANSSON[2]

This paper develops a sequential model of the individual's economic decision problem under risk. On the basis of this model, optimal consumption, investment, and borrowing-lending strategies are obtained in closed form for a class of utility functions. For a subset of this class the optimal consumption strategy satisfies the permanent income hypothesis precisely. The optimal investment strategies have the property that the optimal mix of risky investments is independent of wealth, noncapital income, age, and impatience to consume. Necessary and sufficient conditions for long-run capital growth are also given.

1. INTRODUCTION AND SUMMARY

THIS PAPER presents a normative model of the individual's economic decision problem under risk. On the basis of this model, optimal consumption, investment, and borrowing-lending strategies are obtained in closed form for a class of utility functions. The model itself may be viewed as a formalization of Irving Fisher's model of the individual under risk, as presented in *The Theory of Interest* [**4**]; at the same time, it represents a generalization of Phelps' model of personal saving [**10**].

The various components of the decision problem are developed and assembled into a formal model in Section 2. The objective of the individual is postulated to be the maximization of expected utility from consumption over time. His resources are assumed to consist of an initial capital position (which may be negative) and a noncapital income stream which is known with certainty. The individual faces both financial opportunities (borrowing and lending) and an arbitrary number of productive investment opportunities. The returns from the productive opportunities are assumed to be random variables, whose probability distributions satisfy the "no-easy-money condition." The fundamental characteristic of the approach taken is that the portfolio composition decision, the financing decision, and the consumption decision are all analyzed simultaneously in *one* model. The vehicle of analysis is discrete-time dynamic programming.

In Section 3, optimal strategies are derived for the class of utility functions $\sum_{j=1}^{\infty} \alpha^{j-1}u(c_j)$, $0 < \alpha < 1$, where c_j is the amount of consumption in period j, such that either the relative risk aversion index, $-cu''(c)/u'(c)$, or the absolute risk

[1] This paper was presented at the winter meeting of the Econometric Society, San Francisco, California, December, 1966.

[2] This article is based on my dissertation which was submitted to the Graduate School of Business Administration of the University of California, Los Angeles, in June, 1966. I am greatly indebted to Professors George W. Brown (committee chairman), Jacob Marschak, and Jacques Drèze for many valuable suggestions and comments and to Professors Jack Hirshleifer, Leo Breiman, James Jackson, and Fred Weston for constructive criticisms. I am also grateful to the Ford Foundation for financial support over a three-year period.

aversion index, $-u''(c)/u'(c)$, is a positive constant for all $c \geqslant 0$, i.e., $u(c) = c^{\gamma}$, $0 < \gamma < 1$, $u(c) = -c^{-\gamma}$, $\gamma > 0$, $u(c) = \log c$, and $u(c) = -e^{-\gamma c}$, $\gamma > 0$.

Section 4 is devoted to a discussion of the properties of the optimal consumption strategies, which turn out to be linear and increasing in wealth and in the present value of the noncapital income stream. In three of the four models studied, the optimal consumption strategies precisely satisfy the properties specified by the consumption hypotheses of Modigliani and Brumberg [**9**] and of Friedman [**5**]. The effects of changes in impatience and in risk aversion on the optimal amount to consume are found to coincide with one's expectations. In response to changes in the "favorableness" of the investment opportunities, however, the four models exhibit an exceptionally diverse pattern with respect to consumption behavior.

The optimal investment strategies have the property that the optimal mix of risky (productive) investments in each model is *independent* of the individual's wealth, noncapital income stream, and impatience to consume. It is shown in Section 5 that the optimal mix depends in each case only on the probability distributions of the returns, the interest rate, and the individual's one-period utility function of consumption. This section also discusses the properties of the optimal lending and borrowing strategies, which are linear in wealth. Three of the models always call for borrowing when the individual is poor while the fourth model always calls for lending when he is sufficiently rich. The effect of differing borrowing and lending rates is also examined.

Necessary and sufficient conditions for capital growth are derived in Section 6. It is found that when the one-period utility function of consumption is logarithmic, the individual will always invest the capital available after the allotment to current consumption so as to maximize the expected growth rate of capital plus the present value of the noncapital income stream. Finally, Section 7 indicates how the preceding results are modified in the nonstationary case and under a finite horizon.

2. THE MODEL

In this section we shall combine the building blocks discussed in the previous section into a formal model. The following notation and assumptions will be employed:

c_j: amount of consumption in period j, where $c_j \geqslant 0$ (decision variable).

$U(c_1, c_2, c_3, \ldots)$: the utility function, defined over all possible consumption programs $(c_1, c_2, c_3, \ldots)$. The class of functions to be considered is that of the form

$$(1) \qquad U(c_1, c_2, c_3, \ldots) = u(c_1) + \alpha U(c_2, c_3, c_4, \ldots) = \sum_{j=1}^{\infty} \alpha^{j-1} u(c_j), \quad 0 < \alpha < 1.$$

It is assumed that $u(c)$ is monotone increasing, twice differentiable, and strictly concave for $c \geqslant 0$. The objective in each case is to maximize $E[U(c_1, c_2, \ldots)]$, i.e., the expected utility derived from consumption over time.[3]

x_j: amount of capital (debt) on hand at decision point j (the beginning of the jth period) (state variable).

y: income received from noncapital sources at the *end* of each period, where $0 \leqslant y < \infty$.

[3] While we make use of the expected utility theorem, we assume that the von Neumann-Morgenstern postulates [**12**] have been modified in such a way as to permit unbounded utility functions.

M: the number of available investment opportunities.

S: the subset of investment opportunities which it is possible to sell short.

z_{ij}: amount invested in opportunity i, $i = 1, \ldots, M$, at the beginning of the jth period (decision variable).

$r - 1$: rate of interest, where $r > 1$.

β_i: transformation of each unit of capital invested in opportunity i in any period j (random variable); that is, if we invest an amount θ in i at the beginning of a period, we will obtain $\beta_i\theta$ at the end of that period (stochastically constant returns to scale, no transaction costs or taxes). The joint distribution functions of the β_i, $i = 1, \ldots, M$, are assumed to be known and independent with respect to time j. The $\{\beta_i\}$ have the following properties:

(2) $$\beta_1 = r,$$

(3) $$0 \leqslant \beta_i < \infty \qquad (i = 2, \ldots, M),$$

(4) $$\Pr\left\{\sum_{i=2}^{M} (\beta_i - r)\theta_i < 0\right\} > 0,$$

for all finite θ_i such that $\theta_i \geqslant 0$ for all $i \notin S$ and $\theta_i \neq 0$ for at least one i.

$f_j(x_j)$: expected utility obtainable from consumption over all future time, evaluated at decision point j, when capital at that point is x_j and an optimal strategy is followed with respect to consumption and investment.

Y: present value at any decision point of the noncapital income stream capitalized at the rate of interest, i.e., $Y = y/(r - 1)$.

$\bar{v} \equiv (v_2, \ldots, v_M)$: a vector of real numbers.

$$h(\bar{v}) \equiv E\left[u\left(\sum_{i=2}^{M} (\beta_i - r)v_i + r\right)\right].$$

k: maximum of $h(\bar{v})$ subject to (27) and (28) (see (26)).

$\bar{v}^*$: vector $\bar{v}$ which gives maximum k of $h(\bar{v})$ (see (26)).

$$v^* \equiv \sum_{i=2}^{M} v_i^*.$$

$c^*(x)$: an optimal consumption strategy.

$z_1^*(x)$: an optimal lending strategy.

$z_i^*(x)$: an optimal investment strategy for opportunity i, $i = 2, \ldots, M$.

$s_j \equiv x_j + Y$.

The limitations of utility functions of the form (1) are well known and need not be elaborated here. Condition (4) will be referred to as the "no-easy-money condition." In essence, this condition states (i) that no combination of productive investment opportunities exists which provides, with probability 1, a return at least as high as the (borrowing) rate of interest; (ii) that no combination of short sales exists in which the probability is zero that a loss will exceed the (lending) rate of interest; (iii) that no combination of productive investments made from the proceeds of any short sale can guarantee against loss. For these reasons, (4) may be viewed as a condition that the prices of the various assets in the market must satisfy in equilibrium.

Consumption and investment decisions are assumed to be made at the beginning of each period. The amount allocated to consumption is assumed to be spent immediately or, if spent gradually over the period, to be set aside in a nonearning account. We also assume that any debt incurred by the individual must at all times be fully secured, i.e., that the individual must be solvent at each decision point. In view of the "no-easy-money condition" (4), this implies that his (net) debt cannot exceed the present value, on the basis of the (borrowing) rate of interest, of his noncapital income stream at the end of any period.

We shall now identify the relation which determines the amount of capital (debt) on hand at each decision point in terms of the amount on hand at the previous decision point. This leads to the difference equation

$$x_{j+1} = rz_{1j} + \sum_{i=2}^{M} \beta_i z_{ij} + y \qquad (j = 1, 2, \ldots) \tag{5}$$

where

$$\sum_{i=1}^{M} z_{ij} = x_j - c_j \qquad (j = 1, 2, \ldots). \tag{6}$$

The first term of (5) represents the payment of the debt or the proceeds from savings, the second term the proceeds from productive investments, and the third term the noncapital income received. Combining (5) and (6) we obtain

$$x_{j+1} = \sum_{i=2}^{M} (\beta_1 - r)z_{ij} + r(x_j - c_j) + y \qquad (j = 1, 2, \ldots). \tag{7}$$

This is the difference equation, then, which governs the process we are about to study.

The definition of $f_j(x_j)$ may formally be written

$$f_j(x_j) \equiv \max E[U(c_j, c_{j+1}, c_{j+2}, \ldots)] | x_j. \tag{8}$$

From (1) we obtain, by the principle of optimality,[4] for all j,

$$f_j(x_j) = \max E[u(c_j) + \alpha\{\max E[U(c_{j+1}, c_{j+2}, \ldots)] | x_{j+1}\}] | x_j, \tag{9}$$

since we have assumed the $\{\beta_i\}$ to be independently distributed with respect to time j. By (8), (9) reduces to

$$f_j(x_j) = \max \{u(c_j) + \alpha E[f_{j+1}(x_{j+1})]\}, \quad \text{all } j. \tag{10}$$

Since by our assumptions we are faced with exactly the same problem at decision point $j + 1$ as when we are at decision point j, the time subscript may be dropped. Using (7), (10) then becomes

$$f(x) = \max_{c,\{z_i\}} \left\{ u(c) + \alpha E\left[f\left(\sum_{i=2}^{M} (\beta_i - r)z_i + r(x - c) + y \right) \right] \right\} \tag{11}$$

subject to

$$c \geqslant 0, \tag{12}$$

$$z_i \geqslant 0, \quad i \notin S, \tag{13}$$

and

$$\Pr\left\{ \sum_{i=2}^{M} (\beta_i - r)z_i + r(x - c) + y \geqslant -Y \right\} = 1 \tag{14}$$

at each decision point. Expression (14), of course, represents the solvency constraint.

[4] The principle of optimality states that an optimal strategy has the property that whatever the initial state and the initial decision, the remaining decisions must constitute an optimal strategy with regard to the state resulting from the first decision [2, p. 83].

For comparison, the model studied by Phelps [10] is given by the functional equation

$$(15)\qquad f(x) = \max_{0 \leqslant c \leqslant x} \{u(c) + \alpha E[f(\beta(x - c) + y)]\}.$$

In this model, all capital not currently consumed obeys the transformation β, which is identically and independently distributed in each period. Since the amount invested, $x - c$, is determined once c is known, (15) has only one decision variable (c).[5]

Since x represents capital, $f(x)$ is clearly the utility of money at any decision point j. Instead of being assumed, as is generally the case, the utility function of money has in this model been induced from inputs which are more basic than the preferences for money itself. As (11) shows, $f(x)$ depends on the individual's preferences with respect to consumption, his noncapital income stream, the interest rate, and the available investment opportunities and their riskiness.

3. THE MAIN THEOREMS

We shall now give the solution to (11) for the class of one-period utility functions

$$(16)\qquad u(c) = \frac{1}{\gamma}c^{\gamma}, \qquad 0 < \gamma < 1 \quad \text{(Model I)};$$

$$(17)\qquad u(c) = \frac{1}{\gamma}c^{\gamma}, \qquad \gamma < 0 \quad \text{(Model II)};$$

$$(18)\qquad u(c) = \log c, \qquad \text{(Model III)};$$

$$(19)\qquad u(c) = -e^{-\gamma c}, \qquad \gamma > 0 \quad \text{(Model IV)}.$$

[5] Phelps gives the solution to (15) for the utility functions $u(c) = c^{\gamma}$, $0 < \gamma < 1$, $u(c) = -c^{-\gamma}$, $\gamma < 0$, and for $u(c) = \log c$ when $y = 0$. Unfortunately, this solution is incorrect in the general case, i.e., whenever $y > 0$ *and* the distribution of β is nondegenerate. For example, when $u(c) = -c^{-\gamma}$, the solution is asserted to be, letting $\bar{\beta} \equiv E[\beta^{-\gamma}]$,

$$(15a)\qquad f(x) = -\left[\frac{(\alpha\bar{\beta})^{-1/(\gamma+1)}}{(\alpha\bar{\beta})^{-1/(\gamma+1)} - 1}\right]^{\gamma+1}\left[x + \frac{y}{\bar{\beta}^{-1/\gamma} - 1}\right]^{-\gamma},$$

$$(15b)\qquad c(x) = [1 - (\alpha\bar{\beta})^{1/(\gamma+1)}]\left[x + \frac{y}{\bar{\beta}^{-1/\gamma} - 1}\right],$$

whenever $\alpha\bar{\beta} < 1$. But for this to be a solution, it would be necessary that one be able to write

$$(15c)\qquad E[(\beta(x - c) + y)^{-\gamma}] = E[\beta^{-\gamma}]\left[x - c + \frac{y}{E[\beta^{-\gamma}]^{-1/\gamma}}\right]^{-\gamma}$$

which is clearly impossible unless the distribution of β is degenerate or $y = 0$ or both. The right side of (15c) may, of course, be regarded as a first-order approximation of the left side when the variance of β is small, but this negates the presence of uncertainty. In fact, the preceding solution holds even under certainty only when $\alpha\beta \geqslant 1$ and $x \geqslant [(\alpha\bar{\beta})^{-1/(\gamma+1)} - 1]y/(\bar{\beta} - 1)$, i.e., when $c(x)$ is less than or equal to x in *all* future periods.

It appears that an analytic solution to (15) does not exist when $y > 0$ and the distribution of β is nondegenerate. It is ironic, therefore, that when one generalizes Phelps' problem by introducing the possibility of *choice* among risky investment opportunities *and* the opportunity to borrow and lend (see (11)), an analytic solution does exist (as will be shown). It is the second of these generalizations which guarantees the solution in closed form.

Pratt [11] notes that (16)–(18) are the only monotone increasing and strictly concave utility functions for which the relative risk aversion index

$$(20)\qquad q^*(c) \equiv -\frac{u''(c)c}{u'(c)}$$

is a positive constant and that (19) is the only monotone increasing and strictly concave utility function for which the absolute risk aversion index

$$(21)\qquad q(c) \equiv -\frac{u''(c)}{u'(c)}$$

is a positive constant.[6]

THEOREM 1: *Let $u(c)$, α, y, r, $\{\beta_i\}$, and Y be defined as in Section 2. Then, whenever $u(c)$ is one of the functions (16)–(18) and $k\gamma < 1/\alpha$ in Model I, a solution to (11) subject to (12)–(14) exists for $x \geqslant -Y$ and is given by*

$$(22)\qquad f(x) = Au(x + Y) + C,$$

$$(23)\qquad c^*(x) = B(x + Y),$$

$$(24)\qquad z_1^*(x) = (1 - B)(1 - v^*)(x + Y) - Y,$$

$$(25)\qquad z_i^*(x) = (1 - B)v_i^*(x + Y) \qquad (i = 2, \ldots, M)$$

where the constants v_i^* ($v^* \equiv \sum_{i=2}^M v_i^*$) and k are given by

$$(26)\qquad k \equiv E\left[u\left(\sum_{i=2}^{M} (\beta_i - r)v_i^* + r\right)\right]$$

$$= \max_{\{v_i\}} E\left[u\left(\sum_{i=2}^{M} (\beta_i - r)v_i + r\right)\right],$$

subject to

$$(27)\qquad v_i \geqslant 0, \quad i \notin S,$$

and

$$(28)\qquad Pr\left\{\sum_{i=2}^{M} (\beta_i - r)v_i + r \geqslant 0\right\} = 1,$$

and the constants A, B, and C are given by

(i) *in the case of Models I–II,*

$$A = (1 - (\alpha k\gamma)^{1/(1-\gamma)})^{\gamma - 1},$$

$$(29)\qquad B = 1 - (\alpha k\gamma)^{1/(1-\gamma)},$$

$$C = 0;$$

[6] The underlying mathematical reason why solutions are obtained in closed form (Theorems 1 and 2) for the utility functions (16)–(19) is that these functions are also the only (monotone increasing and strictly concave utility function) solutions (see [8]) to the functional equations $u(xy) = v(x)w(y)$, $u(xy) = v(x) + w(y)$, $u(x + y) = v(x)w(y)$, and $u(x + y) = v(x) + w(y)$, which are known as the generalized Cauchy equations [1, p. 141].

(ii) *in the case of Model III,*

$$(30)\qquad \begin{aligned} A &= \frac{1}{1-\alpha}, \\ B &= 1-\alpha, \\ C &= \frac{1}{1-\alpha}\log(1-\alpha) + \frac{\alpha\log\alpha}{(1-\alpha)^2} + \frac{\alpha k}{(1-\alpha)^2}. \end{aligned}$$

Furthermore, the solution is unique.

In proving this theorem, we shall make use of the following lemma and corollaries.

LEMMA: *Let $u(c)$, $\{\beta_i\}$, and r be defined as in Section 2 and let $\bar{v} \equiv (v_2, \ldots, v_M)$ be a vector of real numbers. Then the function*

$$(31)\qquad h(v_2, v_3, \ldots, v_M) \equiv E\left[u\left(\sum_{i=2}^{M}(\beta_i - r)v_i + r\right)\right]$$

subject to the constraints

$$(27)\qquad v_i \geqslant 0, \quad i \notin S,$$

and

$$(28)\qquad Pr\left\{\sum_{i=2}^{M}(\beta_i - r)v_i + r \geqslant 0\right\} = 1,$$

has a maximum and the maximizing v_i ($\equiv v_i^*$) are finite and unique.

PROOF: Let D be the $(M-1)$-dimensional space defined by the set of points $\bar{v}$ which satisfy (27) and (28). We shall first prove that the set D is nonempty, closed, bounded, and convex, and that h is strictly concave on D.[7]

The nonemptiness of D follows trivially from the observation that $\bar{v}^0 \equiv (0, 0, \ldots, 0)$ is a member of D. By the boundedness of the β_i's and of r ((2) and (3)), there exists a neighborhood of $\bar{v}^0$ in relation to D. That is, there is a neighborhood of points $\bar{v}'$ such that

$$\Pr\left\{\sum_{i=2}^{M}(\beta_i - r)v_i' + r \geqslant 0\right\} = 1$$

where $v_i' \geqslant 0$ for all $i \notin S$.

Now consider the point $\bar{v}^\lambda \equiv \bar{v}^0 + \lambda\bar{v}' = \lambda\bar{v}'$ where $\lambda \geqslant 0$ and $\bar{v}'$ is one of the points in this neighborhood. Let $b(\bar{v})$ be the greatest lower bound on b such that

$$\Pr\left\{\sum_{i=2}^{M}(\beta_i - r)v_i < b\right\} > 0.$$

[7] The author gratefully acknowledges a debt to Professor George W. Brown for several valuable suggestions concerning the proof of the closure and the boundedness of D.

By the "no-easy-money condition" (4), $b(\bar{v}') \geqslant -r$ for $\bar{v}' \in D, b(\bar{v}^0) = 0$, and $b(\bar{v}) < 0$ for all $\bar{v} \neq \bar{v}^0$. Applying the "no-easy-money condition" with respect to the point $\bar{v}^\lambda$ and using the inequality

$$\Pr\left\{\sum_{i=2}^{M} (\beta_i - r)\lambda v_i < \lambda b\right\} > 0,$$

we obtain that $\lambda b(\bar{v}') = b(\lambda\bar{v}')$. But when $\lambda b(\bar{v}') < -r$, or $\lambda > -r/b(\bar{v}')$, the point $\bar{v}^\lambda$ cannot lie in D since $\lambda > -r/b(\bar{v}')$ implies that

$$\Pr\left\{\sum_{i=2}^{M} (\beta_i - r)\lambda v_i' + r \geqslant 0\right\} < 1.$$

Thus, $\lambda_0 \equiv -r/b(\bar{v}')$ is the greatest lower bound on λ such that $\bar{v}^\lambda \notin D$. Since $\lambda_0 b(\bar{v}') = -r$, $\bar{v}^{\lambda_0} \in D$ and is in fact the point farthest from $\bar{v}^0$ lying on the line through $\bar{v}^0$ and $\bar{v}'$ and belonging to D.

We shall only sketch the remainder of the proof establishing the closure and boundedness of D. Let $\bar{v} \neq \bar{v}^0$ be the limit of a sequence of points $\bar{v}^{(n)} \in D$. Since each point in the sequence belongs to D, $b(\bar{v}^{(n)}) \geqslant -r$ for all n. It can now be shown, by utilizing the fact that $\Sigma_{i=2}^M (\beta_i - r)\bar{v}_i$ is continuous at any $\bar{v} \neq \bar{v}^0$, uniformly with respect to the β_i's on any bounded set, that $\overline{\lim}_{n\to\infty} b(\bar{v}^{(n)}) \leqslant b(\bar{v})$, which implies that $\bar{v} \in D$. Consequently, D must be closed.

The boundedness of D is established as follows. Let S_R be the set of points $\bar{v}$ such that $|\bar{v}| = R > 0$. S_R is then clearly both closed and bounded. If $D' \equiv D \cap S_R$ is empty, the boundedness of D follows immediately. Let us therefore assume that D' is nonempty; in this case D' is also bounded and closed since D is closed and S_R is bounded and closed. If $\bar{v}$ is a limit point of the sequence $\langle\bar{v}^{(n)}\rangle$ such that $\bar{v}^{(n)} \in D'$, we must have that $\bar{v} \in D'$ since D' is closed. But $b(\bar{v}) < 0$ by the "no-easy-money condition" (4), since $\bar{v} \neq \bar{v}^0$ by assumption. Therefore, since we already have that $\overline{\lim}_{n\to\infty} b(\bar{v}^{(n)}) \leqslant b(\bar{v})$, 0 cannot be a limit point to the sequence $\langle b(\bar{v}^{(n)})\rangle$, $\bar{v}^{(n)} \in D'$. Consequently, $b(\bar{v})$ for $\bar{v} \in D'$ is bounded away from zero, which implies that D must be bounded.

To prove convexity, let $\bar{v}''$ and $\bar{v}'''$ be two points in D. Then, for any $0 \leqslant \lambda \leqslant 1$,

$$\Pr\left\{\sum_{i=2}^{M} (\beta_i - r)\lambda v_i'' + \lambda r \geqslant 0\right\} = 1,$$

and

$$\Pr\left\{\sum_{i=2}^{M} (\beta_i - r)(1 - \lambda)v_i''' + (1 - \lambda)r \geqslant 0\right\} = 1,$$

which implies

$$\Pr\left\{\sum_{i=2}^{M} (\beta_i - r)(\lambda v_i'' + (1 - \lambda)v_i''') + r \geqslant 0\right\} = 1,$$

so that $\lambda\bar{v}'' + (1 - \lambda)\bar{v}''' \in D$. Thus, D is convex.

Let

$$\tilde{w}_n = \sum_{i=2}^{M} (\beta_i - r)v_i^n + r \qquad (n = 1, 2).$$

Then

(32) $$h(\lambda\bar{v}^1 + (1-\lambda)\bar{v}^2) = E[u(\lambda\tilde{w}_1 + (1-\lambda)\tilde{w}_2)]$$

and

(33) $$\lambda h(\bar{v}^1) + (1-\lambda)h(\bar{v}^2) = \lambda E[u(\tilde{w}_1)] + (1-\lambda)E[u(\tilde{w}_2)].$$

For every pair of values $w_1 \neq w_2$ of the random variables $\tilde{w}_1$ and $\tilde{w}_2$ such that $\bar{v}^1$ and $\bar{v}^2 \in D$, we obtain, by the strict concavity of u,

(34) $$u(\lambda w_1 + (1-\lambda)w_2) > \lambda u(w_1) + (1-\lambda)u(w_2), \quad 0 < \lambda < 1.$$

Consequently, (34) implies

$$E[u(\lambda\tilde{w}_1 + (1-\lambda)\tilde{w}_2)] > \lambda E[u(\tilde{w}_1)] + (1-\lambda)E[u(\tilde{w}_2)], \quad \bar{v}_1^1 \neq \bar{v}_2^2 \in D, \quad 0 < \lambda < 1,$$

which, by (32) and (33), in turn implies that h is strictly concave on D.

Since our problem has now been shown to be one of maximizing a strictly concave function over a nonempty, closed, bounded, convex set, it follows directly that the function h has a maximum and that the v_i^* are finite and unique.

A number of corollaries obtain from this lemma which we shall also require in the proof of Theorem 1.

COROLLARY 1: *Let $u(c)$, $\{\beta_i\}$, and r be defined as in the Lemma. Moreover, let $u(c)$ be such that it has no lower bound. Then the v_i^* which maximize* (31) *subject to* (27) *and* (28) *are such that*

$$Pr\left\{\sum_{i=2}^{M} (\beta_i - r)v_i^* + r > 0\right\} = 1.$$

The proof is immediate from the observation that $h \to -\infty$ as the greatest lower bound on b such that $\Pr \Sigma_{i=2}^{M} (\beta_i - r)v_i + r < b\} > 0$ approaches 0 from above.

COROLLARY 2: *Let $u(c)$, $\{\beta_i\}$, and r be defined as in the Lemma. Then the maximum of the function* (31) *subject to the constraints* (27) *and* (28) *is greater than or equal to $u(r)$.*

PROOF: When $v_i = 0$ for all i, which is always feasible, we obtain by (31) that $h = u(r)$.

COROLLARY 3: *Let $u(c)$, $\{\beta_i\}$, and r be defined as in the Lemma. Moreover, let $u(c)$ be such that $u(c) \leqslant b$. Then the vectors $\bar{v}$ which satisfy* (27) *and* (28) *are such that*

$$h(\bar{v}) \equiv E\left[u\left(\sum_{i=2}^{M} (\beta_i - r)v_i + r\right)\right] < b.$$

The proof is immediate from the observation that $u(c)$ is monotone increasing and that r, $\{\beta_i\}$, and the feasible v_i are bounded.

We are now ready to prove the theorem. The method of proof will be to verify that (22)–(25) is the (only) solution to (11).[8]

PROOF OF THEOREM 1 FOR MODELS I–II: Denote the right side of (11) by $T(x)$ upon inserting (22) for $f(x)$. This gives, for all decision points j,

$$T(x) = \max_{c,\{z_i\}} \left\{ \frac{1}{\gamma} c^\gamma + \alpha(1 - (\alpha k \gamma)^{1/(1-\gamma)})^{\gamma-1} E\left[\frac{1}{\gamma} \left(\sum_{i=2}^{M} (\beta_i - r) z_i + r(x - c) + y + Y \right)^\gamma \right] \right\} \tag{35}$$

subject to

$$c \geqslant 0, \tag{12}$$

$$z_i \geqslant 0, \quad i \notin S, \tag{13}$$

and

$$\Pr\left\{ \sum_{i=2}^{M} (\beta_i - r) z_i + r(x - c) + y + (y/(r-1)) \geqslant 0 \right\} = 1. \tag{14}$$

Since (14) may be written

$$\Pr\left\{ \sum_{i=2}^{M} (\beta_i - r) z_i + r(x + Y - c) \geqslant 0 \right\} = 1,$$

it follows from the "no-easy-money condition" (4) that (14) is satisfied if and only if *either*

$$s - c = 0 \tag{36}$$

and

$$z_i = 0 \qquad (i = 2, \ldots, M), \tag{37}$$

or

$$s - c > 0 \tag{38}$$

and

$$\Pr\left\{ \sum_{i=2}^{M} (\beta_i - r) z_i/(s - c) + r \geqslant 0 \right\} = 1, \tag{39}$$

where $s \equiv x + Y$.

Under feasibility with respect to (14), we then obtain

$$T(x) = \begin{cases} \max\left\{ \frac{1}{\gamma} s^\gamma, \overline{T}(x) \right\}, & 0 < \gamma < 1, \\ \max\{-\infty, \overline{T}(x)\}, & \gamma < 0, \end{cases} \tag{40}$$

[8] A proof based on the method of successive approximations may be found in [7].

where

(41) $$\bar{T}(x) = \sup_{c,\{z_i\}}\left\{\frac{1}{\gamma}c^\gamma + \alpha(1-(\alpha k\gamma)^{1/(1-\gamma)})^{\gamma-1}(s-c)^\gamma \times E\left[\frac{1}{\gamma}\left(\sum_{i=2}^{M}(\beta_i - r)z_i/(s-c) + r\right)^\gamma\right]\right\}$$

subject to (12), (38), (39), and

(42) $$z_i/(s-c) \geqslant 0, \quad i \notin S,$$

since (42) is equivalent to (13) in view of (38). But by (31) the expectation factor in (41) may be written

(43) $$h(z_2/(s-c), \ldots, z_M/(s-c))$$

and (26), the Lemma, and Corollary 2 give

(44) $$k\gamma \geqslant r^\gamma > 0 \quad \text{(Model I)},$$

(45) $$k\gamma \leqslant r^\gamma < 1 \quad \text{(Model II)},$$

while (26), the Lemma, and Corollary 3 give

(46) $$k\gamma > 0 \qquad \text{(Model II)}.$$

Thus, $\partial\bar{T}/\partial h > 0$ always in Model II and in Model I whenever

(47) $$k\gamma < \frac{1}{\alpha}$$

under feasibility. When $k\gamma > 1/\alpha$ in Model I, $\bar{T}(x)$ does not exist; when $k\gamma = 1/\alpha$, (41) and (40) give $T(x) = (1/\gamma)s^\gamma \neq f(x)$. Consequently, it remains to consider the case when $\partial\bar{T}/\partial h > 0$.

Since the maximum of (43) subject to (42) and (39) is k by (26) and the Lemma, we obtain by the Lemma that the strategy

$$\frac{z_i}{s-c} = v_i^* \qquad (i = 2, \ldots, M)$$

or

(48) $$z_i^* = v_i^*(s-c) \qquad (i = 2, \ldots, M)$$

is optimal and unique for *every* c which satisfies (12) and (38) when (38) holds. It is clearly also optimal when (36) and (37) hold. Consequently, (40) reduces to

(49) $$T(x) = \max_{0 \leqslant c \leqslant s}\left\{\frac{1}{\gamma}c^\gamma + \alpha k(1-(\alpha k\gamma)^{1/(1-\gamma)})^{\gamma-1}(s-c)^\gamma\right\}.$$

Since $u(c)$ is strictly concave and $u'(0) = \infty$ in Models I and II, $T(x)$ is strictly concave and differentiable with an "interior" unique solution $c^*(x)$ whenever

(50) $$\alpha k(1-(\alpha k\gamma)^{1/(1-\gamma)})^{\gamma-1}\begin{cases} > 0 & \text{(Model I)}, \\ < 0 & \text{(Model II)}, \end{cases}$$

and $s \geqslant 0$. In this case, setting $dT/dc = 0$ and solving for c, we get,

$$c^{\gamma-1} - \alpha k\gamma(1 - (\alpha k\gamma)^{1/(1-\gamma)})^{\gamma-1}(s - c)^{\gamma-1} = 0$$

or

$$c^*(x) = (1 - (\alpha k\gamma)^{1/(1-\gamma)})(x + Y). \tag{51}$$

In Model I, (50) is satisfied whenever (44) and (47) hold; as noted earlier, no solution exists in Model I for those cases in which $k\gamma \geqslant 1/\alpha$. In Model II, (50) is always satisfied as seen from (45) and (46).

Inserting (51) in (48) we obtain

$$z_i^*(x) = (\alpha k\gamma)^{1/(1-\gamma)}v_i^*(x + Y) \qquad (i = 2, \ldots, M)$$

and (24) follows from (6) upon insertion of $c^*(x)$ and the $z_i^*(x)$. $T(x)$ now becomes, upon insertion of $c^*(x)$ in (49),

$$\begin{aligned} T(x) &= \frac{1}{\gamma}(1 - (\alpha k\gamma)^{1/(1-\gamma)})^{\gamma}s^{\gamma} + \alpha k(1 - (\alpha k\gamma)^{1/(1-\gamma)})^{\gamma-1}s^{\gamma}(\alpha k\gamma)^{\gamma/(1-\gamma)} \\ &= \frac{1}{\gamma}(1 - (\alpha k\gamma)^{1/(1-\gamma)})^{\gamma-1}s^{\gamma} \\ &= f(x) \end{aligned}$$

and the solution clearly exists for $s \geqslant 0$ or

$$x_j \geqslant -Y. \tag{52}$$

Since (52) is an induced constraint with respect to period $j - 1$, it remains to be verified that (52) is either redundant or not effective in period $j - 1$. Because (52) is already present in period $j - 1$ through (14), the induced constraint (52) is redundant, which completes the proof.

Proof of Theorem 1 for Model III: Denote the right side of (11) $T(x)$ upon inserting (22) for $f(x)$. This gives, for all decision points j,

$$T(x) = \max_{c,\{z_i\}} \left\{ \log c + \frac{\alpha}{1-\alpha} E\left[\log\left(\sum_{i=2}^{M} (\beta_i - r)z_i + r(x - c) + y + Y \right)\right] + K \right\}$$

where

$$K \equiv \frac{\alpha}{1-\alpha}\log(1-\alpha) + \frac{\alpha^2 \log\alpha}{(1-\alpha)^2} + \frac{\alpha^2 k}{(1-\alpha)^2}$$

subject to (12), (13), and (14). By the reasoning for Models I and II, we obtain

$$T(x) = \max\{-\infty, \overline{T}(x)\} \tag{53}$$

where

$$\overline{T}(x) = \sup_{c,\{z_i\}} \left\{ \log c + \frac{\alpha}{1-\alpha} \log (s - c) + \frac{\alpha}{1-\alpha} E\left[\log \left(\sum_{i=2}^{M} (\beta_i - r) z_i/(s - c) + r \right) \right] \right\} + K \tag{54}$$

subject to (12),

$$s - c > 0, \tag{38}$$

$$z_i/(s - c) \geqslant 0 \qquad (i = 2, \ldots, M), \tag{42}$$

and

$$\Pr \left\{ \sum_{i=2}^{M} (\beta_i - r) z_i/(s - c) + r \geqslant 0 \right\} = 1. \tag{39}$$

By (31), the next to last term in (54) can be written

$$\frac{\alpha}{1-\alpha} h(z_2/(s - c), \ldots, z_M/(s - c)) \tag{55}$$

where $\partial \overline{T}/\partial h > 0$. Since the maximum of (55) subject to (42) and (39) is $(\alpha k/1 - \alpha)$ by (26) and the Lemma, we obtain from the Lemma that

$$z_i^*(x) = v_i^*(x + Y - c) \qquad (i = 2, \ldots, M) \tag{48}$$

is optimal and unique for *every* c which satisfies (12) and (38). Thus, (53) reduces, in analogy with Models I and II, to

$$T(x) = \max_{0 \leqslant c \leqslant s} \left\{ \log c + \frac{\alpha}{1-\alpha} \log (s - c) + \frac{\alpha k}{1-\alpha} + K \right\} \tag{56}$$

where $T(x)$ always exists since $0 < \alpha < 1$; furthermore, $T(x)$ is strictly concave and differentiable. Setting $\partial T/\partial c = 0$ we obtain

$$c^*(x) = (1 - \alpha)(x + Y), \tag{57}$$

$$z_i^*(x) = \alpha v_i^*(x + Y) \qquad (i = 2, \ldots, M),$$

and (24), all unique. Inserting (57) into (56) gives

$$\begin{aligned} T(x) &= \log (1 - \alpha) + \log s + \frac{\alpha}{1-\alpha} \log \alpha + \frac{\alpha}{1-\alpha} \log s \\ &\quad + \frac{\alpha k}{1-\alpha} + \frac{\alpha}{1-\alpha} \log (1 - \alpha) + \frac{\alpha^2 \log \alpha}{(1-\alpha)^2} + \frac{\alpha^2 k}{(1-\alpha)^2} \\ &= f(x). \end{aligned}$$

Since $f(x)$ exists for $x_j \geqslant -Y$, which as an induced constraint with respect to period $j - 1$ is made redundant by (14) for that period, the proof is complete.

When $y = 0$, the solution to (11) reduces to

$$f(x) = Au(x) + C,$$
$$c^*(x) = Bx,$$
$$z_1^*(x) = (1 - B)(1 - v^*)x,$$
$$z_i^*(x) = (1 - B)v_i^* x \qquad (i = 2, \ldots, M).$$

But then, letting $s \equiv x + Y$,

$$f(s) = Au(s) + C,$$
$$c^*(s) = Bs,$$
$$z_1^*(s) = (1 - B)(1 - v^*)s,$$
$$z_i^*(s) = (1 - B)v_i^* s \qquad (i = 2, \ldots, M).$$

As a result, except for $z_1^*(x + Y)$, the solution to the original problem is not altered when the individual, instead of receiving the noncapital income stream in installments, is given its present value Y in advance. Thus, instead of letting x be the state variable when there is a noncapital income, one could let $x + Y$ be the state variable (pretending there is no income), as long as Y is deducted from $z_1^*(x + Y)$.

Note that it is sufficient, though not necessary, for a solution *not* to exist in Model I that $r^\gamma \geqslant 1/\alpha$ (Corollary 2).

THEOREM 2: *Let α, $\{\beta_i\}$, r, y, and Y be defined as in Section 2. Moreover, let $u(c) = -e^{-\gamma c}$ for $c \geqslant 0$ where $\gamma > 0$. Then a solution to* (11) *subject to* (12)–(14) *exists for $x \geqslant -Y + [r/(\gamma(r - 1)^2)] \log(-\alpha kr)$ and is given by*

$$(58) \qquad f(x) = -\frac{r}{r-1}(-\alpha kr)^{1/(r-1)} e^{-[\gamma(r-1)/r](x+Y)},$$

$$(59) \qquad c^*(x) = \frac{r-1}{r}(x + Y) - \frac{1}{\gamma(r-1)} \log(-\alpha kr),$$

$$(60) \qquad z_1^*(x) = \frac{x}{r} - \frac{y}{r} + \frac{\log(-\alpha kr) - rv^*}{\gamma(r-1)},$$

$$(61) \qquad z_i^*(x) = \frac{r}{\gamma(r-1)} v_i^* \qquad (i = 2, \ldots, M),$$

where the constants k and v_i^* ($v^* \equiv \Sigma_{i=2}^M v_i^*$) are given by

$$(62) \qquad k \equiv E[-e^{-\Sigma_{i=2}^M (\beta_i - r)v_i^*}] = \max_{\{v_i\}} E[-e^{-\Sigma_{i=2}^M (\beta_i - r)v_i}] \quad \textit{subject to } (27)$$

provided that

$$(63) \qquad \log(-\alpha kr) + b(\bar{v}^*) \geqslant 0$$

where $b(\bar{v}^*)$ is the greatest lower bound on b such that

$$Pr\left\{\sum_{i=2}^{M}(\beta_i - r)v_i^* < b\right\} > 0$$

and $\bar{v}^* \equiv (v_2^*, \ldots, v_M^*)$. Moreover, the solution is unique.

Since the conditions under which Theorem 2 holds are quite restrictive, the reader is referred to [7] for the proof. Condition (63) insures that the individual's capital position x is nondecreasing over time with probability 1; it must hold for a solution to exist in closed form. The condition $\alpha r \geqslant 1$ is a necessary, but not sufficient, condition for (63) to be satisfied.[9]

4. PROPERTIES OF THE OPTIMAL CONSUMPTION STRATEGIES

In each of the four models we note that the optimal consumption function $c^*(x)$ is linear increasing in capital x and in noncapital income y. Whenever $y > 0$, positive consumption is called for even when the individual's net worth is negative, as long as it is greater than $-Y$ in Models I–III and greater than $-Y + [r/(\gamma(r-1)^2)] \log(-\alpha kr)$ in Model IV. Only *at* these end points would the individual consume nothing.

Since $x + Y$ may be viewed as permanent (normal) income and consumption is proportional $(0 < B < 1)$ to $x + Y$ in Models I–III, we see that the optimal consumption functions in these models satisfy the permanent (normal) income hypotheses precisely [**9**, **5**, **3**].

In each model, $c^*(x)$ is decreasing in α. Thus, the greater the individual's impatience $1 - \alpha$ is, the greater his present consumption would be. This, of course, is what we would expect.

By (20) and (21), the relative and absolute risk aversion indices of Models I–IV are as follows:

$$q^*(c) = 1 - \gamma \quad \text{(Models I–II)},$$

$$q^*(c) = 1 \quad \text{(Model III)},$$

$$q(c) = \gamma \quad \text{(Model IV)}.$$

[9] For example, when $u(c) = -e^{-.0001c}$, $\alpha = .99$, $y = \$10{,}000$, $r = 1.06$, $M = 2$, and β_2 assumes each of the values .96 and 1.17 with probability .5, a solution exists for $x \geqslant \$-22{,}986$. For selected capital positions, the optimal amounts to consume, lend, and invest in this case are as follows:

x	$c^*(x)$	$z_1^*(x)$	$z_2^*(x)$
\$ −22,986	0	\$ −102,488	\$79,502
0	\$ 1,301	− 80,803	79,502
50,000	4,131	− 33,633	79,502
100,000	6,961	13,537	79,502
500,000	29,601	390,897	79,502
1,000,000	57,901	862,597	79,502

The maximum loss in each period from risky investment is \$3,180.

In Models I–II, we obtain

$$\frac{\partial B}{\partial(1-\gamma)} = -(\alpha k\gamma)^{1/(1-\gamma)}\left\{\frac{[d(k\gamma)/d(1-\gamma)]}{k\gamma(1-\gamma)} - \frac{\log(\alpha k\gamma)}{(1-\gamma)^2}\right\} \tag{64}$$

where $d(k\gamma)/d(1-\gamma)$ is negative whenever $b(\bar{v}^*) \geqslant 1 - r$; otherwise the sign is ambiguous. Since $k\gamma > 0$ and $\alpha k\gamma < 1$, the sign of (64) is ambiguous in both cases; i.e., a change in relative risk aversion may either decrease or increase present consumption. In Model IV, on the other hand, $c^*(x)$ is increasing in γ; i.e., a more risk averse individual consumes more, ceterus paribus.

From (26) and (62) we observe that k is a natural measure of the "favorableness" of the investment opportunities. This is because k is a maximum determined by (the one-period utility function and) the distribution function (F); moreover, F is reflected in the solution only through k, and $f(x)$ is increasing in k. Let us examine the effect of k on the marginal propensities to consume out of capital and non-capital income.

Equation (29) gives

$$\frac{\partial B}{\partial k} = \frac{\alpha\gamma}{\gamma-1}(\alpha k\gamma)^{\gamma/(1-\gamma)}\begin{cases} < 0 & \text{(Model I)}, \\ > 0 & \text{(Model II)}. \end{cases}$$

Thus, we find that the propensity to consume is *decreasing* in k in the case of Model I. This phenomenon can at least in part be attributed to the fact that the utility function is bounded from below but not from above; the loss from postponement of current consumption is small compared to the gain from the much higher rate of consumption thereby made possible later. In Model II, on the other hand, where the utility function has an upper bound but no lower bound, the optimal amount of present consumption is *increasing* in k, which seems more plausible from an intuitive standpoint.

In Model III, we observe from (30) the curious phenomenon that the optimal consumption strategy is independent of the investment opportunities in every respect. While the marginal propensity to consume is independent of k in Model IV also, the *level* of consumption in this case is an increasing function of k as is apparent from (59). We recall that the utility function in Model III is unbounded while that in Model IV is bounded both from below and from above. Thus, the class of utility functions we have examined implies an exceptionally rich pattern of consumption behavior with respect to the "favorableness" of the investment opportunities.

5. PROPERTIES OF THE OPTIMAL INVESTMENT AND BORROWING-LENDING STRATEGIES

The properties exhibited by the optimal investment strategies are in a sense the most interesting. Turning first to Model IV, we note that the portfolio of productive investments is constant, both in mix and amount, at all levels of wealth. The optimal portfolio is also independent of the noncapital income stream and the level of impatience $1 - \alpha$ possessed by the individual, as shown by (61) and (62).

Similarly, we find in Models I–III that, since for all $i, m > 1, z_i^*(x)/z_m^*(x) = v_i^*/v_m^*$ (which is a constant), the *mix* of risky investments is independent of wealth, noncapital income, and impatience to spend. In addition, the *size* of the total investment commitment in each period is proportional to $x + Y$. We also note that when $y = 0$, the ratio that the risky portfolio $\Sigma_{i=2}^{M} z_i^*(x)$ bears to the total portfolio $\Sigma_{i=1}^{M} z_i^*(x)$ is independent of wealth in each model.

In summary, then, we have the surprising result that the optimal mix of risky (productive) investments in each of Models I–IV is independent of the individual's wealth, noncapital income stream, and rate of impatience to consume; the optimal mix depends in each case only on the probability distributions of the returns, the interest rate, and the individual's one-period utility function of *consumption*.

In each case, we find that lending is linear in wealth. Turning first to Models I–III, we find that borrowing always takes place at the lower end of the wealth scale; (24) evaluated at $x = -Y$ gives $-Y < 0$ as the optimal amount to lend. From (24) we also find that $z_1^*(x)$ is increasing in x if and only if $1 - v^* > 0$ since $1 - B$ is always positive. As a result, the models always call for borrowing at least when the individual is poor; whenever $1 - v^* > 0$, they also always call for lending when he is sufficiently rich.

In Model IV, we observe that lending is always increasing in x. Thus, when an individual in this model becomes sufficiently wealthy, he will always become a lender. At the other extreme, when x is at the lower boundary point of the solution set, he will generally be a borrower, though not necessarily, since $z_1^*(x)$ evaluated at $x = -Y + [r/(\gamma(r-1)^2)] \log(-\alpha k r)$ gives

$$-Y + \frac{r \log(-\alpha k r)}{\gamma(r-1)^2} - \frac{r v^*}{\gamma(r-1)}$$

which may be either negative or positive.

We shall now consider the case when the lending rate differs from the borrowing rate as is usually the case in the real world. Let $r_B - 1$ and $r_L - 1$ denote the borrowing and lending rates, respectively, where $r_B > r_L$. Unfortunately, the sign of dv^*/dr is not readily determinable. However, since $f(x)$ is increasing in k, the analysis is straight-forward.[10]

[10] When $r_B > r_L$, the "no-easy-money condition" requires that the joint distribution function of $\beta_2, \ldots, \beta_M$ satisfies

$$\text{(4a)} \qquad \Pr\left\{\sum_{i=2}^{M} (\beta_i - r_B)\theta_i < 0\right\} > 0$$

for all finite numbers $\theta_i \geqslant 0$ such that $\theta_i > 0$ for at least one i;

$$\text{(4b)} \qquad \Pr\left\{\sum_{i \notin S} (\beta_i - r_L)\theta_i < 0\right\} > 0$$

for all finite numbers $\theta_i \leqslant 0$ such that $\theta_i < 0$ for at least one i; and

$$\text{(4c)} \qquad \Pr\left\{\sum_{\substack{i=2 \\ i \notin S^*}}^{M} \beta_i\theta_i - \sum_{k \in S^*} \beta_k\theta_k < 0\right\} > 0$$

for all finite numbers $\theta_i, \theta_k \geqslant 0$ and all $S^* \subseteq S$ such that

$$\sum_{\substack{i=2 \\ i \notin S^*}}^{M} \theta_i = \sum_{k \in S^*} \theta_k,$$

and $\theta_i > 0$ for at least one i. When $r_B = r_L$, 4(a)–4(c) reduce to (4).

Consider first Models I–III when noncapital income $y = 0$. In that case, it is apparent from (24) that when the individual is not in the trapping state (i.e., $x > -Y$), he either always borrows, always lends, or does neither, depending on whether $1 - v^*$ is negative, positive, or zero. Let k_L denote the maximum of (31) when the lending rate is used and the constraint

$$(65) \qquad \sum_{i=2}^{M} v_i \leqslant 1$$

is added to constraints (27) and (28). Since the set of vectors $\bar{v}$ which satisfy (65) is convex and includes $\bar{v} = (0, \ldots, 0)$, the Lemma still holds when (65) is added to the constraint set. Analogously, let k_B denote the maximum of (31) under the borrowing rate r_B subject to (27), (28), and

$$(66) \qquad \sum_{i=2}^{M} v_i \geqslant 1.$$

Again, the Lemma holds since the set of $\bar{v}$ satisfying (66) is convex and any $\bar{v}$ such that $\Sigma_{i=2}^{M} v_i = 1, v_i \geqslant 0$, for example, satisfies all constraints. Setting $k \equiv \max\{k_B, k_L\}$, Theorem 1 holds as before when $y = 0$.

When $y > 0$ in Models I–III and in the case of Model IV, no "simple" solution appears to exist when $r_B > r_L$.

6. THE BEHAVIOR OF CAPITAL

We shall now examine the behavior of capital implied by the optimal investment and consumption strategies of the different models. According to one school, capital growth is said to exist whenever

$$(67) \qquad E[x_{j+1}] > x_j \qquad (j = 1, 2, \ldots),$$

that is, capital growth is defined as expected growth [**10**]. We shall reject this measure since under this definition, as $j \to \infty$, x_j may approach a value less than x_1 with a probability which tends to 1. We shall instead define growth as asymptotic growth; that is, capital growth is said to exist if

$$(68) \qquad \lim_{j \to \infty} \Pr\{x_j > x_1\} = 1.$$

When the $>$ sign is replaced by the $\geqslant$ sign, we shall say that we have capital nondecline. If there is statistical independence with respect to j, (67) is implied by (68) but the converse does not hold, as noted.

Model IV will be considered first. From (63) it follows that nondecline of capital is always implied (in fact, the solution to the problem is contingent upon the condition that capital does not decrease, as pointed out earlier). It is readily seen that a sufficient, but not necessary, condition for growth is that there be a nonzero investment in at least one of the risky investment opportunities since in that case $\Pr\{x_{j+1} > x_j\} > 0, j = 1, 2, \ldots$, by (63). A necessary and sufficient condition for asymptotic capital growth is $\alpha r > 1$, which is readily verified by reference to (62), (63), and the foregoing statement.

Let us now turn to Models I–III and let, as before, $s_j \equiv x_j + Y$. From (7), (23), and (25) we now obtain

$$s_{j+1} = s_j(1 - B)\left[\sum_{i=2}^{M} (\beta_i - r)v_i^* + r\right] \tag{69}$$

$$= s_j W \qquad (j = 1, 2, \ldots)$$

where W is a random variable. By (28), $W \geqslant 0$. Attaching the subscript n to W for the purpose of period identification, we note that since

$$s_j = s_1 \prod_{n=1}^{j-1} W_n, \tag{70}$$

(70) verifies that

$$s_j \geqslant 0 \quad \text{for all } j \text{ whenever } s_1 \geqslant 0 \quad \text{(Models I–III)}.$$

Moreover, since $\Pr\{W > 0\} = 1$ in Models II and III by Corollary 1, it follows that

$$s_j > 0 \quad \text{whenever } s_1 > 0 \text{ for all finite } j \quad \text{(Models II–III)}. \tag{71}$$

From (70) we also observe that $s_j = 0$ whenever $s_k = 0$ for all $j > k$. Consequently, $x = -Y$ is a trapping state which, once entered, cannot be left. In this state, the optimal strategies in each case call for zero consumption, no productive investments, the borrowing of Y, and the payment of noncapital income y as interest on the debt. In Models II and III, it follows from (71) that the trapping state will never be reached in a finite number of time periods if initial capital is greater than $-Y$.

Equation (70) may be written

$$s_j = s_1 e^{\Sigma_{n=1}^{j-1} \log W_n}.$$

The random variable $\Sigma_{n=1}^{j-1} \log W_n$ is by the Central Limit Theorem asymptotically normally distributed; its mean is $(j - 1)E[\log W]$. By the law of large numbers,

$$\frac{\sum_{n=1}^{j-1} \log W_n}{j - 1} \to E[\log W] \quad \text{as } j \to \infty.$$

Thus, since $s_j > s_1$ if and only if $x_j > x_1$, it is necessary and sufficient for capital growth to exist that $E[\log W] > 0$.

It is clear that μ given by $\mu \equiv e^{E[\log W]}$ may be interpreted as the mean growth rate of capital. By (69), we obtain

$$E[\log W] = \log(1 - B) + E\left[\log\left\{\sum_{i=2}^{M} (\beta_i - r)v_i^* + r\right\}\right].$$

For Model III, this becomes, by (30) and (26),

$$E[\log W] = \log \alpha + \max_{\{v_i\}} E\left[\log\left\{\sum_{i=2}^{M} (\beta_i - r)v_i + r\right\}\right]$$

subject to (27) and (28). Thus, a person whose one-period utility function of consumption is logarithmic will always invest the capital available after the allotment to current consumption so as to maximize the mean growth rate of capital plus the present value of the noncapital income stream.

7. GENERALIZATIONS

We shall now generalize the preceding model to the nonstationary case. We then obtain, by the same approach as in the stationary case, for all j,

$$f_j(x_j) = \max_{c_j, \{z_{ij}\}} \left\{u(c_j) + \alpha_j E\left[f_{j+1}\left(\sum_{i=2}^{M_j} (\beta_{ij} - r_j)z_{ij} + r_j(x_j - c_j) + y_j\right)\right]\right\} \tag{72}$$

subject to

$$c_j \geqslant 0, \tag{73}$$

$$z_{ij} \geqslant 0, \quad i \notin S_j, \tag{74}$$

and

$$\Pr\{x_{j+1} \geqslant -Y_{j+1}\}, \tag{75}$$

where the patience factor α, the number of available investment opportunities M and S and their random returns $\beta_i - 1$, the interest rate r, and the noncapital income y may vary from period to period; this, of course, requires that they be time identified through subscript j. Time dependence on the part of any *one* of the preceding parameters also requires that $f(x)$ be subscripted.

As shown in [**7**], the solution to the nonstationary model is qualitatively the same as the solution to the stationary model.

In the case of a finite horizon, the problem again reduces to (72)–(75) with $f_{n+1}(x_{n+1}) \equiv 0$ if the horizon is at decision point $n + 1$. In this case, $f(x)$, x, c, z_i, and Y must clearly be time identified through subscript j even in the stationary model. Under a finite horizon, a solution always exists even for Model I. Again, the solution is qualitatively the same as in the infinite horizon case except that the constant of consumption proportionality B_j increases with time j, $B_n = 1$, and $z_{in}^* = 0$ for all i.[11]

University of California, Berkeley

Manuscript received September, 1966; *revision received January*, 1969.

[11] The implications of the results of the current paper with respect to the theory of the firm may be found in [**6**].

REFERENCES

[1] ACZÉL, J.: *Lectures on Functional Equations and Their Applications.* New York, Academic Press, 1966.

[2] BELLMAN, RICHARD: *Dynamic Programming.* Princeton, Princeton University Press, 1957.

[3] FARRELL, M. J.: "The New Theories of the Consumption Function," *Economic Journal*, December, 1959.

[4] FISHER, IRVING: *The Theory of Interest.* New York, MacMillan, 1930; reprinted, Augustus Kelley, 1965.

[5] FRIEDMAN, MILTON: *A Theory of the Consumption Function.* Princeton, Princeton University Press, 1957.

[6] HAKANSSON, NILS: "An Induced Theory of the Firm Under Risk: The Pure Mutual Fund," *Journal of Financial and Quantitative Analysis*, June 1970.

[7] ———: "Optimal Investment and Consumption Strategies for a Class of Utility Functions," Ph.D. Dissertation, University of California at Los Angeles, 1966; also, Working Paper No. 101, Western Management Science Institute, University of California at Los Angeles, June, 1966.

[8] ———: "Risk Disposition and the Separation Property in Portfolio Selection," *Journal of Financial and Quantitative Analysis*, December, 1969.

[9] MODIGLIANI, F., AND R. BRUMBERG: "Utility Analysis and the Consumption Function: An Interpretation of Cross-Section Data," *Post-Keynesian Economics* (ed. K. Kurihara), New Brunswick, Rutgers University Press, 1954.

[10] PHELPS, EDMUND: "The Accumulation of Risky Capital: A Sequential Utility Analysis," *Econometrica*, October, 1962.

[11] PRATT, JOHN: "Risk-Aversion in the Small and in the Large," *Econometrica*, January–April, 1964.

[12] VON NEUMANN, JOHN, and OSKAR MORGENSTERN: *Theory of Games and Economic Behavior.* Princeton University Press, 1947.

3. MODELS OF OPTION STRATEGY

Reprinted from THE JOURNAL OF POLITICAL ECONOMY
Vol. LXXIV, No. 2, April 1966

THE VALUE OF THE CALL OPTION ON A BOND*

GORDON PYE
University of California, Berkeley

ALMOST all corporate bonds and some government and municipal bonds have call provisions. This call provision gives the issuer the option of buying back his bonds from whomever is holding them at a stated time and price. The call price is typically above par and declines as maturity approaches. Frequently, call is not permitted until several years after the bond has been issued.

The call option has value to the issuer for several reasons. In the future the borrower may wish to remove restrictions placed on him by the bond indenture. For a corporation these might be restrictions on merger, the sale of assets, or the payment of dividends. Without the call provision, the bondholders by hard bargaining might be able to utilize their monopoly position to extract a large premium from the issuer before selling back the bonds or agreeing to change these clauses.

A second source of value is that the borrower may find that he wants to decrease the amount of his borrowing before the bonds mature. Essentially the same effect as retiring his own bonds could be obtained by buying on the market similar bonds issued by someone else. However, because of the transactions costs involved in making interest payments, the cost of bonds to the issuer will always be somewhat greater than their value on the market. There will therefore be some saving to the issuer in retiring his own bonds rather than buying someone else's.

The third and probably most significant source of value of the option to the issuer is the ability it gives him to refinance the issue in the future if interest rates should fall. The analysis of this paper will be confined to considering this aspect of the value of the option. Two models are considered. In one the future price of the bond is assumed to be known and in the other the probability distribution of the future price is assumed to be known and to obey certain conditions. Optimal policies for exercising the option are given for both cases and expressions for the value of the option are derived.[1]

VALUE OF THE OPTION UNDER CERTAINTY

First, consider the case where future bond prices and interest rates are known with certainty. Suppose that p is the market price of a non-callable issue with the same coupon and maturity as a callable issue. Suppose the call price of the latter is c. Suppose that p is greater than c, that transactions costs can be neglected, and that the issuer is too small to affect the market. If the issuer calls the bond at c and issues the otherwise identical non-callable bond at p, he makes a profit of $p - c$. His contractual obligations with respect to the future payment of interest and principal remain unchanged. Therefore, the present value of the interest and principal payments are unchanged and independent of when the option is exercised. The value of the option to the issuer is, of course, dependent on when it is exercised.

The optimal time to exercise the option will be the time at which the present value of $p - c$ is at a maximum. The value of the option to the issuer will be this maximum

* Financial support was provided by the Institute of Business and Economic Research of the University of California, Berkeley.

[1] A technical discussion of the value of the call option on a bond has previously been given by Crockett (1962). The approach taken in this paper is different from hers and additional results are obtained.

present value. Let t be time, r the interest rate, and V the present value of the option. V will be given by

$$V=[p(t)-c(t)]\exp\left[-\int_0^t r(\tau)d\tau\right]. \quad (1)$$

Setting the derivative of V with respect to t equal to zero gives as a necessary condition for the optimal time to call the bond:

$$r(t)[p(t) - c(t)] = p'(t) - c'(t) . \quad (2)$$

In other words, the bond should be called only if the relative rate of growth of the profit from doing so is equal to the market rate of interest. This condition should not be surprising because formally the bond problem is just like the classical capital theory problem of when to cut down a tree or sell a cask of wine.

This optimization condition can be used to evaluate the effect of a shift in the schedule of call prices on the optimum time for call. In particular, assume that the schedule of call prices is shifted up by the factor λ so that it becomes $\lambda c(t)$. Replacing c in the optimization condition by λc, implicitly differentiating, and assuming $c'(t) < 0$, gives

$$\frac{dt}{d\lambda}=\frac{c'-rc}{\partial[p'-\lambda c'-r(p-\lambda c)]/\partial t}>0. \quad (3)$$

The denominator is negative from the second-order condition on the maximum. Thus, shifting up the call price schedule by a constant factor will postpone call of the bond.

The analysis so far has been from the viewpoint of the issuer rather than that of the bondholder. The other side of the profit of $p - c$ obtained by the issuer when he calls the bond is a loss of $p - c$ inflicted on the bondholder. When call occurs, the bondholder receives only c for a future cash flow of interest and principal which is worth p.

The bondholder will assume that the issue will be called at the optimal time from the issuer's point of view. Thus, at any point of time the present value of the interest and principal payments to a bondholder on a callable bond are reduced by the maximum present value of $p - c$. If p^* is the market price of the callable bond and V the value of the option, this means that $p^* = p - V$.

The issuer of the bond will not call the bond as long as the value of the option is greater than the profit of exercising it immediately. Thus, before the option is exercised, it follows that $V > p - c$. Combining this inequality with the fact that $p^* = p - V$ gives $p - p^* > p - c$ or $c > p^*$. Thus, a callable bond will sell below its call price before it is called. When it is called, its price just becomes equal to its call price. In fact, it follows that callable bonds will not sell above their call price in any case. If a callable bond were to sell above its call price, the issuer could make a profit indefinitely by calling the bond and reissuing an identical callable issue. Bondholders realizing the bond would be called at a loss to them if its price were above the call price would never pay more than the call price for the bond.

A MODEL UNDER UNCERTAINTY

It will now be assumed that the future interest rates and bond prices are uncertain. However, their probability distribution will be assumed to be known. Let time be divided into periods and assume that the one-period interest rate can take on a finite number of possible values, $\rho_1, \rho_2, \ldots, \rho_n$. In particular, assume that the probability distribution of the one-period rate in any period depends only on the one-period rate in the previous period.[2] The probability that the one period interest rate is ρ_j given that it was ρ_i in the previous period will be denoted by q_{ij}.

The value of future cash flows will be assumed to be their expected present value. Let p_{ti} be the price of a bond in the tth period when ρ in the tth period is ρ_i. Let V_{ti} be the maximum expected present value of the call option in period t when ρ in t is ρ_i and when the option is required to be exercised subsequent to period t.

[2] Elsewhere it has been shown that this assumption is rich enough to imply under suitable restrictions several observable features of the term structure of interest rates (Pye, 1966).

Given that ρ_j is the one-period interest rate in $t+1$, the maximum expected present value of the option in $t+1$ will be the larger of the following two quantities: (*a*) the value of exercising the option in $t+1$, which is $p_{t+1j} - c_{t+1}$; or (*b*) the maximum expected present value of exercising the option in a period subsequent to $t+1$ which is V_{t+1j}. If ρ_i is the one-period rate in t, the probability that ρ in $t+1$ is ρ_j is q_{ij}. Therefore, given that ρ in t is ρ_i, the expected maximum present value of the option in $t+1$ is

$$\sum_{j=1}^{n} q_{ij} \max(p_{t+1j} - c_{t+1}, V_{t+1j}).$$

The maximum expected present value in t of exercising the option subsequent to t which is V_{ti} must be the maximum expected present value of the option in $t+1$ discounted for the one-period rate of interest in t. Therefore, V_{ti} satisfies the relation

$$V_{ti} = \frac{1}{1+\rho_i} \sum_{j=1}^{n} q_{ij} \times \max(p_{t+1j} - c_{t+1}, V_{t+1j}) \quad (i = 1, 2, \ldots, n). \tag{4}$$

For a bond maturing in T periods the value of exercising the option subsequent to T is clearly zero. Therefore, one has that $V_{Tj} = 0$ for all j. Given the p_{ti}, the V_{ti} can be calculated by iterating backward using equation (4) and this initial condition.[3]

The p_{ti} may be calculated iteratively in a similar manner. If ρ turns out to be ρ_j in $t+1$, a bondholder in t will have in $t+1$ an asset worth $r + p_{t+1j}$ where r is the coupon payment. Given that ρ in t is ρ_i, the expected value of the bond in $t+1$ will be

$$r + \sum_{j=1}^{n} q_{ij} p_{t+1j}.$$

Discounting this for the interest rate between t and $t+1$ gives the expected present value of the bond in t or p_{ti}:

$$p_{ti} = \frac{1}{1+\rho_i}\left(r + \sum_{j=1}^{n} q_{ij} p_{t+1j}\right) \quad (i=1, 2, \ldots, n). \tag{5}$$

If the bond matures in period T, p_{Tj} will be independent of j and equal to the par value of the bond. The p_{ti} can be calculated by iterating backward in time using expression (5) and this initial condition.

In this model the bond should be called as soon as $p_{ti} - c_t > V_{ti}$. This merely says that the bond should be called as soon as the profit from doing so is greater than the value of doing so later. Unlike the certainty case, under uncertainty the optimal time to call the bond cannot be specified initially. Whether or not the bond should be called in any particular period depends on the ρ that happens to occur in that period. When the bond is issued, this ρ is a random variable. All that can be done initially is to specify a rule for calling the bond in any period as a function of the ρ that happens to occur in that period. When the decision about calling the bond actually has to be made in any period, the value of ρ for that period will be known.

Under fairly reasonable conditions the rule for calling the bond as a function of ρ takes a simple form. The bond should be called in a particular period if and only if ρ is less than some critical value for that period. These critical values will increase or at least remain constant as time increases. This characterization of the optimal policy will hold if the following conditions are satisfied: (*a*) the probability that ρ next period is less than or equal to any given value decreases or remains constant as the current value of ρ increases; (*b*) the call prices are

[3] A dynamic programing approach similar to that used here has previously been used by Karlin (1962) in studying the optimal time to sell an asset. However, he has not applied his model to options nor considered uncertain discount rates which are basic to the bond option problem.

Bachelier (1900), Kruizenga (1956), and Boness (1964) have studied the value of a call option on stock. However, they have not studied the multiperiod, sequential decision case or the case of uncertain discount rates which is relevant for bonds.

constant or decrease at a constant or increasing rate (i.e., $\Delta c_t \leq 0$, $\Delta^2 c_t \geq 0$). Interested readers can find a formal statement and proof of this proposition in the Appendix.

No reason has yet been provided to explain why borrowers bother to include a call option on their bonds. If borrowers actually had to pay a price for the option just equal to its value to them as in the certainty case discussed above, they would be just indifferent as to whether or not to make their bonds callable. Under uncertainty, however, the introduction of either probability differences or risk aversion makes it possible to explain why some borrowers will definitely want their bonds callable.

Suppose that individual borrowers and lenders have different probability distributions. Suppose further that the number of bonds to be issued or purchased by each individual is given and that aggregate purchases and sales are equal so that only the proportion of the bonds that are callable is left to be determined. If on the market the price differential between non-callables and callables is small, all the borrowers except the few with the lowest expectations of future bond prices will want to issue callables since to them the value of the option is greater than its price. On the other hand, all the lenders except the few with the lowest expectations of future bond prices will want to buy non-callables since to them the cost of the option is greater than the price they receive for it. Therefore, the demand for callables will be less than the supply. When the price differential is large, the reverse will be true. At an intermediate price differential the supply and demand of callables will be equal. In this equilibrium only the marginal borrower will be just indifferent as to whether he makes his bonds callable or not.

Now assume that borrowers and lenders are averse to risk. This means that they will value a security at more or less than its expected value as it decreases or increases the variability of the net return on their portfolios. Assume that borrowers are corporations which plan on remaining in debt until the bonds mature. If the interest rate falls in the future, competitors with lower capital costs can enter the businesses of these corporations. This low-cost competition will depress product prices and reduce profits. If their bonds are callable, however, the corporations will make profits on their options from the fall in the interest rate which will just offset these losses. Therefore, the call option will be worth more than its expected value to them because it insures against a fall in the interest rate.

Assume that lenders can be divided into two groups: those who plan on holding their bonds until maturity and those who plan on selling their bonds at a point well before maturity. Long-term lenders face the risk of receiving lower yields to maturity if their bonds are called. There seems to be no particular source of gain on other securities they might be holding that would offset such losses. Therefore, the call option will increase the variability of the net return on their portfolios and so its cost to them will be somewhat greater than its expected value.

On the other hand, short-term lenders in holding callables give up a lottery of possible capital gains for a certain payment. If lower interest rates in the near future are associated with unemployment and reduced profit expectations, gains on the bonds will offset losses on any stocks in their portfolios. However, lower interest rates need not be associated with reduced profit expectations. If they are not, lower interest rates will mean higher, not lower, stock prices. Unless lower interest rates are highly associated with lower stock prices, the call option will reduce the variability of the net return for the short-term lender. The cost of the option for him will then be somewhat less than its expected value.

Suppose the price differential between non-callables and callables is equal to the expected value of the option. According to the foregoing analysis all corporations will want to issue callables but only short-term lenders will want to hold them. As the differential rises, some of the less risk-averse corporations will be coaxed into issuing non-callables and some of the less risk-averse long-term lenders will be coaxed into holding callables. Equilibrium will be reached when the differential has risen far enough above expected value.

An interesting fact to explain is why most corporate bonds are callable and most public bonds are not. To public officials callable bonds mean higher taxes now with the possibility of tax savings in the future. The officials might be expected to discount the future savings by the probability that they will no longer be in office when these savings accrue. The effect would be much the same as if public officials had lower expectations of future bond prices than most private borrowers and lenders in the model discussed above. In equilibrium most private bonds would be callable and most public bonds would not.

The phenomenon might also be explained by risk aversion. For state and local governments there seems to be no particular reason to suspect that the tax savings would offset any tax losses or increased expenditures. For the federal government the tax savings would offset higher interest expenditures from deficits to the extent that lower interest rates were associated with unemployment. However, calling bonds during periods of unemployment would be disadvantageous from the point of view of countercyclical policy. Private wealth would be reduced which would reduce spending. This adverse effect would cancel any benefit from stabilized expenditures. Thus, the value of the option to either state and local or federal authorities will tend to be less than its expected value.

Suppose again that the differential between non-callables and callables is equal to the expected value of the option. Public authorities will want to issue non-callables and corporations will want to issue callables. Long-term lenders will want to hold non-callables and short-term lenders will want to hold callables. Equilibrium will be attained at a differential above or below expected value as the proportion of public borrowing is less or greater than the proportion of long-term lending. Unless the proportion of public borrowing is much greater than the proportion of long-term lending, in equilibrium most public bonds will not be callable.

APPENDIX

To prove the proposition stated in the text, the following notation will be used. Let Q be the $n \times n$ probability matrix $[q_{ij}]$. Let D be an $n \times n$ diagonal matrix with the discount factors $1/(1+\rho_i)$ as the diagonal elements. Let p_t and V_t be the $n \times 1$ column vectors of p_{ti} and V_{ti}, respectively. Let c_t be the $n \times 1$ column vector whose elements are all identical and equal to the call price in the tth period. Let r be the $n \times 1$ column vector whose elements are all identical and equal to the coupon payment on the bond. Using this notation the relations (4) and (5) can be written

$$V_t = DQ \max\,[p_{t+1} - c_{t+1}, V_{t+1}]\,, \quad (4')$$

$$p_t = D[r + Qp_{t+1}]\,. \quad (5')$$

A probability distribution $f(x)$ is said to dominate another distribution $g(x)$ if the cumulative distributions satisfy $F(x) \leq G(x)$ for all x. With this definition, the proposition to be proved may be stated as follows:

Proposition: If the probability distribution of each row of Q dominates that of the preceding row and if $\Delta c_t \leq 0$ and $\Delta^2 c_t \geq 0$, then (*a*) the bond should be called in any period if and only if ρ is less than some critical value for that period; (*b*) these critical values will be non-decreasing with time.

Proof: (*a*) For proof it suffices to show that the vector $p_t - c_t - V_t$ is monotone-decreasing for all t. Assume $p_{t+1} - c_{t+1} - V_{t+1}$ is monotone-decreasing.

$$\begin{aligned} p_t - c_t - V_t &= D(r + Qp_{t+1}) - c_t \\ &\quad - DQ \max\,[p_{t+1} - c_{t+1}, V_{t+1}] \\ &= Dr - c_t + DQ(p_{t+1} \\ &\quad - \max\,[p_{t+1} - c_{t+1}, V_{t+1}])\,. \end{aligned} \quad (6)$$

Denote the vector in the parentheses premultiplied by DQ by z. The elements of z are equal to c_{t+1} if ρ_i is less than the critical value for $t+1$ and $p_{t+1i} - V_{t+1i}$ otherwise. In the latter case, $c_{t+1} > p_{t+1i} - V_{t+1i}$. From this and the fact that $p_{t+1} - c_{t+1} - V_{t+1}$ is decreasing, it follows that z must be non-increasing. The expected value with respect to $f(x)$ of a non-increasing function cannot be greater than the expected value of this function with respect to $g(x)$ if $g(x)$ dominates $f(x)$. Therefore, since each row of Q dominates the rows which precede it, the vector Qz must be non-increasing. Multiplication of a non-increasing, positive vector by D gives a decreasing vector. Therefore, DQz and Dr must be decreasing, so from equation (6) $p_t - c_t - V_t$ must be decreasing. For a bond maturing in T periods, $p_T = c_T$ so that

$$p_{T-1} - c_{T-1} - V_{T-1} = D(r + p_T) - c_{T-1} - DQ \max [p_T - c_T, 0]$$

is decreasing. Therefore, the desired result follows for any t by induction.

(*b*) If for any t the critical ρ, say ρ_{i+1}, is less than in $t-1$, then $p_{t-1i} - c_{t-1} - V_{t-1i} > 0$ and $p_{ti} - c_t - V_{ti} \leq 0$. This will be impossible if $\Delta p_{ti} - \Delta c_t - \Delta V_{ti} \geq 0$. Therefore, to show that the critical ρ's are non-decreasing, it suffices to show that $\Delta p_t - \Delta c_t - \Delta V_t \geq 0$. This can be shown by induction as follows:

If $\Delta p_{t+1} - \Delta c_{t+1} - \Delta V_{t+1} \geq 0$, then

$$\Delta V_t = DQ [\max (p_{t+1} - c_{t+1}, V_{t+1}) - \max (p_t - c_t, V_t)] \leq DQ[\Delta p_{t+1} - \Delta c_{t+1}] = \Delta p_t - D\Delta c_{t+1} \leq \Delta p_t - \Delta c_t ,$$

which, upon rearrangement, gives $\Delta p_t - \Delta c_t - \Delta V_t \geq 0$. If the bond matures in T periods, $c_T = p_T$, so that

$$\Delta p_{T-1} - \Delta c_{T-1} - \Delta V_{T-1} = DQ\Delta p_T - \Delta c_{T-1} + DQ \max (p_{T-1} - c_{T-1}, 0) = DQ [\max (p_{T-1} - c_{T-1}, 0) - (p_{T-1} - c_T)] - \Delta c_{T-1} \geq 0$$

and the desired result follows by induction as stated.

REFERENCES

Bachelier, L. *Théorie de la spéculation.* Paris: Gauthier-Villars, 1900.

Boness, A. J. "Elements of a Theory of Stock Option Value," *J.P.E.*, April, 1964.

Crockett, Jean. "A Technical Note on the Value of a Call Privilege," A. Hess and W. Winn, in *The Value of the Call Privilege*, Philadelphia: University of Pennsylvania Press, 1962.

Karlin, S. "Stochastic Models and Optimal Policy for Selling an Asset," in *Studies in Applied Probability and Management Science*, ed. K. Arrow, S. Karlin, and H. Scarf. Stanford, Calif.: Stanford University Press, 1962.

Kruizenga, R. "Put and Call Options: A Theoretical and Market Analysis." Unpublished Ph.D. dissertation, Economics Department, Massachusetts Institute of Technology, 1956.

Pye, G. "A Markov Model of the Term Structure," *Q.J.E.*, February, 1966.

MANAGEMENT SCIENCE
Vol. 14, No. 1, September, 1967
Printed in U.S.A.

EVALUATING A CALL OPTION AND OPTIMAL TIMING STRATEGY IN THE STOCK MARKET*

HOWARD M. TAYLOR†

Cornell University

The optimal strategy for the holder of a "put" or "call" option contract in the stock market is studied under the random walk model for stock prices. Some results are distribution-free in that they depend only on the mean price change. Other results are derived under the assumption that price changes have a normal or Gaussian distribution. Under this assumption explicit results are obtained for the limiting case where the expiration date of the contract is indefinitely far in the future. Knowing the optimal strategy it is possible to evaluate whether the purchase of a given option can be expected to be profitable.

1. Introduction and Summary

A call is an option entitling the holder to buy a block of shares in a given company at a stated price at any time during a stated interval. Thus the call listed in the financial section of the newspaper as:

U. S. Steel...... $49\frac{3}{8}$ 6 mos. $565.50

means that for a price of $565.50 one may purchase the privilege (option) of buying 100 shares of the stock of U. S. Steel at a price of $49.375 per share at any time during the next six months. (Call prices listed in the newspaper are "asking" prices, and bargaining is often possible.)

Of the many reasons for purchasing a call, two are studied here, referred to as the investor's problem (insurance problem [2]) and the speculator's problem. In the first case an investor may have decided to add a block of the stock of a particular company to his portfolio, leaving only the question of the timing of the purchase. He feels that in the near future the price of the stock may drop, enabling a more advantageous buy. However, not being sure and to guard against a near future price rise he purchases a call which guarantees that he, at least, need never pay more than the price stated in the call. Of course, should the price of the stock drop in the near future, he will ignore the call option and make his purchase at the lower price on the open market. As a current example (1966) consider the stock of American Telephone and Telegraph, one of the bluest of the blue chips, which is now selling at or near its lowest price in two years. The price is depressed because of a pending investigation of the company by the Federal Communications Commission, according to the financial newspapers. There is no doubt that the company is sound and soundly managed, and it is

* Received May 1966 and revised January 1967.

† In refereeing an earlier draft of this paper, Arthur Veinott made many detailed and helpful suggestions which resulted in simpler proofs under weaker assumptions with stronger conclusions. I thank him for his conscientious help.

easy to picture an investor having decided to add AT and T to his portfolio as a long run commitment. However, more news of the pending investigation may depress the price further and thus perhaps a call should be considered, rather than immediate purchase.

A speculator, on the other hand, may purchase a call if he feels a sharp price rise in the offing. Again he's not sure and rather than make an outright purchase, for much less money (or more stock controlled) he purchases a call. Should the price rise during the period of the call he will exercise the option to buy at the stated price and immediately resell in the open market at a profit. Should the price fall, he will not exercise his option.

A put is a stock option contract which obligates the writer, or seller, to accept delivery of 100 shares of a particular stock at a set price within a specified period of time. Puts are bought and sold for reasons similar to those underlying calls, and the analysis of puts is similar to the subsequent analysis of calls. Both puts and calls are often purchased for other reasons, usually of a tax planning nature, but our analyses are confined to the investor's and the speculator's problem as outlined above.

The central problem studied is the call owner's optimal strategy in deciding between purchasing on the market or waiting one more day, based on the current market price of the stock and the number of days left in the option contract. Once the optimal strategy is known, it is relatively easy to decide whether or not the purchase of a call could be expected to yield a profit.

Before summarizing the results the fundamental assumptions will be presented so that a reader not willing to make these assumptions need not read further. Most important is the model chosen to represent short term fluctuations in the market price of a stock. We have assumed these prices to be described by a random walk, and examined two models which, following Samuelson [20], we term the absolute random walk and the geometric random walk. Let Y_n be today's market price, Y_{n-1} tomorrow's price, and so on, down to Y_0, the price on the last day the call is effective. Most of our results are for the absolute random walk model where we assume $Y_{k-1} = Y_k + Z_k$ where $\{Z_1, Z_2, \cdots\}$ are independent identically distributed random variables with $E[|Z_k|] < \infty$ for all k. In the geometric model we assume $Y_{k-1} = Y_k Z_k$ where $\{Z_1, Z_2, \cdots\}$ are independent identically distributed positive random variables with $E[Z_k] \leqq 1$. The sequence of prices is a Markov chain under both models, and we let $F(t \mid y) = \Pr(Y_{k-1} \leqq t \mid Y_k = y)$. Some of our results are obtained for the absolute model with the further assumption that the daily price changes Z_j have a Gaussian or normal distribution with mean μ and variance σ^2. That is, the common probability density function $f_Z(\cdot)$ of each Z_k is given by $f_Z(z) = ((2\pi)^{1/2}\sigma)^{-1} \exp\{-\frac{1}{2}(z - \mu)^2/\sigma^2\}$. Many studies, theoretical and empirical, have discussed both random walk models and the Gaussian assumption. Under the absolute random walk model negative prices are possible, a distinct unpleasantness, and in long run problems, a severe weakness. This problem is avoided in the geometric model. Put and call options are typically short run propositions, occasionally running six months or longer but more often having expiration dates

60 to 90 days hence. The range of price changes is relatively small and there should be little difference between the models [20]. Recently there have been some conjectures on martingale models but little published as yet. For further discussion on random walk models for stock prices the reader should consult references [3, 6, 7, 8, 10, 14, 16, 17, 18, 19, 20].

Let p_0 be the price guaranteed in the call.

For the *investor* who assumes an *absolute* random walk model the main results are:

(a1) For the case where $E[Z_k] \leqq 0$, the optimal strategy for the call holder is to do nothing until the last day the call is effective. On that day the purchase is made at the lower of the call price or the market price. (For the geometric case the same result holds if $E[Z_k] \leqq 1$.)

(a2) If $E[Z_k] > 0$ then there exists a decreasing sequence of numbers $a_0 = p_0 \geqq a_1 \geqq a_2 \geqq \cdots$ such that if Y_n is the market price with n days to go to the expiration date of the call, then an open market purchase should be made if $Y_n < a_n$. Otherwise the investor should wait one more day before deciding. The critical numbers a_n may be found as $a_n = \sup \{y: \int P_{n-1}(t)\, dF(t \mid y) > y\}$ where the functions $P_n(\cdot)$ are defined recursively as $P_0(y) = \min \{p_0, y\}$, and

$$P_n(y) = \min \{y, \int P_{n-1}(t)\, dF(t \mid y)\}$$

for all y.

(a3) If the price changes are normally distributed with mean $\mu \geqq 0$ and variance σ^2, and if today's market price is y, and a call guaranteeing a price of p_0 and with an expiration date n days hence is purchased then the minimum expected price paid for the stock is

$$P(y, p_0, n, \mu, \sigma) = p_0 - \sigma(n)^{1/2}[\varphi(\alpha) + \alpha\Phi(\alpha)]$$

where

$$\alpha = (p_0 - y - n\mu)/\sigma(n)^{1/2}$$

and

$$\varphi(t) = (2\pi)^{-1/2} \exp \{-\tfrac{1}{2}t^2\}$$

and

$$\Phi(x) = \int_{-\infty}^{x} \varphi(t)\, dt,$$ the standard normal density and distribution functions, respectively.

(a4) If the price changes are normally distributed with mean $\mu > 0$ and variance σ^2, then the critical numbers $a_0 = p_0 > a_1 \geqq a_2 \geqq \cdots$ referred to in (a2) above converge monotonically to a limit denoted by a which may be found as the solution to the equation $P(y) - y = 0$ where $P_n(y) \to P(y)$ as $n \to \infty$ for all y and $P(\cdot)$ is a solution to the functional equation $P(y) = \min \{y, \int P(t)\, dF(t \mid y)\}$. There are many solutions to this functional equation. It is shown formally that $a = p_0 - \sigma^2/2\mu$. If at any time the market price is less than a, it pays to buy on the market and ignore the call. In particular, a call offered for sale for which a exceeds the current market price of the stock cannot be profitable. Furthermore the quantity $P(y, p_0, n, \mu, \sigma)$ for $\mu = 0$ provides a lower limit on the expected price for the case where $\mu > 0$, thus affording another means of evaluating a call.

The *speculator's* problem is a mirror image of the investor's problem. For the absolute random walk model we say to him:

(b1) For the case where $E[Z_k] \geqq 0$, the optimal strategy is again to do nothing until the last day the call is effective. On that day the call is exercised if the call price is lower than the market price. (Again, the same result holds for the geometric model if $E[Z_k] \geqq 1$.)

(b2) If $E[Z_k] < 0$, then there exists an increasing sequence of numbers $b_0 = p_0 \leqq b_1 \leqq \cdots$ such that if Y_n is the market price with n days to go, then the call should be exercised should $Y_n \geqq b_n$. Otherwise, the speculator should wait one more day before deciding. These critical numbers may be determined by $b_n = \sup \{y: \int Q_{n-1}(t)\, dF(t \mid y) > y\}$, where the functions $Q_n(\cdot)$ are defined recursively by $Q_0(y) = \max\{y, p_0\}$ and $Q_n(y) = \max\{y, \int Q_{n-1}(t)\, dF(t \mid y)\}$ for $n = 1, 2, \cdots$.

(b3) If the price changes are normally distributed with mean $\mu \geqq 0$ and variance σ^2, and if today's market price is y and a call guaranteeing a price of p_0 with an expiration date of n days hence is purchased, then the maximum expected profit the speculator can assume is:

$$Q(y, p_0, n, \mu, \sigma) = \sigma(n)^{1/2}[\varphi(\beta) + \beta\Phi(\beta)]$$

where
$$\beta = (y + n\mu - p_0)/\sigma(n)^{1/2}$$

which must be weighed against the price of the call. The quantity $Q(y, p_0, n, 0, \sigma)$ provides a lower limit on the expected profit.

(b4) If the price changes are normally distributed with mean $\mu < 0$ and variance σ^2, then the critical numbers $b_0 = p_c \leqq b_1 \leqq b_2 \leqq \cdots$ referred to in (b2) above converge monotonically to a limit denoted by b which may be found as the solution to the equation $Q(y) - y = 0$ where $Q(\cdot)$ satisfies the functional equation:

$$Q(y) = \max\{y, \int Q(t)\, dF(t \mid y)\}.$$

If at any time, the price on the market exceeds b it pays to exercise the call and resell on the market.

The motivation for the present work comes from Karlin [9] where similar techniques are used in a different context. References [4, 5, 12, 15] consider variants of this model and obtain closely related results.[1]

[1] These references, which helped considerably to simplify an earlier draft of this paper, were brought to my attention by Arthur Veinott.

Other authors have considered problems similar to ours.[2] Boness [2] correctly states our conclusion (b1), but in a comment appropriate to (b2) he says: "Conversely the holder of a put would exercise his option at the earliest profitable opportunity, since in this case the trend works against further price declines." As our analysis shows this is either a tautology or is incorrect, depending on the sophistication used in the definition of "profitable."

Samuelson [20] has studied the warrant option, an option similar to a call, but usually with an expiration date further in the future. For long-lived warrants he points out why it is important to assume the geometric random walk model, and this is the assumption he predominantly follows: His study concerns "rational" pricing wherein a holder of a warrant will convert or not so as to maximize the expected rate of return. Samuelson gives formulas for the rational price under a variety of assumptions, including the perpetual warrant. Our somewhat simpler formulation and assumptions allow simpler derivations and results.

Finally, Bierman [1] has recently formulated the bond refunding decision in a similar manner.

2. Derivation of Results

Let

p_0 be the price guaranteed in the call,

n, j be subscripts usually indicating the number of periods to go before the option expires,

Y (with or without subscripts) be the market price of the stock (random).

A subscript $j = 0$ refers to the last day in which the call may be exercised, and time is measured backward from this point.

We suppose that:

(2.1) the prices $\{\cdots, Y_2, Y_1, Y_0\}$ form a Markov process and $E[|Y_n|] < \infty$ for all n,

(2.2) $F(t \mid y) = \Pr(Y_{n-1} \leqq t \mid Y_n = y)$ is independent of n and is nonincreasing in y for all t and $n = 1, 2, \cdots$, and

(2.3) $E[Y_{n-1} - Y_n \mid Y_n = y]$ is nonincreasing in y for $n = 1, 2, \cdots$.

Assumptions (2.1) and (2.2) are satisfied in both the absolute and geometric random walk models. In the absolute case one has $F(t \mid y) = F_Z(t - y)$, while in the geometric case $F(t \mid y) = F_Z(t/y)$, both of which are nonincreasing in y for fixed t. Assumption (2.3) is always satisfied in the absolute case, while in the geometric model it is satisfied whenever $E[Z_k] \leqq 1$. From assumption (2.2), if $g(y)$ is a noncreasing function of y, then so also is $\int g(t)\, dF(t \mid y)$ ([11], p. 400).

The Investor's Porblem

Let $P_j(y)$ be the minimum expected purchase price after observing the current price of $Y_j = y$, given that there are $j \geqq 0$ periods to go. Proceeding optimally,

[2] The author is indebted to Eugene Fama for pointing out these references.

the investor would purchase under these conditions with $j \geqq 1$ periods to go if $y < E[P_{j-1}(Y_{j-1}) \mid Y_j = y]$; i.e. if the price y of buying immediately were less than the expected minimum price if an immediate purchase is not made. One has $P_0(y) = \min\{y, p_0\}$ for all y, and, knowing the optimal procedure in any period, we have $P_j(y) = \min\{y, E[P_{j-1}(Y_{j-1}) \mid Y_j = y]\}$, for all y and $j \geqq 1$. Let $H_j(y) = P_j(y) - y$ so that $H_j(y) = \min\{0, E[H_{j-1}(Y_{j-1}) \mid Y_j = y] + E[Y_{j-1} - Y_j \mid Y_j = y]\}$. In terms of the new function, if $Y_j = y$ and $H_j(y) = 0$, we would immediately purchase on the market. More exactly, if $J_j(y) = E[H_{j-1}(Y_{j-1}) \mid Y_j = y] + E[(Y_{j-1} - Y_j) \mid Y_j = y]$, then we would purchase immediately if $J_j(y) > 0$; we are indifferent if $J_j(y) = 0$; and we wait at least one more day if $J_j(y) < 0$.

By assumption (2.3) $E[Y_{j-1} - Y_j \mid Y_j = y]$ is a nonincreasing function of y. Also, $H_0(y) = \min\{0, p_0 - y\}$ is a nonincreasing function of y, and by assumption (2.2) and subsequent remarks, if $H_{j-1}(y)$ is nonincreasing in y, then so also is $E[H_{j-1}(Y_{j-1}) \mid Y_j = y]$. Since the sum of nonincreasing functions and the minimum of 0 and a nonincreasing function are also nonincreasing, we have $J_j(y)$ and $H_j(y)$ are nonincreasing in y for all $j \geqq 0$. There is an additional monotonicity. For $y \leqq p_0$, $H_0(y) = 0$ and $H_1(y) \leqq 0$ so that $H_1(y) \leqq H_0(y)$ in this case. For $y > p_0$,

$$\begin{aligned} H_1(y) &\leqq E[H_0(Y_0) \mid Y_1 = y] + E[Y_0 - Y_1 \mid Y_1 = y] \\ &= \int_{p_0}^{\infty} (p_0 - t)\, dF(t \mid y) + \int_{-\infty}^{+\infty} (t - y)\, dF(t \mid y) \\ &= (p_0 - y) \int_{p_0}^{\infty} dF(t \mid y) + \int_{-\infty}^{p_0} (t - y)\, dF(t \mid y) \leqq (p_0 - y) = H_0(y). \end{aligned}$$

Thus $H_1(y) \leqq H_0(y)$ for all y. Now assume $H_j(y) \leqq H_{j-1}(y)$ for all y. Then

$$\begin{aligned} E[H_j(Y_j) \mid Y_{j+1} = y] &= \int H_j(t)\, dF(t \mid y) \leqq \int H_{j-1}(t)\, dF(t \mid y) \\ &= E[H_{j-1}(Y_{j-1}) \mid Y_j = y]. \end{aligned}$$

Similarly,

$$E[Y_j - Y_{j+1} \mid Y_{j+1} = y] = \int (t - y)\, dF(t \mid y) = E[Y_{j-1} - Y_j \mid Y_j = y].$$

By addition $J_{j+1}(y) \leqq J_j(y)$, and then $H_{j+1}(y) = \min\{0, J_{j+1}(y)\} \leqq \min\{0, J_j(y)\} = H_j(y)$. Thus by induction, $H_j(y)$ and $J_j(y)$ are nonincreasing in j for all y.

We are now ready to draw our first two conclusions. Let $a_0 = p_0$, and for $j \geqq 1$ let $a_j = \sup\{y: J_j(y) > 0\}$ if the set is non-empty, and $a_j = -\infty$ otherwise. With j periods to go and a current price of Y_j, an optimal strategy is to purchase immediately if $Y_j < a_j$ and to wait otherwise. From the monotonic properties of $J_j(y)$ in j and y we have $a_0(= p_0) \geqq a_1 \geqq a_2 \geqq \cdots$. Suppose

$$E[Y_{j-1} - Y_j \mid Y_j = y] \leqq 0$$

for all y. Then, since $E[H_0(Y_0) \mid Y_1 = y] \leqq 0$ for all y, one has $H_1(y) \leqq 0$ for all y and $a_1 = -\infty \geqq a_2 \geqq \cdots$. Thus for the absolute random walk where $E[Z_k] \leqq 0$ or the geometric case where $E[Z_k] \leqq 1$, either of which implies $E[Y_{j-1} - Y_j \mid Y_j = y] \leqq 0$, one would never purchase on the market but would always wait one more period, until the last period. Thus, conclusions (a1) and (a2) have been reached.

Under the absolute random walk assumption, with a current market price of y, and with n days to go, the price of the stock on the last day is $X = y + Z_1 + \cdots + Z_n$ where the changes Z_i are independent and identically distributed. If each Z_i has a normal distribution with mean μ and variance σ^2, then X has a normal distribution with mean $y + n\mu$ and variance $n\sigma^2$. If $\mu \leqq 0$, the optimal policy is to wait until the last day, and then if $X \leqq p_0$, the purchase is made on the market; otherwise the call is exercised. Thus the expected price paid for the stock is

$$\begin{aligned} P(y, p_0, n, \mu, \sigma) &= E[\min\{X, p_0\}] \\ &= y + n\mu + \sigma(n)^{1/2} E[\min\{Z, \alpha\}] \\ &= p_0 - \sigma(n)^{1/2}[\varphi(\alpha) + \alpha\Phi(\alpha)] \end{aligned}$$

where $\alpha = (p_0 - y - n\mu)/\sigma(n)^{1/2}$.

We have assumed $\mu \leqq 0$. Should a call not be purchased, an investor presumably would not purchase the stock until some future time when he believed the expected price change had become positive. Thus to evaluate a call, one would compute $P(y, p_0, n, \mu, \sigma)$ and compare this to $y + n\mu$, the expected market price in n days assuming the call is not purchased. Suppose d is the per share price of the call option itself. Elementary algebra yields

$$[P(y, p_0, n, \mu, \sigma) - (y + n\mu)]/(\sigma n^{1/2}) = \alpha - \varphi(\alpha) - \alpha\Phi(\alpha).$$

Let $g(\alpha) = \alpha - \varphi(\alpha) - \alpha\Phi(\alpha)$. Then one purchases the call provided

$$g(\alpha) + d/(\sigma n^{1/2}) \leqq 0.$$

Since $g'(\alpha) = dq/d\alpha = 1 - \Phi(\alpha) > 0$ for all α, and since $\lim_{\alpha \to -\infty} g(\alpha) = -\infty$, $\lim_{\alpha \to +\infty} g(\alpha) = 0$, a unique solution to $g(\alpha) = -d/(\sigma(n)^{1/2})$ exists.

If $p_0 - (y + n\mu) = 0$, perhaps a rough approximation to the real situation (though it requires that $p_0 \leqq y$ since $\mu \leqq 0$), then

$$P(y, p_0, n, \mu, \sigma) = p_0 - \sigma(n/2\pi)^{1/2}$$

which gives a rough indication of how the price may be expected to vary with σ and n, and might be compared with the present rules of thumb used in pricing call options.

For the investor, the case where $\mu > 0$ is the mcre realistic since most likely he would want to purchase a stock which he thought had an underlying positive bias in its price changes. However, the price paid where $\mu > 0$ must be higher on the average than the price paid where $\mu = 0$, and thus our previous results give a lower bound on the expected price paid under the call option. A call option which appears unfavorable under the assumption $\mu = 0$, cannot be favorable for the *investor* if $\mu > 0$.

In the absolute random walk model where the price changes $Z_1, Z_2, \cdots$ are normally distributed with mean $\mu > 0$ and variance σ^2, one may show that the sequence $a_0 = p_0 \geqq a_1 \geqq a_2 \geqq \cdots$ are bounded below, hence converge to a limit denoted by a, that the functions $H_0(\cdot), H_1(\cdot), \cdots$ converge monotonically to a limit function $H(\cdot)$, and that $a = \sup \{y: H(y) = 0\}$. The recursion relationship for $H_j(\cdot)$ in this case is $H_j(y) = \min \{0, E[H_{j-1}(y + Z)] + \mu\}$, where Z is normally distributed with mean μ and variance σ^2. This leads us to examine the functional equation $H(y) = \min \{0, E[H(y + Z)] + \mu\}$. The solution to this equation is not unique. One set of solutions is given by $H^*(y) = c_1 + \exp \{-2\mu(y + c_2)/\sigma^2\} - y$, for any constants c_1 and c_2 such that $c_1 + \exp \{-2\mu(y + c_2)/\sigma^2\} - y \leqq 0$ for all y. For, in this case, since $E[\exp(-2\mu Z/\sigma^2)] = 1$ we have

$$E[H^*(y + Z)] + \mu = c_1 + \exp \{-2\mu(y + c_2)/\sigma^2\} - y = H^*(y) \leqq 0.$$

Let $a^* = -\sigma^2/2\mu$ and $H^*(y) = p_0 + a^* \exp \{(y - p_0)/a^* - 1\} - y$. For $y = p_0 + a^*$, $H^*(y) = 0$. Also $H^{*\prime}(y) = \exp \{(y - p_0)/a^* - 1\} - 1$, $H^{*\prime}(p_0 + a^*) = 0$ and $H^{*\prime\prime}(y) = a^{*-1} \exp \{(y - p_0)/a^* - 1\} < 0$, so that $H^*(y)$ reaches a maximum of 0 at $y = p_0 + a^*$ and $H^*(y) \leqq 0$ for all y. Now $H^*(y) = \min \{0, E[H^*(y + Z)] + \mu\}$ and $H^*(y) \leqq H_0(y) = \min \{0, p_0 - y\}$ implies that $H^*(y) \leqq H_n(y)$ for all n and all y. In particular $H^*(p_0 + a^*) = 0 \leqq H_n(p_0 + a^*) \leqq 0$ for all n which yields that the sequence $a_0 \geqq a_1 \geqq \cdots$ is bounded below and hence convergent. We denote the limit by a and later will show (somewhat heuristically) that $a = p_0 + a^*$. We have also shown in the above proof that $H_n(y)$ is bounded below by $H^*(y)$ for all n, and, since $H_n(y) \geqq H_{n+1}(y)$, it follows that $H_n(y) \to H(y)$ for some limit function $H(y)$ as $n \to \infty$. By the monotone convergence theorem $H(y) = \min \{0, E[H(y + Z)] + \mu\}$. Since $a \leqq a_n$, $H_n(a) = 0$ and thus $H(a) = 0$.

By reformulating the problem slightly and by examining a continuous time stochastic process we may, at least formally, derive a more explicit formula for the limiting critical value a. We assume that the sequence of prices is described by a Brownian motion or Wiener process superimposed on an increasing mean price trend. More explicitly we assume t time units to go to the expiration date of the call, that the current price is y and that the price with $t - s$ time units to go is given by $Y(t - s) = y + \mu \cdot s + W(s)$ where $\{W(\tau); \tau \leqq 0\}$ is a Gaussian stochastic process with stationary independent increments such that $W(0) = 0$, $E[W(\tau)] = 0$ and $E[W^2(\tau)] = \sigma^2\tau$. Such a model is the continuous time equivalent of the discrete time model previously considered. Now let $R_t(y)$ be the minimum expected price paid given a current price of y and t time units to go. In continuous time the optimality equation becomes

$$R_t(y) = \min \{y, E[R_{t-\Delta t}(y + \Delta Y)]\},$$

where $\Delta Y = \mu \cdot \Delta t + \Delta W$. We've seen that when t is large, the action taken, and hence the expected price, is independent of t and in this limiting case the equation may be written

$$\begin{aligned} R(y) &= \min\{y, E[R(y + \Delta Y)]\} \\ &= y \qquad\qquad \text{for } y \leqq a \\ &= E[R(y + \Delta Y)] \qquad \text{for } y > a. \end{aligned}$$

We expand in a Taylor series and take expectation to get:

$$R(y + \Delta Y) = R(y) + \Delta Y \cdot R'(y) + \tfrac{1}{2}(\Delta Y)^2 R''(y) + \cdots$$

and

$$E[R(y + \Delta Y)] = R(y) + \mu R'(y) \cdot \Delta t + (\sigma^2/2)R''(y) \cdot \Delta t + \cdots.$$

For the region $y > a$ we thus arrive at the differential equation

$$0 = \mu R'(y) + (\sigma^2/2)R''(y).$$

Solving initially with the boundary condition $R'(a) = 1$ yields $R'(y) = \exp\{-(2\mu/\sigma^2)(y - a)\}$ and then solving for $R(y)$ with the boundary condition $R(a) = a$ yields $R(y) = a + (\sigma^2/2\mu)(1 - \exp\{-(2\mu/\sigma^2)(y - a)\})$ for $y > a$. But we know $\lim_{y\to\infty} R(y) = p_0$. Thus $a = p_0 - \sigma^2/2\mu$.

The Speculator's Problem

The speculator's problem is a reflection, in some sense, of the investor's problem. Using the same notation as before, and again assuming $p_0 = 0$ for the derivation, let $Q_n(y)$ be the maximum expected profit assuming n periods to go and a current market price $Y_n = y$. The price of the option itself is not included in $Q_n(y)$ so that a call is profitable only if $Q_N(Y_N)$ exceeds this price where N is the length of the option period. With n periods to go, the optimal strategy is to exercise the option if $Y_n \geqq E[Q_{n-1}(Y_{n-1}) \mid Y_n]$; and otherwise wait at least one more day. Hence $Q_j(y) = \max\{y, E[Q_{j-1}(Y_{j-1}) \mid Y_j = y]\}$. Let $X_j = -Y_j$, $x = -y$, and $P_j(x) = Q_j(-x) = -Q_j(y)$. Then

$$-Q_j(y) = \min\{-y, E[-Q_{j-1}(Y_{j-1}) \mid Y_{j-1} = y]\} \quad \text{and}$$

$$P_j(x) = \min\{x, E[P_{j-1}(X_{j-1}) \mid X_{j-1} = x]\}.$$

This, of course, is the equation that has been analyzed in detail for the investor. We, thus, can map all the known properties of $P_n(\cdot)$ into properties of $Q_n(\cdot)$ and draw the conclusions given for the speculator in the Introduction and Summary section.

It should be mentioned that the analysis of puts, options guaranteeing a market at a stated price, may be made similarly to the previous analysis of calls.

References

1. Bierman, H., "The Bond Refunding Decision as a Markov Process," *Management Science*, 12, 1966, B545–551.
2. Boness, A. J., "Elements of a Theory of Stock Option Value," *J. Political Economy*, 72, 1965, 163–175.

3. Brada, J. C., Ernst, H. and Van Tassel, J., "The Distribution of Stock Price Differences—Gaussian After All?", *Operations Research*, 14, 1966, 334–340.
4. Breiman, L., "Stopping Rule Problems," Chapter 10 in *Applied Combinatorial Mathematics* (E. Beckenbach, ed. Wiley, New York, 1964.
5. Chow, Y. S. and Robbins, H., "A Martingale System Theorem and Applications," In *Proceedings of the Fourth Berkeley Symposium on Mathematical Statistics and Probability*, J. Neyman, ed., University of California Press, 1, 1961, 93–104.
6. Cootner, P. H., *The Random Nature of Stock Market Prices*, The M.I.T. Press, Cambridge, Massachusetts, 1964.
7. Fama, E. F., "Mandelbrot and the Stable Paretian Hypothesis, *J. Business of the University of Chicago*, 36, 1963, 420–429.
8. ——, "The Behavior of Stock Market Prices, *J. Business of the University of Chicago*, 38, 1965, 34–105.
9. Karlin, S., "Stochastic Models and Optimal Policy for Selling on Asset," Chapter 9 in *Studies in Applied Probability and Management Science*, Arrow, Karlin and Scarf, eds., Stanford University Press, Stanford, California, 1962.
10. Laurent, A. G., "Comments on 'Brownian Motion in the Stock Market,' " *Operations Research*, 7, 1957, 806.
11. Lehman, E. L., "Ordered Families of Distribution," *Ann. Math. Statist.* 26, 1955, 399–419.
12. MacQueen, J. and Miller, R. G., "Optimal Persistence Policies, *Operations Research*, 8, 1960, 362–380.
13. McKean, H. P., "Appendix: A Free Boundary Problem for the Heat Equation Arising from a Problem in Mathematical Economies, *Industrial Management Review*, 6, 1965, 32–39.
14. Mandelbrot, B., "The Variation of Certain Speculative Prices," *J. Business of the University of Chicago*, 36, 1963, 394–419.
15. Morris, W. T., "Some Analysis of Purchasing Policy," *Management Science*, 5, 1949, 443–452.
16. Niederhoffer, V., "Clustering of Stock Prices," *Operations Research*, 13, 1959, 258–265.
17. Osborne, M. F. M., "Brownian Motion in the Stock Market," *Operations Research*, 7, 1959, 145–173.
18. ——, 'Reply to "Comments on 'Brownian motion in the stock market' " ', *Operations Research*, 7, 1959, 807–810.
19. ——, "Periodic Structure in the Brownian Motion of Stock Prices," *Operations Research*, 10, 1962, 345–379.
20. Samuelson, P. A., "Rational Theory of Warrant Pricing," *Industrial Management Review*, 6, 1965, 13–31.

MANAGEMENT SCIENCE
Vol. 18, No. 3, November, 1971
Printed in U.S.A.

BOND REFUNDING WITH STOCHASTIC INTEREST RATES*†

BASIL A. KALYMON

University of California, Los Angeles

The bond refunding problem is formulated as a multiperiod decision process in which future interest rates are determined by a Markovian stochastic process. It is assumed that a single bond is to be outstanding at a given time. Given the future requirements for debt financing, the decision maker must decide whether to keep his current bond or to refund by issuing a new bond at the current market interest rates. Over a finite planning horizon, the structure of policies which minimize expected total discounted costs is studied.

Introduction

The class of models considered in this paper is concerned with the bond refunding problem formulated as a multiperiod decision process. The bond refunding problem to be studied assumes that in each of a number of discrete future periods there is a net requirement[1] for financing which is to be met by having outstanding a single bond the original term of which is optional. The bond is callable before maturity, and set-up costs are charged for refunding and issuing a new bond. Given that interest rates payable on new bonds vary with term to maturity and change through time, the decision maker must decide in each period whether to keep his current bond (if unexpired), or to refund and issue a new bond at current market rates. A penalty cost shall be charged for not meeting fund requirements, and a bonus shall be allowed for any excess funds during a given period. The objective shall be to minimize the expected total discounted cost over a finite planning horizon.

Since only a single bond is allowed to be outstanding at any given time, the applicability of the model is restricted to situations in which either the funding requirement is unchanged from present levels, or the future requirement is constant but higher than the currently outstanding bond or there is envisioned a slow increase in the funding requirement but it is desirable to maintain only a single major debt outstanding at a time. Thus, the penalty cost or bonus may be used to model the cost of minor short-term borrowings or deposits which result. As such, this penalty/bonus would reflect the secondary motivation (besides interest payment savings) cited in bond refunding, namely the readjustment of debt level (see [6]).

Over the finite horizon, a nonstationary per-period discount rate shall be assumed, which might represent a single discount factor based on the long-run cost of capital or vary with the expected single-period rates. Such a discounting scheme represents a

* Received February 1970; revised August 1970.

† This paper is based on a section of the author's Ph.D. dissertation [3] in the Department of Administrative Sciences at Yale University and was partially supported by the National Science Foundation. The author is indebted to Harvey M. Wagner for his guidance and encouragement in this research.

[1] For example, if an investment of $1 million is to be made in the first period which provides no return for the first 3 periods, and provides a return of $250 thousand in each of the next 7 periods, then the required financing for this project would be $1 million for each of the first 3 periods, $750 thousand in period 4, and decreasing by $250 thousand in each period thereafter. The net requirement would represent the total requirement over all projects of the given firm.

more general position compared to the extremes of the Weingartner [6] discounting method which uses the interest rate the firm will be paying on issued bonds, and the Pye [4] discounting scheme in which the precise realized single-period market rate is used. The discount factor as used in our model reflects the time preferences of the decision maker, recognizing the imperfections of the market which do not allow him complete flexibility in borrowing and lending. Such a discount factor would be affected by the market interest rate in as much as the latter contains information on future inflation, but would be principally determined by internal considerations. Note that the importance of this factor is deemphasized in the model in that it is used merely to compare different streams of interest payments and issuing costs and not to evaluate the utility of repayment of the principal. As noted above in this regard, the model assumes that the firm has already decided on the appropriate funding requirement in accordance with its investment opportunities. Current practice employs a single discount rate representing either a cost of capital or some type of single-period or multiperiod market rate.

There has been little previous work in bond refunding as a multiperiod decision process. Weingartner [6] is the only one to explicitly formulate the problem as a multiperiod process. His model is completely deterministic, and assumes that a level debt will be maintained, with only a single bond outstanding at a given time. Pye has studied the use of a Markov process to represent interest rate fluctuation in [5], and has used such a representation in [4] to determine the value of the call option by assuming that a single refunding decision will be made in the most favorable period over the term of the issued bond. Also, Bierman in [1] has proposed using a Markovian representation of interest fluctuation to improve short-run timing when refunding a bond, but does not consider the full multiperiod implications. References for work on the term structure and fluctuations of interest rates may be found in Pye [5]. A summary of current practice, (using essentially single decision approaches) as well as a discussion of appropriate discounting rates to be used, can be found in Bowlin [2].

The general model to be discussed is more precisely defined in §1, and the existence of optimal policies of special structure is proved in §2 under differing sets of assumptions. Calculation of optimal debt size after a decision to refund has been made is discussed in §3. In §4, simple necessary conditions for refunding are given for two level debt models.

1. Model Formulation

In this section, the basic model to be studied is formally defined. Let D_i represent the net requirement for debt financing in period i, for $i = 1, 2, \cdots, n$, with $i = n$ representing the chronologically first period of the n-period planning horizon. The D_i's are assumed to be known. For additional clarification we stress that the financing requirement as used in our model represents the exogenously determined amount of debt financing which the firm envisions requiring. It is based on projected cash flows aggregated over all the major projects/investments of the firm. (See Footnote 1.) Let t_i, x_i and r_i represent, respectively, the time to maturity, the size, and the coupon rate of the bond outstanding in period i. Let $\rho_{i,m}$ represent the rate of interest that must be paid on a bond issued in period i with time to maturity (or term) m. Defining ρ_i by $\rho_i = \rho_{i,\infty}$, we shall define

$$\rho_{i,m} = v_i(m, \rho_i), \tag{1}$$

where $v_i(m, \rho_i)$ is an increasing function in ρ_i. Thus $v_i(m, \rho_i)$ represents the term-curve in period i given that the long-term interest rate is ρ_i. The form of $v_i(m, \rho_i)$ is unspecified, permitting $v_i(m, \rho_i)$ to be either decreasing or increasing in the term m, situations which both occur in practice. By (1), we are modeling the term structure of interest rates as a nonstationary function which depends on the current long-term interest rate ρ_i. The long-term interest rate ρ_i is assumed to be determined by a Markovian stochastic process. The distribution function of ρ_i is given by $H(\cdot \mid \rho_{i+1}, i)$. Finally, let α_i represent the firm's time value of money factor for period i.

The cost structure of the model assumes that if, in period i, the decision is to retain the currently outstanding bond then an interest payment of $r_i x_i$ is paid. If the current bond has not reached maturity ($t_i > 0$) and a call option is exercised, a refunding cost of $R_i + c_i x_i$ is incurred. A new callable bond of size y may be issued at an issuing cost of $K_i + d_i y$. In the models studied, it is assumed that only a single bond is outstanding at a given time.

For $\rho_i = \rho$, a penalty cost of $h_i^+(D_i - x_i \mid \rho)$ which is concave in $(D_i - x_i)$ and zero for x_i outside of $[0, D_i]$ is charged if the funds requirement D_i is not met in period i. It shall be assumed that $h_i^+(D_i - x_i \mid \rho)$ is increasing in the long-term interest rate ρ. Since the applicability of the model is restricted to level, or near level, funding requirement, the range of $(D_i - x_i)$ will not be extensive, and is meant to reflect the refunding pressure when the current debt x_i is partly inadequate. The concavity assumption is justifiable in this context since, given that we know the amount of the deficiency, the cost of secondary financing, though increasing in absolute magnitude, should be decreasing per dollar borrowed over the limited range envisioned. For example, a fixed cost plus a constant percentage per unit borrowed would result in a concave penalty cost. Also, a bonus of $-k_i$ is charged per unit of excess borrowing denoted by $(x_i - D_i)^+$. Since $k_i \geqq 0$ may be thought of as interest earnings on short-term deposits, it will be further postulated that $k_i \leqq r$, where r represents any feasible 1-period interest rate on issued bonds. This assumption rules out the possibility of borrowing for future lending. Further assumptions that will be indicated pertain only to the particular case then under consideration.

Finally under the above cost structure we define the optimal cost function $f_i(t, x, r \mid \rho)$ by

$$f_i(t, x, r \mid \rho) = \inf_{y^i} f_i(t, x, r \mid \rho, y^i),$$

where $y^i \equiv \{y_i, y_{i-1}, \cdots, y_1\}$ with y_j a measurable policy for period j that is a function of the state $(t_j, x_j, r_j \mid \rho_j)$, and where $f_i(t, x, r \mid \rho, y^i)$ represents the expected discounted cost over the i remaining periods given that $t_i = t$, $x_i = x$, $r_i = r$, $\rho_i = \rho$ and that policy y^i is to be followed. Assume $f_0(t, x, r \mid \rho) = 0$, for all t, x, r, and ρ. Define y^{i*} as being an optimal policy if $f_i(t, x, r \mid \rho, y^{i*}) = f_i(t, x, r \mid \rho)$.

2. Form of Optimal Policies

The results of this and subsequent sections shall be classified as pertaining to one of the following three cases:

(2)

Case (i). Assumes that the debt financing requirement, D_i, is to be constant, and that only bonds of infinite length (essentially, maturing past the horizon) are to be considered.

Case (ii). Assumes only that the debt financing requirement, D_i, is to be constant, permitting bonds of varying maturity.

Case (iii). Makes no further assumptions.

The three cases vary in the degree of freedom permitted the decision maker, and hence in the form and nature of their solution. In this section, however, the existence of an optimal solution of a basic form shall be proved for Case (iii), and the corresponding forms for Cases (i) and (ii) shall be deduced by noticing that they are special restrictions of the general Case (iii). Thus, we now state and subsequently prove the following

THEOREM 1. *For the general Case* (iii), *given* $\rho_i = \rho$, $t_i = t$ *and* $x_i = x$, *there exists an optimal policy such that for every period i the form of the policy is*:

$$\textit{refund}: \quad \textit{if} \quad r_i \geqq r_i^*(t, x \mid \rho)$$

$$\textit{keep}: \quad \textit{if} \quad r_i < r_i^*(t, x \mid \rho).$$

Such a policy shall be designated as an $r_i^*(t, x \mid \rho)$ *policy.*

Theorem 1 implies that for any given set of values for the term remaining to maturity, the size of the current bond, the current long-term interest rate, and the current period number there exists a critical coupon-rate such that if we are paying more than this rate we refund and if less than we keep our current bond. Before proving Theorem 1 we first establish

LEMMA 1. *Under the assumptions of Case* (iii), $f_i(t, x, r \mid \rho)$ *is an increasing*[2] *function of r for fixed t, x and* ρ.

PROOF. It can be seen by a simple inductive argument that the optimal cost function $f_i(t, x, r \mid \rho)$ satisfies the following recursion:

$$\begin{aligned} f_i(t, x, r \mid \rho) &= \min\{f_i(t, r \mid \rho, Y_K), f_i(t, x, r \mid \rho, Y_R)\} \quad \text{if} \quad t > 0 \\ &= \min_{m,y}[K_i + d_i y + \rho_{i,m} y + h_i^+(D_i - y \mid \rho) - k_i(y - D_i)^+ \\ &\qquad + \alpha_i E_{\mid\rho,i} f_{i-1}(m - 1, y, \rho_{i,m} \mid q)] \quad \text{if} \quad t = 0, \end{aligned} \tag{3}$$

where

$$Y_K \equiv \text{the "keep" strategy is used in the current period,}$$

$$Y_R \equiv \text{the "refund" strategy is used in the current period;}$$

and thus

$$\begin{aligned} f_i(t, x, r \mid \rho, Y_K) \equiv {} & [rx + h_i^+(D_i - x \mid \rho) \\ & - k_i(x - D_i)^+ + \alpha_i E_{\mid\rho,i} f_{i-1}(t - 1, x, r \mid q)] \end{aligned} \tag{4}$$

$$\begin{aligned} f_i(t, x, r \mid \rho, Y_R) \equiv {} & R_i + c_i x + \min_{m,y}[K_i + d_i y + \rho_{i,m} y + h_i^+(D_i - y \mid \rho) \\ & - k_i(y - D_i)^+ + \alpha_i E_{\mid\rho,i} f_{i-1}(m - 1, y, \rho_{i,m} \mid q)], \end{aligned} \tag{5}$$

and where the symbol $E_{\mid\rho,i}$ denotes expectation over q, representing the long-term interest rate in the next period, given that the current interest rate is ρ, and the period is i. The components of the cost function $f_i(t, x, r \mid \rho, Y_K)$ are the interest payment rx that must be made, the shortage penalty $h_i^+(D_i - x \mid \rho)$ which is charged if the size x of the currently outstanding bond is insufficient to meet the current net fund requirements D_i, the bonus $-k_i(x - D_i)^+$ which is allowed if x is greater than D_i

[2] The terms "increasing," "decreasing" shall be used in the nonstrict sense, and shall be qualified otherwise.

and the expected costs in the future given our current decision not to refund represented by $\alpha_i E_{|\rho,i} f_{i-1}(t-1, x, r \mid q)$. The components of $f_i(t, x, r \mid \rho, Y_R)$ are the early refunding penalty cost $R_i + c_i x$ plus the cost in the current and future periods assuming that a new bond with an optimal term m and size y is issued.

Now, notice that $f_0(t, x, r \mid \rho)$ is a constant and, hence, is increasing in r. Next, make the induction hypothesis that $f_{i-1}(t, x, r \mid \rho)$ is increasing in r for all fixed t, x and ρ. This implies that

$$(6) \qquad \begin{aligned} f_i(t, x, r \mid \rho, Y_K) &= rx + h_i^+(D_i - x \mid \rho) \\ &\qquad - k_i(x - D_i)^+ + \alpha_i E_{|\rho,i} f_{i-1}(t-1, x, r \mid q) \end{aligned}$$

is also an increasing function of r since $x \geqq 0$, and since a positively weighted sum of increasing functions is also increasing. Also, from (5)

$$(7) \qquad f_i(t, x, r \mid \rho, Y_R) = (\text{constant with respect to } r),$$

and thus may be considered "increasing" in r.

Thus, for $t > 0$,

$$(8) \qquad f_i(t, x, r \mid \rho) = \min \{f_i(t, x, r \mid \rho, Y_K), f_i(t, x, r \mid \rho, Y_R)\}$$

and is an increasing function of r since the minimum of two functions, both increasing in r, must also be increasing in r. Since, for $t = 0$, $f_i(t, x, r \mid \rho)$ is a constant with respect to r and hence is an increasing function, the induction is complete, and the lemma has been established. ||

We now give the

Proof of Theorem 1. By Lemma 1, $f_i(t, x, r \mid \rho)$ and hence also $f_i(t, x, r \mid \rho, Y_K)$ are increasing functions of r, while $f_i(t, x, r \mid \rho, Y_R)$ is a constant with respect to r. Hence, if we define $r_i^*(t, x \mid \rho)$ as the smallest rate $r_i(t, x \mid \rho)$ for which

$$(9) \qquad f_i(t, x, r_i(t, x \mid \rho) \mid \rho, Y_K) \geqq f_i(t, x, r_i(t, x \mid \rho) \mid \rho, Y_R),$$

then it is immediate that for $r \geqq r_i^*(t, x \mid \rho)$,

$$(10) \qquad \begin{aligned} f_i(t, x, r \mid \rho, Y_K) &\geqq f_i(t, x, r_i^*(t, x \mid \rho) \mid \rho, Y_K) \\ &\geqq f_i(t, x, r_i^*(t, x \mid \rho) \mid \rho, Y_R) \\ &= f_i(t, x, r \mid \rho, Y_R), \end{aligned}$$

and that an optimal policy at r is to refund. Also, for $r < r_i^*(t, x \mid \rho)$,

$$(11) \qquad f_i(t, x, r \mid \rho, Y_K) < f_i(t, x, r \mid \rho, Y_R)$$

by the definition (9) of $r_i^*(t, x \mid \rho)$, and hence the optimal policy is to "keep". Thus, Theorem 1 has been established. ||

One can note that dropping the assumption that fund requirements D_i are deterministic, and assuming instead that the D_i's are independent random variables whose value is known for the current and any future period only up to a probability distribution, does not affect the form of the optimal solution derived above, and merely requires a replacement of $h_i^+(D_i - x \mid \rho)$ by $Eh_i^+(D_i - x \mid \rho)$ and $-k_i(x - D_i)^+$ by $E[-k_i(x - D_i)^+]$, where the expectation is over D_i. If, alternatively, the current requirement D_i becomes known only just before the decision is to be made, or if the distribution of D_i depends on the previous requirement D_{i+1}, then the problem requires an additional state space variable representing either the current or previous requirement and the optimal policy will now also depend on this additional variable.

From Theorem 1, we can readily deduce the following two corollaries:

COROLLARY 1. *Under the assumptions of Case* (ii), $D_i = D$, *level financing requirement, and given that* $t_i = t$ *and* $\rho_i = \rho$, *there exists an optimal policy for period* i *of the form*

$$\begin{aligned} &\textit{refund:} && \textit{if} \quad r_i \geqq r_i^*(t \mid \rho) \\ &\textit{keep:} && \textit{if} \quad r_i < r_i^*(t \mid \rho). \end{aligned}$$

and

COROLLARY 2. *Under the assumptions of Case* (i) (*infinite length bonds, level debt financing requirement*), *and given that* $\rho_1 = \rho$, *there exists an optimal policy for period* i *of the form*

$$\begin{aligned} &\textit{refund:} && \textit{if} \quad r_i \geqq r_i^*(\rho) \\ &\textit{keep:} && \textit{if} \quad r_i < r_i^*(\rho). \end{aligned}$$

PROOF OF COROLLARIES 1 AND 2. Notice that Cases (i) and (ii) are merely specializations of Case (iii). The assumptions of Case (i) imply that $t_i = \infty$, and $x_i = x$ for a fixed x for all i, and thus $r_i^*(t, x \mid \rho)$ may be denoted as $r_i^*(\rho)$ since t and x are constants. Similarly, under the assumptions of Case (ii), $x_i = x$ for a fixed x, for all i, and hence $r_i(t, x \mid \rho)$ may be denoted by $r_i^*(t \mid \rho)$. This completes the proof of Corollaries 1 and 2. ||

The following theorem established the optimality of an even simpler form of policy when interest rates are determined by a random walk.

THEOREM 2. *For Case* (i), *under the assumption that interest rates are determined by a nonstationary additive random walk*[3] (*that is* $\rho_i = \rho_{i+1} + \epsilon_i$ *where* ϵ_i *is a random variable with a distribution that is a function of period* i *only*), *the following hold*

$$(12) \qquad \begin{aligned} &(\text{a}) \ r_i^*(\rho') \leqq r_i^*(\rho) \quad \textit{if} \quad \rho' < \rho \\ &(\text{b}) \ r_i^*(\rho) \ = \rho + \delta_i \quad \textit{with} \quad \delta_i \geqq 0. \end{aligned}$$

Theorem 2 states that in each period i there is a fixed premium δ_i such that if our current coupon rate exceeds the market rate δ by δ_i, it is optimal to refund.

PROOF. First, (b) implies (a).

To prove (b), let the coupon rate to be paid for period i, given $r_i = r$ and $\rho_i = \rho$, be represented by $d_i(r, \rho)$. Thus,

$$(13) \qquad \begin{aligned} d_i(r, \rho) &= r \quad \text{if decision is to keep} \\ &= \rho \quad \text{if decision is to refund.} \end{aligned}$$

We must prove that for some constant $\delta_i \geqq 0$

$$(14) \qquad \begin{aligned} d_i(r, \rho) &= r \quad \text{if} \quad r - \rho < \delta_i \\ &= \rho \quad \text{if} \quad r - \rho \geqq \delta_i. \end{aligned}$$

Let $L_i = R_i + K_i$ and $e_i = c_i + d_i$. For $i = 1$ we have

$$(15) \qquad \begin{aligned} d_1(r, \rho) &= r \quad \text{if} \quad rx < L_1 + e_1x + \rho x \\ &= \rho \quad \text{if} \quad rx \geqq L_1 + e_ix + \rho x, \end{aligned}$$

[3] Under such an assumption of interest rate determination, the magnitude of permissible values for ϵ_i and/or the length of the horizon must be restricted to ensure that interest rates do not become negative.

since $f_1(r \mid \rho, Y_K) = rx$ and $f_1(r \mid \rho, Y_R) = L_1 + e_1x + \rho x$, given that $f_0(r \mid \rho) = 0$ for all r and ρ. Thus, $d_1(r, \rho)$ is of the required form, since (15) may be re-written as

$$(16)\qquad \begin{aligned} d_1(r, \rho) &= r \quad \text{if} \quad r - \rho < \frac{1}{x}(L_1 + e_1 x) \equiv \delta_1 \\ &= \rho \quad \text{if} \quad r - \rho \geqq \frac{1}{x}(L_1 + e_1 x) = \delta_1 . \end{aligned}$$

Next, we make the induction hypothesis that $d_j(r, \rho)$ is of form (14) for all $j \leqq i - 1$. By the definition (13), we know that

$$(17)\qquad \begin{aligned} d_i(r, \rho) &= r \quad \text{if} \quad f_i(r \mid \rho, Y_R) - f_i(r \mid \rho, Y_K) > 0 \\ &= \rho \quad \text{if} \quad f_i(r \mid \rho, Y_R) - f_i(r \mid \rho, Y_K) \leqq 0. \end{aligned}$$

Thus, to show that $d_i(r, \rho)$ is of form (14), it is sufficient to show that $f_i(r \mid \rho, Y_R) - f_i(r \mid \rho, Y_K)$ is a decreasing function of $(r - \rho)$. Also, since

$$(18)\qquad \begin{aligned} f_i(r \mid \rho, Y_R) - f_i(r \mid \rho, Y_K) &= L_i + e_i x + \rho x + \alpha_i E_{\mid i} f_{i-1}(\rho \mid \rho + \epsilon) \\ &\quad - (rx + \alpha_i E_{\mid i}(f_{i-1}(r \mid \rho + \epsilon))) \\ &= (L_i + e_i x) - (r - \rho)x + \alpha_i E_{\mid i}(f_{i-1}(\rho \mid \rho + \epsilon) \\ &\quad - f_{i-1}(r \mid \rho + \epsilon)), \end{aligned}$$

it is sufficient to show that for any fixed outcome of ϵ, say ϵ_i, $\Delta_{i-1}(\rho, r)$ defined by

$$(19)\qquad \Delta_{i-1}(\rho, r) = f_{i-1}(\rho \mid \rho + \epsilon_i) - f_{i-1}(r \mid \rho + \epsilon_i)$$

is of the form

$$(20)\qquad \Delta_{i-1}(\rho, r) = \Delta_{i-1}(r - \rho),$$

with $\Delta_{i-1}(r - \rho)$ being a decreasing function of $(r - \rho)$.

But,

$$(21)\qquad f_{i-1}(r \mid \rho + \epsilon_i) = \textstyle\sum_{j=1}^{i-1} \beta_j E[\delta(d_j - d_{j+1})(L_j + e_j x) + d_j x],$$

where

$$(22)\qquad \begin{aligned} \beta_j &\equiv \textstyle\prod_{k=j}^{i-1} \alpha_k \quad \text{for} \quad j \leqq i - 2 \\ &\equiv 1 \quad \text{for} \quad j = i - 1, \\ \delta(y) &\equiv 0 \quad \text{if} \quad y = 0, \\ &\equiv 1 \quad \text{if} \quad y \neq 0, \\ d_i &\equiv r, \\ d_j &\equiv d_j(d_{j+1}, \rho + \hat{\epsilon}_j) \quad \text{for} \quad j \leqq i - 1, \\ \text{and} \qquad \hat{\epsilon}_j &\equiv \textstyle\sum_{k=j+1}^{i} \epsilon_k . \end{aligned}$$

Note that if we subtract any constant k_1 from both r and $\rho + \epsilon_i$, the resulting new sequence, denoted d'_j, is related to d_j by

$$(23)\qquad d'_j = d_j - k_1 .$$

This is so because, by definition (22),

(24) $$d'_i = r - k_1 = d_i - k_1$$

and supposing that $d'_j = d_j - k_1$, then

(25)
$$\begin{aligned} d'_{j-1} &= d_{j-1}(d'_j, (\rho - k_1) + \hat{\epsilon}_j) = d_{j-1}(d_j - k_1, \rho - k_1 + \hat{\epsilon}_j) \\ &= d_j - k_1 \quad \text{if} \quad (d_j - k_1) - (\rho - k_1 + \hat{\epsilon}_j) < \delta_{j-1} \\ &= \rho - k_1 + \hat{\epsilon}_j \quad \text{if} \quad (d_j - k_1) - (\rho - k_1 + \hat{\epsilon}_j) \geqq \delta_{j-1} \\ &= d_{j-1} - k_1 . \end{aligned}$$

Thus, (23) holds for all $j \leqq i$.

By (23), we see that for all $j \leqq i - 1$

(26) $$\delta(d'_j - d'_{j+1}) = \delta(d_j - d_{j+1})$$

and

(27) $$d'_i x = d_i x - k_1 x.$$

But

(28) $$f_{i-1}(r - k_1 \mid \rho - k_1 + \epsilon_i) = \textstyle\sum_{j=1}^{i-1} \beta_j E[\delta(d'_j - d'_{j+1})(L_j + e_j x) + d'_i x]$$

so that by (26) and (27), we have

(29) $$f_{i-1}(r \mid \rho + \epsilon_i) = f_{i-1}(r - k_1 \mid \rho - k_1 + \epsilon_i) + k_1 x \textstyle\sum_{j=1}^{i-1} \beta_j .$$

Similarly, for any k_2,

(30) $$f_{i-1}(\rho \mid \rho + \epsilon_i) = f_{i-1}(\rho - k_2 \mid \rho - k_2 + \epsilon_i) + k_2 x \textstyle\sum_{j=1}^{i-1} \beta_j .$$

Now, letting

(31) $$k_1 = k_2 = \rho$$

we obtain by definition (19) and Equations (29) and (30) that

(32)
$$\begin{aligned} \Delta_{i-1}(\rho, r) &= f_{i-1}(\rho \mid \rho + \epsilon_i) - f_{i-1}(r \mid \rho + \epsilon_i) \\ &= f_{i-1}(0 \mid \epsilon_i) - f_{i-1}((r - \rho)\epsilon_i). \end{aligned}$$

But, by Lemma 1, $f_{i-1}(s \mid q)$ is an increasing function in s for any q, so that $\Delta_{i-1}(\rho, r)$ must be a decreasing function in $(r - \rho)$ as required. Thus, $d_i(r, \rho)$ is of the required form (14). Finally, $\delta_i \geqq 0$ since otherwise $r_i^*(\rho)$ would be less than ρ, and to refund when the market rate is higher than your current cost can be readily shown to be suboptimal. This completes the induction and establishes Theorem 2. ||

Counterexamples can be constructed showing that a nonadditive interest rate movement structure can destroy monotonicity in ρ of an optimal $r_i^*(\rho)$ policy. Such examples are based on the possibility of delaying refunding at a low rate ρ' in anticipation of refunding at a still lower rate in the next period; at the high rate ρ it is optimal to plan to refund immediately and *also* a few periods later when the much lower rate is reached. Necessary conditions for such monotonicity have so far eluded us.

3. Calculation of Optimal Debt Level

In this section, we derive computational schemes for finding optimal debt level when a refunding decision has been made and a new bond is issued. We first prove a lemma concerning the optimal cost function which is useful in subsequent discussions.

LEMMA 2. *For the general Case* (iii), *the optimal cost function* $f_i(t, x, r \mid \rho)$ *is an increasing function in* x *for* $x \geqq \bar{D}_i$, *where* $\bar{D}_i \equiv \max_{1 \leqq j \leqq i}\{D_j\}$.

PROOF. First note that the hypothesis of the lemma is true for $i = 0$, since

$$f_0(t, x, r \mid \rho) = 0.$$

Next, make the induction assumption that the lemma is true for $i - 1$. Then,

$$E_{|\rho,i} f_{i-1}(t - 1, x, r \mid q)$$

is an increasing function in x for $x \geqq \bar{D}_{i-1}$. Also, for $x \geqq \bar{D}_i \geqq D_i$, $h_i^+(D_i - x \mid \rho) = 0$, so that

$$xr + h_i^+(D_i - x \mid \rho) - k_i(x - D_i)^+ = x(r - k_i) + k_i D_i \tag{33}$$

and is an increasing function of x, since $k_i \leqq r$ by assumption. Since

$$\begin{aligned} \bar{D}_i &= \max_{1 \leqq j \leqq i}\{D_j\} \\ &= \max\{D_i, \bar{D}_{i-1}\} \geqq \bar{D}_{i-1}, \end{aligned} \tag{34}$$

we get that, for $x \geqq \bar{D}_i$, $f_i(t, x, r \mid \rho, Y_K)$ defined by

$$\begin{aligned} &f_i(t, x, r \mid \rho, Y_K) = \\ &\qquad xr + h_i^+(D_i - x \mid \rho) - k_i(x - D_i)^+ + \alpha_i E_{|\rho,i} f_{i-1}(t - 1, x, r \mid q) \end{aligned} \tag{35}$$

is an increasing function of x, by (33) and the induction hypothesis.

Also, it is immediate that $f_i(t, x, r \mid \rho, Y_R)$ defined by

$$\begin{aligned} f_i(t, x, r \mid \rho, Y_R) &= R_i + c_i x + \min_{m,y}\{K_i + d_i y + \rho_{i,m} y + h_i^+(D_i - y \mid \rho) \\ &\qquad - k_i(y - D_i)^+ + \alpha_i E_{|\rho,i} f_{i-1}(m - 1, y, r \mid q)\} \\ &= c_i x + (\text{constant}) \end{aligned} \tag{36}$$

is an increasing function for all x since $c_i \geqq 0$.

From these facts we have that, for $t > 0$,

$$f_i(t, x, r \mid \rho) = \min\{f_i(t, x, r \mid \rho, Y_K), f_i(t, x, r \mid \rho, Y_K)\} \tag{37}$$

is an increasing function of x for $x \geqq \bar{D}_i$.

For $t = 0$,

$$f_i(0, r, x \mid \rho) = f_i(t, r, x \mid \rho, Y_R) - (R_i + c_i x) = (\text{constant with respect to } x) \tag{38}$$

and thus is increasing in x.

This completes the induction and proves the lemma. ||

The following theorem establishes the main result used in computing optimal debt levels. Note that the theorem assumes that some term m is being considered. The determination of an optimal term m would require the consideration of some set of alternative feasible terms, with the one producing the minimum total cost being chosen.

THEOREM 3. *Without loss of optimality* $y_i(m)$, *satisfying the equation*

$$\begin{aligned} &[K_i + (d_i + \rho_{i,m}) y_i(m) + h_i^+(D_i - y_i(m) \mid \rho) - k_i(y_i(m) - D_i)^+ \\ &\quad + \alpha_i E_{|\rho,i} f_{i-1}(m - 1, y_i(m), \rho_{i,m} \mid q)] = \min_y\{K_i + (d_i + \rho_{i,m}) y \\ &\quad + h_i^+(D_i - y \mid \rho) - k_i(y - D_i)^+ + \alpha_i E_{|\rho,i} f_{i-1}(m - 1, y_i(m), \rho_{i,m} \mid q)\}, \end{aligned} \tag{39}$$

can be restricted to the set of values $\{0, D_i, D_{i-1}, \cdots, D_{i-m}\}$.

The above result proves that we need look only at the debt financing requirements over the proposed term m in deciding on an optimal bond size. The proof depends on the concavity of the cost structure, and would fail to hold under general penalty costs.

PROOF. In the proof below we use the notation $\pi^{j(t)}$ to represent the partition of the positive real numbers into the set of intervals

$$[0, D_1^{j(t)}], [D_1^{j(t)}, D_2^{j(t)}], \cdots, [D_{t-1}^{j(t)}, D_t^{j(t)}], [D_t^{j(t)}, \infty] \tag{40}$$

where $D_k^{j(t)}$ is defined by

$$\{D_k^{j(t)} : 1 \leqq k \leqq t\} = \{D_h : j - t + 1 \leqq h \leqq j\} \text{ and } D_1^{j(t)} \leqq D_2^{j(t)} \leqq \cdots \leqq D_t^{j(t)}. \tag{41}$$

We now prove that $f_i(t, x, r \mid \rho)$ is concave in x over each interval of the partition $\pi^{i(t)}$. This property is referred to by calling $f_i(t, x, r \mid \rho)$ piecewise concave over the partition $\pi^{i(t)}$. Properties of piecewise concave functions are given by Zangwill [7].

Again, proof is by induction. For $i = 0$, $f_0(t, x, r \mid \rho) = 0$ is concave in x over $\pi^{0(t)} \equiv [0, \infty]$. Now make the hypothesis that $f_{i-1}(t, x, r \mid \rho)$ is piecewise concave over $\pi^{(i-1)(t)}$ for all values of t, r, and ρ.

Then, for $t > 0$, since $f_{i-1}(t - 1, x, r \mid q)$ is piecewise concave over $\pi^{(i-1)(t-1)}$ by assumption, it is also piecewise concave over $\pi^{i(t)}$ since $\pi^{i(t)}$ is a refinement[4] of $\pi^{(i-1)(t-1)}$. To see that $\pi^{i(t)}$ is a refinement of $\pi^{(i-1)(t-1)}$, observe that the set

$$\{D_n : i - t + 1 \leqq n \leqq i - 1\}$$

defining the possible end points (exclusive of 0 and ∞) of $\pi^{(i-1)(t-1)}$ is contained in the set $\{D_n : i - t + 1 \leqq n \leqq i\}$ defining the end points (exclusive of 0 and ∞) of $\pi^{i(t)}$. Hence, $E_{|\rho,i}f_{i-1}(t - 1, x, r \mid q)$ is piecewise concave on $\pi^{i(t)}$, since over each interval of the partition, $E_{|\rho,i}f_{i-1}(t - 1, x, r \mid q)$ represents a positively weighted sum of concave functions. We also have that

$$xr + h_i^+(D_i - x \mid \rho) - k_i(x - D_i)^+ = xr - k_i(x - D_i) \quad \text{for} \quad x \quad \text{in} \quad [D_i, \infty] \tag{42}$$

and therefore $xr + h_i^+(D_i - x \mid \rho) - k_i(x - D_i)^+$ is piecewise concave over $\pi^{i(t)}$, a refinement of $\{[0, D_i], [D_i, \infty]\}$, by the concavity in x of $-k_i(x - D_i)^+$ over $[D_i, \infty]$ and of $h_i(D_i - x \mid \rho)^+$ over $[0, D_i]$. Thus, we get that

$$\begin{aligned} f_i(t, x, r \mid \rho, Y_K) = {} & xr + h_i^+(D_i - x \mid \rho) \\ & - k_i(x - D_i)^+ + \alpha_i E_{|\rho,i}f_{i-1}(t - 1, r, x \mid q) \end{aligned} \tag{43}$$

is piecewise concave over $\pi^{i(t)}$ since it is a sum of functions piecewise concave over $\pi^{i(t)}$.

Furthermore, from the definition (5), we see that

$$f_i(t, x, r \mid \rho, Y_R) = c_i x + (\text{constant with respect to } x) \tag{44}$$

so that $f_i(t, x, r \mid \rho, Y_R)$ is concave in x and hence also piecewise concave in x over $\pi^{i(t)}$.

But, by Zangwill [7], we know that the minimum of functions piecewise concave over the same partition must also be piecewise concave over that partition. Thus, for $t > 0$, $f_i(t, x, r \mid \rho)$ is piecewise concave in x over $\pi^{i(t)}$ since

$$f_i(t, x, r \mid \rho) = \min \{f_i(t, x, r \mid \rho, Y_K), f_i(t, x, r \mid \rho, Y_R)\}, \tag{45}$$

[4] To say that a partition π' is a refinement of a partition π means that each interval in the partition π' is contained in some interval of the partition π.

where $f_i(t, x, r \mid \rho, Y_K)$ and $f_i(t, x, r \mid \rho, Y_R)$ have both been shown above to be piecewise concave over $\pi^{i(t)}$.

Also, for $t = 0$, from (3), we have

$$f_i(0, x, r \mid \rho) = (\text{constant with respect to } x) \tag{46}$$

so that $f_i(0, x, r \mid \rho)$ is piecewise concave over $\pi^{i(t)}$. This completes the induction, proving that $f_i(t, x, r \mid \rho)$ is piecewise concave over $\pi^{i(t)}$ for all i and t.

From this property as applied to $f_{i-1}(m - 1, y, \rho_{i,m} \mid q)$ and from (42), we can readily deduce by similar arguments as above that the function $g(y \mid m)$ defined by

$$\begin{aligned} g(y \mid m) \equiv K_i + d_i y + \rho_{i,m} y + h_i^+(D_i - y \mid \rho) \\ - k_i(y - D_i)^+ + \alpha_i E_{\mid \rho, i} f_{i-1}(m - 1, y, \rho_{i,m} \mid q) \end{aligned} \tag{47}$$

is piecewise concave in y over $\pi^{i(m)}$. But, by Zangwill [7], the minimum over y of $g(y \mid m)$ is achieved at an extreme point of an interval of the partition $\pi^{i(t)}$. Thus, there exists a $y_i(m)$ in the set $\{0, D_i, D_{i-1}, \cdots, D_{i-m}, \infty\}$, the set of extreme points of $\pi^{i(m)}$, such that

$$g(y_i(m) \mid m) = \min_y \{g(y \mid m)\}. \tag{48}$$

The value of ∞ may be excluded from the set of possible values by noting that, by Lemma 2, $f_{i-1}(m - 1, y, \rho_{i,m} \mid q)$ [and hence $E_{\mid \rho, i} f_{i-1}(m - 1, y, \rho_{i,m} \mid q)$] is an increasing function in y for $y \geqq \bar{D}_i < \infty$, and also that for $y \geqq \bar{D}_i \geqq D_i$,

$$d_i y + \rho_{im} y + h_i^+(D_i - y \mid \rho) - k_i(y - D_i)^+ = (d_i + \rho_{i,m} - k_i)y \tag{49}$$

is an increasing function in y, since by assumption $d_i \geqq 0$ and $\rho_{i,m} = v_i(m \mid \rho) \geqq k_i$. Thus, the minimum of $g(y \mid m)$ must occur at a finite value, and the proof of Theorem 3 is complete. ||

Having established theorems which enable us to compute optimal debt levels, we have now completed our study of the properties of optimal policies and optimal cost function.

4. Necessary Conditions for Refunding

In this final section, we give necessary conditions for Case (i) and Case (ii) for the optimality of the refund decision. These necessary (though not sufficient) conditions are simple to compute, and hence provide a simple test for ruling out the optimality of a refund decision without the extensive calculations involved in calculating the complete optimal policy. Furthermore, these necessary conditions are similar to the criteria used in current practice (see Bowlin [2]) to determine the value of refunding.

THEOREM 4. *For the Case* (i), (*infinite length bonds, level debt financing requirement*), *for* $r \geqq \rho$ *the value* S *defined by*

$$S = \left(\textstyle\sum_{j=1}^{i} \beta_j^{\,i}\right)(r - \rho)x - (L_i + e_i x), \tag{50}$$

where $L_i = R_i + K_i$, $e_i = c_i + d_i$, $\beta_j^{\,i} = \prod_{n=j+1}^{i} \alpha_n$ *for* $j \leqq i - 1$ *and* $\beta_i^{\,i} = 1$, *satisfies the relationship*

$$f_i(r \mid \rho, Y_K) - f_i(r \mid \rho, Y_R) \leqq S. \tag{51}$$

In effect, S represents a simple to compute value which provides a necessary condition for refunding. Namely, if S is negative, then refunding in the current period must be suboptimal.

PROOF. First we prove that for all l and q, if $r \geqq \rho$ then

$$f_l(r \mid q) - f_l(\rho \mid q) \leqq \left(\textstyle\sum_{j=1}^{l} \beta_j^{\,l}\right)(r - \rho)x. \tag{52}$$

Note that if $r \geqq \rho \geqq r_l^*(q)$ for any $l > 0$, then

$$f_l(r \mid q) - f_l(\rho \mid q) = f_l(r \mid q, Y_R) - f_l(\rho \mid q, Y_R) = 0 \tag{53}$$

so that (52) is satisfied. Also, if $r_l^*(q) > \rho$, then

$$\begin{aligned} f_l(r \mid q) - f_l(\rho \mid q) &= f_l(r \mid q) - f_l(\rho \mid q, Y_K) \\ &\leqq f_l(r \mid q, Y_K) - f_l(\rho \mid q, Y_K) \\ &= (r - \rho)x + \alpha_l E_{|q,l}[f_{l-1}(r \mid q_1) - f_{l-1}(\rho \mid q_1)]. \end{aligned} \tag{54}$$

But by Lemma 1, $f_{l-1}(r \mid q_1) - f_{l-1}(\rho \mid q_1) \geqq 0$ for all realizations of q_1 since $r \geqq \rho$, so that by (54) and (53), we have that for $r \geqq \rho$ and any q, $l > 0$,

$$f_l(r \mid q) - f_l(\rho \mid q) \leqq (r - \rho)x + \alpha_l E_{|q,l}[f_{l-1}(r \mid q_1) - f_{l-1}(\rho \mid q_1)]; \tag{55}$$

applying (55) recursively we get

$$\begin{aligned} f_l(r \mid q) - f_l(\rho \mid q) &\leqq (r - \rho)x + \alpha_l E_{|q,l}[(r - \rho)x + \alpha_{l-1} E_{|q_1,l-1}[f_{l-2}(r \mid q_2) \\ &\qquad - f_{l-2}(\rho \mid q_2)]] \\ &= (1 + \alpha_l)(r - \rho)x + \alpha_l \alpha_{l-1} E_{|q,l}[f_{l-2}(r \mid q_2) - f_{l-2}(\rho \mid q_2)] \\ &\ \ \vdots \qquad\qquad \cdot \quad \cdot \quad \cdot \quad \cdot \\ &\leqq \left(\textstyle\sum_{j=1}^{l} \beta_j^{\,l}\right)(r - \rho)x + E_{|q,l}[f_0(r \mid q_l) - f_0(\rho \mid q_l)] \\ &= \textstyle\sum_{j=1}^{l} \beta_j^{\,l}(r - \rho)x, \end{aligned} \tag{56}$$

where the final equality is due to the fact that $f_0(r \mid q_l) = 0 = f_0(\rho \mid q_l)$.

Since (56) was established for any q, $l > 0$, we have

$$f_{i-1}(r \mid q) - f_{i-1}(\rho \mid q) \leqq \left(\textstyle\sum_{j=1}^{i-1} \beta_j^{\,i-1}\right)(r - \rho)x \tag{57}$$

and hence

$$\begin{aligned} f_i(r \mid \rho, Y_K) - f_i(r \mid \rho, Y_R) &= (r - \rho)x - (L_i + e_i x) + \alpha_i E_{|\rho,i}[f_{i-1}(r \mid q) \\ &\qquad - f_{i-1}(\rho \mid q)] \\ &\leqq (r - \rho)x - (L_i + e_i x) + \alpha_i\left(\textstyle\sum_{j=1}^{i-1} \beta_l^{\,i-1}\right)(r - \rho)x \\ &= \left(\textstyle\sum_{j=1}^{i} \beta_l^{\,i}\right)(r - \rho)x - (L_i + e_i x) = S \end{aligned} \tag{58}$$

as required to prove Theorem 4. ||

The final theorem of this section states a similar result for Case (ii) (where bonds are of predetermined finite length) and can be proven by similar reasoning.

THEOREM 5. *For Case* (ii) (*level debt financing requirement, finite length bonds*) *under the assumption that only bonds of predetermined term to maturity m shall be issued, if* $(r - \rho_{i,m})x \geqq R_j + c_j x$ *for all* $j \leqq i$, $t < m < i$ *where* $t_i = t$, $\rho_i = \rho$, $r_i = r$, *then the value S defined by*

$$\begin{aligned} S = {} & \left(\textstyle\sum_{j=i}^{i=t+1} \beta_j^{\,i}\right)(r - \rho_{i,m})x - (R_i + K_i + (c_i + d_i)x) + \beta_{i-t}^{i}(K_{i-t} + d_{i-t}x) \\ & + \beta_{i-t}^{i}\left(\textstyle\sum_{j=i-t}^{i-m+1} \beta_j^{\,i-t}\right)(E[\rho_{j-t,m} \mid \rho_j = \rho] - \rho_{i,m})^+ x + \beta_{i-m}^{i}(R_{i-m} + d_{i-m}x) \end{aligned} \tag{59}$$

satisfies the relationship

$$f_i(t, r \mid \rho, Y_K) - f_i(t, r \mid \rho, Y_R) \leqq S. \tag{60}$$

Under the restriction that only bonds of a fixed length shall be used, the value S is simple to compute and provides a necessary condition for refunding. Specifically, if S turns out to be negative then we know by (60) that to refund is a suboptimal action.

References

1. Bierman, H., Jr., "The Bond Refunding Decision as a Markov Process," *Management Science*, Vol. 12, No. 12 (August 1966), pp. 545–552.
2. Bowlin, O. D., "The Refunding Decision," *J. of Fin.*, Vol. 21, No. 1 (March 1966), pp. 55–68.
3. Kalymon, B. A., "Stochastic Costs in Multi-period Decision Models," Ph.D. Dissertation, Yale University, New Haven, Connecticut, 1970.
4. Pye, G., "The Value of the Call Option on a Bond," *J. of Pol. Econ.* (April 1966), pp. 200–205.
5. ——, "A Markov Model of the Term Structure," *Q. J. of Econ.* (February 1966), pp. 60–72.
6. Weingartner, H. M., "Optimal Timing of Bond Refunding," *Management Science*, Vol. 13, No. 7 (March 1967), pp. 511–524.
7. Zangwill, W. I., "The Piecewise Concave Function," *Management Science*, Vol. 13, No. 11 (July 1967), pp. 900–912.

MANAGEMENT SCIENCE
Vol. 17, No. 7, March, 1971
Printed in U.S.A.

MINIMAX POLICIES FOR SELLING AN ASSET AND DOLLAR AVERAGING*†

GORDON PYE

University of California, Berkeley

The results of this paper fall into two areas. First, it is shown that the conventional wisdom of dollar averaging is related to hedging against large regrets rather than unfavorable outcomes. Suppose that a given sum of dollars is to be irreversibly converted into stock within a given number of periods. Suppose that the share price of dollars follows an arithmetic random walk. It is shown that dollar averaging is a nonsequential minimax strategy, if the largest possible price increase in each period is equal to the largest possible decrease. On the other hand, it is shown that dollar averaging cannot be a nonsequential expected utility maximizing strategy for any strictly concave utility function and any arithmetic random walk. Second, sequential minimax strategies are examined. Here the analysis may be viewed as an extension of the literature on the optimal time to sell an asset or the optimal stopping problem. The minimax sequential strategy is shown to be of the following form. At any time there exists a critical value which depends only on n, the difference between the maximum price since conversions began and the current price. Funds held in excess of this value are converted into stock, otherwise no funds are converted. The critical values are shown to be decreasing functions of n. When the share price of dollars rises (i.e., stock prices fall) few funds are converted while decreases produce large conversions. These results seem to correspond to the second thoughts of those embarking on nonsequential policies of the dollar averaging type. Buying and selling minimax policies are shown to be symmetric. Sequential minimaxing will tend to have a reinforcing effect on price movements.

1. Introduction and Summary

A familiar technique for coping with uncertainty on financial markets is that of dollar averaging. However, no rationalization of this technique based on current decision theory under uncertainty has appeared. One objective of this paper is to provide such a rationalization. The unconventional result emerges that dollar averaging is related to hedging against large regrets or opportunity losses rather than hedging against unfavorable outcomes. Since dollar averaging is a nonsequential strategy the question next arises as to the effect of a regret criterion on a sequential strategy. The second objective of this paper is to characterize such a strategy. Here the work may be viewed as an extension to a minimax criterion of the literature on the optimal time to sell an asset or the optimal stopping problem. An example is provided where such a criterion may be relevant even if one is not willing to consider regret in personal decision making.

A recent statement from the financial press [*Magazine of Wall Street*, January 17, 1970, p. 11] will serve to present the dollar averaging technique and the conventional wisdom for following it:

> Dollar averaging appears to be particularly appropriate as an essential technique in investing in common stocks under present uncertain economic conditions. Under a dollar averaging program if the sum to be invested is \$120,000, twelve units of \$10,000 each might

* Received January 1969; revised March 1970.

† This work was supported in part by NSF Research Grant GS-2225 administered through the Center for Research in Management Science at the University of California, Berkeley. Helpful comment by Robert S. Kaplan on an earlier draft is gratefully acknowledged.

> be invested monthly for a year. With such a program it would be possible to avoid investing the whole fund at the year's high. Dollar averaging as so applied produces a diversification of timing which when accompanied by a diversification between issues and between classes of investment makes for cautious risk spreading.

Similar statements about dollar averaging can be found elsewhere; for example, in Cottle and Whitman [3, pp. 166–168, 180–182]. Examination of such statements indicates the following general features. A given amount of a divisible asset is to be irreversibly converted into another asset over a fixed number of periods. Usually, as in the quotation, a given amount of dollars is to be sold for shares of stock. No firm convictions are held that any available selling strategy will give a significantly higher expected amount of the second asset at the end than any other. Instead, concern with risk seems paramount. None of the possible outcomes is conceived of as a disaster. The policy to be followed is nonsequential since it is not dependent on future price behavior. The seller is too small to affect the market price of the asset. A fairly short period of time is available to complete the conversion.

The fact that the sale of an asset whose price follows a stochastic process is involved suggests that the literature on this topic may be relevant. Karlin [6], Samuelson [10], and Taylor [11] have all studied this problem. A recent contribution is that of Hayes [4]. Additional references may be found there, including references to the closely related optimal stopping problem. In general, in this work the price of the asset has been expected to rise with time and then fall to a given value at some horizon. The criterion adopted has been to maximize the expected discounted value of the price obtained. In the dollar averaging problem differences in the expected value of the alternatives do not seem to be the significant factor. Instead, it is differences in the risk of the alternatives.

The significance of risk suggests a relationship to the portfolio selection problem. Under the nonsequential constraint there is a close similarity to the static portfolio problem first considered by Markowitz [7]. If a sequential policy is allowed there is a similarity to the sequential portfolio problems studied by Tobin [12] and by Mossin [8]. An essential difference, however, between the dollar averaging problem or the asset sale problem and that of portfolio selection is the irreversible nature of the asset transfer. In the asset sale problem the amount of the risky asset which may be held in later periods depends on the amount which was held earlier. In the portfolio problem asset transfers are fully reversible and constrained only by over-all budget requirements. Another difference concerns the appropriate objective. Hedging against large regrets appears to be a significant consideration in the dollar averaging problem and sometimes in other asset sale problems. In previous work on portfolio selection this has not been the case.

The possible significance of regret in the dollar averaging problem is foreshadowed in the quotation. The author justifies the technique in part by stating that in this way it is "possible to avoid investing the whole fund at the year's high". Consideration of regret, however, has largely disappeared from current decision theory in favor of expected utility maximization. Yet, if one suffers regret it seems no less rational to consider it in decision making than it is to consider tomorrow's hangover when consuming alcohol. Though one would like to purge regret from his psyche, it may be no easier to eliminate than the hangover. In cases where the expected utility of the alternatives are not too different, hedging against regret may attain overriding significance. Situations where dollar averaging has been recommended may well fit such a case.

Even if one is not willing to consider regret in personal decision making another situation occurs on financial markets where a regret criterion may be applicable. Frequently, a customer gives a broker an asset to sell for him by a certain date at the broker's "discretion". This means the customer sets the time by which sales must be completed but leaves to the broker when within this time sales are actually to be made. The broker may well have no particular feeling as to whether the price of the asset is going to move up or down over the relatively short time available for selling it. He has a strong feeling, however, that the customer will interpret a failure to get a good price as a lack of skill on his part. In fact, if the price obtained by the broker is far from the best price which could have been obtained, the customer may well take his business elsewhere the next time. Suppose the broker were to sell all of the asset initially. In this case it may be rational to assume that the customer will not be consoled much ex post by the fact that he was guaranteed a certain outcome ex ante, if the price rises significantly. Hence, it will be rational for the broker to hedge against large regrets. The applicability of a regret criterion in situations in which the decision maker is subject to blame has been mentioned by other writers, for instance, Borch [2, p. 82].

In general one might posit the existence of a strictly convex disutility of regret function. The regret criterion would then require the minimization of the expected value of this function. However, a minimax criterion will be adopted here. For a symmetric type of stochastic price behavior such a criterion may give a good approximation to the policies arising from many risk averse disutility functions. The minimax criterion, though, has the distinct advantage of being more tractable. The price at which the asset may be converted will be assumed to follow an arithmetic random walk. Use of such an assumption in an asset sale problem goes back as far as the early work of Bachelier [1]. Such an assumption has been criticized by Samuelson [10] and others in favor of a geometric random walk. In the geometric random walk changes in the logarithm of price follow an arithmetic walk. In the problem here, the horizon is short, and for short periods of time an arithmetic random walk is a good approximation to the geometric walk. This same justification for using an arithmetic walk has been given by Taylor [11, p. 12].

The motivation of the problem is now complete. The plan of the remainder of the paper is as follows. In §2 it is shown that dollar averaging is a nonsequential minimax strategy if the largest possible price increase in each period is equal to the largest possible decrease. It is shown that dollar averaging cannot be a multiperiod, nonsequential, expected utility maximizing strategy for any strictly concave utility function and any arithmetic random walk. In §3 the minimax sequential strategy is shown to be of the following form. At any point of time there exists a critical value which depends only on n, the difference between the current price and the maximum price since sales began. Any asset held in excess of the critical value will be sold. If less than the critical value is held none will be sold. The critical values are shown to be decreasing functions of n. When n becomes so large that it is impossible to exceed the previous maximum before time is up, the critical value becomes equal to zero. Under a sequential policy more will be sold when the price decreases than when it increases. This seems in accord with the second thoughts of those embarking on nonsequential plans of the dollar averaging type. These second thoughts might, of course, also be explained by extrapolative expectations. The variable minimax regret at any point in the sale is shown to be a convex and decreasing function of the two state variables, n and z. The variable z is the amount of the asset left to be sold. In §4 the minimax sequential policies are calculated explicitly for a simple example. In §5 it is shown that the minimax policies for buying a given amount of an asset are symmetric with the selling strategies. Minimax policies will tend to have a reinforcing effect on prive movements. In §6 the sequential and nonsequential minimax policies are derived for the case where repurchases of the asset are allowed so that the problem becomes like portfolio selec-cion.

2. Nonsequential Policies

Assume that there are T points of time at which some of the asset can be sold. The first such point is numbered zero and the last $T - 1$. After each of these points and before the next, the price of the asset changes according to the assumed stochastic process. When the price has changed after $T - 1$, any of the asset remaining is automatically sold at time T. In the nonsequential case a decision is made at time zero as to how much to sell not only at time zero but at each future point of time as well. In the sequential case a decision need not be made as to how much to sell at a particular point until the preceding realizations of the stochastic process are known.

Let the amount of the asset held before sales are made at t be z_t. After sales are made at t the amount held is z_{t+1}. Therefore, the amount sold at t is $z_t - z_{t+1}$ or $-\Delta z_{t+1}$. Since none of the asset can be repurchased the z_t must be nonincreasing with t (i.e., $z_{t+1} \leqq z_t$). For convenience the amount of the asset to be sold initially will be set equal to one (i.e., $z_0 = 1$). Since all of the asset must be sold by T, $z_{T+1} = 0$. The asset is taken to be perfectly divisible.

The regret for a given policy and time series of prices is equal to the difference of two quantities. The first is the largest amount which could be obtained using the optimal feasible policy. Since purchases are prohibited, this is simply to sell all of the asset at the highest price. Let p_t be the price at which the asset can be sold at time t. Let t^* be the time at which the maximum price occurs. Below it will be convenient to make use of the following identity for p_{t^*}

$$p_{t^*} = p_0 + \textstyle\sum_1^{t^*} \Delta p_t \,. \tag{1}$$

When $t^* = 0$ in (1) the summation on the right-hand side is taken to be equal to zero. To obtain regret, the amount obtained using the given policy must be subtracted from this maximum price. The amount obtained is given by the following expression:

$$\textstyle\sum_1^{T+1} - p_{t-1}\Delta z_t = p_0 + \sum_1^{T} \Delta p_t z_t \,. \tag{2}$$

The regret, R, is then equal to (1)–(2) as given in (3).

$$R = \textstyle\sum_1^{t^*} \Delta p_t - \sum_1^{T} \Delta p_t z_t = \sum_1^{t^*} \Delta p_t(1 - z_t) - \sum_{t^*+1}^{T} \Delta p_t z_t \,. \tag{3}$$

The nonsequential minimax policy is calculated in the following way. First, given the values of the z_t, the maximum value of R is calculated over all possible values of the Δp_t. Then, the minimum of this maximum R is calculated for all possible values of the z_t. Consider first the maximization problem. Let the largest possible price increase in any period be equal to a and let the largest possible decrease be equal to b. The maximization will be carried out in two stages. In the first, consider the maximum of R over all possible values of the Δp_t such that the maximum price occurs at t^*. Inspection of (3) indicates that for any feasible values of the z_t (i.e., $0 \leqq z_t \leqq 1$) the maximum of R will occur under the following conditions. The value of Δp_t must take on its largest possible value for all t from 1 through t^* and its lowest possible value for all t larger than t^*. Thus, for all possible price series whose maximum price occurs at t^*, the maximum value of R is given by the expression in (4)

$$a \textstyle\sum_1^{t^*} (1 - z_t) + b \sum_{t^*+1}^{T} z_t \,. \tag{4}$$

The second stage of the maximization procedure is to maximize (4) with respect to t^*. This gives the maximum of R for all possible Δp_t. It is shown by the following argument that this maximum occurs either at $t^* = 0$ or $t^* = T$. The proof is by contradiction. Suppose that a unique maximum with respect to t^* is attained for some t^* greater than zero and less than T. Suppose further that at this value of t^*,

$$a(1 - z_{t^*}) \geqq bz_{t^*}.$$

Since z_t is nonincreasing, it must then be true that $a(1 - z_t) \geqq bz_t$ for all t such that $t^* \leqq t \leqq T$. Inspection of (4) shows that the assumed value of t^* cannot give a unique maximum then, since increasing its value cannot lower the value of R under these conditions. Suppose next that the inequality is reversed at the assumed maximum value of t^* so that $a(1 - z_{t^*}) < bz_{t^*}$. From the fact that z_t is nonincreasing with t it follows that $a(1 - z_t) < bz_t$ for all t such that $0 \leqq t \leqq t^*$. Inspection of (4) indicates that R will increase as t^* is decreased under these conditions. Thus, the assumed value for t^* could not give a maximum. Since it has been shown that a value of t^* other than $t^* = 0$ or $t^* = T$ cannot give a unique maximum to R it follows that the maximum must occur at either $t^* = 0$ or $t^* = T$. Thus, substituting these values for t^* in (4) gives that the maximum of R for all possible values of the Δp_t, given any feasible values of the z_t, must be equal to Max $[a(T - \sum_1^T z_t), b\sum_1^T z_t]$. To obtain the minimax nonsequential strategy the last step is to minimize this maximum over all feasible values of the z_t. This is easily done. Any feasible set of z_t which satisfy the relation in (5) will minimize Max $[a(T - \sum_1^T z_t), b\sum_1^T z_t]$ and give the required strategy.

$$\sum_1^T z_t = aT/(a + b). \tag{5}$$

The minimax value of R using this strategy is given in (6):

$$\text{Min}_z \text{Max}_{\Delta p} R = \text{Min}_z \text{Max} [a(T - \textstyle\sum_1^T z_t), b\sum_1^T z_t] = abT/(a + b). \tag{6}$$

The policy in (5) is of particular interest in the case where the largest possible price increase is equal to the largest possible decrease so that $a = b$. In this case a nonsequential minimax policy is to sell equal amounts of the asset at each point. If equal amounts are sold at each of the $T + 1$ points then z_1 must be equal to $T/(T + 1)$ and, in general, z_t will be equal to $(T - t + 1)/(T + 1)$. The sum of the z_t will thus be equal to the sum of the first T integers divided by $T + 1$. This is equal to $T/2$ which is the policy in (5) when $a = b$.

$$\sum_1^T z_t = \sum_1^T (T - t + 1)/(T + 1) = T(T + 1)/2(T + 1) = T/2.$$

When the asset to be sold is a sum of dollars, equal dollar amounts of stock are bought in each period. The condition for dollar averaging to be a nonsequential minimax policy is that the price of a dollar in terms of shares of stock (i.e., the reciprocal of the stock's price) follow an arithmetic random walk in which the largest possible increases and decreases are equal. For short periods of time this may be approximately satisfied. It is to be noted that dollar averaging is not the only nonsequential minimax policy. For instance, selling half at the beginning and half at the end will also be such a policy under these conditions.

It will now be shown that dollar averaging cannot be a multiperiod, nonsequential, expected utility maximizing strategy for any strictly concave utility function and any arithmetic random walk. Using the expected utility principle, the argument of the utility function is the realized outcome given in (2). The optimal, nonsequential, expected utility maximizing policy will be the solution to the problem in (7).

$$(7) \qquad \mathrm{Max}_z\, E\{U(p_0 + \textstyle\sum_1^T \Delta p_t z_t)\} \qquad 0 \leqq z_T \leqq z_{T-1} \leqq, \cdots, \leqq z_1 \leqq 1.$$

Let the expected utility of the objective in (7) be denoted by $V(z)$ where z is the vector, $z = (z_1, z_2, \cdots, z_T)$. The value of V is invariant to any renumbering of the z_t in z. This follows because given that the Δp_t are independently and identically distributed and that the argument of U is additive in the $\Delta p_t z_t$, any renumbering of the z_t can be written as a renumbering of the variables of integration (i.e., the Δp_t). The latter renumbering has no effect on the value of the expected utility integral. Furthermore, if U is strictly concave, V is a strictly concave function of z. The proof is similar to one provided in [9, p. 112].

It will now be shown that the solution to (7) must have all the values of the z_t equal to each other. Suppose that $z = z'$ maximizes $V(z)$ for all z such that $0 \leqq z \leqq 1$. Suppose, furthermore, that at least two of the elements of z' are not equal to each other. Select two such elements and interchange their values giving z''. Since z'' is a renumbering of z' it follows that $V(z'') = V(z')$ and also that $0 \leqq z'' \leqq 1$. Consider the vector $z = \lambda z' + (1 - \lambda)z''$ where $0 < \lambda < 1$. It follows that $0 \leqq \lambda z' + (1 - \lambda)z'' \leqq 1$. Also, since $z' \neq z''$, it follows from the strict concavity of $V(z)$ that

$$V(\lambda z' + (1 - \lambda)z'') > \lambda V(z') + (1 - \lambda)V(z'') = V(z').$$

Thus, $\lambda z' + (1 - \lambda)z''$ gives a higher value to $V(z)$ than z'. Thus, z' cannot maximize $V(z)$ on $0 \leqq z \leqq 1$ unless its elements are all equal to each other. Having all the z_t equal satisfies the additional constraints in (7) that no value of z_t have a larger value than its predecessor. The maximum of $V(z)$ subject to these additional constraints must therefore also have the values of the z_t equal to each other. This is a necessary condition on the solution to (7). The argument holds for any strictly concave utility function and any arithmetic random walk.

Since the values of the z_t must be equal for an optimum, dollar averaging cannot be an optimal strategy in a multiperiod case. Dollar averaging requires that the z_t strictly decrease over time. Dollar averaging could occur only in a one period case, and then only if by coincidence the value of z_1 turned out to be $\frac{1}{2}$. On the basis of this evidence it seems safe to conclude that dollar averaging is a multiperiod, nonsequential strategy based on hedging against large regrets rather than on risk averse expected utility maximization.

3. Sequential Policies

In general, of course, the seller is perfectly free to let his future decisions depend on the prices which will be observed before the decision actually has to be made. It is therefore of interest to develop the sequential minimax policies. It seems to be common experience to have second thoughts once one has embarked on a nonsequential policy of the dollar averaging type. If the price rises subsequently one has a strong inclination to sell less, while if the price falls one wants to sell more. This might be explained on the basis of extrapolative expectations. Interestingly enough, however, it is also in accord with a sequential minimax policy.

Let p_t^* be defined as the maximum price which occurs up through time t (i.e., $p_t^* = \mathrm{Max}_{0 \leqq i \leqq t}\, [p_i]$). In terms of the notation of the last section then, $p_T^* = p_{t^*}$. Let n_t be defined as the difference between the largest price which has occurred up through t and the price at time t (i.e., $n_t = p_t^* - p_t$). Using this notation, the maximum price p_T^* can be written as in (8)

$$p_T^* = p_t^* + \text{Max}\,[0, -n_t + \textstyle\sum_{t+1}^{i} \Delta p_j\,(t+1 \leqq i \leqq T)]. \tag{8}$$

Subtracting (2) from (8) gives the regret in (9):

$$\begin{aligned} R = {} & p_t^* - p_0 - \textstyle\sum_1^t \Delta p_i z_1 \\ & + \text{Max}\,[0, -n_t + \textstyle\sum_{t+1}^{i} \Delta p_j\,(t+1 \leqq i \leqq T)] - \textstyle\sum_{t+1}^{T} \Delta p_i z_i . \end{aligned} \tag{9}$$

From (9) it follows that, given any time t, R is separable into two sets of terms; one of which is fixed and the other variable. In the fixed set (i.e., $p_t^* - p_0 - \sum_1{}^t \Delta p_i z_i$) the terms depend only on the values of $\Delta p_1, \Delta p_2, \cdots, \Delta p_t$ and $z_1, z_2, \cdots, z_t$. At time t the contribution of these terms to R is fixed independently of subsequent events and policy. Therefore, these terms may be neglected in formulating the optimal policy at time t. The terms in the variable set will be denoted by R_t and called variable regret. These terms are given in (10).

$$R_t = \text{Max}\,[0, -n_t + \textstyle\sum_{t+1}^{i} \Delta p_j\,(t+1 \leqq i \leqq T)] - \textstyle\sum_{t+1}^{T} \Delta p_j z_j, \quad R_T = 0. \tag{10}$$

The terms in R_t depend on $\Delta p_{t+1}, \Delta p_{t+2}, \cdots, \Delta p_T;\ z_{t+1}, z_{t+2}, \cdots, z_T$ and n_t. At time t these price changes are independent of prior price behavior since the price series has been assumed to be generated by an arithmetic random walk. The values of z following t will depend on the past only to the extent that they cannot exceed z_t in value. Thus, in determining an optimal policy at t, prior behavior need be considered only to the extent that it affects the amount of the asset which has not been sold (i.e., z_t) and the value of n_t. Otherwise, the optimal policy is independent of the past and depends only on the future through the contribution to regret of R_t.

The relation for R_t in (10) can be written in recursive form. The expression

$$\text{Max}\,[0, -n_t + \textstyle\sum_{t+1}^{i} \Delta p_j\,(t+1 \leqq i \leqq T)]$$

can be written as

$$\text{Max}\,[0, -n_{t+1} + \textstyle\sum_{t+2}^{i} \Delta p_j\,(t+2 \leqq i \leqq T)], \quad \text{if} \quad \Delta p_{t+1} \leqq n_t .$$

This follows because under this condition, $n_{t+1} = n_t - \Delta p_{t+1}$. If $\Delta p_{t+1} > n_t$, $n_{t+1} = 0$, and the expression can be written as $\text{Max}\,[0, \sum_{t+2}^{i} \Delta p_j\,(t+2 \leqq i \leqq T)] - n_t + \Delta p_{t+1}$. Using this and (10) gives (11)

$$\begin{aligned} R_t(n_t) &= R_{t+1}(n_t - \Delta p_{t+1}) - \Delta p_{t+1} z_{t+1} \qquad && \Delta p_{t+1} \leqq n_t \\ &= R_{t+1}(0) + \Delta p_{t+1}(1 - z_{t+1}) - n_t && \Delta p_{t+1} > n_t . \end{aligned} \tag{11}$$

Let $L_t(n_t, z_t)$ be the value of R_t given n_t, z_t, and that a minimax sequential policy is followed at t and subsequently. Suppose such a policy is followed at $t+1$ and subsequently. The maximum possible value of R_{t+1} is then L_{t+1}. From (11) it follows that any other policy at $t+1$ and subsequently will not be part of a minimax sequential policy at t. Any other policy will give a higher possible value of R_t, given any value of z_{t+1} and Δp_{t+1}. The value of z_{t+1} in a minimax sequential policy for R_t at t can therefore be determined by considering the right-hand side of (11) when R_{t+1} is replaced by $L_{t+1}(n_t - \Delta p_{t+1}, z_{t+1})$. Thus, L_t and z_{t+1} can be determined from the relation in (12) given L_{t+1}. In (12) time subscripts on n_t, Δp_{t+1}, and z_t have been dropped, and z_{t+1} has been replaced by x. Also, to further simplify the notation, $L_t(n, z)$ for negative values of n has been defined as equal to $L_t(0, z) - n$.

$$(12)\quad \begin{aligned}&\text{(a)}\ L_t(n, z) = \operatorname{Min}_{0 \leq x \leq z} \operatorname{Max}_{-b \leq \Delta p \leq a} [L_{t+1}(n - \Delta p, x) - \Delta p x] \\ &\text{(b)}\qquad\qquad\ \ = L_t(0, z) - n \quad \text{if} \quad n < 0.\end{aligned}$$

Since $R_T(n) = 0$ for $n \geqq 0$ it follows that $L_T(n, z) = 0$ for $n \geqq 0$. From (12) for $n < 0$ it follows that $L_T(n, z) = L_T(0, z) - n = -n$. Therefore, $L_T(n, z)$ is given by (13).

$$(13)\quad \begin{aligned} L_T(n, z) &= 0 &\quad n \geqq 0,\ z \geqq 0 \\ &= -n &\quad n < 0,\ z \geqq 0. \end{aligned}$$

With the value of L_T given in (13) the optimal policy and value of $L_t(n, z)$ in any period can be obtained recursively using (12) by working backwards from T. The minimax regret of selling the asset under a sequential policy will be given by $L_0(0, 1)$.

The next step is to show that the recursion relation for calculating L_t in (12) can be substantially simplified. This is true because it turns out that the maximum with respect to Δp in (12) must occur at one of the two boundary values of Δp. Let the function to be maximized in the brackets in (12) be denoted $f(n, x, \Delta p)$. If f, given n and x, is convex in Δp it follows that the maximum will be assumed at one of the boundary values of Δp as stated. Inspection of (12) shows that f, given n and x, will be convex in Δp as required if L_{t+1} is convex in its first argument given the value of its second. It will next be proved that this is the case and, in fact, more generally that L_t is convex in both its arguments for all t.

THEOREM 1. *The minimax variable regret, $L_t(n, z)$, is a convex function of n and z for all t.*

PROOF. The proof is by induction. Assume that $L_{t+1}(n, z)$ is convex. It then follows that $f(n, x, \Delta p)$ is convex in Δp given n and x. Let $h(n, x)$ be the maximum of f with respect to Δp. From the convexity of f with respect to Δp it follows that

$$h(n, x) = \operatorname{Max}_{-b \leq \Delta p \leq a} f(n, x, \Delta p) = \operatorname{Max} [f(n, x, -b), f(n, x, a)].$$

Moreover, $h(n, x)$ is convex in n and x because the maximum of a set of convex functions is itself convex. Also, $g(n, z, x) \equiv h(n, x)$ is a convex function of n, z, and x. Now, a convex function minimized with respect to a subset of its variables on a convex set is also a convex function (Iglehart, [5, p. 199]). Since from (12),

$$L_t(n, z) = \operatorname{Min}_{0 \leq x \leq z} g(n, z, x),$$

it follows that $L_t(n, z)$ is a convex function of n and z. Since L_t has been shown to be convex if L_{t+1} is convex, and since from (13) L_T is convex, it follows by induction that L_t is convex for all t. This completes the proof.

As indicated previously, an immediate corollary of this theorem is that $f(n, x, \Delta p)$ given n and x is convex in Δp. The maximum with respect to Δp in (12) will then occur at one of the boundary values of Δp. Hence, the recursion relation for L_t can be simplified to that given in (14).

$$(14)\quad L_t(n, z) = \operatorname{Min}_{0 \leq x \leq z} \operatorname{Max} [L_{t+1}(n - a, x) - ax, L_{t-1}(n + b, x) + bx].$$

The dependence of the minimax variable regret on n and z is further characterized in Theorem 2. In general $L_t(n, z)$ is shown to be a decreasing function of n or z, given the value of the other variable. The most that the price can rise before time is up is $(T - t)a$. If $n \geqq (T - t)a$ so that the future price cannot exceed its previous maximum before time is up, it is shown that the minimax variable regret is zero. This is achieved by selling all of the asset which remains (i.e., $x = 0$). It is also shown that an upper bound on $L_t(n, z)$ is $a(T - t) - n$ for $n \leqq (T - t)a$. In particular it is shown that $L_t(n, 0) = a(T - t) - n$. That this is an upper bound follows from the fact that L_t is shown also to be a decreasing function of z.

THEOREM 2. (a) *$L_t(n, z)$ is positive for $n < (T - t)a$ and zero for $n \geqq (T - t)a$.* (b) *$L_t(n, z)$ is a strictly decreasing function of n, given z, for $n < (T - t)a$.* (c) *$L_t(n, z)$ is a decreasing function of z, given n, and $L_t(n, 0) = a(T - t) - n$ if $n < (T - t)a$.*

PROOF. (a) The proof is by induction. Suppose that (a) holds for L_{t+1}. Using (14) it then follows that (a) holds for L_t:

$$\begin{aligned} L_t(n, z) &= \mathrm{Min}_{0 \leqq x \leqq z} \mathrm{Max}\, [L_{t+1}(n - a, x) - ax, L_{t+1}(n + b, x) + bx] \\ &= \mathrm{Min}_{0 \leqq x \leqq z}\, bx = 0 \qquad\qquad n \geqq (T - t)a \\ &\qquad h(n, x) \geqq bx > 0 \qquad x > 0 \quad n < (T - t)a \\ &\qquad\qquad \geqq L_{t+1}(n - a, 0) > 0 \quad x = 0. \end{aligned}$$

Inspection of (13) shows that (a) holds for T. Therefore, by induction it follows that (a) holds for all t as was to be proved.

(b) Pick any n' and n'' such that $0 \leqq n' < n'' < a(T - t)$. Set

$$\lambda = [a(T - t) - n'']/[a(T - t) - n']$$

so that $0 < \lambda < 1$ and $\lambda n' + (1 - \lambda)a(T - t) = n''$. From the convexity of L proved in Theorem 1 it follows that $\lambda L_t(n', z) + (1 - \lambda)L_t(a(T - t), z) \geqq L_t(n'', z)$ From (a) $L_t(a(T - t), z) = 0$ and $L_t(n'', z) > 0$. Using these facts, $0 < \lambda < 1$, and the just stated convexity inequality gives:

$$L_t(n', z) \geqq \lambda^{-1}L_t(n'', z) > L_t(n'', z).$$

Since $L_t(n', z) > L_t(n'', z)$ for any n' and n'' such that $0 \leqq n' < n'' < a(T - t)$ and for $n < 0$ from (12) it follows that L_t is a strictly decreasing function of n for all $n < a(T - t)$ as was to be proved.

(c) It follows immediately from (14) that L_t is a decreasing function of z, given n, because increasing the value of z enlarges the set of values on which the minimum with respect to x can occur. L_t need not be strictly decreasing in z. To see that

$$L_t(n, 0) = (T - t)a - n \quad \text{for} \quad n < (T - t)a,$$

first note from part (b) above and (14) that:

$$L_t(n, 0) = \mathrm{Max}\, [L_{t+1}(n - a, 0), L_{t+1}(n + b, 0)] = L_{t+1}(n - a, 0).$$

Using this recursion relation and (13) gives:

$$L_t(n, 0) = L_T(n - (T - t)a, 0) = (T - t)a - n.$$

This completes the proof of the theorem.

It remains to characterize the minimax variable regret with respect to time. In the next theorem it is shown that $L_t(n, z)$ is a decreasing function of time, given n and z. It is a strictly decreasing function of time as long as the value of n is not so large as to preclude the previous maximum value of the price being exceeded before time is up. The proof requires the following trivial lemma.

LEMMA 1. *Let $f(x) = \mathrm{Max}_{0 \leqq i \leqq I}\, [f_i(x)]$ and $g(x) = \mathrm{Max}_{0 \leqq i \leqq I}\, [g_i(x)]$. If $f_i(x) > g_i(x)$ for all i and x, then $f(x) > g(x)$ for all x.*

Proof. Suppose $f(x) \leqq g(x)$ for $x = x'$. Then,

$$g_j(x') = \text{Max}_{0 \leqq i \leqq I} [g_i(x')] \geqq \text{Max}_{0 \leqq i \leqq I} [f_i(x')] \geqq f_j(x')$$

which is impossible. Therefore $f(x) > g(x)$ for all x as stated.

Theorem 3. *$L_t(n, z)$ is a strictly decreasing function of t, given n and z, if*

$$n < (T - t)a.$$

Proof. The proof is by induction. Assume that the result holds for $t + 1$. Let

$$h_{t+1}(n, x) = \text{Max} [L_{t+1}(n - a, x) - ax, L_{t+1}(n + b, x) + bx].$$

Let $L_t = \text{Min}_{0 \leqq x \leqq z} h_{t+1}(n, x) = h_{t+1}(n, x')$. If $n + b < [T - (t + 1)]a$, the conditions of Lemma 1 are satisfied for h_{t+1} and h_{t+2}. Therefore, $h_{t+1}(n, x') > h_{t+2}(n, x')$. Since $L_{t+1}(n, z)$ cannot exceed $h_{t+2}(n, x')$ it follows that $L_t(n, z) > L_{t+1}(n, z)$. The case must now be considered where $n + b \geqq [T - (t + 1)]a$. In this case Lemma 1 cannot be strictly applied because $L_{t+1}(n + b, x) = L_{t+2}(n + b, x) = 0$. However, it does follow by similar reasoning that $h_{t+1}(n, x') \geqq h_{t+2}(n, x')$. Suppose that equality holds. Then $h_{t+1} = h_{t+2} = bx'$, and furthermore, since bx' is a minimum, inspection of h_{t+1} and h_{t+2} shows that $L_{t+1}(n - a, x') - ax' = bx'$ and $L_{t+2}(n - a, x') - ax' = bx'$. The latter equalities, however, imply that $L_{t+1}(n - a, x') = L_{t+2}(n - a, x')$ which contradicts the assumption that $L_{t+1} > L_{t+2}$ for $n - a < [T - (t + 1)]a$. This is impossible because the latter condition must be satisfied if $n < (T - t)a$. Therefore, it must be true for this case as well that $h_{t+1}(n, x') > h_{t+2}(n, x')$ since equality is impossible. Thus, $L_t > L_{t+1}$, if $n < (T - t)a$. From part (a) of Theorem 2 and (12) it follows that $L_{T-1} > L_T$ for $n < (T - t)a = a$ because $L_{T-1} > 0$, $L_T = 0$ if $0 \leqq n < a$ and $L_{T-1} = L_{T-1}(0, z) - n > L_T = -n$ if $n < 0$. Therefore, the result holds for $T - 1$ and hence by induction for all t. This completes the proof.

The characterization of the dependence of the minimax variable regret on t, n and z is now complete. It remains to characterize the optimal policy with respect to sale of the asset. First it will be shown that the optimal policy is of the following form. At any time there exists a critical value which depends on n. All of the asset which is held above this critical value will be sold. If less than this critical value is held, none will be sold.

Theorem 4. *At any time the minimax policy is of the form:* $x^* = z$, *if* $z \leqq z_t^*(n)$, $x^* = z_t^*(n)$, *if* $z > z_t^*(n)$.

Proof. Using (14) and the notation introduced in the proof of Theorem 3, the minimax policy at t minimizes $h_{t+1}(n, x)$ with respect to x on the range $0 \leqq x \leqq z$. Moreover, from Theorem 1 and the fact that the maximum of a set of convex functions is convex it follows that $h_{t+1}(n, x)$ is convex in x given n. Let $z_t^*(n)$ be the x which minimizes $h_{t+1}(n, x)$ on the range, $0 \leqq x \leqq 1$. It follows immediately that $z_t^*(n)$ minimizes $h_{t+1}(n, x)$ on $0 \leqq x \leqq z$ for all $z \geqq z_t^*(n)$ as required. It remains to show that z minimizes $h_{t+1}(n, x)$ on $0 \leqq x \leqq z$ for all $z < z_t^*(n)$. The proof of this is by contradiction. Suppose for some z such that $x' < z < z_t^*(n)$, that $h_{t+1}(n, x') < h_{t+1}(n, z)$. Set $\lambda = (z - z^*)/(x' - z^*)$ so that $0 < \lambda < 1$ and $\lambda x' + (1 - \lambda)z^* = z$. It follows that $\lambda h_{t+1}(n, x') + (1 - \lambda)h_{t+1}(n, z^*) < h_{t+1}(n, z)$, but this is impossible because $h_{t+1}(n, x)$ is convex. This completes the proof. Next, the critical values defined in the last theorem are characterized with respect to their dependence on n. In particular, it is shown that they are decreasing functions of n. It is already known from

Theorem 2(a) that the critical value becomes zero when n becomes so large that exceeding the previous maximum price before time is up is precluded.

THEOREM 5. *The critical value, $z_t^*(n)$, is a decreasing function of n.*

PROOF. The proof is by induction. Suppose that $z_{t+1}^*(n)$ is decreasing in n. Furthermore, suppose that $L_{t+2}[n - a, z] - az \geqq L_{t+2}[n + b, z] + bz$ for all n and all z such that $0 \leqq z \leqq z_{t+1}^*(n)$. For $n < 0$, it is defined that $z_t^*(n) = z_t^*(0)$. For z such that

$$0 \leqq z \leqq z_{t+1}^*(n + b)$$

and all n:

$$L_{t+2}[n - 2a, z] - az = L_{t+2}[n + b - a, z] + bz,$$

$$L_{t+1}[n - a, z] \geqq L_{t+1}[n + b, z] + (a + b)z,$$

$$L_{t+1}[n - a, z] - az \geqq L_{t+1}[n + b, z] + bz.$$

The first inequality follows because by assumption $z_{t+1}^*(n + b) \leqq z_{t+1}^*(n - a)$; the second follows by applying (14) to both sides and using (12b) if $n - a < 0$ or $n + b < 0$. Suppose that the last inequality holds with equality for some

$$z \geqq z_{t+1}^*(n + b).$$

This z must then be $z_t^*(n)$ since Max $[L_{t+1}(n - a, z) - az, L_{t+1}(n + b, z) + bz]$ must strictly increase with z for larger z; $L_{t+1}[n + b, z]$ being invariant with z for $z \geqq z_{t+1}^*(n + b)$. Also, Max $[L_{t+1}(n - a, z) - az, L_{t+1}(n + b, z) + bz]$ must strictly decrease with z for smaller z because $L_{t+1}(n - a, z) - az$ is strictly decreasing in z. If equality does not hold for some z such that $z_{t+1}^*(n + b) \leqq z \leqq 1$ then $z_t^*(n) = 1$. In any case, $L_{t+1}[n - a, z] - az \geqq L_{t+1}(n + b, z) + bz$ for all n and $z: 0 \leqq z \leqq z_t^*(n)$ as required. It will next be shown that $z_t^*(n)$ cannot increase with n. This is true immediately, if $z_t^*(n) = 1$. Suppose next that z such that $z_{t+1}^*(n + b) \leqq z < 1$ satisfies the equation $L_{t+1}(n - a, z) - az = L_{t+1}(n + b, z) + bz$. Take $n' > n$ and let z' satisfy $L_{t+1}(n' - a, z') - az' \geqq L_{t+1}(n' + b, z') + bz'$ with equality unless $z' = 1$. It will next be shown that having $z' > z$ is impossible. First, note that if $z' > z$, then $L_{t+1}(n' - a, z) \geqq L_{t+1}(n' - a, z')$ because from Theorem 2(c) L_{t+1} is a decreasing function of z. Also, $L_{t+1}(n' + b, z') = L_{t+1}(n' + b, z)$ because

$$z' > z \geqq z_{t+1}^*(n + b) \geqq z_{t+1}^*(n' + b).$$

Using these relations and the definitions of z and z' above gives:

$$\begin{aligned} L_{t+1}(n' - a, z) - L_{t+1}(n' + b, z) &\geqq L_{t+1}(n' - a, z') - L_{t+1}(n' + b, z') \\ &\geqq (a + b)z' \\ &> L_{t+1}(n - a, z) - L_{t+1}(n + b, z). \end{aligned}$$

Having the left-hand expression greater than the last term on the right-hand side, however, contradicts the fact established in Theorem 1 that L_{t+1} is convex. Therefore, it must be true that $z' \leqq z$. Thus, it has been shown that $z_t^*(n)$ is a decreasing function of n. To complete the proof by induction it is necessary to show that $z_{T-1}^*(n)$ is a decreasing function of n and that $L_T(n - a, z) - az \geqq L_T(n + b, z) + bz$ for all z such that $0 \leqq z \leqq z_{T-1}^*(n)$ and all n.

$$\begin{aligned} L_{T-1}(n, z) &= \operatorname{Min}_{0 \leqq x \leqq z} \operatorname{Max}\, [L_T(n - a, x) - ax, L_T(n + b, x) + bx] \\ &= \operatorname{Min}_{0 \leqq x \leqq z} \operatorname{Max}\, \begin{cases} [a - n - ax, bx] & 0 \leqq n \leqq a \\ [-ax, bx] & n > a. \end{cases} \end{aligned}$$

Inspection of this last expression shows that $z^*_{T-1}(n)$ is given by

$$z^*_{T-1}(n) = (a - n)/(a + b) \qquad 0 \leqq n \leqq a$$
$$= 0 \qquad n > a.$$

Thus, $z^*_{T-1}(n)$ is a decreasing function of n as required. Using this it is also easily verified that $L_T(a - n, z) - az \geqq L_T(n + b, z) + bz$ for all n and all z: $0 \leqq z \leqq z^*_{T-1}(n)$ as required. This completes the proof.

It might be conjectured that the critical values are a decreasing function of t as well as n. This, however, is not true. A counterexample is provided by the minimax policy calculated for the example in the following section. On the basis of examples it appears that the critical values are much less sensitive to t than n except near T. This, taken together with Theorem 5, has the following implication.

After some of the asset is sold initially, little more will be sold if the price rises or stays the same. In this case n will remain equal to zero. On the other hand, more of the asset will be sold as soon as the price drops because then n will increase. In general, for a sequential policy, more is sold on price dips than rises because the former increase n while the latter do not. An example for a simple case is provided in the next section.

4. An Example of a Sequential Policy

As a simple example of an arithmetic random walk consider the following process. Suppose that each period the price can either go up by one unit, down by one unit or stay the same. This means that $a = b = 1$ and that n can take on only integer values. Suppose furthermore that there are three points of time available to sell the asset before it must all be sold so that $T = 3$. When $t = 2$ and $n \geqq 1$ it is already known from Theorem 2(a) that the minimax policy is to sell all of the asset which is left. For $t = 2$ and $n = 0$, (13) and (14) give the following:

$$\begin{aligned} L_2(0, z) &= \text{Min}_{0 \leqq x \leqq z} \text{Max}\, [L_3(-1, x) - x, L_3(1, x) + x] \\ &= \text{Min}_{0 \leqq x \leqq z} \text{Max}\, [1 - x, x] \\ &= 1 - z \quad z \leqq \tfrac{1}{2} \qquad x^* = z \quad z \leqq \tfrac{1}{2} \\ &= \tfrac{1}{2} \qquad z > \tfrac{1}{2} \qquad\quad = \tfrac{1}{2} \quad z > \tfrac{1}{2}. \end{aligned}$$

Thus, the critical value, $z_2^*(0)$, is $\tfrac{1}{2}$. Any asset which is held above this amount should be sold. If less than this amount is held none should be sold.

Using this result, the minimax strategies for $t = 1$ and $n = 0, 1$ can next be derived.

$$\begin{aligned} L_1(0, z) &= \text{Min}_{0 \leqq x \leqq z} \text{Max}\, [L_2(0, x) + 1 - x, L_2(1, x) + x] \\ &= \text{Min}_{0 \leqq x \leqq z} \text{Max} \left[\begin{cases} 2(1 - x) & x \leqq \tfrac{1}{2} \\ \tfrac{3}{2} - x & y > \tfrac{1}{2} \end{cases}, x \right] \\ &= 2(1 - z) \quad z \leqq \tfrac{1}{2} \qquad\qquad x^* = z \quad z \leqq \tfrac{3}{4} \\ &= \tfrac{3}{2} - h \quad \tfrac{1}{2} < z \leqq \tfrac{3}{4} \qquad\quad = \tfrac{3}{4} \quad z > \tfrac{3}{4} \\ &= \tfrac{3}{4} \qquad z > \tfrac{3}{4} \end{aligned}$$

$$\begin{aligned} L_1(1, z) &= \text{Min}_{0 \leqq x \leqq z} \text{Max}\, [L_2(0, x) - x, L_2(2, x) + x] \\ &= \text{Min}_{0 \leqq x \leqq z} \text{Max} \left[\begin{cases} 1 - 2x & x \leqq \tfrac{1}{2} \\ \tfrac{1}{2} - x & x > \tfrac{1}{2} \end{cases}, x \right] \\ &= 1 - 2z \quad z \leqq \tfrac{1}{3} \qquad x^* = z \quad z \leqq \tfrac{1}{3} \\ &= \tfrac{1}{3} \qquad z > \tfrac{1}{3} \qquad\quad = \tfrac{1}{3} \quad z > \tfrac{1}{3}. \end{aligned}$$

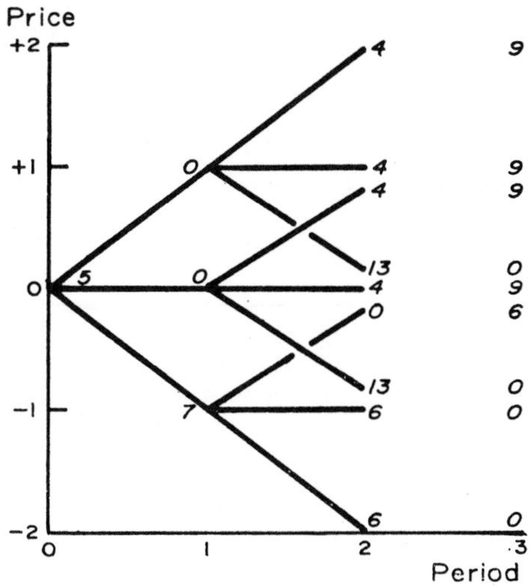

FIGURE 1

Similarly one can calculate $L_0(0, z)$ using the $L_1(0, z)$ and $L_1(1, z)$ just obtained. It is easily found that $L_0(0, 1) = \frac{19}{18}$ and $z_0^*(0) = \frac{13}{18}$. Thus, the minimax loss in this case is $\frac{19}{18}$. This compares with a minimax loss for a nonsequential policy of $T/2$ or $\frac{3}{2}$ as may be seen from (6). The sequential minimax strategy is to sell $\frac{5}{18}$ of the asset initially. The minimax strategies in the following period depend, of course, on subsequent price behavior. They are shown in Figure 1. The number next to each node gives the amount of the asset measured in eighteenths to be sold at that point. The example demonstrates that the critical values need not decrease with time. The critical value initially is $\frac{13}{18}$ or $\frac{26}{36}$ for $n = 0$ while in the next period it is $\frac{3}{4}$ or $\frac{27}{36}$ for $n = 0$.

5. Sequential Policies for Buying an Asset

It is now of interest to consider the problem where a given amount of an asset is to be bought rather than sold. The same formal relations and propositions hold for buying as selling if the variables are reinterpreted. For buying let n be the number of units by which the current price exceeds the previous minimum price. Also, let z be the fraction of the asset left to be purchased. With this reinterpretation, the relation given by (14) and the propositions which follow from (14) apply to buying.

Since the complement of the fraction left to be bought is the amount currently held, a buyer will hold the complement of the amount held by a seller, given t and n. The differences in the amounts held are the amounts bought or sold. The difference of the complement of a variable is equal to the negative of the difference of the variable itself. Therefore, the amount sold by a seller for some t and n must be equal to the amount bought by a buyer for that same t and n. However, for any given price series the n for a seller at any point will not in general be equal to the n for a buyer. In fact, it is true that the price series with the same n for a buyer as a seller in any period is the series which is a mirror image of the seller's series about a horizontal axis equal to the initial price.

This last proposition can be seen as follows. Let the vector $p = (p_0, p_1, \cdots, p_t)$

be the vector of price differences between the actual price in period 0, 1, $\cdots$, t and the initial price. Let $n_s(p)$ be the n for the seller in period t. Let $n_b(p)$ be the n of the buyer in period t. From the definition of n for the seller and buyer it follows that:

$$n_b(p) = p_t - \text{Min}_{0 \leqq \tau \leqq t} [p_\tau] = \text{Max}_{0 \leqq \tau \leqq t} [-p_\tau] - (-p_t) = n_s(-p).$$

Since $-p$ is the mirror image of the price differentials about zero, the stated proposition has been proved.

It has now been established that the amount sold by a seller and the amount bought by a buyer for any n and t are equal, and that $n_b(p) = n_s(-p)$. Initially, a buyer will want to buy the same amount of the asset as a seller. Subsequently, however, if the price rises n_s will continue to equal zero while n_b will rise. Thus, a buyer will want to buy more while a seller will not want to sell more. The converse will hold if the price falls. Behavior of buyers and sellers will be symmetrical. In the example in Figure 1 the minimax behavior of a buyer will be given by the mirror image of each of the series. In other words, the amounts by the nodes when the figure is rotated 180° about the horizontal axis will give the quantities which a buyer will purchase. It is interesting to note that sequential, minimax policies will tend to have a reinforcing effect on price movements. As has been seen, price rises will make buyers want to buy more without causing any increase in sales by sellers.

6. Policies under Reversibility

So far it has been assumed that the seller is prohibited from repurchasing any of the asset. This assumption will now be relaxed to permit repurchases at any time before T. Margin purchases and short sales, however, will continue to be prohibited. As discussed in the introduction this has the effect of changing the flavor of the problem to one of multiperiod portfolio selection. However, it is unconventional in that the objective continues to be one of minimaxing regret. Under these conditions the problem does not seem to replicate situations in which dollar averaging has been recommended. Nevertheless, the problem is of interest because the answer is simple and forms an interesting contrast to the results obtained previously.

The possibility of repurchase changes the maximum amount which can be obtained for the asset, given the price series which occurs. Without repurchase the best one can do is sell out everything at the highest price which occurs. In general, with repurchase one can do better. The best that can be done is to be fully invested during every price increase and completely disinvested during every price decrease. Thus, the amount obtained for the asset will be equal to the initial price plus the sum of all the price increases which occur. The expression for regret in (3) must be changed to read as follows:

$$R = \textstyle\sum_+ \Delta p_t(1 - z_t) - \sum_- \Delta p_t z_t \tag{15}$$

where $\sum_+$ indicates that the summation is taken over all price increases while $\sum_-$ indicates that the summation is over all decreases. Each z_t can now take on any value on the range $0 \leqq z_t \leqq 1$. Also, as before the Δp_t are taken to be independently distributed. Inspection of (15) indicates that the contribution of any z_t and Δp_t to R is independent of the other z_t and Δp_t. The maximum possible contribution of Δp_t to R will be minimized if $z_t = a/(a + b)$. This holds for all t. Thus, the minimax strategy is to sell $b/(a + b)$ of the asset initially and $a/(a + b)$ of it at T. This is both the sequential and nonsequential minimax strategy. Reference to (5) indicates that it also happens to be a nonsequential minimax strategy when repurchases are not allowed.

The question arises as to why sellers might not act as if they can repurchase. One answer is that they may not consider taking advantage of every uptick and downtick in price as a realistic alternative. Therefore, not having done so causes little regret. On the other hand, missing a major turn in the market, given a decision to convert an asset, is cause for considerable unhappiness. Another answer may be transaction costs which might substantially lower the expected value of such a policy. Regret remains an elusive psychological fact. The hope is that this paper has shown that it can be neglected only at the peril of an incomplete understanding of behavior under risk.

References

1. Bachelier, L., *Theory of Speculation* (translation of 1900 French edition), in Cootner, P. H., Editor, *The Random Character of Stock Market Prices*, MIT Press, Cambridge, 1964.
2. Borch, K., *The Economics of Uncertainty*, Princeton Press, Princeton, N. J., 1968.
3. Cottle, S. C. and Whitman, W. T., *Investment Timing: The Formula Plan Approach*, McGraw-Hill, New York, 1953.
4. Hayes, R. H., "Optimal Strategies for Divestiture," *Journal of the Operations Research Society of America*, (March-April 1969), pp. 292–310.
5. Iglehart, D. L., "Capital Accumulation and Production for the Firm: Optimal Dynamic Policies," *Management Science*, Vol. 12, No. 3 (November 1965), pp. 193–205.
6. Karlin, S., "Stochastic Models and Optimal Policy for Selling an Asset," Chapter 9 in *Studies in Applied Probability and Management Science*, Arrow, Karlin and Scarf, Eds., Stanford University Press, Stanford, California, 1962.
7. Markowitz, H. M., *Portfolio Selection: Efficient Diversification of Investments*, John Wiley & Sons, Inc., New York, 1959.
8. Mossin, J., "Optimal Multiperiod Portfolio Policies," *Journal of Business*, (April 1968), pp. 215–229.
9. Pye, G., "Portfolio Selection and Security Prices," *Review of Economics and Statistics*, (February 1967), pp. 111–115.
10. Samuelson, P., "Rational Theory of Warrant Pricing," *Industrial Management Review*, Vol. 6, (1965), pp. 13-31.
11. Taylor, H., "Evaluating a Call Option and Optimal Timing Strategy in the Stock Market," *Management Science*, Vol. 14, No. 1 (September 1967), pp. 111–120.
12. Tobin, J., "The Theory of Portfolio Selection," in Hahn and Brechling (Eds.), *The Theory of Interest Rates*, Macmillan, London, 1965.

4. THE CAPITAL GROWTH CRITERION AND CONTINUOUS-TIME MODELS

Reprinted from *Naval Research Logistics Quarterly* 7, 647-651 (1960).

INVESTMENT POLICIES FOR EXPANDING BUSINESSES OPTIMAL IN A LONG-RUN SENSE

Leo Breiman

University of California

An entrepeneur has a given initial fortune and faces the following situation: during any time period he may invest various amounts of his available fortune in various alternatives (differing from interval to interval) and keep the remainder of his money as a reserve fund. The alternatives materialize and pay off according to a probability distribution in which occurrences in different time intervals are independent. We analyze the problem of finding the investment policies (that is, the division of funds during every time interval between the investment alternatives and the reserve) that will make the long-run growth of the entrepeneur's fortune as rapid as possible.

INTRODUCTION AND SUMMARY

The situation we treat is as follows: at the beginning of the Nth investment period, an entrepeneur has fortune S_N and is prescribed with investment alternatives 1, 2, ..., n. Associated with these alternatives is a random pay-off vector $\overline{r}_N = (r_1, \ldots, r_n)$, $r_i \geq 0$, having some distribution in n-dimensional space, so that if the investor puts amounts $S_i, \ldots, S_n$ of his fortune into alternatives 1, ..., n, then his return at the end of the investment period from alternative i is $r_i\ S_i$. The question is whether there is any optimal investment policy, that is, any optimal way of dividing the available investment fund between the various alternatives. (We note that one of the alternatives could be to put so much money into a reserve fund which then draws a given interest rate.)

Of course, the optimal policies will vary with our criterion of optimality. One classical criterion is to maximize our expected return; that is, invest $S_i, \ldots, S_n, \sum_1^n S_r = S_N$ such that $E\left(\sum_1^n S_i\ r_i\right)$ = maximum. This criterion leads to the highly speculative policy: invest all of S_N in that alternative i such that $Er_i = \max E\ r_j$. A policy of this type is clearly untenable to any continuing investor or firm, since it usually leads to quick bankruptcy proceedings. For example, the alternative with maximum expected return may have $r = 0$ with high probability and r large with small probability. A repeated gambling of all available funds on this alternative will, with probability one, lead to ultimate ruin. One might say that, in this situation, the right thing to do during every investment period is to hold part of one's capital in reserve and invest the remainder in the alternative with the maximum expected return. The obvious objection to this is that the investor is no longer following his optimum policy; besides, what criterion is being used to decide what funds to hold in reserve?

We have discussed the undesirability of the above criterion from the point of view of ultimate ruin; that is, from the long-run point of view. In our opinion, most non-speculative investment policies implicitly take this view. In order to formalize this criterion, we assume that we start with an initial fortune S_0, and that S_{N+1}, our fortune at the beginning of the $N+1^{st}$ period, consists of the total returns from our investments of the Nth period; i.e. everything that we make during the Nth period is plowed back into investments during the $N+1^{st}$ period. This assumption is made in order to be able to compare the asymptotic behavior of S_N under different investment policies, and is approximately valid during the initial expansion period of business firms and in investment holdings where the dividends taken out of the system are small compared with the total dividends. Any well-formulated rule for the removal of funds from the system will evidently enable an asymptotic analysis of S_N; we use the above rule, however, for simplicity and illustration.

Concerning the investment alternatives, we assume very little; they may vary arbitrarily from period to period and the distribution of the pay-off vector r_N for the Nth period may be conditioned in any way by the past. Perhaps our most drastic assumption is that any division of funds between the various alternatives is an allowable investment policy. This is unrealistic in the many situations where there is either an upper limit to investment, or a lower limit, or both.

Naturally, our investment policy for the Nth period may depend both on S_N and on the outcomes of the past. Now let S_N be the fortune of the beginning of the Nth period under one sequence of investment policies and S'_N the fortune of the same time under a competing sequence of investment policies. We will say that the former sequence is inadmissible if there is a fixed number $\alpha > o$ such that for every $\epsilon > o$ there exists a competing sequence such that on a set of probability greater than $\alpha - \epsilon$

$$\lim_N \sup \frac{S_N}{S'_N} = 0$$

and, except for a set of probability at most ϵ,

$$\lim_N \sup \frac{S_N}{S'_N} \leq 1 .$$

Roughly, the sequence leading to S_N is inadmissible if for every $\epsilon > o$ there is a competing sequence that is asymptotically infinitely better with probability at least $\alpha-\epsilon$ and with, at most, probability ϵ of being worse.

Restricting ourselves to admissible sequences of policies, we domonstrate that there is a unique admissible sequence which is distinguished by the fact that every admissible sequence must be asymptotically close to it. This sequence is found as follows: let $\lambda_1 \ldots, \lambda_n$ be the proportions of S_N that we invest in alternatives 1, ..., N, then choose the λ_i so as to maximize

$$E \log \left(\sum_1^n \lambda_i \, r_i \right) , \quad \lambda_i > o, \quad \sum_1^n \lambda_i = 1 .$$

It is easily shown that this is equivalent to choosing the λ_i so that they satisfy

$$\lambda_i = E\left(\frac{\lambda_i r_i}{\sum_i \lambda_i r_i}\right).$$

Hence, we may verbally state this investment policy as: let the investment proportion in alternative i be equal to the expected proportion of the return coming from alternative i. It is interesting, as well as somewhat gratifying, that this policy is one of diversification; that is, in general, this policy is one of dividing S_N between a number of alternatives.

Finally, as far as the mathematics of the situation is concerned, our main tool is the ever-reliable martingale theorem.

MATHEMATICAL FORMULATION AND PROOFS

Let $\bar{\lambda}_N = (\lambda_1, \ldots, \lambda_n)$ be the investment proportion vector for the Nth period, i.e. λ_i is the proportion of S_N invested in alternative i; so using the notation $V_N = \sum_1^n \lambda_i r_i$ we have

$$S_{N+1} = S_N V_N .$$

Now let R_{N-1} be the outcomes during the first N-1 investment periods, $R_{N-1} = (\bar{r}_{N-1}, \ldots, \bar{r}_1)$. We consider two competing sequences of investment policies $(\bar{\lambda}_1, \bar{\lambda}_2, \ldots)$ and $(\bar{\lambda}_1^*, \bar{\lambda}_2^*, \ldots)$ such that under each we start with the same initial fortune S_0. Now λ_N, λ_N^* may depend in an arbitrary manner on R_{N-1} as may the distribution of r_N. Let S_N be the fortune under $(\bar{\lambda}_1, \ldots)$ and S_N^* the fortune under $(\bar{\lambda}_1^*, \ldots)$ where we take $\bar{\lambda}_N^*$ to be defined as that investment proportion vector which maximizes

$$E(\log V_N | R_{N-1}), \qquad V_N = \sum_i \lambda_i r_i, \ \lambda_i \geq 0, \sum_i \lambda_i = 1 .$$

We need some uniform integrability condition, and we will assume that there are constants α, β such that

$$0 < \alpha < r_i < \beta < \infty , \text{ all i, N.}$$

The theorem is true under much weaker conditions, but with the above condition, it reduces to a standard martingale theorem and thus keeps the technicalities at a minimum.

THEOREM 1: Under (A), limit of S_N/S_N^* exists almost surely and $E(\lim S_N/S_N^*) \leq 1$.

PROOF: We have

(A) $$E\left(\frac{S_N}{S_N^*} \,\middle|\, R_{N-1}\right) = E\left(\frac{V_N}{V_N^*} \,\middle|\, R_{N-1}\right) \frac{S_{N-1}}{S_{N-1}^*} .$$

Hence, it is sufficient to prove that $E\left(V_N/V_N^* \mid R_{N-1}\right) \leq 1$, for then S_N/S_N^* is a decreasing semimartingale with $E(\lim S_N/S_N^*) \leq E(S_0/S_0^*) = 1$ (see [1]). Now, for any $\varepsilon > 0$, by the maximizing property of $\bar{\lambda}_N^*$ and by (A), we have

$$E\left\{\log\left[(1-\varepsilon)V_N^* + \varepsilon V_N\right] - \log V_N^* \mid R_{N-1}\right\} < 0,$$

so that

$$\frac{1}{\varepsilon} E\left[\log\left(1 + \frac{\varepsilon}{1-\varepsilon}\frac{V_N}{V_N^*}\right)\Big|R_{N-1}\right] < \frac{1}{\varepsilon}\log\frac{1}{1-\varepsilon}.$$

By (A), V_N/V_N^* is bounded, and we let $\varepsilon \to 0$ to get the desired result.

THEOREM 2: The set on which $\lim_N S_N/S_N^* = 0$ is almost surely equal to the set on which

$$\sum_{N=1}^{\infty}\left[E\left(\log V_N^*|R_{N-1}\right) - E\left(\log V_N|R_{N-1}\right)\right] = \infty.$$

PROOF: The sequence

$$\log\frac{S_N}{S_N^*} - E\left(\log\frac{S_N}{S_N^*}\Big|R_{N-1}\right) = \sum_{k=1}^{N}\left[\log\frac{V_k}{V_k^*} - E\left(\log\frac{V_k}{V_k^*}\Big|R_{k-1}\right)\right]$$

is a martingale sequence with

$$\sup_{k \geq 1}\left|\log\frac{V_k}{V_k^*} - E\left(\log\frac{V_k}{V_k^*}\Big| R_{k-1}\right)\right| \leq \gamma < \infty.$$

Therefore (see [1], pp. 319-320), the sequence converges to a finite value almost surely, which yields theorem 2

THEOREM 3: Under (A) the sequence $(\lambda_1, \lambda_2, \ldots)$ is admissible if, and only if, almost surely,

(B) $$\sum_1^{\infty}\left[E(\log V_k^*|R_{k-1}) - E(\log V_k|R_{k-1})\right] < \infty.$$

PROOF: Suppose (B) is violated, then there is a set E of positive probability (i.e. $P(E) = \alpha > 0$) such that on E, $\lim_N S_N/S_N^* = 0$. For any $\varepsilon > 0$, there exists a set E_M, measurable with respect to R_M such that $P(E_M \,\Delta\, E) < \varepsilon$, where Δ denotes the symmetric set difference. We define a competing sequence of policies $(\bar{\lambda}_1', \bar{\lambda}_2' \ldots)$ as follows: if

$N-1 < M$, $\bar{\lambda}_N = \bar{\lambda}_N$, if R_{N-1}, $N-1 \geqslant M$, is such that $(\bar{r}_1, \ldots, \bar{r}_M)$ is not in E_M, then $\bar{\lambda}'_N = \bar{\lambda}_N$, otherwise $\bar{\lambda}'_N = \bar{\lambda}^*_N$. Now, on E_M,

$$\sum_1^\infty \left[E(\log V^*_k|R_{k-1}) - E(\log V'_k|R_{k-1})\right] < \infty$$

hence, $\lim S'_N/S^*_N > 0$ which implies that $\lim S_N/S'_N = 0$ on $E_M \cup E$. And we have that $P(E_M \cup E) > P(E) - \epsilon = \alpha - \epsilon$, on the complement of E_M, $S_N = S'_N$, so that $\lim S_N/S'_N \leqslant 1$ except with at most probability ϵ. Therefore, $(\bar{\lambda}_1, \bar{\lambda}_2 \ldots)$ is an inadmissible sequence.

Conversely, suppose (B) holds, but $(\bar{\lambda}_1, \ldots)$ is inadmissible with respect to $(\bar{\lambda}'_1 \ldots)$, and that $\lim S_N/S'_N = 0$ on E with $P(E) > 0$. Since (B) holds, $\lim S_N/S^*_N > 0$ almost surely, which implies that $\lim S^*_N/S'_N = 0$ almost surely on E. This implies, in turn, that $\lim S'_N/S^*_N = \infty$ with positive probability, which violates Theorem 1.

The above Thoerem is the result referred to in the introduction, for the essential content is, that unless $\bar{\lambda}_N$ is usually close to $\bar{\lambda}^*_N$ in the sense that $E(\log V^*_N|R_{N-1}) - E(\log V_N|R_{N-1})$ is small, then $(\bar{\lambda}_1, \ldots)$ is not admissible.

REFERENCE

[1] J. L. Doob, Stochastic Processes, John Wiley and Sons, New York, 1953.

* * *

ADDENDUM*

Dr. Breiman was kind enough to submit the following amended proof of Theorem 2. The proof on page 596 is incomplete because it does not consider A^c (see below).

Proof of Theorem 2 By definition

$$E(\log V_k - \log V_k^* \mid R_{k-1}) \leqq 0.$$

Define

$$X_N \equiv \sum_{k=1}^{N} \{\log V_k - \log V_k^* - E(\log V_k - \log V_k^* \mid R_{k-1})\};$$

then,

$$X_{N+1} - X_N = \log V_{N+1} - \log V_{N+1}^* - E(\log V_{N+1} - \log V_{N+1}^* \mid R_N)$$
$$\leqq \gamma < \infty \qquad \text{by assumption,}$$

and $\{X_N\}$ is a martingale.

Applying Theorem 4.1 (iv) (on pp. 319–320 of reference [1]) to the sequences $\{X_N\}$ and $\{-X_N\}$, we have that $\lim_{N\to\infty} X_N$ exists and is finite almost surely

* Adapted by S. Larsson and W. T. Ziemba from personal correspondence with Dr. Leo Breiman.

(a.s.) on the set

$$\left\{\overline{\lim_N}\, X_N < \infty\right\} \cup \left\{\underline{\lim_N}\, X_N > -\infty\right\}. \tag{E1}$$

By Theorem 1, $0 \leqq S_N/S_N{}^*$ converges a.s. to a finite value, and hence $\log S_N/S_N{}^*$ converges a.s. to a finite value or to $-\infty$.

Let A be the set on which $\lim_N (S_N/S_N{}^*) > 0$. We are then required to prove that

$$\sum_{N=1}^{\infty} E(\log V_N \mid R_{N-1}) - E(\log V_N{}^* \mid R_{N-1}) \neq -\infty.$$

Now, $\lim_N (S_N/S_N{}^*) > 0$ implies that $\lim_N \log(S_N/S_N{}^*) > -\infty$, or, equivalently, $\lim_N \sum_{k=1}^N \log(V_k/V_k{}^*) > -\infty$, and since $X_N \geqq \sum_{k=1}^N \log(V_k/V_k{}^*)$, it follows that $\underline{\lim}_N X_N > -\infty$, and by (E1) $\lim X_N$ exists and is finite. Hence

$$\sum_{N=1}^{\infty} E(\log V_N \mid R_{N-1}) - E(\log V_N{}^* \mid R_{N-1}) \neq -\infty.$$

On A^c either (1) $\overline{\lim}\, X_N < \infty$ or (2) $\lim X_N = \infty$. Now (1) means that $\lim_N (S_N/S_N{}^*) = 0$, or equivalently, that $\lim_N \sum_{k=1}^N \log(V_k/V_k{}^*) = -\infty$ but $\lim X_N$ exists and is finite; hence

$$\sum_{n=1}^{\infty} E(\log V_N \mid R_{N-1}) - E(\log V_N{}^* \mid R_{N-1}) = -\infty.$$

In the second case

$$\lim_N \frac{S_N}{S_N{}^*} = 0 \qquad \text{and} \qquad \sum_{k=1}^{\infty} E(\log V_k{}^* \mid R_{k-1}) - E(\log V_k \mid R_{k-1}) = \infty.$$

Hence in all cases $\lim_N (S_N/S_N{}^*) = 0$ iff

$$\sum_{k=1}^{\infty} E(\log V_k{}^* \mid R_{k-1}) - E(\log V_k \mid R_{k-1}) = \infty.$$

We would also like to record the following typographical and other minor errors:

Page[a]	Line	In paper	Change to
647 (593)	6 in abstract	are independent	may be dependent
	5, 12	S_t	S_1
648 (594)	27	$\leqq 1$	$\geqq 1$
	last	$> o$	$\geqq 0$
649 (595)	10	λ_i	λ_1
	last	$V_N/V_N{}^*$	V_{N-1}/V_{N-1}^*
650 (596)	8 from bottom	Theorem 3	Theorem 2
651 (597)	1	$\bar{\lambda}_N =$	$\bar{\lambda}_N{}' =$

[a] Numbers in parentheses indicate page numbers in this volume.

*Portfolio Choice and the Kelly Criterion**

Edward O. Thorp

UNIVERSITY OF CALIFORNIA AT IRVINE

I. Introduction

The Kelly (–Breiman–Bernoulli–Latané or capital growth) criterion is to maximize the expected value $E \log X$ of the logarithm of the random variable X, representing wealth. Logarithmic utility has been widely discussed since Daniel Bernoulli introduced it about 1730 in connection with the St. Petersburg game [3, 28]. However, it was not until certain mathematical results were proved in a limited setting by Kelly in 1956 [14] and then in an expanded and much more general setting by Breiman in 1960 and 1961 [5, 6] that logarithmic utility was clearly distinguished *by its properties* from other utilities as a guide to portfolio selection. (See also Bellman and Kalaba [2], Latané [15], Borch [4], and the very significant paper of Hakansson [11].)

Suppose for each time period ($n = 1, 2, \ldots$) there are k investment opportunities with results per unit invested denoted by the family of random variables $X_{n,1}, X_{n,2}, \ldots, X_{n,k}$. Suppose also that these random variables have only finitely many distinct values, that for distinct n the families are independent of each other, and that the joint probability distributions of distinct families (as subscripted) are identical. Then Breiman's results imply that portfolio strategies Λ which maximize $E \log X_n$, where X_n is the wealth at the end of the nth time period, have the following properties:

Property 1 (Maximizing $E \log X_n$ asymptotically maximizes the rate of asset growth.) If, for each time period, two portfolio managers have the same family of investment opportunities, or investment universes, and one uses a strategy Λ^* maximizing $E \log X_n$ whereas the other uses an "essentially different" [i.e., $E \log X_n(\Lambda^*) - E \log X_n(\Lambda) \to \infty$] strategy Λ, then $\lim X_n(\Lambda^*)/X_n(\Lambda) \to \infty$ almost surely (a.s.).

Property 2 The expected time to reach a fixed preassigned goal x is, asymptotically as x increases, least with a strategy maximzing $E \log X_n$.

The qualification "essentially different" conceals subtleties which are not generally appreciated. For instance, Hakansson [11], which is very close in

* This research was supported in part by the Air Force Office of Scientific Research under Grant AF-AFOSR 1870A. An expanded version of this paper will be submitted for publication elsewhere.

† Reprinted from the 1971 Business and Economics Statistics Section Proceedings of the American Statistical Association.

method to this article, and whose conclusions we heartily endorse, contains some mathematically incorrect statements and several incorrect conclusions, mostly from overlooking the requirement "essentially different." Because of lack of space we only indicate the problem here: If X_n is capital after the nth period, and if $x_j = X_j/X_0$, even though $E \log x_j > 0$ for all j, it need not be the case that $P(\lim x_j = \infty) = 1$. In fact, we can have (just as in the case of Bernoulli trials and $E \log x_j = 0$; see Thorp [26]) $P(\limsup x_j = \infty) = 1$ and $P(\liminf x_j = 0) = 1$ (contrary to Hakansson [11, p. 522, Eq. (18) and following assertions]). Similarly, when $E \log x_j < 0$ for all j we can have these alternatives instead of $P(\lim x_j = 0) = 1$ (contrary to Hakansson [11, p. 522, Eq. (17) and the following statements; footnote 1 is also incorrect]).

We note that with the preceding assumptions, there is a fixed fraction strategy Λ which maximizes $E \log X_n$. A fixed fraction strategy is one in which the fraction of wealth $f_{n,j}$ allocated to investment $X_{n,j}$ is independent of n.

We emphasize that Breiman's results can be extended to cover many if not most of the more complicated situations which arise in real-world portfolios. Specifically, the number and distribution of investments can vary with the time period, the random variables need not be finite or even discrete, and a certain amount of dependence can be introduced between the investment universes for different time periods.

We have used such extensions in certain applications (e.g., Thorp [25; 26, p. 287]).

We consider almost surely having more wealth than if an "essentially different" strategy were followed as the desirable objective for most institutional portfolio managers. (It also seems appropriate for wealthy families who wish mainly to accumulate and whose consumption expenses are only a small fraction of their total wealth.) Property 1 tells us that maximizing $E \log X_n$ is a recipe for approaching this goal asymptotically as n increases. This is to our mind the principal justification for selecting $E \log X$ as the guide to portfolio selection.

In any real application, n is finite, the limit is not reached, and we have $P(X_n(\Lambda^*)/X_n(\Lambda) > 1 + M) = 1 - \varepsilon(n, \Lambda, M)$, where $\varepsilon \to 0$ as $n \to \infty$, $M > 0$ is given, Λ^* is the strategy which maximizes $E \log X_n$, and Λ is an "essentially different" strategy. Thus in any application it is important to have an idea of how rapidly $\varepsilon \to 0$. Much work needs to be done on this in order to reduce $E \log X$ to a guide that is useful (not merely valuable) for portfolio managers. Some illustrative examples for $n = 6$ appear in the work of Hakansson [11].

Property 2 shows us that maximizing $E \log X$ also is appropriate for individuals who have a set goal (e.g., to become a millionaire).

Appreciation of the compelling properties of the Kelly criterion may have been impeded by certain misunderstandings about it that persist in the literature of mathematical economics.

The first misunderstanding involves failure to distinguish among kinds of utility theories. We compare and contrast three types of utility theories:

(1) *Descriptive*, where data on observed behavor are fitted mathematically. Many different utility functions might be needed, corresponding to widely varying circumstances, cultures, or behavior types.[1]

(2) *Predictive*, which "explains" observed data: Hypotheses are formulated from which fits for observed data are deduced; hopefully future data will also be found to fit. Many different utility functions may be needed, corresponding to the many sets of hypotheses that may be put forward.

(3) *Prescriptive* (also called *normative*), which is a guide to behavior, i.e., a recipe for optimally achieving a stated goal. It is not necessarily either descriptive or predictive nor is it intended to be so.

We use logarithmic utility in this last way, and much of the misunderstanding of it comes from those who think it is being proposed as a descriptive or a predictive theory. The $E \log X$ theory is a prescription for allocating resources so as to (asymptotically) maximize the rate of growth of assets. We assert that this is the appropriate goal in important areas of human endeavor and the theory is therefore important.

Another "objection" voiced by some economists to $E \log X$ and, in fact, to all unbounded utility functions, is that it does not resolve the (generalized) St. Petersburg paradox. The rebuttal is blunt and pragmatic: The generalized St. Petersburg paradox does not arise in the real world because any one real-world random variable is bounded (as is any finite collection). Thus in any real application the paradox does not arise.

To insist that a utility function resolves the paradox is an artificial requirement, certainly permissible, but obstructive and tangential to the goal of building a theory that is also a practical guide.

II. Samuelson's Objections to Logarithmic Utility

Samuelson [21, pp. 245–246; 22 pp. 4–5] says repeatedly, authorities (Williams [30], Kelly [14], Latané [15], Breiman [5, 6]) "have proposed a

[1] Information on descriptive utility is sparce; how many writers on the subject have even been able to determine for us their own personal utility?

drastic simplification of the decision problem whenever T (the number of investment periods)[2] is large.

"*Rule* Act in each period to maximize the geometric mean or the expected value of $\log x_t$.

"The plausibility of such a procedure comes from the recognition of the following valid asymptotic result.

"*Theorem* Acting to maximize the geometric mean at every step will, if the period is 'sufficiently long,' 'almost certaintly' result in higher terminal wealth and terminal utility than from any other decision rule....

"From this indisputable fact, it is apparently tempting to believe in the truth of the following false corollary:

"*False Corollary* If maximizing the geometric mean almost certainly leads to a better outcome, then the expected value utility of its outcomes exceeds that of any other rule, provided T is sufficiently large."

Samuelson then gives examples to show that the corollary is false. We heartily agree that the corollary is false. In fact, Thorp [26] had already shown this for one of the utilities Samuelson uses, for he noted that in the case of Bernoulli trials with probability $\frac{1}{2} < p < 1$ of success, one should commit a fraction $w = 1$ of his capital at each trial to maximize expected final gain EX_n ([26, p. 283]; the utility is $U(x) = x$) whereas to maximize $E \log X_n$ he should commit $w = 2p - 1$ of his capital at each trial [26, p. 285, Theorem 4].

The statements which we have seen in print supporting this "false corollary" are by Latané [15, p. 151, footnote 13] as discussed by Samuelson [21, p. 245, footnote 8] and Markowitz [16a, pp. ix and x]. Latané may not have fully supported this corollary for he adds the qualifier "... (in so far as certain approximations are permissible) ...."

That there were (or are?) adherents of the "false corollary," seems puzzling in view of the following formulation. Consider a T-stage investment process. At each stage we allocate our resources among the available investments. For each sequence A of allocations which we choose, there is a corresponding terminal probability distribution F_T^A of assets at the completion of stage T. For each utility function $U(\cdot)$, consider those allocations $A^*(U)$ which

[2] Parenthetical explanation added since we have used n.

maximize the expected value of terminal utility $\int U(x)\, dF_T{}^A(x)$. Assume sufficient hypotheses on U and the set of $F_T{}^A$ so that the integral is defined and that furthermore the maximizing allocation $A^*(U)$ exists. Then Samuelson says that $A^*(\log)$ is not in general $A^*(U)$ for other U. This seems intuitively evident.

Even more seems strongly plausible: that if U_1 and U_2 are inequivalent utilities, then $\int U_1(x)\, dF_T{}^A(x)$ and $\int U_2(x)\, dF_T{}^A(x)$ will in general be maximized for different $F_T{}^A$. (Two utilities U_1 and U_2 are equivalent if and only if there are constants a and b such that $U_2(x) = aU_1(x)+b$, $a > 0$; otherwise U_1 and U_2 are inequivalent.) In this connection we have proved [27a].

Theorem Let U and V be utilities defined and differentiable on $(0, \infty)$, with $0 < U'(x)$, $V'(x)$, and $U'(x)$ and $V'(x)$ strictly decreasing as x increases. Then if U and V are inequivalent, there is a one-period investment setting such that U and V have distinct optimal strategies.[3]

All this is in the nature of an aside, for Samuelson's correct criticism of the "false corollary" does not apply to our use of logarithmic utility. Our point of view is this: If your goal is Property 1 or Property 2, then a recipe for achieving either goal is to maximize $E \log X$. It is these properties which distinguish log for us from the prolixity of utility functions in the literature. Furthermore, we consider these goals appropriate for many (*but not all*) investors. Investors with other utilities, or with goals incompatible with logarithmic utility, will of course find it inappropriate for them.

Property 1 imples that if Λ^* maximizes $E \log X_n(\Lambda)$ and Λ' is "essentially different," then $X_n(\Lambda^*)$ tends almost certainly to be better than $X_n(\Lambda')$ as $n \to \infty$. Samuelson says [21, p. 246] after refuting the "false corollary": "Moreover, as I showed elsewhere [20, p. 4], the ordering principle of selecting between two actions in terms of which has the greater probability of producing a higher result does not even possess the property of being transitive. ... we could have w^{***} better than w^{**}, and w^{**} better than w^*, and also have w^* better than w^{***}."

For some entertaining examples, see the discussion of nontransitive dice by Gardner [9]. [Consider the dice with equiprobable faces numbered as follows: $X = (3, 3, 3, 3, 3, 3)$, $Y = (4, 4, 4, 4, 1, 1)$, $Z = (5, 5, 2, 2, 2, 2)$. Then $P(Z > Y) = \frac{5}{9}$, $P(Y > X) = \frac{2}{3}$, $P(X > Z) = \frac{2}{3}$.] What Samuelson does not tell

[3] We have since generalized this to the case where $U''(x)$ and $V''(x)$ are piecewise continuous.

us is that the property of producing a higher result *almost certainly*, as in Property 1, *is* transitive. If we have $w^{***} > w^{**}$ almost certainly, and $w^{**} > w^{*}$ almost certainly, then we must have $w^{***} > w^{*}$ almost certainly.

One might object [20, p. 6] that in a real investment sequence the limit as $n \to \infty$ is not reached. Instead the process stops at some finite N. Thus we do not have $X_N(\Lambda^*) > X_n(\Lambda')$ almost surely. Instead we have $P(X_n(\Lambda^*) > X_n(\Lambda')) = 1 - \varepsilon_N$ where $\varepsilon_N \to 0$ as $N \to \infty$, and transitivity can be shown to fail.

This is correct. But then an approximate form of transitivity does hold: Let X, Y, Z be random variables with $P(X > Y) = 1 - \varepsilon_1$, $P(Y > Z) = 1 - \varepsilon_2$. Then $P(X > Z) \geqq 1 - (\varepsilon_1 + \varepsilon_2)$. To prove this, let A be the event $X > Y$, B the event $Y > Z$, and C the event $X > Z$. Then $P(A) + P(B) = P(A \cup B) + P(A \cap B) \leqq 1 + P(A \cap B)$. But $A \cap B \subset C$ so $P(C) \geqq P(A \cap B) = P(A) + P(B) - 1$, i.e., $P(X > Z) \geqq 1 - (\varepsilon_1 + \varepsilon_2)$.

Thus our approach is not affected by the various Samuelson objections to the uses of logarithmic utility.

Markowitz [16a, p. ix and x] says "... in 1955–1956, I concluded... that the investor who is currently reinvesting everything for 'the long run' should maximize the expected value of the logarithm of wealth." (This assertion seems to be regardless of the investor's utility and so indicates belief in the false corollary.) "Mossin [18] and Samuelson [20] have each shown that this conclusion is not true for a wide range of [utility] functions .... The fascinating Mossin–Samuelson result, combined with the straightforward arguments supporting the earlier conclusions, seemed paradoxical at first. I have since returned to the view of Chapter 6 (concluding that: for large T, the Mossin–Samuelson man acts absurdly, ..." Markowitz says here, in effect, that alternate utility functions (to $\log x$) are absurd. This position is unsubstantiated and unreasonable.

He continues "...like a player who would pay an unlimited amount for the St. Petersburg game...." If you agree with us that the St. Petersburg game is not realizable and may be ignored when fashioning utility theories for the real world, then his continuation "...the terminal utility function must be bounded to avoid this absurdity; ..." does not follow.

Finally, Markowitz says "...and the argument in Chapter 6 applies when utility of terminal wealth is bounded." If he means by this that the false corollary holds if we restrict ourselves to bounded utility functions, then he is mistaken. Mossin [18] already showed that the optimal strategies for $\log x$ and x^γ/γ, $\gamma \neq 0$, are a fixed fraction for these and only these utilities. Thus any bounded utility besides x^γ/γ, $\gamma < 0$, will have optimal strategies which are *not* fixed fraction, hence not optimal for $\log x$. Samuelson [22] gives counterexamples including the bounded utilities x^γ/γ, $\gamma < 0$. Also the counterexamples satisfying the hypotheses of our theorem include many bounded utilities [e.g., $U(x) = \tan^{-1} x$, $x > 0$, and $V(x) = 1 - e^{-x}$, $x \geqq 0$].

III. An Outline of the Theory of Logarithmic Utility

The simplest case is that of Bernoulli trials with probability p of success, $0 < p < 1$. The unique strategy which maximizes $E \log X_n$ is to bet at trial n the fixed fraction $f^* = p - q$ of total current wealth X_{n-1}, if $p > \frac{1}{2}$ (i.e., if the expectation is positive), and to bet nothing otherwise.

To maximize $E \log X_n$ is equivalent to maximizing $E \log[X_n/X_0]^{1/n} \equiv G(f)$, which we call the (exponential) rate of growth (per time period). It turns out that for $p > \frac{1}{2}$, $G(f)$ has a unique positive maximum at f^* and that there is a critical fraction f_c, $0 < f^* < f_c < 1$, such that $G(f_c) = 0$, $G(f) > 0$ if $0 < f < f_c$, $G(f) < 0$ if $f_c < f \leqq 1$ (we assume "no margin"; the case with margin is similar). If $f < f_c$, $X_n \to \infty$ a.s.; if $f = f_c$, $\limsup X_n = +\infty$ a.s. and $\liminf X_n = 0$ a.s.; if $f > f_c$, $\lim X_n = 0$ a.s. ("ruin").

The Bernoulli trials case is particularly interesting because it exhibits many of the features of the following more general case. Suppose we have at each trial $n = 1, 2, \ldots$ the k investment opportunities $X_{n,1}, X_{n,2}, \ldots, X_{n,k}$ and that the conditions of Property 1 (Section I) are satisfied. This means that the joint distributions of $\{X_{n,i_1}, X_{n,i_2}, \ldots, X_{n,i_j}\}$ are the same for all n, for each subset of indices $1 \leqq i_1 < i_2 < \cdots < i_j \leqq k$. Furthermore $\{X_{m,1}, \ldots, X_{m,k}\}$ and $\{X_{n,1}, \ldots, X_{n,k}\}$ are independent, and all random variables $X_{i,j}$ have finite range. Thus we have in successive time periods repeated independent trials of "the same" investment universe.

Since Breiman has shown that there is for this case an optimal fixed fraction strategy $\Lambda^* = (f_1^*, \ldots, f_k^*)$, we will have an optimal strategy if we find a strategy which maximizes $E \log X_n$ in the class of fixed fraction strategies.

Let $\Lambda = (f_1, \ldots, f_k)$ be any fixed fraction strategy. We assume that $f_1 + \cdots + f_k \leqq 1$ so there is no borrowing, or margin. The margin case is similar (the approach resembles that of Schrock [23]). Using the concavity of the logarithm, it is easy to show (see below) that the exponential rate of growth $(1/n) E \log X_n(\Lambda) = G(f_1, \ldots, f_k)$ is a concave function of $(f_1, \ldots, f_k)$, just as in the Bernoulli trials case. The domain of $G(f)$ in the Bernoulli trials case was the interval $[0, 1)$ with $G(f) \downarrow -\infty$ as $f \to 1$. The domain in the present instance is analogous. First, it is a subset of the k-dimensional simplex $S_k = \{(f_1, \ldots, f_k): f_1 + \cdots + f_k \leqq 1; f_1 \geqq 0, \ldots, f_k \geqq 0\}$.

To establish the analogy further, let $R_j = X_{n,j} - 1$, $j = 1, \ldots, k$, be the return per unit on the ith investment opportunity at an arbitrary time period n.

Let the range of R_j be $\{r_{j,1};\ldots;r_{j,i_j}\}$ and let the probability of the outcome $[R_1 = r_{1,m_1}$ and $R_2 = r_{2,m_2}$ and so on, up to $R_k = r_{k,m_k}]$ be $p_{m_1 m_2,\ldots,m_k}$. Then

$$\begin{aligned} &E \log X_n/X_{n-1} \\ &\quad = G(f_1,\ldots,f_k) \\ &\quad = \sum \{p_{m_1,\ldots,m_k} \log(1+f_1 r_{1,m_1}+\cdots+f_k r_{k,m_k}) \colon 1 \leqq m_1 \leqq i_1;\ldots;1 \leqq m_k \leqq i_k\}, \end{aligned}$$

from which the concavity of $G(f_1,\ldots,f_k)$ can be shown. Note that $G(f_1,\ldots,f_k)$ is defined if and only if $1+f_1 r_{1,m_1}+\cdots+f_k r_{k,m_k} > 0$ for each set of indices $m_1,\ldots,m_k$. Thus the domain of $G(f_1,\ldots,f_k)$ is the intersection of all these open half-spaces with the k-dimensional simplex S_k. Note that the domain is convex and includes all of S_k in some neighborhood of the origin. Note too that the domain of G is all of S_k if (and only if) $R_j > -1$ for all j, i.e., if there is no probability of total loss on any investment. The domain of G includes the interior of S_k if $R_j \geqq -1$. Both domains are particularly simple and most cases of interest are included.

If $f_1,\ldots,f_k$ are chosen so that $1+f_1 r_{1,m_1}+\cdots+f_k r_{k,m_k} \leqq 0$ for some $m_1,\ldots,m_k$, then $P(f_1 X_{n,1}+\cdots+f_k X_{n,k} \leqq 0) = \varepsilon > 0$ for all n and ruin occurs with probability 1.

Computational procedures for finding an optimal fixed fraction strategy (generally unique in our present setting) are based on the theory of concave (dually, convex) functions [29] and will be presented elsewhere. (As Hakansson [11, p. 552] has noted, "...the computational aspects of the capital growth model are [presently] much less advanced" than for the Markowitz model.) A practical computational approach for determining the $f_1,\ldots,f_k$ to good approximation is given in [15a].

The theory may be extended to more general random variables and to dependence between different time periods. Most important, we may include the case where the investment universe changes with the time period, provided only that there be some mild regularity condition on the $X_{i,j}$, such as that they be uniformly a.s. bounded. (See the discussion by Latané [15], and the generalization of the Bernoulli trials case as applied to blackjack betting by Thorp [26].) The techniques rely heavily on those used to so generalize the law of large numbers.

Transaction costs, the use of margin, and the effect of taxes can be incorporated into the theory. Bellman's dynamic programming method is used here.

The general procedure for developing the theory into a practical tool imitates Markowitz [16]. Markowitz requires as inputs estimates of the expectations, standard deviations, and covariances of the $X_{i,j}$. We require joint probability distributions. This would seem to be a much more severe requirement, but in practice does not seem to be so [16, pp. 193–194, 198–201).

Among the actual inputs that Markowitz [16] chose were (1) past history [16, Example, pp. 8–20], (2) probability beliefs of analysts [16, pp. 26–33], and (3) models, most notably regression models, to predict future performance from past data [16, pp. 33, 99–100]. In each instance one can get enough additional information to estimate $E\log(X_n/X_{n-1})$.

There are, however, two great difficulties which all theories of portfolio selection have, including ours and that of Markowitz. First, there seems to be no established method for generally predicting security prices that gives an edge of even a few percent. The random walk is the best model for security prices today (see Cootner [7] and Granger and Morgenstern [10]).

The second difficulty is that for portfolios with many securities the volume of inputs called for is prohibitive: For 100 securities, Markowitz requires 100 expectations and 4950 covariances; and our theory requires somewhat more information. Although considerable attention has been given to finding condensed inputs that can be used instead, this aspect of portfolio theory still seems unsatisfactory.

In Section V we show how both these difficulties were overcome in practice by an institutional investor. That investor, guided by the Kelly criterion, then outperformed for the year 1970 every one of the approximately 400 Mutual Funds listed by the S & P stock guide!

But first we relate our theory to that of Markowitz.

IV. Relation to the Markowitz Theory; Solution to Problems Therein

The most widely used guide to portfolio selection today is probably the Markowitz theory. The basic idea is that a portfolio P_1 is superior to a portfolio P_2 if the expectation ("gain") is at least as great, i.e., $E(P_1) \geqq E(P_2)$, and the standard deviation ("risk") is no greater, i.e., $\sigma(P_1) \leqq \sigma(P_2)$, with at least one inequality. This partially orders the set $\mathscr{P}$ of portfolios. A portfolio such that no portfolio is superior (i.e., a maximal portfolio in the partial ordering) is called *efficient*. The goal of the portfolio manager is to determine the set of efficient portfolios, from which he then makes a choice based on his needs.

This is intuitively very appealing: It is based on standard quantities for the securities in the portfolio, namely expectation, standard deviation, and covariance (needed to compute the variance of the portfolio from that of the component securities). It also gives the portfolio manager "choice."

As Markowitz [16, Chapter 6] has pointed out, the optimal Kelly portfolio is approximately one of the Markowitz efficient portfolios under certain circumstances. If $E = E(P)$ and $R = P-1$ is the return per unit of the portfolio P, let $\log P = \log(1+R) = \log((1+E)+(R-E))$. Expanding in Taylor's series about $1+E$ gives

$$\log P = \log(1+E) + \frac{R-E}{1+E} - \frac{1}{2}\frac{(R-E)^2}{(1+E)^2} + \text{higher order terms.}$$

Taking expectations and neglecting higher order terms gives

$$E \log P = \log(1+E) - \frac{1}{2}\frac{\sigma^2(P)}{(1+E)^2}.$$

This leads to a simple pictorial relationship with the Markowitz theory. Consider the E–σ plane, and plot $(E(P), \sigma(P))$ for the efficient portfolios. The locus of efficient portfolios is a convex nondecreasing curve which includes its endpoints (Fig. 1).

Then constant values of the growth rate $G = E \log P$ approximately satisfy

$$G = \log(1+E) - \frac{1}{2}\frac{\sigma^2(P)}{(1+E)^2}.$$

This family of curves is illustrated in Fig. 1 and the (efficient) portfolio which maximizes logarithmic utility is approximately the one which lies on the greatest G curve. Because of the convexity of the curve of efficient portfolios and the concavity of the G curves, the (E, σ) value where this occurs is unique.

The approximation to G breaks down badly in some significant practical settings, including that of the next section. But for portfolios with large numbers of "typical" securities, the approximation for G will generally provide an (efficient) portfolio which approximately maximizes asset growth. This solves the portfolio manager's problem of which Markowitz-efficient portfolio to choose. Also, if he repeatedly chooses his portfolio in successive time periods by this criterion, he will tend to maximize the rate of growth of his assets, i.e., maximize "performance." We see also that in this instance the problem is reduced to that of finding the Markowitz-efficient portfolios plus the easy step of using Fig. 1. Thus if the Markowitz theory can be applied in practice in this setting, so can our theory. We have already remarked on the ambiguity of the set of efficient portfolios, and how our theory resolves them. To illustrate further that such ambiguity represents a defect in the Markowitz theory, let X_1 be uniformly distributed over [1, 3], let X_2 be uniformly distributed over [10, 100], let $\operatorname{cor}(X_1, X_2) = 1$, and suppose these are the only securities. Then X_1 and X_2 are both efficient with $\sigma_1 < \sigma_2$ and $E_1 < E_2$ so the Markowitz theory does not choose between them. Yet "everyone" would choose X_2 over X_1 because the worst outcome with X_2 is far

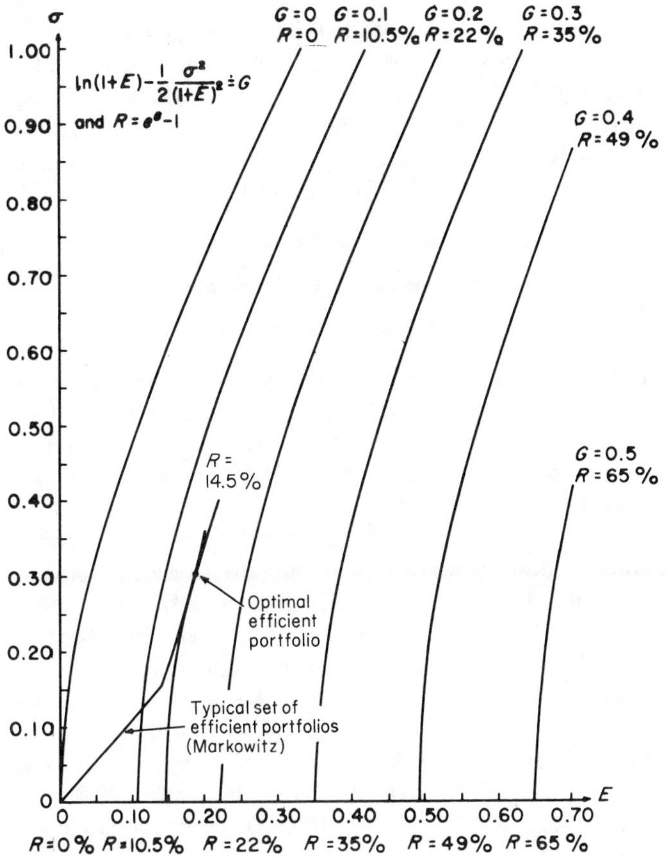

Fig. 1. Growth rate G (return rate R) in the E–σ plane assuming the validity of the power series approximation.

better than the best outcome from X_1. (We presented this example in Thorp [26]. Hakansson [11] presents further examples and extended analysis. He formalizes the idea by introducing the notion of stochastic dominance: X stochastically dominates Y if $P(X \geqq Y) = 1$ and $P(X > Y) > 0$. It is easy to see that

Lemma An $E \log X$ optimal portfolio is never stochastically dominated.

Thus our portfolio theory does not have this defect.)

There are investment universes $(X_1, \ldots, X_n)$ such that a unique portfolio P maximizes $E \log P$, yet P is not efficient in the sense of Markowitz. Then choosing P in repeated independent trials will outperform any strategy limited to choosing efficient portfolios. In addition, the optimal Kelly strategy gives positive growth rate, yet some of the Markowitz-efficient strategies give negative growth rate and ruin after repeated trials. Thorp [26] gave such an example and another appears in the work of Hakansson [11]. See also Hakansson [11, pp. 553–554] for further discussion of defects in the Markowitz model.

V. The Theory in Action: Results for a Real Institutional Portfolio

The elements of a practical profitable theory of convertible hedging were published by Thorp and Kassouf [27]. Thorp and Kassouf indicated an annualized return on investments of the order of 25% per year. Since then the theory has been greatly extended and refined [26] with most of these new results thus far unpublished.

The historical data which have been used to develop the theory include for warrants monthly observations for about 20 years, averaging about 20 warrants, or about 4800 observations, plus weekly observations for three years on an average of about 50 warrants, another 7500 observations. Each of these more than 12,000 observations is an n-tuple including price of common, price of warrant, dividend, dilution, time to expiration, and several other quantities.

The studies also used weekly data for three years on an average of 400 convertible bonds and 200 convertible preferreds, or about 90,000 observations. Including still other data, well over 100,000 observations have been incorporated into the study.

A convertible hedge transaction generally involves *two* securities, one of which is convertible into the other. Certain mathematical price relationships exist between pairs of such securities. When one of the two is underpriced, compared to the other, a profitable convertible hedge may be set up by buying the relatively underpriced security and selling short an appropriate amount of the relatively overpriced security.

The purpose of selling short the overpriced security is to reduce the risk in the position. Typically, one sells short in a single hedge from 50 to 125% as much stock (in "share equivalents") as is held long. The exact proportions depend on the analysis of the specific situation; the quantity of stock sold short is selected to minimize risk. The risk (i.e., change in asset value with fluctuations in market prices) in a suitable convertible hedge should be much less than in the usual stock market long positions.

TABLE I

PERFORMANCE RECORD

Date	Chg. to Date (%)[a]	Growth rate to date[b]	Elapsed time (months)	Closing DJIA[c]	DJIA Chg. (%)[a]	Starting even with DJIA[c]	Gain over DJIA (%)[a]
11-3-69	0.0	—	0	855	0.0	855	0.0
12-31-69	+4.0	+26.8	2	800	−6.3	889	+10.3
9-1-70	+14.0	+17.0	10	758	−11.3	974	+25.3
12-31-70	+21.0	+17.7	14	839	−1.8	1034	+22.8
3-31-71	+31.3	+21.2	17	904	+5.8	1123	+25.5
6-30-71	+39.8	+22.3	20	891	+4.2	1196	+35.6
9-30-71	+49.4	+23.3	23	887	+3.7	1278	+45.7
12-31-71	+61.3	+24.7	26	890	+4.1	1379	+57.2
3-31-72	+69.5	+24.4	29	940	+9.9	1449	+59.6
6-30-72	+75.0	+23.3	32	934	+9.3	1496	+65.7
9-30-72	+78.9	+22.1	35	960	+12.3	1526	+66.6
12-31-72	+84.5	+21.3	38	1020	+19.3	1577	+65.2
3-31-73	+90.0	+20.7	41	957	+11.9	1625	+78.1
6-30-73	+91.6	+19.4	44	891	+4.2	1638	+87.4
9-30-73	+100.3	+19.4	47	947	+10.7	1713	+89.6
12-31-73	+102.9	+17.7	52	851	−0.5	1734	+103.4

[a] Round to nearest 0.1%.
[b] Compound growth rate, annualized.
[c] Round to nearest point; DJIA = Dow Jones industrial average.

The securities involved in convertible hedges include common stock, convertible bonds, convertible preferreds, and common stock purchase warrants. Options such as puts, calls, and straddles may replace the convertible security. For this purpose, the options may be either written or purchased.

The reader's attention is directed to three fundamental papers on the theory of options and convertibles which have since been published by Black and Scholes [3a, 3b] and by Merton [16b].

The theory of the convertible hedge is highly enough developed so that the probability characteristics of a single hedge can be worked out based on an assumption for the underlying distribution of the common. (Sometimes even this can almost be dispensed with! See Thorp and Kassouf [27, Appendix C].) A popular and plausible assumption is the random walk hypothesis: that the future price of the common is log-normally distributed about its current price, with a trend and a variance proportional to the time. Plausible estimates of these parameters are readily obtained. Furthermore, it turns out that the return from the hedge is comparatively insensitive to changes in the estimates for these parameters.

Thus with convertible hedging we fulfill two important conditions for the practical application of our (or any other) theory of portfolio choice: (1) We have identified investment opportunities which are markedly superior to the usual ones. Compare the return rate of 20–25% per year with the long-term rate of 8% or so for listed common stocks. Further, it can be shown that the risks tend to be much less. (2) The probability inputs are available for computing $G(f_1, \ldots, f_n)$.

On November 3, 1969, a private institutional investor decided to commit all its resources to convertible hedging and to use the Kelly criterion to allocate its assets. Since this article was written, this institution has continued to have a positive rate of return in every month; on 12–31–73 the cumulative gain reached +102.9% and the DJIA equivalent reached 1734, whereas the DJIA was at 851, off 0.5%. The performance record is shown in Table I.

The market period covered included one of the sharpest falling markets as well as one of the sharpest rising markets (up 50% in 11 months) since World War II. The gain was +16.3% for the year 1970, which outperformed all of the approximately 400 Mutual Funds listed in the S & P stock guide. Unaudited figures show that gains were achieved during every single month.

(*Note added in proof:* Gains continued consistent in 1974. On 12–31–74 the cumulative gain reached +129%, the DJIA equivalent was 1960, and the DJIA reached +218%. Proponents of efficient market theory, please explain.)

The unusually low risk in the hedged positions is also indicated by the results for the 200 completed hedges. There were 190 winners, 6 break-evens, and 4 losses. The losses as a percentage of the long side of the specific investment ranged from 1 to 15%.

A characteristic of the Kelly criterion is that as risk decreases and expectation rises, the optimal fraction of assets to be invested in a single situation may become "large." On several occasions, the institution discussed above invested up to 30% of its assets in a single hedge. Once it invested 150% of its assets in a single arbitrage. This characteristic of Kelly portfolio strategy is not part of the behavior of most portfolio managers.

To indicate the techniques and problems, we consider a simple portfolio with just one convertible hedge. We take as our example Kaufman and Broad common stock and warrants. A price history is indicated in Fig. 2.

The figure shows that the formula $W = 0.455S$ is a reasonable fit for $S \leqq 38$ and that $W = S - 21.67$ is a reasonable fit for $S \geqq 44$. Between $S = 38$ and $S = 44$ we have the line $W = 0.84x - 15.5$. For simplicity of calculation we

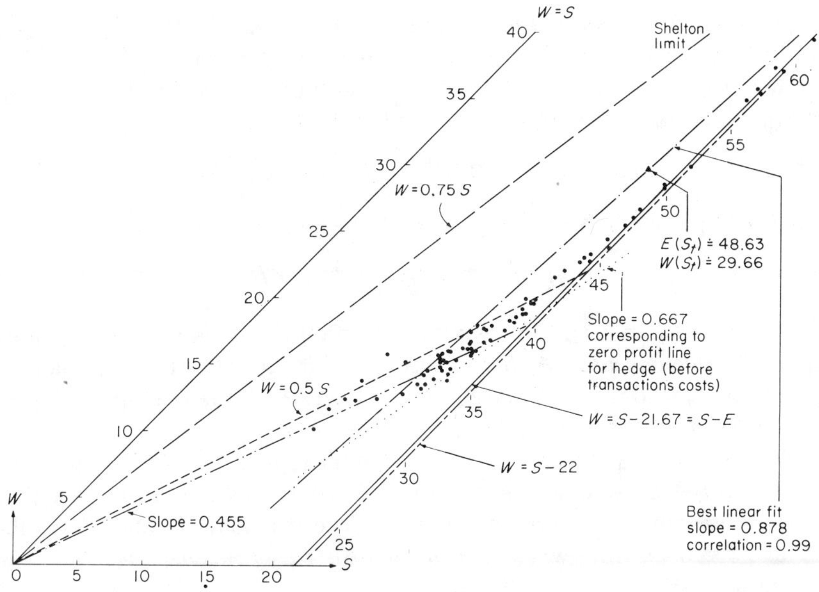

Fig. 2. Price history of Kaufman and Broad common S versus the warrants W. The points moved up and to the right until they reached the neighborhood of (38, 17). At this point a 3 : 2 hedge (15,000 warrants long, 11,200 common short) was instituted. As the points continued to move up and to the right during the next few months, the position was closed but in stages with the final liquidation at about (58, 36). Terms of warrant: one warrant plus \$21.67 → one share common stock until 3–1–74. Full protection against dilution. The company has the right to reduce the exercise price for temporary periods; 750,000 warrants and 5,940,000 common outstanding. Common dividends Q.05ex 10–26–70.

replace this in our illustrative analysis by $W = 0.5S$ if $S \leqq 44$ and $W = S - 22$ if $S \geqq 44$. The lines are also indicated in Fig. 2.

Past history at the time the hedge was instituted in late 1970 supported the fit for $S \leqq 38$. The conversion feature of the warrant ensures $W \geqq S - 21.67$ until the warrant expires. Thus $W = S - 21.67$ for $S \geqq 44$ underestimates the price of the warrant in this region. Extensive historical studies of warrants [12, 13, 24, 27] show that the past history fit would probably be maintained until about two years before expiration, i.e., until about March 1972. Thus it is plausible to assume that for the next 1.3 years, the S may be roughly approximated by $W = 0.5S$ for $S \leqq 44$ and $W = S-22$ for $S \geqq 44$.

Next we assume that S_t, the stock price at time $t > 0$ years after the hedge was initiated, is log-normally distributed with density

$$f_{S_t}(x) = \left(x\sigma\sqrt{2\pi}\right)^{-1} \exp[-(\ln x-\mu)^2/2\sigma^2],$$

mean $E(S_t) = \exp(\mu+\sigma^2/2)$, and standard deviation

$$\sigma(S_t) = E(S_t)(\exp(\sigma^2)-1)^{1/2}.$$

The functions $\mu \equiv \mu(t)$ and $\sigma \equiv \sigma(t)$ depend on the stock and on the time t. If t is the time in years until S_t is realized, we will see below that it is plausible to assume $\mu_t = \log S_0 + mt$ and $\sigma_t^2 = a^2 t$, where S_0 is the present stock price and m and a are constants depending on the stock.

Then $E(S_t) = S_0 \exp[(m+a^2/2)t]$ and a mean increase of 10% per year is approximated by setting $m+a^2/2 = 0.1$. If we estimate a^2 from past price changes, we can solve for m. In the case of Kaufman and Broad it is plausible to take $\sigma \doteq 0.45$, from which $a^2 = \sigma^2 \doteq 0.20$. This yields $m = 0$.

The standard deviation is then

$$\sigma(S_t) = S_0 \exp[(m+a^2/2)t][\exp(a^2 t)-1]^{1/2} \doteq 0.52 S_0.$$

It is generally agreed that the serial correlation of stock-price changes is very weak, and that changes in a stock-price series are approximately independent if the time intervals are nonoverlapping [7, 10]. If the changes in unit time for a stock were bounded independent identically distributed random variables, the central limit theorem would lead to the normal approximation with the mean and variance of a change proportional to the time. But this has difficulties. For instance, the change is bounded below because stock prices are nonnegative. Also, the magnitude of the price change per unit time is in fact dependent on the current price, increasing as the price increases.

A more realistic model which eliminates these difficulties and seems more plausible, is to assume that the price changes are proportional to the current price. This leads to the hypothesis that the changes in the logarithms of the prices are bounded identically distributed independent random variables, i.e., that $\log S_t - \log S_0$ is normally distributed with $\mu(t) = mt$ (hence $\log S_t$ has mean $\log S_0 + mt$) and $\sigma(t)^2 = a^2 t$. This is a sketch of the thinking behind our assumptions in the present example. For a detailed discussion, see Osborne [19] and Ayres [1a].

It is by no means established that the log-normal model is the appropriate one for stock-price series [10, Chapter 7]. However, once we clarify certain general principles by working through our example on the basis of the log-normal model, it can be shown that the results are substantially unchanged by choosing instead any distribution that roughly fits observation!

For a time of one year, a computation shows the return $R(S)$ on the stock to be $+10.5\%$, the return $R(W)$ on the warrant to be $+34.8\%$, $\sigma(S) = 0.52$, $\sigma(W) = 0.92$, and the correlation coefficient $\text{cor}(S, W) = 0.99$. The difference in $R(S)$ and $R(W)$ shows that the warrant is a much better buy than the common. Thus a hedge of long warrants and short common has a substantial positive expectation. The value $\text{cor}(S, W) = 0.99$ shows that a hedge corresponding to the best linear fit of W to S has a standard deviation of approximately $(1-0.99)^{1/2} = 0.1$, which suggests that $\sigma(P)$ for the optimal hedged portfolio is probably going to be close to 0.1. The high return and low risk for the hedge will remain, it can be shown, under wide variations in the choice of m and a.

To calculate the optimal mix of warrants long to common short we maximize $G(f_1, f_2) = E \log(1 + f_1 S + f_2 W)$. The detailed computational procedures are too lengthy and involved to be presented here. We hope to publish them elsewhere.

The actual decision made by our institutional investor took into consideration other positions already held, some of which might have to be closed out to release assets, and also those which were currently candidates for investment. The position finally taken was to short common and buy warrants in the ratio of three shares to four. The initial market value of the long side was about 14% of assets and the initial market value of the short side was about 20% assets. The profit realized was, in terms of the initial market value of the long side, about 20% in six months. This resulted from a move in the common from about 40 to almost 60.

VI. Concluding Remarks

As we remarked above, we do not propose logarithmic utility as descriptive of actual investment behavior, nor do we believe any one utility function could suffice. It would be of interest, however, to have empirical evidence showing areas of behavior which are characterized adequately by logarithmic utility. Neither do we intend logarithmic utility to be predictive; again, it would be of interest to know what it does predict.

We only propose the theory to be normative or prescriptive, and only for those institutions, groups or individuals whose overriding current objective is maximization of the rate of asset growth. Those with a different "prime directive" may find another utility function which is a better guide.

We remark that $E \log X$ has in our experience been a valuable qualitative guide and we suggest that this could be its most important use. Once familiarity with its properties is gained, our experience suggests that many investment decisions can be guided by it without complex supporting calculations.

It is interesting to inquire into the sort of economic behavior is to be expected from followers of $E \log X$. We find that insurance is "explained," i.e., that even though it is a negative expectation investment for the insured and we assume both insurer and insured have the same probability information, it is often optimal for him (as well as for the insurance company) to insure [3]. It usually turns out that insurance against large losses is indicated and insurance against small losses is not. (Do not insure an old car for collision, take $200 deductible, not $25, etc.)

We find that if all parties to a security transaction are followers of $E \log X$, they will often find it mutually optimal to make securities transactions. This may be true whether the transactions be two party (no brokerage), or three party (brokerage), and whether or not they have the same probability information about the security involved, or even about the entire investment universe.

Maximizing logarithmic utility excludes portfolios which have positive probability of total loss of assets. Yet it can be argued that an impoverished follower of $E \log X$ might in some instances risk "everything." This agrees with some observed behavior, but is not what we might at first expect in view of the prohibition against positive probability of total loss. Consider each individual as a piece of capital equipment with an assignable monetary value. Then if he risks and loses all his cash assets, he hasn't really lost everything [3].

All of us behave as though death itself does not have infinite negative utility. Since the risk of death, although generally small, is ever present, a negative infinite utility for death would make all expected utilities negative infinite and utility theory meaningless. In the case of logarithmic utility as applied to the extended case of the (monetized) individual plus all his resources, death has a finite, though large and negative, utility. The value of this "death constant" is an additional arbitrary assumption for the enlarged theory of logarithmic utility.

In the case of investors who behave according to $E \log X$ (or other utilities unbounded below), it might be possible to discover their tacit "death constants."

Hakansson [11, p. 551] observes that logarithmic utility exhibits decreasing absolute risk aversion in agreement with deductions of Arrow [1] and others on the qualities of "reasonable" utility functions. Hakansson says, "What the relative risk aversion index [given by $-xU''(x)/U'(x)$] would look like for a meaningful utility function is less clear.... In view of Arrow's conclusion that '...broadly speaking, the relative risk aversion must hover around 1, being, if anything, somewhat less for low wealths and somewhat higher for high wealths...' the optimal growth model seems to be on safe ground." As he notes, for $U(x) = \log x$, the relative risk aversion is precisely 1. However, in both the extension to valuing the individual as capital equipment, and the further extension to include the death constant, we are led to $U(x) = \log(x+c)$, where c is positive. But then the relative risk-aversion index is $x/(x+c)$, which behaves strikingly like Arrow's description. See also the discussion of $U(x) = \log(x+c)$ by Freimer and Gordon [8, pp. 103, 112].

Morgenstern [17] has forcefully observed that assets are random variables, not numbers, and that economic theory generally does not incorporate this. To replace assets by *numbers* having *the same expected utility* in valuing companies, portfolios, property, and the like, allows for comparisons when asset values are given as random variables. We, of course, think logarithmic utility will often be the appropriate tool for such valuation.

Acknowledgment

I wish to thank James Bicksler for several stimulating and helpful conversations.

References

1. Arrow, K. J., *Aspects of the Theory of Risk-Bearing*. Yrjö Jahnssonin Säätiö, Helsinki, 1965. Reprinted in K. J. Arrow, *Essays in the Theory of Risk-Bearing*. Markham, Chicago and North-Holland, London (1970).

1a. Ayres, H. F., "Risk Aversion in the Warrant Markets." S. M. Thesis, MIT, Cambridge, Massachusetts; *Industrial Management Review* **5**: 1 (1963), 45–53; reprinted by Cootner [7, pp. 479–505].

2. Bellman, R., and Kalaba, R., "On the Role of Dynamic Programming in Statistical Communication Theory." *IRE Transactions of the Professional Group on Information Theory* **IT-3**: 3 (1957), 197–203.

3. Bernoulli, D., "Exposition of a New Theory on the Measurement of Risk." *Econometrica* **22** (1954), 23–36 (translated by Louise Sommer).

3a. Black, F., and Scholes, M., "The Valuation of Option Contracts and a Test of Market Efficiency." *Journal of Finance* **27** (1972), 399–417.

3b. Black, F., and Scholes, M., "The Pricing of Options and Corporate Liabilities." *Journal of Political Economy* **81** (1973), 637–654.

4. Borch, K. H., *The Economics of Uncertainty*. Princeton Univ. Press, Princeton, New Jersey, 1968.

5. Breiman, L., "Investment Policies for Expanding Businesses Optimal in a Long Run Sense." *Naval Research Logistics Quarterly* **7**: 4 (1960), 647–651.

6. BREIMAN, L., "Optimal Gambling Systems for Favorable Games." *Symposium on Probability and Statistics 4th, Berkeley, 1961*, **1** pp. 65–78.
7. COOTNER, P. H. (ed.), *The Random Character of Stock Market Prices.* MIT Press, Cambridge, Massachusetts, 1964.
8. FREIMER, M., and GORDON, M. S., "Investment Behavior with Utility a Concave Function of Wealth." In *Risk and Uncertainty* (K. Borch and J. Mossin, eds.), pp. 94–115. St. Martin's Press, New York, 1968.
9. GARDNER, M., "Mathematical Games: The Paradox of the Non-Transitive Dice and the Elusive Principle of Indifference." *Scientific American* **223** (December 1970), 110.
10. GRANGER, C., and MORGENSTERN, O., *Predictability of Stock Market Prices.* Heath, Lexington Books, Lexington, Massachusetts, 1970.
11. HAKANSSON, N., "Capital Growth and the Mean-Variance Approach to Portfolio Selection." *Journal of Finance and Quantitative Analysis* **6** (1971), 517–557.
12. KASSOUF, S. T., "A Theory and an Econometric Model for Common Stock Purchase Warrants." Ph.D. Thesis, Columbia Univ. New York, 1965; published by Analytic Publ., New York, 1965. [A regression model statistical fit of normal price curves for warrants. There are large systematic errors in the model due to faulty (strongly biased) regression techniques. The average mean square error in the fit is large. Thus, it is not safe to use the model in practice as a predictor of warrant prices. However, the model and the methodology are valuable as a first *qualitative* description of warrant behavior and as a guide to a more precise analysis.]
13. KASSOUF, S. T., "An Econometric Model for Option Price with Implications for Investors' Expectations and Audacity." *Econometrica* **37**: 4 (1969), 685–694. [Based on the thesis. The variance of residuals is given as .248 for a standard deviation of about .50 in y, the normalized warrant price, and a standard error of about .34. The mid-range of y varies from 0 to .5 and is never greater, thus the caveat about not using the model for practical predictions!]
14. KELLY, J. L., "A New Interpretation of Information Rate." *Bell System Technical Journal* **35** (1956), 917–926.
15. LATANÉ, H. A., "Criteria for Choice Among Risky Ventures." *Journal of Political Economy* **67** (1959), 144–155.
15a. MAIER, S. F., PETERSON, D. W., and VANDER WEIDE, J. H., "A Monte Carlo Investigation of Characteristics of Optimal Geometric Mean Portfolios." *Duke Working Paper* #120. Presented to session on Multiperiod Portfolio Theory, American Financial Association Annual Meeting, San Francisco (December, 1974).
16. MARKOWITZ, H. M., *Portfolio Selection: Efficient Diversification of Investments.* Wiley, New York, 1959.
16a. MARKOWITZ, H. M., *Portfolio Selection: Efficient Diversification of Investments*, 2nd printing, Preface. Yale Univ. Press, New Haven, Connecticut, 1970, reprint.
16b. MERTON, R. C., "Theory of Rational Option Pricing." *Bell Journal of Economic and Management Science* **4** (1973), 141–183.
17. MORGENSTERN, O., *On the Accuracy of Economic Observations*, 2nd ed., revised. Princeton Univ. Press, Princeton, New Jersey, 1963.
18. MOSSIN, J., "Optimal Multiperiod Portfolio Policies." *Journal of Business* **41** (1968), 215–229.
19. OSBORNE, M. F. M., "Brownian Motion in the Stock Market." *Operations Research* **7** (1959), 145–173, reprinted in Cootner [7, pp. 100–128].

20. Samuelson, P. A., "Risk and Uncertainty: A Fallacy of Large Numbers." *Scienta (Milan)* **57** (1963), 153–158.
21. Samuelson, P. A., "Lifetime Portfolio Selection by Dynamic Stochastic Programming." *Review of Economics and Statistics* **51** (1969), 239–246.
22. Samuelson, P. A., "The 'Fallacy' of Maximizing the Geometric Mean in Long Sequences of Investing or Gambling." *Proceedings of the National Academy of Sciences of the United States* **68** (1971), 2493–2496.
23. Schrock, N. W., "The Theory of Asset Choice: Simultaneous Holding of Short and Long Positions in the Futures Market." *Journal of Political Economy* **79**: 2 (1971), 270–293.
24. Shelton, J. P., "The Relation of the Price of a Warrant to its Associated Common Stock." *Financial Analysts Journal* **23**: 3 (1967), 143–151; **23**: 4 (1967), 88–99.
25. Thorp, E., "A Winning Bet in Nevada Baccarat." *Journal of the American Statistical Association* **61** Pt. I (1966), 313–328.
26. Thorp, E., "Optimal Gambling Systems for Favorable Games." *Review of the International Statistical Institute* **37**: 3 (1969), 273–293.
27. Thorp, E., and Kassouf, S., *Beat the Market.* Random House, New York, 1967.
27a. Thorp, E., and Whitley, R. J., "Concave Utilities are Distinguished for Their Optimal Strategies." *Colloquia Mathematica Societatis Janos Bolyai* **9**. European Meeting of Statisticians, Budapest (Hungary), 1972, in press.
28. Todhunter, I., *A History of the Mathematical Theory of Probability*, 1st edition. Cambridge, 1865, as reprinted by Chelsea, New York, 1965. (See pp. 213 ff. for details on Daniel Bernoulli's use of logarithmic utility.)
29. Wagner, H. M., *Principles of Operations Research, with Application to Managerial Decisions.* Prentice-Hall, Englewood Cliffs, New Jersey, 1969.
30. Williams, J. B., "Speculation and the Carryover." *Quarterly Journal of Economics* **50** (1936), 436–455.

Reprinted from JOURNAL OF ECONOMIC THEORY
Vol. 3, No. 4, December 1971

Printed in Belgium

Optimum Consumption and Portfolio Rules in a Continuous-Time Model*

ROBERT C. MERTON

Sloan School of Management, Massachusetts Institute of Technology, Cambridge, Massachusetts 02139

Received September 30, 1970

1. INTRODUCTION

A common hypothesis about the behavior of (limited liability) asset prices in perfect markets is the random walk of returns or (in its continous-time form) the "geometric Brownian motion" hypothesis which implies that asset prices are stationary and log-normally distributed. A number of investigators of the behavior of stock and commodity prices have questioned the accuracy of the hypothesis.[1] In particular, Cootner [2] and others have criticized the independent increments assumption, and Osborne [2] has examined the assumption of stationariness. Mandelbrot [2] and Fama [2] argue that stock and commodity price changes follow a stable-Paretian distribution with infinite second moments. The nonacademic literature on the stock market is also filled with theories of stock price patterns and trading rules to "beat the market," rules often called "technical analysis" or "charting," and that presupposes a departure from random price changes.

In an earlier paper [12], I examined the continuous-time consumption-portfolio problem for an individual whose income is generated by capital gains on investments in assets with prices assumed to satisfy the "geometric Brownian motion" hypothesis; i.e., I studied Max $E \int_0^T U(C, t)\,dt$

* I would like to thank P. A. Samuelson, R. M. Solow, P. A. Diamond, J. A. Mirrlees, J. A. Flemming, and D. T. Scheffman for their helpful discussions. Of course, all errors are mine. Aid from the National Science Foundation is gratefully acknowledged. An earlier version of the paper was presented at the second World Congress of the Econometric Society, Cambridge, England.

[1] For a number of interesting papers on the subject, see Cootner [2]. An excellent survey article is "Efficient Capital Markets: A Review of Theory and Empirical Work," by E. Fama, *Journal of Finance*, May, 1970.

where U is the instantaneous utility function, C is consumption, and E is the expectation operator. Under the additional assumption of a constant relative or constant absolute risk-aversion utility function, explicit solutions for the optimal consumption and portfolio rules were derived. The changes in these optimal rules with respect to shifts in various parameters such as expected return, interest rates, and risk were examined by the technique of comparative statics.

The present paper extends these results for more general utility functions, price behavior assumptions, and for income generated also from non-capital gains sources. It is shown that if the "geometric Brownian motion" hypothesis is accepted, then a general "Separation" or "mutual fund" theorem can be proved such that, in this model, the classical Tobin mean-variance rules hold without the objectionable assumptions of quadratic utility or of normality of distributions for prices. Hence, when asset prices are generated by a geometric Brownian motion, one can work with the two-asset case without loss of generality. If the further assumption is made that the utility function of the individual is a member of the family of utility functions called the "HARA" family, explicit solutions for the optimal consumption and portfolio rules are derived and a number of theorems proved. In the last parts of the paper, the effects on the consumption and portfolio rules of alternative asset price dynamics, in which changes are neither stationary nor independent, are examined along with the effects of introducing wage income, uncertainty of life expectancy, and the possibility of default on (formerly) "risk-free" assets.

2. A Digression on Itô Processes

To apply the dynamic programming technique in a continuous-time model, the state variable dynamics must be expressible as Markov stochastic processes defined over time intervals of length h, no matter how small h is. Such processes are referred to as infinitely divisible in time. The two processes of this type[2] are: functions of Gauss–Wiener Brownian motions which are continuous in the "space" variables and functions of Poisson processes which are discrete in the space variables. Because neither of these processes is differentiable in the usual sense, a more general type of differential equation must be developed to express the dynamics of such processes. A particular class of continuous-time

[2] I ignore those infinitely divisible processes with infinite moments which include those members of the stable Paretian family other than the normal.

Markov processes of the first type called Itô Processes are defined as the solution to the stochastic differential equation[3]

$$dP = f(P, t)\,dt + g(P, t)\,dz, \tag{1}$$

where P, f, and g are n vectors and $z(t)$ is an n vector of standard normal random variables. Then $dz(t)$ is called a multidimensional Wiener process (or Brownian motion).[4]

The fundamental tool for formal manipulation and solution of stochastic processes of the Itô type is Itô's Lemma stated as follows[5]

LEMMA. *Let $F(P_1, \ldots, P_n, t)$ be a C^2 function defined on $R^n X[0, \infty)$ and take the stochastic integrals*

$$P_i(t) = P_i(0) + \int_0^t f_i(P, s)\,ds + \int_0^t g_i(P, s)\,dz_i, \qquad i = 1, \ldots, n;$$

then the time-dependent random variable $Y \equiv F$ is a stochastic integral and its stochastic differential is

$$dY = \sum_1^n \frac{\partial F}{\partial P_i}\,dP_i + \frac{\partial F}{\partial t}\,dt + \frac{1}{2}\sum_1^n \sum_1^n \frac{\partial^2 F}{\partial P_i\,\partial P_j}\,dP_i\,dP_j,$$

where the product of the differentials $dP_i\,dP_j$ are defined by the multiplication rule

$$dz_i\,dz_j = \rho_{ij}\,dt, \qquad i, j = 1, \ldots, n,$$
$$dz_i\,dt = 0, \qquad i = 1, \ldots, n,$$

[3] Itô Processes are a special case of a more general class of stochastic processes called Strong diffusion processes (see Kushner [9, p. 22]). (1) is a short-hand expression for the stochastic integral

$$P(t) = P(0) + \int_0^t f(P, s)\,ds + \int_0^t g(P, s)\,dz,$$

where $P(t)$ is the solution to (1) with probability one.

A rigorous discussion of the meaning of a solution to equations like (1) is not presented here. Only those theorems needed for formal manipulation and solution of stochastic differential equations are in the text and these without proof. For a complete discussion of Itô Processes, see the seminal paper of Itô [7], Itô and McKean [8], and McKean [11]. For a short description and some proofs, see Kushner [9, pp. 12–18]. For an heuristic discussion of continuous-time Markov processes in general, see Cox and Miller [3, Chap. 5].

[4] dz is often referred to in the literature as "Gaussian White Noise." There are some regularity conditions imposed on the functions f and g. It is assumed throughout the paper that such conditions are satisfied. For the details, see [9] or [11].

[5] See McKean [11, pp. 32–35 and 44] for proofs of the Lemma in one and n dimensions.

where ρ_{ij} is the instantaneous correlation coefficient between the Wiener processes dz_i and dz_j.[6]

Armed with Itô's Lemma, we are now able to formally differentiate most smooth functions of Brownian motions (and hence integrate stochastic differential equations of the Itô type).[7]

Before proceeding to the discussion of asset price behavior, another concept useful for working with Itô Processes is the differential generator (or weak infinitesimal operator) of the stochastic process $P(t)$. Define the function $\mathring{G}(P, t)$ by

$$\mathring{G}(P, t) \equiv \lim_{h \to 0} E_t \left[\frac{G(P(t+h), t+h) - G(P(t), t)}{h} \right], \tag{2}$$

when the limit exists and where "E_t" is the conditional expectation operator, conditional on knowing $P(t)$. If the $P_i(t)$ are generated by Itô Processes, then the differential generator of P, $\mathscr{L}_P$, is defined by

$$\mathscr{L}_P \equiv \sum_1^n f_i \frac{\partial}{\partial P_i} + \frac{\partial}{\partial t} + \frac{1}{2} \sum_1^n \sum_1^n a_{ij} \frac{\partial^2}{\partial P_i \, \partial P_j},$$

where $f = (f_1, ..., f_n)$, $g = (g_1, ..., g_n)$, and $a_{ij} \equiv g_i g_j \rho_{ij}$. Further, it can be shown that

$$\mathring{G}(P, t) = \mathscr{L}_P[G(P, t)]. \tag{4}$$

$\mathring{G}$ can be interpreted as the "average" or expected time rate of change of

[6] This multiplication rule has given rise to the formalism of writing the Wiener process differentials as $dz_i = \mathscr{S}_i \sqrt{dt}$ where the $\mathscr{S}_i$ are standard normal variates (*e.g.*, see [3]).

[7] Warning: derivatives (and integrals) of functions of Brownian motions are similar to, but different from, the rules for deterministic differentials and integrals. For example, if

$$P(t) = P(0)\, e^{\int_0^t dz - \frac{1}{2}t} = P(0)\, e^{z(t) - z(0) - \frac{1}{2}t},$$

then $dP = P dz$. Hence

$$\int_0^t \frac{dP}{P} = \int_0^t dz \neq \log(P(t)/P(0)).$$

Stratonovich [15] has developed a symmetric definition of stochastic differential equations which formally follows the ordinary rules of differentiation and integration. However, this alternative to the Itô formalism will not be discussed here.

the function $G(P, t)$ and as such is the natural generalization of the ordinary time derivative for deterministic functions.[8]

3. Asset Price Dynamics and the Budget Equation

Throughout the paper, it is assumed that all assets are of the limited liability type, that there exist continuously-trading perfect markets with no transactions costs for all assets, and that the prices per share, $\{P_i(t)\}$, are generated by Itô Processes, i.e.,

$$\frac{dP_i}{P_i} = \alpha_i(P, t)\, dt + \sigma_i(P, t)\, dz_i\,, \tag{5}$$

where α_i is the instantaneous conditional expected percentage change in price per unit time and σ_i^2 is the instantaneous conditional variance per unit time. In the particular case where the "geometric Brownian motion hypothesis is assumed to hold for asset prices, α_i and σ_i will be constants. For this case, prices will be stationarily and log-normally distributed and it will be shown that this assumption about asset prices simplifies the continuous-time model in the same way that the assumption of normality of prices simplifies the static one-period portfolio model.

To derive the correct budget equation, it is necessary to examine the discrete-time formulation of the model and then to take limits carefully to obtain the continuous-time form. Consider a period model with periods of length h, where all income is generated by capital gains, and wealth, $W(t)$ and $P_i(t)$ are known at the *beginning* of period t. Let the decision variables be indexed such that the indices coincide with the period in which the decisions are implemented. Namely, let

$N_i(t) \equiv$ number of shares of asset i purchased during period t, i.e., between t and $t + h$

and (6)

$C(t) \equiv$ amount of consumption per unit time during period t.

[8] A heuristic method for finding the differential generator is to take the conditional expectation of dG (found by Itô's Lemma) and "divide" by dt. The result of this operation will be $\mathscr{L}_P[G]$, i.e., formally,

$$\frac{1}{dt} E_t(dG) = \overset{\circ}{G} = \mathscr{L}_P[G].$$

The "$\mathscr{L}_P$" operator is often called a Dynkin operator and is often written as "D_P".

The model assumes that the individual "comes into" period t with wealth invested in assets so that

$$W(t) = \sum_1^n N_i(t-h)\,P_i(t). \tag{7}$$

Notice that it is $N_i(t-h)$ because $N_i(t-h)$ is the number of shares purchased for the portfolio in period $(t-h)$ and it is $P_i(t)$ because $P_i(t)$ is the *current* value of a share of the i-th asset. The amount of consumption for the period, $C(t)\,h$, and the new portfolio, $N_i(t)$, are simultaneously chosen, and if it is assumed that all trades are made at (known) current prices, then we have that

$$-C(t)\,h = \sum_1^n [N_i(t) - N_i(t-h)]\,P_i(t). \tag{8}$$

The "dice" are rolled and a new set of prices is determined, $P_i(t+h)$, and the value of the portfolio is now $\sum_1^n N_i(t)\,P_i(t+h)$. So the individual "comes into" period $(t+h)$ with wealth $W(t+h) = \sum_1^n N_i(t)\,P_i(t+h)$ and the process continues.

Incrementing (7) and (8) by h to eliminate backward differences, we have that

$$\begin{aligned} -C(t+h)\,h &= \sum_1^n [N_i(t+h) - N_i(t)]\,P_i(t+h) \\ &= \sum_1^n [N_i(t+h) - N_i(t)][P_i(t+h) - P_i(t)] \\ &\quad + \sum_1^n [N_i(t+h) - N_i(t)]\,P_i(t) \end{aligned} \tag{9}$$

and

$$W(t+h) = \sum_1^n N_i(t)\,P_i(t+h). \tag{10}$$

Taking the limits as $h \to 0$,[9] we arrive at the continuous version of (9) and (10),

$$-C(t)\,dt = \sum_1^n dN_i(t)\,dP_i(t) + \sum_1^n dN_i(t)\,P_i(t) \tag{9'}$$

[9] We use here the result that Itô Processes are right-continuous [9, p. 15] and hence $P_i(t)$ and $W(t)$ are right-continuous. It is assumed that $C(t)$ is a right-continuous function, and, throughout the paper, the choice of $C(t)$ is restricted to this class of functions.

and

$$W(t) = \sum_{1}^{n} N_i(t)\, P_i(t). \tag{10'}$$

Using Itô's Lemma, we differentiate (10′) to get

$$dW = \sum_{1}^{n} N_i\, dP_i + \sum_{1}^{n} dN_i P_i + \sum_{1}^{n} dN_i\, dP_i\,. \tag{11}$$

The last two terms, $\sum_1^n dN_i P_i + \sum_1^n dN_i\, dP_i$, are the net value of additions to wealth from sources other than capital gains.[10] Hence, if $dy(t)$ = (possibly stochastic) instantaneous flow of noncapital gains (wage) income, then we have that

$$dy - C(t)\, dt = \sum_{1}^{n} dN_i P_i + \sum_{1}^{n} dN_i\, dP_i\,. \tag{12}$$

From (11) and (12), the budget or accumulation equation is written as

$$dW = \sum_{1}^{n} N_i(t)\, dP_i + dy - C(t)\, dt. \tag{13}$$

It is advantageous to eliminate $N_i(t)$ from (13) by defining a new variable, $w_i(t) \equiv N_i(t)\, P_i(t)/W(t)$, the percentage of wealth invested in the i-th asset at time t. Substituting for dP_i/P_i from (5), we can write (13) as

$$dW = \sum_{1}^{n} w_i W \alpha_i\, dt - C\, dt + dy + \sum_{1}^{n} w_i W \sigma_i\, dz_i\,, \tag{14}$$

where, by definition, $\sum_1^n w_i \underset{t}{\equiv} 1$.[11]

Until Section 7, it will be assumed that $dy \equiv 0$, i.e., all income is derived from capital gains on assets. If one of the n-assets is "risk-free"

[10] This result follows directly from the discrete-time argument used to derive (9′) where $-C(t)\, dt$ is replaced by a general $dv(t)$ where $dv(t)$ is the instantaneous flow of funds from all noncapital gains sources.

It was necessary to derive (12) by starting with the discrete-time formulation because it is not obvious from the continuous version directly whether $dy - C(t)\, dt$ equals $\sum_1^n dN_i P_i + \sum_1^n dN_i\, dP_i$ or just $\sum_1^n dN_i P_i$.

[11] There are no other restrictions on the individual w_i because borrowing and short-selling are allowed.

(by convention, the n-th asset), then $\sigma_n = 0$, the instantaneous rate of return, α_n, will be called r, and (14) is rewritten as

$$dW = \sum_1^m w_i(\alpha_i - r) W\,dt + (rW - C)\,dt + dy + \sum_1^m W_i \sigma_i\,dz_i, \quad (14')$$

where $m \equiv n - 1$ and the $w_1, \ldots, w_m$ are unconstrained by virtue of the fact that the relation $w_n = 1 - \sum_1^m w_i$ will ensure that the identity constraint in (14) is satisfied.

4. Optimal Portfolio and Consumption Rules: The Equations of Optimality

The problem of choosing optimal portfolio and consumption rules for an individual who lives T years is formulated as follows:

$$\max E_0 \left[\int_0^T U(C(t), t)\,dt + B(W(T), T)\right] \quad (15)$$

subject to: $W(0) = W_0$; the budget constraint (14), which in the case of a "risk-free" asset becomes (14′); and where the utility function (during life) U is assumed to be strictly concave in C and the "bequest" function B is assumed also to be concave in W.[12]

To derive the optimal rules, the technique of stochastic dynamic programming is used. Define

$$J(W, P, t) = \max_{\{C,w\}} E_t \left[\int_t^T U(C, s)\,ds + B(W(T), T)\right], \quad (16)$$

where as before, "E_t" is the conditional expectation operator, conditional on $W(t) = W$ and $P_i(t) = P_i$. Define

$$\phi(w, C; W, P, t) \equiv U(C, t) + \mathscr{L}[J], \quad (17)$$

[12] Where there is no "risk-free" asset, it is assumed that no asset can be expressed as a linear combination of the other assets, implying that the $n \times n$ variance-covariance matrix of returns, $\Omega = [\sigma_{ij}]$, where $\sigma_{ij} \equiv \rho_{ij}\sigma_i\sigma_j$, is nonsingular. In the case when there is a "risk-free" asset, the same assumption is made about the "reduced" $m \times m$ variance-covariance matrix.

given $w_i(t) = w_i$, $C(t) = C$, $W(t) = W$, and $P_i(t) = P_i$.[13] From the theory of stochastic dynamic programming, the following theorem provides the method for deriving the optimal rules, C^* and w^*.

THEOREM I.[14] *If the $P_i(t)$ are generated by a strong diffusion process, U is strictly concave in C, and B is concave in W, then there exists a set of optimal rules (controls), w^* and C^*, satisfying $\sum_1^n w_i^* = 1$ and $J(W, P, T) = B(W, T)$ and these controls satisfy*

$$0 = \phi(C^*, w^*; W, P, t) \geqslant \phi(C, w; W, P, t)$$

for $t \in [0, T]$.

From Theorem I, we have that

$$0 = \max_{\{C,w\}}\{\phi(C, w; W, P, t)\} \tag{18}$$

In the usual fashion of maximization under constraint, we define the Lagrangian, $L \equiv \phi + \lambda[1 - \sum_1^n w_i]$ where λ is the multiplier and find the extreme points from the first-order conditions

$$0 = L_C(C^*, w^*) = U_C(C^*, t) - J_W, \tag{19}$$

$$\begin{aligned} 0 = L_{w_k}(C^*, w^*) &= -\lambda + J_W \alpha_k W + J_{WW} \sum_1^n \sigma_{kj} w_j^* W^2 \\ &+ \sum_1^n J_{jW} \sigma_{kj} P_j W, \qquad k = 1,..., n, \end{aligned} \tag{20}$$

$$0 = L_\lambda(C^*, w^*) = 1 - \sum_1^n w_i^*, \tag{21}$$

[13] "$\mathscr{L}$" is short for the rigorous $\mathscr{L}_{P,W}^{w,C}$, the Dynkin operator over the variables P and W for a given set of controls w and C.

$$\begin{aligned} \mathscr{L} \equiv \frac{\partial}{\partial t} &+ \left[\sum_1^n w_i \alpha_i W - C\right] \frac{\partial}{\partial W} + \sum_1^n \alpha_i P_i \frac{\partial}{\partial P_i} \\ &+ \frac{1}{2} \sum_1^n \sum_1^n \sigma_{ij} w_i w_j W^2 \frac{\partial^2}{\partial W^2} + \frac{1}{2} \sum_1^n \sum_1^n P_i P_j \sigma_{ij} \frac{\partial^2}{\partial P_i \partial P_j} \\ &+ \sum_1^n \sum_1^n P_i W w_j \sigma_{ij} \frac{\partial^2}{\partial P_i \partial W}. \end{aligned}$$

[14] For an heuristic proof of this theorem and the derivation of the stochastic Bellman equation, see Dreyfus [4] and Merton [12]. For a rigorous proof and discussion of weaker conditions, see Kushner [9, Chap. IV, especially Theorem 7].

where the notation for partial derivatives is $J_W \equiv \partial J/\partial W$, $J_t \equiv \partial J/\partial t$, $U_C \equiv \partial U/\partial C$, $J_i \equiv \partial J/\partial P_i$, $J_{ij} \equiv \partial^2 J/\partial P_i\, \partial P_j$, and $J_{jW} \equiv \partial^2 J/\partial P_j\, \partial W$.

Because $L_{CC} = \phi_{CC} = U_{CC} < 0$, $L_{Cw_k} = \phi_{Cw_k} = 0$, $L_{w_k w_k} = \sigma_k{}^2 W^2 J_{WW}$, $L_{w_k w_j} = 0$, $k \neq j$, a sufficient condition for a unique interior maximum is that $J_{WW} < 0$ (i.e., that J be strictly concave in W). That assumed, as an immediate consequence of differentiating (19) totally with respect to W, we have

$$\frac{\partial C^*}{\partial W} > 0. \tag{22}$$

To solve explicitly for C^* and w^*, we solve the $n+2$ nondynamic implicit equations, (19)–(21), for C^*, and w^*, and λ as functions of J_W, J_{WW}, J_{jW}, W, P, and t. Then, C^* and w^* are substituted in (18) which now becomes a second-order partial differential equation for J, subject to the boundary condition $J(W, P, T) = B(W, T)$. Having (in principle at least) solved this equation for J, we then substitute back into (19)–(21) to derive the optimal rules as functions of W, P, and t. Define the inverse function $G \equiv [U_C]^{-1}$. Then, from (19),

$$C^* = G(J_W, t). \tag{23}$$

To solve for the $w_i{}^*$, note that (20) is a linear system in $w_i{}^*$ and hence can be solved explicitly. Define

$$\begin{aligned} \Omega &\equiv [\sigma_{ij}], \qquad \text{the } n \times n \text{ variance-covariance matrix,} \\ [v_{ij}] &\equiv \Omega^{-1},^{15} \\ \Gamma &\equiv \sum_1^n \sum_1^n v_{ij}. \end{aligned} \tag{24}$$

Eliminating λ from (20), the solution for $w_k{}^*$ can be written as

$$w_k{}^* = h_k(P, t) + m(P, W, t)\, g_k(P, t) + f_k(P, W, t), \qquad k = 1, \ldots, n, \tag{25}$$

where $\sum_1^n h_k \equiv 1$, $\sum_1^n g_k \equiv 0$, and $\sum_1^n f_k \equiv 0$.[16]

[15] Ω^{-1} exists by the assumption on Ω in footnote 12.

[16]
$$h_k(P, t) \equiv \sum_1^n v_{kj}/\Gamma; \; m(P, W, t) \equiv -Jw/WJww;$$
$$g_k(P, t) \equiv \frac{1}{\Gamma} \sum_1^n v_{kl} \left(\Gamma \alpha_l - \sum_1^n \sum_1^n v_{ij} \alpha_j \right); \; f_k(P, W, t)$$
$$\equiv \left[\Gamma J_{kw} P_k - \sum_1^n J_{iw} P_i \sum_1^n v_{kj} \right] \Big/ \Gamma WJww.$$

Substituting for w^* and C^* in (18), we arrive at the fundamental partial differential equation for J as a function of W, P, and t,

$$
\begin{aligned}
0 = {} & U[G, t] + J_t + J_W \left[\frac{\sum_1^n \sum_1^n v_{kj} \alpha_k W}{\Gamma} - G\right] \\
& + \sum_1^n J_i \alpha_i P_i + \frac{1}{2} \sum_1^n \sum_1^n J_{ij} \sigma_{ij} P_i P_j + \frac{W}{\Gamma} \sum_1^n J_{jW} P_j \\
& - \frac{J_W}{\Gamma J_{WW}} \left(\sum_1^n \Gamma J_{kW} P_k \alpha_k - \sum_1^n J_{jW} P_j \sum_1^n \sum_1^n v_{kl} \alpha_l \right) \\
& + \frac{J_{WW} W^2}{2\Gamma} - \frac{1}{2\Gamma J_{WW}} \left[\sum_1^n \sum_1^n J_{jW} J_{mW} P_j P_m \sigma_{mj} \Gamma - \left(\sum_1^n J_{iW} P_i \right)^2 \right] \\
& - \frac{J_W^2}{2\Gamma J_{WW}} \left[\sum_1^n \sum_1^n v_{kl} \alpha_k \alpha_l \Gamma - \left(\sum_1^n \sum_1^n v_{kl} \alpha_k \right)^2 \right]
\end{aligned} \tag{26}
$$

subject to the boundary condition $J(W, P, T) = B(W, T)$. If (26) were solved, the solution J could be substituted into (23) and (25) to obtain C^* and w^* as functions of W, P, and t.

For the case where one of the assets is "risk-free," the equations are somewhat simplified because the problem can be solved directly as an unconstrained maximum by eliminating w_n as was done in (14′). In this case, the optimal proportions in the risky assets are

$$
w_k^* = -\frac{J_W}{J_{WW} W} \sum_1^m v_{kj}(\alpha_{j-r}) - \frac{J_{kW} P_k}{J_{WW} W}, \qquad k = 1, \ldots, m. \tag{27}
$$

The partial differential equation for J corresponding to (26) becomes

$$
\begin{aligned}
0 = {} & U[G, T] + J_t + J_W[rW - G] + \sum_1^m J_i \alpha_i P_i \\
& + \frac{1}{2} \sum_1^m \sum_1^m J_{ij} \sigma_{ij} P_i P_j - \frac{J_W}{J_{WW}} \sum_1^m J_{jW} P_j (\alpha_j - r) \\
& - \frac{J_W^2}{2 J_{WW}} \sum_1^m \sum_1^m v_{ij} (\alpha_i - r)(\alpha_j - r) - \frac{1}{2 J_{WW}} \sum_1^m \sum_1^m J_{iW} J_{jW} \sigma_{ij} P_i P_i
\end{aligned} \tag{28}
$$

subject to the boundary condition $J(W, P, T) = B(W, T)$.

Although (28) is a simplified version of (26), neither (26) nor (28) lend themselves to easy solution. The complexities of (26) and (28) are caused by the basic nonlinearity of the equations and the large number of state variables. Although there is little that can be done about the non-linearities, in some cases, it may be possible to reduce the number of state variables.

5. Log-Normality of Prices and the Continuous-Time Analog to Tobin–Markowitz Mean-Variance Analysis

When, for $k = 1,..., n$, α_k and σ_k are constants, the asset prices have stationary, log-normal distributions. In this case, J will be a function of W and t only and not P. Then (26) reduces to

$$0 = U[G, t] + J_t + J_W \left[\frac{\sum_1^n \sum_1^n v_{kj}\alpha_k}{\Gamma} W - G\right] + \frac{J_{WW} W^2}{2\Gamma} - \frac{J_W^2}{2\Gamma J_{WW}}\left[\sum_1^n \sum_1^n v_{kl}\alpha_k\alpha_l \Gamma - \left(\sum_1^n \sum_1^n v_{kl}\alpha_k\right)^2\right]. \tag{29}$$

From (25), the optimal portfolio rule becomes

$$w_k{}^* = h_k + m(W, t)\, g_k\,, \tag{30}$$

where $\sum_1^n h_k \equiv 1$ and $\sum_1^n g_k \equiv 0$ and h_k and g_k are constants.

From (30), the following "separation" or "mutual fund" theorem can be proved.

Theorem II.[17] *Given n assets with prices P_i whose changes are log-normally distributed, then (1) there exist a unique (up to a nonsingular transformation) pair of "mutual funds" constructed from linear combinations of these assets such that, independent of preferences (i.e., the form of the utility function), wealth distribution, or time horizon, individuals will be indifferent between choosing from a linear combination of these two funds or a linear combination of the original n assets. (2) If P_f is the price per share of either fund, then P_f is log-normally distributed. Further, (3) if δ_k = percentage of one mutual fund's value held in the k-th asset and if λ_k = percentage of the other mutual fund's value held in the k-th asset, then one can find that*

$$\delta_k = h_k + \frac{(1-\eta)}{\nu} g_k\,, \qquad k = 1,..., n,$$

[17] See Cass and Stiglitz [1] for a general discussion of Separation theorems. The only degenerate case is when all the assets are identically distributed (i.e., symmetry) in which case, only one mutual fund is needed.

and

$$\lambda_k = h_k - \frac{\eta}{\nu} g_k , \qquad k = 1,\ldots, n$$

where ν, η *are arbitrary constants* ($\nu \neq 0$).

Proof. (1) (30) is a parametric representation of a line in the hyperplane defined by $\sum_1^n w_k^* = 1$.[18] Hence, there exist two linearly independent vectors (namely, the vectors of asset proportions held by the two mutual funds) which form a basis for all optimal portfolios chosen by the individuals. Therefore, each individual would be indifferent between choosing a linear combination of the mutual fund shares or a linear combination of the original n assets.

(2) Let $V \equiv N_f P_f =$ the total value of (either) fund where $N_f =$ number of shares of the fund outstanding. Let $N_k =$ number of shares of asset k held by the fund and $\mu_k \equiv N_k P_k / V =$ percentage of total value invested in the k-th asset. Then $V = \sum_1^n N_k P_k$ and

$$\begin{aligned} dV &= \sum_1^n N_k\, dP_k + \sum_1^n P_k\, dN_k + \sum dP_k\, dN_k \\ &= N_f\, dP_f + P_f\, dN_f + dP_f\, dN_f . \end{aligned} \tag{31}$$

But

$$\begin{aligned} \sum_1^n P_k\, dN_k + \sum_1^n dP_k\, dN_k &= \text{net inflow of funds from non-capital-gain sources} \\ &= \text{net value of new shares issued} \\ &= P_f\, dN_f + dN_f\, dP_f . \end{aligned} \tag{32}$$

From (31) and (32), we have that

$$N_f\, dP_f = \sum_1^n N_k\, dP_k . \tag{33}$$

By the definition of V and μ_k, (33) can be rewritten as

$$\begin{aligned} \frac{dP_f}{P_f} &= \sum_1^n \mu_k \frac{dP_k}{P_k} \\ &= \sum_1^n \mu_k \alpha_k\, dt + \sum_1^n \mu_k \sigma_k\, dz_k . \end{aligned} \tag{34}$$

[18] See [1, p. 15].

By Itô's Lemma and (34), we have that

$$P_f(t) = P_f(0) \exp\left[\left(\sum_1^n \mu_k \alpha_k - \tfrac{1}{2} \sum_1^n \sum_1^n \mu_k \mu_j \sigma_{kj}\right) t + \sum_1^n \mu_k \sigma_k \int_0^t dz_k\right]. \quad (35)$$

So, $P_f(t)$ is log-normally distributed.

(3) Let $a(W, t; U) \equiv$ percentage of wealth invested in the first mutual fund by an individual with utility function U and wealth W at time t. Then, $(1 - a)$ must equal the percentage of wealth invested in the second mutual fund. Because the individual is indifferent between these asset holdings or an optimal portfolio chosen from the original n assets, it must be that

$$w_k^* = h_k + m(W, t)\, g_k = a\delta_k + (1 - a)\,\lambda_k, \qquad k = 1,\ldots, n. \quad (36)$$

All the solutions to the linear system (36) for all W, t, and U are of the form

$$\begin{aligned} \delta_k &= h_k + \frac{(1-\eta)}{\nu} g_k, && k = 1,\ldots, n, \\ \lambda_k &= h_k - \frac{\eta}{\nu} g_k, && k = 1,\ldots, n, \\ a &= \nu m(W, t) + \eta, && \nu \neq 0. \end{aligned} \quad (37)$$

Note that

$$\sum_1^n \delta_k = \sum_1^n \left(h_k + \frac{(1-\eta)}{\nu} g_k\right) \equiv 1$$

and

$$\sum_1^n \lambda_k = \sum_1^n \left(h_k - \frac{\eta}{\nu} g_k\right) \equiv 1. \qquad \text{Q.E.D.}$$

For the case when one of the assets is "risk-free," there is a corollary to Theorem II. Namely,

COROLLARY. *If one of the assets is "risk-free," then the proportions of each asset held by the mutual funds are*

$$\delta_k = \frac{\eta}{\nu} \sum_1^m v_{kj}(\alpha_j - r), \qquad \lambda_k = \frac{(\eta - 1)}{\nu} \sum_1^m v_{kj}(\alpha_j - r),$$

$$\delta_n = 1 - \sum_1^m \delta_k, \qquad \lambda_n = 1 - \sum_1^m \lambda_k.$$

Proof. By the assumption of log-normal prices, (27) reduces to

$$w_k^* = m(W, t) \sum_1^m v_{kj}(\alpha_j - r), \qquad k = 1,..., m, \tag{38}$$

and

$$w_n^* = 1 - \sum_1^m w_k^* = 1 - m(W, t) \sum_1^m \sum_1^m v_{kj}(\alpha_j - r). \tag{39}$$

By the same argument used in the proof of Theorem II, (38) and (39) define a line in the hyperplane defined by $\sum_1^n w_i^* = 1$ and by the same technique used in Theorem II, we derive the fund proportions stated in the corollary with $a(W, t; u) = \nu m(W, t) + \eta$, where ν, η are arbitrary constants ($\nu \neq 0$). Q.E.D.

Thus, if we have an economy where all asset prices are log-normally distributed, the investment decision can be divided into two parts by the establishment of two financial intermediaries (mutual funds) to hold all individual securities and to issue shares of their own for purchase by individual investors. The separation is complete because the "instructions" given the fund managers, namely, to hold proportions δ_k and λ_k of the k-th security, $k = 1,..., n$, depend only on the price distribution parameters and are independent of individual preferences, wealth distribution, or age distribution.

The similarity of this result to that of the classical Tobin–Markowitz analysis is clearest when we choose one of the funds to be the risk-free asset (i.e., set $\eta = 1$), and the other fund to hold only risky assets (which is possible by setting $\nu = \sum_1^m \sum_1^m v_{ij}(\alpha_j - r)$, provided that the double sum is not zero). Consider the investment rule given to the "risky" fund's manager when there exists a "risk-free" asset (money) with zero return ($r = 0$). It is easy to show that the δ_k proportions prescribed in the corollary are derived by finding the locus of points in the (instantaneous) mean-standard deviation space of composite returns which minimize variance for a given mean (i.e., the efficient risky-asset frontier), and then by finding the point where a line drawn from the origin is tangent to the locus. This point determines the δ_k as illustrated in Fig. 1.

Given the α^*, the δ_k are determined. So the log-normal assumption in the continuous-time model is sufficient to allow the same analysis as in the static mean-variance model but without the objectionable assumptions of quadratic utility or normality of the distribution of absolute price changes. (Log-normality of price changes is much less objectionable, since this does invoke "limited liability" and, by the central limit theorem

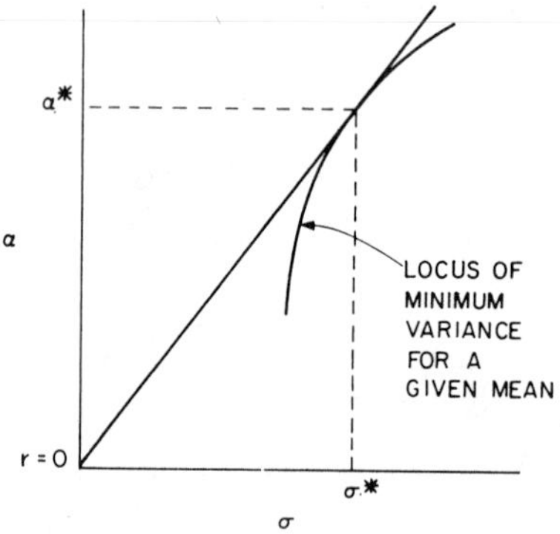

FIG. 1. Determination of the optimal combination of risky assets.

is the only regular solution to any continuous-space, infinitely-divisible process in time.)

An immediate advantage for the present analysis is that whenever log-normality of prices is assumed, we can work, without loss of generality, with just two assets, one "risk-free" and one risky with its price log-normally distributed. The risky asset can always be thought of as a composite asset with price $P(t)$ defined by the process

$$\frac{dP}{P} = \alpha\, dt + \sigma\, dz, \tag{40}$$

where

$$\begin{aligned} \alpha &\equiv \sum_1^m \sum_1^m v_{kj}(\alpha_j - r)\, \alpha_k \Big/ \sum_1^m \sum_1^m v_{ij}(\alpha_j - r), \\ \sigma^2 &\equiv \sum_1^m \sum_1^m \delta_k \delta_j \sigma_{kj}\,, \\ dz &\equiv \sum_1^m \delta_k \sigma_k\, dz_k / \sigma. \end{aligned} \tag{41}$$

6. Explicit Solutions for a Particular Class of Utility Functions

On the assumption of log-normality of prices, some characteristics of the asset demand functions were shown. If a further assumption about the preferences of the individual is made, then Eq. (28) can be solved in closed form, and the optimal consumption and portfolio rules derived explicitly. Assume that the utility function for the individual, $U(C, t)$, can be written as $U(C, t) = e^{-\rho t}V(C)$, where V is a member of the family of utility functions whose measure of absolute risk aversion is positive and hyperbolic in consumption, i.e.,

$$A(C) \equiv -V''/V' = 1\Big/\left(\frac{C}{1-\gamma} + \eta/\beta\right) > 0,$$

subject to the restrictions:

$$\gamma \neq 1; \quad \beta > 0; \quad \left(\frac{\beta C}{1-\gamma} + \eta\right) > 0; \quad \eta = 1 \text{ if } \gamma = -\infty. \qquad (42)$$

All members of the HARA (hyperbolic absolute risk-aversion) family can be expressed as

$$V(C) = \frac{(1-\gamma)}{\gamma}\left(\frac{\beta C}{1-\gamma} + \eta\right)^{\gamma}. \qquad (43)$$

This family is rich, in the sense that by suitable adjustment of the parameters, one can have a utility function with absolute or relative risk aversion increasing, decreasing, or constant.[19]

[19]

TABLE I

Properties of HARA Utility Functions

$A(C) = \dfrac{1}{\dfrac{C}{1-\gamma} + \dfrac{\eta}{\beta}} > 0$	(implies $\eta > 0$ for $\gamma > 1$)
$A'(C) = \dfrac{-1}{(1-\gamma)\left(\dfrac{C}{1-\gamma} + \dfrac{\eta}{\beta}\right)^2}$	< 0 for $-\infty < \gamma < 1$ > 0 for $1 < \gamma < \infty$ $= 0$ for $\gamma = +\infty$
Relative risk aversion $R(C) \equiv -V''C/V' = A(C)C$	
$R'(C) = \dfrac{\eta/\beta}{\left(\dfrac{C}{1-\gamma} + \dfrac{\eta}{\beta}\right)^2}$	> 0 for $\eta > 0$ $(-\infty \leqslant \gamma \leqslant \infty, \gamma \neq 1)$ $= 0$ for $\eta = 0$ < 0 for $\eta < 0$ $(-\infty < \gamma < 1)$

Note that included as members of the HARA family are the widely used isoelastic (constant relative risk aversion), exponential (constant absolute risk aversion), and quadratic utility functions. As is well known for the quadratic case, the members of the HARA family with $\gamma > 1$ are only defined for a restricted range of consumption, namely $0 < C < (\gamma - 1)\eta/\beta$. [1, 5, 6, 10, 12, 13, 16] discuss the properties of various members of the HARA family in a portfolio context. Although this is not done here, the HARA definition can be generalized to include the cases when γ, β, and η are functions of time subject to the restrictions in (42).

Without loss of generality, assume that there are two assets, one "risk-free" asset with return r and the other, a "risky" asset whose price is log-normally distributed satisfying (40). From (28), the optimality equation for J is

$$0 = \frac{(1-\gamma)^2}{\gamma} e^{-\rho t} \left[\frac{e^{\rho t} J_W}{\beta}\right]^{\frac{\gamma}{\gamma-1}} + J_t + [(1-\gamma)\,\eta/\beta + rW]\,J_W - \frac{J_W^2}{J_{WW}} \frac{(\alpha - r)^2}{2\sigma^2}. \tag{44}$$

subject to $J(W, T) = 0$.[20] The equations for the optimal consumption and portfolio rules are

$$C^*(t) = \frac{(1-\gamma)}{\beta} \left[\frac{e^{\rho t} J_W}{\beta}\right]^{\frac{1}{\gamma-1}} - \frac{(1-\gamma)\,\eta}{\beta} \tag{45}$$

and

$$w^*(t) = -\frac{J_W}{J_{WW} W} \frac{(\alpha - r)}{\sigma^2}, \tag{46}$$

where $w^*(t)$ is the optimal proportion of wealth invested in the risky asset at time t. A solution[21] to (44) is

$$J(W, t) = \frac{\delta \beta^{+\gamma}}{\gamma} e^{-\rho t} \left[\frac{\delta\left(1 - e^{-\left(\frac{\rho - \gamma\nu}{\delta}\right)(T-t)}\right)}{\rho - \gamma\nu}\right]^{\delta} \left[\frac{W}{\delta} + \frac{\eta}{\beta r}\left(1 - e^{-r(T-t)}\right)\right]^{\gamma}, \tag{47}$$

where $\delta \equiv 1 - \gamma$ and $\nu \equiv r + (\alpha - r)^2/2\delta\sigma^2$.

From (45)–(47), the optimal consumption and portfolio rules can be written in explicit form as

$$C^*(t) = \frac{[\rho - \gamma\nu]\left[W(t) + \dfrac{\delta\eta}{\beta r}\left(1 - e^{r(t-T)}\right)\right]}{\delta\left(1 - \exp\left[\dfrac{(\rho - \gamma\nu)}{\delta}(t - T)\right]\right)} - \frac{\delta\eta}{\beta}. \tag{48}$$

and

$$w^*(t)\,W(t) = \frac{(\alpha - r)}{\delta\sigma^2} W(t) + \frac{\eta(\alpha - r)}{\beta r \sigma^2}\left(1 - e^{r(t-T)}\right). \tag{49}$$

[20] It is assumed for simplicity that the individual has a zero bequest function, i.e., $B \equiv 0$. If $B(W, T) = H(T)(aW + b)^\gamma$, the basic functional form for J in (47) will be the same. Otherwise, systematic effects of age will be involved in the solution.

[21] By Theorem I, there is no need to be concerned with uniqueness although, in this case, the solution is unique.

The manifest characteristic of (48) and (49) is that the demand functions are linear in wealth. It will be shown that the HARA family is the only class of concave utility functions which imply linear solutions. For notation purposes, define $I(X, t) \subset \text{HARA}(X)$ if $-I_{XX}/I_X = 1/(\alpha X + \beta) > 0$, where α and β are, at most, functions of time and I is a strictly concave function of X.

THEOREM III. *Given the model specified in this section, then* $C^* = aW + b$ *and* $w^*W = gW + h$ *where* a, b, g, *and* h *are, at most, functions of time if and only if* $U(C, t) \subset \text{HARA}(C)$.

Proof. "If" part is proved directly by (48) and (49). "Only if" part: Suppose $w^*W = gW + h$ and $C^* = aW + b$. From (19), we have that $U_C(C^*, t) = J_W(W, t)$. Differentiating this expression totally with respect to W, we have that $U_{CC}\, dC^*/dW = J_{WW}$ or $aU_{CC} = J_{WW}$ and hence

$$\frac{-U_{CC}a}{U_C} = \frac{-J_{WW}}{J_W}. \tag{50}$$

From (46), $w^*W = gW + h = -J_W(\alpha - r)/J_{WW}\sigma^2$ or

$$-J_{WW}/J_W = 1\Big/\left[\left(\frac{\sigma^2 g}{(\alpha - r)}\right) W + \frac{\sigma^2 h}{(\alpha - r)}\right]. \tag{51}$$

So, from (50) and (51), we have that U must satisfy

$$-U_{CC}/U_C = 1/(a'C^* + b'), \tag{52}$$

where $a' \equiv \sigma^2 g/(\alpha - r)$ and $b' \equiv (a\sigma^2 h - b\sigma^2 g)/(\alpha - r)$. Hence $U \subset \text{HARA}(C)$. Q.E.D.

As an immediate result of Theorem III, a second theorem can be proved.

THEOREM IV. *Given the model specified in this section,* $J(W, t) \subset \text{HARA}(W)$ *if and only if* $U \subset \text{HARA}(C)$.

Proof. "If" part is proved directly by (47). "Only if" part: suppose $J(W, t) \subset \text{HARA}(W)$. Then, from (46), w^*W is a linear function of W. If (28) is differentiated totally with respect to wealth and given the specific price behavior assumptions of this section, we have that C^* must satisfy

$$C^* = rW + \frac{J_{tW}}{J_{WW}} + \frac{rJ_W}{J_{WW}} - w^*W\frac{d(w^*W)}{dW}\sigma^2 - \frac{J_{WWW}}{2J_{WW}}\left(\frac{J_W}{J_{WW}}\right)^2\frac{(\alpha - r)^2}{\sigma^2}. \tag{53}$$

But if $J \subset \mathrm{HARA}(W)$, then (53) implies that C^* is linear in wealth. Hence, by Theorem III, $U \subset \mathrm{HARA}(C)$. Q.E.D.

Given (48) and (49), the stochastic process which generates wealth when the optimal rules are applied, can be derived. From the budget equation (14′), we have that

$$\begin{aligned} dW &= [(w^*(\alpha - r) + r) W - C^*] \, dt + \sigma w^* W \, dz \\ &= \left\{ \left[\frac{(\alpha - r)^2}{\sigma^2 \delta} - \frac{\mu}{1 - e^{\mu(t-T)}} \right] dt + \frac{(\alpha - r)}{\sigma\delta} dz \right\} X(t) \qquad (54) \\ &\quad + r \left[W + \frac{\delta\eta}{\beta r} \right] dt, \end{aligned}$$

where $X(t) \equiv W(t) + \delta\eta/\beta r(1 - e^{r(t-T)})$ for $0 \leqslant t \leqslant T$ and $\mu \equiv (\rho - \gamma\nu)/\delta$. By Itô's Lemma, $X(t)$ is the solution to

$$\frac{dX}{X} = \left[\delta - \frac{\mu}{(1 - e^{\mu(t-T)})} \right] dt + \frac{(\alpha - r)}{\sigma\delta} dz. \qquad (55)$$

Again using Itô's Lemma, integrating (55) we have that

$$\begin{aligned} X(t) = X(0) \exp \left\{ \left[\delta - \mu - \frac{(\alpha - r)^2}{2\sigma^2\delta^2} \right] t + \frac{(\alpha - r)}{\sigma\delta} \int_0^t dz \right\} \\ \times (1 - e^{\mu(t-T)})/(1 - e^{-\mu T}) \end{aligned} \qquad (56)$$

and, hence, $X(t)$ is log-normally distributed. Therefore,

$$W(t) = X(t) - \frac{\delta\eta}{\beta r} (1 - e^{r(t-T)})$$

is a "displaced" or "three-parameter" log-normally distributed random variable. By Itô's Lemma, solution (56) to (55) holds with probability one and because $W(t)$ is a continuous process, we have with probability one that

$$\lim_{t \to T} W(t) = 0. \qquad (57)$$

From (48), with probability one,

$$\lim_{t \to T} C^*(t) = 0. \qquad (58)$$

Further, from (48), $C^* + \delta\eta/\beta$ is proportional to $X(t)$ and from the definition of $U(C^*, t)$, $U(C^*, t)$ is a log-normally distributed random variable.[22] The following theorem shows that this result holds only if $U(C, t) \subset \text{HARA}(C)$.

THEOREM V. *Given the model specified in this section and the time-dependent random variable* $Y(t) \equiv U(C^*, t)$, *then* Y *is log-normally distributed if and only if* $U(C, t) \subset \text{HARA}(C)$.

Proof. "If" part: it was previously shown that if $U \subset \text{HARA}(C)$, then Y is log-normally distributed. "Only if" part: let $C^* \equiv g(W, t)$ and $w^*W \equiv f(W, t)$. By Itô's Lemma,

$$\begin{aligned} dY &= U_C\, dC^* + U_t\, dt + \tfrac{1}{2}U_{CC}(dC^*)^2, \\ dC^* &= g_W\, dW + g_t\, dt + \tfrac{1}{2}g_{WW}(dW)^2, \\ dW &= [f(\alpha - r) + rW - g]\, dt + \sigma f\, dz. \end{aligned} \tag{59}$$

Because $(dW)^2 = \sigma^2 f^2\, dt$, we have that

$$dC^* = [g_W f(\alpha - r) + g_W rW - gg_W + \tfrac{1}{2}g_{WW}\sigma^2 f^2 + g_t]\, dt + \sigma f g_W\, dz \tag{60}$$

and

$$\begin{aligned} dY = \{&U_C[g_W f(\alpha - r) + rg_W W - gg_W + \tfrac{1}{2}g_{WW}\sigma^2 f^2 + g_t] + U_t \\ &+ \tfrac{1}{2}U_{CC}\sigma^2 f^2 g_W{}^2\}\, dt + \sigma f g_W U_C\, dz. \end{aligned} \tag{61}$$

A necessary condition for Y to be log-normal is that Y satisfy

$$\frac{dY}{Y} = F(Y)\, dt + b\, dz, \tag{62}$$

where b is, at most, a function of time. If Y is log-normal, from (61) and (62), we have that

$$b(t) = \sigma f g_W U_C / U. \tag{63}$$

From the first-order conditions, f and g must satisfy

$$U_{CC} g_W = J_{WW}, \qquad f = -J_W(\alpha - r)/\sigma^2 J_{WW}. \tag{64}$$

[22] $$U = \frac{(1-\gamma)}{\gamma} e^{-\rho t}\left[\frac{\beta C}{1-\gamma} + \eta\right]^{\gamma}$$

and products and powers of log-normal variates are log normal with one exception: the logarithmic utility function ($\gamma = 0$) is a singular case where $U(C^*, t) = \log C^*$ is normally distributed.

But (63) and (64) imply that

$$bU/\sigma U_C = fg_W = -(\alpha - r)\, U_C/\sigma^2 U_{CC} \tag{65}$$

or

$$-U_{CC}/U_C = \eta(t)\, U_C/U, \tag{66}$$

where $\eta(t) \equiv (\alpha - r)/\sigma b(t)$. Integrating (66), we have that

$$U = [(\eta + 1)(C + \mu)\, \zeta(t)]^{\frac{1}{n+1}}, \tag{67}$$

where $\zeta(t)$ and μ are, at most, functions of time and, hence, $U \subset \mathrm{HARA}(C)$.
Q.E.D.

For the case when asset prices satisfy the "geometric" Brownian motion hypothesis and the individual's utility function is a member of the HARA family, the consumption-portfolio problem is completely solved. From (48) and (49), one could examine the effects of shifts in various parameters on the consumption and portfolio rules by the methods of comparative statics as was done for the isoelastic case in [12].

7. Noncapital Gains Income: Wages

In the previous sections, it was assumed that all income was generated by capital gains. If a (certain) wage income flow, $dy = Y(t)\, dt$, is introduced, the optimality equation (18) becomes

$$0 = \max_{\{C, w\}}[U(C, t) + \mathscr{L}(J)], \tag{68}$$

where the operator $\mathscr{L}$ is defined by $\mathscr{L} \equiv \mathscr{L} + Y(t)\, \partial/\partial W$. This new complication causes no particular computational difficulties. If a new control variable, $\tilde{C}(t)$, and new utility function, $V(\tilde{C}, t)$ are defined by $\tilde{C}(t) \equiv C(t) - Y(t)$ and $V(\tilde{C}, t) \equiv U(\tilde{C}(t) + Y(t), t)$, then (68) can be rewritten as

$$0 = \max_{\{\tilde{C}, w\}}[V(\tilde{C}, t) + \mathscr{L}[J]], \tag{69}$$

which is the same equation as the optimality equation (18) when there is no wage income and where consumption has been re-defined as consumption in excess of wage income.

In particular, if $Y(t) \equiv Y$, a constant, and $U \subset \mathrm{HARA}(C)$, then the optimal consumption and portfolio rules corresponding to (48) and (49) are

$$C^*(t) = \frac{[\rho - \gamma\nu]\left[W + \dfrac{Y(1 - e^{r(t-T)})}{r} + \dfrac{\delta\eta}{\beta r}(1 - e^{r(t-T)})\right]}{\delta(1 - \exp[(\rho - \gamma\nu)(t - T)/\delta])} - \frac{\delta\eta}{\beta} \tag{70}$$

and

$$w^* W = \frac{(\alpha - r)}{\delta\sigma^2}\left(W + \frac{Y(1 - e^{r(t-T)})}{r}\right) + \frac{(\alpha - r)\,\eta}{\beta r\sigma^2}(1 - e^{r(t-T)}). \tag{71}$$

Comparing (70) and (71) with (48) and (49), one finds that, in computing the optimal decision rules, the individual capitalizes the lifetime flow of wage income at the market (risk-free) rate of interest and then treats the capitalized value as an addition to the current stock of wealth.[23]

The introduction of a stochastic wage income will cause increased computational difficulties although the basic analysis is the same as for the no-wage income case. For a solution to a particular example of a stochastic wage problem, see example two of Section 8.

8. Poisson Processes

The previous analyses always assumed that the underlying stochastic processes were smooth functions of Brownian motions and, therefore, continuous in both the time and state spaces. Although such processes are reasonable models for price behavior of many types of liquid assets, they are rather poor models for the description of other types. The Poisson process is a continuous-time process which allows discrete (or discontinuous) changes in the variables. The simplest independent Poisson process defines the probability of an event occuring during a time interval of length h (where h is as small as you like) as follows:

$$\begin{aligned}
&\text{prob}\{\text{the event does not occur in the time interval } (t, t+h)\} \\
&\qquad = 1 - \lambda h + O(h), \\
&\text{prob}\{\text{the event occurs once in the time interval } (t, t+h)\} \\
&\qquad = \lambda h + O(h), \\
&\text{prob}\{\text{the event occurs more than once in the time interval} \\
&\qquad (t, t+h)\} = O(h),
\end{aligned} \tag{72}$$

[23] As Hakansson [6] has pointed out, (70) and (71) are consistent with the Friedman Permanent Income and the Modigliani Life-Cycle hypotheses. However, in general, this result will not hold.

where $O(h)$ is the asymptotic order symbol defined by

$$\psi(h) \text{ is } O(h) \qquad \text{if } \lim_{h \to 0}(\psi(h)/h) = 0 \tag{73}$$

and $\lambda =$ the mean number of occurrences per unit time.

Given the Poisson process, the "event" can be defined in a number of interesting ways. To illustrate the degree of latitude, three examples of applications of Poisson processes in the consumption-portfolio choice problem are presented below. Before examining these examples, it is first necessary to develop some of the mathematical properties of Poisson processes. There is a theory of stochastic differential equations for Poisson processes similar to the one for Brownian motion discussed in Section 2. Let $q(t)$ be an independent Poisson process with probability structure as described in (72). Let the event be that a state variable $x(t)$ has a jump in amplitude of size $\mathscr{S}$ where $\mathscr{S}$ is a random variable whose probability measure has compact support. Then, a Poisson differential equation for $x(t)$ can be written as

$$dx = f(x, t)\, dt + g(x, t)\, dq \tag{74}$$

and the corresponding differential generator, $\mathscr{L}_x$, is defined by

$$\mathscr{L}_x[h(x, t)] \equiv h_t + f(x, t)\, h_x + E_t\{\lambda[h(x + \mathscr{S}g, t) - h(x, t)]\}, \tag{75}$$

where "E_t" is the conditional expectation over the random variable $\mathscr{S}$, conditional on knowing $x(t) = x$, and where $h(x, t)$ is a C^1 function of x and t.[24] Further, Theorem I holds for Poisson processes.[25]

Returning to the consumption-portfolio problem, consider first the two-asset case. Assume that one asset is a common stock whose price is log-normally distributed and that the other asset is a "risky" bond which pays an instantaneous rate of interest r when not in default but, in the event of default, the price of the bond becomes zero.[26]

From (74), the process which generates the bond's price can be written as

$$dP = rP\, dt - P\, dq, \tag{76}$$

[24] For a short discussion of Poisson differential equations and a proof of (75) as well as other references, see Kushner [9, pp. 18–22].

[25] See Dreyfus [4, p. 225] and Kushner [9, Chap. IV].

[26] That the price of the bond is zero in the event of default is an extreme assumption made only to illustrate how a default can be treated in the analysis. One could made the more reasonable assumption that the price in the event of default is a random variable. The degree of computational difficulty caused by this more reasonable assumption will depend on the choice of distribution for the random variable as well as the utility function of the individual.

where dq is as previously defined and $\mathscr{S} \equiv 1$ with probability one. Substituting the explicit price dynamics into (14′), the budget equation becomes

$$dW = \{wW(\alpha - r) + rW - C\}\, dt + w\sigma W\, dz - (1 - w)\, W\, dq. \quad (77)$$

From (75), (77), and Theorem I, we have that the optimality equation can be written as

$$\begin{aligned} 0 = {} & U(C^*, t) + J_t(W, t) + \lambda[J(w^* W, t) - J(W, t)] \\ & + J_W(W, t)[(w^*(\alpha - r) + r)\, W - C^*] + \tfrac{1}{2} J_{WW}(W, t)\, \sigma^2 w^{*2} W^2, \end{aligned} \quad (78)$$

where C^* and w^* are determined by the implicit equations

$$0 = U_C(C^*, t) - J_W(W, t) \quad (79)$$

and

$$0 = \lambda J_W(w^* W, t) + J_W(W, t)(\alpha - r) + J_{WW}(W, t)\, \sigma^2 w^* W. \quad (80)$$

To see the effect of default on the portfolio and consumption decisions, consider the particular case when $U(C, t) \equiv C^\gamma/\gamma$, for $\gamma < 1$. The solutions to (79) and (80) are

$$C^*(t) = AW(t)/(1 - \gamma)(1 - \exp[A(t - T)/1 - \gamma]), \quad (79')$$

where

$$A \equiv -\gamma \left[\frac{(\alpha - r)^2}{2\sigma^2(1 - \gamma)} + r\right] + \lambda \left[1 - \frac{(2 - \gamma)}{\gamma}\, w^{*\gamma} - \frac{\gamma(\alpha - r)}{2\sigma^2(1 - \gamma)}\, w^{*\gamma - 1}\right]$$

and

$$w^* = \frac{(\alpha - r)}{\sigma^2(1 - \gamma)} + \frac{\lambda}{\sigma^2(1 - \gamma)}\, (w^*)^{\gamma - 1}. \quad (80')$$ [27]

As might be expected, the demand for the common stock is an increasing function of λ and, for $\lambda > 0$, $w^* > 0$ holds for all values of α, r, or σ^2.

For the second example, consider an individual who receives a wage, $Y(t)$, which is incremented by a constant amount ϵ at random points in time. Suppose that the event of a wage increase is a Poisson process with parameter λ. Then, the dynamics of the wage-rate state variable are described by

$$dY = \epsilon\, dq, \qquad \text{with } \mathscr{S} \equiv 1 \text{ with probability one.} \quad (81)$$

[27] Note that (79′) and (80′) with $\lambda = 0$ reduce to the solutions (48) and (49) when $\eta = \rho = 0$ and $\beta = 1 - \gamma$.

Suppose further that the individual's utility function is of the form $U(C, t) \equiv e^{-\rho t}V(C)$ and that his time horizon is infinite (i.e., $T = \infty$).[28] Then, for the two-asset case of Section 6, the optimality equation can be written as

$$\begin{aligned} 0 = {} & V(C^*) - \rho I(W, Y) + \lambda[I(W, Y + \epsilon) - I(W, Y)] \\ & + I_W(W, Y)[(w^*(\alpha - r) + r) W + Y - C^*] \\ & + \tfrac{1}{2} I_{WW}(W, Y)\, \sigma^2 w^{*2} W^2, \end{aligned} \tag{83}$$

where $I(W, Y) \equiv e^{\rho t}J(W, Y, t)$. If it is further assumed that $V(C) = -e^{-\eta C}/\eta$, then the optimal consumption and portfolio rules, derived from (83), are

$$C^*(t) = r\left[W(t) + \frac{Y(t)}{r} + \frac{\lambda}{r^2}\left(\frac{1 - e^{-\eta\epsilon}}{\eta}\right)\right] + \frac{1}{\eta r}\left[\rho - r + \frac{(\alpha - r)^2}{2\sigma^2}\right] \tag{84}$$

and

$$w^*(t)\, W(t) = \frac{(\alpha - r)}{\eta \sigma^2 r}. \tag{85}$$

In (84), $[W(t) + Y(t)/r + \lambda(1 - e^{-\eta\epsilon})/\eta r^2]$ is the general wealth term, equal to the sum of present wealth and capitalized future wage earnings. If $\lambda = 0$, then (84) reduces to (70) in Section 7, where the wage rate was fixed and known with certainty. When $\lambda > 0$, $\lambda(1 - e^{-\eta\epsilon})/\eta r^2$ is the capitalized value of (expected) future increments to the wage rate, *capitalized at a somewhat higher rate than the risk-free market rate reflecting the risk-aversion of the individual.*[29] Let $X(t)$ be the "Certainty-equivalent wage rate at time t" defined as the solution to

$$U[X(t)] = E_0 U[Y(t)]. \tag{86}$$

[28] I have shown elsewhere [12, p. 252] that if $U = e^{-\rho t}V(C)$ and U is bounded or ρ sufficiently large to ensure convergence of the integral and if the underlying stochastic processes are stationary, then the optimality equation (18) can be written, independent of explicit time, as

$$0 = \max_{\{C, w\}} [V(C) + \bar{\mathscr{L}}[I]], \tag{82}$$

where $\bar{\mathscr{L}} \equiv \mathscr{L} - \rho - \dfrac{\partial}{\partial t}$ and $I(W, P) \equiv e^{\rho t}J(W, P, t)$.

A solution to (82) is called the "stationary" solution to the consumption-portfolio problem. Because the time state variable is eliminated, solutions to (82) are computationally easier to find than for the finite-horizon case.

[29] The usual expected present discounted value of the increments to the wage flow is

$$E_t \int_t^\infty e^{-r(s-t)}\,[Y(s) - Y(t)]\, ds = \int_t^\infty \lambda\epsilon\, e^{-r(s-t)}(s - t)\, ds = \lambda\epsilon/r^2,$$

which is greater than $\lambda(1 - e^{-\eta\epsilon})/\eta r^2$ for $\epsilon > 0$.

For this example, $X(t)$ is calculated as follows:

$$-\frac{e^{-\eta X(t)}}{\eta} = -\frac{1}{\eta} E_0 e^{-\eta Y(t)}$$
$$= -\frac{1}{\eta} e^{-\eta Y(0)} \sum_{k=0}^{\infty} \frac{(\lambda t)^k}{k!} e^{-\lambda t} e^{-\eta k\epsilon} \tag{87}$$
$$= -\frac{1}{\eta} e^{-\eta Y(0)-\lambda t+\lambda e^{-\eta\epsilon}t}.$$

Solving for $X(t)$ from (87), we have that

$$X(t) = Y(0) + \lambda t(1 - e^{-\eta\epsilon})/\eta. \tag{88}$$

The capitalized value of the Certainty-equivalent wage income flow is

$$\int_0^\infty e^{-rs} X(s)\, ds = \int_0^\infty Y(0)\, e^{-rs}\, ds + \int_0^\infty \frac{\lambda(1 - e^{-\eta\epsilon})}{\eta} s e^{-rs}\, ds$$
$$= \frac{Y(0)}{r} + \frac{\lambda(1 - e^{-\eta\epsilon})}{\eta r^2}. \tag{89}$$

Thus, for this example,[30] the individual, in computing the present value of future earnings, determines the Certainty-equivalent flow and then capitalizes this flow at the (certain) market rate of interest.

The third example of a Poisson process differs from the first two because the occurrence of the event does not involve an explicit change in a state variable. Consider an individual whose age of death is a random variable. Further assume that the event of death at each instant of time is an independent Poisson process with parameter λ. Then, the age of death, τ, is the first time that the event (of death) occurs and is an exponentially distributed random variable with parameter λ. The optimality criterion is to

$$\max E_0 \left\{ \int_0^\tau U(C, t)\, dt + B(W(\tau), \tau) \right\} \tag{90}$$

and the associated optimality equation is

$$0 = U(C^*, t) + \lambda[B(W, t) - J(W, t)] + \mathscr{L}[J]. \tag{91}$$

[30] The reader should not infer that this result holds in general. Although (86) is a common definition of Certainty-equivalent in one-period utility-of-wealth models, it is not satisfactory for dynamic consumption-portfolio models. The reason it works for this example is due to the particular relationship between the J and U functions when U is exponential.

To derive (91), an "artificial" state variable, $x(t)$, is constructed with $x(t) = 0$ while the individual is alive and $x(t) = 1$ in the event of death. Therefore, the stochastic process which generates x is defined by

$$dx = dq \qquad \text{and} \qquad \mathscr{S} \equiv 1 \text{ with probability one} \tag{92}$$

and τ is now defined by x as

$$\tau = \min\{t \mid t > 0 \text{ and } x(t) = 1\}. \tag{93}$$

The derived utility function, J, can be considered a function of the state variables W, x, and t subject to the boundary condition

$$J(W, x, t) = B(W, t) \qquad \text{when} \quad x = 1. \tag{94}$$

In this form, example three is shown to be of the same type as examples one and two in that the occurrence of the Poisson event causes a state variable to be incremented, and (91) is of the same form as (78) and (83).

A comparison of (91) for the particular case when $B \equiv 0$ (no bequests) with (82) suggested the following theorem.[31]

THEOREM VI. *If τ is as defined in* (93) *and U is such that the integral $E_0[\int_0^\tau U(C, t)\, dt]$ is absolutely convergent, then the maximization of $E_0[\int_0^\tau U(C, t)\, dt]$ is equivalent to the maximization of $\mathscr{E}_0[\int_0^\infty e^{-\lambda t} U(C, t)\, dt]$ where "E_0" is the conditional expectation operator over all random variables including τ and "$\mathscr{E}_0$" is the conditional expectation operator over all random variables excluding τ.*

Proof. τ is distributed exponentially and is independent of the other random variables in the problem. Hence, we have that

$$\begin{aligned} E_0\left[\int_0^\tau U(C, t)\, dt\right] &= \int_0^\infty \lambda e^{-\lambda\tau}\, d\tau\, \mathscr{E}_0 \int_0^\tau U(C, t)\, dt \\ &= \int_0^\infty \int_0^\tau \lambda g(t)\, e^{-\lambda\tau}\, dt\, d\tau, \end{aligned} \tag{95}$$

where $g(t) \equiv \mathscr{E}_0[U(C, t)]$. Because the integral in (95) is absolutely con-

[31] I believe that a similar theorem has been proved by J. A. Mirrlees, but I have no reference. D. Cass and M. E. Yaari, in "Individual Saving, Aggregate Capital Accumulation, and Efficient Growth," in "Essays on the Theory of Optimal Economic Growth," ed., K. Shell, (M.I.T. Press 1967), prove a similar theorem on page 262.

vergent, the order of integration can be interchanged, i.e., $\mathscr{E}_0 \int_0^T U(C,t)\,dt = \int_0^T \mathscr{E}_0 U(C, t)\,dt$. By integration by parts, (95) can be rewritten as

$$\int_0^\infty \int_0^T e^{-\lambda\tau} g(t)\,dt\,d\tau = \int_0^\infty e^{-\lambda s} g(s)\,ds = \mathscr{E}_0 \int_0^\infty e^{-\lambda t} U(C, t)\,dt. \qquad \text{Q.E.D.} \tag{96}$$

Thus, an individual who faces an exponentially-distributed uncertain age of death acts as if he will live forever, but with a subjective rate of time preference equal to his "force of mortality," i.e., to the reciprocal of his life expectancy.

9. Alternative Price Expectations to the Geometric Brownian Motion

The assumption of the geometric Brownian motion hypothesis is a rich one because it is a reasonably good model of observed stock price behavior and it allows the proof of a number of strong theorems about the optimal consumption-portfolio rules, as was illustrated in the previous sections. However, as mentioned in the Introduction, there have been some disagreements with the underlying assumptions required to accept this hypothesis. The geometric Brownian motion hypothesis best describes a stationary equilibrium economy where expectations about future returns have settled down, and as such, really describes a "long-run" equilibrium model for asset prices. Therefore, to explain "short-run" consumption and portfolio selection behavior one must introduce alternative models of price behavior which reflect the dynamic adjustment of expectations.

In this section, alternative price behavior mechanisms are postulated which attempt to capture in a simple fashion the effects of changing expectations, and then comparisons are made between the optimal decision rules derived under these mechanisms with the ones derived in the previous sections. The choices of mechanisms are not exhaustive nor are they necessarily representative of observed asset price behavior. Rather they have been chosen as representative examples of price adjustment mechanisms commonly used in economic and financial models.

Little can be said in general about the form of a solution to (28) when α_k and σ_k depend in an arbitrary manner on the price levels. If it is specified that the utility function is a member of the HARA family, i.e.,

$$U(C, t) = \frac{(1-\gamma)}{\gamma} F(t) \left(\frac{\beta C}{1-\gamma} + \eta\right)^\gamma \tag{97}$$

subject to the restrictions in (42), then (28) can be simplified because $J(W, P, t)$ is separable into a product of functions, one depending on W and t, and the other on P and t.[32] In particular, if we take $J(W, P, t)$ to be of the form

$$J(W, P, t) = \frac{(1-\gamma)}{\gamma} H(P, t)\, F(t) \left(\frac{W}{1-\gamma} + \frac{\eta}{\beta r}\,[1 - e^{r(t-T)}]\right)^{\gamma}, \quad (98)$$

substitute for J in (28), and divide out the common factor

$$F(t)\left(\frac{W}{1-\gamma} + \frac{\eta}{\beta r}\left[1 - e^{r(t-T)}\right]\right)^{\gamma},$$

then we derive a "reduced" equation for H,

$$\begin{aligned} 0 = {} & \frac{(1-\gamma)^2}{\gamma}\left(\frac{H}{\beta}\right)^{\frac{\gamma}{\gamma-1}} + \frac{(1-\gamma)}{\gamma}\left(\frac{\dot{F}}{F} + H_t\right) + (1-\gamma)\, rH \\ & + \frac{(1-\gamma)}{\gamma}\sum_1^m \alpha_i P_i H_i + \frac{(1-\gamma)}{2\gamma}\sum_1^m\sum_1^m \sigma_{ij} P_i P_j H_{ij} \\ & + \sum_1^m (\alpha_i - r)\, P_i H_i + \frac{H}{2}\sum_1^m\sum_1^m v_{ij}(\alpha_i - r)(\alpha_j - r) \\ & + \frac{1}{2H}\sum_1^m\sum_1^m \sigma_{ij} P_i P_j H_i H_j \end{aligned} \quad (99)$$

and the associated optimal consumption and portfolio rules are

$$C^*(t) = \frac{(1-\gamma)}{\beta}\left[\left(\frac{H}{\beta}\right)^{\frac{1}{\gamma-1}}\left(\frac{W}{1-\gamma} + \frac{\eta}{\beta r}\,[1 - e^{r(t-T)}]\right) - \eta\right] \quad (100)$$

and

$$w_k^*(t)\, W = \left[\sum_1^m v_{jk}(\alpha_j - r) + \frac{H_k P_k}{H}\right]\left(\frac{W}{1-\gamma} + \frac{\eta}{\beta r}\,[1 - e^{r(t-T)}]\right), \quad (101)$$

$$k = 1, \ldots, m.$$

Although (99) is still a formidable equation from a computational point of view, it is less complex than the general equation (28), and it is possible

[32] This separability property was noted in [1, 5, 6, 10, 12, and 13]. It is assumed throughout this section that the bequest function satisfies the conditions of footnote 20.

to obtain an explicit solution for particular assumptions about the dependence of α_k and σ_k on the prices. Notice that both consumption and the asset demands are linear functions of wealth.

For a particular member of the HARA family, namely the Bernoulli logarithmic utility ($\gamma = 0 = \eta$ and $\beta = 1 - \gamma = 1$) function, (28) can be solved in general. In this case, J will be of the form

$$J(W, P, t) = a(t) \log W + H(P, t) \qquad \text{with} \quad H(P, T) = a(T) = 0, \quad (102)$$

with $a(t)$ independent of the α_k and σ_k (and hence, the P_k). For the case when $F(t) \equiv 1$, we find $a(t) = T - t$ and the optimal rules become

$$C^* = \frac{W}{T - t} \quad (103)$$

and

$$w_k^* = \sum_1^m v_{kj}(\alpha_j - r), \qquad k = 1, \ldots, m. \quad (104)$$

For the log case, the optimal rules are identical to those derived when α_k and σ_k were constants, with the understanding that the α_k and σ_k are evaluated at current prices. Hence, although we can solve this case for general price mechanisms, it is not an interesting one because different assumptions about price behavior have no effect on the decision rules.

The first of the alternative price mechanisms considered is called the "asymptotic 'normal' price-level" hypothesis which assumes that there exists a "normal" price function, $\bar{P}(t)$, such that

$$\lim_{t \to \infty} E_T[P(t)/\bar{P}(t)] = 1, \qquad \text{for} \quad 0 \leqslant T < t < \infty, \quad (105)$$

i.e., independent of the current level of the asset price, the investor expects the "long-run" price to approach the normal price. A particular example which satisfies the hypothesis is that

$$\bar{P}(t) = \bar{P}(0)\, e^{\nu t} \quad (106)$$

and

$$\frac{dP}{P} = \beta[\phi + \nu t - \log(P(t)/P(0))]\, dt + \sigma\, dz, \quad (107)$$

where $\phi \equiv k + \nu/\beta + \sigma^2/4\beta$ and $k \equiv \log[\bar{P}(0)/P(0)]$.[33] For the purpose of analysis, it is more convenient to work with the variable

[33] In the notation used in previous sections, (107) corresponds to (5) with $\alpha(P, t) \equiv \beta[\phi + \nu t - \log(P(t)/P(0))]$. Note: "normal" does not mean "Gaussian" in the above use, but rather the normal long-run price of Alfred Marshall.

$Y(t) \equiv \log[P(t)/P(0)]$ rather than $P(t)$. Substituting for P in (107) by using Itô's Lemma, we can write the dynamics for Y as

$$dY = \beta[\mu + \nu t - Y]\,dt + \sigma\,dz, \tag{108}$$

where $\mu \equiv \phi - \sigma^2/2\beta$. Before examining the effects of this price mechanism on the optimal portfolio decisions, it is useful to investigate the price behavior implied by (106) and (107). (107) implies an exponentially-regressive price adjustment toward a normal price, adjusted for trend. By inspection of (108), Y is a normally-distributed random variable generated by a Markov process which is not stationary and does not have independent increments.[34] Therefore, from the definition of Y, $P(t)$ is log-normal and Markov. Using Itô's Lemma, one can solve (108) for $Y(t)$, conditional on knowing $Y(T)$, as

$$Y(t) - Y(T) = \left(k + \nu T - \frac{\sigma^2}{4\beta} - Y(T)\right)(1 - e^{-\beta\tau}) + \nu\tau + \sigma e^{-\beta t}\int_T^t e^{\beta s}\,dz, \tag{109}$$

where $\tau \equiv t - T > 0$. The instantaneous conditional variance of $Y(t)$ is

$$\operatorname{var}[Y(t) \mid Y(T)] = \frac{\sigma^2}{2\beta}(1 - e^{-2\beta\tau}). \tag{110}$$

Given the characteristics of $Y(t)$, it is straightforward to derive the price behavior. For example, the conditional expected price can be derived from (110) and written as

$$\begin{aligned} &E_T(P(t)/P(T)) \\ &\quad = E_T \exp[Y(t) - Y(T)] \\ &\quad = \exp\left[\left(k + \nu T - \frac{\sigma^2}{4\beta} - Y(T)\right)(1 - e^{-\beta\tau}) + \nu\tau + \frac{\sigma^2}{4\beta}(1 - e^{-2\beta\tau})\right]. \end{aligned} \tag{111}$$

It is easy to verify that (105) holds by applying the appropriate limit process to (111). Figure 2 illustrates the behavior of the conditional expectation mechanism over time.

For computational simplicity in deriving the optimal consumption and portfolio rules, the two-asset model is used with the individual having an infinite time horizon and a constant absolute risk-aversion utility

[34] Processes such as (108) are called Ornstein–Uhlenbeck processes and are discussed, for example, in [3, p. 225].

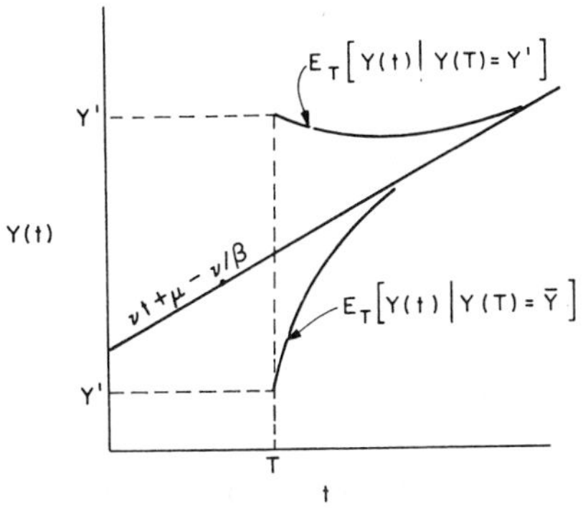

FIG. 2. The time-pattern of the expected value of the logarithm of price under the "normal" price-level hypothesis.

function, $U(C, t) = -e^{-\eta C}/\eta$. The fundamental optimality equation then is written as

$$0 = -e^{-\eta C^*}/\eta + J_t + J_W[w^*(\beta(\phi + \nu t - Y) - r)\, W + rW - C^*] + \tfrac{1}{2}J_{WW}w^{*2}W^2\sigma^2 + J_Y\beta(\mu + \nu t - Y) + \tfrac{1}{2}J_{YY}\sigma^2 + J_{YW}w^*W\sigma^2 \quad (112)$$

and the associated equations for the optimal rules are

$$w^*W = -J_W[\beta(\phi + \nu t - Y) - r]/J_{WW}\sigma^2 - J_{YW}/J_{WW} \quad (113)$$

and

$$C^* = -\log(J_W)/\eta. \quad (114)$$

Solving (112), (113), and (114), we write the optimal rules in explicit form as

$$w^*W = \frac{1}{\eta r\sigma^2}\left[\left(1 + \frac{\beta}{r}\right)(\alpha(P, t) - r) + \frac{\beta^2}{r^2}\left(\frac{\sigma^2}{2} + \nu - r\right)\right] \quad (115)$$

and

$$C^* = rW + \frac{\beta^2}{2\sigma^2\eta r}\, Y^2 - \frac{\beta}{\eta r\sigma^2}\left(\beta\nu t + \beta\phi - r + \beta\left(\nu + \frac{\sigma^2}{2} - r\right)\right) Y + a(t), \quad (116)$$ [35]

[35] $$a(t) \equiv \frac{1}{\eta}\left\{\frac{r}{2\sigma^2} - 1 + \frac{\beta}{\sigma^2}\left(\phi - 1 - \frac{\sigma^2}{2r}\right) + \frac{\beta^2}{r\sigma^2}\left[\left(1 - \frac{\sigma^2}{2r}\right)\left(\phi + \frac{\nu}{r} + \frac{\sigma^2}{2r} - 1\right) - \frac{\phi^2}{2} - \frac{\sigma^2}{2r}\right] + \frac{\beta\nu}{\sigma^2 r^2}\left(r + \beta - \beta\phi - \frac{\beta\sigma^2}{2r}\right) - \frac{\beta^2\nu^2}{2\sigma^2 r^3} + \frac{\beta\nu t}{r\sigma^2}\left[r + \beta - \beta\phi - \frac{\beta\sigma^2}{2r}\right] - \frac{\beta^2\nu^2 t^2}{2\sigma^2 r}\right\}.$$

where $\alpha(P, t)$ is the instantaneous expected rate of return defined explicitly in footnote 33. To provide a basis for comparison, the solutions when the geometric Brownian motion hypothesis is assumed are presented as[36]

$$w^*W = \frac{(\alpha - r)}{\eta r \sigma^2} \tag{117}$$

and

$$C^* = rW + \frac{1}{\eta r}\left[\frac{(\alpha - r)^2}{2\sigma^2} - r\right]. \tag{118}$$

To examine the effects of the alternative "normal price" hypothesis on the consumption-portfolio decisions, the (constant) α of (117) and (118) is chosen equal to $\alpha(P, t)$ of (115) and (116) so that, in both cases, the *instantaneous* expected return and variance are the same at the point of time of comparison. Comparing (115) with (117), we find that the proportion of wealth invested in the risky asset is always larger under the "normal price" hypothesis than under the geometric Brownian motion hypothesis.[37] In particular, notice that even if $\alpha < r$, unlike in the geometric Brownian motion case, a positive amount of the risky asset is held. Figures 3a and 3b illustrate the behavior of the optimal portfolio holdings.

The most striking feature of this analysis is that, despite the ability to make continuous portfolio adjustments, a person who believes that prices satisfy the "normal" price hypothesis will hold more of the risky asset than one who believes that prices satisfy the geometric Brownian motion hypothesis, even though they both have the same utility function and the same expectations about the instantaneous mean and variance.

The primary interest in examining these alternative price mechanisms is to see the effects on portfolio behavior, and so, little will be said about the effects on consumption other than to present the optimal rule.

The second alternative price mechanism assumes the same type of price-dynamics equation as was assumed for the geometric Brownian motion, namely,

$$\frac{dP}{P} = \alpha\, dt + \sigma\, dz. \tag{119}$$

However, instead of the instantaneous expected rate of return α being a

[36] For a derivation of (117) and (118), see [12, p. 256].

[37] It is assumed that $\nu + \sigma^2/2 > r$, i.e., the "long-run" rate of growth of the "normal" price is greater than the sure rate of interest so that something of the risky asset will be held in the short and long run.

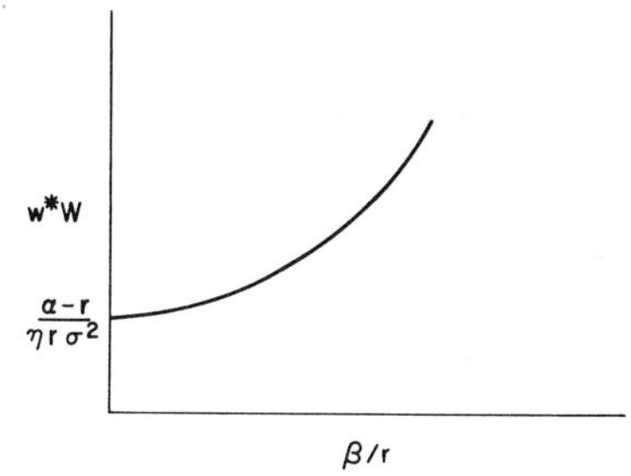

FIG. 3a. The demand for the risky asset as a function of the speed of adjustment

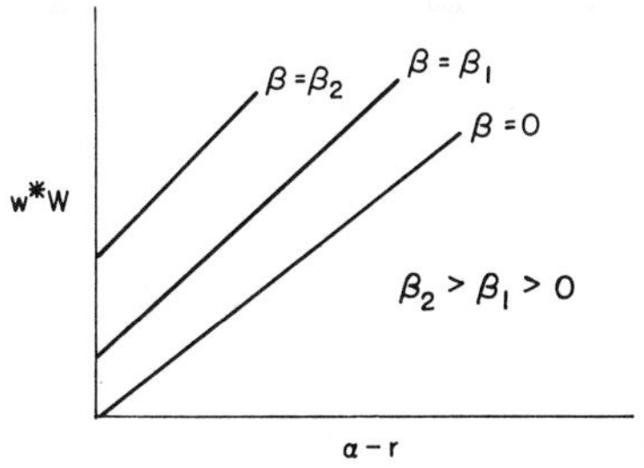

FIG. 3b. The demand for the risky asset as a function of the expected return.

constant, it is assumed that α is itself generated by the stochastic differential equation

$$\begin{aligned} d\alpha &= \beta(\mu - \alpha)\,dt + \delta\left(\frac{dP}{P} - \alpha\,dt\right) \\ &= \beta(\mu - \alpha)\,dt + \delta\sigma\,dz. \end{aligned} \tag{120}$$

The first term in (120) implies a long-run, regressive adjustment of the expected rate of return toward a "normal" rate of return, μ, where β is the speed of adjustment. The second term in (120) implies a short-run, extrapolative adjustment of the expected rate of return of the "error-learning" type, where δ is the speed of adjustment. I will call the assumption of a price mechanism described by (119) and (120) the "De Leeuw" hypothesis for Frank De Leeuw who first introduced this type mechanism to explain interest rate behavior.

To examine the price behavior implied by (119) and (120), we first derive the behavior of α, and then P. The equation for α, (120), is of the same type as (108) described previously. Hence, α is normally distributed and is generated by a Markov process. The solution of (120), conditional on knowing $\alpha(T)$ is

$$\alpha(t) - \alpha(T) = (\mu - \alpha(T))(1 - e^{-\beta\tau}) + \delta\sigma e^{-\beta t} \int_T^t e^{\beta s}\, dz, \qquad (121)$$

where $\tau \equiv t - T > 0$. From (121), the conditional mean and variance of $\alpha(t) - \alpha(T)$ are

$$E_T(\alpha(t) - \alpha(T)) = (\mu - \alpha(T))(1 - e^{-\beta\tau}) \qquad (122)$$

and

$$\text{var}[\alpha(t) - \alpha(T) \mid \alpha(T)] = \frac{\delta^2\sigma^2}{2\beta}(1 - e^{-2\beta\tau}). \qquad (123)$$

To derive the dynamics of P, note that, unlike α, P is not Markov although the joint process $[P, \alpha]$ is. Combining the results derived for $\alpha(t)$ with (119), we solve directly for the price, conditional on knowing $P(T)$ and $\alpha(T)$,

$$Y(t) - Y(T) = (\mu - \tfrac{1}{2}\sigma^2)\,\tau - \frac{(\mu - \alpha(T))}{\beta}(1 - e^{-\beta\tau}) + \sigma\delta \int_T^t \int_T^s e^{-\beta(s-s')}\, dz(s')\, ds + \sigma \int_T^t dz, \qquad (124)$$

where $Y(t) \equiv \log[P(t)]$. From (124), the conditional mean and variance of $Y(t) - Y(T)$ are

$$E_T[Y(t) - Y(T)] = (\mu - \tfrac{1}{2}\sigma^2)\,\tau - \frac{(\mu - \alpha(T))}{\beta}(1 - e^{-\beta\tau}) \qquad (125)$$

and

$$\text{var}[Y(t) - Y(T) \mid Y(T)] = \sigma^2\tau + \frac{\sigma^2\delta^2}{2\beta^3}[\beta\tau - 2(1 - e^{-\beta\tau}) + \tfrac{1}{2}(1 - e^{-2\beta\tau})] + \frac{2\delta\sigma^2}{\beta^2}[\beta\tau - (1 - e^{-\beta\tau})]. \qquad (126)$$

Since $P(t)$ is log-normal, it is straightforward to derive the moments for $P(t)$ from (124)–(126). Figure 4 illustrates the behavior of the expected price mechanism. The equilibrium or "long-run" (i.e., $\tau \to \infty$) distribution for $\alpha(t)$ is stationary gaussian with mean μ and variance $\delta^2\sigma^2/2\beta$, and the equilibrium distribution for $P(t)/P(T)$ is a stationary log-normal. Hence, the long-run behavior of prices under the De Leeuw hypothesis approaches the geometric Brownian motion.

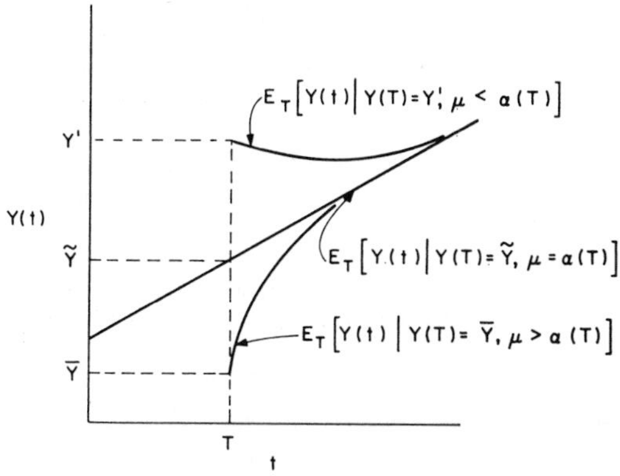

FIG. 4. The time-pattern of the expected value of the logarithm of price under the De Leeuw hypothesis.

Again, the two-asset model is used with the individual having an infinite time horizon and a constant absolute risk-aversion utility function, $U(C, t) = -e^{-\eta C}/\eta$. The fundamental optimality equation is written as

$$0 = -\frac{e^{-\eta C^*}}{\eta} + J_t + J_W[w^*(\alpha - r)\,W + rW - C^*]$$

$$+ \tfrac{1}{2}J_{WW}w^{*2}W^2\sigma^2 + J_\alpha\beta(\mu - \alpha) + \tfrac{1}{2}J_{\alpha\alpha}\delta^2\sigma^2 + J_{W\alpha}\delta\sigma^2 w^* W. \quad (127)$$

Notice that the state variables of the problem are W and α, which are both Markov, as is required for the dynamic programming technique. The optimal portfolio rule derived from (127) is,

$$w^*W = -\frac{J_W(\alpha - r)}{J_{WW}\sigma^2} - \frac{J_{W\alpha}\delta}{J_{WW}}. \quad (128)$$

The optimal consumption rule is the same as in (114). Solving (127) and (128), the explicit solution for the portfolio rule is

$$w^*W = \frac{1}{\eta r\sigma^2(r + 2\delta + 2\beta)}\left[(r + \delta + 2\beta)(\alpha - r) - \frac{\delta\beta(\mu - r)}{r + \delta + \beta}\right]. \quad (129)$$

Comparing (129) with (127) and assuming that $\mu > r$, we find that under the De Leeuw hypothesis, the individual will hold a smaller amount of the risky asset than under the geometric Brownian motion hypothesis. Note also that w^*W is a decreasing function of the long-run normal rate of return μ. The interpretation of this result is that as μ increases for a given α, the probability increases that future "α's" will be more favorable relative to the current α, and so there is a tendency to hold more of one's current wealth in the risk-free asset as a "reserve" for investment under more favorable conditions.

The last type of price mechanism examined differs from the previous two in that it is assumed that prices satisfy the geometric Brownian motion hypothesis. However, it is also assumed that the investor does not know the true value of the parameter α, but must estimate it from past data. Suppose P is generated by equation (119) with α and σ constants, and the investor has price data back to time $-\tau$. Then, the best estimator for α, $\hat{\alpha}(t)$, is

$$\hat{\alpha}(t) = \frac{1}{t+\tau}\int_{-\tau}^{t} \frac{dP}{P}, \tag{130}$$

where we assume, arbitrarily, that $\hat{\alpha}(-\tau) = 0$. From (130), we have that $E(\hat{\alpha}(t)) = \alpha$, and so, if we define the error term $\epsilon_t \equiv \alpha - \hat{\alpha}(t)$, then (119) can be re-written as

$$\frac{dP}{P} = \hat{\alpha}\, dt + \sigma\, d\hat{z}, \tag{131}$$

where $d\hat{z} \equiv dz + \epsilon_t\, dt/\sigma$. Further, by differentiating (130), we have the dynamics for $\hat{\alpha}$, namely

$$d\hat{\alpha} = \frac{\sigma}{t+\tau}\, d\hat{z}. \tag{132}$$

Comparing (131) and (132) with (119) and (120), we see that this "learning" model is equivalent to the special case of the De Leeuw hypothesis of pure extrapolation (i.e., $\beta = 0$), where the degree of extrapolation (δ) is decreasing over time. If the two-asset model is assumed with an investor who lives to time T with a constant absolute risk-aversion utility function, and if (for computational simplicity) the risk-free asset is money (i.e., $r = 0$), then the optimal portfolio rule is

$$w^*W = \frac{(t+\tau)}{\eta\sigma^2}\log\left(\frac{T+\tau}{t+\tau}\right)\hat{\alpha}(t) \tag{133}$$

and the optimal consumption rule is

$$\begin{aligned} C^* = \frac{W}{T-t} - \frac{1}{\eta}\Big[&\log(T+\tau) \\ &+ \frac{2}{T-t}(T - t - (T+\tau)\log(T+\tau) + (t+\tau)\log(t+\tau)) \\ &+ \frac{\hat{\alpha}^2}{2\sigma^2}\left[\frac{(t+\tau)^2}{(T-t)}\log\left(\frac{T+\tau}{t+\tau}\right) - \frac{(T-t)}{t+\tau}\right]\Big]. \end{aligned} \tag{134}$$

By differentiating (133) with respect to t, we find that w^*W is an increasing function of time for $t < \bar{t}$, reaches a maximum at $t = \bar{t}$, and then is a decreasing function of time for $\bar{t} < t < T$, where $\bar{t}$ is defined by

$$\bar{t} = [T + (1 - e)\,\tau]/e. \tag{135}$$

The reason for this behavior is that, early in life (i.e. for $t < \bar{t}$), the investor learns more about the price equation with each observation, and hence investment in the risky asset becomes more attractive. However, as he approaches the end of life (i.e., for $t > \bar{t}$), he is generally liquidating his portfolio to consume a larger fraction of his wealth, so that although investment in the risky asset is more favorable, the absolute dollar amount invested in the risky asset declines.

Consider the effect on (133) of increasing the number of available previous observations (i.e., increase τ). As expected, the dollar amount invested in the risky asset increases monotonically. Taking the limit of (133) as $\tau \to \infty$, we have that the optimal portfolio rule is

$$w^*W = \frac{(T - t)}{\eta\sigma^2}\,\alpha \qquad \text{as} \quad \tau \to \infty, \tag{136}$$

which is the optimal rule for the geometric Brownian motion case when α is known with certainty. Figure 5 illustrates graphically how the optimal rule changes with τ.

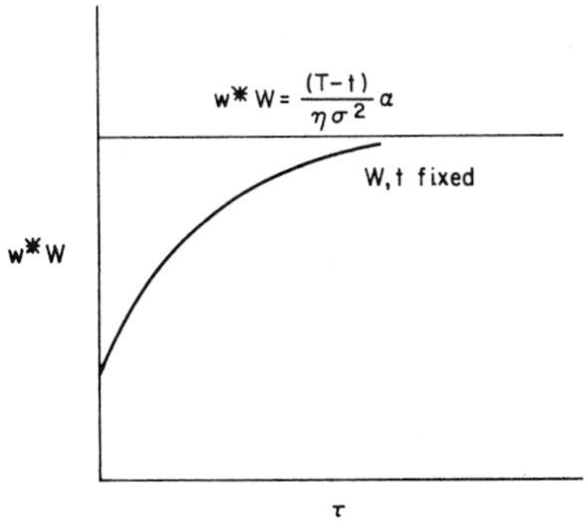

FIG. 5. The demand for the risky asset as a function of the number of previous price observations.

10. Conclusion

By the introduction of Itô's Lemma and the Fundamental Theorem of Stochastic Dynamic Programming (Theorem I), we have shown how to construct systematically and analyze optimal continuous-time dynamic models under uncertainty. The basic methods employed in studying the consumption-portfolio problem are applicable to a wide class of economic models of decision making under uncertainty.

A major advantage of the continuous-time model over its discrete time analog is that one need only consider two types of stochastic processes: functions of Brownian motions and Poisson processes. This result limits the number of parameters in the problem and allows one to take full advantage of the enormous amount of literature written about these processes. Although I have not done so here, it is straightforward to show that the limits of the discrete-time model solutions as the period spacing goes to zero are the solutions of the continuous-time model.[38]

A basic simplification gained by using the continuous-time model to analyze the consumption-portfolio problem is the justification of the Tobin–Markowitz portfolio efficiency conditions in the important case when asset price changes are stationarily and log-normally distributed. With earlier writers (Hakansson [6], Leland [10], Fischer [5], Samuelson [13], and Cass and Stiglitz [1]), we have shown that the assumption of the HARA utility function family simplifies the analysis and a number of strong theorems were proved about the optimal solutions. The introduction of stochastic wage income, risk of default, uncertainty about life expectancy, and alternative types of price dynamics serve to illustrate the power of the techniques as well as to provide insight into the effects of these complications on the optimal rules.

References

1. D. Cass and J. E. Stiglitz, The structure of investor perferences and asset Returns, and Separability in Portfolio Allocation: A Contribution to the Pure Theory of Mutual Funds," *J. Econ. Theory* **2** (1970), 122–160.
2. "The Random Character of Stock Market Prices," (P. Cootner, Ed.), Massachusetts Institute of Technology Press, Cambridge, Mass., 1964.
3. D. A. Cox and H. D. Miller, "The Theory of Stochastic Processes," John Wiley and Sons, New York, 1968.
4. S. E. Dreyfus, "Dynamic Programming and the Calculus of Variations," Academic Press, New York, 1965.

[38] For a general discussion of this result, see Samuelson [14].

5. S. FISCHER, Essays on assets and contingent commodities, Ph.D. Dissertation, Department of Economics, Massachusetts Institute of Technology, August, 1969.
6. N. H. HAKANSSON, Optimal investment and consumption strategies under risk for a class of utility functions, *Econometrica* to appear.
7. K. ITÔ, On stochastic differential equations, *Mem. Amer. Math. Soc.* No. 4 1951.
8. K. ITÔ AND H. P. MCKEAN, JR., "Diffusion Processes and Their Sample Paths, Academic Press, New York, 1964.
9. H. J. KUSHNER, "Stochastic Stability and Control," Academic Press, New York, 1967.
10. H. E. LELAND, Dynamic portfolio theory, Ph.D. Dissertation Department of Economics, Harvard University, May, 1968.
11. H. P. MCKEAN, JR., "Stochastic Integrals," Academic Press, New York, 1969.
12. R. C. MERTON, Lifetime portfolio selection under uncertainty: the continuous-time case, *Rev. Econ. Statist.* LI (August, 1969), 247–257.
13. P. A. SAMUELSON, Lifetime Portfolio Selection by Dynamic Stochastic Programming, *Rev. Econ. Statist.* LI (August 1969), 239–246.
14. P. A. SAMUELSON, The fundamental approximation theorem of portfolio analysis in terms of means, variances and higher moments, *Rev. Econ. Stud.* (October, 1970).
15. R. L. STRATONOVICH, "Conditional Markov Processes and Their Application to the Theory of Optimal Control," American Elsevier, New York, 1968.
16. F. BLACK, "Individual Investment and Consumption Strategies Under Uncertainty," Associates in Finance Financial Note No. 6C, September, 1970.

ERRATA

p. 382 (630). Footnote 16 should read:

$$m(P, W, t) \equiv -J_W/WJ_{ww};$$

$$g(P, t) \equiv \frac{1}{\Gamma}\sum_1^n v_{kl}\left(\Gamma\alpha_1 - \sum_1^n\sum_1^n v_{ij}\alpha_j\right).$$

p. 384 (632). Lines 1 and 2 of Theorem II should read:

THEOREM II.[17] *Given n assets with prices P_i whose changes are stationarily and log-normally distributed, then* (1) *there exists a unique* (*up to a nonsingular*)

p. 386 (634). Last line should read:

$$k = 1, \ldots, m$$

p. 387 (635). Line 8 should read:

the corollary with $a(W, t; U) = vm(W, t) + \eta$, where v, η are arbitrary

Line 21 should read:

asset (i.e., $\eta = 1$), and the other fund to hold only risky assets (which

COMPUTATIONAL AND REVIEW EXERCISES

1. In the Drèze–Modligliani article it is convenient to assume that both present and future consumption are not inferior goods or that the marginal propensity to consume is positive but less than unity.

(a) Show that this condition requires that

$$(U_1/U_2)\,U_{22} - U_{12} < 0 \qquad \text{and} \qquad U_{11} - (U_1/U_2)\,U_{12} < 0,$$

where $r > 0$ is the borrowing–lending rate of interest, U is the utility function over consumption C_1, C_2, and subscripts refer to partial differentiation.

(b) Show that the conditions are satisfied when $U(\cdot) = \log C_1 + \log C_2$ but not when $U(\cdot) = \log(C_1 + C_2)$. What happens when $U(\cdot) = \log(\alpha C_1 + \beta C_2)$ for $\alpha, \beta > 0$?

(c) Show that the conditions are satisfied when $U(\cdot) = C_1^{\alpha} C_2^{1-\alpha}$ for $0 < \alpha < 1$.

2. Consider an investor with \$1000 to invest over two investment periods between a risky and a risk-free asset. The risk-free asset returns 5% per period. In period 1 the risky asset returns 0, 5, or 10% with equal probabilities. The returns in period 2 for the risky asset are

$$p_r[r_2 = 3 \mid r_1 = 0] = \tfrac{1}{2}, \qquad p_r[r_2 = 5 \mid r_1 = 0] = \tfrac{1}{2}, \qquad p_r[r_2 = 3 \mid r_1 = 5] = \tfrac{1}{3},$$
$$p_r[r_2 = 6 \mid r_1 = 5] = \tfrac{2}{3}, \qquad p_r[r_2 = 5 \mid r_1 = 10] = \tfrac{1}{4}, \qquad p_r[r_2 = 8 \mid r_1 = 10] = \tfrac{1}{4},$$
$$p_r[r_2 = 12 \mid r_1 = 10] = \tfrac{1}{2}.$$

Suppose that the goal is to maximize expected wealth at the end of period 2.

(a) Formulate the two-period decision problem.

(b) Show that the problem in (a) is equivalent to a linear program.

(c) Formulate and interpret the dual linear program. (Recall that $\{\min b'y \mid A'y \geqq c,\ y \geqq 0\}$ is the dual to $\{\max c'x \mid Ax \leqq b,\ x \geqq 0\}$ where c, $x \in E^n$, b, $y \in E^m$, and A is an $m \times n$ matrix.)

(d) Solve the linear program.

(e) For what risk-free return rates is it optimal to invest entirely in either the risky or risk-free assets in both periods.

3. Consider an investor who wishes to allocate his funds between assets A and B. Asset A has net returns of 0.01 and 0.04 over one- and two-month horizons, respectively, while asset B returns 0.02 per month over either a one- or two-month horizon.

(a) Assume that there are no transaction costs, that A and B are perfectly liquid assets, and that the investor wishes to maximize his wealth at the end of the second month. Show that it is optimal to hold B in the first month and to sell B and buy A. What is the two-month net rate of return?

(b) Suppose that the cost of switching from B to A at the end of the first month is 1.5% of the value of the portfolio. Show that the investor will hold A during both months. Suppose the investor is only concerned with his wealth at the end of the first month. Show that he will hold B. When will the investor hold a mixture of A and B?

Suppose asset A is illiquid. Over two months one dollar invested in A will increase in value to $(1+r_1)^2$. But A cannot be sold until the end of the second month. However, each dollar invested in the liquid asset B can be realized at the end of the first month for $1+r_2$ or at the end of the second month for $(1+r_2)^2$. The individual expects to receive y_1 and y_2 from exogenous sources at the end of each month. His problem is to decide on the fraction x of his initial wealth that is invested in A to maximize his utility $U(C_1, C_2)$ over consumption in the two months.

(c) What happens if $r_2 \leqq r_1$?

(d) Suppose $r_1 > r_2$. Formulate the optimization problem.
(e) Suppose $r_1 = 0.02$, $r_2 = 0.01$, $y_1 = 1$, $y_2 = 2$, and $U = \log C_1 + \log C_2$. Find x^*.
(f) Graph the situation when $U = \alpha \log C_1 + \beta \log C_2$ for $\alpha, \beta > 0$.
(g) What must α and β be for x^* to be 0 or 1?

4. Suppose an investor holds security A that was purchased last year for one dollar and has appreciated by the percentage g. Let α and β represent the expected percentage (net) returns over a planning horizon for securities A and B, respectively. Suppose $\beta > \alpha$ and that the rate of taxation on capital gains is $0 < w < 1$. Assume that the investor wishes to maximize the expected value of terminal wealth net of capital gains taxes.

(a) Show that it is optimal to switch from A to B whenever

$$\beta/\alpha > \frac{1+g}{1+g(1-w)}.$$

(b) Interpret this condition. What happens in the special cases $w = 0$, $w = 1$, and $g = 0$?
(c) Suppose the brokerage charge is $0 < b < 1$ dollars for switching from A to B. Show that it is optimal to switch whenever

$$\beta/\alpha > \frac{1+g}{(1+g)(1-b)^2 - gw(1-b)} + \frac{b(2-b)}{(1-b)(1-bw)}.$$

(d) Suppose $g = 0.10$, $\alpha = 0.15$, $\beta = 0.20$, $w = 0.5$, and $b = 0.02$. Is it optimal to switch from A to B?

Suppose α and β have the discrete distributions

$$P_r\{\alpha = \alpha_i\} = p_i > 0, \quad i = 1, \ldots, m, \qquad P_r\{\beta = \beta_j\} = q_j > 0, \quad j = 1, \ldots, n.$$

(e) Develop the expressions corresponding to those in (a) and (c).
(f) Is the conclusion in (d) altered if $\alpha_1 = 0.10$, $\alpha_2 = 0.20$, $p_1 = p_2 = \frac{1}{2}$, $\beta_1 = 0$, $\beta_2 = 0.40$, and $q_1 = q_2 = \frac{1}{2}$?

5. Consider an investor whose preference ordering for consumption over two periods is represented by the cardinal utility function $u(c_1, c_2)$. Suppose that his preference ordering is *mutually utility independent*; that is, his preference ordering for gambles on c_1 is independent of the level of c_2 and his preference ordering for gambles on c_2 is independent of the level of c_1.

(a) Show that $u(c_1, c_2)$ may be represented as

$$u(c_1, c_2) = a_1(c_2) + b_1(c_2)\, u(c_1, c_2^{\,0}) \tag{1}$$

and

$$u(c_1, c_2) = a_2(c_1) + b_2(c_1)\, u(c_1^{\,0}, c_2) \tag{2}$$

for fixed but arbitrary $c_1^{\,0}$ and $c_2^{\,0}$.
(b) Show that u may be represented as

$$u(c_1, c_2) = f_1(c_1) + f_2(c_2) + k f_1(c_1) f_2(c_2). \tag{3}$$

[*Hint*: From (1), if $u(c_1^{\,0}, c_2^{\,0}) = 0$, $a_1(c_2) = u(c_1^{\,0}, c_2)$. Then if $u(c_1^{\,1}, c_2^{\,0}) \neq 0$, we have

$$u(c_1, c_2) = u(c_1^{\,0}, c_2) + \frac{u(c_1^{\,1}, c_2) - u(c_1^{\,0}, c_2)}{u(c_1^{\,1}, c_2^{\,0})}\, u(c_1, c_2^{\,0}).$$

Derive a similar expression using (2), then substitute the appropriate values of c_1 and c_2 into the representation given above.]
(c) If $k > 0$ in (3) above, show that the investor's utility function may be transformed

into the equivalent form

$$\bar{u}(c_1, c_2) = g_1(c_1) g_2(c_2). \tag{4}$$

(d) From (3) above, show that $u(c_1, c_2)$ is concave in c_1 for any fixed c_2, and concave in c_2 for any fixed c_1 if f_1, f_2 are nonnegative concave functions and $k \geqq 0$.

(e) Give an example of a function of the form (3) in which f_1, f_2 are positive, strictly increasing, strictly concave functions over $(0, \infty)$, $k \geqq 0$, but $u(c_1, c_2)$ is *not* concave over $E_+^2 = (0, \infty) \times (0, \infty)$.

(f) Give an example of a function $u(c_1, c_2)$ of form (3), strictly increasing and strictly concave in c_1 and c_2 separately, for which there exists a fair gamble on the pair (c_1, c_2) (i.e., a gamble in which c_1, c_2 are correlated) which is *preferred* to the status quo. Note, however, that fair gambles on c_1 and c_2 separately are always less preferred than the status quo.

6. In a two-period consumption–investment problem, let the investor's utility function for two-period consumption be $u(c_1, c_2) = c_1^{\alpha} c_2^{\beta}$.

(a) Show that u is concave over $E_+^2 = [0, \infty) \times [0, \infty)$ if $0 < \alpha, \beta < 1$ and $\alpha + \beta \leqq 1$.

Define first- and second-period absolute risk-aversion measures as

$$A_1(c_1, c_2) = -\frac{u_{11}(c_1, c_2)}{u_1(c_1, c_2)} \quad \text{and} \quad A_2(c_1, c_2) = -\frac{u_{22}(c_1, c_2)}{u_2(c_1, c_2)},$$

respectively, where u_1 and u_{11} are first and second partial derivatives with respect to the first argument, etc.

(b) Show that A_1 is strictly decreasing in c_1, and A_2 is strictly decreasing in c_2 for c_1 and $c_2 > 0$.

Let initial wealth be w, let amount $c_1 \in [0, w]$ be consumed in period 1, with the remainder invested in a risk-free asset returning $r = 1 + i \geqq 1$, and a risky asset returning $\rho \geqq 0$ per dollar invested. Assume that wealth at the end of period 1 is totally consumed in period 2. Suppose that borrowing and short sales are not allowed.

(c) Defining $h \equiv \rho - r$, show that the investor's problem is

$$\max_{c_1, a} Eu(c_1, (w - c_1) r + ah), \qquad \text{s.t.} \quad c_1 \in [0, w], \quad a \in [0, w - c_1].$$

(d) Show that the problem in (c) can be rewritten as

$$\max_{c_1 \in [0, w]} U(c_1, w), \tag{1}$$

$$U(c_1, w) = \max_{a \in [0, w - c_1]} Eu(c_1, (w - c_1) r + ah). \tag{2}$$

(e) If $E[(\rho - r)\rho^{\beta - 1}] < 0$, show that (2) has an interior solution. In the sequel, assume that this is the case.

(f) Show that $U(c_1, w) = Bc_1^{\alpha}(w - c_1)^{\beta}$, where $B = r^{\beta} E(1 + \sigma h)^{\beta}$ and σ is the solution of $Eh(1 + \sigma h)^{\beta - 1} = 0$.

Define the induced first-period absolute risk-aversion measure to be

$$\bar{A}(c_1, w) = -U_{11}(c_1, w) / U_1(c_1, w).$$

(g) Show that $\bar{A}(c_1, w)$ is strictly decreasing in c_1 for $c_1 \in (0, w)$.

(h) Show that the optimal first-period consumption level is

$$c_1 = \frac{\alpha}{\alpha + \beta} w.$$

(i) Show that the induced utility function for wealth, $\bar{U}(w) = \max_{c_1 \in [0, w]} U(c_1, w)$ is given by $\bar{U}(w) = B' w^{\alpha + \beta}$, where

$$B' = B\left[\frac{\alpha}{\alpha+\beta}\right]^{\alpha}\left[\frac{\beta}{\alpha+\beta}\right]^{\beta}.$$

(j) Show that the absolute risk-aversion measure for wealth, $-\bar{U}''(w)/\bar{U}'(w)$, is nonincreasing in w.

7. Refer to the Bradley–Crane paper.

(a) Formulate the general two- and four-period models.

(b) Illustrate the calculations that would be needed to solve the two-period model for some hypothetical numerical data.

(c) Indicate the problem sizes for the two models as a function of the cardinality of the various parameters.

(d) Discuss the computational aspects of these two models. Is it ever easier to solve the two-period model?

(e) What simplificatoins occur if the random events are period-by-period independent?

(f) What would happen to the model if e_1 were a vector rather than a scalar?

(g) Verify that (4.4) is the dual of (4.3).

(h) Attempt to generalize the model to include a nonlinear preference function. [*Hint*: Consult Lasdon (1970).]

8. This problem illustrates a two-period stochastic programming model for commercial bank bond portfolio management. The maximum size of the portfolio as well as certain cash flow constraints are determined outside the model's framework. At the beginning of each period, the portfolio manager has an inventory of bonds and cash. Based upon present credit market conditions and his assessment of future interest rates and cash flows, the manager must decide upon which bonds to hold, sell, and purchase in the next period, subject to the portfolio size constraint.

There are $m = 1, \ldots, M$ bond categories which can be purchased at the beginning of periods $i = 1, 2$. Bonds may be sold at the start of period $j = 2$ and must be sold when $j = 3$. Random events $e = 1, \ldots, E$ and $f = 1, \ldots, F$ occur at the end of periods 1 and 2, respectively. The (nonnegative) decision variables, in terms of book value, are buy (B_{ie}^{m}), sell (S_{ije}^{m}), and hold (H_{ije}^{m}), and the upper limits on portfolio size are (P_{ie}).

(a) Interpret B_{22}^{4}, S_{122}^{2}, H_{123}^{5}, and P_{22}.

(b) Which combinations of subscripts are meaningful for B, S, H, and P?

Because of special accounting practices and tax regulations of commercial banks, the "profits" from these decisions are broken into two components: an income yield (y_{ie}^{m}) stemming from the coupon interest on the bonds and a capital gain or loss rate (g_{ijef}^{m}) which occurs when the bonds are sold. These "profit" coefficients are defined as after-taxes. The gain coefficient also takes into consideration the broker's commission.

(c) Show that purchases in period 1 must satisfy $\sum_m B_1{}^m \leqq P_1$.

(d) Show that, for each event e and bond category m, bonds must be either held or sold so that $-B_1{}^m + S_{12e}^{m} + H_{12e}^{m} = 0$. How many constraints of this form are there?

In the second period, the portfolio consists of bonds held over from the first period plus bonds purchased at the start of the second period. The dollar volume of these assets is constrained by the limit on portfolio size, which resulted from the exogenous random cash inflow or outflow. Also, income earned on bonds held in the first period is available for reinvestment in the second, as is the after-tax gain or tax savings resulting from sales at the end of the first period.

(e) Show that for each event e the portfolio size constraint at the start of period 2 is

$$-\sum_m y_1{}^m B_1{}^m - \sum_m g_{12e}^m S_{12e}^m + \sum_m H_{12e}^m + \sum_m B_{2e}^m \leqq P_{2e}.$$

It is also of interest to add constraints which limit losses on the sale of bonds. These constraints are included for two reasons. First, there is an effective constraint on maximum losses, say L, imposed by the magnitude of the bank's operating earnings. Security losses can be offset against earnings to reduce taxes in the current year, as well as in some past and future years, but there is no tax saving after taxable income is fully absorbed by security losses and other offsets. Also, it is a method used by some portfolio managers to express their risk aversion since too large a loss in any given period may have an adverse effect on the bank's stock price.

A loss constraint is needed for each event at the start of the second period, since the manager may make a sell decision at this point. In addition, all bonds in the portfolio during the second period are sold at the end of that period. To take into account the possible losses that could occur at this point, the constraints assume that the maximum possible loss coefficient occurs, i.e., that the event with the highest interest rates occurs. This event at the end of period 2 is, say, $f = 1$.

(f) Show that the loss constraint is

$$-\sum_m g_{12e}^m S_{12e}^m - \sum_m g_{13e1}^m H_{12e}^m - \sum_m g_{23e1}^m B_{2e}^m \leqq L.$$

(g) Consider some alternative loss constraints.

Let R_{ef} be the total return if events e and f occur.

(h) Show that

$$R_{ef} = \sum_m \{y_1{}^m B_1{}^m + g_{12e}^m S_{12f}^m + (y_1{}^m + g_{13ef}^m) H_{12e}^m + (y_{2e}^m + g_{23ef}^m) B_{2e}^m\}.$$

Suppose that $p_e = \text{Prob}\{\text{event } e \text{ occurs at the end of period 1}\}$ and $p_{f/e} = \text{Prob}\{\text{event } f$ occurs at the end of period 2, given $e\}$. Then the expected return is $\varphi(R) = \sum_e p_e\{\sum_f p_{f/e} R_{ef}\}$.

(i) Discuss the advantages and disadvantages of the linear criterion $\varphi(R)$.

(j) Discuss how the formulation relates to the mean–variance criterion. [*Hint*: Refer to the paper by Pyle and Turnovsky in Part III.]

(k) Discuss solution methods for the linear programming model. [*Hint*: Look carefully at the model's structure.]

9. Refer to Exercise 8. Let $M = 7$ and suppose that the bond maturity classes are 1, 2, 3, 4, 5, 10, and 20 years. Let the time periods be of six months' duration. Suppose $P_1 = \$100{,}000$ and that the initial yield curve is linear over the maturity range rising from 3% after taxes for a one-year maturity bond to 4% for the 20-year rate. Suppose $E = 3$. With probability $\frac{1}{2}$, there will be a uniform increase in rates of 0.4% and there will be a cash outflow of 10% ($10,000). With probability $\frac{1}{4}$, there will be no change in rates and no cash flow. The third event, having probability $\frac{1}{4}$, represents an easing of credit conditions with a uniform decrease in rates of 0.4% and a cash inflow of 10%. Suppose $F = 3$. The second-period events are independent of those in period 1 and consist of a 0.4, 0, and -0.4% change in the yields that prevailed at the end of period 1 with probabilities $\frac{1}{2}$, $\frac{1}{4}$, and $\frac{1}{4}$, respectively. Let L, the maximum security loss, be 2% of the book value of the initial portfolio, and transaction costs be a constant $\frac{2}{32}$ spread between bid and asked prices of the bonds.

(a) Calculate the $\{y_{ie}^m\}$ and the $\{g_{ijef}^m\}$.

(b) Write out the entire linear programming model with numerical coefficients.

The optimal first-period decision and policy are

	1-year maturity ($)	20-year maturity ($)
Buy for initial portfolio	44,900	55,100
Second-period decisions:		
If event I (rates increase):		
Sell	—	55,100
Hold	44,900	—
Buy	45,000	—
If event II (no rate change):		
Sell	—	—
Hold	44,900	55,100
Buy	—	200
If event III (rates fall):		
Sell	—	55,100
Hold	44,900	—
Buy	70,500	—

(c) Verify that the above is an optimal policy by finding a dual optimal solution to the problem in (b).
(d) Investigate the dual multipliers on P_1, L, and the P_{2e}.
(e) Attempt to determine why only the extreme maturity bonds have positive weights. [*Hint*: Investigate the risk-aversion properties of the program along with the structure of returns.]
(f) Attempt to prove mathematically that such an optimal policy will always obtain.
(g) Suppose the loss constraint is deleted. Show that the optimum first-period portfolio is to invest entirely in 20-year bonds. Indicate why intermediate-length bonds are not chosen when the loss constraint is imposed and is binding.
(h) In additional numerical experiments, it has been found that the percentage invested in the longest bond tends to increase with increases in the slope of the yield curve, or the probability of lower rates in the future and decreases in the magnitude of risk aversion and transaction costs. Explain this behavior.

10. (Interest rates and bond prices: Deterministic case) Let ρ_t be the instantaneous (one-period) interest at time $t = 0, 1, \ldots, T$. Let $R(t)$ be the long-term interest rate over the interval $[t, T]$ (i.e., the effective constant one-period interest rate applying in the interval $[t, T]$). Assume that ρ_t is known with certainty.
(a) Using arbitrage arguments in perfect capital markets, show that

$$[1+R(t)]^{T-t} = \prod_{\tau=t}^{T}(1+\rho_\tau).$$

(b) In the continuous time limit, show that (a) becomes

$$R(t) = (T-t)^{-1}\int_t^T \rho(\tau)\, d\tau.$$

Suppose a bond of par value $1, time to maturity T, and (continuous) coupon rate r is issued at time $t = 0$. Let $p(t)$ be market value of the bond at time t.
(c) Show that $p(t)$ satisfies the differential equation

$$p'(t) - \rho(t)p(t) = -r, \qquad \text{and} \qquad p(T) = 1.$$

(d) Show that

$$p(t) = \exp(-(T-t)\,R(t))\left[1 + r\int_t^T \exp((T-\tau)\,R(\tau))\,d\tau\right].$$

Consider the special case in which $\rho(t)$ is linearly strictly decreasing and $\geqq 0$ in $[0, T]$, and suppose that $R(0) = r$. Suppose also that $p(0) = 1$; that is, the bonds are issued at par.

(e) Show that $p(t)$ is strictly increasing in some closed interval $[0, t_1]$, $t_1 > 0$, and strictly decreasing in some closed interval $[t_2, T]$, $t_2 < T$.

(f) Show that any stationary point of $p(t)$ must be a strict local maximum of $p(\cdot)$.

(g) Show that $p(\cdot)$ has one and only one stationary point in $(0, T)$, and this point is the unique global maximum of $p(\cdot)$ in $[0, T]$. Let this point be t_m.

(h) If $p''(0) \leqq 0$, show that $p''(t) < 0$ for all $t \in (0, T]$; thus $p(\cdot)$ is strictly concave in $(0, T]$.

(i) If $p''(0) > 0$, show that there exists $t_0 \in (0, t_m)$ such that $p(\cdot)$ is strictly increasing and strictly convex in $(0, t_0)$, and strictly concave in $(t_0, T]$.

(j) Show that (h) holds iff $T \leqq 2/\rho(0)$.

[*Hint*: In (e)–(j), analyze the differential equation (c) directly, rather than the solution (d).]

Referring to the Pye paper, suppose that $c(t)$ is a nonincreasing convex function, and $c(T) = 1$.

(k) Show that Pye's equation (2) is a sufficient as well as necessary condition for optimality of a call at time t.

11. Referring to the article by Pye concerned with call options on bonds. Suppose $n = 5$, $(\rho_1, \rho_2, \rho_3, \rho_4, \rho_5) = (0.04, 0.045, 0.05, 0.055, 0.06)$, $T = 4$, $r = \$50$, $p_{Tj} = \$1000$, $c = (c_1, c_2, c_3, c_4) = (1080, 1060, 1020, 1000)$, and

$$Q = \begin{pmatrix} 0.6 & 0.3 & 0.1 & 0 & 0 \\ 0.2 & 0.5 & 0.2 & 0.1 & 0 \\ 0 & 0.1 & 0.8 & 0.1 & 0 \\ 0 & 0 & 0.1 & 0.6 & 0.3 \\ 0 & 0 & 0.1 & 0.2 & 0.7 \end{pmatrix}.$$

(a) Utilize Eq. (5) to calculate the p_{ti}.

(b) Calculate the v_{ti} via Eq. (4).

(c) Describe and interpret the optimal policy.

(d) Referring to the Appendix, show that each row of Q dominates the preceding row.

(e) Show that $\Delta c_t \leqq 0$ but $\Delta^2 c_t$ is not nonnegative.

(f) Suppose $c = (1080, 1040, 1020, 1000)$. Show that $\Delta c_t \leqq 0$ and $\Delta^2 c_t \geqq 0$.

(g) For this new value of c, calculate the critical values associated with the proposition in the Appendix.

(h) Describe and interpret the optimal policy using the results in (g).

12. Referring to Exercises ME-7 and 8 formulate the call option problem (Exercise 11):

(a) as a pure entrance-fee problem, and

(b) as a linear programming problem.

13. Referring to Pye's paper in Chapter 2 of this part, suppose that the maximum possible asset price increases and decreases are unequal; that is, $a \neq b$.

(a) Show that an optimal nonsequential minimax policy for asset selling is to sell amounts s_t at $t = 0, 1, \ldots, T$, where

$$s_t = \frac{2b}{(a+b)(T+1)}, \quad 0 \leqq t \leqq T-1, \qquad \text{and} \qquad s_T = \frac{(a-b)\,T + a + b}{(a+b)(T+1)}.$$

(b) Show that the optimal policy in (i) reduces to dollar averaging when $a = b$.
(c) Interpret the policy in (i) when b (the maximum possible price decrease) is very small.
(d) Derive an optimal policy similar to (i) which would be appropriate when a (the maximum possible price increase) is small.

14. Referring to Pye's paper in Chapter 2 of this part, let p_t $(1 \leqq t \leqq T)$ be the asset prices, and suppose the p_t follow an arithmetic random walk: $p_t = p_{t-1} + y_t$, $t = 1, 2, \ldots$. The y_t are independent and identically distributed. Let a and b be the maximum possible price increases and decreases: $-b \leqq y_t \leqq a$. Assume that $b \geqq a$. Consider the case of a nondivisible asset, where the only possible decisions are to sell or not to sell the entire asset. Let $p_t^* = \max_{0 \leqq i \leqq t}(p_i)$, $R_t = p_t^* - p_t$ be the regret at time t, and let P_0 be given.
(a) Show that R_t and p_t^* obey the recursion relations

$$R_{t+1} = \max(0, R_t - y_{t+1}), \qquad p_{t+1}^* = p_t^* + \max(0, -R_t + y_{t+1}).$$

Let $f_n(R)$ be the maximum regret given R, given that the asset has not been sold with n periods to go, and given that a minimax policy is used in the future. Note that $f_0(R) = R$.
(b) Show that

$$f_1(r) = \min\{\underset{\text{(sell)}}{\max(R, a)},\ \underset{\text{(retain)}}{R + b}\} = \max(R, a),$$

and the minimax policy is to sell on the next-to-last period for any R.
(c) Show by induction that

$$f_n(R) = \min_{0 \leqq j \leqq n-1} \{\max(R + jb, (n-j)\,a)\}.$$

(d) Show that an optimal stopping rule is to sell the asset with n periods to go if the current value of R satisfies $R \geqq na - b$, and to wait one more period before deciding if $R < na - b$.
(e) If $ka \leqq b \leqq (k+1)\,a$, show that the asset will always be sold at least k periods prior to the terminal date.
(f) If $R_1 = 0$, show that the asset will not be sold prior to

$$\left[T\frac{a}{a+b} - \frac{b}{a+b}\right],$$

where $[x]$ denotes the largest integer $\leqq x$.

15. Consider a three-state Markov chain with transition matrix

$$P = \begin{bmatrix} \frac{1}{2} & \frac{1}{4} & \frac{1}{4} \\ \frac{1}{2} & 0 & \frac{1}{2} \\ \frac{1}{2} & \frac{1}{2} & 0 \end{bmatrix}.$$

Refer to Exercises ME-15 and 16 for notation and concepts.
(a) Find the limiting distribution π_i, $i = 1, 2, 3$.
(b) Show that the mean first passage times μ_{ij} are

$$[\mu_{ij}] = \begin{bmatrix} 2 & 2 & 2 \\ \frac{10}{3} & 4 & \frac{8}{3} \\ \frac{10}{3} & \frac{8}{3} & 4 \end{bmatrix},$$

where μ_{ii} is the expected recurrence time of state i.

Suppose that a cost $C(i)$ is incurred each time the chain is in state i: $C(1) = 2$, $C(2) = 0$, $C(3) = 1$.

(c) Find the long-run average cost of the chain.
(d) Find the expected discounted total cost of the chain, using an interest rate of 25%.

Suppose that the chain starts in state 1, that you win one dollar every time it passes through state 2 without having returned to state 1, and that the game is over upon returning to 1.
(e) What is the fair price of such a gamble?

16. It is of interest to develop conditions under which an optimal sequence of portfolio choices is stationary. That is, in each period it is optimal to invest a constant proportion of total wealth in each asset. Tobin (1965) developed such a result in a simplified dynamic mean–variance framework. This problem illustrates the result and shows that the result is not generally true.

Suppose that the investor wishes to make his total consumption during period T. Hence, only the expected utility of wealth at time T is valued. Assume that asset returns are statistically independent. Suppose also that expected utility is a function of the mean and variance of wealth.

Let R_t, E_t, and σ_t be the (gross) return, expected value of return, and standard deviation of return in period t. The efficient set of portfolios is given by $\sigma_t^2 = f(E_t)$. The T-period variables are denoted by R, E, and σ. Assume that initial wealth is one dollar. Given reinvestment of intermediate returns $R = \prod E_t$.

(a) Show that $E = \exp R = \prod E_t$ and

$$\sigma^2 + E^2 = \exp R^2 = \prod[\sigma_t^2 + E_t^2] = \prod[f(E_t) + E_t^2].$$

One may explore the properties of efficient sequences of portfolios by minimizing $\sigma^2 + E^2$ subject to $E =$ constant.

(b) Show that the solution to the first-order conditions is a stationary sequence with $E_1 = E_2 = \cdots = E_T$.

(c) Show, however, that this sequence does not necessarily satisfy the second-order conditions. [*Hint*: Let $T = 2$ and $\sigma_t = f(E_t) = (E_t - r_t)^2$, where $r_t > 1$ is a riskless rate of interest.]

(d) Show that the second-order conditions are satisfied if $\sigma_t^2 = \exp[A(E_{t-1})] - E_t^2$.

*(e) Whether or not the second-order conditions are satisfied depends crucially on the generalized convexity properties of the objective function and constraints in the problem

$$\text{minimize} \prod[f_t(E_t) + E_t^2] \qquad \prod E_t \leqq \text{const.}$$

Attempt to utilize the results in Mangasarian's second paper in Part I to determine sufficient conditions on the f_t that lead to the satisfaction of the second-order conditions.

17. Suppose an investor has $100 to invest over three investment periods. Investment 3 yields a certain return of 2% per period. Investments 1 and 2 have the following distribution in each of the three periods

$$p_r\{r_1 = 0.03,\ r_2 = 0.02\} = \tfrac{1}{4}, \qquad p_r\{r_1 = 0.02,\ r_2 = 0.04\} = \tfrac{1}{4},$$
$$p_r\{r_1 = 0.01,\ r_2 = -0.02\} = \tfrac{1}{4}, \qquad p_r\{r_1 = 0.12,\ r_2 = 0.08\} = \tfrac{1}{4}.$$

Suppose that the goal is to maximize the expected utility of terminal wealth, w, where $u(w) = w^{1/2}$.

(a) Formulate the problem as a three-period stochastic program.
(b) Show that the program in (a) is equivalent to a deterministic program that is concave.
(c) Solve the program in (b) by dynamic programming.
(d) Solve the program in (b) by finding a solution to the Kuhn–Tucker conditions.

18. (The Kelly criterion) A gambler makes repeated bets in sequence against an infinitely rich adversary. At time t his wager is β_t dollars out of his total wealth W_t. Suppose that the game is stationary and favorable with the probability of winning and losing being $p > \frac{1}{2}$ and $q = 1-p$, respectively. Suppose that the objective is to maximize the exponential rate of growth of W_t, namely, $G \equiv \lim_{t\to\infty} (1/t)\log(W_t/W_0)$, where W_0 is the gambler's initial wealth. Assume that the gambler bets a constant fraction f of his capital in each period so $\beta_t = fW_t$.

(a) Show that in each period t the gambler's wealth is

$$W_t = (1+f)^v(1-f)^{t-v}W_0$$

if he wins v of the bets.

(b) Show that $G = p\log(1+f)+q\log(1-f)$.

(c) Interpret this result in terms of expected utility theory.

(d) Show that the optimal bet is

$$f^* = p - q \qquad \text{and} \qquad G^* = p\log p + q\log q + \log 2 > 0.$$

(e) Interpret f^* and G^*.

(f) Calculate G, G^*, and f^* if there is a tie with probability $r > 0$.

(g) Show that no nonstationary sequence $\{f_1, f_2, \ldots\}$ is superior to the stationary sequence $\{f^*, f^*, \ldots\}$.

19. (Definition of a martingale) Suppose a sequence of random variables $\{X_i\}$, $i = 1, 2, \ldots$, has the property that $E|X_i| < \infty$ for all i. Then if $E(X_{n+1}\,|\,X_n, \ldots, X_1) = X_n$ a.e. for all n, the sequence is said to be a martingale. If $E(X_{n+1}\,|X_n, \ldots, X_1) \geqq (\leqq)\, X_n$ a.e. for all n, it is said to be a submartingale (supermartingale).

(a) Consider the game of roulette in which 38 numbers are equally likely, 18 of which are red, 18 are black, and 2 are green. Suppose a player utilizes the following strategy:

Step 1: Bet \$1 on black.
Step 2: If black occurs go to 1. If red or green occurs go to 3.
Step 3: Go to step 1 but bet double the previous bet.

Let S_n be his winnings after play n. Is S_n a martingale or a sub- or supermartingale?

(b) More generally, let $\{X_n\}$ be a sequence of random variables where $\Pr\{X_n = +1\} = p \geqq 0$ and $\Pr\{X_n = -1\} = 1-p \geqq 0$, respectively. Suppose on the nth trial we bet $B_n = f(X_1, \ldots, X_{n-1})$ and we receive $2B_n$ if $X_n = +1$ and zero if $X_n = -1$. Let $W_0 > 0$ be our initial fortune and S_n the fortune after n plays. Is S_n a martingale, sub- or supermartingale?

(c) In the proof of Theorem 1 in Breiman's paper the following theorem [see, e.g., Feller (1966)] was used: Let $\{X_n\}$ be a submartingale. If $\sup_n E(X_n^+) < \infty$, where X_n^+ is the nonnegative part of X_n, then there is an integrable random variable X such that $X_n \to X$ a.e. Show that the conditions of the theorem are satisfied. [*Hint*: If $\{X_i\}$ is a supermartingale, then $\{-X_i\}$ is a submartingale.]

20. Refer to Breiman's paper.

(a) Interpret the meaning of an inadmissible investment policy sequence.

(b) Show that the optimal investment proportions satisfy $\lambda_i = E(\lambda_i r_i/\sum \lambda_i r_i)$. [*Hint*: Utilize the Kuhn–Tucker conditions.]

(c) Show that the boundedness assumptions on the r_i may be weakened to the following: Assume that there are constants α, β such that $0 < \alpha < V_N < \beta < \infty$ for all N.

(d) Interpret assumption (A). Is this a strong assumption?

(e) Interpret assumption (B). Is this a strong assumption?

21. Refer to Thorp's paper. Let X_n denote the investor's wealth at the end of period n.

(a) Show that $E\log X_n/X_{n-1} > 0$ for all n does not imply $P(\lim X_n = \infty) = 1$.

(b) Suppose $E\log X_n/X_{n-1} \geqq 0$. Show that it is possible that $P(\limsup X_n = \infty) = P(\liminf X_n = 0) = 1$.

(c) Can the result in (b) be true if $E \log X_n/X_{n-1} < 0$?
(d) Provide a counterexample to the "false corollary": "If maximizing the geometric mean almost certainly leads to a better outcome, then the expected value utility of its outcomes exceeds that of any other rule provided it is sufficiently large."
(e) Attempt to determine a method for constructing utility functions over final wealth that satisfy the false corollary. [*Hint*: Suppose U is bounded.]
(f) Provide an example to verify the theorem in Section II.
(g) Give an example of increasing inequivalent utility functions U and V that have distinct optimal strategies for specific Bernoulli investments.
(h) Can the strictly concave assumption be dropped from the statement of the theorem in Section II?
(i) Show that the almost certainly transitivity result in Section II does not hold if the time horizon is finite.
(j) Refer to Section IV. What are the higher order terms in the Taylor series expansion? When will they be "small"?

Suppose there are three investments having mean returns of (1.1, 1.05, 1.20) per dollar invested and variance–covariance matrix of returns equal to

$$\begin{pmatrix} 0.15 & 0 & -0.2 \\ 0.0 & 0 & 0.0 \\ -0.2 & 0 & 0.5 \end{pmatrix}$$

(k) Compute Markowitz's efficient surface.
(l) Compute Thorp's efficient surface.
(m) Compare the surfaces.
(n) Prove the lemma: An $E \log X$ optimal portfolio is never stochastically dominated.
(o) Comment on the statement: "A characteristic of the Kelly criterion is that as risk decreases and expectation rises, the optimal fraction of assets to be invested in a single situation may become 'large'" in light of the fact that the optimal asset proportions, say λ_i, satisfy $\lambda_i = E(\lambda_i r_i / \sum \lambda_i r_i)$, where r_i is the gross return from investment i.

22. This exercise presents an extension of Breiman's model to allow for more general constraint sets and value functions. The development extends Theorem 1 and indicates that no strategy has higher expected return than the expected log strategy in any period. The model applies to common investment circumstances that include borrowing, transactions and brokerage costs, taxes, possibilities of short sales, and so on. Suppose initially that the value function has stochastic constant returns to scale: $V_N = \sum_i^N \lambda_i^N r_i^N$, as in Breiman's formulation. It will be convenient to utilize the superscripts N to denote the investment period under consideration. Let $\Lambda_N \subset E^{n_N}$ denote the set of feasible investment allocations in period N. Suppose that Λ_N is closed, connected, and nonempty and that $\{\max E \log(\lambda^N r^N) \mid \lambda^N \in \Lambda^N\}$ has a unique maximizer λ^{N*}.

(a) Show that without loss of generality we may set $w = 1$ in any constraint set of the form $\Lambda(w) = \{\lambda \mid \sum \lambda_i = w,\ w > 0,\ g_j(\lambda) \geqq 0,\ j \in J\}$, where the g_j are differentiable, concave, and homogeneous, since the maximizing allocations over $\Lambda(w)$, say $\lambda^*(w)$, equal $w \cdot \lambda^*(1)$. [*Hint*: Use the Kuhn–Tucker conditions.]
(b) Show that the following constriant sets belong to Λ_N:
(i) $\{(\lambda_1, \lambda_2, \lambda_3) \mid \lambda_1 - \lambda_2 + \lambda_3 = 1,\ \lambda_i \geqq 0,\ \lambda_2 \leqq M < \infty\}$;
(ii) $\{(\lambda_1, \lambda_2, \lambda_3, \lambda_4) \mid \lambda_1 + \lambda_2 - \lambda_3 + \lambda_4 = 1,\ 0 \leqq \alpha_i \leqq \lambda_i \leqq \beta_i < \infty\}$;
(iii) $\{\lambda \mid \lambda \in E^n,\ \bar{r}\lambda - k\lambda V\lambda \geqq \alpha,\ \sum \lambda_i = 1,\ \lambda \geqq 0,\ \alpha > 0,\ k > 0\}$, where V is the variance–covariance matrix of r;
(iv) $\{\lambda \mid \sum |\lambda_i| \leqq 1\}$; and

(v) $\{\lambda^N \mid \sum \lambda_i^N - \sum g_i(\lambda_i^N - \lambda_i^{N-1}) \leqq w,\ \lambda_i^N \geqq 0,\ w > 0,\ 0 \leqq g_i \leqq G < \infty\}$.

(c) Interpret the conomic meaning of these constraint sets.

*(d) When is the result in (a) valid for the constraint set in part (v) of (b)?

(e) Suppose $V_N = \sum f_i(\lambda_i^N) r_i^N$ and $\lambda^N \in \Lambda^N$. Show that the following modified value function and constraint set are permissible for the model under consideration: $\hat{V}_N = \sum \hat{\lambda}_i^N r_i^N$, where

$$\hat{\lambda}^N \in \{\hat{\lambda}^N \mid \hat{\lambda}_i^N = f_i(\lambda_i^N),\ \lambda^N \in \Lambda^N\}.$$

(f) Show how the value function in (e) permits certain forms of taxation to be included in the model.

Some of the constraint sets in (b) are intended to include borrowing, in which case it is not appropriate to assume that all $r_i > \alpha > 0$. Hence we will suppose that each V_N satisfies $0 < V_N < \beta < \infty$ and Breiman's condition (A), namely that

$$E\left(\frac{S_N}{S_{N*}} \,\middle|\, R_{N-1}\right) = E\left(\frac{V_N}{V_{N*}} \,\middle|\, R_{N-1}\right) \frac{S_{N-1}}{S_{N-1}^*}.$$

That is, given the wealth levels, S_{N-1} and S_{N-1}^* are a known function of the past outcomes in periods $1, \ldots, N-1$, namely R_{N-1}. The key idea in the proof is to show that

$$E\left(\frac{V_N}{V_N^*} \,\middle|\, R_{N-1}\right) \leqq 1,$$

for then $E(S_N/S_{N*} \mid R_{N-1})$ is a supermartingale; that is,

$$E\left(\frac{S_N}{S_{N*}} \,\middle|\, R_{N-1}\right) \leqq \frac{S_{N-1}}{S_{N-1}^*} \qquad \text{for all} \quad N.$$

But then

$$E\left[E\frac{S_N}{S_{N*}} \,\middle|\, R_{N-1}\right] \leqq E\left(\frac{S_{N-1}}{S_{N-1}^*}\right) \qquad \text{or} \qquad E\left(\frac{S_N}{S_{N*}}\right) \leqq E\left(\frac{S_{N-1}}{S_{N-1}^*}\right).$$

Hence $E(S_N/S_{N*}) \leqq 1$ for all N. [*Note*: This is a stronger statement than $E \lim(S_N/S_N{}^*) \leqq 1$ since it holds for all N.]

(g) Show that $\lim_{n\to\infty}(1+x/(n-1))^n = e^x$ if $|x| < \infty$.

(h) Show that Λ_N is sufficiently regular for there to exist a strategy $\lambda^N(\varepsilon)$ such that $V_N = \lambda^N(\varepsilon) r^N = (1-\varepsilon) V_N{}^* + \varepsilon V_N$ for any $\varepsilon \geqq 0$. [*Hint*: Let $\lambda^N(\varepsilon) = (1-\varepsilon)\lambda^{N*} + \varepsilon\lambda^N$.] Since λ^{N*} is the unique expected log maximizer, it follows that for $\varepsilon \geqq 0$,

$$E \log[(1-\varepsilon) V_N{}^* + \varepsilon V_N] - \log V_N{}^* \leqq 0.$$

(i) Show that this last line is equivalent to

$$E \log\left(1 + \frac{1}{1/\varepsilon - 1} \frac{V_N}{V_{N*}}\right)^{1/\varepsilon} \leqq \frac{1}{\varepsilon} \log \frac{1}{1-\varepsilon}.$$

(j) Show that the assumptions imply that V_N/V_{N*} is bounded. Let $n = 1/\varepsilon$ and $x = V_N/V_{N*}$. Use (g) to show that

$$E\left(\frac{V_N}{V_{N*}} \,\middle|\, R_{N-1}\right) \leqq 1.$$

Hence by the supermartingale theorem, $\lim S_N/S_{N*}$ exists almost surely and $E(\lim S_N/S_{N*}) \leqq 1$.

(k) Show that Theorem 1 is valid under the following relaxed assumptions on the value function and investment allocation sets:

(i) $V_N = f(\lambda^N, r^N)$ is continuous in λ^N for all fixed r^N, where $0 < \alpha < V_N < \beta < \infty$.

(ii) There exists a sequence of strategies $\Delta_N(\varepsilon) \in \Lambda_N$ such that

$$V_N(\Delta_N(\varepsilon)) \equiv f[\Delta_N(\varepsilon), r^N] \to f(\lambda^{N*}, r^N) \equiv V_N{}^*$$

as $\varepsilon \to 0$, with associated ε defined by $V_N(\Delta_N(\varepsilon)) = (1-\varepsilon) V_N{}^* + \varepsilon V_N$.

Exercise Source Notes

Exercise 3 was adapted from Tobin (1965); Exercise 4 was adapted from Holt and Shelton (1961); Exercises 5 and 6 were adapted from Keeney (1972); Exercises 8 and 9 were adapted from Crane (1971); Exercise 10 was based in part on a private communication from G. Pye; Exercise 14 was based on Bawa (1973); Exercise 16 was adapted from Stevens (1972); Exercise 18 was based on Kelly (1956); Exercise 19 was adapted from notes written by S. Larsson; and Exercise 22 was adapted from Brumelle and Larsson (1973).

MIND-EXPANDING EXERCISES

1. Consider a two-period problem and let the decision vectors in periods 1 and 2 be x_1 and x_2, where these vectors must be chosen from the convex sets K_1 and $K_2(\cdot,\cdot)$ respectively. Suppose that ξ_1 and ξ_2 have a known joint distribution and that the preference functions in periods 1 and 2 are f_1 and f_2, where $f_1(x_1,\cdot)$ and $f_2(x_1,x_2,\cdot,\cdot)$ are finite and convex, respectively. Assume that all maxima and expectations indicated below are finite. The problem is

$$\min_{x_1 \in K_1} E_{\xi_1}\left[f_1(x_1,\xi_1) + \min_{x_2 \in K_2(x_1,\xi_1)} \left\{ E_{\xi_2|\xi_1} f_2(x_1,x_2,\xi_1,\xi_2) \right\} \right] \equiv Z^*.$$

Let $\bar{\xi}_1$ be the mean of ξ_1 and $\bar{\xi}_2(\xi_1)$ be the mean of ξ_2 conditional on ξ_1, and denote by $(\bar{x}_1,\bar{x}_2)$ a solution to

$$\min_{x_1 \in K_1, x_2 \in K_2(x_1,\bar{\xi}_1)} [f_1(x_1,\bar{\xi}_1) + f_2(x_1,x_2,\bar{\xi}_1,\bar{\xi}_2(\bar{\xi}_1))] \equiv Z_L.$$

(a) Show that $Z_L \leqq Z^*$. [*Hint*: Utilize Jensen's inequality.]
(b) Under what assumptions is

$$Z^* \leqq E_{\xi_1}\{f_1(\bar{x}_1,\xi_1) + E_{\xi_2|\xi_1} f_2(\bar{x}_1,\bar{x}_2,\xi_1,\xi_2)\}?$$

(c) Compute these bounds when $f_1(\cdot,\cdot) = (x_1-\xi_1)^4$, $f_2(\cdot,\cdot,\cdot,\cdot) = (x_2-\xi_2),^4$ $K_1 = K_2(\cdot,\cdot) = [x \mid 1 \geqq x \geqq 0,\ x \in R]$,

$$P_r[\xi_1 = 0] = \tfrac{1}{4}, \qquad P_r[\xi_1 = 1] = \tfrac{1}{4}, \qquad P_r[\xi_1 = 2] = \tfrac{1}{2},$$
$$P_r[\xi_2 = 2 \mid \xi_1 = 0] = \tfrac{1}{2}, \qquad P_r[\xi_2 = 3 \mid \xi_1 = 0] = \tfrac{1}{2}, \qquad P_r[\xi_2 = 3 \mid \xi_1 = 1] = \tfrac{1}{3},$$
$$P_r[\xi_2 = 4 \mid \xi_1 = 1] = \tfrac{2}{3}, \qquad P_r[\xi_2 = 2 \mid \xi_1 = 2] = \tfrac{2}{5}, \qquad P_r[\xi_2 = 4 \mid \xi_1 = 2] = \tfrac{3}{5}.$$

(d) Suppose that ξ_1 and ξ_2 have general discrete distributions. Show that the random problem is equivalent to a deterministic problem.
(e) Show that the deterministic problem in (d) is convex.
(f) Illustrate the problem of (c) via the result of (d). Solve this problem and indicate how sharp the bounds are.

2. This problem concerns the effect of uncertainty on savings decisions when the consumer has uncertainty concerning either his future income or his yield on capital investments.

Suppose that the consumer's utility function over present and future consumption (C_1, C_2) is $U(C_1, C_2)$, where u is strictly monotone, concave, and three times continuously differentiable. Equation (2.8) in the Drèze–Modigliani article develops the risk-aversion function, which may be written as

$$\frac{2}{h^2}p = \frac{-U_{22}(C_1,C_2)}{U_2(C_1,C_2)}$$

for equally likely gambles (C_1, C_2-h) and (C_1, C_2+h), where $p > 0$ is the risk premium (as discussed in Pratt's paper in Part II). Assume that the risk-aversion function is increasing in C_1 and decreasing in C_2; that is, there is decreasing temporal risk aversion.

(a) Graphically interpret the concept of decreasing temporal risk aversion.
(b) Suppose the consumer is endowed with consumption (C_1, C_2) and is offered a gamble having outcomes $\pm h$ of future consumption. Show that the consumer will accept the gamble only when the probability of a gain of h, say $\pi(h)$, is greater than $\frac{1}{2}$.
(c) Show that π is an increasing function of C_1.
(d) Show that π will fall with a simultaneous increase in C_2 and decrease in C_1.

Consider first the effects of uncertainty regarding the consumer's future income. His first-period budget constraint is $Y_1 = C_1 + S_1$ where Y_1 is his (certain) income in period 1 and S_1 is savings. Future consumption is $C_2 = Y_2 + S_1(1+r)$, where r is the rate of interest, assumed to be *known* in the case of pure income risk, and Y_2 is the uncertain income in period 2.

(e) Show that expected utility is

$$EU = \int u\{C_1, Y_2 + (Y_1 - C_1)(1+r)\}\, dF(Y_2).$$

(f) Show that the first- and second-order conditions for an optimal choice of C_1 are

$$E[U_1 - (1+r)\,U_2] = 0 \qquad \text{and} \qquad D = E[U_{11} - 2(1+r)\,U_{12} + (1+r)^2 U_{22}] < 0,$$

respectively.

(g) Verify that $D < 0$.

(h) Show that the effect of an increase in present income is

$$\partial C_1/\partial Y_1 = -(1+r)\,E[U_{12} - (1+r)\,U_{22}]/D.$$

(i) Assume that the expression in (h) is always positive. Show that this means that both present and future consumption are not inferior goods.

Write future income as $\gamma Y_2 + \theta$, where γ and θ are multiplicative and additive shift factors, respectively.

(j) Interpret the meaning of the two types of shift parameters.

Since $Y_2 \geqq 0$ a multiplicative shift around zero will increase the mean. Hence to maintain a constant expected value, the additive shift must be negative. Taking the differential, the requirement is that $dE(\gamma Y_2 + \theta) = E(Y_2\, d\gamma + d\theta) = 0$ or that $d\theta/d\gamma = -E[Y_2] \equiv -\phi$.

(k) Show that

$$\left(\frac{\partial C_1}{\partial \gamma}\right)_{\partial\theta/\partial\gamma = -\phi} = -\left(\frac{1}{D}\right) E[(U_{12} - (1+r)\,U_{22})(Y_2 - \phi)].$$

(l) Show that decreasing temporal risk aversion is sufficient for the expression in (k) to be < 0. Hence increased uncertainty about future income increases savings.

Consider the case of capital risk. Suppose that in the first period the consumer can allocate his resources (Y_1) between present consumption (C_1) and capital investment K. Capital investment is transformed into resources available for future consumption by the function $f(K, \xi)$, where ξ is random. Consider the simple form $C_2 = K\xi$, where $\xi \geqq 0$.

(m) Show that expected utility is $EU = \int U\{C_1, (Y_1 - C_1)\,\xi\}\, dF(\xi)$.

(n) Show that necessary and sufficient conditions for a maximum of EU are $E[U_1 - \xi U_2] = 0$ and $H = E[U_{11} - 2\xi U_{12} + \xi^2 U_{22}] < 0$.

Let the yield on capital be $\gamma(\xi - 1) + \theta$. For a multiplicative shift around zero to keep the mean constant we must have $dE[\gamma(\xi-1)+\theta] = 0$ or $d\theta/\partial\gamma = -E[\xi - 1] \equiv -\mu$.

(o) Show that

$$(\partial C_1/\partial\gamma)_{\partial\theta/\partial\gamma = -\mu} = -(1/H)\,KE\{(U_{12} - \xi U_{22})(\xi - \mu)\} + (1/H)\,E[U_2(\xi - \mu)].$$

[*Hint*: Differentiate the first-order conditions with respect to γ and evaluate the derivative at $(\gamma, \theta) = (1, 0)$.] We refer to the first term as the income effect [because of its similarity to the expression in (k)] and the second term as the substitution effect.

(p) Show that the existence of risk aversion, i.e., U is concave, is a necessary and sufficient condition for the substitution effect to be positive. Show that the additional assumption of decreasing temporal risk aversion is sufficient for the income effect to be negative. Hence the total effect is ambiguous without further assumptions.

Note the following distinction between income and capital risks. With income risk increased saving raises the expected value of future consumption, while leaving higher moments unaffected. Hence the consumer reacts to increased uncertainty by increasing savings so that the increased uncertainty (variance, etc.) of future consumption is compensated by a higher mean.

However, with capital risk there is also the second effect because savings increases will increase both the mean and uncertainty (i.e., variance) of future consumption, hence the ambiguity.

3. In Exercise II-CR-10 the reader was asked to prove Arrow's result that nonincreasing absolute risk aversion is sufficient for a risky asset not to be an inferior good. In this exercise the reader is asked to attempt to establish a similar result in a generalized two-period model of Fisher's model of saving.

Suppose that the consumer has a thrice continuously differentiable, increasing, concave utility function $u(C_1, C_2)$ over present and future consumption C_1 and C_2, respectively. Assume that present and future income (Y_1, Y_2) are exogenously given and that the consumer has access to a perfect capital market where he can borrow or lend at the (net) risk-free rate r. Let the amount invested in the risk-free asset be m. Suppose that there is a risky asset having random rate of return x and that a is the amount invested in this asset. The budget constraint in the present period is then $C_1 + a + m = Y_1$ and future consumption is $C_2 = Y_2 + a(1+x) + m(1+r)$. Let subscripts on u refer to partial differentiation with respect to the C_t.

(a) Show that the first-order conditions for a maximum are

$$E[u_1 - (1+r)u_2] = 0 \quad \text{and} \quad E[u_2(x-r)] = 0.$$

(b) Discuss the economic interpretation of the conditions in (a).

(c) Develop the second-order conditions.

Assume, as in Exercise 2, that the risk-aversion function $-u_{22}/u_2$ is decreasing in C_2 and increasing in C_1.

(d) Show that

$$\frac{\partial}{\partial C_2}\left[\frac{-u_{22}}{u_2}\right] \leqq 0 \quad \text{implies} \quad E[(x-r)u_{22}] \begin{cases} \geqq 0 & \text{if } a \geqq 0 \\ \leqq 0 & \text{if } a \leqq 0. \end{cases}$$

(e) Show that

$$\frac{\partial}{\partial C_1}\left[\frac{-u_{22}}{u_2}\right] \geqq 0 \quad \text{implies} \quad E[(x-r)u_{12}] \begin{cases} \leqq 0 & \text{if } a \geqq 0 \\ \geqq 0 & \text{if } a \leqq 0. \end{cases}$$

(f) Utilize (d) and (e) to show that

$$E[(x-r)u_{12} - (1+r)(x-r)u_{22}] \begin{cases} \leqq 0 & \text{if } a \geqq 0 \\ \geqq 0 & \text{if } a \leqq 0. \end{cases}$$

(g) Show that necessary conditions for the marginal propensity to consume, $\partial C_1/\partial Y_1$, to be positive and less than unity are

$$E[(1+r)u_{22} - u_{12}] < 0 \quad \text{and} \quad E[u_{11} - (1+r)u_{12}] < 0.$$

(h) Compare the results in (g) with those in Exercise ME-1.

(i) Show that

$$\frac{\partial a}{\partial Y_1} = \frac{-1}{H}\{E[u_{11} - (1+r)u_{12}]E[(1+r)(x-r)u_{22}] + E[(1+r)u_{22} - u_{12}]E[(1+r)(x-r)u_{12}]\},$$

where $H > 0$ is the determinant of the Hessian matrix of second-order partial derivatives developed in (b).
(j) Show that the sign of $\partial a/\partial Y_1$ is, unfortunately, ambiguous for all a.
(k) Give an economic interpretation of why this ambiguity arises. [*Hint*: Note that higher income increases both present consumption and planned future consumption via present savings.]
(l) Suppose $u(\cdot) = v(C_1)+w(C_2)$, where v is strictly concave. Show that $\partial a/\partial Y_1 > 0$ under the decreasing absolute risk-aversion assumption.
(m) Investigate alternative assumptions on u that lead to the result that the risky asset is not an inferior good.

4. This problem concerns the development of relatively easily computable upper and lower bounds on the value of information in uncertain decision problems, such as that illustrated by Eq. (2.5) of the Drèze–Modigliani article.

Suppose that a decision maker must choose a decision vector x from a constraint set K. A choice x will provide him with the payoff $f(x,\xi)$, where ξ is a random vector distributed independently of x. Assume that the decision maker knows the distribution function of ξ, and that his utility function is u. If the decision must be made before ξ is observed, then the decision maker achieves maximum expected utility by solving

$$Z_n = \max_{x \in K} E_\xi u[f(x,\xi)],$$

where E_ξ denotes mathematical expectation with respect to ξ.

Now if a clairvoyant were to tell the decision maker the precise realization of ξ that will occur before the decision x must be chosen, then the decision-maker's maximum expected utility would be

$$Z_p = E_\xi \max_{x \in K} u[f(x,\xi)].$$

(a) Show that $Z_p \geqq Z_n$.

Define the value of information V as the solution to

$$E_\xi \max_{x \in K} u[f(x,\xi)-V] = \max_{x \in K} E_\xi u[f(x,\xi)].$$

Thus V is the cost which would equate the maximum attainable expected utility with and without perfect information.

Assume that u is strictly increasing and that all relevant expectations and maxima exist.
(b) Show that $V \geqq 0$. [*Hint*: Show that the assumption that $V < 0$ leads to a contradiction.]
(c) Let $V(x) \equiv h[a_1(x),\ldots,a_m(x)]$ where $V\colon C \to R$, $h\colon R^m \to R$, $a_i\colon C \to R$, $i = 1,\ldots,m$, and $C \subset R^n$ is a convex set. Show that V is concave if h is concave and nondecreasing and the a_i are concave.
(d) Suppose that u is concave on R and f is concave on the convex set $K \times \Xi$, where Ξ is the domain of ξ. Show that $V \leqq C_1 \leqq C_2$ where

$$C_i \equiv \max_{x \in K} f(x,\bar{\xi}) - u^{-1}(c_i), \qquad i = 1,2,$$

$$c_1 \equiv \max_{x \in K} E_\xi u[f(x,\xi)], \qquad \text{and} \qquad c_2 \equiv E_\xi u\{f[\bar{x},\xi]\},$$

where $\bar{x}$ solves $\max_{x \in K}[f(x,\bar{\xi})]$. [*Hint*: Utilize Jensen's inequality, the result in (c), and recall that the optimal solution to $\max_{x \in L} h(g(x))$ also solves $\max_{x \in L} g(x)$ if h is monotone increasing.]

(e) Discuss the computational aspects of these two bounds.

(f) Suppose that the utility function takes the linear form $aw+b$ where $a > 0$. Show that

$$V/a = E_\xi \max_{x \in K} f(x,\xi) - \max_{x \in K} E_\xi f(x,\xi).$$

(g) Let $v(x) = \max_{y \in D(x)} w(x,y)$ where $w: C \times D \to R$, $v: C \to R$ with $C \subset R^n$ and $D(\cdot) \subset R^l$ being convex sets. Suppose that for each $x \in C$ the maximum is attained for some $y_0(x) \in D(x)$. Show that v is a concave function of x if w is a concave function of (x,y).

(h) Let $a = 1$. Suppose that f is concave on the convex set $K \times \Xi$. Show that $V \leqq f(\bar{x},\bar{\xi}) - E_\xi f(\bar{x},\xi) \equiv A$.

(i) Show that

$$V \leqq \max_{x \in K} \{f(x,\bar{\xi}) + \nabla f(x,\bar{\xi})'(\xi - \bar{\xi})\} - E_\xi f(\tilde{x},\xi) \equiv B$$

where $\bar{\xi}$ is any $\xi \in \Xi$, the gradient operator ∇f is the vector of partial derivatives of f with respect to the components of x, and $\tilde{x}$ solves $\max_{x \in K} f(x,\bar{\xi})$.

(j) Show that $A \leqq B$.

(k) Discuss the computational aspects of the bounds A and B.

5. This problem is concerned with the effect of uncertainty on the savings–consumption–portfolio decision of an expected utility maximizing investor faced with a two-period horizon. The analysis is based on the following result due to Hahn (1970) and Rothschild and Stiglitz (1971).

(a) Suppose that the investor wishes to maximize $Eu(\theta,\alpha)$ with respect to the decision variable α, where θ is a random variable. Show that the first- and second-order conditions for a unique maximum α^* to exist are $Eu_\alpha(\theta,\alpha) = 0$ and $Eu_{\alpha\alpha}(\theta,\alpha) < 0$ (subscripts denote differentiation), assuming that the appropriate integrals are well behaved.

(b) Show that α^* (increases, is constant, decreases) if $u_\alpha(\theta,\alpha)$ is a (concave, linear, convex) function of θ when the risk of θ is increased. [*Hint*: Note that if $u_\alpha(\theta,\alpha)$ is concave, then increasing risk makes $Eu_\alpha(\theta,\alpha^*) < 0$ and to restore the equality we need to decrease α^*.]

Suppose there are two periods, present and future. Let initial wealth be W, and s be the savings rate; then consumption in the present period is $C = (1-s)W$. Suppose r is the risk-free (gross) rate of interest and that the individual will receive the random bequest Y at the end of the present period. Assuming that utility is stationary and additive and that $0 < \beta < 1$ is a discount factor, the problem is to $\max_{0 \leqq s \leqq 1}\{u[(1-s)W] + \beta Eu(Y + rsW)\}$.

(c) Determine when there will be an interior maximum.

(d) Develop the first- and second-order conditions assuming that there is an interior maximum.

(e) Show that the marginal propensity to save with respect to wealth is between 0 and 1. What happens if $s^* = 0$ or 1?

(f) Show that if u' (primes denote differentiation) is a convex (concave) function of consumption, that is, $u''' > 0$ (<0), the investor will increase (decrease) his savings rate when there is increased uncertainty regarding his future income.

(g) Show that $u''' > 0$ for all utility functions displaying decreasing absolute risk aversion.

(h) Show that the investor will increase savings with increasing income uncertainty for the utility functions $[1/(1-\alpha)]\,C^{1-\alpha}$, $\log C$, and $-\phi^{-\gamma C}$.

Suppose now that the investor derives all his income from wealth, that is, $Y \equiv 0$, but that the rate of return on wealth, r, is random.

(i) Show that increasing risk in r increases (decreases) savings if $2u''(C)+Cu'''(C) > (<) 0$.
(j) Show that increased risk increases (decreases) savings if $\alpha > 1$ (<1) when $u(C)=(1-\alpha)^{-1}C^{1-\alpha}$.
(k) Investigate the case when both r and Y are random.

Suppose now that the investor derives all his income from his return on wealth (i.e., $Y \equiv 0$) but that he may allocate his investment between two investments. Let δ and $1-\delta$ be the proportions invested in assets 1 and 2 that have random (gross) rates of return r_1 and r_2, respectively.
(l) Suppose that $u' > 0$ and $u'' < 0$. Show that it is not possible to determine if it is advisable to reduce the proportion invested in an asset that becomes riskier.
(m) Suppose $u(C)=(1-\alpha)^{-1}C^{1-\alpha}$ and $\alpha < 1$. Show that the optimal savings rate and the optimal proportion invested in asset 1 are decreased with an increase in the variability of r_1.
(n) Show that the result is ambiguous if $\alpha > 1$.
(o) What happens if $\alpha = 1$?

***6.** In the dynamic programming approach to the multiperiod consumption–investment problem, it is of interest to examine those properties of the n-period utility function which are preserved under the backward induction procedure. For example, if $U(c_1, c_2, \ldots, c_n)$, the utility function for consumption over n periods, is increasing and concave over the appropriate convex set $K_n \subset E^n$, the Fama paper shows that all the "induced" utility functions $U_k(c_1, c_2, \ldots, c_k, w)$ are also increasing and concave over the appropriate convex set $K_{k+1} \subset E^{k+1}$. Thus, concavity is one such "preserved" property. As another example, the Neave paper shows that if U is additive and the single-period utility functions exhibit nonincreasing absolute risk aversion, then all the induced utility functions for wealth also exhibit nonincreasing absolute risk aversion.
(a) Find the most general class of increasing, concave utility functions $U(c_1, \ldots, c_n)$ such that the induced utility functions $U_k(c_1, \ldots, c_k, w)$ exhibit nonincreasing absolute risk aversion in the variables $c_1, \ldots, c_k$ and w.
(b) As in (a) above, with "nonincreasing absolute risk aversion" replaced by "nondecreasing relative risk aversion."
(c) As in (a) above, with "nonincreasing absolute risk aversion" replaced by "nonincreasing absolute and nondecreasing relative risk aversion."

7. The call option problem as presented in the Pye paper is actually very closely related to a general class of so-called *stopping rule* problems. By applying certain known results from the theory of stopping rule problems, further insight can be gained into the nature of the bond refunding decision. In this approach, the bond problem is characterized in terms of (1) "states" of a probabilistic process (stochastic interest rates), (2) a binary decision (observe the state, then either call the bond or not), and (3) a reward structure. In the bond problem, the states are pairs (i,t), $i=1,\ldots,n$, $t=1,\ldots,T$, and transitions between states are governed by a Markovian interest rate model:

$$q_{jt} \equiv P[j,t \mid i_0, t_0, i_1, t_1, \ldots, i_r, t_r] = P[j,t \mid i_r, t_r] \qquad \text{if} \quad t_0 < t_1 < \cdots < t_r < t.$$

The decision/reward structure given state (i,t) is (a) *stop* (call the issue) and collect $F(i,t) = p_{t,i} - c_t$ or (b) *continue* and pay an *entrance fee* $f(i,t) = r/(1+\rho_i)$ for the privilege of watching another transition. Here r is the coupon rate, assumed to be paid at the *end* of period t; the reward $F(i,t)$ is assumed to be collected at the beginning of period t, at which time the appropriate interest rate ρ_i is assumed to be known. Let $U(i,t)$ be the expected reward obtained by following an optimal policy starting from state (i,t).

(a) Show that U obeys the functional equation

$$U(i,t) = \max\left[F(i,t), \frac{1}{1+\rho_i}\sum_j q_{ij}\,U(j,t+1) - f(i,t)\right]$$

for $t = 1, \ldots, T-1$ and $U(i, T) = 0$ for all i.

It is useful to reformulate the problem as a *pure entrance-fee* problem in which the reward for stopping is $F'(i,t) = 0$ and an entrance fee $f'(i,t)$ must be paid to continue. Note that if $f'(i,t) < 0$, it is actually profitable to continue. In the present case, let

$$f'(i,t) = F(i,t) + f(i,t) - \frac{1}{1+\rho_i}\sum_j q_{ij}\,F(j,t+1)$$

and $F'(i,t) \equiv 0$. Let $U'(i,t)$ be the maximum expected reward for the pure entrance-fee problem, starting from state (i,t).

(b) Show that $U'(i,t)$ obeys the equation

$$U'(i,t) = \max\left[0, -f'(i,t) + \sum_j q_{ij}\,U(j,t+1)\right]$$

for $t = 1, 2, \ldots, T-1$, and $U'(i, T) = 0$ for all i.

(c) Show that $U(i,t) = F(i,t) + U'(i,t)$. [*Hint*: Use backward induction.]

(d) Show that the optimal refunding policy for the original and the entrance-fee versions is the same.

(e) Recalling the functional equation obeyed by the bond prices $p_{t,i}$, show that

$$f'(i,t) = \frac{2r + c_{t+1} - c_t}{1+\rho_i}, \qquad t = 1, \ldots, T-1.$$

(f) Recalling Pye's assumptions $\rho_i < \rho_{i+1}$, $\Delta c_t \leqq 0$, $\Delta^2 c_t \geqq 0$, show that

$$f'(i+1,t) < f'(i,t) \qquad \text{and} \qquad f'(i,t+1) \geqq f'(i,t).$$

(g) Show that the optimal policy is this: Call the bond if the interest rate $\rho(t)$ in period t is less than a critical value $\rho^*(t)$, and don't call it otherwise. Show also that $\rho^*(t+1) \geqq \rho^*(t)$.

8. Using the pure entrance-fee approach of Exercise 7, the call option problem may be formulated as a linear programming problem: Choose $i_1 \in \{1, \ldots, n\}$ and $t_1 \in \{1, \ldots, T-1\}$, and solve

$$\min x_{i_1,t_1}, \qquad \text{s.t.}\quad x_{i,t} \geqq 0, \quad x_{i,t} \geqq -f'(i,t) + \sum_j q_{ij}\,x_{j,t+1}$$

$$\text{for}\quad i = 1, \ldots, n, \quad t = 1, \ldots, T-1, \qquad \text{and} \qquad x_{i,T} = 0 \quad \text{for} \quad i = 1, \ldots, n.$$

Let G be the set of all vectors $(x_{i,t})$ satisfying the inequalities given above.

(a) If $x \in G$ and $y \in G$, show that $z \in G$, where $z = \min(x, y) = (\min(x_{i,t}, y_{i,t}))$.

Let $x^* = (x^*_{i,t})$ be an optimal solution to the linear programming problem, and let

$$y_{i,t} = \max\left[0, -f'(i,t) + \sum_j q_{ij}\,x^*_{j,t+1}\right].$$

(b) Show that $y \in G$.

(c) Assuming that x^* is unique, show that its components satisfy the equation

$$x^*_{i,t} = \max\left[0, -f'(i,t) + \sum_j q_{ij}\,x^*_{j,t+1}\right].$$

(d) Show that $U'(i,t) = x^*_{i,t}$ for all (i,t).

(e) Show that the optimal policy is to call the bond in state (i,t) if $x^*_{i,t} = 0$, and to not call it in state (i,t) otherwise.

9. (Linear programming for Markovian decisions) Let a probabilistic system have at any time a fixed set of possible states $i = 1, 2, \ldots N$. At discrete times $t = 0, 1, 2, \ldots$, a decision is to be made. Given that the state is i at time t, a decision $k \in K_i$ (K_i a finite set, independent of t) is selected and generates:

(1) a transition probability distribution $p_{ij}^k \geqq 0$ $(1 \leqq j \leqq N)$ with $\sum_{j=1}^N p_{ij}^k = 1$ (p_{ij}^k is independent of previous states and decisions);

(2) a cost c_i^k for making decision k in state i.

A *pure* policy is a rule which selects for each time t and state i a unique decision $k \in K_i$. An α-optimal policy is a policy which minimizes total discounted expected cost when a discount factor $\alpha \in (0,1)$ is used. The decision problem can be generalized to include *mixed* (randomized) policies, which select for each t and i a probability distribution $\pi_i^k(t)$ over the decisions $k \in K_i$. Since any pure policy is actually a degenerate mixed policy, minimization of expected discounted cost may be carried out over the set of mixed policies without loss of optimality.

Let $\pi_j(0) \geqq 0$ be the probability that the state is j at $t = 0$. Given a mixed policy, let $\pi_j^k(t)$ be the joint probability that the state is j and decision $k \in K_j$ is made at time t.

(a) Show that $\sum_{k \in K_j} \pi_j^k(0) = \pi_j(0)$ and

$$\sum_{k \in K_j} \pi_j^k(t) = \sum_{i=1}^N \sum_{k \in K_j} p_{ij}^k \pi_i^k(t-1), \qquad t = 1, 2, \ldots, \quad \text{for} \quad 1 \leqq j \leqq N.$$

Now let $\{\pi_j^k(t)\}$, $t = 0, 1, 2, \ldots$, $1 \leqq j \leqq N$, $k \in K_j$, be any nonnegative solution of the equations in (a).

(b) Show that $\pi_j^k(t)$ is a probability distribution. Let

$$V = \sum_{t=0}^{\infty} \alpha^t \sum_{i=1}^N \sum_{k \in K_i} c_i^k \pi_i^k(t)$$

be the total expected discounted cost.

(c) If $c = \min_{i,k} c_i^k$ and $\bar{c} = \max_{i,k} c_i^k$, show that

$$c/(1-\alpha) \leqq V \leqq \bar{c}/(1-\alpha) \qquad \text{for any} \quad \{\pi_j^k(t)\} \quad \text{satisfying (a).}$$

(d) Show that the optimal mixed policy is given by the linear programming problem in infinitely many variables $\{\pi_j^k(t)\}$:

$$\inf_{\{\pi\}} V \qquad \text{s.t.} \quad \{\pi\} \text{ satisfies (a) for all} \quad t = 0, 1, 2, \ldots.$$

(e) Given any feasible sequence $\{\pi_j^k(t)\}$ of (d), show that

$$x_j^k = \sum_{t=0}^{\infty} \alpha^t \pi_j^k(t) \qquad \text{exists for all} \quad \alpha \in (0,1).$$

(f) Show that (d) becomes

$$\min \sum_{j=1}^N \sum_{k \in K_j} c_j^k x_j^k$$

s.t.

$$\sum_{k \in K_j} x_j^k - \alpha \sum_{i=1}^N \sum_{k \in K_i} p_{ij}^k x_i^k = \pi_j(0) \qquad (1 \leqq j \leqq N),$$

$$\text{and} \qquad x_i^k \geqq 0 \quad (1 \leqq i \leqq N; \quad k \in K_i).$$

(g) If $\pi_j(0) > 0$ $(1 \leqq j \leqq N)$, show that a basic feasible solution to (f) will have only one nonzero variable x_i^k for each i $(1 \leqq i \leqq N)$; that is, show that a unique $k \in K_i$ is selected for each i. [*Hint*: Recall that at most N variables x_i^k can be nonzero in a basic feasible solution. If $\{x_i^k\}$ is such a solution, then show that $\sum_{k \in K_j} x_j^k \geqq \pi_j(0) > 0$ $(1 \leqq j \leqq N)$.]

(h) If $\pi_j(0) > 0$ $(1 \leqq j \leqq N)$, show that a basic feasible solution to (f) is nondegenerate, that is, has N positive variables.

(i) Show that an optimal basis to (f) for $\pi_j(0) > 0$ $(1 \leqq j \leqq N)$ is also an optimal basis for *any* nonnegative $\pi_j(0)$.

(j) Given an optimal basic solution to (f) with positive $\pi_i(0)$ for all i, show that an optimal decision policy is to choose decision $k \in K_i$ if $x_i^k > 0$. Note that this policy is stationary, that is, time independent.

10. (Stochastic cash balance problem: Linear programming approach) To meet its daily transaction requirements, a firm keeps a cash balance. Associated with the cash balance are the following costs: (a) holding costs for positive cash balance (since cash has alternative uses); (b) penalty costs for negative cash balance (failure to meet requirements); (c) transfer costs to change the level of the balance. The daily inflows and outflows of cash are stochastic. The goal of the firm is to find an optimal cash balance policy, that is, a decision rule which minimizes the expected present value of all costs associated with its cash balance. The discount factor is $\alpha = 1/(1+c_{\mathrm{h}})$, where c_{h} is the unit holding cost for positive cash balance. Assume that the cash balance states are discrete: $i = 1, 2, \ldots, N$ (e.g., whole dollars). State 1 represents the lowest possible and state N the highest possible level of cash balance. At discrete times $t = 0, 1, 2, \ldots$, decisions are made to (1) move the cash balance from i to $k > i$; (2) move the cash balance from i to $k < i$; or (3) do nothing. Assume that the change of state takes place instantaneously. Thus the firm observes the state i at time t, and then decides to start the period in state k. Assume also that the transfer costs $T(i \to k)$ are linear,

$$t_i^k = T(i \to k) = \begin{cases} c_u(k-i) & \text{if } k \geqq i \\ c_d(i-k) & \text{if } k \leqq i \end{cases}$$

and that the penalty or holding costs are linear,

$$L(k) = \begin{cases} c_{\mathrm{p}}(M-k) & \text{if } k \leqq M, \\ c_{\mathrm{h}}(k-M) & \text{if } k \geqq M, \end{cases}$$

where M is the zero cash balance level. Assume that all the c's are greater than zero. Finally, assume that the stochastic changes in cash balance are independent identically distributed random variables with discrete probability distribution p_k, $k = 0, \pm 1, \ldots, \pm K$ (K finite).

(a) Show that the cash balance problem can be formulated as the linear programming problem:

$$P0: \qquad \min \sum_{j=1}^{N} \sum_{k=1}^{N} [t_j^k + L(k)]\, x_j^k,$$

$$\text{s.t.}$$

$$\sum_{k=1}^{N} x_j^k - \alpha \sum_{i=1}^{N} \sum_{k=1}^{N} p_{ij}^k x_i^k = \pi_j(0) \qquad (1 \leqq j \leqq N),$$

and

$$x_j^k \geqq 0 \qquad (1 \leqq j, k \leqq N), \quad \text{with} \quad \pi_j(0) \geqq 0 \text{ and } \sum_{j=1}^{N} \pi_j(0) = 1.$$

(b) Show that p_{ij}^k (the probability of making a transition to j given initial state i and decision k) is *independent of* i: $p_{ij}^k = p_j^k$ for all i.

(c) Show that the dual of problem $P0$ is

$$D0: \qquad \max \sum_{j=1}^{N} \pi_j(0)\, u_j,$$

s.t.

$$u_i - \alpha \sum_j p_j{}^i u_j \leqq L(i) \qquad (1 \leqq i \leqq N), \tag{1}$$

$$u_i - \alpha \sum_j p_j{}^i u_j \leqq L(k) + c_u(k-i) \qquad (1 \leqq i \leqq N-1;\quad k > i), \tag{2}$$

$$u_i - \alpha \sum_j p_j{}^i u_j \leqq L(k) + c_d(i-k) \qquad (2 \leqq i \leqq N;\quad k < i). \tag{3}$$

(d) Show that problem $D0$ is equivalent to the following problem:

$$D1: \qquad \max \sum_{j=1}^{N} \pi_j(0)\, u_j,$$

s.t.

$$u_i - \alpha \sum_j p_j{}^i u_j \leqq L(i) \qquad (1 \leqq i \leqq N), \tag{1'}$$

$$u_i - w_i \leqq -c_u i \qquad (1 \leqq i \leqq N-1), \tag{2'}$$

$$u_i - v_i \leqq c_d i \qquad (2 \leqq i \leqq N), \tag{3'}$$

$$w_i - w_{i+1} \leqq 0 \qquad (1 \leqq i \leqq N-1;\quad w_N \equiv 0), \tag{4'}$$

$$v_i - v_{i-1} \leqq 0 \qquad (2 \leqq i \leqq N;\quad v_1 \equiv 0), \tag{5'}$$

$$w_i - \alpha \sum_j p_j{}^k u_j \leqq L(k) + c_u k \qquad (1 \leqq i \leqq N-1,\quad k > i), \tag{6'}$$

$$v_i - \alpha \sum_j p_j{}^k u_j \leqq L(k) - c_d k \qquad (2 \leqq i \leqq N,\quad k < i). \tag{7'}$$

[*Hint*: If u', v', w' is any feasible solution of (1′)–(7′), show that u' is also a feasible solution of (1)–(3). Conversely, if u is a feasible solution of (1)–(3), define

$$w_j = \min_{k > i} \left[L(k) + c_u k + \alpha \sum_{i=1}^{N} p_i{}^k u_i \right],$$

etc. Show that u, w, v satisfy (1′)–(7′).]

(e) Show that problem $D1$ is equivalent to problem $D2$, which is obtained from $D1$ by replacing (6′) and (7′) by the conditions for $k = i+1$ and $k = i-1$, respectively [that is, the condition (6′) for all $k > i$ is assumed to be true only for $k = i+1$, etc.]

(f) Show that the dual of problem $D2$ is

$$P2: \qquad \min \sum_{i=1}^{N} L(i)\, x_i{}^i + \sum_{i=1}^{N-1} (-c_u i)\, s_i + \sum_{i=2}^{N} (c_d i)\, r_i \tag{1''}$$

$$+ \sum_{k=2}^{N} [L(k) + c_u k]\, y_k + \sum_{k=1}^{N-1} [L(k) - c_d k]\, q_k,$$

s.t.

$$x_i{}^i - \alpha \sum_{k=1}^{N} p_i{}^k x_k{}^k + s_i - \alpha \sum_{k=2}^{N} p_i{}^k y_k + r_i - \alpha \sum_{k=1}^{N-1} p_i{}^k q_k = a_i \qquad (1 \leqq i \leqq N), \tag{2''}$$

$$-s_j - t_j + t_{j+1} + y_{j+1} = b_j \qquad (1 \leqq j \leqq N-1), \tag{3''}$$

$$-r_j + z_{j-1} - z_j + q_{j-1} = c_j \qquad (2 \leqq j \leqq N), \tag{4''}$$

and r_1, s_N are omitted. The right-hand-side parameters are $a_i = \pi_i(0)$, $b_i = 0$, $c_i = 0$, but it is convenient to consider the case of more general a_i, b_i, and c_i in the sequel.

Note that problem $P2$ has $3(N-1)+1$ constraints and $7(N-1)+1$ variables, while problem $P0$ has N constraints and N^2 variables. Define a *selection basis* of problem $P2$ to be any basic feasible solution of $P2$ such that, for each i, only one of the $x_i{}^i$, r_i, or s_i, only one of the y_i or t_i, and only one of the q_i or z_i are nonzero.

(g) Prove the following proposition: For any selection basis of problem $P2$:

(1) The resulting subsystem has a unique solution;

(2) if $a_i, b_i, c_i \geqq 0$ for all i, then the solution is nonnegative;

(3) if $a_i, b_i, c_i > 0$ for all i, then the solution is strictly positive.

[*Hint*: Define the sets

$I^- = \{i \mid s_i+r_i+x_i{}^i < 0\}$, $\quad J^- = \{i \mid t_{i+1}+y_{i+1} < 0\}$, $\quad K^- = \{i \mid z_{i-1}+q_{i-1} < 0\}$.

For a selection basis, one has

$$\begin{aligned} i \in I^- \quad &\Rightarrow \quad x_i{}^i \geqq 0, \quad s_i \leqq 0, \quad \text{and} \quad r_i \leqq 0; \\ i \in J^- \quad &\Rightarrow \quad y_{i+1} \geqq 0 \quad \text{and} \quad t_{i+1} \leqq 0; \\ i \in K^- \quad &\Rightarrow \quad q_{i-1} \geqq 0 \quad \text{and} \quad z_{i-1} \leqq 0. \end{aligned}$$

Similarly

$$\begin{aligned} i \notin I^- \quad &\Rightarrow \quad x_i{}^i \geqq 0, \quad s_i \geqq 0, \quad \text{and} \quad r_i \geqq 0; \\ i \notin J^- \quad &\Rightarrow \quad y_{i+1} \geqq 0 \quad \text{and} \quad t_{i+1} \geqq 0; \\ i \notin K^- \quad &\Rightarrow \quad q_{i-1} \geqq 0 \quad \text{and} \quad z_{i-1} \geqq 0. \end{aligned}$$

Sum Eqs. (2″), (3″), and (4″) over I^-, J^-, and K^-, respectively, and add. Show that each term on the left-hand side of the result is $\leqq 0$, by showing that $\sum_{i \in J^-}(t_{i+1}-t_i) \leqq 0$, that $\sum_{i \in I^-} r_i - \sum_{i \in K^-} r_i \geqq 0$, etc. Thus if $a_i = b_i = c_i = 0$, then $i \in I^- \Rightarrow x_i{}^i = 0$, $i \in J^- \Rightarrow y_{i+1} = 0$, and $i \in K^- \Rightarrow q_{i-1} = 0$. Thus the x, y, and q variables are $\geqq 0$ for all i. Show that this is true also for the r, s, t, and z variables, so that the sets I^-, J^-, and K^- are empty. By similar reasoning, show that all the variables are $\leqq 0$, so that the homogeneous system has only the null solution. This establishes statement (1) of the proposition; (2) and (3) may be obtained by similar reasoning.]

(h) Show that if a_i, b_i, and c_i are >0 for all i, then any basic feasible solution of (2″)–(4″) is a selection basis; furthermore, this basic solution is nondegenerate. [*Hint*: Show that for all i, $x_i{}^i+r_i+s_i \geqq 0$, etc.]

(i) If a_i, b_i, and $c_i > 0$ for all i, show that problem $P2$ has an optimal basic solution and its dual has a unique optimal solution. Show also that the optimal basis remains optimal for *any* nonnegative a_i, b_i, c_i. [*Hint*: Show that all variables are bounded, by showing that

$$(1-\alpha)\left\{\sum_i x_i{}^i + \sum_i y_i + \sum_i q_i\right\} = \sum_i a_i + \sum_i b_i + \sum_i c_i,$$

etc.]

(j) If $a_i, b_i, c_i > 0$ for all i, show that an optimal solution to problem $P0$ is obtained from an optimal basic solution of problem $P2$ by means of the following rules:

(1) if $x_i{}^i > 0$, let $x_i{}^i$ be in the basis for $P0$;

(2) if $r_i > 0$, let $x_i{}^k$ be in the basis, where

$$k = \max\{j \mid j < i \text{ and } q_j > 0\};$$

(3) if $s_i > 0$, let $x_i{}^k$ be in the basis, where

$$k = \min\{j \mid j > i \text{ and } y_j > 0\}.$$

[*Hint*: Consider the dual problem $D2$, and use complementary slackness. If $s_i > 0$, then $u_i - w_i = -c_u i$; then if $y_i > 0$, show that $u_i - \alpha \sum_k p_k^{i+1} u_k = L(i+1) + c_u(i+1-i)$; however, if $y_i = 0$, then $t_i > 0$ and $w_{i+1} = w_i$, and either $y_{i+1} > 0$ or $y_{i+1} = 0$, etc.]

[*Note*: The stochastic cash balance problem has now been reduced to a linear programming problem of manageable proportions. After solving the linear programming problem $P2$, the optimal cash balance policy is (1) if $x_i^i > 0$, do nothing; (2) if $r_i > 0$, move the cash balance down to $k = \max\{j \mid j < i \text{ and } q_j > 0\}$; (3) if $s_i > 0$, move the cash balance up to $k = \min\{j \mid j > i \text{ and } y_j > 0\}$.]

(k) Formulate a linear programming problem analogous to $P2$ when *fixed* transfer costs are also included, that is, for the case

$$t_i^k = \begin{cases} K_u + c_u(k-i) & \text{if } k > i, \\ 0 & \text{if } k = i, \\ K_d + c_d(i-k) & \text{if } k < i, \end{cases}$$

where K_u and K_d are greater than zero.

11. (Dynamic programming formulation of the cash balance problem) Referring to problem 10, let the cash balance states be renumbered as $i = 0, \pm 1, \pm 2, \ldots$; suppose also that the i are not bounded. Given an n-period problem, let $f_n(i)$ be the minimum expected discounted cost in state i with n periods to go.

(a) Show that f_n satisfies

$$f_n(i) = \min_j \left[t_i^j + L(j) + \alpha \sum_{k=-K}^{K} f_{n-1}(j+k)p_k \right],$$

$$f_0(i) = 0 \qquad \text{for all} \quad i.$$

Define

$$G_n(j) = L(j) + \alpha \sum_{k=-K}^{K} f_{n-1}(j+k)p_k \qquad \text{and} \qquad \Delta G_n(j) = G_n(j+1) - G_n(j).$$

(b) Show that $f_n(i)$ is convex in i for all n. Show also that the optimal policy for an n-period problem is a (U_n, D_n)-policy, defined: (1) if $i > D_n$, decrease the cash balance to D_n; (2) if $U_n \leqq i \leqq D_n$, do nothing; (3) if $i < U_n$, increase the cash balance to U_n. (Possibly $U_n = -\infty$ or $D_n = +\infty$.) [*Hint*: Use induction. Define

$$D_n = \min\{j \mid \Delta G_n(j) > c_d\},$$

and

$$U_n = \max\{j \mid \Delta G_n(j) > -c_u\},$$

and prove that G_n is convex, so ΔG_n is monotone nondecreasing.]

(c) Show that $U_n \leqq 0$ and $D_n \geqq 0$ for all n. [*Hint*: Show that $\Delta G_n(0) > -c_u$ and $\Delta G_n(-1) < c_d$.]

Assume now that $c_u < c_p/(1-\alpha)$ and $c_d < c_h/(1-\alpha)$.

(d) Show that there are finite n_1 and n_2 such that $D_{n_1} < \infty$ and $U_{n_2} > -\infty$. [*Hint*: Let $n_1 = \min\{m \mid \sum_{n=0}^{n_1-1} \alpha^n c_h > c_d\}$. If $D_n = \infty$ for all $n < n_1$, show that $\Delta G_{n_1}(n_1 K) = \sum_{n=0}^{n_1-1} \alpha^n c_h > c_d$, etc.]

(e) Show that $D_{n+1} \leqq D_n + K$, $U_{n+1} \geqq U_n - K$.

Note that (d) and (e) imply the finiteness of U_n and D_n for all n sufficiently large.

(f) Show that the sequence $f_n(i)$ is bounded from above for any finite value of i. [*Hint*: Consider the not necessarily optimal policy of returning to state i at the beginning of each period. For any n, the expected present value of this policy dominates $f_n(i)$.]

(g) Show by induction that $f_{n+1}(i) \geqq f_n(i)$ for all n, i.

Note that (f) and (g) imply the existence of the limit $f(i) = \lim_{n\to\infty} f_n(i)$. Now let

$$G(j) = L(j) + \alpha \sum_{k=-K}^{K} f(j+k)p_k,$$

and define

$$D = \min\{j \mid \Delta G(j) > -c_u\}, \qquad U = \max\{j \mid \Delta G(j) > c_d\}.$$

(h) Show that the return points U_n, D_n converge to U, D and that the latter are finite. [*Hint*: Show that

$$K < j < D_n \quad \Rightarrow \quad \Delta G_n(j) \geqq \Delta L(j) + \alpha\, \Delta G_{n-1}(j-K).$$

Show that

$$0 < j < D_n \quad \Rightarrow \quad G_n(j) \geqq \sum_{k=0}^{M} \alpha^k c_h - \alpha^{M+1} c_u,$$

where

$$M = \max\{m \mid m \leqq j/K\}.$$

Now let

$$\bar{n} = \min\{m \mid \sum_{k=0}^{m} \alpha^k c_h - \alpha^{m+1} c_u > c_d\}.$$

Note that $\bar{n} < \infty$, and show that $D_n \leqq (\bar{n}+2)K$ for $n > \max(n_1, n_2)$.]

(i) Show that $f(i)$ satisfies the functional equation

$$f(i) = \min_j \left[t_i^{\,j} + L(j) + \alpha \sum_{k=-K}^{K} f(j+k)p_k \right].$$

Show also that the optimal cash balance policy is of the form (U, D): (1) if $i > D$, decrease the cash balance to D; (2) if $i < U$, increase the cash balance to U; (3) if $U \leqq i \leqq D$, do nothing.

Assume now that there are also *fixed* transfer costs for changing the cash balance. Thus

$$t_i^{\,j} = \begin{cases} K_u + c_u(j-i) & \text{if } j > i, \\ 0 & \text{if } j = i, \\ K_d + c_d(i-j) & \text{if } j < i, \end{cases}$$

where K_u and K_d are greater than zero. By analogy with inventory theory, it might be suspected that the optimal policy is of the form $(u, U; D, d)$: (1) if $i > d$, decrease the cash balance to D; (2) if $i < u$, increase the cash balance to U; (3) if $u \leqq i \leqq d$, do nothing. (Whether or not this is true in general appears to be unknown.)

Assume now that the only transfer costs are the fixed costs; that is, assume that $c_u = c_d = 0$.

(j) Show that the optimal policy is of the form $(u, U; D, d)$ with $U = D$, that is, show that the two return points are identical.

(k) If in addition $c_h = c_p$ and $K_u = K_d$, and if p_k is symmetric ($p_k = p_{-k}$, $|k| \leqq K$), show that $d = -u$ and $D = U = 0$. [*Note*: This problem says that for a symmetric cash balance problem involving only fixed transfer costs, the optimal policy is symmetric.]

(l) What is the form of the optimal policy for $c_p = \infty$?

12. Refer to the article by Pye concerned with call options on bonds.

(a) Suppose that the ρ's have a general distribution $F(\rho^1, \rho^2, \ldots, \rho^T)$, where ρ^i refers to the value of ρ in period i. Formulate a T-period stochastic program that describes the decision problem, and develop an appropriate dynamic programming functional equation.

Develop solution methods for the cases when:

(b) the ρ^t are independent and identically distributed,

(c) the ρ^t are independent, and
(d) the ρ^t have discrete distributions.
(e) Attempt to develop a result similar to that in the proposition in the Appendix that describes a simple optimal policy for case (d).

13. The proofs of Corollaries 1 and 2 of Theorem 1 in the Kalymon paper implicitly assume that x_i remains constant in an optimal policy. Prove rigorously that this is true.

14. (Elementary properties of renewal processes) Let $\{X_n, n=1,2,\ldots\}$ be a sequence of nonnegative iid random variables with distribution F, such that $P[X_n=0]<1$. Let $\mu = EX_n$ (which exists, but may be ∞). Define $S_0=0$, $S_n=\sum_{j=1}^n X_j$, $n \geqq 1$, and define $N(t)=\sup\{n \mid S_n \leqq t\}$. The stochastic process $\{N(t), t \geqq 0\}$ is called a *renewal process*. Intuitively, the X_n are "interarrival times" of some probabilistic process, and $N(t)$ is the number of "arrivals" in $[0,t]$.
(a) Show that $S_n/n \to \mu$ w.p. 1 as $n \to \infty$, and $N(t) < \infty$ w.p. 1. [*Hint*: Use the strong law of large numbers.]
(b) Show that $\lim_{t\to\infty} N(t)/t = 1/\mu$ w.p. 1. [*Hint*: Show that $S_{N(t)} \leqq t \leqq S_{N(t)+1}$, and $N(t) \to \infty$ w.p. 1 as $t \to \infty$.]

Suppose that a reward is earned by the renewal process. Let $Y(t)$ be the total reward earned by time t, and Y_n the incremental reward earned at the nth renewal. Assume that EX_n and $E|Y_n|$ are finite, and that the pairs (X_n, Y_n) are iid. Assume temporarily that $Y_n \geqq 0$ and $Y(t) \geqq 0$ is nondecreasing in t.
(c) Show that

$$\sum_{n=1}^{N(t)} Y_n/t \to EY_1/EX_1 \text{ w.p. 1} \quad \text{as} \quad t \to \infty.$$

[*Hint*: $Y_n/t = Y_n/N(t) \cdot N(t)/t$, and use the strong law of large numbers.]
(d) Show that $Y_{N(t)+1}/t \to 0$ w.p. 1 as $t \to \infty$.
(e) Show that $Y(t)/t \to EY_1/EX_1$ w.p. 1 as $t \to \infty$.
(f) Prove (e) when the Y_n and $Y(t)$ are not restricted in sign.

Properties (e) and (f) will play a fundamental role in the investigation of time averages for Markov chains in a later exercise. Under the conditions stated above, it is true that (e) and (f) also hold in the expected value sense: $EY(t)/t \to EY_1/EX_1$ as $t \to \infty$. The proof of the latter statement requires the use of more detailed, technical properties of renewal processes [see Ross (1970)].

A simple but important generalization of the results above pertains to delayed renewal processes. Let $\{X_n, n=1,2,\ldots\}$ be nonnegative independent random variables such that X_1 has distribution G and $X_2, X_3, \ldots$ have the common distribution F, $EX_2 = \mu$. Let

$$S_0 = 0, \qquad S_n = \sum_{k=1}^n X_k, \quad n \geqq 1, \qquad \text{and} \qquad N_D(t) = \sup\{n \mid S_n \leqq t\}.$$

The process $\{N_D(t), t \geqq 0\}$ is a *delayed renewal process.*
(g) Show that $N_D(t)/t \to 1/\mu$ w.p. 1 as $t \to \infty$.
(h) Formulate and prove results similar to (e) and (f) for delayed renewal processes.

15. (Properties of finite Markov chains) Let $\{X_n, n=0,1,\ldots\}$ be a stationary Markov chain with finitely many states $i=1,2,\ldots,k$ and stationary one-step transition matrix

$P = (p_{ij})$, where $p_{ij} = P[X_{n+1} = j \mid X_n = i]$. Note that the n-step transition matrix $P^{(n)} = (p_{ij}^{(n)})$, $p_{ij}^{(n)} = P[X_n = j \mid X_0 = i]$ is the nth power of the one-step matrix P: $P^{(n)} = P^n$, $n \geqq 1$. The definitions and results below relate to an important class of Markov chains for financial applications.

A state j is said to be *accessible* from state i, written $i \to j$, if there exists $n \geqq 0$ such that $p_{ij}^n > 0$. States i and j are mutually accessible, written $i \leftrightarrow j$, if $i \to j$ and $j \to i$.

(a) Show that (i) $i \leftrightarrow i$, (ii) $i \leftrightarrow j$ implies $j \leftrightarrow i$, and (iii) $i \leftrightarrow j$ and $j \leftrightarrow k$ implies $i \leftrightarrow k$.

In view of (a)(i)–(iii), the states can be decomposed into disjoint classes of states, each of which is a maximal class of mutually accessible states. A Markov chain is said to be *irreducible* if all states are accessible from each other. For any states i and j, let f_{ij}^n be the probability that starting in state i, the first transition to j occurs at time n, that is:

$$f_{ij}^0 = 0, \qquad f_{ij}^n = P[X_n = j,\ X_k \neq j,\ k = 1, \ldots, n-1 \mid X_0 = i].$$

(b) Show that

$$p_{ij}^n = \sum_{k=0}^{n} f_{ij}^k p_{jj}^{n-k}.$$

Let $f_{ij} = \sum_{n=1}^{\infty} f_{ij}^n = P[X_n = j \text{ for some } n \geqq 1 \mid X_0 = i]$. Clearly f_{ij} is the probability of ever reaching j, starting from i. A state is said to be *recurrent* if $f_{ii} = 1$.

(c) Show that i is recurrent iff $\sum_{n=1}^{\infty} p_{ii}^n = \infty$.

(d) Show that if i is recurrent and $i \leftrightarrow j$, then j is recurrent.

Suppose that i is recurrent, and define the *mean recurrence time* $\mu_{ii} = \sum_{n=1}^{\infty} n f_{ii}^n$. State i is said to be *positive recurrent* if $\mu_{ii} < \infty$, and *null recurrent* if $\mu_{ii} = \infty$.

(e) If $i \leftrightarrow j$, show that

$$\lim_{n \to \infty} \sum_{k=1}^{n} p_{ij}^k / n = 1/\mu_{jj}.$$

[*Hint*: Use limit results for delayed renewal processes.]

(f) Show that no state of a finite Markov chain can be null recurrent. [*Note*: This is false for chains with infinitely many states.]

In (g)–(k) below, let the Markov chain be stationary and irreducible, and suppose that $p_{ii} > 0$ for some i.

(g) Show that all states are positive recurrent.

(h) Show that $\pi_j = \lim_{n \to \infty} p_{ij}^n > 0$ exists, $j = 1, \ldots, K$.

(i) Show that $\pi = \{\pi_i, i = 1, \ldots, K\}$ is a probability distribution on $\{1, \ldots, K\}$, and is the *unique* probability distribution satisfying the equations

$$\pi_j = \sum_{i=1}^{K} \pi_i p_{ij}, \qquad j = 1, \ldots, K. \tag{i1}$$

[*Hint*: $\sum_{j=1}^{K} p_{ij}^n = 1$, and $p_{ij}^{n+1} = \sum_{k=0}^{K} p_{ik}^n p_{kj}$. If π_i' is any probability distribution satisfying (i1), show that $\pi_j' = \sum_{i=1}^{K} \pi_j \pi_i' = \pi_j$.]

The distribution π is the *limiting* or *equilibrium distribution* of the Markov chain. Let $X_0 = i$ and let $v(j)$ be the expected number of visits of $\{X_1, X_2, \ldots\}$ to j before return to i. Define $v(i) = 1$.

(j) Show that $v(j) = \sum_{k=1}^{K} v(k) p_{kj}$, $j = 1, \ldots, K$.

(k) Show that $\sum_{j=1}^{K} v(j) = \mu_{ii}$; thus $v(j)/\mu_{ii} = \pi_j$, by part (i).

16. In many cases it is important to know the limiting behavior of the time average of a stochastic process $\{X_t, t = 1, 2, \ldots\}$, that is, the existence and the value of $S_T = \sum_{t=1}^{T} X_t / T$

as $T \to \infty$. If $\{X_t\}$ is a stationary process, ergodic arguments may often be applied to obtain the limit. If $\{X_t\}$ is nonstationary but is asympotically stationary, ergodic arguments will sometimes be applicable, but may require tedious and technical limiting arguments. In this problem, an important special case is developed using "elementary" arguments based on Exercises 14 and 15.

Suppose that $\{X_n, n = 0, 1, \ldots\}$ is a stationary, irreducible Markov chain with finitely many states $\{1, \ldots, K\}$, all positive recurrent.

Theorem For any finite, real-valued function f on $\{1, \ldots, K\}$, and for any starting state $X_0 = i$,

$$\lim_{N \to \infty} \frac{1}{N} \sum_{n=0}^{N} f(X_n) = E_\pi f(X_1),$$

where E_π is the expected value with respect to the limiting distribution π.

(a) Prove the theorem when $f(k) = 1$ for $i = k$ and $f(k) = 0$ otherwise. [*Hint*: Use Exercises 14(f) and 15(k).]

(b) Prove the theorem for general f.

(c) Prove that the theorem holds if the lower limit $n = 0$ in the sum is replaced by any fixed finite value.

17. Let $\{X_n, n = 0, 1, \ldots\}$ be a stationary Markov chain with finitely many states and transition matrix P. Let p_{ij}^n be the probability that $X_n = j$, given $X_0 = i$, and let f_{ij}^n be the probability that starting in i the first transition to j occurs at time $n \geqq 1$, and let $f_{ij}^0 = 0$. Define

$$P_{ij}(s) = \sum_{n=0}^{\infty} p_{ij}^n s^n \qquad \text{and} \qquad F_{ij}(s) = \sum_{n=0}^{\infty} f_{ij}^n s^n.$$

(a) Show that $P_{ij}(s)$ and $F_{ij}(s)$ exist for all i, j when $|s| < 1$.

(b) Show that

$$F_{ij}(s) = P_{ij}(s)/P_{jj}(s) \quad (i \neq j) \qquad \text{and} \qquad F_{ii}(s) = 1/(1 - P_{ii}(s)).$$

(c) Show that $P_{ij}(s)$ satisfies the equations

$$P_{ij}(s) = \delta_{ij} + s \sum_k p_{ik} P_{kj}(s)$$

where

$$\delta_{ij} = \begin{cases} 1, & i = j \\ 0, & i \neq j. \end{cases}$$

The *expected first passage time* μ_{ij} from i to j is defined to be the expected time of the first visit to j, starting from i; that is,

$$\mu_{ij} = \sum_{n=0}^{\infty} n f_{ij}^n.$$

(d) Show that the μ_{ij} satisfy the equations

$$\mu_{ij} = 1 + \sum_{k \neq j} p_{ik} \mu_{kj}.$$

18. Consider a Bernoulli random walk $\{X_n\}$ which drifts between barriers at 0 and h, such that the process returns instantly to z $(0 < z < h)$ upon hitting either barrier. Let the probabilities of positive and negative unit steps be p and q, respectively, where $p, q > 0$ and $p + q = 1$.

(a) Show that $\{X_n\}$ is a stationary Markov chain with states $\{1, \ldots, h-1\}$ and transition matrix

$$
\begin{aligned}
P[X_1 = i+1 \mid X_0 = i] &= p \qquad (i = 1, \ldots, h-2),\\
P[X_1 = i-1 \mid X_0 = i] &= q \qquad (i = 2, \ldots, h-1),\\
P[X_1 = z \mid X_0 = h-1] &= \mathrm{p},\\
P[X_1 = z \mid X_0 = 1] &= q.
\end{aligned}
$$

(b) If $q \neq p$, show that the equilibrium or limiting distribution is

$$
\pi_i = \begin{cases} A[1-(p/q)^i], & 1 \leqq i \leqq z,\\ C[1-(p/q)^{i-h}], & z \leqq i \leqq h-1, \end{cases}
$$

where

$$
C = A\left[\frac{1-(p/q)^z}{1-(p/q)^{z-h}}\right] \qquad \text{and} \qquad A = \frac{1-(p/q)^{z-h}}{z[1-(p/q)^{z-h}]+(h-z)[1-(p/q)^z]}.
$$

(c) Show that the expected time between hits at 0 is

$$
\mu_0 = \frac{z}{q-p} + \frac{z-h}{q-p} \cdot \frac{(q/p)^z - 1}{(q/p)^h - (q/p)^z}.
$$

[*Hint*: Consider an auxiliary process with 0 added as an absorbing state.]

(d) Show that the expected time between hits at h is

$$
\mu_h = \frac{z-h}{q-p} + \frac{z}{q-p} \cdot \frac{(q/p)^h - (q/p)^z}{(q/p)^z - 1}.
$$

Let μ be the expected time between hits at either 0 or h.

(e) Show that $1/\mu = (1/\mu_0) + (1/\mu_h)$; thus

$$
\mu = \frac{z}{q-p} - \frac{h}{q-p} \cdot \frac{(q/p)^z - 1}{(q/p)^h - 1}.
$$

[*Hint*: Interpret $N(t) = N_0(t) + N_h(t)$, where $N_0(t)$ and $N_h(t)$ are the number of hits at 0 and h in time t.]

(f) Show that

$$
E_\pi X_n = \frac{1}{2}\left\{\frac{1}{q-p} + h + z - \frac{hz[1-(p/q)^{z-h}]}{z[1-(p/q)^{z-h}] + (h-z)[1-(p/q)^z]}\right\}.
$$

(g) Show that for $p = q = \frac{1}{2}$, the corresponding results are

$$
\pi_i = \begin{cases} 2i/hz, & i = 1, \ldots, z,\\ 2(h-i)/h(h-z), & i = z, \ldots, h-1, \end{cases}
$$

$$
\mu_0 = zh, \qquad \mu_h = (h-z)h, \qquad \mu = z(h-z), \qquad \text{and} \qquad E_\pi X_n = \tfrac{1}{3}(h+z).
$$

19. Consider a cash balance which is controlled by a $(0, z, h)$-policy: If the cash balance states are $0, \pm 1, \pm 2, \ldots$, then (i) do nothing if $0 < i < h$, (ii) move the cash balance instantly up to z if $i \leqq 0$, (iii) move the cash balance instantly down to z if $i \geqq h$, where $0 < z < h$. Suppose there is a fixed cost of $c_u > 0$ to move the cash balance up to z, a fixed cost of $c_d > 0$ to move it down to z, and a holding cost of c_h per dollar of cash balance per unit time.

Assume that the changes in cash balance are a Bernoulli random walk with equal probability of unit positive and negative steps. The average cost of the cash balance over a time T is

$$C_T = c_u \frac{N_u(T)}{T} + c_d \frac{N_d(T)}{T} + c_h \frac{\sum_{t=1}^{T} X_t}{T},$$

where X_t is the cash balance at time t and $N_u(T)$, $N_d(T)$ are the number of transfers up and down to z, respectively, in time T.

(a) Show that

$$C_T \to (c_u/\mu_0) + (c_d/\mu_h) + c_h E_\pi X \qquad \text{w.p. 1 as } T \to \infty,$$

where μ_0 and μ_h are the expected times between hits at 0 and h, respectively, and π is the limiting distribution of the Markov chain $\{X_t\}$.

(b) Show that $C(z, h) = \lim_{T\to\infty} C_T$ is given by

$$C(z,h) = \frac{c_u}{zh} + \frac{c_d}{h(h-z)} + \frac{c_h}{3}(h+z).$$

(c) Neglecting the integer restrictions on z and h, show that the z^* and h^* which minimize C are related:

$$h^* - z^* = kz^*, \qquad \text{where} \quad c_u k^3 = 3c_d k + 2c_d.$$

(d) For $c_u = c_d$, show that

$$h^* = 3z^*, \qquad z^* = \left(\frac{3c_u}{4c_h}\right)^{1/3}.$$

(e) Show that k is relatively insensitive to the ratio c_u/c_d.

20. Referring to Exercise 19, suppose that the cash balance policy is of the form $(0, u; d, h)$: When the balance falls to 0, return it to u; when the balance reaches h, return it to d; otherwise, do nothing ($0 \leq u \leq d \leq h$). Assume the cash balance follows a Bernoulli random walk with equal probability of unit steps up and down. Assume there is a unit holding cost c_h per unit time, a fixed cost c_t, and a unit cost τ for transfers to or from the cash balance.

(a) Show that the long-run average cost of the cash balance policy is

$$C(u,d,h) = \frac{c_h}{3}\cdot\frac{h^2+d^2-u^2+hd}{h+d-u} + \frac{c_t+u\tau}{u(h+d-u)} + \frac{c_t+(h-d)\tau}{(h-d)(h+d-u)}.$$

Let u^*, d^*, h^* minimize $C(u,d,h)$.

(b) Show that $u^* = d^*$ if $\tau = 0$.
(c) Show that $u^* < d^*$ if $\tau > 0$.
(d) Show that $u^* = 1$, $d^* = h^* - 1$ if $c_t = 0$ and $\tau > 0$.

21. Consider an investor whose utility for lifetime consumption is $u(c_0, c_1, \ldots, c_T)$. For each i, suppose the elasticity of marginal utility of c_i is constant and equal to $\gamma_i - 1$.

(a) Show that for all i

$$u(c_0, c_1, \ldots, c_T) = \begin{cases} a_i + b_i c_i^{\gamma_i} & \text{if } \gamma_i \neq 0, \\ a_i + b_i \log c_i & \text{if } \gamma_i = 0, \end{cases}$$

where a_i and b_i are independent of c_i (but may be functions of c_j for $j \neq i$).

Note that the additive utility functions of Samuelson and Hakansson are of this form, with γ_i constant over i, and b_i independent of all the c_j. Assume instead that either $\gamma_i \neq 0$ and a_i is independent of the c_j, for all j and i, or $\gamma_i = 0$ and b_i is independent of the c_j, for all j and i.

(b) Show that

$$u = a + b\prod_{i=0}^{T} c_i^{\gamma_i} \quad \text{or} \quad u = a + b\log\prod_{i=0}^{T} c_i^{\gamma_i} \quad \text{for constants } a \text{ and } b.$$

Assume that marginal utility of consumption is strictly positive for all i.

(c) Show that all γ_i must have the same sign, and $\gamma_i \cdot b > 0$.

(d) Show that a necessary and sufficient condition for strict concavity of u is $\sum_{i=0}^{T}\gamma_i < 1$.

Suppose that the γ_i reflect a constant rate of impatience between successive periods: $\gamma_i = \gamma\alpha^i$ for all i, where $\alpha \in (0,1)$ is the impatience factor. By a linear transformation of the utility index, u may thus be chosen as

$$u = \delta\prod_{i=0}^{T} c_i^{\gamma\alpha^i} \quad \text{where} \quad \delta = \begin{cases} 1 & \text{if } \gamma > 0, \\ -1 & \text{if } \gamma < 0, \end{cases}$$

or

$$u = \log\prod_{i=0}^{T} c_i^{\alpha^i}.$$

Let w_t be net worth in t, and let c_t be consumption in t. Suppose there are n securities $1, 2, \ldots, n$ having gross rates of return $\rho_{t1}, \rho_{t2}, \ldots, \rho_{tn}$ in period t. Define:

z_{t0}	Proportion of w_t consumed
$(1-z_{t0})\,w_t$	Amount invested
z_{ti}	Proportion of $(1-z_{t0})\,w_t$ invested in security i
z_t	$(z_{t0}, z_{t1}, \ldots, z_{tn})$
ρ_t	$(\rho_{t1}, \rho_{t2}, \ldots, \rho_{tn})$

Assume that $\rho_t \geqq 0$ (limited liability), and that the ρ_t are intertemporally independent. Suppose that the investor possesses no source of income other than return on investment. Assume further that borrowing or short selling is allowed, with default occurring if $\sum_1^n z_{ti}\rho_{ti} < 0$.

(e) Show that $w_{t+1} = (1-z_{t0})\,w_t R_{t+1}$, where $R_{t+1} = \max\{\sum_1^n z_{ti}\rho_{ti}, 0\}$.

Let $Z = \{z_t \mid 0 \leqq z_t \leqq 1 \text{ and } \sum_0^n z_{ti} = 1\}$ be the constraint set for all t. The investor's problem is thus

$$\max Eu(c_0, c_1, \ldots, c_T), \quad \text{s.t.} \quad z_t \in Z, \quad c_t = z_{t0}\,w_t \quad (t = 0, \ldots, T),$$

$$w_{t+1} = (1-z_{t0})\,w_t R_{t+1} \quad (t = 0, \ldots, T-1), \quad \text{and} \quad R_{t+1} = \max\{\textstyle\sum_1^n z_{ti}\rho_{ti}, 0\}.$$

For the power-law utility function, define $U_{T+1} = \delta$ and $U_t = \delta\prod_{i=t}^{T} c_i^{\gamma\alpha^i} = c_t^{\gamma\alpha^t}U_{t+1}$. Let $f_t(w_t)$ be the maximum value of EU_t given w_t (i.e., given that an optimal sequential policy is followed in t and subsequently).

(f) Show that

$$f_t(w_t) = \max_{z_t \in Z}(z_{t0}\,w_t)^{\gamma\alpha^t}E\rho_{t+1}\,f_{t+1}[(1-z_{t0})\,w_t R_{t+1}],$$

with $f_{T+1} = \delta$.

(g) Show that the optimal policy for the power-law utility function is

$$f_t(w) = A_t w^{\lambda_t}, \qquad \lambda_t = \gamma\sum_{i=t}^{T}\alpha^i; \qquad z_{t0}^* = 1\Big/\sum_{i=0}^{T-t}\alpha^i;$$

z_{ti}^* solves $\max_{z_t \in Z}\delta E[R^{\lambda_t+1}]$ $(i = 1, \ldots, n)$; and

$$A_t = \delta_{\lambda_t}^{-\lambda_t}\prod_{i=t}^{T}(\gamma\alpha^i)^{\gamma\alpha^i}\prod_{i=t}^{T-t}E[R(z_i^*)^{\lambda_i+1}] \qquad \text{for} \quad t = 0, 1, \ldots, T,$$

except that $A_T = \delta$ and $z_{Ti}^* = 0$ $(i = 1, \ldots, n)$. [*Hint*: Use backward induction.]

(h) Show that the optimal policy for the logarithmic utility is

$$f_t(w) = A_t \log w + B_t; \qquad z_{t0}^* = 1 \Big/ \sum_{i=0}^{T-t} \alpha^i;$$

z_{ti}^* solves $\max_{z_t \in Z} E \log R$ $(i = 1, \ldots, n)$; $A_t = \sum_{i=0}^{T-t} \alpha^i$; and

$$B_t = -A_t \log A_t + \log \alpha \sum_{i=t+1}^{T} \alpha^{i-t} A_i + \sum_{i=t+1}^{T} \alpha^{i-t} A_i E \log R(z_{i-1}^*) \qquad \text{for} \quad t = 0, \ldots, T,$$

except that $z_{Ti}^* = 0$ $(i = 1, \ldots, n)$ and $B_T = 0$.

The solutions given above show that the optimal portfolio selection problem in each period requires the expected utility maximization of a single-period utility function with constant relative risk aversion r_t in period t.

(i) Show that $r_t = 1 - \lambda_{t+1} = 1 - \gamma \sum_{i=t+1}^{T} \alpha^i$, with $\gamma = 0$ for the logarithmic case.

Note that risk tolerance increases or decreases with age according as risk tolerance is greater or less than that of the logarithm.

(j) Show that for the power-law utility function, the optimal policy as $T \to \infty$ is

$$f_t(w) = A(\lambda)\, w^\lambda, \qquad \lambda = \gamma\alpha^t/(1-\alpha); \qquad z_{t0}^* = 1 - \alpha;$$

z_{ti}^* solves $\max_{z \in Z} \delta E[R^{\alpha\lambda}]$ $(i = 1, \ldots, n)$; and

$$A(\lambda) = \delta(1-\alpha)^\lambda \alpha^{\alpha\lambda/(1-\alpha)} \prod_{i=1}^{\infty} E[R(z_i^*)^{\lambda\alpha^i}] \qquad \text{for all} \quad t.$$

(k) Derive the optimal policy for the logarithmic case as $T \to \infty$.

22. (The Kelly criterion: Additional properties) Refer to Exercise CR-18: Suppose $r = 0$; i.e., ties are not allowed.

(a) Show that there is a unique fraction $f_c > 0$ such that $G(f_c) = 0$ and $f^* < f_c < 1$.

(b) Show that G is a strictly concave function, and hence is strictly increasing from 0 to $G(f^*)$ on $[0, f^*]$ and strictly decreasing from $G(f^*)$ to $-\infty$ on $[f^*, 1]$.

The proofs of the next four parts may be established using the Borel strong law of large numbers. Let S_t denote the number of successes in t Bernoulli trials. Then Borel's law is that $\Pr\{S_t/t \to p\} = 1$.

(c) Show that $G(f) > 0$ implies $\lim W_t = \infty$ a.s. (almost surely); i.e., for all M, $\Pr\{\liminf W_t > M\} = 1$.

(d) Show that $G(f) < 0$ implies $\lim W_t = 0$ a.s.; i.e., for all $\varepsilon > 0$, $\Pr\{\limsup W_t < \varepsilon\} = 1$.

(e) Show that $G(f) = 0$ implies $\limsup W_t = \infty$ a.s. and $\liminf W_t = 0$ a.s.

(f) Show that $G(f_1) > G(f_2)$ implies $\lim (W_t(f_1)/W_t(f_2)) = \infty$ a.s.

(g) Interpret the results in (c)–(f).

23. (Generalized Kelly criterion) Refer to the Breiman and Thorp papers. Consider two investment strategies θ and θ^* having payoffs V_N and V_N^* after period N. These strategies are said to be significantly different if and only if $\exists \varepsilon > 0, M$ such that for all $N > M$

$$\frac{1}{N} \sum_{n=1}^{N} E|\log V_n - \log V_n^*| > \varepsilon.$$

(a) Interpret this definition.

A set of strategies Ξ is bounded if and only if for all N and $\theta \in \Xi$ there exists $\alpha(\theta) < \infty$ such that $\theta \in \Xi$ implies

$$\frac{1}{N} \operatorname{var} \sum_{n=1}^{N} \log V_n < \alpha(\theta).$$

(b) Interpret this assumption. Is it strong or weak?

(c) Show that Breiman's assumption that the random returns are finite and bounded away from zero is sufficient but not necessary for the satisfaction of the boundedness assumption.

(d) Use the Chebychev inequality to prove that for independent returns, a $\max E \log V_N$ criterion has a modified version of Property 1 (where convergence in probability replaces almost sure convergence) stated in Thorp's paper. Assume that all feasible strategies are bounded.

(e) Use the Chebychev inequality to prove that for dependent returns, a $\max E \log V_N$ criterion has the modified Property 1. Assume that all strategies under consideration are significantly different and bounded, and that the return distributions are independent of the investor's strategy.

(f) What can be said about Property 2 in Thorp's paper?

***24.** Refer to the papers by Thorp and Breiman.

(a) Attempt to prove that Thorp's Property 2 does hold for Breiman's model.

(b) Attempt to prove that for a sequence of investment choices Thorp's Property 1 holds if and only if Property 2 holds.

(c) Assume that the investor's initial wealth is the positive integer M, that only integer-valued returns are possible, and that the λ_i must be integers. Suppose $\bar{\lambda}_N{}^*$ is a maximum expected log strategy. Attempt to show that Theorems 1–3 apply to $\bar{\lambda}_N{}^*$.

(d) Does Property 2 hold for this model?

25. (The Kelly criterion and expected utility) Suppose an investor's utility function is $W_T{}^\alpha$ over period T's wealth W_T, where $\alpha < 1$ and $\alpha \neq 0$. Let the gross rate of return in each period t be r_t, where $r_t \equiv \sum r_i X_{it}$, the X_{it} are the relative investment allocations in period t, and the r_i are the gross rates of return for the individual investments assumed to be independently and identically distributed in time. Suppose $W_0 > 0$ is the investor's initial wealth; then his wealth at time T is

$$W_T = W_0 \prod_{t=1}^{T} r_t.$$

(a) Show that one may maximize W_T for fixed r_t by choosing the X_{it} so that in each period they maximize the expected logarithm of one period return, namely,

$$\max \log \sum_i r_i X_{it}, \qquad \text{s.t.} \quad \sum_i X_{it} = 1, \quad X_{it} \geqq 0.$$

(b) Show that the optimal strategy, say X^l, is stationary in time, i.e., $X^l = (X_{1t}, \ldots, X_{nt})$ for all t.

(c) Show that with probability 1 the *return* under the X^l strategy is at least as high as under the strategy X^P, where X^P is the solution to

$$\max E\left(\sum r_i X_i\right)^\alpha, \qquad \text{s.t.} \quad \sum X_i = 1, \quad X_i \geqq 0.$$

Let us consider now the expected utilities of strategies X^P and X^l.

(d) Find a distribution $F(r)$ and an α such that

$$\frac{E(r)^\alpha}{E \log r} = k > 1 \qquad \text{for some specific} \quad k.$$

Hence the power portfolio is better than the Kelly portfolio in terms of expected utility.

(e) Show that as $T \to \infty$ the superiority of the power portfolio is infinitely better than the Kelly portfolio.

***26.** Exercise ME-13 in Part III shows how the monotone and dominated convergence theorems may be used to verify the validity of an interchange of limit operations associated with expectation and differentiation operations. Such interchanges are useful in many stochastic investment problems. This problem illustrates two common fallacies that occur because of an improper limit exchange.

(a) Consider two investment strategies that produce random total wealth V and W over T periods having distribution functions $F(V;T)$ and $G(W;T)$, respectively. Show that $\lim_{T\to\infty} \text{Prob}(V>W) = 1$ does not imply nor is implied by $\lim_{T\to\infty}(EV-EW) > 0$.

(b) Devise assumptions on F and G so that these two properties are equivalent.

(c) Suppose $F_n(\xi)$, $n = 1, 2, \ldots$, is a sequence of distribution functions for the random vector ξ. Assume that $F_n(\xi) \to F_*(\xi)$ pointwise. Suppose that for each n, λ^n is a unique optimal solution to

$$\max_{\lambda \in M} \int u(\xi'\lambda)\, dF_n(\xi),$$

where u is concave and M is convex and compact. Suppose λ^* is a unique optimal solution to

$$\max_{\lambda \in M} \int u(\xi'\lambda)\, dF_*(\xi).$$

Show that $\lim_{n\to\infty} \lambda^n \neq \lambda^*$ in general.

(d) Devise sufficient conditions on u and F_n such that $\lim_{n\to\infty} \lambda^n = \lambda^*$.

***27.** Refer to Property 1 in Thorp's paper.

(a) Determine whether or not the following conjecture is true. Let $U = \{u \mid u$ is monotone nondecreasing and concave$\}$. Suppose $\max Eu(X_n(\Lambda^*)) - Eu(X_n(\Lambda)) \to \infty$ as $n \to \infty$; then

$$\lim_{n\to\infty} \frac{X_n(\Lambda^*)}{X_n(\Lambda)} = \infty.$$

(b) Can the concavity assumption be deleted in (a)?

(c) If the conjecture in (a) is false, find a subset of U for which it is true.

28. Consider the following continuous-time, deterministic consumption–investment problem. During the planning period $[0, T]$, a stream of income $m(t)$ and rate of interest $r(t)$ will prevail with certainty. A consumption plan $c(t)$ is a nonnegative function on $[0, T]$. Assume that all assets (positive or negative) are held in the form of notes bearing interest at the rate $r(t)$. Assume also that initial asset holdings $S(0)$ are zero.

(a) Show that terminal assets (bequests) S are given by

$$S = \int_0^T \left[\exp \int_t^T r(\tau)\, d\tau\right] [m(t) - c(t)]\, dt.$$

Now assume that $r(t) = j$ for all $t \in [0, T]$. Define lifetime wealth M to be

$$M = \int_0^T e^{j(T-t)} m(t)\, dt.$$

(b) Show that

$$S = M - \int_0^T e^{j(T-t)} c(t)\, dt.$$

Assume that the consumer's preferences are given by a utility function $V = \int_0^T \alpha(t) g[c(t)]\, dt$, where g is the utility associated with the rate of consumption at any time, and $\alpha(t)$ is a subjective discount factor. Assume that g is strictly concave and twice continuously differentiable on $[0, \infty)$.

(c) If terminal wealth S has no utility, that is, if bequests are disregarded, show that the consumer's problem may be written as $\max_{S \geqq 0} V$.

(d) Show that (c) may be written as

$$\max_c \int_0^T \alpha(t) g[c(t)]\,dt, \qquad \text{s.t.} \quad \int_0^T e^{j(T-t)} c(t)\,dt = M, \quad \text{and} \quad c(t) \geqq 0, \quad t \in [0, T].$$

(e) Show that a necessary and sufficient condition for c^* to be a solution to (d) is that

$$e^{j(t-T)} \alpha(t) g'[c^*(t)] = k \qquad \text{for some} \quad k > 0,$$

for all t such that $c^*(t) > 0$. [*Hint*: Consider a discrete time approximation and go to the limit of continuous time. Alternatively, use constrained calculus of variations methods.]

(f) Show that $e^{j(t-T)} \alpha(t) g'[c^*(t)] \leqq k$ for $c^*(t) = 0$.

(g) Show that c^* is differentiable in the interior of the set $\{t \mid c^*(t) > 0\}$ if g'' is continuous and less than zero.

(h) Show that

$$\frac{dc^*(t)}{dt} = -\left\{ j + \frac{1}{\alpha(t)} \frac{d\alpha(t)}{dt} \right\} \frac{g'[c^*(t)]}{g''[c^*(t)]} \qquad \text{for} \quad c^*(t) > 0.$$

Note that $-(\alpha(t))^{-1}\, d\alpha(t)/dt$ is the relative rate of subjective discount at time t.

(i) Interpret the result in (h).

Now assume that bequests S have utility $\phi(S)$, where ϕ is twice differentiable and strictly concave.

(j) Show that the consumer's problem now becomes

$$\max_c \int_0^T \alpha(t) g[c(t)]\,dt + \phi(S), \qquad \text{s.t.} \quad c(t) \geqq 0 \qquad \text{for} \quad t \in [0, T].$$

(k) Show that a necessary and sufficient condition for c^* to be optimal is

$$e^{j(t-T)} \alpha(t) g'[c^*(t)] \leqq \phi'(S^*),$$

and $c^*(t) = 0$ whenever $<$ holds. Here,

$$S^* = M - \int_0^T e^{j(T-t)} c^*(t)\,dt.$$

(l) Formulate an algorithm for solving the problem in (k).

29. This problem is a continuation of Exercise 5 when the investor has an infinite time horizon. Suppose that in period t the investor has the option of either consuming all his wealth W_t or investing $W_t - C_t$ of it, where C_t denotes consumption in period t. Assume that the individual derives all his income from wealth and that his wealth in period $t+1$ is $(W_t - C_t) r_t$, where the r_t are nonnegative iid random variables. Suppose that the investor's instantaneous utility function is $U(C_t)$ where $U' \geqq 0$ and $U'' < 0$. The problem is

$$\max_{\{C_t\}} E \sum_{t=0}^{\infty} \beta^t U(C_t), \qquad \text{s.t.} \quad W_{t+1} = (W_t - C_t) r_t, \quad W_t \geqq C_t \geqq 0,$$

where W_0 and $0 < \beta < 1$ are given.

Let $V(W_0)$ be the expected value of the sum of discounted utilities attainable from initial wealth W_0 following the policy $C = f(W)$. Then a dynamic programming recursion is

$$V(W_0) = U[f(W_0)] + \beta E V[(W_0 - f(W_0)) r_0].$$

(a) Show that

$$U'[f(W)] = \beta \int r U'[f\{(W - f(W)) r\}]\,dF \quad \text{and} \quad E\beta^t U'\{f(W_t) W_t\} \to 0 \quad \text{as } t \to \infty$$

is sufficient for $f(W)$ to be an optimal policy. [*Hint*: Compare with an alternative policy.]

(b) Interpret these conditions.

Suppose that $u(C) = (1-\alpha)^{-1}C^{1-\alpha}$, and that $\beta E r^{1-\alpha} < 1$.

(c) Suppose $\alpha \neq 1$. Show that the optimal policy indicates that the investor should always consume a constant proportion $0 < \lambda < 1$ of his wealth in each period, where $\lambda \equiv 1-(\beta E r^{1-\alpha})^{1/\alpha}$.

(d) Show that $V(W_0) = \lambda^{1-\alpha} W_0^{1-\alpha}/(1-\alpha)$.

(e) Suppose $\alpha = 1$. Show that $\lambda = 1-\beta$ and

$$V(W_0) = \frac{1}{1-\beta}\log W_0 + \frac{1}{1-\beta}\log(1-\beta) + \frac{\beta}{(1-\beta)^2}\log\beta + \frac{\beta}{(1-\beta)^2}E\log r.$$

We now consider the influence of uncertainty on the policy $f(W)$. Let the random variable $q \equiv r + p(r-\bar{r})$ where $p \geqq 0$.

(f) Show that q is riskier than r if $p > 0$.

(g) Show that the proportion consumed, i.e., λ (increases, is constant, decreases) if $\alpha(<, =, >)1$.

(h) Interpret these results.

Suppose now that the investor may allocate his investment between two investments having rates of return r_1 and r_2, respectively. Let δ be the proportion invested in the first asset and suppose $\alpha = 1$.

(i) Show that one may maximize $V(W)$ by maximizing $E\log(\delta r_1 + (1-\delta)r_2)$.

(j) Show that an increase in the riskiness of r_1 reduces the proportion invested in it but leaves the rate of savings unchanged—as in the two-period case (Exercise 5).

(k) Suppose $\alpha \neq 1$, and the optimal $\delta \neq 0$ or 1. Show that an increase in the variability of r_1 (increases, decreases) the investment in this asset if $\alpha\,(>, <)\,1$.

(l) Suppose α is general. Show that $\delta^* = 1$ if $E(r_1 - r_2)r_1^{-\alpha} \geqq 0$.

(m) Show that increasing the variability of r_1 *may* cause investment in both assets. [*Hint*: Suppose r_1 and r_2 have independent log-normal distributions.]

Exercise Source Notes

Exercise 1 was adapted from Ziemba (1974); Exercise 2 was adapted from Sandmo (1970); Exercise 3 was adapted from Sandmo (1969); Exercise 4 was adapted from Avriel and Williams (1970) and Ziemba and Butterworth (1974); Exercise 5 was adapted from Levhari (1972); Exercises 7 and 8 were adapted from Breiman (1964); Exercise 9 was adapted from de Ghellinck and Eppen (1967) and Eppen and Fama (1968); Exercise 10 was adapted from Eppen and Fama (1968); Exercise 11 was adapted from Eppen and Fama (1969); Exercise 14 was adapted from Ross (1970); Exercise 15 was adapted from Breiman (1968) and Ross (1970); Exercise 17 was adapted from Hillier and Lieberman (1967) and Ross (1970); Exercise 18 was adapted from Miller and Orr (1966) and Feller (1966); Exercise 19 was adapted from Miller and Orr (1966) and Weitzman (1968); Exercise 20 was adapted from Miller and Orr (1968); Exercise 21 was adapted from Pye (1972); Exercise 22 was adapted from Thorp (1969); Exercise 23 was adapted from Aucamp (1971); portions of Exercise 24 were conjectured by S. Larsson; Exercise 25 was adapted from Markowitz (1972); Exercise 26 was adapted from Merton and Samuelson (1974); Exercise 28 was adapted from Yaari (1964); and Exercise 29 was adapted from Levhari (1972).

Bibliography

Journal Abbreviations

AER	*American Economic Review*
AISM	*Annals of the Institute of Statistical Mathematics*
AMM	*American Mathematical Monthly*
AMS	*Annals of Mathematical Statistics*
BSTJ	*Bell System Technical Journal*
E	*Econometrica*
EJ	*Economic Journal*
IER	*International Economic Review*
JASA	*Journal of the American Statistical Association*
JB	*Journal of Business*
JBR	*Journal of Bank Research*
JET	*Journal of Economic Theory*
JF	*Journal of Finance*
JFE	*Journal of Financial Economics*
JFQA	*Journal of Financial and Quantitative Analysis*
JIE	*Journal of Industrial Engineering*
JMAA	*Journal of Mathematical Analysis and Applications*
JPE	*Journal of Political Economy*
LAA	*Linear Algebra and its Applications*
MP	*Journal of Mathematical Programming*
MS	*Management Science*
OR	*Journal of the Operations Research Society of America*
PNAS	*Proceedings of the National Academy of Sciences*
QJE	*Quarterly Journal of Economics*
RES	*Review of Economic Studies*
RE Stat.	*Review of Economic and Statistics*
RISI	*Review of the International Statistical Institute*
RSSJ	*Journal of the Royal Statistical Society*
SIAM	*Journal of the Society of Industrial and Applied Mathematics*
SIAM Rev.	*Review of the Society of Industrial and Applied Mathematics*
SJE	*Swedish Journal of Economics*
SOEJ	*Southern Economic Journal*

ABADIE, J. (ed.) (1967). *Nonlinear Programming*. North-Holland Publ., Amsterdam.

ABADIE, J. (ed.) (1970). *Integer and Nonlinear Programming*. North-Holland Publ., Amsterdam.

ADLER, M. (1969). "On the risk-return trade-off in the valuation of assets." *JFQA* **4**, 493–512.

Agnew, N. H. *et al.* (1969). "An application of chance constrained programming to portfolio selection in a casualty insurance firm." *MS* **15**, B-512–B-520.

Ahsan, S. M. (1973). "Chance-constraints, safety-first, and portfolio selection." Working Paper No. 73–22. Econ. Dep., McMaster Univ., Hamilton, Ontario, Canada.

Aitchison, J., and Brown, J. A. C. (1966). *The Lognormal Distribution.* Cambridge Univ. Press, London and New York.

Allen, R. G. D. (1971). *Mathematical Analysis for Economists.* Macmillan, New York.

Archer, S. H., and D'Ambrosio, C. A. (1967). *The Theory of Business Finance: A Book of Readings.* Macmillan, New York.

Arrow, K. J. (1964). "Optimal capital policy, the cost of capital, and myopic decision rules." *AISM* **16**, 21–30.

Arrow, K. J. (1965). *Aspects of the Theory of Risk Bearing.* Yrjö Jahnsson Foundation, Helsinki.

Arrow, K. J. (1971). *Essays in the Theory of Risk Bearing.* Markham Publ., Chicago, Illinois.

Arrow, K. J., (1974). "Unbounded Utility Functions in Expected Utility Maximization: Response." *QJE*, **88**, 136–138.

Arrow, K. J., and Enthoven, A. C. (1961). "Quasi-concave programming." *E* **29**, 779–800.

Aucamp, D. C. (1971). "A new theory of optimal investment." Mimeograph. Southern Illinois Univ.

Avriel, M. (1972). "*r*-convex functions." *MP* **2**, 309–323.

Avriel, M., and Williams, A. C. (1970). "The value of information and stochastic programming." *OR* **18**, 947–954.

Balinsky, M. L., and Baumal, W. J. (1968). "The dual in nonlinear programming and its economic interpretation." *RES* **35**, 237–256.

Baumol, W. J. (1952). "The transactions demand for cash: An inventory theoretic approach." *QJE* **66**, 545–556.

Bawa, V. (1973). "Minimax policies for selling a non-divisible asset." *MS* **19**, 760–762.

Beckmann, M. J. (1968). *Dynamic Programming of Economic Decisions.* Springer-Verlag, Berlin and New York.

Bellman, R. (1957). *Dynamic Programming.* Princeton Univ. Press, Princeton, New Jersey.

Bellman, R. (1961). *Adaptive Control Processes: A Guided Tour.* Princeton Univ. Press, Princeton, New Jersey.

Bellman, R., and Dreyfus, S. (1962). *Applied Dynamic Programming.* Princeton Univ. Press, Princeton, New Jersey.

Ben-Tal, A., and Hochman, E. (1972). "More bounds on the expectation of a convex function of a random variable." *Journal of Applied Probability* **9**, 803–812.

Ben-Tal, A., Huang, C. C., and Ziemba, W. T. (1974). "Bounding the Expectation of a Convex Function of a Random Variable." Mimeograph, Univ. of British Columbia.

Berge, C. (1963). *Topological Spaces.* Macmillan, New York.

Berge, C., and Ghouila-Houri, A. (1962). *Programming, Games and Transportation Networks.* Wiley, New York.

Bergthaller, C. (1971). "Minimum risk problems and quadratic programming." CORE Discuss. Paper No. 7115. Catholic Univ., Louvain, Belgium.

Bernoulli, D. (1954). "Exposition of a new theory on the measurement of risk." *Papers of the Imperial Academy of Science.* St. Petersburg, 1738; reprinted in *E* **22**, 23–36.

Bertsekas, D. P. (1974). "Necessary and Sufficient Conditions for Existence of an Optimum Portfolio." *JET* **8**, 235–247.

Bessler, S. A., and Veinott, A. F., Jr. (1966). "Optimal Policy for Dynamic Multi-Echelon Inventory Models." *Naval Research Logistics Quarterly* **13**, 335–388.

Bharucha-Reid, A. T. (1960). *Elements of the Theory of Markov Processes and Their Application*. McGraw-Hill, New York.

Bierwag, G. O. (1973). "Liquidity preference and risk aversion with an exponential utility function and comment." *RES* **40**, 301–302.

Bierwag, G. O. (1974). "The rationale of mean-standard deviation analysis: Comment. *AER* **64**, 431–433.

Blackwell, D. (1951). "Comparison of experiments." *Proc. Symp. Math. Statist. and Probability, 2nd, Berkeley, 1951* (J. Neyman and L. M. LeCam, eds.), pp. 93–102. Univ. of California Press, Berkeley.

Blackwell, D. (1962). "Discrete dynamic programming." *AMS* **33**, 719–726.

Blackwell, D. (1967). "Positive dynamic programming." *Proc. Symp. Math. Statist. and Probability, 5th, Berkeley, 1967* (L. M. LeCam and J. Neyman, eds.). Univ. of California Press, Berkeley.

Blackwell, D. (1970). "On stationary strategies." *RSSJ Ser. A* **133**, 33–37.

Boltyanskii, V. G. (1971). *Mathematical Methods of Optimal Control*, translated by K. N. Trirogoff. Holt, New York.

Borch, K. H. (1968). *The Economics of Uncertainty*. Princeton Univ. Press, Princeton, New Jersey.

Borch, K. H. (1974). "The rationale of mean-standard deviation analysis: Comment." *AER* **64**, 428–430.

Bradley, S. P., and Crane, D. B. (1973). "Management of commercial bank government security portfolios: An optimization approach under uncertainty." *JBR* **4**, 18–30.

Breen, W. (1968). "Homogeneous risk measures. and the construction of composite assets." *JFQA* **3**, 405–413.

Breiman, L. (1961). "Optimal gambling systems for favorable games." *Symp. Probability and Statist. 4th, Berkeley, 1961*, **1**, 65–78.

Breiman, L. (1964). "Stopping rule problems." In *Applied Combinatorial Mathematics* (E. F. Beckenbach, ed.), pp. 284–319. Wiley, New York.

Breiman, L. (1968). *Probability*. Addison-Wesley, Reading, Massachusetts.

Brennan, M. J., and Kraus, A. (1974). "The Portfolio Separation–Myopia Nexus." Mimeograph, Univ. of British Columbia.

Brumelle, S. L. (1974). "When does diversification between two investments pay?" *JFQA* **9**, 473–483.

Brumelle, S. L., and Larsson, S. (1973). Private communication.

Burrill, C. W. (1972). *Measure, Integration and Probability*. McGraw-Hill, New York.

Cass, D., and Stiglitz, J. E. (1970). "The structure of investor preferences and asset returns, and separability in portfolio allocation: A contribution to the pure theory of mutual funds." *JET* **2**, 122–160.

Cass, D., and Stiglitz, J. E. (1972). "Risk aversion and wealth effects on portfolios with many assets." *RES* **39**, 331–354.

Cass, D. (1974). "Duality: A symmetric approach from the economist's vantage point." *JET* **7**, 272–295.

Chen, A. H. Y., Jen, F. C., and Zionts, S. (1971). "The optimal portfolio revision policy." *JB* **44**, 51–61.

Chen, A. H. Y., Jen, F. C., and Zionts, S. (1972). "Portfolio models with stochastic cash demands." *MS* **19**, 319–332.

Chipman, J. S. (1973). "The ordering of portfolios in terms of mean and variance." *RES* **40**, 167–190.

Chow, Y. S. *et al.* (1971). *Great Expectations: The Theory of Optimal Stopping*. Houghton, Boston, Massachusetts.

Cox, D. R. (1962). *Renewal Theory*. Methuen, London.
Crane, D. B. (1971). "A stochastic programming model for commercial bank bond portfolio management." *JFQA* **6**, 955–976.
Daellenback, H. G., and Archer, S. H. (1969). "The optimal bank liquidity: A multiperiod stochastic model." *JFQA* **4**, 329–343.
Danskin, J. M. (1967). *The Theory of Max-Min*. Springer-Verlag, Berlin and New York.
Davis, M. H. A., and Varaiya, P. (1973). "Dynamic programming conditions for partially observable stochastic systems." *SIAM* **11**, 226–261.
de Ghellinck, G. T., and Eppen, G. D. (1967). "Linear programming solutions for separable markovian decision problems." *MS* **13**, 371–394.
Denardo, E. V., and Mitten, L. G. (1967). "Elements of sequential decision processes." *JIE* **18**, 106–112.
Dexter, A. S., Yu, J. N. W., and Ziemba, W. T. (1975). "Portfolio Selection in a Lognormal Market When the Investor has a Power Utility Function: Computational Results." In *Proceedings of the International Conference on Stochastic Programming* (M. A. H. Dempster, ed.). Academic Press, New York. To be published.
Diamond, P. A., and Stiglitz, J. E. (1974). "Increases in risk and in risk aversion." *JET* **8**, 337–360.
Diamond, P. A., and Yaari, M. E. (1972). "Implications of a theory of rationing for consumer choice under uncertainty." *AER* **62**, 333–343.
Diewert, W. E. (1973). "Separability and a generalization of the Cobb-Douglas cost, production and indirect utility functions." Mimeograph. Dep. of Manpower and Immigration, Ottowa.
Dirickx, Y. M. I., and Jennergren, L. P. (1973). "On the optimality of myopic policies in sequential decision problems." Preprint I/73-13. Int. Inst. of Management, Berlin, Germany.
Doob, J. L. (1953). *Stochastic Processes*. Wiley, New York.
Doob, J. L. (1971). "What is a Martingale?" *AMM* **78**, 451–462.
Dragomirescu, M. (1972). "An algorithm for the minimum-risk problem of stochastic programming." *OR* **20**, 154–164.
Dreyfus, S. E. (1965). *Dynamic Programming and the Calculus of Variations*. Academic Press, New York.
Dubins, L., and Savage, L. (1965). *How to Gamble if You Must*. McGraw-Hill, New York.
Eppen, G. D., and Fama, E. F. (1968). "Solutions for cash balance and simple dynamic portfolio problems." *JB* **41**, 94–112.
Eppen, G. D., and Fama, E. F. (1969). "Optimal policies for cash balance and simple dynamic portfolio models with proportional costs." *IER* **10**, 119–133.
Eppen, G. D., and Fama, E. F. (1971). "Three asset cash balance and dynamic portfolio problems." *MS* **17**, 311–319.
Epstein, R. (1967). *Theory of Gambling and Statistical Logic*. Academic Press, New York.
Fama, E. F., and Miller, M. H. (1972). *The Theory of Finance*. Holt, New York.
Feldstein, M. S. (1969). "Mean-variance analysis in the theory of liquidity preference and portfolio selection." *RES* **36**, 5–12.
Feller, W. (1962). *An Introduction to Probability Theory and its Applications*, Vol. I. Wiley, New York.
Feller, W. (1966). *An Introduction to Probability Theory and its Applications*, Vol. II. Wiley, New York.
Ferland, J. A. (1971). "Quasi-convex and pseudo-convex functions on solid convex sets." Tech. Rep. No. 71-4. Oper. Res. House, Stanford Univ., Stanford, California.

FIACCO, A. V., and MCCORMICK, G. P. (1968). *Non-Linear Programming: Sequential Unconstrained Minimization Technique*. Wiley, New York.

FISHBURN, P. C. (1968). "Utility theory." *MS* **14**, 335–378.

FISHBURN, P. C. (1970). *Utility Theory for Decision Making*. Wiley, New York.

FISHBURN, P. C. (1974a). "Convex stochastic dominance with continuous distribution functions." *JET* **7**, 143–158.

FISHBURN, P. C. (1974b). "Convex stochastic dominance with finite consequence sets." *Theory and Decision*, To be published.

FISHER, J. L. (1961). "A class of stochastic investment problems." *OR* **9**, 53–61.

FLETCHER, R. (1969a). "A review of methods for unconstrained optimization." In *Optimization* (R. Fletcher, ed.). Academic Press, New York.

FLETCHER, R. (ed.) (1969b). *Optimization*. Academic Press, New York.

FREIMER, M., and GORDON, M. J. (1968). "Investment behaviour with utility a concave function of wealth." In *Risk and Uncertainty* (Proc. Conf. held by the Int. Econ. Assoc.) (K. Borch and J. Mossin, eds.). St. Martin's Press, New York.

FREUND, R. J. (1956). "The introduction of risk into a programming model." *E* **24**, 253–263.

FROST, P. A. (1970). "Banking services, minimum cash balances, and the firm's demand for money." *JF* **25**, 1029–1039.

GEOFFRION, A. M. (1971). "Duality in nonlinear programming: A simplified applications-oriented development." *SIAM Rev.* **13**, 1–37.

GINSBERG, W. (1973). "Concavity and quasiconcavity in economics." *JET* **6**, 596–605.

GIRGIS, N. M. (1969). "Optimal cash balance levels." *MS* **15**, 130–140.

GLUSS, B. (1972). *An Elementary Introduction to Dynamic Programming*. Allyn & Bacon, Rockleigh, New Jersey.

GLUSTOFF, E., and NIGRO, N. (1972). "Liquidity preference and risk aversion with an exponential utility function." *RES* **39**, 113–115.

GOLDMAN, M. B. (1974). "A negative report on the 'near-optimality' of the max-expected-log policy as applied to bounded utilities for long-lived programs." *JFE* **1**, 97–103.

GOULD, J. P. (1974). "Risk, stochastic preference, and the value of information." *JET* **8**, 64–84.

GREENBERG, H. J., and PIERSKALLA, W. P. (1971). "A review of quasi-convex functions." *OR* **19**, 1553–1570.

HADAR, J., and RUSSELL, W. R. (1969). "Rules for ordering uncertain prospects." *AER* **59**, 25–34.

HADAR. J., and RUSSELL, W. R. (1971). "Stochastic dominance and diversification." *JET* **3**, 288–305.

HADAR. J., and RUSSELL, W. R. (1974). "Diversification of interdependent prospects." *JET* **7**, 231–240.

HADLEY, G. (1964). *Nonlinear and Dynamic Programming*. Addison-Wesley, Reading, Massachusetts.

HADLEY, G. (1967). *Introduction to Probability and Statistical Decision Theory*. Holden-Day, San Francisco, California.

HAGEN, K. P. (1972). "On the problem of optimal consumption and investment policies over time." *SJE* **74**, 201–219.

HAHN, R. H. (1970). "Savings and uncertainty." *RES* **37**, 21–24.

HAKANSSON, N. H. (1969a). "Risk disposition and the separation property in portfolio selection." *JFQA* **4**, 401–416.

HAKANSSON, N. H. (1969b). "Optimal investment and consumption strategies under risk, an uncertain lifetime, and insurance." *IER* **10**, 443–466.

Hakansson, N. H. (1970). "Friedman-Savage utility functions consistent with risk aversion." *QJE* **84**, 472–487.

Hakansson, N. H. (1971a). "Capital growth and the mean-variance approach to portfolio selection." *JFQA* **6**, 517–557.

Hakansson, N. H. (1971b). "Mean-variance analysis of average compound returns." Mimeograph. Univ. of California, Berkeley.

Hakansson, N. H. (1971c). "Optimal entrepreneurial decisions in a completely stochastic environment." *MS* **17**, 427–449.

Hakansson, N. H. (1971d). "Multi-period mean-variance analysis: Toward a general theory of portfolio choice." *JF* **26**, 857–884.

Hakansson, N. H. (1974). "Convergence to isoelastic utility in multiperiod portfolio selection." *JFE* **1**, 201–224.

Hakansson, N. H., and Miller, B. L. (1972). "Compound-return mean-variance efficient portfolios never risk ruin." Working Paper No. 8. Res. Program in Finance, Graduate School of Bus. Admin., Univ. of California, Berkeley.

Hammer, P. L., and Zoutendijk, G., eds. (1974). Mathematical programming: Theory and Practice. North-Holland Publ., Amsterdam.

Hammond, J. S. (1973). "Simplifying the choice between uncertain prospects where preference is nonlinear." *MS* **20**, 1047–1072.

Hanoch, G., and Levy, H. (1971). "Efficient portfolio selection with quadratic and cubic utility." *JB* **43**, 181–189.

Hart, O. D., and Jaffee, D. M. (1974). "On the application of portfolio theory to depository financial intermediaries." *RES* **42**, 129–147.

Hausman, W. H., and Sanchez-Bell, A. (1973). "The stochastic cash balance problem with average compensating-balance requirements." Working Paper No. 663-73. Sloan School of Management, MIT, Cambridge, Massachusetts.

Henderson, J. M., and Quandt, R. E. (1958). *Microeconomic Theory*. McGraw-Hill, New York.

Hester, D. D. (1967). "Efficient portfolios with short sales and margin holdings." In *Risk Aversion and Portfolio Choice* (D. D. Hester and J. Tobin, eds.). Wiley, New York.

Hester, D. D., and Tobin, J., (eds.) (1967). *Risk Aversion and Portfolio Choice*. Wiley, New York.

Heyman, D. P. (1973). "A model for cash balance management." *MS* **19**, 1407–1413.

Hicks, J. R. (1962). "Liquidity." *EJ* **72**, 787–802.

Hillier, F. S. (1969). *The Evaluation of Risky Interrelated Investments*. North-Holland Publ., Amsterdam.

Hillier, F. S., and Lieberman, G. J. (1967). *Introduction to Operations Research*. Holden-Day, San Francisco, California.

Hogan, W. W., and Warren, J. M. (1972). "Computation of the efficient boundary in the E–S portfolio selection model." *JFQA* **7**, 1881–1896.

Hogan, W. W. (1973). "Directional Derivatives for Extremal-Value Functions with Applications to the Completely Convex Case." *OR* **21**, 188–209.

Holt, C. C., and Shelton, J. P. (1961). "The implications of the capital gains tax for investment decisions." *JF* **16**, 559–580.

Homonoff, R. B., and Mullins, D. W. Jr., (1972). "Cash management: Applications and extensions of the Miller–Orr Control limit approach." Sloan School of Management, MIT, Cambridge, Massachusetts.

Howard, R. A. (1960). *Dynamic Programming and Markov Processes*. Technol. Press of MIT Press, Cambridge, Massachusetts.

Howard, R. A. (1972). *Dynamic Probabilistic Systems*, Vols. I and II. Wiley, New York.

IGLEHART, D. (1965). "Capital accumulation and production for the firm: Optimal dynamic policies." *MS* **12**, 193–205.

IGNALL, E., and VEINOTT, A. F., JR. (1969). "Optimality of myopic inventory policies for several substitute products." *MS* **15**, 284–304.

ITÔ, K., and MCKEAN, H. P., JR. (1964). *Diffusion Processes and their Sample Paths.* Academic Press, New York.

JACOBS, O. L. R. (1967). *An Introduction to Dynamic Programming.* Chapman & Hall, London.

JAZWINSKI, A. H. (1970). *Stochastic Processes and Filtering Theory.* Academic Press, New York.

JEN, F. (1971). "Multi-period portfolio strategies." Working Paper No. 108. State Univ. of New York, Buffalo.

JEN, F. (1972). "Criteria in multi-period portfolio decisions." Working Paper No. 131. State Univ. of New York, Buffalo.

JENSEN, J. L. W. V. (1906). "Sur les fonctions convexes et les inégalites entre les valeurs moyennes." *Acta Math.* **30**, 175–193.

JENSEN, M. C. (ed.) (1972). *Studies in the Theory of Capital Markets.* Praeger, New York.

JENSEN, N. E. (1967a). "An introduction to Bernoullian utility theory: I: Utility Functions." *SJE* **69**, 163–183.

JENSEN, N. E. (1967b). "An introduction to Bernoullian utility theory: II: Interpretation, evaluation and application; a critical survey." *SJE* **69**, 229–247.

JOY, O. M., and PORTER, R. B. (1974). "Stochastic dominance and mutual fund performance." *JFQA* **9**, 25–31.

KARLIN, S. (1958). "Steady state solutions." In *Studies in the Mathematical Theory of Inventory and Production.* (K. J. Arrow, S. Karlin, and H. Scarf, eds.), Chapter XIV. Stanford Univ. Press, Stanford, California.

KARLIN, S. (1962). *Mathematical Methods and Theory in Games, Programming, and Economics,* Vol. I. Addison-Wesley, Reading, Massachusetts.

KATZNER, D. W. (1970). *Static Demand Theory.* Macmillan, New York.

KAUFMANN, A., and CRUON, R. (1967). *Dynamic Programming.* Academic Press, New York.

KEENEY, R. L. (1972). "Utility functions for multi-attributed consequences." *MS* **18**, 276–287.

KELLY, J. L., JR. (1956). "A new interpretation of the information rate." *BSTJ* **35**, 917–926.

KEMENY, J. G., and SNELL, J. L. (1960). *Finite Markov Chains.* Van Nostrand-Reinhold, Princeton, New Jersey.

KIHLSTROM, R. E., and MIRMAN, L. J. (1974). "Risk aversion with many commodities." *JET* **8**, 361–388.

KIM, Y. C. (1973). "Choice in the lottery-insurance situation augmented income approach." *QJE* **87**, 148–156.

KLEVORICK, A. K. (1969). "Risk aversion over time and a capital-budgeting problem." Discuss. Paper No. 268. Cowles Foundation, Yale Univ., New Haven, Connecticut.

KLEVORICK, A. K. (1973). "A note on 'The Ordering of Portfolios in Terms of Mean and Variance.'" *RES* **40**, 293–296.

KLINGER, A., and MANGASARIAN, O. L. (1968). "Logarithmic convexity and geometric programming." *JMAA* **24**, 388–408.

KOPP, R. E. (1962). "Pontryagin maximum principle." In *Optimization Techniques* (G. Leitmann, ed.). Academic Press, New York.

KORTANEK, K. O., and EVANS, J. P. (1968). "Asymptotic Lagrange regularity for pseudoconcave programming with weak constraint qualification." *OR* **16**, 849–857.

KUHN, H. W. (ed.) (1970). *Proc. Symp. Math. Programming, Princeton, 1970.* Princeton Univ. Press, Princeton, New Jersey.

KUSHNER, H. J. (1967). *Stochastic Stability and Control.* Academic Press, New York.

KUSHNER, H. J. (1971). *Introduction to Stochastic Control.* Holt, New York.

LARSON, R. E. (1968). *State Increment Dynamic Programming.* Amer. Elsevier, New York.

LASDON, L. S. (1970). *Optimization Theory for Large Systems.* Macmillan, New York.

LATANÉ, H. A. (1959). "Criteria for choice among risky ventures." *JPE* **67**, 144–155.

LATANÉ, H. A. (1972). "An optimum growth portfolio selection model." In *Mathematical Methods in Investment and Finance* (G. Szegö and K. Shell, eds,) pp. 336–343. North-Holland Publ., Amsterdam.

LEHMANN, E. L. (1955). "Ordered families of distributions." *AMS* **26**, 399–419.

LELAND, H. E. (1968). "Saving and uncertainty: The precautionary demand for saving." *QJE* **82**, 465–473.

LELAND, H. E. (1971). "Optimal forward exchange positions." *JPE* **79**, 257–269.

LELAND, H. E. (1972). "On turnpike portfolios." In *Mathematical Methods in Investment and Finance* (G. Szegö and K. Shell, eds.), pp. 24–33. North-Holland Publ., Amsterdam.

LEPPER, S. J. (1967). "Effects of Alternative Tax Structures on Individual Holdings of Financial Assets." In *Risk Aversion and Portfolio Choice* (D. D. Hester and J. Tobin, eds.), pp. 51–109. Wiley, New York.

LEVHARI, D. (1972). "Optimal savings and portfolio choice under uncertainty." In *Mathematical Methods in Investment and Finance* (G. Szegö and K. Shell, eds.), pp. 34–48. North-Holland Publ., Amsterdam.

LEVHARI, D., and SRINIVASAN, T. N. (1969). "Optimal savings under uncertainty." *RES* **36**, 153–163.

LEVY, H. (1973). "Stochastic dominance among log-normal prospects." Res. Rep. No. 1/1973. Dept. of Bus. Admin., The Hebrew Univ. Jerusalem.

LEVY, H. (1974). "The rationale of mean-standard deviation analysis: Comment." *AER* **64**, 434–441.

LEVY, H. and HANOCH, G. (1970). "Relative effectiveness of efficiency criteria for portfolio selection." *JFQA* **5**, 63–76.

LEVY, H., and SARNAT, M. (1970). "Alternative efficiency criteria: An empirical analysis." *JF* **25**, 1153–1158.

LEVY, H., and SARNAT, M. (1971). "Two period portfolio selection and investors' discount rates." *JF* **26** 757–761.

LEVY, H., and SARNAT, M. (1972a). *Investment and Portfolio Analysis.* Wiley, New York.

LEVY, H., and SARNAT, M. (1972b). "Safety first: An expected utility principle." *JFQA* **7**, 1829–1834.

LIPPMAN, S. A. (1971). *Elements of Probability and Statistics.* Holt, New York.

LIPPMAN, S. A. (1972). "Optimal reinsurance." *JFQA* **7**, 2151–2155.

LOÉVE, M. (1963). *Probability Theory,* 3rd ed. Van Nostrand-Reinhold, Princeton, New Jersey.

LONG, J. B., JR. (1972). "Consumption-investment decisions and equilibrium in the securities market." In *Studies in the Theory of Capital Markets* (M. C. Jensen, ed.), pp. 146–222. Praeger, New York.

LUENBERGER, D. G. (1969). *Optimization by Vector Space Methods.* Wiley, New York.

LUENBERGER, D. E. (1973). *Introduction to Linear and Nonlinear Programming.* Addison-Wesley, Reading, Massachusetts.

LUSTIG, P. A., and SCHWAB, B. (1973). *Managerial Finance in a Canadian Setting.* Holt, New York.

MADANSKY, A. (1962). "Methods of solution of linear programs under uncertainty." *OR* **10**, 165–176.

MANGASARIAN, O. L. (1966). "Sufficient conditions for the optimal control of nonlinear systems." *SIAM* **4**, 139–152.
MANGASARIAN, O. L. (1969). *Nonlinear Programming*. McGraw-Hill, New York.
MANGASARIAN, O. L. (1970). "Optimality and duality in nonlinear programming." *Proc. Symp. Math. Programming, Princeton, 1970* (H. W. Kuhn, ed.), pp. 429–443. Princeton Univ. Press, Princeton, New Jersey.
MANNE, A. (1960). "Linear programming and sequential decisions." *MS* **6**, 259–267.
MAO, J. C. T. (1969). *Quantitative Analysis of Financial Decisions*. Macmillan, New York.
MAO, J. C. T. (1970). "Models of capital budgeting, E–V vs. E–S." *JFQA* **4**, 657–675.
MARGLIN, S. A. (1963). *Approaches to Dynamic Investment Planning*. North-Holland Publ., Amsterdam.
MARKOWITZ, H. M. (1952). "Portfolio selection." *JF* **6** 77–91.
MARKOWITZ, H. M. (1959). *Portfolio Selection: Efficient Diversification of Investments*. Wiley, New York.
MARKOWITZ, H. M. (1972). "Investment for the long run." Working Paper No. 20–72. Rodney L. White Center for Financial Res., Univ. of Pennsylvania, Philadelphia.
MENEZES, C. F., and HANSON, D. L. (1971). "On the theory of risk aversion." *IER* **11**, 481–487.
MERTON, R. C. (1969). "Lifetime portfolio selection under uncertainty: The continuous time case." *RE Stat.* **51**, 247–257.
MERTON, R. C. (1972). "An analytic derivation of the efficient portfolio frontier." *JFQA* **7**, 1851–1872.
MERTON, R. C., and SAMUELSON, P. A. (1974). "Fallacy of the log-normal approximation to optimal portfolio decision-making over many periods." *JFE* **1**, 67–94.
MEYER, P. A. (1966). *Probability and Potential*. Ginn (Blaisdell), Boston, Massachusetts.
MEYER, R. F. (1970). "On the relationship among the utility of assets, the utility of consumption, and investment strategy in an uncertain, but time-invariant world." *Proc. Int. Conf. on Operational Res., 5th, Tavistock, 1970* (J. Lawrence, ed.), pp. 627–648.
MILLER, B. L. (1974a). "Optimal consumption with a stochastic income stream." *E* **42**, 253–266.
MILLER, B. L. (1974b). "Optimal portfolio decision making where the horizon is infinite." Mimeograph. Western Management Sci. Inst., Univ. of California, Los Angeles.
MILLER, M. H., and ORR, D. (1966). "A model of the demand for money by firms." *QJE* **80**, 413–435.
MILLER, M. H., and ORR, D. (1968). "The demand for money by firms: Extensions of analytic results." *JF* **23**, 735–759.
MIRMAN, L. J. (1971). "Uncertainty and optimal consumption decisions." *E* **39**, 179–185.
MIRRLEES, J. A. (1965). "Optimum accumulation under uncertainty." Mimeograph, Cambridge Univ., Cambridge, England.
MITTEN, L. G. (1964). "Composition principles for synthesis of optimal multi-stage processes." *OR* **12**, 610–619.
MOSSIN, J. (1968a). "Aspects of rational insurance purchasing." *JPE* **76**, 553–568.
MOSSIN, J. (1968b). "Taxation and risk-taking: An expected utility approach." *E* **35**, 74–82.
MOSSIN, J. (1968c). "Optimal multiperiod portfolio policies." *JB* **41**, 215–229.
MOSSIN, J. (1973). *Theory of Financial Markets*. Prentice-Hall, Englewood Cliffs, New Jersey.
MURTY, K. G. (1972). "On the number of solutions to the complementary problem and spanning properties of complementary cones." *LAA* **5**, 65–108.
NEAVE, E. H. (1970). "The stochastic cash balance problem with fixed costs for increases and decreases." *MS* **16**, 472–490.

Neave, E. H. (1973). "Optimal Consumption-investment decisions and discrete-time dynamic programming." Mimeograph. Queen's Univ., Kingston, Ontario, Canada.

Nemhauser, G. L. (1966). *Introduction to Dynamic Programming*. Wiley, New York.

Newman, P. (1969). "Some properties of concave functions." *JET* **1**, 291–314.

Ohlson, J. A. (1972a). "Portfolio selection in a log stable market." Res. Paper No. 137. Graduate School of Bus., Stanford Univ., Stanford, California.

Ohlson, J. A. (1972b). "Optimal portfolio selection in a log-normal market when the investor's utility-function is logarithmic." Res. Paper No. 117. Graduate School of Bus., Stanford Univ., Stanford, California.

Ohlson, J. A. (1974). "Quadratic approximations of the portfolio selection problem when the means and variances of returns are infinite. Mimeograph. Graduate School of Bus., Stanford Univ., Stanford, California.

Ohlson, J. A., and Ziemba, W. T. (1974). "Portfolio selection in a lognormal market when the investor has a power utility function." *JFQA*. To be published.

Olsen, P. L. (1973a). "Multistage stochastic programming: The deterministic equivalent problem." Tech. Rep. No. 191. Dept. of Oper. Res., Cornell Univ., Ithaca, New York.

Olsen, P. L. (1973b). "Measurability in stochastic programming." Tech. Rep. No. 196. Dep. of Oper. Res., Cornell Univ., Ithaca, New York.

Orgler, Y. E. (1969). "An unequal-period model for cash management decisions." *MS* **16**, 472–490.

Orgler, Y. E. (1970). *Cash Management: Methods and Models*. Wadsworth, Belmont, California.

Orr, D. (1970). *Cash Management and the Demand for Money*. Praeger, New York.

Paine, N. R. (1966). "A case study in mathematical programming of portfolio selections." *Applied Statistics* **1**, 24–36.

Parikh, S. C. (1968). "Lecture notes for the course in stochastic programming." Dep. of I.E. and O.R., Univ. of California, Berkeley.

Penner, R. G. (1964). "A note on portfolio selection and taxation." *RES* **31**, 83–86.

Phelps, E. S. (1962). "The accumulation of risky capital: A sequential utility analysis." *E* **30**, 729–743.

Phelps, R. R. (1966). *Lectures on Choquet's Theorem*. Van Nostrand-Reinhold, Princeton, New Jersey.

Pogue, G. A. (1970). "An intertemporal model for investment management." *JBR* **1**, 17–33.

Pollak, R. A. (1967). "Additive von Neumann–Morgenstern utility functions." *E* **35**, 485–494.

Ponstein, J. (1967). "Seven kinds of convexity." *SIAM Rev.* **9**, 115–119.

Pontryagin, L. S., Boltyanskii, V. G., Gamkrelidze, R. V., and Mischenko, E. F. (1962). *The Mathematical Theory of Optimal Processes*, translated by K. N. Trirogoff. Wiley (Interscience), New York.

Porter, R. B. (1961). "A model of bank portfolio selection." *Yale Economic Essays*, Vol. I, pp. 323–357.

Porter, R. B. (1972). "A comparison of stochastic dominance and stochastic programming as corporate financial decision models under uncertainty." Mimeograph. School of Business, Univ. of Kansas, Lawrence, Kansas.

Porter, R. B. (1974). "Semivariance and stochastic dominance: A comparison." *AER*. To be published.

Porter, R. B., and Gaumnitz, J. E. (1972). "Stochastic dominance vs. mean-variance portfolio analysis: An empirical evaluation." *AER* **62**, 438–446.

Porter, R. B., Wart, J. R., and Ferguson, D. L. (1973). "Efficient algorithms for conducting stochastic dominance tests on large numbers of portfolios." *JFQA* **8**, 71–81.

PORTEUS, E. L. (1972). "Equivalent formulations of the stochastic cash balance problem." *MS* **19**, 250–253.

PORTEUS, E. L., and NEAVE, E. H. (1972). "The stochastic cash balance problem with charges levied against the balance." *MS* **18**, 600–602.

PRATT, J., RAIFFA, H., and SCHLAIFER, R. (1964). "The foundations of decision under uncertainty: An elementary exposition." *JASA* **59**, 353–375.

PRESS, S. J. (1972). *Applied Multivariate Analysis.* Holt, New York.

PYE, G. (1972). "Lifetime portfolio selection with age dependent risk aversion." In *Mathematical Methods in Investment and Finance* (G. Szegö and K. Shell, eds.), pp. 49–64. North-Holland Publ., Amsterdam.

PYE, G. (1973). "Sequential policies for bank money management." *MS* **20**, 385–395.

PYE, G. (1974). "A note on diversification." *JFQA* **9**, 131–136.

PYLE, D. H. (1971). "On the theory of financial intermediation." *JF* **26**, 737–748.

PYLE, D. H., and TURNOVSKY, S. J. (1971). "Risk aversion in chance constrained portfolio selection." *MS* **18**, 218–225.

QUIRK, J. P., and SAPOSNIK, R. (1962). "Admissibility and measurable utility functions." *RES* **29**, 140–146.

RAMSEY, F. P. (1928). "A mathematical theory of saving." *EJ* **38**, 543–559.

RENTZ, W. F. (1971). "Optimal consumption and portfolio policies." Unpublished dissertation. Univ. of Rochester, Rochester, New York.

RENTZ, W. F. (1972). "Optimal multi-period consumption and portfolio policies with life insurance." Working Paper No. 72–36. Graduate School of Bus., Univ. of Texas, Austin.

RENTZ, W. F. (1973). "The family's optimal multiperiod consumption, term insurance and portfolio policies under risk." Working Paper No. 73-20. Graduate School of Bus., Univ. of Texas, Austin.

RENTZ, W. F., and WESTIN, R. B. (1972). "Optimal portfolio composition and the generalized equivalent-return class of firms." Working Paper No. 72-56. Graduate School of Bus., Univ. of Texas, Austin.

RICHTER, M. K. (1960). "Cardinal utility, portfolio selection, and taxation." *RES* **27**, 152–166.

ROCKAFELLAR, R. T. (1970). *Convex Analysis.* Princeton Univ. Press, Princeton, New Jersey.

ROLL, R. (1972). "Evidence on the 'growth-optimum' model." *JF* **28**, 551–566.

ROSEN, J. B., MANGASARIAN, O. L., and RITTER, K. (eds.) (1970). *Nonlinear Programming.* Academic Press, New York.

ROSS, S. A. (1970). *Applied Probability Models with Optimization Applications.* Holden-Day, San Francisco, California.

ROSS, S. A. (1971). "Portfolio and capital market theory with arbitrary preferences and distributions—The general validity of the mean-variance approach in large markets." Mimeograph. Rodney L. White Center for Financial Res., Univ. of Pennsylvania, Philadelphia.

ROSS, S. A. (1974a). "Portfolio turnpike theorems for constant policies." *JFE* **1**, 171–198.

ROSS, S. A. (1974b). "Some Portfolio Turnpike Theorems." Mimeograph. Univ. of Pennsylvania, Philadelphia.

ROSS, S. (1972). "Dynamic programming and gambling models." Tech. Rep. ORC 72-24. Oper. Res. Center, Univ. of California, Berkeley.

ROTHSCHILD, M., and STIGLITZ, J. E. (1970). "Increasing risk: I. A definition." *JET* **2**, 225–243.

ROTHSCHILD, M., and STIGLITZ, J. E. (1971). "Increasing risk: II. Its economic consequences." *JET* **3**, 66–84.

Roy, A. (1952). "Safety first and the holding of assets." *E* **20**, 431–449.
Royama, S., and Hamada, K. (1967). "Substitution and complementarity in the choice of risky assets." In *Risk Aversion and Portfolio Choice* (D. D. Hester and J. Tobin, eds.). Wiley, New York.
Russell, W. R., and Smith, P. E. (1970). Taxation, risk-taking, and stochastic dominance." *SOEJ* **36**, 425–433.
Ryan, T. M. (1974). "Unbounded utility functions in expected utility maximization: Comment." *QJE* **88**, 133–135.
Samuelson, P. A. (1965). *Foundations of Economic Analysis.* Atheneum Publ., New York.
Samuelson, P. A. (1967). "Efficient portfolio selection for Pareto–Levy investments." *JFQA* **11**, 107–122.
Samuelson, P. A. (1971). "The 'fallacy' of maximizing the geometric mean in long sequences of investing or gambling." *PNAS* **68**, 2493–2496.
Samuelson, P. A., and Merton, R. C. (1973). "Generalized mean-variance tradeoffs for best pertubation corrections to approximate portfolio decisions." Mimeograph. Dep. of Econ., MIT, Cambridge, Massachusetts.
Sandmo, A. (1968). Portfolio choice in a theory of saving." *SJE* **60**, 106–122.
Sandmo, A. (1969). "Capital risk, consumption and portfolio choice." *E* **37**, 586–599.
Sandmo, A. (1970). "The effect of uncertainty on saving decisions." *RES* **37**, 353–360.
Sandmo, A. (1972). "Two-period models of consumption decisions under uncertainty: A survey." Discuss. Papers 04/72. Inst. of Econ., Norwegian School of Econ. and Bus. Admin., Norway.
Sankar, U. (1973). "A utility function for wealth for a risk averter." *JET* **6**, 614–617.
Sharpe, W. F. (1964). "Capital asset prices: A theory of market equilibrium under conditions of risk." *JF* **19**, 425–442.
Sharpe, W. F. (1970). *Portfolio Theory and Capital Markets.* McGraw-Hill, New York.
Shell, K. (1972). "Selected elementary topics in the theory of economic decision-making under uncertainty." In *Mathematical Methods in Investment and Finance* (G. Szegö and K. Shell, eds.), pp. 65–75. North-Holland Publ., Amsterdam.
Sherman, S. (1951). "On a theorem of Hardy, Littlewood, Polya and Blackwell." *PNAC* **37**, 826–831.
Simon, H. A. (1956). "Dynamic programming under uncertainty with a quadratic criterion function." *E* **24**, 74–81.
Smith, K. V. (1967). "A transition model for portfolio revision." *JF* **22**, 425–439.
Smith, K. V. (1971). *Portfolio Management.* Holt, New York.
Stevens, C. V. G. (1972). "On Tobin's multiperiod portfolio theorem." *RES* **39**, 461–468.
Stiglitz, J. D. (1969). "Behavior towards risk with many commodities." *E* **37**, 660–667.
Stiglitz, J. D. (1972). "Portfolio allocation with many risky assets." In *Mathematical Methods in Investment and Finance* (G. Szegö and K. Shell, eds.), pp. 76–125. North-Holland Publ., Amsterdam.
Stoer, J., and Witzgall, C. (1970). *Convexity and Optimization in Finite Dimensions*, Vol. I. Springer-Verlag, Berlin and New York.
Stone, B. K. (1970). *Risk, Return and Equilibrium.* MIT Press, Cambridge, Massachusetts.
Strauch, R. E. (1966). "Negative dynamic programming." *AMS* **37**, 871–889.
Szegö, G., and Shell, K. (eds.) (1972). *Mathematical Methods in Investment and Finance.* North-Holland Publ., Amsterdam.
Telser, L. (1955–1956). "Safety first and hedging." *RES* **23**, 1–6.
Theil, H. (1957). "A note on certainty equivalence in dynamic planning." *E* **25**, 346–349.
Thomas, G. B., Jr. (1953). *Calculus.* Addison-Wesley, Reading, Massachusetts.

THOMASION, A. J. (1969). *The Structure of Probability Theory with Applications.* McGraw-Hill, New York.
THORE, S. (1968). "Programming bank reserves under uncertainty." *SJE* **70**, 123–137.
THORP, E. O. (1969). "Optimal gambling systems for favorable games." *RISI*, **37** 273–293.
THORP, E. O., and WHITLEY, R. (1973). "Concave utilities are distinguished by their optimal strategies." Mimeograph. Math. Dept., Univ. of California, Irvine.
TOBIN, J. (1958). "Liquidity preference as behavior toward risk." *RES* **25**, 65–86.
TOBIN, J. (1965). "The theory of portfolio selection." In *The Theory of Interest Rates* (F. H. Hahn and F. P. R. Brechling, eds.). Macmillan, New York.
TSIANG, S. C. (1972). "The rationale of the mean-standard deviation analysis, skewness preference, and the demand for money." *AER* **62**, 354–371.
TSIANG, S. C. (1974). "The rationale of the mean-standard deviation analysis: Reply and errata for original article." *AER* **64**, 442–450.
VEINOTT, A. F. (1969). "Discrete dynamic programming with sensitive discount optimality criteria." *AMS* **40**, 1635–1660.
VIAL, J. P. (1972). "A continuous time model for the cash balance problem." In *Mathematical Methods in Investment and Finance* (G. Szegö and K. Shell, eds.), pp. 244–291. North-Holland Publ., Amsterdam.
VICKSON, R. G. (1974). "Stochastic dominance for decreasing absolute risk aversion: I: Discrete random variables." Mimeograph. Univ. of Waterloo, Waterloo, Ontario, Canada.
VICKSON, R. G. (1975). "Stochastic dominance for decreasing absolute risk aversion." *JFQA*. To be published.
VON HORNE, J. C. (1968). *Financial Management and Policy.* Prentice Hall, Englewood Cliffs, New Jersey.
VON NEUMANN, J., and MORGENSTERN, O. (1944). *Theory of Games and Economic Behavior.* Princeton Univ. Press, Princeton, New Jersey.
WEITZMAN, M. (1968). "A model of the demand for money by firms: Comment." *QJE* **82**, 161–164.
WETS, R. (1966). "Programming under uncertainty: The solution set." *SIAM* **14**, 1143–1151.
WETS, R. (1972a). "Induced constraints for stochastic optimization problems." In *Techniques of Optimization* (A. V. Balachrician, ed.). Academic Press, New York.
WETS, R. (1972b). "Stochastic programs with recourse: A basic theorem for multistage problems." *Z. Wahrscheinlichkeitstheorie verw. Gebiete* **21**, 201–206.
WHITE, D. J. (1969). *Dynamic Programming.* Holden-Day, San Francisco, California.
WHITMORE, G. A. (1970). "Third-degree stochastic dominance." *AER* **60**, 457–459.
WILDE, D. J., and BEIGHTLER, C. S. (1967). *Foundations of Optimization.* Prentice-Hall, Englewood Cliffs, New Jersey.
WILLIAMS, A. C. (1970). "Nonlinear activity analysis." *MS* **17**, 127–139.
WONG, E. (1971). *Stochastic Processes in Information and Dynamical Systems.* McGraw-Hill, New York.
YAARI, M. E. (1964). "On the consumer's lifetime allocation process." *IER* **5**, 304–317.
YAARI, M. E. (1965a). "Convexity in the theory of choice under risk." *QJE* **79**, 278–290.
YAARI, M. E. (1965b). "Uncertain lifetime, life insurance and the theory of the consumer." *RES* **32**, 137–150.
YAARI, M. E. (1969). "Some remarks on measures of risk aversion and on their uses." *JET* **1**, 315–329.
YOUNG, W. E., and TRENT, R. H. (1969). "Geometric Mean Approximations of Individual Security and Portfolio Performance." *JFQA* **4**, 179–199.
ZABEL, E. (1973). "Consumer choice, portfolio decisions and transaction costs." *E* **44**, 321–335.

ZANGWILL, W. I. (1969). *Nonlinear Programming: A Unified Approach.* Prentice-Hall, Englewood Cliffs, New Jersey.

ZECKHAUSER, R., and KEELER, E. (1970). "Another type of risk aversion." *E* **38**, 661–665.

ZIEMBA, W. T. (1969). "A myopic capital budgeting model." *JFQA* **6**, 305–327.

ZIEMBA, W. T. (1971). "Transforming stochastic dynamic programs into nonlinear programs." *MS* **17**, 450–462.

ZIEMBA, W. T. (1972a). "Solving nonlinear programming problems with stochastic objective functions." *JFQA* **7**, 1809–1827.

ZIEMBA, W. T. (1972b). "Note on optimal growth portfolios when yields are serially correlated." *JFQA* **7**, 1995–2000.

ZIEMBA, W. T. (1974). "Stochastic programs with simple recourse." In P. L. Hammer and G. Zoutendijk, eds., "Mathematical Programming: Theory and Practice," pp. 213–273. North-Holland Publ., Amsterdam.

ZIEMBA, W. T., and BUTTERWORTH, J. E. (1974). "Bounds on the value of information in uncertain decision problems." *Stochastics* **4**, 1–18.

ZIEMBA, W. T., PARKAN, C., and BROOKS-HILL, F. J. (1974). "Calculation of investment portfolios with risk free borrowing and lending." *MS* **21**, 209–222.

INDEX

E

F

G

H

I

J

K

L

V

W

Z

A 5
B 6
C 7
D 8
E 9
F 0
G 1
H 2
I 3
J 4